PHYSICS

Volume I: Mechanics and Heat

Volume II: Electricity, Magnetism
and Optics

Barnes & Noble International
Textbook Series

Chemistry: Structure and Changes of Matter
Uno Kask

Experimental General Chemistry
George G. Hess and Uno Kask

Quantitative Analytical Chemistry
    Volume I: Introduction to Principles
    Volume II: Short Introduction to Practice
H. A. Flaschka, A. J. Barnard, Jr., and P. E. Sturrock

Chemical Analysis: An Intensive Introduction
    to Modern Analysis
W. E. Harris and B. Kratochvil

Elements of Quantum Mechanics with Chemical Applications
Jean Barriol (translated by J. Warren Blaker)

INTERNATIONAL TEXTBOOK SERIES

PHYSICS

VOLUME I: MECHANICS AND HEAT
VOLUME II: ELECTRICITY, MAGNETISM AND OPTICS

ARNOLD L. REIMANN

Research Professor in Physics,
University of Queensland

BARNES & NOBLE, INC., NEW YORK

Publishers · Booksellers · Founded 1873

Library of Congress Catalogue Card No. 73-137431
ISBN 389 00462 6

Distributed:

In Canada by Methuen Publications, Toronto

In Australia and New Zealand by Hicks, Smith & Sons Pty. Ltd.,
    Sydney and Wellington

In the United Kingdom, Europe, and South Africa
    by Chapman & Hall Ltd., London

Printed in the United States of America

FOREWORD

Textbooks, like womens' clothing, conform to different styles. The avowed intention of most texts, generally paralleling female dress, is to be as revealing as possible within dignity. Recitation of only the facts in a physics textbook, like nudity, quickly becomes a bore. The balance between promise of greater adventure ahead and satisfaction in the present is difficult to achieve.

Professor Reimann in his physics textbook has come close to fulfilling this dream. Whereas most physics textbooks lecture *to* the student, this textbook discusses nature and physics *with* the reader. Instead of following the trend of introducing the "big physics" accomplishments of recent years in an attempt to encourage the reader on, Professor Reimann develops physics as a natural unfolding of observations. His texts, particularly volumes I and II, are spiced with examinations of phenomena of everyday interest. These examples bind the reader to the text through a common tie of experience. As merely one example, atmospheric precipitation is treated at some length—the formation of snow, hail, cold rain, and warm rain. By-products of this discourse are such phenomena as nucleation, supersaturation, and supercooling. With a touch of "British humor," cloud seeding is explained as water molecules being "fooled" into freezing onto silver iodide crystals.

The conversation-discourse writing style of this textbook is not new in itself. What is new and imaginative here is the effective employment of this style well beyond the usual arena of highly introductory, elementary textbooks. Professor Reimann's volumes are introductory in that no vast preliminary preparation is assumed. But Reimann does not employ the gambit of many authors by starting these texts with "mathematical preparations"; rather the necessary mathematical techniques are introduced at the most propitious moments within the development of the central theme—physics. Vector analysis arises in the treatment of particle kinetics. Differentiation first appears as a "streamlined" method to obtain velocity from displacement following the more cumbersome, non-calculus approach to this topic. Before the student completes the book, he has been taken well into integral calculus, including line integration. These volumes may not be classified as "elementary," if this term is to imply that no degree of sophistication can be tolerated. The fact is that these volumes are both elementary *and* sophisticated. They are elementary in that they deal with the basic elements of physics, a point worth recognizing. They are sophisticated in that they offer the reader an appreciation and a command of much of the methods and essence of physics.

One example of Reimann's approach and scope is the problem sets that follow each chapter. The problems are proposed in sufficient detail that the reader is not left frustrated and annoyed; they are graded, progressing toward increased difficulty. The reader who successfully works his way to the end of a problem set will know that he has partaken of much of the physics contained in the respective chapter. An instructor, needless to say, is free to select from the extensive problem sets to satisfy the particular needs of his class.

These texts will appeal to a wide range of prospective readers. At the center, the largest group contains those who will gain a knowledge and appreciation of physics that will be of major assistance in their pursuits of related subjects. These texts, while supplying the answer to such demands, will still appeal excellently to those on either side of the center. For the individual who chooses to read further in physics and perhaps specialize in physics, the scope of these texts provides a firm background. For the reader who merely wishes to obtain a perspective of physics, these texts are eminently readable (a novelty in textbook writing).

Since physics is a truly international language, there should be no reason for concern on the reader's part that these texts were written by an author off the American scene. On the contrary, by happenstance or otherwise, Professor Reimann has succeeded in writing a set of books that are in tune with current trends in the United States and elsewhere regarding relevance in teaching. Fortunately, there need be no undue concern for those who care to use an introductory course as a showcase for modern physics. Volume III, which follows the two introductory volumes, is devoted to a description of the history and present status of a wide span of today's "frontier" activities and the previous two volumes draw upon these fields for examples of more classical physics.

A. Sosin
University of Utah

PREFACE

This book is intended for science majors or engineering students entering a physics course who have previously studied physics at a more elementary level and either already have some knowledge of the calculus or are studying this subject concurrently with their university physics.

The book is divided into three approximately equal parts: (1) mechanics, mechanical properties of matter, and heat (chapters 1–18, constituting volume I); (2) electricity and magnetism, followed by optics (chapters 19–40, constituting volume II); and (3) modern physics (chapters 1–14, constituting volume III). The treatment of electricity and magnetism in volume II is predominantly classical, but it includes a few topics which many physicists would consider as belonging more properly to modern physics. Wherever this is done, it is for a special reason. For example, thermionic and photoelectric emission are dealt with in this volume because their discussion leads naturally to the concept of electrons, and so to the picture of an electrical current in a metal flow as a flow of these particles and to the theory of electronic and ionic processes in gas discharges.

The coverage is sufficiently comprehensive to afford teachers a wide choice of topics when planning their courses and at the same time to meet the needs of students who wish to extend their reading beyond the confines of a set course. Among the interesting extensions and applications of basic physics that are dealt with at some length are the theory of plastic deformations, the modern theory of friction, the processes by which snow, rain, and hail are formed, the generation of thunderstorms, electron optics, the theory of discharges in gases, and the self-excited dynamo theory of terrestrial magnetism.

The approach is one of investigation, with emphasis on the use of constructive imagination, experimentation, and inductive reasoning. This approach, being in harmony with the processes by which practically all advances in physics are made, is not only more realistic but must be more interesting and intellectually satisfying to the student than a mere recital of facts, unsupported by the evidence on which our knowledge is based.

Physical principles and theories are not propounded as something ready-made, but are allowed to unfold naturally in their proper logical sequence throughout the text. For example, the laws of force and mass, instead of just being quoted uncritically in their original Newtonian form, are developed from simple experiments, and set out in the revised form due to Mach. It is then shown why

Mach's presentation is preferable to Newton's. In the treatment of electrostatics, positive and negative charges are not introduced at the outset as defects and excesses, respectively, of electrons. Instead, the idea of electrons is made to emerge from a discussion of the appropriate electrical experiments. Only rarely have I resorted to the device of "borrowing from the future," for example, explaining interatomic and intermolecular forces as electrical in nature before electricity and atomic structures have been dealt with; I have done this only in cases where a lucid theoretical exposition has seemed hardly possible without such borrowing; and then I have, if possible within the scope of this book, repaid the debt at the appropriate time by properly establishing the information that has been invoked. Too frequent resort to this device is, I am convinced, not only philosophically to be deprecated, but also educationally misguided; its excessive use engenders in the student a growing confusion between items of knowledge whose experimental and logical basis he understands and those which he has merely been told and asked to accept on trust. It also deprives him of much of the excitement he should experience in his later, more advanced studies, where what he has been told prematurely is finally scientifically established.

New concepts and mathematical techniques are introduced, not in isolation from their application, but functionally, where they are required. For example, vector composition is introduced at the stage where it will be needed for dealing with such topics as centripetal acceleration and simple harmonic motion, dot products where the concept of work is developed, and cross products where the magnetic force per unit length exerted on a current-carrying conductor is related with the field and the current.

If a student is to be prepared for more advanced studies it is necessary not only to hold his interest but also to present him with a continuing challenge. I have therefore not hesitated to make considerable demands on the reader's mathematical ability and on his powers of abstract thinking, both in the main text and in many of the problems set at the ends of chapters. To faciltate his working on problems concurrently with the reading of the text, I have arranged these in approximately the same order as the corresponding material is dealt with in the text. This appears preferable to grading them on the basis of relative difficulty.

The units I have employed are, in the main, metric units. While FPS units are also used to some extent in the chapters on mechanics, these are then faded out. In the nonelectrical part of the book CGS and MKSA units are employed about equally, without bias either way, but in the chapters on electricity rationalized

MKSA units are used exclusively. My object in using more than one system in the nonelectrical chapters is to familiarize the student with unit conversions, which every working physicist has on occasion to carry out. It also serves to impress on the student the fact, of some philosophical importance, that the validity of any theoretical equation in physics is not dependent on the units in which the quantities are expressed, as long as these units are fundamental (that is, not merely empirical) and mutually consistent, and have not been adjusted for some special purpose. The student will be in a position to appreciate this universality of formulas only if he is familiar with at least two systems.

The experimental basis of our knowledge involves instrumentation. While I do not believe that a discussion of the finer details of instrumentation would serve any useful purpose in a book of this kind, I do feel that for complete intellectual satisfaction the thoughtful student will wish to know and understand the broad principles of the measurement of quantities dealt with in theoretical discussions, and that he will be interested in the main techniques used in experimental investigations. Accordingly, I have devoted some space to these matters.

My best thanks are due to Professor H. C. Webster, Head of the Department of Physics in the University of Queensland, for his encouragement in many practical ways in the task of preparing the manuscript and for a number of helpful suggestions. I am greatly indebted to my colleagues Dr. Jim Crouchley, Dr. W. B. Lasich, Dr. W. A. Macky, Professor D. Mugglestone, Dr. J. H. Noon, Professor R. W. Parsons, and Dr. F. D. Stacey for their careful reading and constructive criticism of portions of the manuscript. I also wish to thank Professor Eugene V. Ivash, University of Texas, for some valuable suggestions; Professor Peng Somanabanhu, Chulalongkorn University, Bangkok, Thailand, for his kind cooperation in reading and checking part of an earlier draft of the manuscript; and Professors Richard L. Jacob, Cornell College; Robert Knox, University of Rochester; and A. Sosin, University of Utah for their careful readings and advice on the final manuscript. Professors Ralph Hautau and Neil Goldman receive my thanks for their checking of the problems in volumes I and II, respectively, and for working out their solutions. Finally, I should like to thank my wife for her help in the considerable task of correcting the proofs.

Arnold L. Reimann

ADVICE TO STUDENTS

In offering what I believe will be helpful advice to you, the readers of this book, I want to avoid formality, and so I shall address you directly, in the second person.

Acquisition of Knowledge Physics is, as you know, preeminently a thinking and reasoning subject. The material thought and reasoned about is generally summarized in the form of "laws," based on what we observe in natural phenomena and in the experiments we devise and perform. It is obviously necessary for you, as students of physics, to become familiar with these laws. But no attempt should be made to learn them by rote. Rote learning is a complete waste of time; all it does is to substitute the memory of word sequences for genuine understanding.

People with hobbies are extremely well informed about what they are interested in. It is *through their interest* that all their detailed knowledge is acquired. The same principle applies in the study of physics; you do not have consciously to *learn* its laws, as you cultivate and maintain your interest in the subject.

I do not recommend the rote learning of formulas any more than I do that of laws. If a formula is simple, its frequent use will impress it on your memory. If it is long and complex, you probably will not remember it. But does this really matter? What you presumably *will* remember is the fact that there is a formula connecting the quantities with which you are concerned, and when you need this formula you can look it up.

Consolidation Knowledge gained by reading or by attending lectures remains somewhat superficial and ephemeral until it has been used constructively; to make one's knowledge real and lasting one must learn by doing. Provision is normally made for this in universities by requiring students to perform laboratory experiments and by conducting tutorial classes in which problems are worked by students. In this book there is a set of problems at the end of each chapter. The order in which these are arranged is such that you will not have to wait until you have read the entire chapter before attempting the corresponding problems; you will be able to work on them in stages concurrently with your study of the text.

Difficulties in Understanding Theories The standard of difficulty in this book covers the whole spectrum ranging from passages such as those dealing with the generation of thunderstorms and geomagnetism, which can be read almost as narratives, to quite difficult sections such as the proof of Ampère's line-integral theorem. Most of you will not always at once understand completely the reasoning in these more difficult sections, or grasp the full signif-icance of the methods employed in this reasoning and the conclusions arrived at.

After a first, or perhaps a second, unsuccessful attempt to understand a theory or a derivation, it is generally unprofitable to hammer away at it; at this stage it is advisable to try other means of achieving clarification. It sometimes helps simply to discontinue working on the problem for a day or two, and then to try again. In the interval your subconscious mind will have had an opportunity to sort things out, and you may well find some or all of the initial difficulty has been resolved. Discussion of your problem with some-one else is also worth trying; the very act of explaining your diffi-culty to him often produces clarification. A third course which is sometimes useful is to read related material; this, too, may well clarify the issues in question.

You cannot claim to understand an argument completely unless, besides being able to follow all its steps, you also appreciate fully why the argument proceeds in just the way it does and the reasons for inventing and introducing any new concepts involved in it. It will be well worth your while to devote some thought to these matters.

Problem Working Success in the working of difficult problems is an art, the nature of which is by no means generally appreciated. It is an art in which you can gain competence only by intelligent application and perseverance.

The best problems have as their objective your training in the realization of situations, the excercise of critical thought, and the intelligent choice of a method of attack. Do not rush at such problems. Take your time; look before you leap. Try to see your problem as a whole, in its proper context. Try to think around corners and to realize all the implications that may be relevant. Above all, try to ascertain the essential operative factor or factors on which the problem hinges. If in doubt, search the text for all references to the kind of situation or process with which the problem is concerned, and then try to determine the relevant factors.

When you have done this, consider the various items of information relevant to the problem which you already know or are given and those you are required to find. In the light of this, ascertain what formula or formulas, if any, the problem requires you to use.

A problem in the working of which you can easily become involved in a long and complicated procedure can often, alter-natively, be solved by a much shorter and more elegant method, depending on the realization of some key factor. It is well worth-while, in such cases, to spend time in an attempt to discover what this is.

Many good problems are, in some sense, trick problems. In contemplating your problem, be on the lookout for any trap that may have been set for you.

Some problems may aptly be described as "series problems." In some of these you are specifically asked a number of questions, such that in finding the answers to the later ones you will need to make use of information obtained in answering one or more of the earlier ones. It is important to recognize such series where they occur. In other, more difficult cases, you will have to work out your own series in the total chain of reasoning. It pays to be methodical in doing this, writing down what you think might be the relevant series leading to the final solution, subjecting your list to critical scrutiny, and then if necessary, trying again.

Numerical data in problems are often given in nonabsolute units and/or in units belonging to different systems. In such cases the first thing you should do is to express all your quantities in mutually consistent units, which, if you have to use a theoretical formula, should be in an absolute system. Usually, all your quantities should be expressed, according to preference or convenience, either in CGS or in MKSA absolute units.

Except, possibly, in the simplest and most straightforward problems, you should keep your mathematics in algebraic form right until the end, only then substituting numerical values for symbols. This will help to keep your working tidy, and in compact, easily surveyable form, thus substantially reducing the risk of error. It will also avoid much unnecessary work often carried out by those who substitute numerical values prematurely, such as, for example, multiplying by 2π in one place and then later dividing by this same quantity.

As stated at the outset, the above comments represent what is, in my view, helpful advice. The extent to which you follow this, and any further advice your professor may have to offer, will be up to you.

CONTENTS*

Foreword   v

Preface   vii

Advice to Students   xi

Volume I
Section I

INTRODUCTION   1

1   GENERAL CONSIDERATIONS   3
     Methods of Scientific Advance   3
     Role of Physical Theory in Explanation and
         Unification   4
     Scope of Physics   5
     Role of Mathematics in Physics   9
     Need for Accuracy in Measurement   9
     Degrees of Accuracy   11
     Graphical Representation of Data   11
     Fundamental and Derived Quantities   12
     Dimensions   17
     Problems   20

Section II

KINEMATICS   25

2   CLASSICAL PARTICLE KINEMATICS   27
     Concept of a Particle   27
     Displacements, Velocities, and Accelerations   28
     Motion of a Particle along a Straight Line   28
     Acceleration of Freely Falling Bodies   31
     Composition and Resolution of Directed Quantities   33
     Uniform Circular Motion   37
     Variation of Gravitational Acceleration with
         Distance of Attracted Body   42
     Simple Harmonic Motion   45
     Simple Harmonic Disturbances   52
     Problems   53

3   WAVE MOTION   59
     Waves along a Stretched String or Rope   60
     Longitudinal Waves   67
     Waves in General   68
     Doppler Effect   69

---

*The complete contents of volumes I and II are included; see page xxi
of the Contents for volume II.

Superposition 71
Problems 80

Section III

DYNAMICS 87

4 THE LAWS OF FORCE AND MASS 89
Frames of Reference 89
Force 91
Relation between Force and Acceleration 95
Mass 98
Systems of Units 100
Absolute Units of Force 100
Complete Definition of Force 103
Momentum 104
Interaction between Bodies 105
Conservation of Momentum 110
Laws of Force and Mass 111
Problems 113

5 CENTRAL FORCES 119
Centripetal and Centrifugal Forces 119
Kepler's Laws of Planetary Motion 121
Law of Universal Gravitation 123
Simple Pendulum 125
Oscillating Mass on Spring 129
Problems 132

6 WORK AND ENERGY 137
Physical Definition of Work 137
Energy 141
New Way of Regarding Force 145
Conservation of Energy 147
Units of Work and Energy 152
Power 152
Problems 152

Section IV

RIGID-BODY MECHANICS 157

7 PRINCIPLES OF ROTATIONAL DYNAMICS 159
Torque 159
Rotational Quantities as Vectors 166
Forces on an Extended Body 172
Center of Mass 173
Values of Moment of Inertia 180
Statics 185
Problems 187

8   ROTATIONAL OSCILLATIONS AND
        GYROSCOPIC ACTION   193

     PART I: ROTATIONAL OSCILLATIONS   193
     General Theory   193
     Rigid-Body Pendulum or Compound Pendulum   195
     Determination of the Gravitational Constant G 196

     PART II: GYROSCOPIC ACTION   198
     General Theory   198
     Applications and Consequences   202
     Problems   205

Section V

CONTINUUM MECHANICS   **209**

9   HYDROSTATICS AND HYDRODYNAMICS   211

     PART I: HYDROSTATICS   211
     Density   211
     Pressure   213
     Archimedes' Principle   222
     Pressure–Volume Relation for Gases   224
     Bulk Modulus of Elasticity   226

     PART II: HYDRODYNAMICS   228
     Types of Flow   228
     Transverse Velocity Gradient   228
     Longitudinal Velocity Gradient   230
     Viscosity   230
     Bernoulli's Theorem   236
     Problems   240

10   DEFORMATIONS OF SOLIDS   249

     PART I: ELASTIC DEFORMATIONS   249
     Three Elastic Moduli   249
     Relation among the Elastic Moduli   255

     PART II: PLASTIC DEFORMATIONS   256
     Plastic Flow and Work Hardening   256
     Structural Changes   258

     PART III: FRICTION   266
     Main Phenomena   267
     Laws of Friction   269
     Modern Theory of Friction   272
     Problems   275

11   WAVES IN A CONTINUUM   279
     Wave Propagation along a Stretched String   279

Wave Propagation in a Fluid   281
Velocity of Waves in a Gas   283
Identification with Sound   285
Velocities in Liquids and Solids   286
Effects of Wind and Temperature Gradients on
   Audibility   289
Reflection   292
Pitch and Frequency   293
Resonance   295
Quality   298
Problems   300

12   LIQUID SURFACES   305
Condition of Surface Molecules   306
Surface Energy and Surface Tension   308
Drops and Bubbles   311
Capillarity   313
Measurement of Surface Tension   316
Stability of Liquid Films   318
Applications and Consequences of Surface Tension   319
Problems   322

Section VI

HEAT   **327**

13   TEMPERATURE AND ITS MEASUREMENT   329
Definition of Temperature   329
Standard Celsius Scale   334
Gas Thermometers   337
Kelvin Scale of Temperature   340
Calibration of Other Than Gas Thermometers   342
Practical Thermometers   343
Problems   354

14   HEAT AND THERMAL TRANSPORT   357
Thermal Expansion   357
Heat as a Physical Quantity   358
Specific and Latent Heats   362
Dulong and Petit's Law   364
Heat Transfer   366
Problems   371

15   THE KINETIC THEORY OF GASES   375
Equation of State for a Gas   375
Molecular Picture of a Gas   378
Behavior of a Gas in Terms of Molecules   382
Problems   398

16   THERMAL BEHAVIOR OF MATERIALS   401

Dulong and Petit's Law   401

Changes of State   402

Vapor Pressure   407

Triple Point   411

Unsaturated and Supersaturated Vapors   412

Vapor Pressure and Curvature of Surface   413

Atmospheric Precipitation   416

Boiling   421

Complete Isothermals   422

Problems   424

17   THE FIRST LAW OF THERMODYNAMICS   425

Heat and Conservation of Energy   426

Heat and Internal Energy   430

Joule–Kelvin Experiment   431

Isothermal Volume Change of a Perfect Gas   433

Adiabatic Volume Change   434

Variation of Temperature with Altitude in the Troposphere   438

Liquefaction of Gases   441

Carnot's Cycle   445

Problems   448

18   THE SECOND LAW OF THERMODYNAMICS   451

Degradation of Energy   451

Statement of the Second Law   452

Evidence for the Truth of the Second Law   455

Maxwell's Demon   455

Macroworlds and Microworlds   456

Reversible and Irreversible Processes   458

Carnot's Principle   460

Efficiencies of Practical Heat Engines   463

Thermodynamic Scale of Temperature   464

Clausius–Clapeyron Equation   466

Variation of Vapor Pressure with Temperature   469

Effect of Pressure on the Freezing Point   471

Surface Tension and Surface Energy   473

Problems   476

CONTENTS

Volume II
Section VII

ELECTRIC CHARGES AND CURRENTS   479

19   BASIC PHENOMENA OF ELECTROSTATICS   481
     Electrification by Friction and by Induction   481
     Insulators and Conductors   483
     Classification of Charges   484
     Nature of the Conduction Process   489
     Electroscope   489
     Explanation of Electrostatic Induction   491
     Coulomb's Law   493
     MKSA Unit of Charge   495
     Problems   497

20   ELECTROSTATIC FIELD THEORY   501
     Concept of Electric Field   501
     Further Consideration of Coulomb's Law   507
     Gauss's Theorem   509
     Potential   523
     Problems   530

21   ELECTROSTATIC DEVICES   533
     Van de Graaff Generator   533
     Thunderstorm Cell   538
     Electrostatic Voltmeters   541
     Capacitors (Condensers)   542
     Dielectrics   547
     Capacitors in Parallel and in Series   550
     Energy of a Charged Capacitor   551
     Force on a Charged Surface   552
     Attracted-Disk Electrometer   555
     Problems   557

22   BASIC ELECTRONICS I   563
     Positive and Negative Electricity in Metals   563
     Electrons   568
     Explanation of Surface Potential Barrier   568
     Thermionic Emission   570
     Electronic Charge-to-Mass-Ratio   582
     Millikan's Oil-Drop Experiment   583

Information Deduced from Value of Electronic
Charge 586
Problems 589

23 BASIC ELECTRONICS II 593
Photoelectric Emissions 593
Quantum Theory 597
Photoelectric Cathodes 602
Cathode Rays 603
X Rays 607
Thermionic Triode 610
Modern Cathode-Ray Tube 612
Problems 614

24 DISCHARGES IN GASES 617
Photoionization 618
Ionization by Collision 620
Secondary Electron Emission 624
Processes Involved in Gas Discharges 627
Problems 634

25 DIRECT-CURRENT THEORY 637
Measurement of Current Strength 637
Moving-Coil Galvanometer 640
Potential Difference in Current Electricity 647
Ohm's Law 648
Charge and Discharge of Capacitor in an RC Circuit 651
Conductors in Series and Parallel 656
Electromotive Force 659
Cells in Series and in Parallel 665
Kirchhoff's Laws 666
Problems 669

26 DIRECT-CURRENT INSTRUMENTATION 675
Comparison of Resistances 675
Insulation in Static and Current Electricity 681
Voltmeters 682
Ammeters 684
Ammeter–Voltmeter Combinations 686
Potentiometer 687
Potential Divider 693
Problems 694

27 ELECTROCHEMISTRY 697
Electrolytic Processes 697

Reason for Dissociation 700
Electrolysis of Water 701
Pt–H_2SO_4–Pt Cell as a Generator 701
Primary and Secondary Cells 704
Thermodynamical Theory of Reversible Cells 714
Problems 717

28 THERMOELECTRICITY 719
Seebeck Effect 719
Peltier Effect 721
Application of Thermodynamics 723
Thomson Effect 724
Thermoelectric Power 727
Physical Explanations of Peltier and Thomson
   Effects 729
Problems 731

Section VIII

MAGNETISM 733

29 MAGNETIC EFFECTS OF CURRENTS 735
Magnetic Field and Its Effects 736
Field Produced by a Current Element 749
Force between Current Elements 754
Definition of the Ampere 755
Problems 757

30 MAGNETIC FIELDS 761
Magnetic Permeability 761
Field Due to Infinitely Long Straight Wire 762
Field on Axis of Flat Circular Coil 763
Field Due to Current in Solenoid 764
Field Due to Small Circuit or Coil 766
Magnetic Moment 769
Field Due to Large Circuit 770
Ampère's Line-Integral Theorem 771
Geomagnetism 782
Action of Magnetic Field on Moving Electrons 787
Problems 796

31 MAGNETIC PROPERTIES OF MATERIALS 801
Magnetization and Demagnetization 802
Ferromagnetic Materials 806
Weber–Ewing Theory of Magnetism 808

Theory of Magnetic Domains 810
Curie Temperature 816
Paramagnetism 816
Diamagnetism 820
Equivalence of Bar Magnet and Current-Carrying Solenoid 820
Electromagnets 822
Forces between Magnets 822
Problems 823

32 ELECTROMAGNETIC INDUCTION 825
Law of Electromagnetic Induction 825
Self-Induction 831
Energy of Current-Carrying Inductor 832
Growth and Decay of Current in an Inductive Circuit 833
Mutual Induction 836
Measurement of B–H Characteristic 837
Generators and Motors 839
Damping of Galvanometer 847
Problems 850

33 AC CIRCUITS AND INSTRUMENTS 857

PART I: PEAK, RMS, AND "MEAN" VALUES OF CURRENTS AND VOLTAGES 857
Relation between Peak and RMS Values for Sinusoidal Alternating Current 859
Relation between RMS and "Mean" Values for Sinusoidal Alternating Current 860

PART II: AC MEASURING INSTRUMENTS 861
Rectifier Systems 862
Electrodynamometer 865
Moving-Iron Instruments 867
Hot-Wire Instruments 869
Electrostatic Voltmeters 869

PART III: CURRENT-VOLTAGE RELATIONS IN AC CIRCUITS 870
RLC Circuit 870
Use of Phasors in AC Theory 872
Resonance 878
Problems 881

34   ALTERNATING VOLTAGE
         TRANSFORMATIONS   887

     PART I: TRANSFORMERS   887
     Ideal Transformer   888
     Transformer Design   898
     Autotransformer   899
     Induction Heating   900

     PART II: USE OF THERMIONIC TRIODE
         IN AC CIRCUITRY   901
     Voltage Amplification   901
     LC Oscillator   904
     Problems   905

                                          Section IX

LIGHT   909

35   THE PROPAGATION AND VELOCITY
         OF LIGHT   911
     Concept of a Ray   912
     Huygens's Principle   912
     Laws of Reflection   916
     Laws of Refraction   917
     Total Internal Reflection   921
     Determinations of the Velocity of Light   923
     Problems   928

36   MIRRORS AND PRISMS   931
     Real and Virtual Images and Objects   931
     Reflection by a Spherical Mirror   933
     Aplanatic Reflecting Surfaces   935
     Sign Convention for Algebraical Formulas   936
     Spherical-Mirror Formula   938
     Images of Extended Objects   941
     Refraction by a Prism   943
     Dispersive Power   945
     Problems   947

37   LENSES AND OPTICAL INSTRUMENTS   949
     Refraction at a Spherical Surface   949
     Thin Lens   952
     Two Thin Lenses in Contact   956

Applications of Mirror and Lens Formulas ... 957
Geometrical Constructions ... 959
Optical Instruments ... 961
Problems ... 970

38   INTERFERENCE AND DIFFRACTION I ... 975
Interference Produced by Two Point Sources ... 975
Phasor-Diagram Method of Treatment ... 981
Actual Light Sources ... 986
Effects of Finite Widths and Lengths of Slits ... 988
Problems ... 989

39   INTERFERENCE AND DIFFRACTION II ... 993
Diffraction Grating ... 993
Babinet's Principle ... 997
Fraunhofer Diffraction by a Single Slit ... 998
Fraunhofer Diffraction by a Circular Aperture ... 1001
Interference Due to Thin Films ... 1004
Haidinger's Fringes ... 1013
Michelson's Interferometer ... 1015
X-Ray Spectroscopy ... 1017
Problems ... 1023

40   POLARIZATION ... 1027
Polarized and "Unpolarized" Light ... 1027
Polarization by Reflection ... 1027
Polarization by Scattering ... 1028
Double Refraction ... 1031
Dichroism ... 1036
Production of Circularly and Elliptically
   Polarized Light ... 1037
Photoelasticity ... 1039
Kerr Effects ... 1041
Optical Rotation ... 1042
Problems ... 1044

APPENDIX

GREEK ALPHABET ... A-1

CONVERSION FACTORS ... A-2

FUNDAMENTAL PHYSICAL CONSTANTS ... A-3

NATURAL TRIGONOMETRIC FUNCTIONS ... A-4

COMMON LOGARITHMS — A-8

DEGREES OF ACCURACY — A-11
Estimation of Relative Error — A-11
Relative Error of a Mean — A-14
Errors and Mistakes — A-16
Error of Calculated Quantity in Terms of Errors
    of Separate Items — A-16

ANSWERS TO SELECTED PROBLEMS — A-19

INDEX — A-97

PHYSICS

Volume I: Mechanics and Heat

**Volume II: Electricity, Magnetism
and Optics**

SECTION I

INTRODUCTION

1

GENERAL CONSIDERATIONS

Methods of Scientific Advance

All science is ultimately based on the observation of phenomena and reasoning applied to what is observed. On the basis of this observation and reasoning there is gradually built up a body of organized knowledge, a **science.**

SCIENTIFIC LAWS The first step in organizing observed phenomena into a science is to formulate certain **laws of nature.** It should be particularly noted that such laws are defined simply as generalized statements of experience. A law is sometimes also called a **principle.** It is a statement of a general truth, that is, something that is in complete accord with all experience. To be useful it must also summarize and generalize experience within a certain category. For example, if we observe that at each of a number of temperatures, arbitrarily chosen, the volume of a certain sample of nitrogen is inversely proportional to the pressure exerted on it by its surroundings, and that the same thing is true of oxygen, hydrogen, argon, carbon dioxide, and all other gases, it is useful to summarize this experience in the statement that at any constant temperature the volume of a given sample of gas is inversely proportional to its pressure. This is the statement known as Boyle's law. Here, as in all cases, a law summarizes a certain type of behavior or relationship. This "boiling down" or classification of experience is a necessary preliminary to the building up of our complete organized body of knowledge.

A law, in science, allows no exceptions. Should an exception be found, and its validity firmly established, the law must be restated, and the new statement then supersedes the old.

Science is not static. It is a living, growing thing, which, in the process of its growth, is constantly subject to revision, not only in matters of detail (laws), but even, occasionally, in general plan. And each item of discovery of which a science is built represents the fruits of original research; this is the ultimate origin and basis of every statement that appears in a scientific book or is made by a teacher.

There are exact laws and approximate laws. An example of the former is the inverse-square law of gravitation (see Chapters 2 and 5). An example of an approximate law is that just considered— Boyle's law. This is generally sufficiently well understood not to need specific statement. Quite strictly, a more acceptable statement of Boyle's law would be that at any constant temperature the volume of a given mass of gas is *approximately* inversely proportional to its pressure. How closely the variation approximates to inverse proportionality depends on the nature of the gas, the temperature, and the range of pressures covered.

THEORIES AND HYPOTHESES After having established our laws, the next step is to coordinate them with one another, building them into a thought structure known as a **theory.** In a theory we explain certain things in terms of others, and in its development we are obliged to make use of certain ideas known as **abstractions.** Examples of these are mass, force, temperature, heat, electric field, and potential. These things do not always have the same meanings in physics as in everyday language, and in order to be quite specific it is necessary to define what we mean by them.

The development of a theory usually takes some time, depending on a process which, by its nature, cannot be hurried. This is the process known as **inductive reasoning,** which has been rather aptly described as "inspired guesswork." After pondering a problem for a time, we form an idea of what might be the explanation of the phenomenon in question. Assuming our guess to be correct, we then make certain logical deductions, leading to predictions of new kinds of phenomena that we should be able to observe. We then see if, in fact, these predicted phenomena are observable. If not, we think again. If we do observe them, such observation constitutes strong evidence for the correctness of our guess. Finally, if after exhaustive tests of this kind the assumed explanation proves to be consistent with all the relevant observed phenomena, we accept it as true, by the only criterion of truth science can recognize. Until this stage is reached we do not accept the assumed explanation as necessarily true. It is still on probation, and we do not yet call it a theory, only a **hypothesis.** Only when we are finally satisfied of its truth beyond reasonable doubt is the hypothesis raised to the status and dignity of a theory.

Role of Physical Theory in Explanation and Unification

In physics, as in other sciences, we proceed from the known to the unknown, from the apparently obvious to the unfamiliar and intriguing. In doing this, we must of necessity use the concepts formed in our ordinary lives. Although these are not likely to be the most suitable concepts on which to base the structure of theoretical physics, they are all we have—perhaps all we *can* have. Also, we tend—at least in the first instance—to employ the kinds of reasoning characteristic of our everyday thinking, in the hope that these may serve to explain all the phenomena in which we are interested. In practice we find that although many of the theories we develop along these lines justify this hope, there are cases where "commonsense" modes of thought, however logical, fail us completely. This often happens when phenomena outside the range of ordinary observation are considered. We must then be prepared to relax commonsense arguments, replacing these, if possible, with new kinds of reasoning that can be accommodated in a self-consistent and logical whole, however strange this may seem. Such self-consistency is all we are entitled to demand; the fulfillment of this requirement should completely satisfy us.

Why is it that theoretical physics as we know it today has had to resort to these unfamiliar lines of reasoning? It is because we are gross, macroscopic, slow-moving beings who have no direct experience of the microscopic world of atoms, electrons, and so on, or of objects moving with very high velocities. All our thinking has developed in the environment of the macroscopic and slow-moving world we know. Although within this world it serves us admirably, we cannot expect that it will be equally effective in explaining processes in the world of atoms and subatomic particles or phenomena involving very high velocities.

The usefulness of a theory is not necessarily confined to the phenomenon or group of phenomena for whose explanation it was developed; it often also plays an important role in the development of theories in what appear to be totally different fields of physics. In this way, physical theory constitutes a valuable unifying element, revealing the interrelations among different branches of the subject.

Scope of Physics

Logically, as well as traditionally, the study of physics begins with the consideration of those areas with which we, as thinking beings, have had our longest experience—the various branches of mechanics. In the study of mechanics, we normally deal with matter on a macroscopic scale, but we are nevertheless always conscious of the fact that all gross matter is made up of atoms. Accordingly, we naturally speculate on whether the laws of mechanics as found for ordinary samples of matter might not also apply, at least to some extent, to these atoms, or even to subatomic particles. In the absence of direct experience of atoms we have no right to *assume* that they do, except tentatively, as part of the process of inductive reasoning. The real test of the validity of the ordinary laws of mechanics comes only when we consider processes occurring in atoms and study a branch of physics developed during the last few decades, quantum mechanics. Meanwhile, even in the study of macroscopic mechanics, it is useful on occasion to invoke the concept of atoms and their mechanical interactions in a general way, particularly in the theory of elastic and plastic deformations, and also in the theory of surface tension, which may be considered to belong peripherally to mechanics.

Man has had experience of thermal effects, such as variations of temperature, freezing, melting, and evaporation, for as long as he has of mechanical phenomena, but it was not until the nineteenth century that the existence of a connection between heat and mechanics was recognized. The realization that heat is a form of energy at once suggested an interpretation of heat in terms of molecular motion, and this idea later developed into the kinetic theory of gases, the theory of the atomic heats of solid elements (Dulong and Petit's law), a mechanistic interpretation of temperature, and a molecular explanation of the processes of fusion and evaporation. It also led naturally to the development of the wider disciplines of thermodynamics and statistical mechanics, which are now recognized as different aspects of the same general field of physics. Both of these disciplines have been extraordinarily fruitful in widening our understanding of thermal phenomena.

Historically, the study of electricity began with electrostatics. At the very outset this was based on the observation of the exertion of forces between charges, so establishing a link with mechanics. This link was later strengthened with the development of the concept of potential, defined as potential energy per unit charge. Later still, positive and negative charges were shown to consist of defects and excesses, respectively, of carriers which we now know as electrons and which were found to have mass as well as charge. Finally, electric currents in metals were established as consisting of a flow of electrons within them, and free electrons in a vacuum were shown to respond to electric and magnetic forces according to the laws of mechanics. It was also found that the most convenient way to obtain free electrons in a vacuum was to "boil" them out of a hot metal and, furthermore, that (1) heat was developed by the flow of electrons along a wire in response to a potential gradient, and

(2) electrons constituted the main vehicle for thermal conduction in metals (Wiedemann–Franz law). In these and many other ways, mechanics, heat, and electricity were brought together in a unified physical theory. Somewhat more recently, magnetic fields have been shown to be due to electricity in motion, and so magnetism also has been included in this synthesis.

The application of energy considerations in which a current-carrying conductor moves transversely to a magnetic field leads to the law of electromagnetic induction, a law that was originally established empirically by Faraday, Lenz, and Neumann. Also, an important theorem connecting what is known as the line integral of the magnetic field around a closed path (see Chapter 30) with the current flowing through this was derived from known electromagnetic relations by Ampère. An imaginative mathematical combination of the law of electromagnetic induction with Ampère's line-integral theorem by Maxwell in the nineteenth century led to the prediction of electromagnetic waves in a vacuum traveling with a velocity of 3×10^8 m/sec. This was precisely the velocity with which Fizeau and others had found light to travel. And light was already known, from the evidence of optical interference, to be propagated as waves of some sort. Thus Maxwell's work brought light into the orbit of electromagnetic theory. The electromagnetic wave theory of light so established appeared eminently satisfactory until in 1900 a theoretical investigation by Max Planck of the radiation emitted from hot bodies, and the discovery of the photoelectric effect, both showed conclusively that electromagnetic radiation also has certain properties characteristic of particles. As is explained in books dealing with some of the more sophisticated developments in physics that have taken place in the present century, these radiation particles, or "quanta," have now been shown to have momentum, and to exert pressure on surfaces on which they impinge, establishing a theoretical link between light and mechanics.

The discovery of the dual (wave and particle) nature of radiation was the first great shock sustained by classical physics, a shock from which it has taken several decades to recover. The process of "recovery" consisted of a theoretical reconciliation of the two aspects of radiation. Meanwhile, the fundamental relation found between the energy of a quantum and the frequency of the associated wave train was used by Bohr, Sommerfeld, and others to set up a successful model of the atom which was able to account for all the main features of the line spectra characteristic of different kinds of atoms. The processes envisaged in the absorption and emission of these spectral lines were definitely "nonclassical"; they differed

7

radically from any processes directly observable in the macroscopic world of our ordinary experience.

As a counterpart of the particle aspect of electromagnetic waves came another discovery no less strange, during the 1920s: that particles in motion also have a wave aspect. This discovery led to the development of what is known as "wave mechanics," which, together with other theoretical developments, constitutes the considerable general field of quantum mechanics. This has had considerable success, first in disposing of what had appeared to be a basic inner contradiction in the theory of the Bohr–Sommerfeld atom, and later in many other branches of physics, notably in the theory of solids.

Within five years of the first shock sustained by the structure of physics in the birth of the quantum theory, classical physics was again shaken to its foundations by the development in 1905 of Einstein's special theory of relativity. Einstein showed that as a logical consequence of the absence of any unique frame of reference against which to measure velocities, the concepts of absolute length and of absolute time must be abandoned. He also showed that mass and energy are not independent quantities, separately conserved, but that there is a definite relation between them, and that the two separate conservation laws must be replaced by a single law of conservation of mass–energy. Einstein's mass–energy relation has proved to be of the most fundamental importance in many branches of physics, particularly in accounting for nuclear transmutations such as occur in natural and artificial radioactivity and in the generation of "atomic energy."

The intensive study of nuclear physics within the last few decades has led to the discovery of a surprisingly large number of "elementary particles" in addition to the electron and the proton (the nucleus of the hydrogen atom). Not only are the various families of elementary particles so far discovered of great interest in themselves, but the philosophical implications of particle physics are profound and far-reaching.

As this brief survey indicates, the emphasis in theoretical physics has tended to move away from the familiar macroscopic world of everyday experience toward the unfamiliar world of atoms and subatomic particles. And even where large aggregates of matter are still considered, there is often a fundamental involvement, statistically, of their constituent particles. We have seen also how in the development of physical theory one thing leads to another, with the eventual production of a unified structure of theoretical physics in which there are many connecting links between apparently different branches.

Although in the present book we shall now and again have occasion to refer to modern theoretical developments in physics, we shall in the main confine ourselves to a study of the classical foundations of the subject. Without a reasonable mastery of classical physics a student is in no position to understand and appreciate the more difficult concepts and reasoning of modern physical theories. However, with an adequate background of classical physics, such as this book provides, he may undertake with confidence an introductory course of study in modern physics, covering such subjects as electromagnetic radiation, special relativity, quantum mechanics, atomic and nuclear physics, and the physics of elementary particles; he will find this an intellectually rewarding and exciting experience.

Role of Mathematics in Physics

We have seen that an essential ingredient of inductive reasoning is deduction. Deduction is familiar to us in the various processes of mathematics, which are used extensively in physics, and without which we should not be able to progress very far. Mathematics may be regarded as a highly organized form of reasoning, employing certain agreed-upon symbols and conventions to the end of vastly improving on the reasoning powers with which we are endowed by nature. It represents the same kind of extension of the unaided reasoning powers of the human brain as hammer, chisel, screwdriver, and machine tools are an extension of the powers of one's bare hands, or as the written or printed word is an extension of memory.

Need for Accuracy in Measurement

Because mathematics deals with quantities, we cannot have any basis for its employment until we have measured our quantities and expressed them in numerical form. Only in this way can we make physics an exact science. In many cases a fairly rough measurement suffices, whereas in others we need the greatest accuracy it is possible to attain. It all depends on what we need the measurement for. Where we need great accuracy it is not necessarily for accuracy's sake, in the narrow sense. Metrology, the science of accurate measurement, is not everybody's meat, and appeals only to certain types

9

of mind. For others, the interest in accurate measurement attaches not so much to the measurement itself as to the information it is capable of yielding concerning the nature of some physical process. Thus the quantitative becomes the instrument for furthering our knowledge of the qualitative, and it is generally to the latter that the real interest attaches.

For example, we may make a series of measurements with the greatest possible accuracy, to see how closely they fit a certain hypothesis. If the fit is found not to be satisfactory, it is indicated either that this hypothesis should be replaced by some other, or that the existing hypothesis might still serve if amended by taking into account certain factors whose neglect has been responsible for the discrepancies. In other cases it is possible, on the basis of a definite hypothesis concerning the mechanism of a process, to predict a precise value for a quantity depending on this process. If, upon measurement of this quantity, it is found to agree with the value predicted, we have the strongest possible confirmation of the correctness of the supposed mechanism.

An example of the first of these uses of accurate measurement is concerned with Boyle's law. As we have already noted, sufficiently accurate measurements of pressure and volume show that this is not an exact but only an approximate law. But this is not in itself very interesting. The approximate inverse proportionality is in accord with the predictions of an approximate version of the kinetic theory of gases, in which the molecules are supposed to have negligible volume compared with the volume occupied by the gas, to travel in straight lines between collisions, and to rebound without loss of kinetic energy and without the exertion of appreciable mutual forces except during collisions.

However, this picture could obviously be an oversimplification of the true state of affairs, at least with regard to the first and last items listed. Accordingly, in 1873 J. D. van der Waals worked out a more elaborate version of the kinetic theory of gases, in which he succeeded in making due allowance for the finite sizes of the molecules and the mutual forces exerted between them at times other than during collisions. No measurements were available for either of these items, so this allowance for their existence had to take the form of an elaboration of the equation connecting pressure and volume, the effects of the mutual forces between the molecules and their finite sizes being represented by new algebraic symbols, a and b, in this equation (see Chapter 15). By substituting suitable values for these quantities, it was now found possible to match pressure–volume curves obtained over a wide range of experimental conditions very faithfully indeed. Not only did this indicate the essential correctness of van der Waals's elaborated kinetic theory, but the numerical values found for the constants yielded valuable information, since confirmed by other methods, concerning the sizes of the molecules of different kinds of gas and the forces they exert on each other. It has also proved possible, by using van der Waals's equation with the values of a and b substituted, to account quantitatively, at least reasonably well, for an interesting property of a gas known as its inversion temperature (see Chapter 17). Thus accurate measurement enables us to obtain a much deeper insight into the processes occurring in gases than could otherwise be achieved.

An example of the second of the uses of accurate measurement referred to above concerns the mechanism of the propagation of light. This has already been outlined in our general discussion of the scope of physics. It is considered in detail in Volume III.

Degrees of Accuracy

The only kind of quantitative determination that can be made with absolute accuracy is one of counting a number of separate objects, for example bricks in a pile, where the question of fractional quantities does not arise. Apart from such cases, no measurement is ever completely accurate; there is always "a certain uncertainty," depending on the limitations of the instruments used and on the skill of the observer. And, of course, such error as is incurred in the making of a measurement must infect the result of any calculation in which this measurement enters as one of the data. For a detailed discussion of errors the reader is referred to the Appendix.

Graphical Representation of Data

The great usefulness of representing physical data graphically is often not appreciated as much as it deserves to be. Graphical representation can be a most powerful integrative agency, enabling us to see and understand at a glance information that cannot be nearly so well or so compactly conveyed in tabular form or in words. It is also an important aid in certain types of reasoning about physical phenomena. Graphs lend themselves as do no other means to depicting trends and relationships, and it is to these rather than to mere isolated quantities that nearly all the interest in physics attaches.

Graphs may be either curved or straight. Both have their special fields of usefulness, but straight-line graphs have a particular value of their own in the investigation of quantitative relationships. Accordingly, it is in many cases useful to search for a manner of graphing in which the data can be represented by a straight line. For example, if we plot the volume of a gas at constant temperature against the pressure, we obtain a curve, but the plot of the reciprocal of the volume against the pressure, or that of the product of pressure and volume against the pressure, comes out as a straight line. Again, in a study of the formation of images by a lens, if we plot image distances against object distances, measuring always from the lens, we find that the points representing the data lie on a curve, but if we plot the reciprocals of image distances against the reciprocals of object distances the points all lie on a straight line. Such straight-line graphs lend themselves much more directly to algebraic representation than do curves, and when we have found an algebraic expression to represent our data we are often well on the way toward their physical interpretation. Several examples of this procedure of experimental graphing and of the algebraic representation of rectilinear plots will appear in the following pages.

Fundamental and Derived Quantities

In discussing the great variety of physical phenomena, we shall find that we need a correspondingly large number of quantitative concepts, or "quantities," in terms of which to try to correlate and explain them. Examples of these are force, elastic stress, quantity of heat, luminous intensity, and electric potential. These are only a few, chosen from among a great number. Such quantities are not independent of one another, but are interrelated. In conformity with our basic philosophical principle of exercising the utmost economy of ideas, therefore, we try to find from among them those that are the most fundamental, those in terms of which we may express, as simply as possible, all the others. We find that we cannot manage with less than four, and the four that are chosen by present-day physicists are length, time, mass, and electric current. Although there can be no guarantee that future generations of physicists will necessarily make the same choice, these four quantities do appear to us in the present state of knowledge to be the simplest kinds of "bricks" with which to build our structure of theoretical physics.

In addition to these four fundamental *quantities*, we shall also have need of a certain number of *qualitative* concepts, and these will be introduced as we require them. Sometimes these are combined with quantities to form modified quantities. Thus if a length is given a direction it becomes a displacement, and if a speed is given a direction it becomes a velocity, direction being in both cases the qualitative concept in question. These things will be discussed more fully in their proper context.

LENGTH Length, time, mass, and electric current are not all equally simple and obvious quantities. The simplest of these is length. We cannot, of course, "explain" length in terms of anything simpler. If we could, length would cease to be a fundamental quantity. We feel nevertheless that there is nothing particularly difficult in the idea, and we are all familiar with some of the processes used for measuring length in terms of a chosen standard. And, combining length with itself, we have three simple derived quantities: angle, area, and volume.

The units of length most commonly employed in scientific work are the centimeter and the meter, the former being the one-hundredth part of the latter. The meter was formerly defined as the distance between two fine lines engraved on gold plugs near the ends of a bar of platinum–iridium alloy, known as the standard meter, kept at the International Bureau of Weights and Measures at Sèvres, near Paris, when the bar is at the temperature of melting ice. The present-day definition is 1,650,763.73 vacuum wavelengths of the orange spectral line in the spectrum of the krypton isotope of atomic weight 86; this is the experimentally determined number of wavelengths of this line between the scratches on the standard meter. There is a reliable technique for counting large numbers of such wavelengths.

The unit of length in the British system, the foot, is based on what is now, both in the United Kingdom and in the United States, the legal yard. This is defined as 0.9144 m, the corresponding inch–centimeter relation being that 1 in. is exactly equal to 2.54 cm.

TIME It does not seem possible to think of time except in terms of a sequence of events. This being so, it is only natural that we should define equal intervals of time as those that are "occupied" by equal events, the criterion of equality of events being perforce a negative one—that we are unable to think of any reason why they should not be equal. Thus we might regard as equal intervals of time those "marked off" by successive transits of the pendulum of a clock in a given direction across its midpoint of swing, or as those corresponding to complete rotations of the earth about its axis.

Because the distances of the stars are incomparably greater than the dimensions of the earth's orbit about the sun, successive

13

transits of a selected distant star across the meridian at any point on the earth's surface may be taken as corresponding to complete rotations of the earth about its axis. The interval between two successive such transits is known as a **sidereal day,** and, because the earth rotates in the same direction as it revolves about the sun in describing its orbit, a sidereal day is somewhat shorter than the mean solar day—by about 3 min, 56 sec. The number of sidereal days in 1 year is one greater than the number of solar days.

It is now possible to time sidereal days by extremely accurate clocks, and such timing has revealed that there is an irregular variation in the length of the sidereal day of the order of 1 part in 10^8. Actually this is not altogether surprising, in view of the considerable proportion of the earth that is fluid. In the surface fluid— the oceans and the atmosphere—we have ocean currents and wind systems and seasonal variations in the deposits of ice and snow in regions of high latitude. Also, it is known from the evidence of earthquakes that part of the deep interior of the earth is liquid, and it is possible that there may be disturbances in this. All such effects would be expected to react on the solid part of the earth; they would be quite capable of affecting its rate of rotation.

There is also a long-term effect associated with tides. By converting some of the rotational kinetic energy of the earth into heat, these must tend to slow down the rate of rotation, and it has been estimated that due to this cause the sidereal day increases by about $\frac{1}{1000}$ second per century. There are, however, other factors simultaneously operating, and it is in some doubt whether the overall effect is a retardation or an acceleration of the earth's rate of spin.

From these considerations it must be concluded that the rotating earth can no longer be regarded as a satisfactory clock; the second, 1/86,400 of a mean solar day, is not a perfectly definite quantity. In the same way as the old definition of the meter, as the distance between two scratches on a bar of metal, has been superseded by a definition in terms of an optical wavelength, so the present definition of the unit of time will eventually have to be revised. It seems likely that in the future the unit of time will be defined in terms of some natural atomic or molecular period. Highly reliable "atomic clocks" have already been developed, and could be used for a redefinition of the unit of time.

The first atomic clock to be constructed was based on the vibration of the nitrogen atom in a molecule of ammonia back and forth through the plane of the three hydrogen atoms. The frequency of this vibration is about 2.387×10^{10} per second. If a suitable electric circuit is adjusted to have a natural frequency of oscillation closely approximating this value, it is possible to arrange for it to be controlled by the vibrations of the molecules in a beam of ammonia gas passed through the apparatus, so that it oscillates with a frequency agreeing to a high degree of accuracy with that of the ammonia molecules. An ammonia clock run under carefully controlled conditions has been found to maintain its frequency to within ± 1 in 10^{10} over periods of several hours. In 1948 it was proposed by an international conference on weights and measures that this type of clock be used as a basis for time measurements.

A rival to the ammonia clock is the cesium clock, whose frequency is that of a selected very fine line in the spectrum of cesium. This frequency is $9,192,631,830 \pm 10$ Hz (cycles per second). The best cesium clock now in existence is one operated at the National Physical Laboratory at Teddington, near London, England. If maintained in its least satisfactory condition of operation this clock would give an error of 1 second in just over 31 years. Under actual operating conditions the accuracy must be substantially better than this.

MASS This is an alternative name for **inertia,** the difficulty of changing the state of motion of a body. We shall see in Chapter 4 how it is possible to achieve a quantitative measure of this "difficulty."

ELECTRIC CURRENT We shall deal with this in Chapter 25.

FUNDAMENTAL QUANTITIES IN CLASSICAL AND RELATIVISTIC PHYSICS Prior to 1905 the fundamental quantities of physics were believed to be "absolute," in a sense that will become clear in the light of a few examples. From all ordinary experience, the linear dimensions of a given body appear to be totally independent of its state of motion; they appear to be the same whether the body is at rest or in motion relative to the person observing it. The same seems to apply to its mass; this, like the linear dimensions, appears to be a unique property of the body. Finally, the passage of time seems to be unaffected by state of motion. Classical kinematics and dynamics, which we shall study in Chapters 2 to 12, are based on the assumption of the absoluteness of these quantities, and, as we shall see, the equations developed in these disciplines serve us very well indeed in all the ordinary affairs of everyday life; they appear to describe observed phenomena with complete accuracy. Notwithstanding this, Albert Einstein in 1905 showed the assumption of the absoluteness of these quantities to be untenable; he showed that the classical equations must be replaced by less simple equations, the equations of what is known as "special relativity." However, this does not mean that we have no further use for the classical equations. These are limiting cases of the more sophisticated equations of special relativity, which are valid to an extremely high degree of accuracy in all the ordinary situations of everyday life, where the velocities with which we are concerned are small compared with the velocity of light.

In the following pages we shall begin our study of physics with a consideration of classical kinematics and dynamics. Only when those disciplines have been mastered shall we be properly equipped to understand the more recondite and philosophically sophisticated reasoning that led Einstein to his formulation of the theory of special relativity.

SIMPLE DERIVED QUANTITIES Examples of derived quantities that may profitably be considered briefly at this stage are angle, area, volume, frequency, and speed.

Let us begin with the concept and scientific measure of an angle. If in Fig. 1.1 s is the arc of a circle centered on O which lies between the lines OA and OB, and r is the radius of this circle, then, for a given r, the angle θ between these lines is something we think of as proportional to s. And, for a given angle θ, s is itself proportional to r. Hence the ratio s/r is independent of r; it is simply proportional to θ, and may be taken as the natural measure of this angle. The angle for which this ratio is 1 (that is, for which s is equal to r) is the natural unit of angle, which we call the **radian** (rad). And because for a full circle s is equal to $2\pi r$, the corresponding angle, 360°, is, in terms of the natural unit, 2π rad. A radian is equal to $360°/2\pi$, or $57°17'45.6''$.

Many quantitative relations in physics involving angles can be derived and expressed much more simply in terms of the natural unit, the radian, than in degrees. It is for this reason that the radian is used so much as a unit in theoretical work.

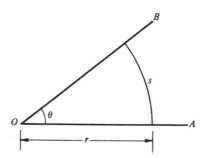

FIGURE 1.1 An angle considered in terms of the arc of a circle and its radius.

In contrast to an angle, a measure of which is a length (arc) divided by a length (radius), an area is expressible as the product of two lengths, and corresponding units of area are square centimeters, square meters, square feet, square miles, and so on. Alternatively, these units may be written cm², m², ft², miles², and so on.

Similarly, a volume may be expressed as the product of three lengths, corresponding units being cm³, m³, ft³, miles³, and so on.

Frequency is a quantity expressing the number of occasions an event occurs in unit time and is found by dividing the number occurring in a certain time by that time. Thus we may express the frequency of electrical supply in the United States as 60 hertz (Hz).*

Velocity involves two different kinds of fundamental quantity, length (distance covered) and time, its value being obtained by dividing the former by the latter. Correspondingly, its units are centimeters per second, meters per second, feet per second, miles per hour, and so on. Alternative ways of expressing these units are cm/sec, m/sec, ft/sec, and miles/hr; or cm sec^{-1}, m sec^{-1}, ft sec^{-1}, and miles hr^{-1}.

Acceleration also involves a length and a time, being the rate of change of velocity. The units in which it is measured are centimeters per second per second, meters per second per second, feet per second per second, and so on. The usual abbreviations for these units are cm/sec², ft/sec², and so on.

ARBITRARY AND NATURAL UNITS The units of length, mass, time, and electric current considered above are clearly arbitrary; their magnitudes have no fundamental scientific significance. In contrast to these, we may note four quantities which are unique, and which could in principle serve as nature's special units in a system replacing the familiar systems of physical quantities used by physicists today—perhaps at some time in the future they actually will. The first of these is the velocity of light in a vacuum, 2.998×10^8 m/sec. The second is the charge of the electron, 1.6021×10^{-19} coulomb (C). Next there is Planck's quantum of action, 6.625×10^{-34} kg-m²/sec. Finally, we have the natural unit of magnetic moment, the so-called Bohr magneton, 1.1667×10^{-29} kg-m³/sec²/A (A for ampere). All these will be discussed later.

Dimensions

All the derived quantities we have considered except the first (angle) are said to have **dimensions,** the dimensions signifying

---

* In Great Britain and a number of other countries it is 50 Hz.

the manner of their derivation from the fundamental quantities. To express dimensions we use the abbreviations L for length, T for time, M for mass, and I for electric current. The two last-named, mass and current, do not, of course, occur in any of the examples just considered, but we shall have occasion to use them later.

The meaning of dimensions and the customary method of signifying the dimensions of a derived quantity will be understood from the following examples. An area, which is expressed as so many cm², m², ft², and so on, is said to have the dimensions L^2. A volume has dimensions L^3, a frequency T^{-1}, a velocity LT^{-1}, and an acceleration LT^{-2}.

A measure of an angle is, as we have seen, the ratio of a length (arc) to a length (radius). Its dimensions are therefore LL^{-1}, or L^0. The power to which L is raised is zero, and similarly for T and M, so we may say that an angle is a dimensionless quantity, or that it has no dimensions.

A knowledge of the dimensions of quantities is useful in various ways. For example, a partial check on the validity of an equation may be made by seeing if the dimensions of the two sides are the same. If they are not, the equation is certainly wrong. If they are, it is likely to be correct—at least apart from numerical factors or other dimensionless quantities such as angles, concerning which this check gives no information. Conversely, except for such dimensionless quantities, theoretical relationships may often actually be *established* by assuming that the relevant quantities enter as powers and matching the dimensions on the two sides of an equation. A third important use of dimensions is that of helping us to understand processes and situations. The following examples will serve to illustrate these three kinds of applications.

CHECKING EQUATIONS   In Chapter 2 the formula

$$v^2 = v_0{}^2 + 2as$$

will be derived, expressing the final velocity v attained by a uniformly accelerated body in terms of the initial velocity v_0, the acceleration a, and the distance traveled s. The dimensions of velocity are, as we have already noted, LT^{-1}; acceleration (rate of change of velocity) has the dimensions LT^{-2}; and, of course, a distance has dimensions L. Thus, writing down the dimensions of the left side of the above equation and of the two items on the right side, we have

$$L^2T^{-2} = L^2T^{-2} + LT^{-2}L$$
$$= L^2T^{-2} + L^2T^{-2}.$$

We see that the dimensions are consistent throughout, and this is our check.

ESTABLISHING FORMULAS A particle traveling with uniform speed v around a circle of radius r may be shown to have an acceleration a directed toward the center of the circle. This must surely depend on both v and r; and it is difficult to see how it could be affected by anything else. If then we assume that

$$a = kv^\alpha r^\beta,$$

where k is a numerical constant, this leads to the dimensional equation

$$LT^{-2} = (LT^{-1})^\alpha L^\beta.$$

We see at once that for this to balance with respect to T, α must be equal to 2. Hence, considering L, we have, for β,

$$1 = 2 + \beta,$$

which means that β must be equal to -1. Consequently, we may write

$$a = k\frac{v^2}{r},$$

where k is a dimensionless constant. Actually, as we shall see in Chapter 2, k has the value 1.

UNDERSTANDING SITUATIONS Unicellular organisms, such as amebas, are always exceedingly small. Why is this? Let us consider such organisms of linear dimensions (for example, diameter, if spherical) l. These organisms obtain their nourishment through their cell walls, so we may set the rate of food intake as proportional to surface area, that is, to l^2. The bodily "activity" of the organism— the sum total of its life processes—can take place only by virtue of the conversion of chemical energy, derived from food, into other forms of energy, for example mechanical energy. Such activity is distributed through the volume of the organism, which is proportional to l^3. Hence for what we might call the "level of bodily activity," or rate of energy conversion per unit volume of the organism, we must have

$$\text{level of activity} \propto \frac{\text{rate of nourishment}}{\text{volume}}$$

$$\propto \frac{l^2}{l^3};$$

that is,

$$\propto l^{-1}.$$

Any given level of bodily activity is therefore realizable only below a certain limit of size; small size favors a high level of activity.

Problems

It is assumed that students using this book will already have some knowledge of elementary physics, so that they will understand the terminology (pressure, density, acceleration, temperature, and so on) used in these problems. Included are several relating to errors, which are discussed in the Appendix.

1.1   The diameter of a sphere is measured as 4.37 cm, using a pair of calipers. The volume is calculated on the basis of this figure. What is the accuracy of the result? What effect on the final result would there be if later examination showed that a zero error of 0.10 cm had not been allowed for?

1.2   The mass of a certain body is found to be 6.742 g, and the volume (found by displacement) is measured as 3.4 cm³. Calculate the density and estimate the accuracy of the result. Can you find anything to criticize concerning the two measurements on which the density determination is based?

1.3   Given the quantities $x = 5.7$ cm and $y = 4.2$ cm, calculate (a) the proportional error, and (b) the absolute error, for the quantities x/y and $x - y$. Hence find the proportional and absolute errors for $(x/y)(x - y)$.

1.4   The pressure of an enclosed gas is found from the reading of a U-tube manometer in conjunction with that of a barometer. The barometer reads 75.98 cm Hg. The U-tube manometer is read with the aid of a cathetometer, a traveling telescope that measures heights above a certain horizontal plane. The height of the mercury in the open arm is measured as 8.43 cm; that in the other arm is 7.21 cm. Find (a) the pressure in the enclosure, and (b) the pressure difference between the inside and outside of the enclosure. Estimate both the possible absolute errors and the possible relative errors in these quantities.

1.5   The density of a liquid is determined by a method based on Archimedes' principle, allowance being made for "air buoyancy"; and the formula applying to this experiment is

$$\rho = \rho_0 \frac{m_1 - m_2}{m_1 - m_3} + \rho_a \frac{m_2 - m_3}{m_1 - m_3},$$

where ρ, ρ_0, and ρ_a are the densities of the liquid, water, and air, respectively. Given that

$$\rho_0 = 0.99754 \text{ g/cm}^3,$$
$$\rho_a = 0.0012 \text{ g/cm}^3,$$
$$m_1 = 28.273 \text{ g},$$
$$m_2 = 20.587 \text{ g},$$

and

$$m_3 = 18.662 \text{ g},$$

find ρ and estimate the accuracy of the determination.

1.6   The theoretical formula for the period of oscillation T of a certain system is

$$T = 2\pi \sqrt{\frac{b^2 + 3a^2}{2ag}},$$

where g is the acceleration due to gravity. Find the value of g and estimate the accuracy of the result obtained, given that the measured values of a, b, and T are as follows:

$$a = 10.27 \text{ cm,}$$
$$b = 12.54 \text{ cm,}$$
$$T = 0.964 \pm 0.002 \text{ sec.}$$

1.7   The related quantities x and y measured in a certain experiment have the following values:

x	12	15	20	30	60
y	60	30	20	15	12

Find a method of graphing these results that will give a straight line, and write down the corresponding algebraic formula.

1.8   A thin wooden beam is supported horizontally on two knife-edges equidistant from its midpoint, at which a fixed load is suspended. The separation x of the knife-edges is varied, and the corresponding depression y of the midpoint is observed. The following values are obtained:

x (cm)	40	50	60	70	80
y (cm)	0.32	0.63	0.99	1.72	2.56

Given that the relation between the two quantities is of the form $y = Ax^n$, find, by employing a graphical method, the values of the constants A and n.

1.9   A meter stick is used as a rigid pendulum and swung on pivots at different distances h from the center of mass, and the following values of the period T are obtained, corresponding to a series of values of h:

h (cm)	5	10	20	30	40	50
T (sec)	2.76	2.03	1.63	1.57	1.60	1.66

Given that the relevant formula is

$$T = 2\pi \sqrt{\frac{k^2 + h^2}{gh}},$$

where k is a constant for the stick, determine by means of a suitable linear graph the values of k and g.

1.10   From measurements made with tungsten filaments mounted in vacuum lamps, it has been found that the rate of loss of energy by radiation varies with the absolute temperature as follows:

Temp. (°K)	Radiation (W/cm²)	Temp. (°K)	Radiation (W/cm²)
400	1.99×10^{-3}	1600	7.72
600	3.04×10^{-2}	1800	14.19
800	1.69×10^{-1}	2000	24.04
1000	6.02×10^{-1}	2200	38.2
1200	1.66	2400	57.7
1400	3.86	2600	83.8

Plot these results in such a manner that a good approximation to a straight line is obtained, and hence find an approximate algebraical expression for power radiated per unit area as a function of temperature. From your graph estimate what would be the power radiated at 3655°K, the melting point of tungsten.

1.11 The following are figures that have been obtained for the pressure p of the atmosphere as a function of height h above sea level:

h (m)	p (mm Hg)	h (m)	p (mm Hg)
0	760	6000	354
1000	674	7000	308
2000	596	8000	268
3000	526	9000	231
4000	462	10,000	199
5000	405		

Find a simple manner of graphing these results that gives a reasonably close approximation to a straight line, and hence write an approximate formula for the variation of pressure with height within the range of the observations. Extrapolating from your graph, estimate the pressure at an altitude of 15,000 m.

1.12 An equation for the range d of a projectile fired horizontally with velocity v from the top of a cliff of height h appears as

$$d = v\frac{2h}{g},$$

g being the acceleration due to gravity. Check by the method of dimensions whether this equation can be right. If it is found to be in error, try to suggest how it should be amended.

1.13 What are the dimensions of (a) angular velocity, defined as the rate of description of an angle, and (b) the cosine of an angle? In a theoretical discussion of simple harmonic motion, the equation

$$v = \omega r^2 \cos \theta$$

appears, relating the velocity v of the particle executing simple harmonic motion to the motion of a "point of reference" around a circle of radius r with the angular velocity ω. It is suspected that this equation contains a misprint. If it is in error, can you suggest how it should be corrected?

1.14   Physical considerations indicate that the time t taken by a body to fall freely from rest through a height h must depend solely on this height and the acceleration g due to gravity. Using the method of dimensions, and assuming that these quantities enter as powers, find (apart from a numerical factor) the equation relating t with h and g.

1.15   A stretched string has a tension F and a mass per unit length μ. Show by the method of dimensions that the velocity v with which waves are propagated along the string is given by the formula

$$v = k \sqrt{\frac{F}{\mu}},$$

where k is a dimensionless constant.

1.16   Ship designers make measurements on models in water tanks to predict the performance of a full-sized ship at various speeds. Waves generated by the ship carry away energy, so work has to be done by the ship, and the corresponding drag due to the action of the water on the model can be ascertained. It is known that the retarding force depends on the mass and length of the ship or model, its velocity, and the acceleration g due to gravity. Show by dimensional analysis that the retarding force per unit mass for a model of length l moving with velocity v is the same as for a ship of length Nl moving with velocity $\sqrt{N} \cdot v$. Hence show that

$$\left(\frac{p}{mv}\right)_{\text{model}} = \left(\frac{P}{MV}\right)_{\text{ship}},$$

where p and P are powers, m and M are masses, and v and V are velocities. (This is the statement of Froude's law of corresponding speeds.)

SECTION II

KINEMATICS

2

CLASSICAL PARTICLE KINEMATICS

Down through the ages the contemplation of movement in all its rich variety of forms and changes has aroused men's curiosity and inspired speculation, but it was not until the time of Isaac Newton (1643–1727) that some order began to emerge from the preexisting chaos of ideas and a real science of motion was born.

It is proposed here to present the salient features of this science of motion in two main parts, each of which will be appropriately subdivided. In the present chapter and in Chapter 3 we shall simply analyze motion itself. Then, in Chapters 4 to 9 and in Chapter 11 we shall also take into account the role played by the properties of the bodies that move.

The study of motion without regard to any property of the body that moves is called **kinematics;** the complete science of motion, in which the relevant properties of the body or bodies concerned *are* taken into account, is known as **dynamics.**

Motion may be steady or it may vary with time. It may be along a single straight line or it may change its direction. The motion with which we are concerned may be that of a single particle or of a system of particles. And in the latter case the system may be a rigid body; a deformable body such as, for example, a string; or it may be a fluid (liquid or gas).

All these cases have a peculiar interest of their own, which will appear as each is dealt with in its turn. The broad plan of study will be to lead naturally from one to the next, from the simple to the complex, from particle kinematics, through particle dynamics, to the dynamics of systems of particles.

Concept of a Particle

The simplest kind of motion to consider is that in which the body concerned moves as a whole without simultaneous change of orientation. Such motion is known as **pure translation.** In cases where the body also rotates, the rotation need not always be taken into account; it can be ignored in discussions where it is unnecessary to specify to which particular point we are referring, the dimensions

27

of the body being small compared with the distances with which we are otherwise concerned. Such bodies are known as **particles.** A particle need not always be very small, as it is in the everyday meaning of the term. Thus, in considering the mechanics of the solar system, even bodies as large as the planets may for most purposes be treated as particles, their dimensions being negligible in comparison with their distances from the sun, the body to which their positions and motions are normally referred.

Displacements, Velocities, and Accelerations

When we use the terms **displacement, velocity,** and **acceleration** in everyday language, we naturally think not only of quantities having a certain magnitude but also of the associated directions. If we move an object from one place to another, the displacement is fully specified only if, in addition to stating how far we have moved it, for example 10 m, we also say in what direction, for example horizontally and northeast.

In their physical meanings, also, these terms are generally understood to include direction. Thus, if a car travels at 60 kilometers per hour (km/hr) in a northerly direction it is said to have a different velocity from one traveling at 60 km/hr in a southeasterly direction. If we do not wish to concern ourselves with direction, we should, strictly, use the word "speed" rather than "velocity." However, this rule is not always adhered to in physical literature. Although "speed" never includes the idea of direction, the use of the word "velocity" does not *always* imply that we are concerned with direction. Thus it is customary to speak of the "velocity of sound" or the "velocity of light" without any reference to the direction of travel. In this book we shall conform to usage in this matter; it will always be clear from the context whether or not reference to direction as well as to magnitude is intended.

Motion of a Particle along a Straight Line

The position of a particle on a straight line is most conveniently specified in relation to an agreed reference point in the line that we consider as the origin in one-dimensional coordinate geometry, with one direction from this point regarded as positive and the other negative. We shall now consider the fundamental equations of motion based on this system, writing x for the coordinate of the moving particle, this being positive if the particle is on one side of the origin and negative if on the other.

CONSTANT VELOCITY   Let the velocity of the particle be v, this being reckoned as positive if the particle is moving in one direction ("positive x direction") and negative if moving in the other. Let the coordinate of the particle at time t, counted from some given moment, be x, and let the value of x at the chosen zero for time ($t = 0$) be x_0. Then, because v is the time rate of increase (regarded algebraically) of the coordinate, that is, the increase occurring per unit of time, the increase that occurs in t units must be vt. Hence at the time t the coordinate x must be the algebraic sum of the initial coordinate x_0 and the change vt it has undergone in t units of time (see Fig. 2.1):

$$x = x_0 + vt. \tag{2.1}$$

CONSTANT ACCELERATION   Just as velocity is the rate of change of position, so acceleration is the rate of change of velocity. If distances are measured in meters and time in seconds, so that velocity is measured in meters per second, then the acceleration is the number of meters per second by which the velocity increases every second. Thus a freely falling body increases its velocity downward by 9.8 m/sec every second, and we express this by saying that its acceleration downward is 9.8 meters per second per second, or 9.8 m/sec².

If at time $t = 0$ the velocity is v_0 and the constant acceleration is a, then in t units of time the velocity added on is, by definition of acceleration, at, so that the final velocity v at time t is the initial velocity v_0 plus what has been added on, at; that is,

$$v = v_0 + at. \tag{2.2}$$

This is completely analogous to Eq. (2.1), and, as we have just seen, its derivation follows the same lines.

If we wish to find the distance the particle travels in the positive x direction in time t, this being the difference between the coordinates x_0 and x at the beginning and end, respectively, of this time interval, we may proceed as in the derivation of Eq. (2.1), except that the mean velocity replaces the constant velocity in this equation. The mean of the initial velocity v_0 and the final velocity $v_0 + at$ is $v_0 + \frac{1}{2}at$, and substituting this for v in Eq. (2.1) we obtain

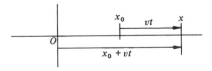

FIGURE 2.1   Position of a particle having constant velocity as a function of time.

the equation

$$x = x_0 + v_0 t + \tfrac{1}{2} a t^2,$$

or, writing s for the distance $x - x_0$ traveled in the positive x direction,

$$s = v_0 t + \tfrac{1}{2} a t^2. \tag{2.3}$$

Although this simple reasoning does, as it happens, lead to the right answer, it is somewhat wanting in rigor; its weakness lies in the assumption that the effective mean velocity for the whole time interval considered is necessarily half the sum of the initial and final velocities. A more acceptable derivation of Eq. (2.3) is the following.

Because the acceleration is constant, the velocity added on is proportional to time; hence a plot of velocity against time will be a straight line. Let AB in Fig. 2.2 be such a plot. Then the number of scale units in OA is numerically equal to v_0; in this sense we may write $OA = v_0$. Similarly, $CB = v = v_0 + at$ and $OC = t$. Let us consider times t_p and t_q, represented by OP and OQ, respectively, such that the interval $t_q - t_p$ is very small. At the beginning of this time interval the velocity is represented by the ordinate PR drawn through P, and at the end the velocity is QS. If the time interval is short enough, we may take PR and QS as sufficiently nearly equal and either of them as representing the velocity during the whole of this interval. And if we multiply the velocity during a short interval of time by this interval we obtain the corresponding distance covered. Hence the shaded area represents the distance covered during the time interval $t_q - t_p$. But we may cut up the whole of the time t into a number of short intervals like this one, and for each of these the distance covered is by similar reasoning given by the area of the corresponding vertical strip. All these areas added together are equal to the area of the figure $OABC$, so the whole distance covered in the time t must be equal to this area, which is

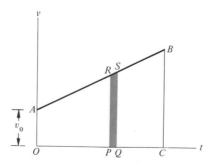

FIGURE 2.2 Graph of velocity versus time for constant acceleration.

$\frac{1}{2}(OA + CB) \cdot OC$. But $OA = v_0$, $CB = v_0 + at$, and $OC = t$. Hence

$$s = v_0 t + \tfrac{1}{2}at^2. \qquad (2.3)$$

The process of reasoning just considered is one of graphical integration. It has the advantage of helping one to visualize the physical situation throughout the whole period of time contemplated. However, if the reader is familiar with the calculus, he may prefer to carry out an algebraic integration, not only to obtain Eq. (2.3) but also Eqs. (2.1) and (2.2). Thus, since

$$\frac{dx}{dt} = v,$$

we have, integrating this,

$$x = x_0 + vt, \qquad (2.1)$$

where the constant of integration x_0 is the value of x corresponding to $t = 0$. Similarly, because

$$\frac{dv}{dt} = a,$$

$$v = v_0 + at, \qquad (2.2)$$

where v_0 is the velocity at zero time. And finally, writing ds/dt for v in Eq. (2.2) and integrating, we have

$$s = v_0 t + \tfrac{1}{2}at^2, \qquad (2.3)$$

the constant of integration in this case obviously being zero.

We sometimes wish to know the velocity of the particle in terms of the initial velocity, the acceleration, and the distance covered, that is, v in terms of v_0, a, and s. We may obtain the equation for this by eliminating t between Eqs. (2.2) and (2.3). This at once gives us

$$v^2 = v_0{}^2 + 2as. \qquad (2.4)$$

Acceleration of Freely Falling Bodies

An important case of uniform acceleration in a straight line is that which occurs at the earth's surface due to gravity. Since the time of Galileo Galilei (1564–1642) we have known that at any particular locality on the earth's surface all freely falling bodies have the same acceleration downward. Here "free" fall is to be understood as that which occurs when air resistance is negligible or is eliminated.

31

The acceleration with which bodies fall toward the earth is conventionally denoted by the symbol g. Up to the present time, or at least until very recently, the most accurate method of determining g has been by the measurement of the period of oscillation of a pendulum, a somewhat crude version of which will be considered in Chapter 4. When this method is carried out with all possible refinements, the relative accuracy attainable is to within about 1 part in a million. It now appears likely, however, that this method will presently be superseded by the direct observation of the free fall of a body in an evacuated glass tube. So great have been the advances made in recent years in the techniques of automatic optical and electrical timing that the direct method incorporating the use of these promises to yield results of even greater accuracy than are attainable by the older pendulum method. This direct determination of g is based on measurements of the times at which the body passes three points, A, B, and C, in the course of its fall, beginning at a point higher than any of these. Its velocity as it passes A being v_0 (unknown), the times t_1 and t_2, which it takes to fall through the distances s_1 and s_2 equal to AB and AC, respectively, are measured. By substituting these times and distances in equations of the type (2.3), and subtracting, we eliminate the unknown velocity v_0, and we may then solve for a $(=g)$.

Precision determinations have shown that although g is identical for all bodies at a particular locality, it varies by a few parts in a thousand over the surface of the globe, its value at sea level increasing with latitude from the equator to the poles. It also decreases appreciably with increasing height above sea level.

It is convenient to refer certain quantities, for example the "standard atmosphere" of pressure, to a "standard" value of g. This is defined as that found at latitude 45° and sea level. Its value is 9.80616 m/sec², or 980.616 cm/sec².

For calculation purposes it will in most cases be sufficiently accurate to use for g the value 9.81 m/sec², or 981 cm/sec². The corresponding value of g in the British system of units—now seldom used except by engineers in English-speaking countries—is 32.2 ft/sec².

Denoting the distance fallen through from rest $(v_0 = 0)$ by h (height), Eqs. (2.3) and (2.4) as applied to freely falling bodies become

$$h = \tfrac{1}{2}gt^2 \qquad (2.5)$$

and

$$v^2 = 2gh, \qquad (2.6)$$

respectively.

Composition and Resolution of Directed Quantities

Having discussed the derivation of the relevant equations for motion along a straight line, it will now be necessary for us to consider what mathematical procedure might be used for compounding displacements, velocities, and accelerations in the general case, where these quantities do not necessarily have the same direction.

COMPOSITION OF DISPLACEMENTS   To fix the ideas, let us consider a man walking from P to Q on the deck of a ship (Fig. 2.3), so that when he has arrived at Q his displacement relative to the ship is PQ. Let the ship also be in motion, so that the simultaneous displacement of the ship relative to the water is given in magnitude and direction by PP' or QQ', P' and Q' being the positions reached by the points P and Q at the end of the time interval in question. Then obviously the man's final position relative to the water is Q', and the resultant displacement of the man relative to the water, due to the combined motions of the man on the deck of the ship and the ship in the water, is PQ'.

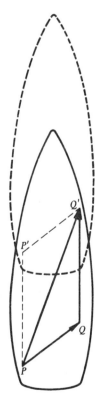

FIGURE 2.3   Composition of two simultaneous displacements.

The displacements PQ, QQ', and PQ' are examples of **vector quantities,** or **vectors,** and the triangle PQQ' is known as a **vector triangle.** The precise definition of a vector will be given later. Meanwhile we may note that vectors are quantities that have both magnitude and direction, although the converse is not necessarily true—not all directed quantities are vectors. On the other hand, quantities that either have no direction (for example, time or mass) or of whose associated directions we take no account (for example, speed) are known as **scalars.**

If in the case of the man walking on the deck of a ship we represent the component displacements in the reverse order, it obviously makes no difference; we still obtain the same resultant. If we use both orders and combine the corresponding triangles, we have the so-called **vector parallelogram,** $PQQ'P'$. The vector-parallelogram construction is sometimes given as an alternative to the triangle procedure for finding the resultant of two vectors. According to the parallelogram procedure, lines representing the two vectors are both drawn from the same point, the parallelogram is completed, and then the diagonal through this same point represents the resultant. Although either construction serves equally well for ascertaining the resultant, the validity of the parallelogram procedure is not immediately self-evident as is that of the vector triangle.

If more than two vectors are to be compounded, this may be done by constructing a vector polygon, an extension of the vector-triangle procedure. The polygon construction is indicated in Fig. 2.4. The quantities are represented to an agreed scale by directed straight lines, which are joined end to end, and then a line is drawn from the starting point O to the final point reached, R. This line OR represents both to scale and in direction the resultant of all the component vectors. That this must be so is self-evident in the case where the vectors are displacements. It is also obvious that, as in the case of the vector triangle, the order in which the separate vectors are dealt with is immaterial.

COMPOSITION OF VELOCITIES In the case of the man walking on the deck of the ship, let both the velocity of the man relative to the ship and that of the ship relative to the water be uniform, and let the displacements represented by the lines PQ and QQ' occur in the time t. Then each velocity is in magnitude $1/t$ of the corresponding displacement, and, of course, the velocities have the same directions as the displacements. Hence on some scale the sides of this same triangle PQQ' represent the velocities, PQ and QQ' the velocities to be compounded, and PQ' the resultant. This triangle, or

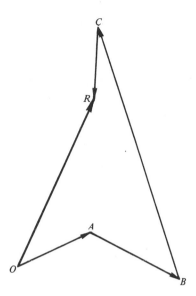

FIGURE 2.4 Vector polygon.

any similar and similarly orientated triangle, is therefore also a vector triangle of velocities. Extending this reasoning to the case where more than two velocities are to be compounded, we see that here the vector-polygon construction gives us the resultant of these velocities.

COMPOSITION OF ACCELERATIONS Earlier in this chapter acceleration was defined as the rate of change of velocity. We were there concerned with motion in a single direction, but it is not intended that the definition should be confined to this case; it is to be taken as having *general* validity. *Any* change in velocity, whether in magnitude alone, direction alone, or both together, is to be regarded as giving a particle an acceleration, in accordance with the definition.

Just as we have derived the procedure for compounding velocities from that for displacements, so, by again dividing by the scalar *t* throughout, we could derive the procedure for compounding accelerations from that for velocities, the velocities with which we are concerned being those by which the original velocity is changed in the time *t*. In this way it may readily be shown, by reasoning closely analogous to that just given, that accelerations can be compounded by the same kind of procedure as applies in the case of displacements and velocities.

DEFINITION OF A VECTOR We have just seen that displacements, velocities, and accelerations may all be compounded by the triangle or polygon construction. These are only a few of a large number

35

of directed quantities, all of which conform to this procedure for composition. All such quantities are conveniently classed together, and we do this by defining them as **vectors.**

Vectors so defined form an important group of quantities, not only because of the possibility of compounding them in the manner referred to but also because we can perform many other mathematical operations with them which are of the very greatest physical interest and importance. These operations are dealt with in a special kind of algebra, expressly developed for its physical applications, known as **vector algebra.**

The reason the proviso concerning conformity to the composition procedure is included in the definition is that only those directed quantities that do conform (and not all do) can be dealt with by the methods of vector algebra. It would be pointless and confusing to class as vectors quantities to which the rules of vector algebra do not apply.

RESOLUTION OF VECTORS Resolution of vectors is the inverse of composition. Any vector, such as is represented by PQ' in Fig. 2.3, is equivalent to a pair of vectors such that the lines representing them make with PQ' a complete triangle. There are obviously an infinite number of pairs of vectors having different directions that are equivalent to the original single vector, and these are in all cases called the **components** of the vector in the chosen directions. A case of particular importance is that in which the components are at right angles to one another, for example having the directions of the x and y axes of a Cartesian coordinate system as in Fig. 2.5, which refers to the case where the vector is a velocity v. If we speak simply of "the component" of a certain vector in a specified direction, the

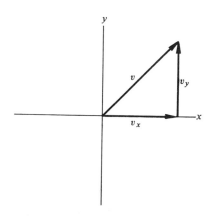

FIGURE 2.5 Components of a velocity v in x and y directions.

other component is generally understood to be the one at right angles to this.

As an interesting example of resolution, let us consider separately the horizontal and vertical motions of a particle in free flight. To take a simple case, let us suppose a stone is projected horizontally with a known velocity v outward from the edge of a vertical cliff of height h, and let it be required to find the distance from the foot of the cliff at which the stone will strike the ground. As far as the vertical components of motion (displacement, velocity, and acceleration) are concerned, the stone falls from rest with uniform acceleration g, covering a total distance h. Using Eq. (2.5) we can therefore calculate the time t that elapses before the stone strikes the ground,

$$t = \sqrt{\frac{2h}{g}},$$

and multiplying this by the horizontal component of velocity, v, which is constant throughout (there being no horizontal component of acceleration), we obtain the horizontal distance covered, d. Thus

$$d = v \sqrt{\frac{2h}{g}}.$$

If, for example, the cliff is 100 m high and v is 10 m/sec, the distance d from the foot of the cliff at which the stone strikes the ground will be $10 \sqrt{2 \times 100/9.8}$ m, that is, 45.1 m.

Quite generally problems on projectiles may be worked out by treating the horizontal and vertical components of velocity as independent of one another, the horizontal component remaining constant throughout, while the vertical component varies with time because of the gravitational acceleration g downward. Strictly, of course, this simple procedure is valid only for the ideal case where there is no air resistance. Actually such resistance must always have some effect on the flight of the projectile.

Uniform Circular Motion

Having discussed motion in a constant direction but with uniformly varying speed (constant acceleration), let us now consider the converse of this—the motion of a particle having constant speed but moving in a uniformly varying direction, that is, in a circular path. This will be our first important application of the composition of vectors.

Because the *velocity* of the particle (as distinct from speed) is constantly changing, the particle must be accelerated. We wish to find the acceleration in both magnitude and direction corresponding to any particular position of the particle in its circular path. To do this we must find the change of velocity occurring in a short interval of time and divide it by this time interval. This will give us the mean acceleration during this interval. The shorter the interval taken, the less will be the change in the acceleration that occurs during it (actually it is constantly changing in direction), and the more nearly therefore will the mean acceleration approximate to the actual acceleration at a particular moment. When the time interval taken is made infinitesimally small, the mean and instantaneous accelerations become identical.

Let us first consider a finite, but short, interval of time δt during which the particle moves from P to Q (Fig. 2.6). We shall find the mean acceleration during this interval and then see what will be the magnitude and direction of this in the limiting case where δt is made infinitesimally small.

Let AB and AC be drawn to represent the velocity of the particle when at P and at Q, respectively. Let the constant speed of the particle be v. Then AB and AC will be equal and of length to represent v to scale. And since AB and AC are in the directions of the tangents at P and Q, respectively, $\angle BAC$ must be equal to $\angle POQ$. Let this angle be denoted by $\delta\theta$.

If we join B to C we have a vector triangle in which AB represents the velocity at the beginning of the time interval considered and AC that at the end. We now have to find what velocity δv compounded with the initial velocity will produce the final velocity.

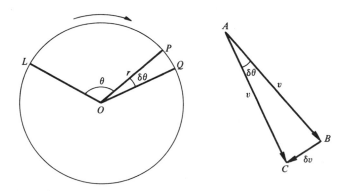

FIGURE 2.6 Vector construction for derivation of centripetal acceleration.

From the vector triangle we see at once that this is the velocity represented by BC. This is the change of velocity we are seeking, and by dividing it by δt we shall obtain the mean acceleration during this time interval.

Now, with $\delta\theta$ sufficiently small, we may regard BC as very closely approximating the arc of a circle of radius AB, so that, to this degree of approximation,

$$BC = AB \cdot \delta\theta.$$

The smaller the time interval δt, and with it the angle $\delta\theta$, the closer this approximation is, and, in the limit, where these quantities become infinitesimally small, the equation becomes exact.

We must now make use of the fact that the sides of the vector triangle represent the corresponding velocities to scale; that is, $v = k \cdot AB$, and $\delta v = k \cdot BC$, where k is a constant. Hence also

$$\delta v = v \cdot \delta\theta.$$

We have now merely to divide both sides of this last equation by δt to obtain the acceleration we seek. Thus we have

$$a = \frac{\delta v}{\delta t} = v\frac{\delta\theta}{\delta t}.$$

The quantity $\delta\theta/\delta t$ is the rate of increase of the angle θ made by the radius OP with an agreed reference line OL; it is the rate of description of angle, the **angular velocity** with which the particle executes its circular orbit. If we denote this angular velocity by ω, our equation for acceleration becomes

$$a = v\omega.$$

With $\delta\theta$ infinitesimally small, BC becomes perpendicular to AB, so that the direction of the acceleration is radial and toward the center of the circle. The acceleration is accordingly known as **centripetal,** that is, "center-seeking."

Because

$$\delta\theta = \frac{PQ}{OP},$$

and therefore

$$\frac{\delta\theta}{\delta t} = \omega = \frac{PQ/\delta t}{OP} = \frac{v}{r},$$

where r is the radius of the circle, we may substitute either v/r for ω or ωr for v in the expression for a. In this way we obtain the three alternative expressions for the centripetal acceleration,

$$a = v\omega = \frac{v^2}{r} = \omega^2 r. \qquad (2.7)$$

EFFECT OF EARTH'S ROTATION ON g As an interesting application of this result, and at the same time as an example of the composition of accelerations, let us now investigate the effect of the earth's rotation on the observed values of g. In doing this, let us idealize the situation by imagining the earth to be a true sphere, with spherically symmetrical distribution of material.

Let us consider a point P on the earth's surface, at latitude θ, as shown in Fig. 2.7(a). Let the earth's radius and its angular velocity of rotation be r and ω, respectively. Then the radius of the circle in which this point rotates is $r \cos \theta$, and its centripetal acceleration, directed along the radius PQ of this circle, toward Q, is $\omega^2 r \cos \theta$.

Now, the observed acceleration g of a freely falling body is the acceleration relative to the earth's surface, which is itself being accelerated in the direction PQ with the acceleration $\omega^2 r \cos \theta$. The "absolute" acceleration, as we may call it, that is, the acceleration relative to a nonrotating system of reference, such as would be seen by an observer outside the earth and not partaking of its rotation, must be the vector resultant of these two components. This situation is analogous to that of the man walking on the deck of the ship, considered earlier, with accelerations replacing velocities and the deck of the ship corresponding to the rotating surface of the earth. In the present problem, let us denote the absolute acceleration by g_0. From considerations of symmetry it is evident that this

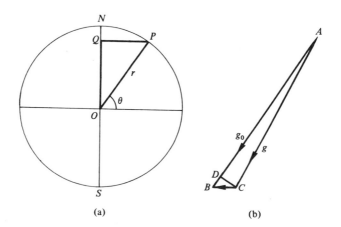

(a)   (b)

FIGURE 2.7   Effect of earth's rotation on g.

must be in a direction through the center of the earth. Hence, if in Fig. 2.7(b) AB is drawn parallel to PO, this may be taken to represent vectorially, on some scale, the absolute acceleration g_0, and a line such as CB will represent the centripetal acceleration of the earth's surface, of magnitude $\omega^2 r \cos \theta$. Then the third side of the triangle, AC, must represent g. From Table 2.1 it will be seen that at the equator, where the effect of the earth's rotation is greatest, g is only 5.2 cm/sec² less than at the poles, and this is only a fraction of 1 percent of the value of g anywhere on the earth's surface. Consequently, in a vector triangle of the kind shown in Fig. 2.7(b) drawn properly to scale, the side BC would be very small indeed compared with the other two sides. Hence if we drop a perpendicular CD from C on AB it is permissible to regard AC and AD as equal, so that the effect of rotation is to make g less than g_0 by $\omega^2 r \cos \theta \cdot \cos \theta$, or $\omega^2 r \cos^2 \theta$.

The period of rotation of the earth is a sidereal day, or 86,164 sec. The value of ω is therefore $2\pi/86,164$, and ω^2 is $\pi^2/(43,082)^2$. For r we may take the mean radius, which is one third of the sum of the three mutually perpendicular semiaxes of the ellipsoid, that is, $\frac{1}{3}(6378 + 6378 + 6357)$ km, or 6.371×10^8 cm. Hence the value of $\omega^2 r$ is 3.388 cm/sec². Applying the correction term $\omega^2 r \cos^2 \theta$ to find g_0 from g, we then obtain the values given in Table 2.1.

It is interesting to note that the extreme spread of values of g_0, from the equator to the poles, is only 1.8 cm/sec², as against 5.2 cm/sec² for g. The spread of values of g_0 is due to the lack of spherical symmetry in the distribution of the earth's material, including its ellipsoidal external form. This, in turn, is due to the earth's rotation. At any point on the earth's surface, what we call the horizontal plane (for example, the plane of the free surface of a body of water) must necessarily be perpendicular to the direction of g. There can be no component of g along this surface; for, if there were, water would flow in the direction of this component, and there

TABLE 2.1   Values of g and g_0

Location	Latitude	g (cm/sec²)	g_0 (cm/sec)
Equator	0°0′	978.03	981.42
Bangkok	13°54.7′N	978.33	981.53
Brisbane	27°28′S	979.16	981.82
New York	40°38.6′N	980.23	982.19
London	51°28.1′N	981.20	982.52
Reykjavik	64°1.7′N	982.28	982.93
North Pole	90°0′N	983.22	983.22

would be a corresponding readjustment in the orientation of the free surface. From Fig. 2.7(b) we see that at all points on the earth's surface except at the poles and at the equator the angle between the direction of g and the plane of the equator is greater than the latitude. It is for this reason that the earth has adjusted itself to its ellipsoidal figure.

An acceptable mean value of g_0, corresponding to the mean radius of the earth, is obtained by dividing the sum of the polar value and twice the equatorial value by 3. This is 982.0 cm/sec².

Variation of Gravitational Acceleration with Distance of Attracted Body

This is most conveniently found by considering the accelerations of the planets toward the sun, or those of the satellites of any planet having more than one of these toward the planet. In these systems the orbits are very nearly circular, and if we treat them as such it is a simple matter to calculate the centripetal accelerations of the orbiting bodies from their orbital radii and their periods of revolution. Also from the simple (almost circular) nature of the orbits, and from the simple law of variation of the periods of revolution of the orbiting bodies with their distances from the attracting body at the center of the orbits, it may be inferred that the orbiting bodies do not affect one another's motions to any appreciable extent. Indeed, from the smallness of the planets compared with the sun, or, for example, of the satellites of Jupiter compared with that planet, it would be surprising if it were otherwise.

To ascertain how the acceleration of an orbiting body varies with its distance from the attracting body, a useful first step would appear to be to try to find how the period T of revolution in the orbit varies with the orbital radius r. For this purpose we might try the method of experimental graphing referred to in Chapter 1. We might try plotting T against r, T^2 against r, T^3 against r^2, T^2 against r^3, and so on, hoping that eventually we should find a method of graphing that gives a straight line. If we did this, we should find that the plot of T^2 against r^3 is rectilinear and passes through the origin. This means that T^2 is proportional to r^3.

There is, however, a much more elegant and less time-consuming experimental graphing procedure by which we may ascertain whether one quantity is proportional to a power of another, and, if so, finding what that power is. Because of its general interest and usefulness it will be worthwhile to consider this method here.

TABLE 2.2 Data for Planets

Planet	Mean distance from sun (millions of kilometers)	Period (years)
Mercury	57.9	0.2408
Venus	108.1	0.6152
Earth	149.5	1.0000
Mars	227.8	1.8809
Jupiter	778	11.86
Saturn	1426	29.46
Uranus	2869	84.01
Neptune	4496	164.8
Pluto	5899	248.4

If two variables x and y are related by the equation

$$y = kx^n,$$

where k and n are constants, then

$$\log y = \log k + n \log x.$$

Hence if we plot $\log y$ against $\log x$, we obtain a straight line whose slope is n.

Let us now apply this method to the data for the planets of our solar system, which are given in Table 2.2.

In Fig. 2.8 $\log T$ is plotted against $\log r$ for all these planets except Pluto, the outermost planet, whose orbit is rather eccentric.

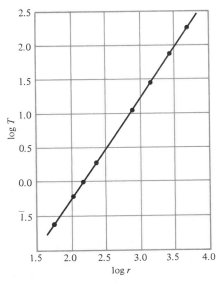

FIGURE 2.8 Plot of $\log T$ against $\log r$ for planets of the solar system.

We see that all the points lie on a straight line, and from this we may infer that T is proportional to a power of r. The slope tells us what this power is. It is $\frac{3}{2}$; that is,

$$\log T = \tfrac{3}{2} \log r + C,$$

or

$$\log T = \tfrac{3}{2} \log r + \log C',$$

where C and C' are constants. Hence

$$T = C'r^{3/2},$$

or

$$T \propto r^{3/2};$$

that is,

$$T^2 \propto r^3.$$

Having found how T varies with r, let us now work out the corresponding relation between the centripetal acceleration a and r. Because

$$T^2 \propto r^3,$$

and the angular velocity ω of a planet describing its orbit about the sun is inversely proportional to its period of revolution T, we must have

$$\omega^2 \propto \frac{1}{r^3}$$

and hence

$$a = \omega^2 r \propto \frac{1}{r^2};$$

that is, the gravitational acceleration of a planet toward the sun is inversely proportional to the square of its distance from it.

Had we examined in the same way the corresponding data for the satellites of Mars, Jupiter, or Saturn, we should have obtained the same result. Also, the inverse-square law of gravitational acceleration may be obtained by comparing the centripetal acceleration of the moon in its orbit around the earth with the value of g_0 at the earth's surface, assuming that the relevant distances determining these accelerations are those from the center of the earth to the center of the moon and to the attracted object at the earth's surface, respectively.

The fact that the same law of variation of acceleration with distance applies in all these cases is highly significant, suggesting the existence of a *universal* law of gravitation between bodies. This will be studied in full in Chapter 5.

Simple Harmonic Motion

In our study of particle kinematics so far we have considered the two important special cases of (1) uniform rate of change of speed of a particle traveling in a constant direction, and (2) uniform rate of change of direction of motion of a particle traveling with constant speed. We are now ready to consider a slightly more complex and extremely important type of motion known as **simple harmonic motion** (SHM). One of two alternative definitions of this—the other will be stated presently—is the projection of a "point of reference" moving with constant speed around a "circle of reference" on a diameter of this circle or on any other line in its plane. In Fig. 2.9, if the point P moves around the circle shown with constant angular velocity, then its projection Q on the diameter AB executes simple harmonic motion. The distance between the midpoint O of Q's oscillation and one of the extremities, A or B, is called the **amplitude** of the SHM. This is, of course, the same as the radius r of the circle of reference.

Let OP momentarily make the angle θ with the line OC drawn perpendicular to AB, and let the corresponding displacement OQ of Q from O be denoted by x, this being reckoned positive when Q is to the right of O and negative when it is to the left. Then this displacement is given by the equation

$$x = r \sin \theta. \tag{2.8}$$

Next, let us find Q's velocity. The speed of the point P in its circular path is ωr and is in a direction tangential to the circle; that is, the momentary velocity has the magnitude ωr and the direction

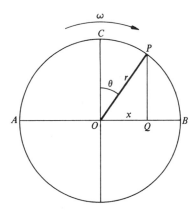

FIGURE 2.9 Simple harmonic motion as a projection of uniform circular motion.

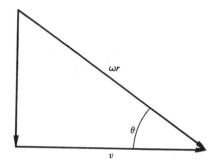

FIGURE 2.10 Velocity of particle executing simple harmonic motion.

perpendicular to OP. In Fig. 2.10 are shown the components of this velocity perpendicular and parallel to AB. Only the latter of these interests us, this being the required projection of the velocity of P along AB, and as we see at once from the figure the magnitude of this component is $\omega r \cos \theta$. Accordingly, if the projected velocity is denoted by v, we have for this the general equation

$$v = \omega r \cos \theta, \tag{2.9}$$

v being reckoned positive when the motion is from left to right and negative when it is from right to left.

Finally, let us investigate Q's acceleration. P's acceleration is $\omega^2 r$, directed from P toward O along the line PO. From the vector diagram of accelerations shown in Fig. 2.11, we see that the component of this in the direction AB (that is, the acceleration of the projection Q of P on this line) has the magnitude $\omega^2 r \sin \theta$ and is directed from Q toward O. Thus, reckoning accelerations from left to

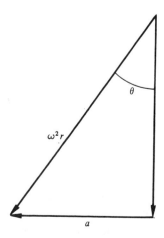

FIGURE 2.11 Acceleration of particle executing simple harmonic motion.

right as positive and from right to left as negative, we may express the acceleration a of Q by the equation

$$a = -\omega^2 r \sin \theta. \tag{2.10}$$

If time is counted from the moment P is at C, then θ is equal to ωt, and Eqs. (2.8), (2.9), and (2.10) may be written

$$x = r \sin \omega t, \tag{2.11}$$
$$v = \omega r \cos \omega t, \tag{2.12}$$

and

$$a = -\omega^2 r \sin \omega t, \tag{2.13}$$

respectively. These are useful alternative expressions for the three quantities.

Those familiar with the calculus will realize that Eqs. (2.12) and (2.13) may be derived directly from Eq. (2.11) by simple differentiation. Differentiating Eq. (2.11) with respect to t, we have

$$\frac{dx}{dt} = v = \omega r \cos \omega t,$$

and by again differentiating we obtain

$$\frac{d^2 x}{dt^2} = \frac{dv}{dt} = a = -\omega^2 r \sin \omega t.$$

This is a quick, "streamlined," and obviously sound method of obtaining the equations for the velocity and acceleration, even though it does not give the physical insight into the behavior of the oscillating particle that is given by the vector diagrams of Figs. 2.10 and 2.11.

From Eqs. (2.11) and (2.13) we see that

$$a = -\omega^2 x; \tag{2.14}$$

that is, the acceleration is proportional to the displacement by the proportionality factor $-\omega^2$. The minus sign means that the direction of the acceleration is always opposite to that of the displacement; it is always toward the midpoint of swing.

From Eq. (2.14) we can at once find the period of the oscillation in terms of the ratio of the acceleration to the displacement. For the **period** T, defined as the time occupied by one complete oscillation, we have

$$T = \frac{2\pi}{\omega}, \tag{2.15}$$

47

and since, from Eq. (2.14),

$$\frac{1}{\omega} = \sqrt{\frac{|x|}{|a|}},$$

signs being disregarded, this is the same as

$$T = 2\pi \sqrt{\frac{|x|}{|a|}} = 2\pi \sqrt{\frac{\text{displacement}}{\text{acceleration}}}. \tag{2.16}$$

The alternative definition of SHM referred to above is the motion of a point along a straight line such that its acceleration is always proportional to its distance from some fixed point in the line and directed toward it. Of the two definitions, this is the one that is usually preferred. The two definitions will now be shown to be equivalent to one another.

Expressed mathematically, the alternative definition is

$$a = -kx,$$

or

$$\frac{d^2x}{dt^2} = -kx, \tag{2.17}$$

where k is a positive constant. The most general solution of this differential equation is

$$x = A \sin (\sqrt{k} \cdot t + \phi), \tag{2.18}$$

where A and ϕ are constants of integration. The reader who is not familiar with differential equations may readily verify for himself that this is indeed a solution by differentiating the right side of Eq. (2.18) with respect to t twice in succession. The resulting expression, $-kA \sin (\sqrt{k} \cdot t + \phi)$, is seen to be $-k$ times the expression (2.18) for x.

The equivalence of the two definitions of SHM follows at once from the geometrical interpretation of Eq. (2.18), which is the projection on a diameter of a point moving around a circle of radius A with uniform angular velocity $\sqrt{k}$. Writing ω in place of $\sqrt{k}$, we may express Eq. (2.18) in the alternative form

$$x = A \sin (\omega t + \phi),$$

which differs from Eq. (2.11) only in regard to the constant angle ϕ.

The angle represented by the quantity in parentheses in the last equation is known as the **phase** of the SHM; this increases linearly with time. The constant angle ϕ appearing in the expression for the phase is called the **phase angle.**

If we have two simple harmonic motions of equal period (or frequency) expressed mathematically by the equations

$$x_1 = A_1 \sin (\omega t + \phi_1)$$

and

$$x_2 = A_2 \sin (\omega t + \phi_2),$$

in which the phase angles ϕ_1 and ϕ_2 have different values, the difference between these angles is commonly referred to as the **phase difference** between the two simple harmonic motions, expressed as an angle. The phase difference may alternatively be expressed in terms of the period T of the simple harmonic motions. Thus, if $\phi_1 - \phi_2$ is $\pi/2$, we may either say that the first SHM leads the second in phase by $\pi/2$ (or 90°), or that it leads by $T/4$. In general, we may convert from an angle expressing a phase difference to the equivalent time by multiplying the angle by $T/2$.

The particular importance of simple harmonic motion derives from the fact that in so many periodic motions along a straight line occurring in nature the acceleration is proportional to the displacement from a fixed point by a negative proportionality factor. This is, for example, the case for a system subjected to an elastic "control," the negative proportionality factor corresponding to Hooke's law (see Chapter 10). Many other, nonmechanical, systems behave in a corresponding manner; for example, in an electrical circuit in which the plates of a capacitor, initially charged, are connected via an inductor (see Chapter 32), the charge of the capacitor varies sinusoidally with time. All these oscillatory phenomena may be classed together as simple harmonic *disturbances*. As we shall see later, the observation of a simple harmonic disturbance often enables us to obtain valuable quantitative information concerning the control (gravitational, elastic, electrical, and so on) under which it operates. Simple harmonic disturbances also constitute the basis of all accurate timing mechanisms, for example, the pendulum or balance wheel of a clock, or a quartz oscillator. Finally, simple harmonic disturbances are of fundamental importance in the study of the propagation of almost all kinds of waves.

CONICAL PENDULUM A very direct and convenient way to observe simple harmonic motion, defined as a projection of uniform circular motion, is to set up and operate what is known as a **conical pendulum.** This is indicated in Fig. 2.12. It consists of a bob suspended from a point S by a light string, which describes a circular orbit in a horizontal plane. In the figure the circle PQ described by the bob is seen edge on. If viewed in this way, the bob, which actually moves with constant speed around its circular orbit, *appears* merely to move back and forth in a straight line between P

49

FIGURE 2.12   Apparent motion of the bob of a conical pendulum when viewed in the plane of motion.

and Q. And if it is viewed from a sufficient distance, the line of sight may be regarded as always sufficiently nearly at right angles to the diameter PQ. Accordingly, the apparent motion of the bob is the projection of its uniform circular motion on this diameter.

SIMPLE PENDULUM   The simple pendulum is shown in Fig. 2.13. It consists of a bob supported by a light string from a fixed point S, which executes oscillations back and forth along an arc AB of a vertical circle in a plane defined by the points A, B, and S. It moves fastest when passing through its midpoint of swing O, vertically below S, and is momentarily at rest at the two turning points A and B at the extremities of its swing. Provided the arc AB is not too long relative to the length SO of the pendulum, the motion of the bob approximates very closely indeed to simple harmonic motion.

It may be objected that simple harmonic motion has been defined as the projection of uniform circular motion on a diameter, a *straight line*, whereas here we have motion along the arc of a circle, a *curve*. However, there is no reason why we should not imagine a straight line along which a particle is executing SHM to be bent into a curve. It is in this sense that the motion of the bob of the simple pendulum along the arc AB may be regarded as simple harmonic.

That this motion is indeed simple harmonic, at least to a close approximation, may be verified experimentally as follows. A simple pendulum and a conical pendulum are set up adjacent to one another, as indicated in Fig. 2.14, and set swinging with the ampli-

50

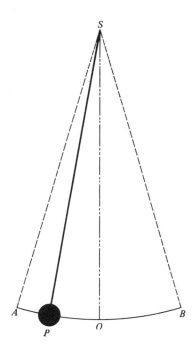

FIGURE 2.13 Simple pendulum.

tude of the simple pendulum small compared with the distance of the bob below the support. The lengths of the two supporting strings are adjusted so that the periods are as nearly as possible equal. It is also arranged that (1) the diameter of the circle described by the bob of the conical pendulum and the chord of the circular arc described by that of the simple pendulum are equal, and (2) when the two pendula are viewed from a distant point to the right (indicated by the arrow) and in the plane of the orbit of the conical pendulum, the two bobs pass across O simultaneously and in the same direction. In practice it is difficult to adjust the periods to exact equality. How-

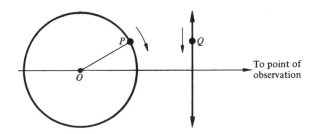

FIGURE 2.14 Correspondence between the projected motion of the bob of a conical pendulum and the actual motion of the bob of a simple pendulum oscillating with small amplitude.

51

ever, this does not matter; a reasonably close approximation is all that is required. Actually, a slight inequality is an advantage, as it causes one pendulum to gain on the other, and it is then only a matter of time before they are "in step," and continue to be so—at least sufficiently nearly—for a number of successive oscillations, so satisfying the second condition.

When these conditions are fulfilled, it will be observed that there is a detailed correspondence between the two motions seen; throughout the oscillations the two bobs always appear to be in line with one another; that is, their displacements from the line joining O to the eye of the distant observer are always the same. (Here we need not distinguish between the horizontal displacement of the bob of the simple pendulum and the displacement along the arc of swing; with a small amplitude the curvature of the arc is not important.) From this it may be inferred that the position Q of the bob of the simple pendulum is the projection of the position P of the bob of the conical pendulum on the line of oscillation of the simple pendulum. This oscillation is, therefore, at least to a close approximation, simple harmonic.

This is an important result, which we shall have occasion to make use of in establishing the laws of particle dynamics in Chapter 4.

Simple Harmonic Disturbances

As we have already noted, simple harmonic motion as executed by a particle is only one example of a whole class of oscillations known as **simple harmonic disturbances.** A simple harmonic disturbance may be defined mathematically as any disturbance s (for example, linear displacement, angular displacement, pressure change, temperature change, or change in electric field strength) whose variation with time t can be put into the form

$$s = A \sin (\omega t + \phi), \tag{2.19}$$

where A, ω, and ϕ are constants.

The constant A in this equation is called the **amplitude** of the periodic disturbance, while ω is related to the period T by the equation

$$T = \frac{2\pi}{\omega}. \tag{2.15}$$

The **frequency** f is defined as the number of periods in a unit of time (for example, in a second) and is thus equal to $1/T$. The frequency is therefore related to ω by the equation

$$f = \frac{\omega}{2\pi}. \tag{2.20}$$

As in the case of linear SHM, so with simple harmonic disturbances generally, there are two alternative definitions that are mathematically equivalent. One of these we have already expressed in Eq. (2.15). The mathematical form of the other is

$$\frac{d^2s}{dt^2} = -ks, \tag{2.21}$$

where k is a constant. Corresponding to the case of a linear simple harmonic motion, the most general solution of Eq. (2.21) is Eq. (2.15), with ω equal to $\sqrt{k}$.

Problems

2.1 A car, uniformly accelerated from rest, acquires a velocity of 40 miles/hr in 12 sec. What is the acceleration, and what is the distance covered in this time?

2.2 A train decreases speed uniformly from 60 to 30 miles/hr in 15 sec. Find how far it travels during this time.

2.3 A shell fired from a gun whose barrel is 5 m long leaves the muzzle with a velocity of 150 m/sec. Assuming a constant acceleration of the shell within the barrel, find the magnitude of this acceleration.

2.4 A truck traveling at 30 miles/hr sets out on a journey of 400 miles, and 3 hr later a car starts from the same place for the purpose of overtaking the truck before it arrives at its destination. If the car travels at 50 miles/hr, how far along the road will it overtake the truck? At what speed would the car have to travel to arrive at the destination simultaneously with the truck?

2.5 A man in a motorboat traveling along a river drops a bottle overboard. Half an hour later he decides that he needs the bottle after all, turns back, and retrieves the bottle 4 km downstream from the point where he discarded it. Assuming the speed of the motorboat relative to the water to be constant, find the speed of the current.

2.6 A ship travels between two points on a river whose distance apart is 10 km. The journey takes 40 min downstream and 1 hr and 20 min upstream. Find the speed v_0 of the current and the speed v of the ship relative to the water.

2.7 A car A is 200 m behind a car B and both are traveling in the same direction at 90 km/hr. Both cars begin at the same moment to accelerate at steady rates, but A has the greater acceleration and passes B in 20 sec, by which time B is doing 99 km/hr. Neglecting the lengths of the cars, find A's acceleration, its speed at the moment of passing B, and the distance it has traveled in the 20 seconds.

2.8 The measured distances through which a body falls from rest in various times are as follows:

Time of fall (sec)	$\frac{1}{2}$	$\frac{3}{4}$	1	$1\frac{1}{4}$	$1\frac{1}{2}$	$1\frac{3}{4}$
Distance (m)	1.2	3.0	4.5	7.2	10.8	15.0

Plot these measurements in an appropriate manner, and from your graph calculate the value of g and estimate the accuracy of your result.

2.9 Two objects are shot simultaneously with the same initial velocity v_0 from a platform projecting from the roof of a tall building, one vertically upward and the other vertically downward. Neglecting air resistance, find how the distance between the bodies will vary with time.

2.10 A passenger on a ship sailing due north at 15 miles/hr observes a wind apparently blowing from the northeast at 30 miles/hr. In what direction and with what velocity is the wind actually blowing?

2.11 A car traveling along a straight road at 60 km/hr passes through an intersection with another straight road at right angles, and 20 min later another car, traveling at 100 km/hr along the other road, passes through the same intersection. How does the distance between the two cars vary with time from the moment of the second crossing? What will this distance be when the second car is 20 km from the intersection?

2.12 A circular hoop in a vertical plane has a number of stiff straight wires joining its highest point to various other points on the hoop, the wires thus constituting chords of the circle. Prior to attachment to the hoop, each wire has been threaded through a small ring. Show that if these rings are released simultaneously from the top of the hoop and slide frictionlessly down their wires, they will all arrive simultaneously at the other ends.

2.13 An airplane flies at a height h along a horizontal straight line with speed v toward a target which the pilot intends to bomb. Neglecting air resistance, find at what angle to the horizontal he must see the target at the moment of dropping the bomb.

2.14 A particle, initially at rest, slides frictionlessly down a curve of arbitrary form. Show that after having lost height by an amount h its velocity is $\sqrt{2gh}$, irrespective of the shape of the curve.

2.15 Two ships are traveling in opposite directions along parallel paths whose distance apart is d, one with velocity v_1, the other with velocity v_2; and, at the moment when the line joining them is perpendicular to their directions of travel, one ship fires on the other. Assuming the horizontal component v_0 of the shell velocity relative to the ship firing it to be constant, find, in terms of v_1, v_2, v_0, and d, what should be the angle θ between the direction of the target ship and the vertical plane through the axis of the gun at the moment of firing for a hit to be made.

2.16 Repeat Problem 2.15 for the case where the ships are both traveling in the same direction instead of in opposite directions.

2.17 A motorboat is headed at right angles across a stream which is 200 m wide and which flows at 15 km/hr. The boat reaches the opposite bank at a point 400 m downstream. What is the velocity of the boat (a) relative to the water, and (b) relative to the bed of the stream?

The boat is now required to return across the stream at right angles to the bank. In what direction must it be headed if the engine gives the boat a speed four times as great as in the first crossing?

2.18 A passenger in a train traveling over a viaduct drops a heavy object into the valley below. What will be the path described by the object in its fall (a) as seen by the passenger, and (b) as seen by a stationary outside observer? Is either description *in principle* preferable to the other?

2.19 A pebble is embedded in the tread of one of the front wheels of a car, and its distance from the axis of the wheel is 20 cm. When the car travels forward at 30 m/sec the pebble flies horizontally off the wheel from a point directly above the axle. What will be the horizontal distance between the axle and the pebble at the instant when the pebble strikes the ground? (Assume that the car has no mudguards to interfere with the flight of the pebble and that air resistance may be neglected.)

2.20 A ship at sea fires a shell with a muzzle velocity of 150 m/sec from a gun elevated at an angle of 30° to the horizontal. Neglecting air resistance, calculate (a) the maximum height attained by the shell, (b) the duration of its flight, and (c) its range.

2.21 A shell fired from a gun travels in a curved path toward a distant target. At what point will the component of the acceleration of the shell normal to its trajectory be a maximum? (Neglect air resistance.)

2.22 Show that for each horizontal range less than the maximum of a shell fired from a gun with a given initial velocity there are two angles of elevation of the gun which give this range. Show also that for the maximum range the gun must be elevated at an angle of 45°. (Neglect air resistance.)

2.23 The acceleration due to gravity on the surface of the moon is 0.164 of that on the earth's surface. How would (a) the time of flight, (b) the range, and (c) the maximum height of a body projected with a given velocity at a given angle to the horizontal on the moon's surface compare with the corresponding quantities on the surface of the earth?

2.24 A gun fires a shell at a target situated a horizontal distance d away and at a vertical height h (which may be either positive or negative) above the gun. The velocity of the shell as it leaves the gun is v, and the angle θ of elevation of the gun when it is fired is such that on its way to the target the shell reaches a height greater than that of the target. Neglecting air resistance, find answers to the following questions: (a) What is the maximum height above the gun reached by the shell? (b) How long does the shell take to reach this height? (c) How long does it take the shell to fall from its maximum height to the level of the target? (d) What is the total time taken from the moment of firing to the moment the shell is at the level of the target on its way down? (e) What is the horizontal distance covered by the shell during this time? (f) What is the algebraical equation expressing the condition that the shell hits the target? (g) What angle ϕ does the direction of travel of the shell make with the horizontal just before it strikes the target, and with what velocity is it traveling? (h) If the locations of the gun and target are interchanged, and the gun is elevated at an angle ϕ to the horizontal, with what velocity must the shell be fired to enable it to hit the target?

2.25   A stone is released from rest at the top of a vertical cliff of height h, and at the same moment another stone is projected vertically upward from the foot of the cliff with an initial speed equal to that with which the first stone will eventually strike the ground (assuming that the stones do not collide). Neglecting air resistance, calculate how far down the cliff the two stones will pass each other and at what time this will occur.

2.26   A number of small projectiles are launched with the same initial speed v simultaneously in different directions from the top of a tower. Show that in the absence of air resistance these will subsequently all lie on the surface of a sphere whose center falls vertically with the acceleration g and whose radius is vt, where t is the time that has elapsed since the moment of launching.

2.27   A bucket is tied to the end of a rope which unwinds without slipping from a drum of radius 20 cm as the bucket falls from rest with acceleration 3 m/sec² down a well. Find the angular acceleration $d\omega/dt$* of the wheel. Write an equation expressing the total angle turned through by the wheel as a function of time.

2.28   The velocity of a point on the circumference of a wheel rolling along a plane surface without slip may be regarded as the resultant of two separate velocities: the translational velocity of the wheel as a whole, and the velocity due to rotation of the wheel about an axis through its center with angular velocity ω. Find the components of this resultant velocity parallel and at right angles to the plane, and hence show that it is the same in both direction and magnitude as the velocity the point would have if it were rotating momentarily with angular velocity ω about a stationary axis through the point of contact of the wheel with the plane.

2.29   Supposing Galileo had realized the gravitational nature of the relevant accelerations, could he have predicted the result of his Pisa experiment simply from a knowledge of the sizes of the planets (which are all small compared with the sun and vary considerably, in a somewhat irregular manner) and of their orbital motions about the sun as expressed by Kepler's laws? If so, how?

2.30   Given that the radius of the earth is 6.37×10^6 m and that the height of Pike's Peak above sea level is 4293 m, calculate by how much the value of g at the summit of this mountain would be less than the sea-level value at the same latitude if there were no complications due to local irregularities in the distribution of material and to the attraction of the mountain itself.

2.31   An artificial satellite is to be put into a circular orbit such that it will always appear vertically above a certain fixed point on the earth's surface. Can this be achieved for any point one wishes to select or only for points having a certain latitude? What will be the plane of the orbit, and its radius? (Radius of earth = 6371 km.)

2.32   Given, in addition to information which is well known, that the radii of the earth's orbit about the sun and the sun itself are 1.495 ×

---

* Those readers who are not yet familiar with the calculus may simply accept $d\omega/dt$ as a symbol for the rate of increase of angular velocity with time.

10^8 km and 6.95×10^5 km, respectively, calculate the value of g_0 at the surface of the sun.

2.33 A particle executes simple harmonic motion with amplitude A. At what displacement from the midpoint of its oscillation is its speed one half of its maximum speed?

2.34 A simple pendulum has a period of 3 sec and an amplitude of 40 cm. Find the velocity of the bob at the midpoint of its swing. Find also the acceleration (a) at one extremity, and (b) 20 cm from the midpoint of swing.

2.35 The equation of motion of a particle is

$$x = 6.5 + 8.4 \sin 20t,$$

where x is in centimeters and t in seconds. State what kind of motion this is, and give full quantitative particulars concerning its various features.

2.36 A particle is executing simple harmonic motion between the limits -2 cm and $+8$ cm, and its speed at the midpoint of its motion is 150 cm/sec. What are (a) the amplitude, (b) the frequency, and (c) the maximum acceleration of the particle? What is the equation of motion referred to the midpoint of swing as origin?

2.37 A particle executing simple harmonic motion travels a total distance of 32 cm during each cycle, and its maximum acceleration is 12 m/sec². What are (a) the frequency of the oscillation, and (b) the maximum velocity of the particle? What is the equation of motion referred to the midpoint of swing as origin?

2.38 A particle has x and y coordinates varying with time according to the equations

$$x = 10 \sin 18t$$

and

$$y = 10 \sin \left(18t + \frac{\pi}{2}\right),$$

x and y being in centimeters and t in seconds. Give full particulars concerning the nature of the motion of this particle and the way in which the distance and direction of the particle from the origin, its velocity, and its acceleration vary with time.

2.39 A particle has x and y coordinates varying with time according to the equations

$$x = 15 + 10 \sin 15t$$

and

$$y = 5 \sin \left(15t - \frac{\pi}{2}\right),$$

where x and y are in centimeters and t in seconds. Describe, as fully as you can, the various features of this motion.

2.40 Show that two uniform circular motions in which the circles, of equal radii, are described in equal times but in opposite directions, combine vectorially to give simple harmonic motion. What is the amplitude of the simple harmonic motion, and what determines its direction?

3

WAVE MOTION

The study of waves is of fundamental importance in physics. Not only do we have water waves and waves traveling along strings, which are familiar to everybody, but sound, light, and radio signals are also propagated as waves. The latter two are members of a family known as electromagnetic waves, which are of basic significance in physical theory. Since the mid-1920s we also know of waves, called "matter waves," which are much more difficult to visualize than any of those mentioned above, but which play a fundamental role in the branch of modern physics known as quantum mechanics.

Strictly speaking, not all waves come under the heading of wave *motion*. For example, although there certainly is motion involved in the propagation of water waves, waves traveling along strings, and sound waves, there is no motion associated with the propagation of electromagnetic waves. Nevertheless, electromagnetic waves are easier to visualize and to understand if we have first studied waves propagated mechanically in material media, such as waves in a stretched string, whose particles move rhythmically as the waves travel along it; there are several useful points of analogy between the two kinds of wave propagation. We shall therefore give special consideration to mechanically propagated waves.

In the present chapter we shall analyze wave motion kinematically and develop mathematical equations that describe it. In Chapter 11 we shall deal with the dynamics of wave propagation in material media.

In the mechanical propagation of waves, the form of the waves at any particular instant is determined by the displacements of all the particles of the medium, and their motion is determined by the motions of all these particles. The situation is an orderly one, the displacements and motions of all the separate particles being related to one another in a perfectly definite manner. Accordingly, we shall find that the theoretical elucidation and mathematical representation of wave motions are not nearly as difficult as we might have anticipated.

59

Waves along a Stretched String or Rope

If a long rope is fixed at one end, and the other end is shaken rhythmically up and down in a direction at right angles to the rope, waves will be seen to travel along it. These may be observed without any complication due to reflection at the fixed end if an appreciable length of the rope at this end is arranged to be under water. The energy of the oncoming waves will then be dissipated in the water, and a continuous train of waves will be seen to travel in a single direction along the rope.

Let us consider the case in which the motion imparted to the free end of the rope where the waves are generated is simple harmonic. The form of the wave train traveling along it at any particular instant will then be such as is indicated in Fig. 3.1. This is a **sine wave,** so called because the displacement y of any point on the rope from its undisturbed position may, for the instant in question, be expressed as a sine function of the coordinate x of this point measured from a "zero" point or origin. What the precise nature of this sine function is we shall see later. Meanwhile, we may take note of two important parameters of such a wave train. These are indicated in the figure. One is the **amplitude** A of the waves, defined as the maximum distance a point is displaced from its undisturbed position as the wave train travels through it. The other is the **wavelength** λ. This is the smallest distance between pairs of corresponding points on the wave train; corresponding, in the sense that not only are the displacements of the two points at the particular instant the same but also the slopes of the curve defining the wave train at these points.

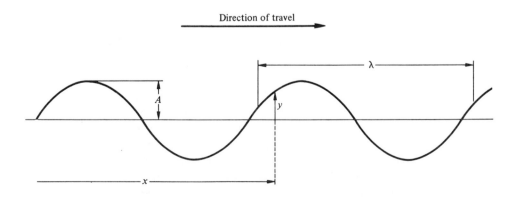

FIGURE 3.1   Transverse sine wave and associated quantities.

The nature of the motion of the individual elements, or "particles," of the rope may be observed by suitably marking one of these, with chalk, for example. The point marked will then be seen to execute simple harmonic motion along a line perpendicular to the undisturbed direction of the rope. Thus we may think of the rope as consisting of a string of particles, each oscillating with simple harmonic motion transversely to its undisturbed direction. Accordingly, the displacement y of a selected element of the rope from its undisturbed position may be expressed as a function of time t by the equation

$$y = A \sin (\omega t + \phi), \tag{3.1}$$

where A is the amplitude of the SHM (the same as the amplitude of the wave train, defined earlier); ω is $2\pi/T$, where T is the period of the oscillation; and ϕ is a constant angle whose magnitude is characteristic of the element of the rope under consideration.

All the different elements of the rope oscillate in the same direction with the same period T, so ω is the same for all of them. Also, the amplitude of the oscillation is everywhere the same. But these items are all that the elements have in common. Obviously they do not all do the same thing (for example, pass in a certain direction through their midpoints of swing) at the same time; they do not vibrate in step with one another. This means that each element has its own characteristic ϕ; the value of ϕ, which, as we have seen, is constant *with respect to time* for each element, must vary from one element along the rope to the next. If we can find the nature of this variation, and make the corresponding substitution for ϕ in Eq. (3.1), we shall have our complete **wave equation,** an equation that gives y as a function of both t and position along the rope for the whole wave train.

We obviously need to specify, mathematically, the position of an element along the rope. This we have, in fact, already done (see Fig. 3.1) by assigning it a Cartesian coordinate x, measured along the rope from a fixed origin, with an agreed direction from this origin counting as positive. The value (including sign) of x then identifies the particular element of rope whose motion we are considering. Our problem is now to express ϕ as a function of x.

Let us consider three points on the rope, A, B, and C, indicated in Fig. 3.2, such that the length of rope between B and C is the same as that between A and B. Then if the rope is uniform and the tension is the same throughout its length, the oscillation of C must be related in time to that of B in exactly the same way as B's oscillation is related to that of A. With a noncapricious progression of the wave train it could, indeed, hardly be otherwise; only on the basis of

61

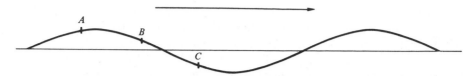

FIGURE 3.2 Instantaneous configuration of a transverse wave train.

equality of the two time relations would it be possible for the waves to travel with uniform velocity and not in jerks. And because these time relations depend on the values of ϕ for the three points, we must have

$$\phi_B - \phi_C = \phi_A - \phi_B,$$

where ϕ_A, ϕ_B, and ϕ_C are the values in question. In other words, ϕ varies linearly with x.

We now have to ascertain whether ϕ increases or decreases with increasing x for waves traveling in the positive x direction. In Fig. 3.3 let the solid curve represent the wave train at time t and the dashed curve that at the later time $t + \delta t$, when the waves have progressed a short distance to the right. Let A be a point on the former and B a corresponding point on the latter; that is, A and B are two points on the rope such that not only is the displacement y of B at time $t + \delta t$ the same as that of A at time t, but also the rate at which the displacement varies with time is the same for B at time $t + \delta t$ as it is for A at time t. These quantities (the displacements and their time variations) will be the same for both points if the angles whose sines appear in equations of the type (3.1) are the same in both cases. If ϕ_A is the value of ϕ characteristic of the point A and ϕ_B that characteristic of B, the condition to be fulfilled is that

$$\omega t + \phi_A = \omega(t + \delta t) + \phi_B.$$

Obviously, the two sides of this equation can be equal only if, to compensate for $\omega(t + \delta t)$ being greater than ωt, ϕ_B is less than ϕ_A. This means that for a forward-traveling wave train ϕ must decrease

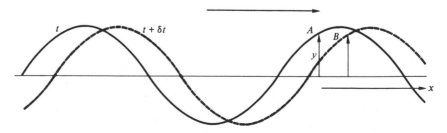

FIGURE 3.3 Two consecutive configurations of a transverse wave train.

with increasing x; that is, ϕ may be represented as a function of x by an equation of the type

$$\phi = \phi_0 - kx, \tag{3.2}$$

where ϕ_0 (the value of ϕ at the origin) and k are constants, k more particularly a *positive* constant.

Substituting the value of ϕ as given by Eq. (3.2) in Eq. (3.1), we now have the wave equation for a forward-traveling wave train,

$$y = A \sin (\omega t - kx + \phi_0), \tag{3.3}$$

an equation giving y in terms of both variables, t and x.

Although this is indeed the wave equation we seek, it is not yet in a form we generally find convenient. In a more usual form the variation of y with t and x is expressed, not in terms of ω and k, but of the period T of oscillation of each particle and the wavelength λ. The relation between ω and T for a particle executing simple harmonic motion is expressed by Eq. (2.15). This (rearranged) is

$$\omega = \frac{2\pi}{T}.$$

And we can see how k may be expressed in terms of λ as follows. Let us consider two points on the rope, one of coordinate x, and the other, one wavelength farther along in the direction in which the waves are traveling, of coordinate $x + \lambda$. At any particular instant of time, both y and its rate of variation with time are the same for these two points. This means that the angles in terms of which the values of y for the two points are expressed must differ by 2π; only so can the conditions for the two points being one wavelength apart, as expressed by the definition of a wavelength, be satisfied. Moreover, the angle for the point of coordinate $x + \lambda$ must be less than that for the point of coordinate x—less, not greater, because of the minus sign in Eq. (3.3). Accordingly,

$$\omega t - k(x + \lambda) + \phi_0 = \omega t - kx + \phi_0 - 2\pi,$$

or

$$k = \frac{2\pi}{\lambda};$$

thus k is related to λ in the same way as ω is related to T. Hence, making the corresponding substitutions in Eq. (3.3), we have the wave equation in one of its more usual forms,

$$y = A \sin 2\pi \left(\frac{t}{T} - \frac{x}{\lambda} + \epsilon \right), \tag{3.4}$$

where ϵ is written for $\phi_0/2\pi$.

63

Obviously, with a suitable adjustment of zero for either t or x, that is, with a suitable choice of either the instant from which time is counted or the position of the origin from which the coordinate x is measured, the constant ϵ can be made to vanish, the wave equation then taking the somewhat simpler form

$$y = A \sin 2\pi \left(\frac{t}{T} - \frac{x}{\lambda}\right), \qquad (3.5)$$

or, in terms of frequency instead of period,

$$y = A \sin 2\pi \left(ft - \frac{x}{\lambda}\right). \qquad (3.6)$$

With a further adjustment in one or other of the zeros, the order of the quantities in parentheses may be reversed, the equation appearing as

$$y = A \sin 2\pi \left(\frac{x}{\lambda} - \frac{t}{T}\right),$$

or

$$y = A \sin 2\pi \left(\frac{x}{\lambda} - ft\right)$$

We may also express y in terms of A, t, x, λ, and the velocity v of propagation of the waves. The distance traveled by the wave train during one complete period is a wavelength, so we must have

$$v = \frac{\lambda}{T},$$

or

$$v = f\lambda. \qquad (3.7)$$

Hence, substituting for f in the last equation its value, v/λ, we obtain the wave equation in yet another form,

$$y = A \sin \frac{2\pi}{\lambda} (x - vt).$$

All these are common forms in which to represent the progression of a wave train mathematically.

POLARIZATION Waves such as we have been considering, in which the motions of the particles are entirely normal to the direction of travel of the waves, are known as **transverse waves.** There are, of course, an infinite number of directions normal to the direction of travel, and the particles may move along any one of these or along a combination of them. If, as in the case considered, their motion is confined to a single direction, the waves are said to be **plane-polarized,** and their **plane of polarization** is the plane defined by the direction of travel of the waves and the direction of oscillation of the particles. The plane of polarization of plane-polarized transverse waves may be any one of the planes containing the direction in which the waves are traveling.

Transverse waves need not necessarily be plane-polarized. Thus, if in the rope experiment the hand holding the free end of the rope, instead of being shaken to and fro, is turned around in circles, the plane of motion being at right angles to the undisturbed direction of the rope, waves of a different form will be generated, each particle describing a circular orbit in a plane parallel to this. Again, there will be the usual phase relation between the motions of the various particles, and the configuration of the rope will be that of a helix, the turns of which move from the hand where they are generated, along the rope, and into the water at the far end (Fig. 3.4(a)). Such waves are said to be **circularly polarized.** Circularly polarized waves are of two kinds, right-handed and left-handed, according to whether, as viewed in the direction of travel of the waves, the orbits of the particles appear to be described in a clockwise or in an anti-clockwise direction (Fig. 3.4(b)). Because of the progressively increasing phase lag with increasing x which applies here also, the helical wave form generated in a right-handed circularly polarized wave train is left-handed, like the turns of a left-handed screw,

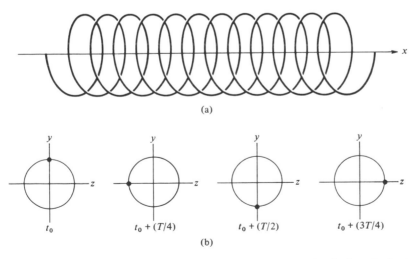

(a)

(b)

FIGURE 3.4   (a) Wave form of a train of left-handed circularly polarized waves.
(b) Position of a particle at four successive instants of time in a plane normal to the x axis.

65

while in the case of a left-handed circularly polarized wave train the helix is right-handed.

It is useful to think of circularly polarized waves in terms of a series of clocks each having one hand only, whose dials are all at right angles to the direction of propagation of the waves and face the observer when he is looking in the direction of propagation. If the circular polarization is right-handed, the hands turn in their usual direction, and at any particular moment the clocks tell progressively earlier times with increasing x. This is because of the progressively increasing phase lag. And the wave train itself will be in the form of the curve that goes through the ends of the hands of all the clocks. Clearly, this will be a left-handed helix. In the case of left-handed circular polarization we may use the same picture of a succession of clocks, but with all of them going backward, and with the times they tell (on *conventional* clock faces) at any particular instant being progressively later with increasing x. To this must correspond a right-handed helical wave form.

In either of the two cases of circular polarization, the position of the hand of any clock may be specified in terms of an angle θ, the angle turned through from an arbitrary starting position. And, the radius of the circles being given, the rest of the information concerning the waves is contained in a wave equation of the form

$$\theta = 2\pi \left(\frac{t}{T} - \frac{x}{\lambda} + \epsilon \right), \tag{3.8}$$

or

$$\theta = 2\pi \left(ft - \frac{x}{\lambda} + \epsilon \right). \tag{3.9}$$

In addition to plane-polarized and circularly polarized waves we may have transverse waves that are **elliptically polarized.** In the rope experiment elliptically polarized waves may be generated by moving the hand that holds the free end of the rope in an elliptical path, again, of course, in a plane at right angles to the undisturbed direction of the rope.

It may be shown that circularly polarized and elliptically polarized waves may be regarded as compounded of two plane-polarized waves whose planes of polarization are mutually at right angles. This follows from the possibility of resolving the circular or elliptical motion of the individual particles into two simple harmonic motions at right angles to one another. This may be seen in the case of circular motion from the definition of simple harmonic motion as the projection of uniform circular motion along a straight line. We have merely to resolve the displacement of a particle

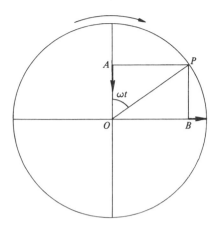

FIGURE 3.5 Equivalence of uniform circular motion to two simple harmonic motions of equal amplitude (radius of circle) differing in phase by $\pi/2$.

executing uniform circular motion, measured from the center of the circle, along two mutually perpendicular diameters, as indicated in Fig. 3.5, OA and OB being the components along these diameters of the displacement OP. In this way we see that the uniform circular motion of the point P may be resolved into the two simple harmonic motions of the points A and B, with phase difference $\pi/2$. Conversely, these two simple harmonic motions may be compounded into uniform circular motion.

Longitudinal Waves

Waves in which the particles execute simple harmonic motions in the same direction as that in which the waves travel are known as **longitudinal waves.** An example of longitudinal wave motion may be seen when one end of a long helical spring is fixed, and the other, held in one's hand, is moved rapidly back and forth in a direction coinciding with the axis of the spring. Some cotton wool or other suitable material may be inserted in the fixed end to absorb the energy of the oncoming waves and so prevent the waves from being reflected. It will be seen that a regular succession of "compressions" and "rarefactions" of the coils of the spring travels from the hand, where they are generated, along the spring, toward the fixed end.

The most important example of longitudinal waves is the series of compressions and rarefactions by which sound is propagated.

67

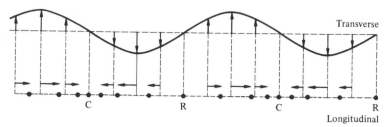

FIGURE 3.6 Relation between plane-polarized transverse waves and longitudinal waves.

The relation between transverse and longitudinal waves may be seen from Fig. 3.6, the arrows representing the displacements in the two cases. In the lower part of the figure the positions of the compressions and rarefactions are indicated by the letters C and R, respectively. We see that there is complete correspondence between the two types of waves except that the displacement y in longitudinal wave motion has the same direction as x, instead of being at right angles to it. This being understood, all the equations describing plane-polarized transverse waves also apply to longitudinal waves whose amplitude is constant with respect to x. It is, of course, important not to confuse the quantities x and y, which now have the same direction. As in the case of plane-polarized transverse waves, x is the distance of the midpoint of swing of a particle from the fixed reference point or plane and y its instantaneous displacement from this midpoint.

Waves in General

Not all mathematical representations of the propagation of material waves need be explicitly in terms of the motions of particles. As we have just noted, sound is propagated in the form of longitudinal waves, and these may be represented by the same equations as apply to plane-polarized transverse waves, in which the displacement y of a particle of the medium is expressed as a function of t and x. There are, however, alternative ways of representing sound waves. One of these, for example, is to express it in terms of the pressure above or below the mean pressure prevailing in the medium. The equation for sound waves then takes the form

$$p = p_0 \sin 2\pi \left(\frac{t}{T} - \frac{x}{\lambda} \right),$$

where p denotes the pressure increase (positive or negative) at the point x at time t and p_0 is a constant which we may call the pressure amplitude.

Not all waves are *material* waves (that is, motion of particles of matter). Light waves are an important example of nonmaterial waves. In the propagation of light there is a variation of both electric and magnetic field transverse to the direction of propagation, the directions of the electric and magnetic fields also being at right angles to one another. Light is accordingly said to be propagated in the form of electromagnetic waves, and the variation of electric and magnetic fields, E and B, in a plane-polarized wave train may be expressed by equations such as

$$E = E_0 \sin 2\pi \left(\frac{t}{T} - \frac{x}{\lambda} \right)$$

and

$$B = B_0 \sin 2\pi \left(\frac{t}{T} - \frac{x}{\lambda} \right),$$

respectively.

Strictly speaking, the equations given above for sound waves and for electromagnetic waves refer only to so-called "plane waves" —those in which at any given instant the disturbance is everywhere the same in any plane perpendicular to the direction of propagation and in which the amplitude does not fall off with increasing distance. By contrast, an equation representing spherical waves— actually sound waves—spreading out uniformly in all directions from a small spherical source would be

$$p = \frac{p_0}{r} \sin 2\pi \left(\frac{t}{T} - \frac{r}{\lambda} \right),$$

where p is the pressure (above normal) at the distance r from the center of the oscillating sphere at which the sound is generated and p_0 is a constant. The amplitude factor p_0/r falls away with distance according to an inverse-proportionality law, as may be shown to be required by energy considerations.

Doppler Effect

Every motorist is familiar with the sudden drop in pitch of the sound of a passing car's horn at the moment of passing, that is, at the moment when relative approach between the source of the sound and the observer changes to relative recession. This is an example of the **Doppler effect.** It will be shown in Chapter 11 that the pitch of a sound depends uniquely on its frequency, becoming higher with increasing frequency. Thus the phenomenon just cited may be interpreted in terms of frequency as a fall in frequency of the sound heard accompanying the transition from relative approach to relative recession.

Although an effect essentially of this kind was first observed by Doppler with electromagnetic waves in the optical range of frequencies, it is best studied in the first instance in relation to sound. We shall accordingly now consider the most general one-dimensional case of the acoustical Doppler effect by supposing that not only are the source and the observer in motion, but also the medium.

Let us define the direction from the source to the observer as the positive x direction. Let the velocities in this direction of the source S, the observer O, and the medium be u, v, and w, respectively (Fig. 3.7). We shall consider these velocities algebraically; they may have either positive or negative values, or one or more of them may be zero. And let V be the velocity of sound in still air, that is, the velocity relative to the medium. This is superimposed on the wind velocity w, so that the velocity relative to the ground with which the sound waves travel toward the observer is $V + w$. And, finally, let the frequency of vibration of the source be f.

The frequency f' which the observer attributes to the sound he hears is the number of waves that he intercepts per second. To find this number we shall have to know not only the velocity of the wave train relative to the observer (that is, the length of wave train intercepted by him in a second) but also the wavelength λ. The former divided by the latter must then necessarily be f'.

Let us first find the wavelength, the distance between the centers of two successive compressions. For this we shall need to

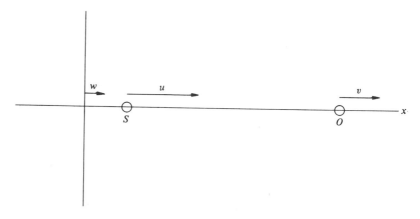

FIGURE 3.7 Diagrammatic representation of medium, source of sound (S), and observer (O), with their respective velocities.

know the time interval between the emission of two successive compressions by the source, the distance the first compression has moved during this time, and the distance moved by the source. The time interval in question is, of course, the period of vibration, $1/f$. During this time the first compression, traveling with the velocity $V + w$, has moved forward a distance $(V + w)/f$, while the source, traveling with the velocity u, has moved forward the distance u/f. And at the end of the time interval the second compression is just leaving the source. The distance between two successive compressions must be the difference between the two distances just quoted, $(V + w - u)/f$. This is the wavelength λ:

$$\lambda = \frac{V + w - u}{f}.$$

The velocity of the wave train relative to the observer is obviously $V + w - v$; this is the length of wave train that he intercepts in a second. In this length there are $(V + w - v)/\lambda$ waves. Hence for the frequency f' heard by the observer we have

$$f' = \frac{V + w - v}{\lambda} = \frac{V + w - v}{V + w - u}f.$$

Superposition

It sometimes happens that two or more wave trains simultaneously traverse the same region of a medium. In such a case the question arises of what is the nature and magnitude of the resulting disturbance at each point. Close observation of a number of cases suggests a simple answer: Each wave train retains its identity and travels independently of the other(s); and the disturbance at every point is at all times the algebraic sum, or the vectorial resultant, of the separate disturbances. This is what is known as the **principle of superposition.** For small-amplitude waves* of all kinds (waves traveling along strings, sound waves, water waves, electromagnetic waves, and so on) the validity of this principle is confirmed by a rigorous mathematical analysis of the mechanism of propagation of the waves.

In physical literature it is customary to refer to the algebraic or vectorial addition of the separate disturbances as "interference" rather than "superposition." This is perhaps a little unfortunate,

---

* In the great majority of cases that concern us in physics, the amplitudes are, in fact, small enough for us to consider this condition to be, for all practical purposes, satisfied.

as the uninitiated might interpret this expression as implying mutual cancellation of the separate disturbances or some tendency in this direction. Actually, although mutual cancellation does occur when the separate disturbances add up, algebraically or vectorially, to zero, there can equally well be reinforcement, when the separate disturbances are in the same direction and so give a resultant greater than either. It is important to realize that in physics the term "interference" is used in this wider sense, covering both "destructive interference" (total or partial mutual cancellation) and "reinforcement."

In this chapter we shall consider two special cases of interference, under the headings "Standing Waves" and "Beats." In Chapters 38 and 39 we shall analyze in considerable detail various cases of optical interference, that is, interference of electromagnetic waves. We may note here that interference phenomena constitute the main, and in many cases the only, basis for the determination of wavelengths. We owe our knowledge of the wavelengths of light of different colors entirely to observations of optical interference phenomena.

STANDING WAVES   In the rope and helical-spring experiments designed to demonstrate transverse and longitudinal wave motions, means were employed to suppress reflected waves. It is now of interest to inquire what happens when reflection from the fixed end is not suppressed, and the reflected and incident waves are allowed to interfere. In order to study this case experimentally, let a rope be fastened at one end as before, but this time not under water, and let the free end be held in the hand.

As a preliminary, some important facts about reflection may be learned by carrying out some very simple experiments. The free end held in the hand is first simply flicked sharply, and the pulse so generated is watched as it progresses along the rope. It will be observed that when the pulse reaches the far end of the rope it is reflected back again and appears to travel on its return journey with the same speed as it did on the forward journey. There is, indeed, no apparent reason why it should return at a different speed from that with which it previously advanced, and we may therefore assume that the speeds in the two directions are the same.

Next, let the rope be flicked, say, three times in rapid succession, but at unequal intervals, so as to generate a series of pulses with a characteristic pattern, as depicted in Fig. 3.8(a). It will be observed that in due course this series returns with its characteristic pattern reversed in direction, but otherwise unchanged, as indicated in Fig. 3.8(b). This is highly significant, because during the process of reflection the beginning of the reflected series must travel on its

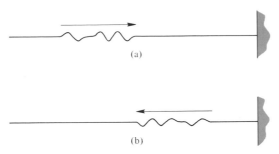

(a)

(b)

FIGURE 3.8 Incident and reflected pulses traveling along a stretched string.

return journey through that part of the series which is still approaching the end in the forward direction. Yet, despite this mixing up of the incident and reflected parts of the disturbance, the reflected series eventually emerges from the region of the fixed end with its character unchanged. This is one of the cases referred to above where close observation of an interference phenomenon suggests that the interfering wave trains travel independently of one another, each as if it were the only one present.

We have now to consider the additional question of what, precisely, the reflecting agency, for example, the post to which the far end is tied, does to the waves. Does it *merely* turn the waves back on their path, or does it do something else besides? In Fig. 3.9 let B be the reflecting "boundary," I the incident wave train, and I' an imaginary continuation of this beyond the boundary. The question now is whether the action of the boundary on the waves is simply to cause I' to be reversed in direction as from the boundary B, so that the reflected waves are as depicted by the dashed curve R. It is at once evident that this cannot be so, for, if it were, the rope at the boundary would be displaced by twice the amount corresponding to the incident waves. This is clearly impossible; the boundary condition imposed by the attachment of the rope at O is that at this point it does not move.

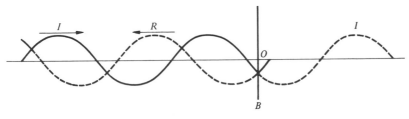

FIGURE 3.9 Diagram showing impossibility of zero phase change on reflection.

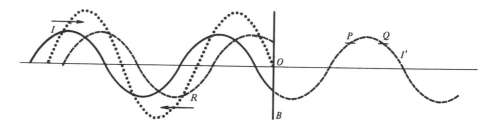

FIGURE 3.10 Incident and reflected wave trains, showing phase reversal on reflection.

For this boundary condition to be fulfilled, the displacement at the boundary due to the reflected waves must at all times be exactly equal and opposite to that due to the incident waves. The two will then cancel each other out. This implies a phase change on reflection. What is this phase change? Do the return waves start from B as at P in the imaginary prolongation I' shown in Fig. 3.10 or as at Q? A moment's consideration will show that the waves cannot be reflected from B with their phase as at Q, for, if they were, then (1) there would be no consistent phase relation between the incident and reflected waves, and (2) the resultant displacement, the algebraic sum of the separate displacements associated with the incident and reflected waves, would be zero everywhere and at all times, and this is obviously not the case. The return waves must therefore begin with their phase as at P, which may be regarded as one half of a wavelength ahead of the point O. This is sometimes also expressed by saying that there is a *reversal* of phase on reflection. The reflected waves with their phase correspondingly adjusted are drawn in as the dashed curve R on the left of B in Fig. 3.10 and the resultant of the incident and reflected waves is indicated by the dotted curve.

The dotted curve represents the resultant of the incident and reflected waves at a particular instant. Of course, this resultant varies with time: Twice in each period the resultant disturbance passes through a maximum value, and twice the separate disturbances due to the two waves just cancel each other out. This is shown in Fig. 3.11, where the conditions for these four special cases are shown in chronological order.

In Fig. 3.12 the two maximum resultant disturbances corresponding to the second and fourth series are shown together. These are limiting configurations of the combined waves. At all times the disturbances lie somewhere between these limits, and at any point, such as P, a particle executes simple harmonic oscillations between

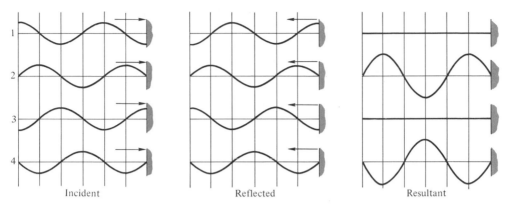

Incident   Reflected   Resultant

FIGURE 3.11 Four special cases of incident and reflected wave trains, and the corresponding resultant disturbances.

corresponding limits, that is, between the points Q and R. The resultant motion as a whole constitutes what are known as **standing waves.** By reference to Fig. 3.11 it is easily seen that the line QR has its minimum length (zero) at the points marked N and its maximum at those marked A. These points of zero and maximum disturbance are known as **nodes** and **antinodes,** respectively. It is easily seen that the distance between two successive nodes or two successive antinodes is $\lambda/2$ and that the distance between a node and an adjacent antinode is $\lambda/4$. If l_1 is the former distance and l_2 the latter (see Fig. 3.12),

$$\lambda = 2l_1 = 4l_2, \tag{3.10}$$

and hence, from Eq. (3.2), we see that

$$v = 2fl_1 = 4fl_2, \tag{3.11}$$

or

$$f = \frac{v}{2l_1} = \frac{v}{4l_2}. \tag{3.12}$$

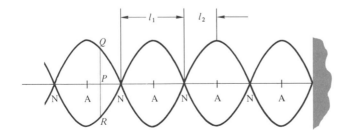

FIGURE 3.12 Limiting configurations of a standing wave disturbance. **75**

We may now inquire whether there is any optimum length of rope, or series of lengths, between the fixed end and that at which motion is imparted to it, corresponding to a given velocity of the waves and a given impressed frequency. The answer to this question cannot be obtained from kinematical considerations alone, but it will nevertheless be convenient to deal with it here. When once the motion is well established, there is no need to feed energy (see Chapter 6) into the rope at any considerable rate, because there is little loss. Correspondingly, it is found that the most favorable position of the hand or other disturbing agency is near a node, where its motion is relatively slight. There is then reflection from this point in much the same way as from the fixed end, only under slightly modified conditions with regard to phase relation.

Standing waves in a rope or stretched string may also be generated in cases where both ends are fixed, as in that of a violin string. When a violin is being played, the bow, which generates the vibration of a string, must be applied at some distance from the bridge, which is the location of one of the fixed ends. The finite range of movement of the string at the point where the bow is applied is needed for feeding in energy at an appreciable rate, to compensate for that dissipated as heat and radiated in the form of sound.

It is instructive to consider the vibration of a string fixed at both ends mathematically.

The motion of the string is regarded as the resultant of two wave motions, one traveling in the positive and the other in the negative x direction, so we must first find an expression for the latter.

As we have seen, the equation for plane-polarized waves traveling in the positive x direction may be written

$$y = A \sin 2\pi \left(ft - \frac{x}{\lambda} \right). \tag{3.6}$$

We should expect the corresponding expression for waves traveling in the negative x direction to be obtained by replacing x in Eq. (3.6) by $-x$; that is, we should expect it to be

$$y = A \sin 2\pi \left(ft + \frac{x}{\lambda} \right). \tag{3.13}$$

That this equation does indeed represent backward-traveling waves may be seen by finding the coordinate $x + \delta x$ of the point where, at the time $t + \delta t$, the value of y in the equation will be the same as it is at x at the time t. The condition for this is that

$$f(t + \delta t) + \frac{x + \delta x}{\lambda} = ft + \frac{x}{\lambda},$$

whence

$$f\,\delta t + \frac{\delta x}{\lambda} = 0,$$

so that

$$\frac{\delta x}{\delta t} = -f\lambda.$$

But, as is obvious physically, $\delta x/\delta t$ is the velocity of the waves represented by Eq. (3.13). And, comparing our last equation with Eq. (3.7), we see that this is equal in magnitude but opposite in sign to the velocity of the forward-traveling waves represented by Eq. (3.6). Hence Eq. (3.13) does, as expected, represent waves traveling in the backward direction. More generally, using the same zeros for x and t as have been selected for the forward-traveling waves, we may represent waves traveling in the backward direction by the equation

$$y = A \sin 2\pi \left(ft + \frac{x}{\lambda} + \epsilon \right), \tag{3.14}$$

where ϵ is a constant.

It should be pointed out that Eqs. (3.6) and (3.14), which correspond to the waves we have labeled as forward-traveling and backward traveling, respectively, are really, physically, on an equal footing. We might equally well change our point of view and regard the negative x direction as the forward direction and the positive x direction as the negative. In that case we should have to interchange our labels for the wave trains corresponding to the two equations. We shall, in fact, have occasion to do this presently, when dealing with the reflection of waves traveling from right to left at the point $x = 0$, the reflected waves then traveling from left to right.

Let us now combine the equation for waves of amplitude A_1 traveling from left to right with that for waves of amplitude A_2 traveling from right to left, both wave trains having the same frequency; that is, we combine

$$y_1 = A_1 \sin 2\pi \left(ft - \frac{x}{\lambda} \right)$$

with

$$y_2 = A_2 \sin 2\pi \left(ft + \frac{x}{\lambda} + \epsilon \right),$$

y_1 and y_2 being the displacements for the separate wave trains. Then for the resultant displacement $y\ (=y_1 + y_2)$ at the time t of a point having the coordinate x we have

$$y = A_1 \sin 2\pi \left(ft - \frac{x}{\lambda} \right) + A_2 \sin 2\pi \left(ft + \frac{x}{\lambda} + \epsilon \right). \quad (3.15)$$

Let one end of the string be at $x = 0$. Then at this fixed end y must be zero at all times; that is, irrespective of the value of t, we must have

$$A_1 \sin 2\pi ft + A_2 \sin 2\pi \left(ft + \epsilon \right) = 0. \quad (3.16)$$

To satisfy this condition, it is necessary that A_1 and A_2 shall be equal and that ϵ shall have the value $\pm n/2$, where n is an odd number. Let us set $A_1 = A_2 = A$ and $\epsilon = \frac{1}{2}$. Then Eq. (3.15) may be written

$$y = A \left[\sin 2\pi \left(ft - \frac{x}{\lambda} \right) + \sin 2\pi \left(ft + \frac{x}{\lambda} + \tfrac{1}{2} \right) \right],$$

or, because for any angle α,

$$\sin(\alpha + \pi) = -\sin \alpha,$$

$$y = A \left[\sin 2\pi \left(ft - \frac{x}{\lambda} \right) - \sin 2\pi \left(ft + \frac{x}{\lambda} \right) \right].$$

This may be simplified to

$$y = -2A \sin \frac{2\pi x}{\lambda} \cos 2\pi ft. \quad (3.17)$$

From this equation we see that each point in the string executes simple harmonic motion, but that the amplitude, given by the factor $2A \sin (2\pi x/\lambda)$, varies from point to point along the string. This amplitude has a maximum value of $2A$ at the points where the values of x are $\lambda/4$, $3(\lambda/4)$, $5(\lambda/4)$, $7(\lambda/4)$, ... These are the positions of the antinodes. The nodes, on the other hand, occur at the points where x has the values $\lambda/2$, $2(\lambda/2)$, $3(\lambda/2)$, ...

If the string is also fixed at the point $x = l$, then, as we see by substituting l for x in Eq. (3.17), it is necessary that

$$\sin \frac{2\pi l}{\lambda} = 0.$$

This means that there can now only be certain quite definite values of λ. These are

$$2l, \frac{2l}{2}, \frac{2l}{3}, \frac{2l}{4}, \cdot \cdot \cdot \cdot$$

It may be shown (see Chapter 6) that for a given stretched string there is only one velocity v with which waves can travel along it. This being so, it follows from

$$v = f\lambda \tag{3.7}$$

that only certain definite values of the frequency are possible. These are

$$\frac{v}{2l}, 2\frac{v}{2l}, 3\frac{v}{2l}, 4\frac{v}{2l}, \cdots$$

The frequency $v/2l$ is known as the fundamental, or first harmonic. The higher frequencies, in the order in which they appear above, are called the second, third, fourth, . . . harmonics.

BEATS The phenomenon of beats is best known in acoustics. When two notes of nearly but not quite the same frequency are sounded together, a rhythmic rise and fall in intensity is observed, the intensity being a maximum when the interfering waves reaching the observer reinforce one another, while it is a minimum when they interfere destructively. The maxima are known as "beats." We shall now analyze this phenomenon mathematically.

For simplicity let us consider the special case where the amplitudes of the two disturbances at the location of the observer are equal. Let the two simple harmonic disturbances at this point be

$$s_1 = A \sin 2\pi f_1 t$$

and

$$s_2 = A \sin 2\pi f_2 t.$$

Then, writing ω_1 for $2\pi f_1$ and ω_2 for $2\pi f_2$, we have for the algebraic sum of the two disturbances,

$$s = s_1 + s_2 = A(\sin \omega_1 t + \sin \omega_2 t)$$

$$= 2A \cos \frac{\omega_1 - \omega_2}{2} t \sin \frac{\omega_1 + \omega_2}{2} t,$$

or, if we write ω for $(\omega_1 + \omega_2)/2$,

$$s = 2A \cos \frac{\omega_1 - \omega_2}{2} t \sin \omega t.$$

We may regard $2A \cos[(\omega_1 - \omega_2)/2]t$ as a slowly and periodically changing amplitude factor in a simple harmonic disturbance. The fact that for half a period this amplitude factor is positive and for the other half negative does not affect this situation; for a transition from a positive to a negative amplitude factor or vice versa is equivalent merely to a reversal in the phase of the oscillation

79

represented by the factor sin ωt, and there are many periods of this oscillation during each half-period of $\cos[(\omega^2 - \omega^2)/2]t$. Thus a half-period of this function, $2\pi/(\omega_1 - \omega_2)$, represents the beat period, that is, the interval of time between zero and zero, or between one maximum and the next, in the sound heard. The beat frequency is, therefore, $(\omega_1 - \omega_2)/2\pi$, or $f_1 - f_2$. That the beat frequency must be equal to the difference of the frequencies of the components is also obvious physically.

Problems

Problems 3.37, 3.38, 3.39, and 3.40 are for readers who are familiar with differential calculus.

3.1 Sketch graphs showing the variation with time of the displacement y, velocity v, and acceleration a of a particle on a stretched string along which a train of plane-polarized waves is traveling. State the phase relationship among y, v, and a.

3.2 The displacements from their undisturbed positions of the crests and troughs of a sinusoidal wave train are each 6 cm, and from each crest or trough to the nearest point whose displacement is 3 cm the distance is 50 cm. What is the wavelength?

3.3 Waves of frequency 1 Hz travel along a rope at a speed of 10 m/sec. What is the difference in phase between the motions of points on the rope (a) 1 m apart, and (b) 10 m apart?

3.4 The figure shows a sinusoidal wave train traveling along a rope as it appears at times zero, 0.1 sec, and 0.5 sec. What are (a) the amplitude, (b) the wavelength, (c) the minimum possible wave speed, and (d) the minimum frequency? In what direction do the waves travel? Write the equation for this wave motion.

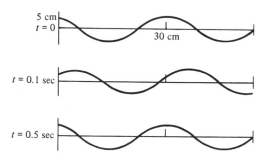

Figure for problem 3.4.

3.5 The figure shows the configuration of a wave pulse at a certain instant as it travels from right to left along a rope. Sketch curves showing how the transverse displacement, velocity, and acceleration along the rope vary with (a) time for a given particle, and (b) distance along the rope at a given instant.

Figure for problem 3.5.

3.6 The point of maximum slope on a wave pulse traveling along a string has a slope of 0.2, and the speed with which the pulse travels is 5 m/sec. What is the maximum transverse speed of a particle of the string?

3.7 Given that the equation of a transverse wave disturbance traveling along a stretched string is

$$y = \sin (10t - x),$$

find the ratio between the velocity of a particle in the string at its midpoint of swing and the velocity of the wave train.

3.8 The equation of a plane-polarized transverse wave train traveling along a rope is

$$y - 0.1 \sin (12t - 0.8x),$$

x and y being in meters and t in seconds. What are (a) the amplitude, (b) the period, (c) the maximum velocity of the particles of the rope, (d) the wavelength, and (e) the velocity of the waves?

3.9 A sinusoidal wave train traveling along a stretched string in the positive x direction with a velocity of 10 m/sec has a wavelength of 1 m and an amplitude of 5 cm. What is the equation of this wave disturbance?

3.10 P_1 and P_2 are points on a stretched string, less than one wavelength apart, which execute simple harmonic motions according to the equations

$$y_1 = A \sin 10\pi t$$

and

$$y_2 = A \sin (10\pi t + 10°),$$

respectively, in consequence of the passage through them of a wave train. What are the direction of travel and the velocity of the wave train, and what are the wavelength and frequency? What is the wave equation?

3.11 If the wave train of Problem 3.9 is a right-handed circularly polarized wave disturbance, instead of being plane-polarized and sinusoidal, what two equations give full information concerning it? To what would one of these equations have to be changed if the circularly polarized wave train were left-handed and traveled in the negative x direction?

3.12 What adjustments would have to be made in the zero for time or distance to convert the "cosinusoidal" form of the wave equation

$$y = A \cos 2\pi \left(\frac{t}{T} - \frac{x}{\lambda} \right)$$

81

to the sinusoidal form

$$y = A \sin 2\pi \left(\frac{t'}{T} - \frac{x}{\lambda} \right),$$

or

$$y = A \sin 2\pi \left(\frac{t}{T} - \frac{x'}{\lambda} \right),$$

t' or x' denoting time or distance measured from the new zero?

3.13 Show that, with a suitable adjustment in the zero for either t or x in the equation

$$y = A \sin 2\pi \left(\frac{t'}{T'} - \frac{x}{\lambda} \right)$$

applying to plane-polarized transverse or longitudinal waves, the equation may be written in the form

$$y = A \sin 2\pi \left(\frac{x}{\lambda} - \frac{t}{T} \right).$$

3.14 In the progression of a circularly polarized wave train along a string, does the helical wave form move like the corresponding corkscrew when in use, may the turns be thought of as rotating in the opposite direction without progression, or does the wave form simply move forward without rotating in either direction?

3.15 How must the angle θ in the equation

$$\theta = 2\pi \left(\frac{x}{\lambda} - \frac{t}{T} \right)$$

be defined in order that this equation, instead of Eq. (3.5), shall represent a forward-traveling train of right-handed circularly polarized waves?

3.16 A rope is stretched along the z direction of a Cartesian coordinate system, and a train of waves travels along it. Given that at a certain point

$$x = A \sin \omega t$$

and

$$y = 2A \sin (\omega t + \phi),$$

draw the ellipses traced out by this point for $\phi = 0$, $\phi = \pi/4$, and $\phi = \pi/2$. Are the figures traced out by other points on the rope the same or different?

3.17 A sound is propagated by longitudinal waves having a frequency of 500 Hz, a velocity of 350 m/sec, and an amplitude of 10^{-5} mm.* What is the maximum velocity of the particles of the air, and how does this compare with the velocity of the waves? How does the amplitude compare with the wavelength?

3.18 Show that two circularly polarized wave trains of equal amplitude and wavelength, one of which is right-handed and the other left-handed, may be compounded vectorially to give a plane-polarized trans-

---

* This would be a sound of medium loudness.

verse wave train of twice the amplitude. What determines the plane of polarization of this resultant wave train?

3.19 After a whistling railway locomotive has passed an observer standing close to the track, the pitch of the note heard is found to have dropped from its original value by one full tone. Given that this drop in pitch corresponds to a fall in frequency in the ratio 1.1226:1, and that under the prevailing atmospheric conditions the speed of sound is 340 m/sec, calculate the speed of the train.

3.20 Show that the equation for the Doppler frequency change is the same as

$$f' = \frac{V - v_r}{V - u_r} f,$$

where u_r and v_r are the velocities of the source and the observer, respectively, relative to the medium. Show that this reduces to

$$f' = f\left(1 + \frac{u_r}{v_r}\right)$$

in the limiting case where u_r and v_r are both small compared with V.

3.21 In the propagation of waves on the surface of deep water, each particle moves with uniform speed in a circular path, in such a direction that the particle moves forward when on the crest of the wave and backward when in a trough. The radius of the circle in which each particle moves is half the vertical distance (difference in height) between a crest and a trough. The radius of this circle being given, the remaining information concerning the wave motion may be stated in the form of a single equation, which we may appropriately refer to as the wave equation for water waves. In this wave equation the direction of each particle from the center of the circle in which it moves is expressed in terms of the period T, the wavelength λ, the coordinate x of the center of the circle with respect to some agreed reference point, and the time t. What is this equation?

3.22 Show that a train of water waves may be represented by a pair of equations, one giving the ordinate y of a particle describing a circular path in terms of T, λ, x (the x coordinate of the center of the circle), and t, and the other giving x' (the x component of the displacement of the particle from the point x) in terms of the same parameters. What kinds of wave motions do these equations *separately* represent?

3.23 Construct a series of clocks, each with only a single hand, whose centers all lie on a straight line, showing times such as would simultaneously be shown by clocks in a series of places such as Sydney, Delhi, London, New York, San Francisco, and Honolulu; show that the ends of the hands of such a series of clocks (not necessarily these particular ones), appropriately spaced apart, lie on the profile of a train of water waves.

3.24 Show that if all the clocks of Problem 3.23 are rotated through a right angle in one direction, the (three-dimensional) curve passing through the ends of all the hands will give the wave form of a right-handed circularly polarized wave train, while if they are rotated through a right angle in the other direction the corresponding curve will be the wave form of a left-handed circularly polarized wave train. Which are these two directions of rotation?

3.25 How are (a) a longitudinal sinusoidal wave train and (b) a transverse sinusoidal wave train related geometrically to a train of water waves?

3.26 A condition for the applicability of the principle of superposition is that the amplitude of the waves is small compared with the wavelength. Show that, in the case of sinusoidal waves, this means that the velocity of propagation of the waves must be large compared with the maximum transverse speed of any particle.

3.27 A stretched string of length 70 cm vibrates in a plane in its fundamental mode with a frequency of 250 vibrations per second. What is the velocity with which waves travel along this string? What is the maximum velocity of the string itself (a) at the center, and (b) at a point one eighth of the length of the string from one end, if the amplitude at the center is 5 cm?

3.28 Show that a right-handed circularly polarized wave train is reflected as a left-handed circularly polarized wave train, and vice versa.

3.29 Two longitudinal wave trains travel simultaneously in opposite directions along a tube, the separate air displacements being represented by the equations

$$y = A \sin 2\pi \left(ft - \frac{x}{\lambda} \right)$$

and

$$y = A \sin 2\pi \left(ft + \frac{x}{\lambda} \right).$$

What is the equation representing the resulting standing waves set up in the tube, and at what points will the nodes and the antinodes occur? What is the amplitude of the simple harmonic disturbance at an antinode, and what is it halfway between a node and an antinode?

3.30 A string is clamped at each end and vibrates according to the equation

$$y = \cos \frac{\pi x}{3} \sin 40\pi t,$$

where x and y are in centimeters and t in seconds. (a) Calculate the amplitude and velocity of two superimposed wave trains that can produce this vibration. (b) What is the distance between nodes? (c) Calculate the velocity of a particle of the string at the position $x = 6$ cm when $t = 1$ sec. (d) Describe the motion of this particle.

3.31 A string stretched between two points is pulled aside at its center so that it has the configuration of two straight lines meeting at a point. It is then released. At what point would you expect the subsequent disturbance of the vibrating string to be a maximum? What harmonics would you expect in the oscillation?

3.32 The equation of a wave disturbance is $y = A \sin 2\pi[ft - (x/\lambda)] + (2A/3) \sin 4\pi[ft - (x/\lambda)] + (A/3) \sin 6\pi[ft - (x/\lambda)]$. Plot y against x

between $x = 0$ and $x = \lambda$ for the times (a) $t = 0$, (b) $t = T/4$, and (c) $t = T/2$. What does a comparison among these plots suggest?

3.33   Beats are heard at the rate of 12 every 5 seconds when two open organ pipes, of lengths 84 cm and 85 cm, are sounded together in their fundamental mode. Find the velocity of sound in the air at the time of the observation. (An open organ pipe has an antinode at each end.)

3.34   A person standing between a loudspeaker fed from an oscillator and an organ pipe hears 5 beats per second, then walks slowly toward the speaker and hears 2 beats per second. Explain what has happened, and deduce which source of sound has the higher frequency.

3.35   A whistle emits sound waves at a frequency f, and a stationary reflector is placed behind this so that an observer hears the reflected and direct sounds from the whistle simultaneously. When the whistle is moved with a velocity of 1 m/sec toward the observer he hears 5 beats per second. Given that the velocity of sound under the prevailing conditions is 340 m/sec, find f.

3.36   A man is standing on a straight road between two factories. It is a windless day, and the velocity of sound under the prevailing conditions is 340 m/sec. Simultaneously, sirens are sounded at the two factories. One has a frequency of 352 Hz, while that of the other is slightly lower, and as a result the man hears 4 beats per second. He immediately mounts his bicycle, and as he accelerates to his final speed he notices that the beat frequency drops to zero and then rises again to 6 per second. Toward which factory is he riding, and at what speed? What beat frequency will he hear if he continues riding past the factory without slowing down?

3.37   The equation of a wave disturbance is

$$y = A \sin 2\pi \left(ft - \frac{x}{\lambda} \right).$$

First differentiate twice in succession treating x as a constant, so obtaining an expression for the acceleration of a particular point. Express this acceleration in the notation of a "partial" second derivative as $\partial^2 y / \partial t^2$. Then carry out a partial differentiation twice in succession treating t as a constant, so obtaining an expression for $\partial^2 y / \partial x^2$. Show that

$$\frac{\partial^2 y}{\partial t^2} = v^2 \frac{\partial^2 y}{\partial x^2},$$

where v is the velocity of the wave train.

3.38   Show that the standing-wave disturbance

$$y = A \sin \frac{2\pi x}{\lambda} \cos 2\pi ft$$

is compatible with the wave equation in its differential form,

$$\frac{\partial^2 y}{\partial t^2} = v^2 \frac{\partial^2 y}{\partial x^2}.$$

85

3.39 Show that at any point x of a stretched string along which a wave disturbance

$$y = f(x - vt)$$

is traveling, f being any continuous function of the variable $x - vt$, the tangent of the angle the string makes with its undisturbed direction at the time t is minus the ratio of the transverse velocity of the string at x and t and the velocity v of the wave disturbance.

3.40 Show that for a wave disturbance conforming to the equation

$$y = f_1(x - vt) + f_2(x + vt),$$

in which f_1 and f_2 are any two continuous functions of the variables $x - vt$ and $x + vt$, respectively,

$$\frac{\partial^2 y}{\partial t^2} = v^2 \frac{\partial^2 y}{\partial x^2}.$$

Interpret physically the two terms in the above equation for y.

SECTION III

DYNAMICS

THE LAWS OF FORCE AND MASS

In order to make further progress in our study of motion and its changes it now becomes necessary to consider not merely motion itself but also how changes of motion are conditioned by the interactions between bodies and by the dynamical properties of these bodies. The interactions between bodies we call **forces** and the relevant dynamical properties their **masses.**

In this chapter we shall consider the basic principles governing the interactions between particles, the definitions of force and mass, the units in which they are expressed, and their measurement. We shall also introduce an important new concept, that of momentum. Then, in the two following chapters, we shall deal with extensions of these basic principles—their application to systems in which forces are always directed toward a fixed center, in Chapter 5, and, in Chapter 6, the introduction of the concepts of work and energy.

Frames of Reference

When in Chapter 2 we spoke of displacements, velocities, and accelerations, it was implied that these quantities were measured relative to some agreed "frame of reference," for example, the axes of a Cartesian coordinate system. The fact that we did not define this frame, or consider whether or not it was accelerated (relative to some other frame) or rotated, did not seem to matter; from everyday experience one might argue that, because all motion is relative, the validity of kinematical equations cannot be affected by any motion of the frame of reference.

However, before embarking on a discussion of particle dynamics in the present chapter and of rigid-body dynamics in Chapter 5, we may well pause to consider what effect, if any, our choice of a frame of reference will have on dynamical laws. The importance of this choice will at once be realized when we contemplate the effects we observe in an airplane while it is gathering speed prior to takeoff or, again, during the period of slowing down after landing. On both occasions extra forces seem to be exerted on ourselves and other objects, and if we attempted to set up a system of laws applying to these, with velocities and accelerations measured *relative to the aircraft,* we might well expect these laws to be less simple than those applying in relation to a frame fixed with respect to the earth's surface. Actually, even the latter would really not be the best choice it is possible to make in our search for an ideal frame, because the earth rotates on its axis, and, as we have seen, with this is associated an acceleration. Although the complications arising from this acceleration are much less important, quantitatively, than are those experienced in an airplane just prior to takeoff or just after landing, they nevertheless exist, and in measurements of the highest accuracy they have to be taken into account.

Isaac Newton, near the beginning of the eighteenth century, established certain simple "laws of motion" using the earth's surface as his frame of reference. Newton himself recognized the limitations of this frame and realized that the simplicity of the laws of motion he obtained was due to a combination of fortunate circumstances: (1) that the earth's rotation is relatively slow, and (2) that his measurements were not of the highest accuracy. He believed that his laws would be strictly valid only in relation to a nonrotating frame which is also not accelerated as a whole.

INERTIAL FRAMES OF REFERENCE The specification of an ideal frame as one that neither rotates nor is accelerated raises the question: rotates or is accelerated with respect to what? In our search for a satisfactory frame of reference we might think of the sun as a body that is presumably not accelerated, and of the system of stars, taken as a whole, as not rotating with any appreciable angular velocity. We should be encouraged in making our choice of a frame of reference accordingly by the fact that within the limits of accuracy attainable by present-day measuring techniques Newton's laws apply precisely—at least if we exclude from consideration bodies having velocities comparable with the speed of light—with respect to a frame of reference that is stationary relative to the sun and in relation to which distant stars have fixed directions.

The simplicity of Newton's laws suggests that any frame with respect to which they apply precisely is something special; it is preferable to a frame in relation to which the laws of mechanics are less simple. The "astronomical" frame just referred to appears to qualify for acceptance by this criterion. However, the question of whether even this frame in fact rotates or is accelerated is still unanswered. Actually it must forever remain unanswered, because the question is meaningless; we can in the nature of things never find an "absolute" frame against which to test for rotation or acceleration. This being so, we are obliged to define our ideal frame on some other basis; thus we define it as one with respect to which

Newton's laws apply precisely, at least for bodies having not-too-high velocities.

As Newton recognized, his laws are valid not only with respect to our astronomical frame anchored to the sun, but also in relation to all frames moving with constant velocity relative to it and not rotating. All such frames with respect to which Newton's laws are valid are known as **inertial frames of reference.** The whole of our discussion of dynamics on which we are now embarking is to be understood as relating to the use of such an inertial frame.

Force

In physics the idea of a force is always associated in our minds with some agency for its exertion. This might, for example, be a stretched or compressed spring, or it might be one's hand pulling or pushing an object. But the agency is not always a material or tangible one; thus we also have gravitational, electric, and magnetic forces.

Forces are conveniently divided into two categories: "active forces" and "passive forces." The force exerted by a spring is an example of an active force. Passive forces are those which by their nature resist relative motion between two bodies or systems. The most familiar example of a passive force is the force of friction, exerted tangentially between two surfaces in contact and either preventing relative motion ("static" friction) or, if motion has been established, resisting it ("kinetic" friction). Frictional forces will be discussed at some length in Chapter 10. Another example of a passive force, which is often referred to, somewhat misleadingly, as frictional, is the viscous resistive force exerted by a liquid or a gas on a body moving through it. Viscous forces will be considered in Chapter 9.

A scientific definition of force must include all cases, tangible and intangible, active and passive forces, within its scope.

In our search for a suitable definition, let us think, in the first instance, of the simplest possible situation, where there is, actually or effectively, only a single force acting on a body. Since all bodies on the surface of the earth are subjected to the earth's gravitational pull, the only possibility of having only a single force acting is for this force to be gravitational (see Fig. 4.1). The elimination of other forces simultaneously acting means that the body in question must not be prevented from falling by any support, and it must not be exposed to any other agency resisting its fall. Strictly, these con-

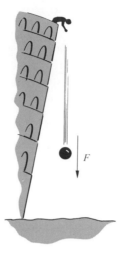

FIGURE 4.1 Free fall of a body produced by gravitational force.

ditions can be satisfied only by allowing the body to fall freely in a vacuum. However, the absence of nongravitational forces may be approximated by allowing the body to fall in air in such cases where the air resistance is small. As we have already seen, a freely falling body falls with the constant acceleration g, directed vertically downward. We may regard the gravitational force as the cause of this acceleration.

The gravitational force* exerted on a body by the earth is given a special name, the **weight** of the body; the earth's gravitational force and weight are interchangeable terms, meaning the same thing. In an extension of the idea, we may also speak of the weight of a body on some other planet, satellite, or star; thus by the weight of a body on the moon we mean the gravitational force exerted by the moon on this body.

BALANCING OUT THE WEIGHT OF A BODY If, instead of being allowed to fall freely, a body is supported, for example by a table, spring, string, or floor, it is no longer accelerated, and it seems

---

* Strictly speaking, we should say *apparent* gravitational force, or the force as measured by an observer on the surface of the earth and rotating with it. As will be appreciated later, this is not quite the same thing as the earth's gravitational pull uncorrected for rotation, just as the observed acceleration g is not quite the same as g_0. This is an unfortunate complication to have to bear in mind in the discussion that is to follow. For simplicity, we shall disregard the distinction between apparent and true gravitational force, imagining that we are on a nonrotating earth. Later it will be a useful exercise for the student to amend the discussion to take account of the rotation the earth actually has.

reasonable to assume that there is no longer any effective or resultant force acting on it. The presence of the table or whatever cannot prevent the gravitational force from acting, but it does provide a neutralizing force, one that is equal and opposite to the weight. This is the "reaction" exerted on the body by the table or floor on which it rests, or the tension of the spring or string.

EFFECT OF APPLIED FORCE IN GENERAL CASE The possibility of balancing out the weight of a body provides the opportunity of applying, in effect, a single force to it other than its weight and observing the effect this force produces. For example, let us suppose we have a body suspended by a long string, and let the body be held stationary while a spring is attached to it and held out horizontally. If the body is now released it will move off with acceleration in the same horizontal direction as the effectively single applied force provided by the spring.

This experiment could equally well be carried out in a moving train or inside an airplane, traveling with uniform velocity, or within a uniformly moving elevator, and the result observed would be precisely the same as before.

By such experiments we find that in all cases where (in effect) a single force is applied to a body, the body is accelerated in the direction of the force. A corollary of this is that in the absence of any resultant force the body will either remain at rest or continue moving with uniform velocity; that is, it will not be accelerated relative to any inertial frame.

QUALITATIVE DEFINITION OF FORCE We have hitherto spoken of force as we understand it in everyday life, but the term has not yet been precisely defined. We are now in a position to define it, qualitatively, as **that which produces, or tends to produce, an acceleration in the body to which it is applied, the direction of the force being that of the acceleration produced.** The term "tends to produce" is included to cater for those cases where two or more forces are applied simultaneously. This partial definition, being consistent with all our experience of forces as ordinarily thought of, holds corresponding promise of scientific usefulness.

To advance beyond this merely qualitative definition, we have now to ascertain (1) how the acceleration of any particular body varies with the applied force as we are accustomed to measure it in our everyday lives, and (2) how, for a particular applied force, the acceleration varies from one body to another. When the answers to these questions have been obtained we shall be in a position to formulate a completely quantitative definition of force that will be

93

in harmony with everyday practice as regards its measurement. However, before we can enter into a discussion of these questions, there are certain preliminaries to be dealt with. We shall first have to develop a method for applying measured forces in any required directions, and then, using this technique, investigate whether or not force is a vector, so that, if it is, we may make use of its vector properties.

MEASUREMENT OF FORCE In everyday life it is customary to measure all forces, directly or indirectly, in terms of weight. Thus, an agreed-upon unit might be a "pound weight," the weight of a mass of metal we call a pound. This being agreed upon, a number of other bodies could be assembled, each of which has been adjusted (by trimming or by addition) to have the same weight, the criterion of equality being, for example, that when suspended from a given spring they all extend it by the same amount. This assemblage of bodies, all having the same weight, could then be used to calibrate a spring balance. Finally, this spring balance could be used to apply known forces in any desired direction, or, alternatively, to measure tensions of strings, and so on, having any orientations in space. As is indicated in Fig. 4.2, a calibrated spring balance could be used to check that the tension of a string on one side of a well-lubricated pulley over which it passes is equal to the weight of the combination of standard masses suspended from it on the other side.

FORCE AS A VECTOR To ascertain whether force is a vector, the method indicated in Fig. 4.3 may be used. This method is self-explanatory. The junction of the three strings is at rest, so the forces applied there must be in equilibrium. If force is a vector, then, when a body is in equilibrium under the action of three forces, the

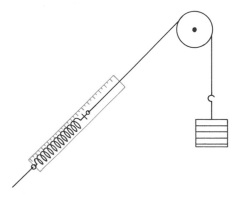

FIGURE 4.2 Use of pulley for applying a force, known in terms of weight, in any desired direction.

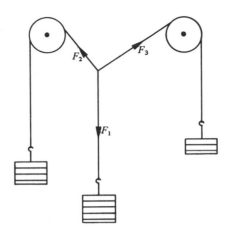

FIGURE 4.3 Three simultaneously applied forces with zero resultant.

resultant of any two of these as found by the vector-triangle construction (see Fig. 4.4) should be equal and opposite to the third force. It is found in all cases that this is, indeed, the case. From this it may be inferred that **force is a vector.**

Relation between Force and Acceleration

We are now ready to address ourselves to the question of how the acceleration produced in a given body varies with the magnitude of the force applied to it. The answer to this question is provided in the most elegant manner by studying the component of the gravitational force acting on the bob of a simple pendulum tan-

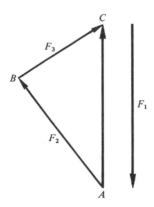

FIGURE 4.4 Vector triangle of forces.

95

gentially to the arc of swing and of the acceleration in the same direction for different positions of the bob.

Let the bob of a simple pendulum, supported from the point S, swing backward and forward along the arc AB, as indicated in Fig. 4.5. At any point P of its swing, where the angular displacement of the bob from its rest position O is ϕ, it is acted upon by two forces: its weight W and the tension of the string. Let the weight be resolved in two mutually perpendicular directions, along the string and tangentially to the arc of swing, as indicated. We see that the component F along the direction of motion is $W \sin \phi$. The other component, $W \cos \phi$, along the direction of the string, is at right

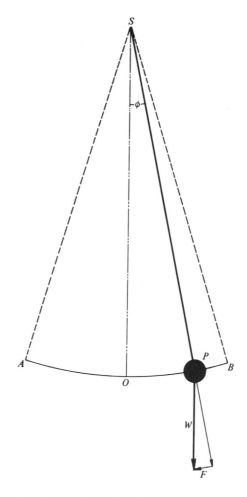

FIGURE 4.5   Resolution of weight of bob of simple pendulum along and at right angles to direction of motion.

angles to the direction of motion, as is also the tension of the string. Having no component in the direction of motion, neither of these forces concerns us in the present discussion. And W is constant, so we have for the force in the direction of motion

$$F \propto \sin \phi.$$

How the acceleration of the bob along the direction of motion varies with ϕ we have, in effect, already ascertained in Chapter 2, in our discussion of simple harmonic motion in general and of the motion of a simple pendulum in particular. There we saw (1) that a characteristic of SHM is that the acceleration of the oscillating particle is proportional to its displacement from the midpoint of swing, and (2) that the motion of the bob of a simple pendulum is simple harmonic, or at least very nearly so. Hence, because angular and linear displacements are proportional to one another, we have, at least to a close approximation,

$$a \propto \phi,$$

ϕ being the angle referred to in Fig. 4.5.

A condition adhered to in our experimental study in Chapter 2 of the detailed correspondence, in projection, between the simple and conical pendula was that the amplitudes were always kept small, and so there was always a close approximation to equality between ϕ and $\sin \phi$. In view of this, and of the experimental limitations of the method, it would be unrealistic to attempt to discriminate between ϕ and $\sin \phi$. Just as acceptable an inference from our observations as proportionality between a and ϕ would be proportionally between a and $\sin \psi$; that is, within the limits of experimental uncertainty it may, alternatively, be inferred that

$$a \propto \sin \phi.$$

This, considered in conjunction with the conclusion reached in our analysis of the component of the weight of the bob of a simple pendulum in the direction of motion,

$$F \propto \sin \phi,$$

means that, within these same limitations, it has been established that

$$a \propto F. \tag{4.1}$$

There remains for consideration the question of how, for a particular applied force, the acceleration varies from one body to another. The discussion of this question can only be profitably embarked upon after we have first considered the relevant property

97

that determines the response of a body to a given force—the mass of the body.

Mass

In Chapter 1 mass was defined, qualitatively, as the difficulty of changing the state of motion of a body. We have since seen that any such change in state of motion must be ascribed to the application of a force. An obvious measure of the difficulty of changing the state of motion of a body would be the magnitude of the force that is required to produce in it a given acceleration. It is therefore indicated that we now, tentatively, set up a quantitative definition of mass as something proportional to the force necessary to produce a given acceleration. If a force F_1 applied to the first of two bodies produces in it the same acceleration as does the force F_2 when applied to the second body, then, according to this definition,

$$\frac{m_1}{m_2} = \frac{F_1}{F_2}.$$

(4.2)

Alternatively, mass might be defined as inversely proportional to the acceleration produced in a body by a given force. According to this definition, if a given force, when applied in turn to two bodies, produces in them the accelerations a_1 and a_2, then their masses, m_1 and m_2, are such that

$$\frac{m_1}{m_2} = \frac{a_2}{a_1}.$$

(4.3)

Actually, from the fact that for a given body the acceleration is proportional to the force applied, it follows at once that these two definitions are equivalent to one another. It is left as an exercise for the reader to show that this is so.

We must now ascertain whether these definitions are mathematically useful, that is, whether, irrespective of the acceleration produced in the two bodies if we use the first definition, or of the force applied if we use the second, the ratio m_1/m_2 is always the same for a given pair of bodies, and, furthermore, whether the ratios found between the masses of pairs selected from a number of bodies are mutually consistent mathematically.

In view of the proportionality between force and acceleration for a given body, we may at once answer the first of these questions, in either of the alternative forms, in the affirmative.

We may obtain the answer to the second question from the fact that at a given locality g is the same for all bodies. A freely falling body has only one force acting on it, its weight. Hence, if we consider a number of freely falling bodies, we see that the forces which produce in all of these the same acceleration g are simply their weights. And, according to the first of the two alternative definitions, the masses are proportional to these forces. Thus we have the important result that **the masses of bodies are proportional to their weights:**

$$m_1 : m_2 : m_3 \cdot \cdot \cdot = W_1 : W_2 : W_3 \cdot \cdot \cdot \qquad (4.4)$$

Because the weight of a given body at a given place is a definite quantity, not subject to variation, and experience shows that weights are additive, this must also be true of masses: **Masses, as defined, conform to the laws of mathematics.** This is all we can ask for; our definition is eminently useful.

It should particularly be noted that it is only the *ratio* between masses that has been defined. If we could define a mass in terms of other quantities not involving mass, the concept of mass would not be fundamental. This, however, we cannot do. In the same way as length and time, so also mass is regarded as a fundamental quantity. Accordingly, we can only express the mass of a body in terms of some arbitrarily chosen unit; all that we can determine experimentally is the ratio of the mass of the body to that of the unit.

UNITS OF MASS The unit of mass in the older of the two current versions of the metric system is the **gram,** defined as the one-thousandth part of the mass of a cylinder of platinum–iridium alloy known as the **international kilogram,** kept at the International Bureau of Weights and Measures at Sèvres, near Paris. The original intention was that the international kilogram should have the mass of 1000 cm³ of pure water at 3.98°C, the temperature at which water has its greatest density (see Chapter 6). Actually, 1 kg of water at 3.98°C occupies 1000.028 cm³; that is, 1 g of water at this temperature has a volume of 1.000028 cm³. However, except in cases where we are concerned with extreme accuracy, it is not necessary to make any distinction between the volume of 1 g of water at 3.98°C and 1 cm³.

The unit of mass in the British system is the **pound.** Since 1959 in America and since 1960 in the United Kingdom the pound has been defined in terms of the kilogram. According to this definition

$$1 \text{ lb} = 0.45359237 \text{ kg.}$$

This means that

$$1 \text{ kg} = 2.2046226 \text{ lb.}$$

Systems of Units

We noted in Chapter 1 that present-day physicists use as "building bricks" for the structure of theoretical physics four fundamental quantities: length, mass, time, and electric current. Until comparatively recently the necessity for the fourth of these was not realized; it was believed possible to define all electrical quantities solely in terms of measurements of length, mass, and time. However, physicists are now generally agreed that this is not feasible, and that a fourth fundamental quantity, electrical in nature, is also required. It is nevertheless still customary, as formerly, to refer only to the first three quantities in specifying systems of units, at least in the British system and in the older of the two metric systems. Although this is hardly logical, it does at least suffice for the discussion of those branches of physics that are not concerned with electricity.

The original metric system, based on the centimeter, gram, and second, is known as the CGS (centimeter-gram-second) system. The British system, now falling into disuse among physicists, is known as the FPS (foot-pound-second) system.

An alternative metric system to the CGS which is coming into increasing use is that based on the meter, kilogram, and second. This is known as the MKSA system (A for ampere). The units of most quantities in this system differ from those in the CGS system only by factors of powers of 10, but in electrical units there are also factors of 4μ and the numerical value (in meters per second) of the velocity of light in empty space, which have been introduced for reasons of convenience. In this way it has been contrived to make MKSA electrical units coincide with those used in practice (amperes, volts, ohms, and so on), and accordingly the MKSA system has an important advantage over the CGS in the field of electricity.

Absolute Units of Force

The units of force we have used thus far, the pound weight, the gram weight, and the kilogram weight, vary slightly from one locality on the earth's surface to another. This follows from the fact that the acceleration produced in a given mass is proportional to the force applied, or the force is proportional to the acceleration. The weight of a body is proportional to g, and, as we saw in Chapter 2, g varies somewhat with locality on the earth's surface. This difficulty would be aggravated if we were no longer earthbound.

100

For example, a kilogram weight on the surface of the moon would be only about one sixth of what it is on the earth's surface.

But scientifically the objections to expressing forces in gravitational units go deeper than this. Physics is a universal science, spatially as in other respects, and it is therefore in principle unsatisfactory to have the unit of force, one of its most basic concepts, specifically associated with the gravitational attraction of the earth or any other body. It is not even *possible* in all cases to measure forces in gravitational units. For example, the fact that the earth and the moon describe orbits about a common center can only be ascribed to a gravitational attraction between them; yet there is no possibility of directly measuring this mutual force in pounds weight or any other gravitational units. Again, on what basis are we to calculate the propulsive thrust in pounds weight exerted on a rocket by the ejected stream of high-velocity gas? Although gravitational units of force are extremely convenient in engineering and in the ordinary affairs of everyday life, they sometimes fail us even in engineering problems.

This difficulty is avoided by basing the unit of force on the definition of mass, in conjunction with the law connecting the acceleration produced in a given mass with the force applied to it. In this, to be sure, gravitational attraction is still involved; thus we compare *masses* by comparing weights. However, the ratio found must be the same, irrespective of the locality where the comparison is made. This represents some progress. Another difficulty might appear to be the fact that the law of variation of force with acceleration is based on reasoning which makes use of the fact that force *measured as a weight* is a vector. This, again, does not really matter; for the nature of the law of variation in no way depends on the value of g at the locality where the vector experiment is carried out: The same result would be found whether one worked at the equator, at the North Pole, or on the surface of the moon. Actually, in a way, the measurement of forces as weights as discussed above is advantageous, for in whatever alternative manner we may eventually define units of force, we must require these to be applicable in those cases where our forces *are* weights. Our use of forces as weights ensures that this shall be so.

Our definition of mass may be stated in the form

(1) For a given a, $F \propto m$;

and the law connecting force and acceleration is

(2) For given m, $F \propto a$.

101

Obviously these two statements are merely special cases of the more general statement

$$F \propto ma,$$

or

$$F = kma, \tag{4.5}$$

where k is a constant.

This equation suggests that we might decide on a convenient proportionality factor and then specify forces in terms of the accelerations they would, if acting alone, produce in bodies of known mass.

Before definitely deciding in favor of this course, let us consider the question of the reliability of the proposed defining equation (4.5). Can we be sure that the accelerations produced in a given body are *quite strictly* proportional to the forces applied to them? It must be admitted that the experiments from which this has been inferred do not lend themselves to the attainment of high accuracy. Hence, *as far as these experiments alone are concerned*, the answer to our question must be in the negative. The proposed defining equation for force is correspondingly an extrapolation on the experimental accuracy available. There is, however, both a priori and a posteriori justification for such extrapolation. The proposed defining equation represents an extremely simple relationship mathematically, and it is correspondingly attractive. It is a good principle in science to try simple things first, and so it is indicated that we should at least investigate the suitability of this definition. But its final justification is that the scheme of theoretical physics—at least of *classical* theoretical physics—in which this is one of the fundamental equations, is completely self-consistent. Hence we have every reason for adopting this definition, at least until such time as it may become necessary to modify it.

For simplicity, the unit of force, both in the two metric systems and in the British system, is so chosen that the constant k in Eq. (4.5) has the value 1. The equation thus becomes

$$F = ma. \tag{4.6}$$

The unit of force is then such as will impart unit acceleration to unit mass, or, more generally, such as will impart to a mass m an acceleration numerically equal to $1/m$. In the CGS system the unit so defined is called the **dyne** (dyn), in the MKSA system it is the **newton** (N), and the FPS absolute unit is known as the **poundal.** Thus a dyne is such a force as imparts to a mass of 1 g an acceleration of 1 cm/sec², and the magnitude of a force in dynes is equal to the number of grams constituting the mass on which it acts multiplied by the number of centimeters per second per second representing the acceleration produced. By substituting newtons for dynes, kilograms for grams, and meters for centimeters in this last sentence, we have the corresponding situation in the MKSA system. And by substituting, instead, poundals, pounds, and feet, we have it in the FPS system.

Incidentally, we may note that a newton is 1000 (number of grams in a kilogram) times 100 (number of centimeters in a meter) dynes, that is, 10^5 dyn. It is roughly 100 grams weight (more nearly 102).

In Chapter 1 we saw that the dimensions of an acceleration are LT^{-2}. Because force is equal, or at least proportional by a pure number (according to definition), to the product of mass and acceleration, its dimensions are MLT^{-2}, the dimensional symbol M denoting mass.

A freely falling body has only one force acting on it—that exerted by the earth gravitationally. This force is thus measurable in absolute units as the product of the mass acted upon and the acceleration produced in it. But, as we have seen, the acceleration is the same at any given locality for all freely falling bodies—g. The weight W of a body in absolute units is therefore given by the equation

$$W = mg, \qquad (4.7)$$

which is a special case of Eq. (4.6), W being written for F in this equation and g for a.

From Eq. (4.7) we see that the number of absolute units representing the weight of a body is g times the number of gravitational units. Thus 1 gram weight is equal to g dyn (about 981 dyn, varying with locality), 1 kilogram weight is equal to g N (about 9.81 N), and 1 pound weight is equal to g (about 32.2) poundals.

Complete Definition of Force

Equation (4.5) may be regarded as a quantitative definition of force. It is important to realize that this equation is not a *complete* statement of what we mean by this term. This becomes evident when we consider how absurd it would be to say that a mass multiplied by an acceleration, or something proportional to it, produces a certain acceleration in a body. Force is not merely a name given to k times the product of mass and acceleration. Equation (4.5) represents only the quantitative side of the matter. In addition we

103

always think, rightly or wrongly, of something qualitative, as is implied when we speak of a force being "exerted." We suppose that there is always some agency for the exertion of a force, whether this is something tangible, in the material sense, such as the tension of a string or of a stretched spring, or otherwise. The qualitative aspect of force was referred to when we first introduced the idea, and, incorporating the statement then made with Eq. (4.5), we may give the following complete definition of force: **A force is that which produces, or tends to produce, an acceleration in a body, and its measure is the product of the mass of the body and the acceleration produced in it when no other forces are acting.**

FURTHER CONSIDERATION OF FORCE AS A VECTOR Let us now again briefly turn our attention to the question of whether force is a vector. As we have seen, an experimental test indicates that force measured empirically in terms of weight *is* a vector. It will now be of interest to discuss the question also from the theoretical point of view, in terms of our new definition. It will be recalled that in Chapter 2, beginning with a consideration of displacement as a vector, we divided twice in succession by the scalar time and so inferred that velocity and acceleration must also be vectors. We may now multiply each side of a vector triangle or polygon of accelerations by another scalar, the mass of the body accelerated, and so show that the product of mass and acceleration is a vector. From this we may infer that forces as now defined may be compounded vectorially when they act simultaneously on a body, each force being thought of as producing its own acceleration.

Momentum

Yet another important concept in mechanics is the product of the velocity and the mass of a body, known as its **momentum.** Denoting this quantity by the symbol p, we may express its definition mathematically by the equation

$$p = mv. \tag{4.8}$$

By reasoning completely analogous to that employed to show that force is a vector, it follows that momentum, also, is a vector. From

$$F = ma = \text{mass} \times \text{rate of change of velocity}$$
$$= \text{rate of change of (mass} \times \text{velocity)}$$
$$= \text{rate of change of momentum,}$$

or, in calculus notation,

$$F = ma = m\frac{dv}{dt} = \frac{d}{dt}(mv) = \frac{dp}{dt}, \qquad (4.9)$$

which must hold if we may assume that the mass of a body is independent of its velocity and time, we see that a quantitative definition of force alternative to the product of mass and acceleration is the rate of change of momentum.

In classical mechanics, which we are studying in this chapter, these two definitions of force, expressed by Eqs. (4.6) and (4.9), may be regarded as completely equivalent to one another and so equally acceptable. However, classical mechanics does not have general validity, even with respect to inertial frames of reference. Classical mechanics represents a limiting case of a system of mechanics that we may conveniently refer to as **relativistic mechanics,** which was worked out by Einstein early in the present century. For all practical purposes classical and relativistic mechanics are indistinguishable where we are concerned only with velocities that are small compared with the velocity of light, and this applies to all ordinary situations in everyday life. It is in these situations that we may regard the two definitions of force as equivalent. But in cases where we have to deal with particles having velocities comparable with the velocity of light, such as, for example, beta rays (fast electrons emitted by certain radioactive materials) or positive ions accelerated to extra high energies in modern particle accelerators, this is no longer the case. As the speed of a particle approaches the speed of light $(3 \times 10^8 \text{ m/sec})$ its mass becomes substantially greater than its "rest mass," asymptotically approaching infinity, and correspondingly its momentum increases at a faster rate than merely in proportion to its velocity. Hence for bodies moving at such speeds the two definitions of force are no longer equivalent. As Einstein has shown, the only *generally* acceptable definition of force is the time rate of change of momentum.

Interaction between Bodies

If a body is supported against gravity, for example by resting on a table, then, as we have seen, the reaction of the table on the body balances its weight. At the same time we must surely assume that the body presses down on the table with a force equal to its weight. This is a simple case of the exertion of mutual forces between bodies, one of which we may regard as the action of one body on the other, while the other force is the reaction of the other body on the

105

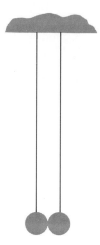

FIGURE 4.6 Double-pendulum system for investigating interaction between colliding bodies.

one. We find that, in all situations in statics that we are able to analyze, the forces that pairs of bodies exert on one another are equal in magnitude and opposite in direction: "Action and reaction are equal and opposite."

Let us now investigate whether this law applies to interacting bodies in motion as well as to bodies at rest. For this purpose we shall use the system indicated in Fig. 4.6. Here two pendulum bobs are suspended by threads so that they just touch one another when at rest. If in such a system one bob is drawn aside and then released, it will collide with the stationary bob, after which both bobs will move in the same manner as does the bob of a simple pendulum, until they again collide. By measuring how far and in what direction each bob moves after the first collision it is possible to determine whether the law that action and reaction are equal and opposite applies to interacting bodies in motion relative to one another as well as to those at rest.

The test must in the nature of things be an indirect one. What we do is to *assume* that at each stage of the collision action and reaction are equal and opposite, and on the basis of this assumption predict how the maximum excursions of the two bobs after the collision, considered algebraically, should be related to the displacement of the "striker" bob at the moment of release. We then carry out the relevant measurements to see if this prediction is borne out in fact. If it is, we infer that our assumption must be right, for surely if it were otherwise we should have obtained a different result.

We may regard the colliding pendulum bobs, during the brief

time interval of the collision, as constituting an almost ideal "two-body system," that is, a system of two bodies each of which exerts a force on the other but neither of which is acted on by any "outside" body. At least this is the case as far as forces and velocities *in the horizontal direction* are concerned. For, during the collision, the strings are to all intents and purposes vertical, and such slight horizontal components of their tensions as there may be (at such times when they are not quite vertical) are as nothing compared with the very considerable forces that the bobs exert on one another.

During the period of the collision the force between the bobs varies considerably from one moment to another. We can imagine this period to be divided into a large number of much smaller intervals, during each one of which the ratio of the velocity changes is equal to the ratio of the accelerations, which should, in turn (according to our assumption), be equal to minus the inverse ratio of the masses; the minus sign takes account of the fact that the forces are in opposite directions. Accordingly, integrating over the whole period of the collision, we should have

$$\frac{\delta v_2}{\delta v_1} = -\frac{m_1}{m_2}, \qquad (4.10)$$

where δv_1 and δv_2 denote the changes in the velocities of the two bodies, of masses m_1 and m_2, respectively. Experimentally, one would find the ratio of masses by weighing, for, as we have seen, masses are proportional to weights. If, then, the ratio of the changes in velocity is found to be equal to minus the inverse ratio of the weights, we must infer that here, as in statics, action and reaction are equal and opposite.

In performing the experiment, we find, on some scale, the velocities before and after the collision in the following manner. First we measure the distance the striking bob is drawn aside before being released, and then we observe the distances both bobs travel in their respective directions after the collision, to the point where they turn back. All three of these distances are amplitudes of simple-pendulum oscillations, observed during a quarter of a complete period. And these amplitudes are a measure of the corresponding velocities at the midpoint of swing. This follows at once from the detailed correspondence between the motion of a simple pendulum and the projected motion of a conical pendulum of equal period and having an orbit of radius A equal to the amplitude of the simple pendulum; for at the midpoint of swing the velocity of the bob is momentarily equal to the orbital velocity ωA of the bob of the conical pendulum. Hence the amplitudes, or "extreme elongations," A, are simply proportional to the velocities at the midpoint of swing, by the proportionality factor ω. By measuring these elongations,

107

and treating them algebraically, we can at once find the ratio of the velocity changes.

If we do this, using each bob in turn as the striker, and for a series of distances to which the striker bob is drawn aside, it is found in every case that the ratio of velocity changes so determined is minus the inverse ratio of the masses. Hence, as explained above, it is established that here, as in statics, action and reaction are equal and opposite.

Among the many applications of this principle in dynamics, it is proposed here to consider only two: the theory of action of a simple kind of water turbine, known as a Pelton wheel, and the action of a sailing ship, close-hauled for sailing into the wind.

PELTON WHEEL The Pelton wheel is shown in Fig. 4.7. A number of buckets are attached at regular intervals to the rim of a large wheel mounted on a horizontal axle, and a stream of water issuing from a nozzle is directed against the bucket momentarily at the bottom of the wheel. As the wheel rotates, each bucket in turn comes into position to receive the stream of water. Each bucket has two compartments, as shown enlarged in section below

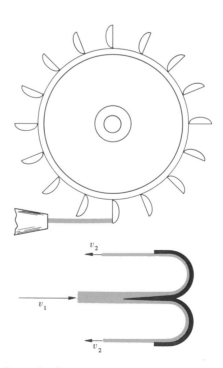

FIGURE 4.7 Pelton wheel.

the diagram of the wheel, these being joined by a common wall with a sharp dividing edge onto which the jet is directed. Here the water, divided by the edge, slides along the surfaces of the two compartments and finally emerges on the outside with its direction of motion reversed.

Let the velocity of the jet impinging on the wheel from left to right relative to the earth (not to the moving bucket) be v_1, its area of cross section A, and its density (mass per unit volume) ρ. Then the volume of water impinging on the wheel per second will be Av_1, the mass impinging per second ρAv_1, and the momentum directed toward the wheel per second $\rho Av_1{}^2$. Similarly, if the velocity of emergence from the buckets (again relative to the earth) is v_2 from right to left, the momentum of the water leaving the wheel per second is ρAv_1v_2. The two momenta are in opposite directions, so the rate of change of momentum must be equal to the sum of these quantities, $\rho Av_1(v_1 + v_2)$. This is the force exerted by the wheel on the water from right to left. The force exerted by the water on the wheel is, of course, of the same magnitude but is directed from left to right.

If it were not for viscosity (see Chapter 9), the velocities of impact and emergence *relative to the bucket* would be equal, and for simplicity let us idealize the situation by ignoring the effects of viscosity and supposing that these relative velocities *are equal.* Then if the wheel is stationary v_2 will be equal to v_1, and the force, which will have its maximum value, will be equal to $2\rho Av_1{}^2$. But to be stationary is hardly the purpose of a waterwheel! If now we gradually allow the peripheral velocity of the wheel (velocity of the buckets) to increase from zero, the two velocities relative to the bucket will continue to be equal, but the velocity v_2 relative to the earth will gradually become less, until, when the peripheral velocity of the wheel is $\frac{1}{2}v_1$, v_2 will be zero. The force exerted on the wheel will now be $\rho Av_1{}^2$. Although this is only half as large as the force exerted when the wheel is stationary, the condition where v_2 is zero is nevertheless the most favorable from the point of view of power developed, as will be appreciated when we have studied energy and its transformations and transferences.

In an actual turbine the effects of viscosity are never negligible, and the optimum peripheral velocity of the wheel, for which v_2 is zero, will be somewhat less than $\frac{1}{2}v_1$.

SAILING SHIP A ship sailing into the wind is represented schematically in Fig. 4.8. Only one sail is shown in the diagram, the configuration and orientation of all of them being similar. Let the ship be moving in the direction of its length, and let the wind be blowing in the direction OA. Then, if the sail is set correctly, the

109

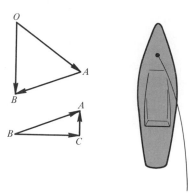

FIGURE 4.8   Forward and sideways components of force exerted by wind on close-hauled sailing ship.

horizontal tangent to it at its leading edge will have a direction coinciding with the direction of the wind *relative to the ship*. The air then flows around, following the curve of the sail, and finally leaves it tangentially at its after edge. This deflection of the wind by the sail is not confined to the windward side; some also occurs on the lee side.

Let OA represent the initial momentum of the air deflected by the sail per second and OB its final momentum. Then AB represents the rate of change of momentum of the air, that is, the force the ship exerts on it due to the action of the sail. The force exerted on the sail is equal and opposite to this; it is represented by BA. This in turn may be resolved into two components, BC at right angles to the ship and CA in the direction of its length. The former component causes the ship to make a certain amount of leeway, but because of the resistance the water offers to lateral motion this is generally small, particularly if the ship has a keel of large area. The latter component CA drives the ship forward.

Conservation of Momentum

From Eq. (4.10) it follows at once that the total momentum of the two bobs in the double-pendulum experiment is the same after the collision as before. For, cross-multiplying, we have

$$m_1\, \delta v_1 + m_2\, \delta v_2 = 0,$$

or

$$\delta p_1 + \delta p_2 = 0,$$

where δp_1 and δp_2 denote the changes in momentum suffered by the two bodies; that is, there is no change in the algebraic sum of the two momenta.

The momentum of a single, isolated body, on which no force acts, does not change. And, as we have seen, there is also conservation of momentum for a two-body system. The fact that momentum is conserved in both these cases suggests that it is also conserved for a many-body system, such as, for example, the solar system—if we may regard this system as not acted on by appreciable forces from outside. On the assumption that all forces in nature are ultimately exerted between pairs of particles, these forces being equal and opposite, the rate of change of momentum associated with each pair must be zero, and so the sum total of all such rates taken vectorially must also be zero. Observation shows that the total momentum of an isolated complex system of bodies is, indeed, conserved; and from this **law of conservation of momentum** it seems reasonable to infer that forces are, as we have imagined, exerted between pairs of particles, being equal and opposite for the members of each such pair. This is an important principle, which is made use of in the dynamics of systems of particles.

Laws of Force and Mass

In science our aim must always be to achieve the greatest possible degree of generalization; we do not wish to be involved unnecessarily in special cases. Hence, before proceeding to formulate our findings in the form of laws, it will be desirable to set up a new definition of the ratio of two masses which does not depend on the use of forces measured as weights. We can do this by considering a two-body system—a system of two bodies isolated from (not acted upon by) other bodies—and referring only to *what we observe to happen* in such systems. What we observe is that the accelerations of the components of a two-body system are in a constant ratio to one another and are in opposite directions along the line joining them. Thus we may replace our earlier definition of mass by one in which **the ratio of the masses of the components of a two-body system is defined as minus the inverse ratio of their accelerations:**

$$\frac{m_1}{m_2} = -\frac{a_2}{a_1}. \tag{4.11}$$

Not only is the constancy of the ratio of accelerations on which this definition is based something we observe in such approximations to two-body systems as we are able to devise experimentally (for example, the double pendulum during the period of collision), but it is also found in all other cases, including, for example, binary stars. Furthermore, we find that, for all combinations of two-body systems we choose to consider, **the mass ratios so defined are**

mathematically consistent. Our new definition is therefore not only universal in scope but mathematically useful.

We may now formulate the laws of force and mass in the following statements, which, for convenience, also incorporate our key definitions:

1. In any two-body system the accelerations a_1 and a_2 of the components have opposite directions along the line joining the bodies, and bear a constant ratio to one another. If a quantity, mass, is defined by the equation

$$\frac{m_1}{m_2} = -\frac{a_2}{a_1},$$

then the masses of bodies so defined are additive.

2. Force, in addition to certain qualitative considerations, is defined quantitatively by the equation

$$F \propto ma,$$

or, alternatively, by

$$F \propto \text{rate of change of momentum.}$$

3. All forces are exerted mutually between pairs of particles, and for each such pair of forces action and reaction are equal in magnitude and opposite in direction.

It will be observed that there is no specific mention here of force being a vector. That it is a vector is a mathematical consequence of the definition of force as proportional to the product of mass (a constant scalar quantity) and acceleration (a vector).

NEWTON'S "LAWS OF MOTION" The first scientific attempt to formulate the laws of force and mass was made near the beginning of the eighteenth century by Isaac Newton, the great English physicist and mathematician, in his *Philosophiae Naturalis Principia Mathematica.*

Newton defined mass, not as we do today, but as "quantity of matter," and his laws of motion should be considered against the background of this definition. They may be rendered as follows:

1. Every body continues in its state of rest or of uniform motion in a straight line, except insofar as it is compelled by forces to change that state.

2. Rate of change of momentum is proportional to the applied force and takes place in the direction in which the force acts.

3. To every action there is always an equal and opposite reaction; or, the mutual actions of any two bodies are always equal and oppositely directed.

If we could accept Newton's definition of mass as quantity of matter, and if we could take force as something measurable in its own right, without reference to accelerations produced in known masses, the three laws enunciated by him could all be taken as true laws of observation and would be unexceptionable. However, although we can measure weights (gravitation forces) or volumes, we cannot measure a "quantity of matter"—the term is meaningless. Hence we cannot accept Newton's definition of mass. Also there are difficulties in setting up anything better than a makeshift definition of force if it is to be independent of considerations of mass and acceleration.

Newton's first statement is formally unnecessary, being merely a special case (zero force) of the second. Nevertheless, it was well justified in Newton's time when everybody firmly believed the application of a forward force to be required merely to sustain a body in motion; the true state of affairs certainly needed emphasizing.

The second statement is the one most open to criticism. Newton intended this as the statement of a law, mass and force being independently measurable. As we have seen, this is not true of mass, and it is true of force only in a limited sense. The statement could, instead, be regarded as a quantitative definition of force, mass being known; or, alternatively, of mass, force being known. In the former case it would then correspond completely to the quantitative part of our statement 2. However, Newton's definition of mass as quantity of matter not being acceptable, the former alternative, regarded as based on *this* definition, must be ruled out. As regards the second alternative, Newton considered mass as already having been defined; he certainly did not intend this statement as an independent definition of this quantity.

Newton's third statement corresponds, of course, to our statement 3.

Our own three statements may be regarded as a tidied-up version of Newton's laws, amended in the light of criticisms of these laws advanced by Mach toward the end of the nineteenth century.

Problems

Problems 4.22 and 4.30 are for readers who are familiar with the calculus.

4.1   A rope, suspended from a hook at its upper end, supports a 10-kg mass at its lower end. The mass of the rope is 100 g. Find, in newtons, the tension of the rope (a) at its lower end, (b) at its upper end, and (c) at its center.

4.2   A bird of weight 200 g lands on a telephone wire of negligible mass, causing the wire to be depressed into two straight segments which make angles of 3° and 5° with the horizontal. What are the tensions in the two segments of the wire?

4.3   A body of mass 1 kg is attached to the lower end of a string of negligible mass. The string is held fixed at its upper end. Find what horizontal force must be applied to the body to deflect the string from the vertical by 30°. When so deflected, what is the tension in the string?

4.4   A car of mass 1000 kg, starting from rest, acquires a velocity of 60 km/hr in 12 sec. What mean forward force is exerted on the car? By what is this force exerted, and how?

4.5   The classical problem of the horse and cart is usually stated in this way: If the cart pulls on the horse with the same force as the horse pulls on the cart, then how do the horse and cart move forward? Explain.

4.6   A car weighing 1000 kg and traveling along a horizontal road at 20 m/sec is braked suddenly and comes to rest in a distance of 40 m. Calculate the horizontal force, expressed in kilograms weight, exerted frictionally by the road on the tires. (Assume this force to be constant.)

4.7   What must be the minimum value of the coefficient of limiting static friction (maximum ratio of tangential frictional force to normal force in the absence of sliding) exerted between the tires of a car and the road to enable the car, starting from rest, to develop a speed of 100 km/hr in a distance of 200 m on a horizontal road?

4.8   A body slides down a plane inclined at an angle θ to the horizontal. The frictional force exerted by the plane on the body is a constant fraction μ of the normal force. What must be the angle θ such that, with sliding once established, it continues without acceleration? Obtain an expression for the acceleration in terms of μ and θ generally.

4.9   A block sliding up an inclined plane of vertical height 80 m and horizontal length 150 m has an initial velocity of 20 m/sec. The coefficient of kinetic friction (ratio of tangential frictional force to normal force exerted during sliding) is 0.2. Find (a) how far the block goes up the plane, (b) what time elapses before it reaches the bottom again, and (c) what velocity it has when it reaches the bottom.

4.10   On the ends of a light string hung over a light frictionlessly mounted pulley masses of 1 kg and 2 kg are suspended. Find (a) the tension of the string, and (b) the acceleration of the masses.

4.11   A light flexible rope hangs over a light frictionlessly mounted pulley, with two monkeys, each of the same mass, and each suspended from one end at the same height above the ground. One monkey begins to climb up the rope and, observing that the other is also rising, decides to try to race it to the pulley. Can he do so? Give reasons for your answer.

4.12   It is observed that, while an airplane is gathering speed preparatory to takeoff, objects suspended from hooks within it swing backward through an angle of 30°. What is the acceleration, and with what proportion of their normal weights do the passengers feel themselves to be pressed backward in their seats?

4.13 A mass m_1 on the surface of a frictionless inclined plane is attached by a light string passing over a light frictionless pulley to a mass m_2, as shown in the figure. Obtain an expression for the acceleration of the masses in terms of m_1, m_2, and the angle θ of inclination of the plane to the horizontal. Find also the tension of the string.

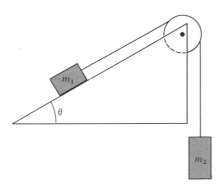

Figure for problem 4.13.

4.14 To the 1-kg bob of a very long simple pendulum is attached a light string which passes over a light frictionless pulley and supports a mass of 2 kg at its other end (see the figure). The portion of this string between the bob and the pulley is at right angles to the supporting string of the simple pendulum, which latter makes an angle of 45° with the vertical. What is the acceleration of the bob? What is the tension in the string supporting the 2-kg mass?

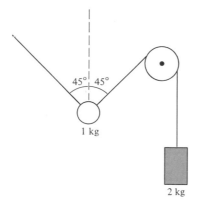

Figure for problem 4.14.

4.15 Three equal cubical blocks, each of mass 1 kg and supported by long strings of negligible mass, are in contact, as shown in the figure. A force F equal to 1 kg weight is applied horizontally to the first block in the direction toward the others. Find the force exerted by the second block (a) on the first, and (b) on the third. What is the acceleration of the blocks consequent on the application of the force?

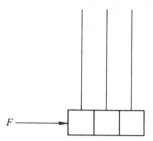

Figure for problem 4.15.

4.16 A rectangular frame from one corner of which a small ball of mass m is suspended by a light string is held near the top of a plane inclined at an angle θ to the horizontal, as shown in the figure. It is then released and slides frictionlessly down the plane. What is the direction of the string supporting the ball during this sliding? What is the tension of the string?

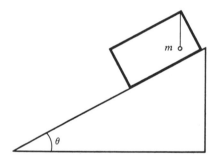

Figure for problem 4.16.

4.17 A "rough-and-ready" rule for road safety is that one should keep at least one car length behind the car in front for each 15 km/hr of one's speed. Assuming that the car in front stops suddenly without warning, and that the driver of the rear car takes immediate action, investigate whether this rule gives as good a margin of safety at high speeds as at low. (Assume constant available braking force.) Are all these assumptions realistic? If not, reassess the usefulness of the rule in the light of any amendments to the assumptions which you feel should be made.

4.18 A body of mass 100 g, in flight upward in a direction momentarily making an angle of 30° to the horizontal, encounters a resistive force of 0.2 N due to the air through which it is traveling. What is its acceleration, in magnitude and direction?

4.19 A chain of length l and weight W is suspended from one end so that the other (lower) end just touches the floor. It is then released. What will be the maximum downward force exerted on the floor during its fall?

4.20 A chain of length l is held on a table of height l in such a way that a negligibly short section hangs through a hole in the table. It is then released. Find the speed at which a link reaches the floor as a function of its position on the chain. (Assume that friction is negligible, and that on passing through the hole a link has the direction of its motion changed from horizontal to vertical with no change in speed.)

4.21 Small droplets of oil, falling under the combined forces of their weight acting downward and the viscous resistance of the air acting upward, soon acquire a constant speed downward, their "terminal velocity." For a given liquid, this terminal velocity is found to be proportional to the cube of the diameter of a droplet. Assuming that the resistive force is proportional to the product of the superficial area of a droplet and some function of its speed, find what this function is.

4.22 A body of mass m, initially moving with velocity u, is acted upon by a force, of magnitude kv^2, which opposes the motion, k being a constant and v the velocity at any instant. Find how the velocity varies with time and with distance covered.

4.23 A man weighing 80 kg stands on a spring balance in an elevator. What scale reading does the balance show if the elevator (a) ascends at a constant velocity of 3 m/sec? (b) has an upward acceleration of 1 m/sec²? (c) has a downward acceleration of 2 m/sec²? (d) falls freely in consequence of the cable breaking?

4.24 A man throws a ball of mass m upward and catches it as it descends. What is the force that he exerts upward on the ball, averaged over the total time of throwing, free flight, and catching? (Neglect air resistance.)

4.25 A 5-kg shell is fired horizontally with a velocity of 300 m/sec from a 1000-kg gun. What is the recoil velocity of the gun? The gun is mounted on a track provided with a mechanism for bringing it to rest 50 cm behind the point of firing. Assuming the retarding force to be constant, calculate the magnitude of this force.

4.26 An empty coal truck weighing 15,000 kg is moving along a horizontal track at a speed of 4 km/hr. A load of 5000 kg of coal is suddenly dumped, at zero horizontal velocity, into the truck from above. What is the new speed of the truck and its contents? (Neglect rotational dynamical effect on the wheels.)

4.27 A particle of mass m_1 traveling with velocity v_1 collides elastically* with a stationary particle of mass m_2, after which the first particle moves with velocity v_1' in a direction making an angle θ with its original direction of travel, and the second particle moves with velocity v_2 in another direction. Find this other direction. Also find how v_1' and v_2 are related to m_1, m_2, v_1, and θ.

4.28 At a certain instant in its flight, a rocket has a mass of 2500 kg and burns fuel at a rate of 20 kg/sec, with the resulting gas issuing from the rear of the rocket having a velocity relative to the rocket of 3 km/sec.

---

* An elastic collision is one in which the component of the velocity of one particle relative to the other in any direction after the collision is minus that before the collision.

If the rocket travels vertically upward, what is its acceleration at this instant? (Use the value 9.8 m/sec² for g and ignore air resistance.)

4.29   The action of a rocket is usually explained in terms of the conservation of momentum, or the equality and oppositeness of action and reaction. However, there must be some *mechanism* for the exertion of the forward force on the rocket. Can you explain what it is that exerts this force, and how?

4.30   A rocket of mass M contains a mass m of fuel (additional to its own mass) at blastoff. Ignoring the effect of gravity during the burning period and the viscous drag of the atmosphere, show that the velocity of the rocket when the fuel is completely burnt is

$$v = u \ln \left(1 + \frac{m}{M} \right),$$

where u is the velocity of the ejected gases relative to the rocket.*

---

* In mathematical symbolism "ln" stands for "natural logarithm of," that is, $\log_e$, where e is the base of natural logarithms.

5

CENTRAL FORCES

Centripetal and Centrifugal Forces

We saw in Chapter 3 that a particle moving in a circular orbit of radius r with uniform angular velocity ω has a centripetal acceleration of $\omega^2 r$. Accordingly, if the mass of the particle is m, there must, by the definition of force, be a **centripetal force** acting on the particle (Fig. 5.1), a force directed toward the center of the circle, given by the equation

$$F = m\omega^2 r = mv\omega = m\frac{v^2}{r}. \tag{5.1}$$

This centripetal force must be exerted by another body or system of bodies, which we may call the "constraining system," the system that constrains the particle to remain in its circular path. And because forces act in pairs, the particle in question must exert on the constraining system a force that is equal and opposite to that which the constraining system exerts on the particle. The particle thus exerts a force in a radial direction, *away from the center*, a **centrifugal force,** of magnitude $m\omega^2 r$, on the constraining system. This centrifugal force may be regarded as the reaction of the centripetal force, and, for emphasis, let it again be stated that it is exerted *by* the particle in question *on* the constraining system.

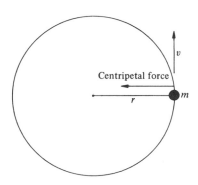

FIGURE 5.1   Particle of mass m describing a circular orbit of radius r with speed v.

There is a popular fallacy according to which a centrifugal force (away from the center) is supposed to act *on* the body moving in the circular path, and it is important that this idea be dispelled at the outset. It arises from a tendency to regard the situation as one in statics, with the body in equilibrium under the action of a pair of equal and opposite forces, instead of what it really is, a dynamical situation. A passenger in a car, for example, tends to take his bearings from the car rather than from the landscape through which he is traveling, and if the car rounds a curve, the passenger, feeling himself impelled toward the outer side of the car, says he is being "centrifuged." He feels that because of the motion around the curve a force has been brought into being, acting on him, in a direction away from the center. Actually, he simply tends to continue moving in a straight line, along the tangent to the circle, and if he and the car are not to part company he must be constrained to travel around the curve with the car by a centripetal force which the car exerts on him, for example frictionally, or by the outer side of the car pressing him inward. At the same time he exerts on the car (the constraining system) an equal and opposite centrifugal force.

Now let us consider the car as the body going around a circular path and the road as the constraining system. If the car is to keep

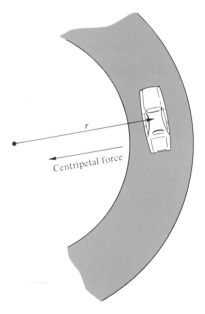

FIGURE 5.2 Centripetal force, provided frictionally, acting on a car traveling along a curved road.

to the road, the latter must exert on it, frictionally if the road is horizontal, the necessary centripetal force; and the car exerts on the constraining system, the road, again frictionally, an equal and opposite centrifugal force (Fig. 5.2). But if friction is insufficient to supply the necessary force, the car goes into a skid.

Perhaps the action generally found most difficult to explain otherwise than by invoking a fictitious centrifugal force exerted on the rotating body is that of a "centrifuge," a familiar example of which is the cream separator. In this, a liquid containing small particles of a solid or other liquid in suspension is rotated at high speed and this causes a separation of the less dense from the denser constituents, the former moving toward the axis, the latter away from it. Let us consider a small volume of the composite liquid. This can rotate about the axis only if acted upon by the requisite centripetal force, and this force is supplied by a pressure gradient (see Chapter 9) which is set up in the liquid, the pressure increasing with increasing distance from the axis. Now, with regard to the individual constituents of the liquid we have the following situation. At a given distance from the axis there can only be one pressure gradient, determined by the angular velocity and the *mean* density of the liquid. For the denser-than-average particles this will provide insufficient centripetal force, and consequently these particles will increase their distance from the axis. On the other hand, for the less dense particles the centripetal force provided by the pressure gradient will be too large. Hence these particles will move toward the axis.

Kepler's Laws of Planetary Motion

In Chapter 2 we saw that the periods of the motions of the planets around the sun are related to the radii of the orbits by the law

$$T^2 \propto r^3,$$

and that a similar law holds for the orbits of the satellites about a planet in cases where there is more than one satellite. Actually this is one of three laws of planetary motion enunciated by Johannes Kepler (1571–1630), a pupil of the Danish astronomer, Tycho Brahe (1546–1601), the last great astronomical observer before the invention of the telescope. After an enormous amount of calculation based on Tycho Brahe's observations, Kepler satisfied himself that these could all be brought together in the form of the following three laws:

1. Every planet moves so that the line joining it to the sun sweeps out equal areas in equal intervals of time (see Fig. 5.3). **121**

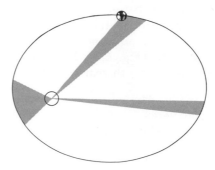

FIGURE 5.3   Kepler's "equal-areas" law illustrated. The three shaded areas shown, swept out in equal times by the line joining the sun to a planet, are all equal. (The eccentricity of the elliptical orbit and the sizes of the sun and planet are greatly exaggerated.)

2. The orbit of every planet is an ellipse, with the sun at one of its foci.

3. The squares of the periods of the planets are proportional to the cubes of their mean distances from the sun.*

From the second law in conjunction with the first it may be shown that this acceleration varies with distance according to an inverse-square law. And, as we saw in Chapter 2, this same result also comes out quite simply from the third law if we idealize the orbits to circles about the sun as center.

Let the mass of the sun be denoted by m_1 and that of a planet by m_2. If the planet's acceleration toward the sun is a and the distance between them is r, then, as we have just seen,

$$a \propto \frac{1}{r^2},$$

and hence, because the force F exerted by the sun on the planet is $m_2 a$,

$$F \propto \frac{m_2}{r^2}.$$

---

* We now know that the third law should be amended to read: The squares of the periods of the planets are proportional to the cubes of the major axes of their orbits. In other words, they are proportional to the cubes of the means of the greatest and least distances of the planets from the sun. However, for the nearly circular orbits with which we are here concerned, the differences between these means and the mean distances from the sun are much smaller than it would have been possible to detect in Tycho Brahe's and Kepler's times.

The mutual force between sun and planet is thus proportional to the mass of the attracted body as well as being inversely proportional to the square of the distance.

Clearly, the distinction between attract*ed* and attract*ing* is merely a matter of point of view; we may equally well think of the planet as attracting the sun with the force F. The sun then becomes the attracted body, and accordingly F must be proportional to the mass of the sun, m_1. From this requirement of physical reciprocity and corresponding mathematical symmetry it must be inferred that F is simultaneously proportional to m_1, to m_2, and to $1/r^2$. Accordingly, we may write

$$F \propto \frac{m_1 m_2}{r^2},$$

or

$$F = G\frac{m_1 m_2}{r^2}, \tag{5.2}$$

where G is the proportionality factor.

Law of Universal Gravitation

It would be almost inconceivable that the constant G in Eq. (5.2) should have one value for the sun and planets and another for, say, the system of Jupiter and its satellites—this both for a general and for a particular reason. Our whole experience is that natural laws are universal in their scope—we do not, for example, have ice melting at one temperature in the Northern Hemisphere and at another in the Southern Hemisphere. Also, if the value were different for the two systems, it would not apply to the case where, for example, Jupiter is regarded as the attracting body and the sun and a satellite of Jupiter as two bodies attracted by it. From such considerations we must infer that G is a *universal* constant, having the same value for all pairs of gravitationally attracted bodies, whatever or wherever they might be. This is known as the **constant of gravitation.** How the value of this constant may be determined we shall see in Chapter 8.

The law of gravitation expressed by Eq. (5.2) was first deduced from Kepler's laws of planetary motion by Newton, who then applied it to a comparison between g and the centripetal acceleration of the moon in its orbit. He found it to apply here, too, showing that the fall of a freely falling body at the earth's surface, for example, of an apple from a tree (if we may believe the legend), the revolution of the moon in its orbit about the earth, and the revolution of

123

the planets about the sun are all related phenomena—truly an epoch-making discovery.

Newton was at first not at all happy about the assumption he had to make that the earth's attraction is the same as if the whole of its mass were concentrated at its center. It was not until 18 years after he had formulated his law of gravitation that he finally succeeded in proving, on the basis of the assumed validity of this law, that it must be so; or, more generally, that the resultant gravitational attraction exerted on an external particle by a body having a spherically symmetrical distribution of mass is the same as would be exerted by a particle located at the center having the same mass.*

GRAVITATIONAL FIELD   A useful concept in physics is that of a field, a condition existing at each particular point in space which, in conjunction with some relevant property of a body or system located there, for example the mass or electric charge of the body, or the magnitude and orientation of a "current element" (see Chapter 29), determines the force experienced by this body or system. We shall introduce the idea in this chapter with a brief discussion of a **gravitational field.**

A gravitational field at a point is defined as the gravitational force exerted on a body located there per unit mass of this body. Thus if the magnitude of the gravitational force exerted on a body of mass m located at a particular point is F, the gravitational field at this point has the magnitude F/m; in other words, the "gravitational field strength" is F/m. And, since force is a vector quantity, a gravitational field is also a vector.

We have already dealt with one important example of a gravitational field in our discussion of the gravitational acceleration g of a body at the earth's surface.† This, being the gravitational force exerted on the body per unit mass, is, according to our definition, the gravitational field strength at the earth's surface. In our discussion of Kepler's laws of planetary motion we have, in effect, again been concerned with a gravitational field—that due to the sun as a function of distance from it. It is this field that produces the centripetal accelerations of the planets toward the sun. We may

---

* It is possible to see that this must be so by reasoning analogous to that employed in electrostatic field theory for showing that a uniformly electrically charged sphere produces external effects identical with what would be produced by the same total charge concentrated at the center of the sphere. This will be shown in Chapter 20.

† Strictly g_0, but let us now drop the subscript and use the symbol g quite generally for gravitational field.

express the magnitude g of this field as a function of distance r from the sun by dividing both sides of Eq. (5.2) by m_2 and writing m instead of m_1 for the mass of the sun. Thus

$$g = G \frac{m}{r^2}.$$

Clearly, the validity of this equation is not confined to the case where the sun is the attracting body. The gravitational field strength due to *any* isolated body of mass m at a distance r from it is given by the same equation. And, of course, the direction of the field is everywhere that toward the attracting body.

In considering the field at the location of any one planet of the solar system the contributions of all the other planets, satellites, and so on, should, strictly, be taken into account and the vectorial resultant of all the separate fields found. Although this may in general be omitted in the case of the solar system, because of the extreme smallness of the contributions of these lesser bodies compared with the field due to the sun, cases do obviously arise in which a vectorial combination of contributions due to a number of separate bodies or elements of mass must be carried out. Reference has already been made to one example of this: In the evaluation of g at the earth's surface from first principles all the contributions due to the elements of mass constituting the earth must be combined.

Simple Pendulum

In Chapter 2 it was shown experimentally that the motion of the bob of a simple pendulum is, at least to a close approximation, simple harmonic. This motion will now be investigated analytically and an expression derived for the period of oscillation.

As we have already seen, the component of the weight of the bob tangential to the arc of swing is $W \sin \phi$, where ϕ is the angle shown in Fig. 5.4. The other component of the weight is $W \cos \phi$. The tension of the string just balances out this latter component when the bob is at either extremity of its swing and momentarily at rest. At other times, when the bob is in motion, the tension of the string must exceed $W \cos \phi$ by the centripetal force that is necessary to keep the bob traveling in its circular path. We merely note this in passing; what mainly concerns us is the tangential force, acting on the bob in the direction of motion. Since W is equal to mg, this is $mg \sin \phi$.

125

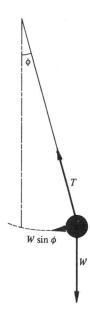

FIGURE 5.4 The forces exerted on the bob of a simple pendulum are its weight W acting vertically downward and the tension T of the string. The component of W along the arc of swing is $W \sin \phi$.

It will be convenient to consider both the angle ϕ and the tangential force algebraically, ϕ being reckoned positive or negative according as the bob is to the right or to the left of the midpoint of swing, O, while the tangential force is taken as positive or negative according as its direction along the arc AB is from left to right or from right to left. On this basis the tangential force becomes $-mg \sin \phi$, and the corresponding tangential acceleration a (also reckoned algebraically) must therefore be $-g \sin \phi$.

In Chapter 2 we saw that in simple harmonic motion the acceleration is proportional to the displacement by a negative proportionality factor. We have just seen that the tangential acceleration of the bob of a simple pendulum is always directed toward the midpoint of swing, O, and so is of opposite sign to the displacement from this point, both quantities being reckoned algebraically. We have now to ascertain whether the acceleration is also *proportional* to the displacement; if we find that it is, the motion must be simple harmonic.

The displacement x of the bob along the arc of swing, measured from O, is $l\phi$, where l is the distance from the point of support to the center of the bob, known as the "length of the pendulum."

For the ratio of the acceleration (a) to the displacement we have, therefore,

$$\frac{a}{x} = -\frac{g}{l} \frac{\sin \phi}{\phi}.$$

We see that this is not strictly constant, and accordingly the motion of the bob of a simple pendulum is not strictly simple harmonic. However, for sufficiently small angles, $\sin \phi$ approximates very closely indeed to ϕ itself, and so, to this degree of approximation, we may write

$$\frac{a}{x} = -\frac{g}{l}. \tag{5.3}$$

Thus, for arcs of swing that are small compared with l, the motion approximates closely to simple harmonic motion.

If now we substitute for the quantity under the square-root sign in Eq. (2.16),

$$T = 2\pi \sqrt{\frac{\text{displacement}}{\text{acceleration}}},$$

we at once obtain the formula for the period of the simple pendulum,

$$T = 2\pi \sqrt{\frac{l}{g}}. \tag{5.4}$$

The student who has the necessary mathematical background will be able to establish that the motion is (to a close approximation) simple harmonic and obtain the expression for the period without reference to Eq. (2.16) as follows: Eq. (5.3) may be written in the form

$$\frac{d^2x}{dt^2} + \frac{g}{l} x = 0.$$

This is a differential equation, whose most general solution is

$$x = A \sin\left(\sqrt{\frac{g}{l}} t + \epsilon\right), \tag{5.5}$$

where A and ϵ are constants of integration. That this is so may readily be checked by working backward and differentiating with respect to time twice in succession. Obviously, an equation of this form is a mathematical representation of simple harmonic motion, defined, as in Chapter 2, as the projection of uniform circular motion, A being the radius of the circle and $\sqrt{g/l}$ the angular velocity of the point of reference. Either by using this angular velocity, or directly from inspection of the equation, we see at once that x must have the period $2\pi \sqrt{l/g}$.

From Eq. (5.4), or its equivalent,

$$g = \frac{4\pi^2 l}{T^2},$$

we see that by setting a simple pendulum of measured length oscillating with small amplitude and ascertaining the period of oscillation we may determine the value of g with quite reasonable accuracy. Rather than perform the experiment with only a single length of the pendulum, it would be better to measure the values of T corresponding to a series of values of l and then plot T^2 against l. This would give a straight line passing through the origin, of slope $4\pi^2/g$, and from the measured value of the slope g could accordingly be calculated. Such a determination of g would be accurate to within about 0.1 percent.

The accuracy of determinations of g attainable with a simple pendulum is limited mainly by the finite mass of the string or thread and by the want of complete definiteness in its effective length. The most accurate determinations are made with a rigid-body pendulum developed by Kater (Fig. 5.5), which is provided with two knife-edges, one at either end, the positions of these being adjusted so that the period of oscillation is the same from whichever of the two knife-edges the pendulum is suspended. The theory of Kater's pendulum is given in many of the older textbooks. In its use, cor-

FIGURE 5.5   Kater's pendulum.

rections are applied for the curvature of the knife-edge, air buoyancy and resistance, and the finiteness of the arc of swing. It is by such means that the values of g quoted in Table 2.1 were obtained; these are accurate to about 1 part in a million.

Oscillating Mass on Spring

Let us now consider a mass suspended from a light helical spring and set into vertical oscillations under the joint action of the tension of the spring and the earth's gravitation. Such oscillations are of particular interest for the information they give concerning the elastic properties of the spring, and also as providing a means for determining g alternative to, although somewhat less accurate than, the use of a simple pendulum.

Let the mass of the suspended body be m, that of the spring being negligible in comparison. The gravitational force acting on the body (its weight) is, of course, constant, and equal to mg. On the other hand, the force exerted by the spring varies with the extent to which it is pulled out. The nature of this variation may readily be established experimentally; it is found, by suspending, statically, different masses from the spring, and noting the corresponding extensions, that the extension is simply proportional to the tension (weight of the suspended mass). This is a special case of Hooke's law, which will be considered in Chapter 10.

Let the tensile force that the spring exerts upward on the body of mass m when it is at rest be F_l. Because the body *is* at rest, this must just balance the weight of the body, mg; that is,

$$F_l = mg.$$

Let us write l for the observed extension of the spring under these conditions. Then for the ratio of the tension F_l to extension we have

$$\frac{F_l}{l} = \frac{mg}{l}.$$

But we know that this ratio is a constant for an ideal spring: It is a unique property of the spring and so must apply irrespective of whether the mass is in equilibrium or not. When, in the course of the oscillation of this mass, its displacement downward from the equilibrium position is x (see Fig. 5.6) so that the total extension of the spring is $l + x$, the corresponding tension must be $mg[(l + x)/l]$, or the sum of the two terms mg and mgx/l. The first is equal to the weight of the body, balancing this out. The second therefore

129

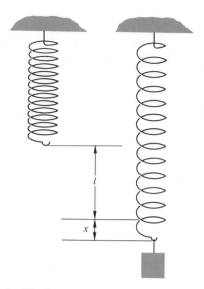

FIGURE 5.6 Oscillating mass on a spring.

represents the resultant force acting on the body, this force being exerted in the negative direction (upward) when x is positive (displacement downward), and vice versa. The magnitude of the acceleration corresponding to this force is gx/l. Taking both acceleration and displacement algebraically, the ratio of the former to the latter is $-g/l$, a negative constant. From this it follows that the oscillation must be simple harmonic, the period being expressed by the same formula,

$$T = 2\pi \sqrt{\frac{l}{g}},$$

as applies to the simple pendulum, except that here l stands for the extension of the spring under equilibrium conditions, whereas in the case of the simple pendulum it stands for the length of this pendulum.

This formula shows that g could be evaluated from the extension produced by a given suspended mass when at rest and the corresponding period of oscillation. However, in practice, owing to the finite mass of the spring, which in our discussion we have idealized as negligible, such a determination would tend to be less accurate than one made with the simple pendulum.

In the cases of both the simple pendulum and the oscillating mass on the spring we have found the period from the ratio of the displacement to the acceleration, using Eq. (2.16),

$$T = 2\pi \sqrt{\frac{\text{displacement}}{\text{acceleration}}}.$$

It is worth noting that a useful alternative formula for the period may be obtained by multiplying both the numerator and the denominator under the square-root sign by the mass of the oscillating body. Thus

$$T = 2\pi \sqrt{\frac{\text{mass} \times \text{displacement}}{\text{mass} \times \text{acceleration}}}.$$

From this the alternative formula follows at once,

$$T = 2\pi \sqrt{\frac{\text{mass}}{\text{restoring force per unit displacement}}},$$

or, in symbols,

$$T = 2\pi \sqrt{\frac{m}{|F/x|}}. \tag{5.6}$$

In the present case $|F/x|$ is mg/l, the ratio of tension to extension. It is interesting to note that this ratio may be evaluated either statically, being g times the reciprocal of the slope of the plot of the extension l against the mass m of the suspended body, or dynamically. In the latter case, one would, for convenience, rewrite Eq. (5.6) in the form

$$\left| \frac{F}{x} \right| = \frac{4\pi^2 m}{T^2}.$$

For best accuracy, one would evaluate $|F/x|$ from the slope of a plot of T^2 against m.

Of the two methods for determining the ratio of tension to extension, the static is the more direct, and at least as accurate as the dynamic. Consideration of the dynamic method is nevertheless important, because it serves as a useful introduction to a whole class of problems in which the timing of oscillations is used for the quantitative determination of the "control" of an oscillating system. We have already had two examples of this. Thus, in the simple pendulum, whose control is gravitational, the period in conjunction with the length of the pendulum gives us the value of g. And in the present system the ratio of tension to extension, which is an elastic property of the spring, may be calculated from the mass of the suspended body and the corresponding period of oscillation. In Chapter 8 we shall consider another important example, where the method of oscillations is employed to evaluate the gravitational constant G.

Problems

Problems 5.10 and 5.11 are for readers familiar with the calculus.

5.1 A 1-kg mass is whirled at the end of a string 1 m long at uniform speed in a vertical circle at 2 m/sec. What is the tension in the string (a) when the mass is at the top of the circle, (b) when it is at the bottom, and (c) when the string is horizontal?

5.2 A humpbacked bridge over a stream has a radius of curvature of 20 m. What is the maximum speed with which a car may travel over this bridge without "taking off"?

5.3 A body A of mass 0.5 kg rests on a horizontal turntable with its center at a distance of 20 cm from the axis of rotation, and the coefficient of limiting static friction between the body and the turntable is 0.4 (see the figure). A light string attached to A passes over a frictionless pulley mounted on the turntable and hangs through a hole in the center as shown. A body B of mass 0.6 kg is tied to the lower end of the string. Between what limits must the number n of revolutions of the turntable per second lie if B is not to move either up or down?

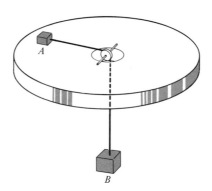

Figure for problem 5.3.

5.4 A passenger in an artificial earth satellite observes that he and all objects surrounding him appear to be devoid of weight. Yet the earth certainly exerts on him and on these objects a substantial gravitational pull. What, then, is the explanation of the apparent weightlessness? The sensation of "weightlessness" is not confined to passengers in space vehicles; most of us have experienced it during brief intervals of time. Can you think of an example of this?

5.5   Find the period of oscillation of the bob of a conical pendulum which describes its circular orbit in a plane whose distance below the point of support is h.

5.6   The bob of a conical pendulum has a mass of 5 g and the distance of its center from the point of suspension of the string is 1 m. Find (a) the tension of the string, and (b) the period of the pendulum, when the bob revolves in a circle of radius 50 cm.

5.7   A curve in a road has a radius of 200 m and is intended for use by traffic traveling at 60 km/hr. At what angle should it be banked? If a car travels around this curve at 100 km/hr, what is the ratio of tangential to normal force exerted on it by the road?

5.8   A thin hoop of steel ($\rho = 7.8$ g/cm^3), whose radius is 1 m, rotates about the axis through its center perpendicular to its plane at 1500 rpm. Find the resultant of the tensile forces exerted at the two ends of a short element of the hoop, and hence determine the stress (force per unit cross-sectional area) developed in the hoop. (Use must be made of the condition for the element to go around in a circular path.) If the radius of the hoop were halved and the angular velocity doubled, how would the stress be affected?

5.9   Two small balls are attached to a light string of length l supported at one end from a ceiling, one, of mass m_1, halfway down, and the other, of mass m_2, at the bottom. The system is now set into rotation about a vertical axis with the upper ball rotating in a horizontal circle of radius r_1 and the other in a circle of radius r_2. Investigate how r_1, r_2, and l are related.

5.10   Show by direct integration based on the inverse-square law of gravitation that at any point outside a uniform spherical shell whose thickness is small compared with its radius the gravitational field is the same as it would be if the whole mass of the shell were concentrated at its geometric center. Show also that at any point inside such a shell the field is zero. Extend these results to any body having a spherically symmetrical distribution of mass.

5.11   Given that the moon's radius is 1740 km and that the acceleration due to gravity at the surface of the moon is 161 cm/sec^2, calculate the "velocity of escape" from the moon; that is, find the minimum velocity with which a particle must be projected vertically upward from a point on the moon's surface in order that it shall escape from the moon's gravitational attraction altogether.

5.12   Mars and Neptune each have two small satellites describing approximately circular orbits. One of Mars's satellites revolves in an orbit of radius 23,400 km with a period of 30 hr, 18 min, and one of Neptune's revolves in an orbit of radius 354,000 km with a period of 5 days, 21 hr. Compare the masses of the two planets.

5.13   The distance apart of the two components of the binary star Capella is 114.6 million km and the radii of their (circular) orbits are 54.3 and 60.3 million km. Their period of revolution about the common center of their orbits is 0.285 yr. Find the masses of the components in terms of that of the sun, given that the earth's distance from the sun is 149.5 million km.

5.14 Estimate the maximum disturbance to the value of g at the earth's surface that can be produced by the gravitational attractions of the sun and moon, given the following information:

mean radius of earth = 6371 km,
mass of sun = 333,000m_e, where m_e is the mass of the earth,
mass of moon = 0.01222m_e,
distance of sun from earth = 1.495×10^8 km,
distance of moon from earth = 3.84×10^5 km.

(*N.B.*: Think well before embarking on your calculations. The "obvious" method could well be too naive, giving quite the wrong answer.)

5.15 The following periods T are found for a series of lengths l of a simple pendulum:

l (cm)	22.7	28.4	35.5	42.3	51.1	59.4
T (sec)	0.955	1.070	1.197	1.307	1.434	1.546

Plot these results in a suitable manner, and from your graph calculate the value of g.

5.16 A simple pendulum is suspended from the ceiling of a room and the time t for 50 oscillations is measured for different values of the height h of the center of the bob above the floor. Values of h and t are as follows:

h (cm)	38.4	72.5	88.3	111.8	130.7	155.7
t (sec)	163	152	146	138	131	121

Using a graphical method, find (a) the height of the ceiling above the floor, and (b) the value of g.

5.17 Is the period of a simple pendulum greater or less when its amplitude is large than when it is small? Explain.

5.18 A simple pendulum 1 m long swings through a total angle of 120°. Find the velocity of the bob at the lowest point and the corresponding tension of the string in terms of the mass m of the bob, the length l of the pendulum, and g. Find also the acceleration of the bob at one of the extreme points of its swing.

5.19 The periods of vertical oscillation T of a series of masses m suspended from a certain spring are as follows:

m (g)	20	40	60	80	100	120
T (sec)	0.475	0.671	0.822	0.950	1.063	1.164

How far will the spring be stretched if a mass of 150 g is suspended from it? What will be the corresponding period of oscillation?

5.20 The tension in stretched india rubber varies with extension in the manner shown. A mass suspended from a strip of india rubber produces an extension corresponding to the point P in the figure. It is then set into

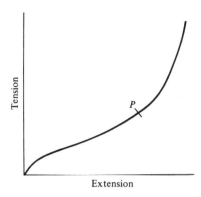

Figure for problem 5.20.

vertical oscillation. Sketch the curve of displacement against time (a) for a small amplitude of oscillation, and (b) for a large amplitude.

5.21   A light helical spring under tension whose axis is vertical is fixed at both ends, and a small mass is attached to its center. This is then set into oscillation along a horizontal straight line. Will this oscillation be simple harmonic? If not, sketch the displacement time curve. Will the period of the oscillation vary with its amplitude? If so, how?

5.22   In the system of Problem 5.21 the mass is set into vertical oscillation. Show that this oscillation is simple harmonic, and investigate whether the period varies with the original extension of the spring and, if so, how.

6

WORK AND ENERGY

One of the most useful and fundamental concepts of physics is that of **energy.** A closely related concept, without which it would hardly be possible to discuss energy intelligently, is that of **work.**

In everyday language we often speak of doing work, or of having or expending energy, and in some of the meanings of these expressions we are directly concerned with physical phenomena. For example, we should say we are doing work if we are drawing water from a well. Or a railway locomotive may be said to do work when it is hauling a train. When a person drawing water has been working for some time he may say he has less energy, less capacity for doing work, than before. Let us see if we can give these terms, work and energy, precise meanings that will be useful in our scheme of physics.

Physical Definition of Work

The man drawing water might naturally measure the amount of work he has performed by the quantity of water he has drawn from the well. But if the depth of the well were doubled and he drew the same quantity of water from it, he could legitimately claim to have performed twice the amount of work; he would, in effect, have carried out the original task twice over. Thus, generalizing, we may consider the work as given by the product of the amount of water drawn and the height through which it has been raised. (For simplicity we may consider the case where the weight of the bucket and rope is negligible.) Applying this to each unit operation, we should say that the work performed in pulling up each bucketful is proportional to the product of the weight of the latter and the height through which it has been raised. With an appropriate choice of units, the proportionality becomes an equality. Accordingly, in the case under consideration, where the displacement of the point of application of the applied force has the same direction as this force, the work may be defined as the product of the force and the displacement.

A rough alternative measure of the work performed by the man would be the fatigue he feels on completion of his task, his loss of energy. But fatigue is not a very tangible thing. In attempting to put loss of energy on a physical basis we might, by way of trial, measure it by the loss of weight the man has sustained in the performance of his task, over and above that which he would have sustained when resting because of his metabolic processes. An experimental test would show that, allowing for the uncertainty in the metabolic rate when resting, this appears satisfactory.

In the most general case we have to take account of the fact that the force and the displacement do not necessarily have the same direction. An example of this would be a horse or a locomotive on the bank of a canal pulling a barge (Fig. 6.1(a)); the towrope by which the force is applied is necessarily inclined at an angle to the direction of the canal. An extreme case is that of a conical pendulum. In the absence of air resistance, no expenditure of energy would be

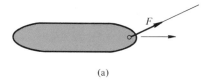

(a)

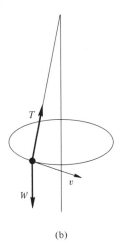

(b)

FIGURE 6.1   (a) A barge being towed along a canal.
(b) A conical pendulum. The tension T of the string, the weight W of the bob, and the resultant of these two forces are all at right angles to the velocity v.

necessary to maintain the motion, yet there is certainly a force exerted by the string, and the point of application of this force moves (Fig. 6.1(b)). The motion is, however, always at right angles to the force. Under these conditions, we can retain our association of the performance of work with loss of energy by the agency performing it only by asserting that, not only is no energy expended, but also no work is done. This now becomes a feature of our *definition* of work.

In considering the general case where the direction of the applied force makes an angle θ with that of the displacement, we make use of the fact that force is a vector quantity. Let the applied force of magnitude F be resolved into two components, $F \cos \theta$ in the direction of the displacement and $F \sin \theta$ in a perpendicular direction. Then only the component $F \cos \theta$ will contribute to the work done, and if s is the displacement of the point of application of the force, the corresponding amount of work W performed is the product of s and $F \cos \theta$:

$$W = s \cdot F \cdot \cos \theta. \tag{6.1}$$

We may associate $\cos \theta$ with either F or s; that is, we may say that the work performed is equal either to the product of the displacement of the point of application of the force and the component of the force in the direction of the displacement or to the product of the force and the component of the displacement in the direction of the force.

WORK AS THE SCALAR PRODUCT OF TWO VECTORS The right side of Eq. (6.1) defines a quantity known in vector algebra as the **scalar product,** or **dot product,** of two vectors, these being in the present case the force and the displacement of its point of application.

By convention, a vector quantity is represented *fully* by a symbol in boldface type; for example, a force considered vectorially is represented by the symbol **F**. The corresponding symbol printed in italics represents merely the magnitude of this vector, without regard to direction; thus the symbol F stands for the magnitude of the force **F**. Accordingly, F is a scalar. In our discussions hitherto we have used only italicized symbols; for example, we have considered

$$F = \frac{dp}{dt} \tag{6.2}$$

as the defining equation for force. Because this equation makes no reference to directions, the additional information that the force and the rate of change of momentum have the same direction has

139

been conveyed in words. An alternative way of defining force in its quantitative aspect *completely*, without resort to words, would have been to print the equation as

$$\mathbf{F} = \frac{d\mathbf{p}}{dt}.$$

This would have signified, in conventional mathematical symbolism, that not only are the magnitudes of the force and the rate of change of momentum the same but also their directions. Similarly, the equations

$$\mathbf{F} = m\mathbf{a}$$

and

$$m_1\,\delta\mathbf{v}_1 = -m_2\,\delta\mathbf{v}_2,$$

corresponding to Eqs. (4.6) and (4.10), respectively, convey the information that the directions, as well as the magnitudes, of the quantities on the two sides are the same.

Let us now consider two vectors **A** and **B** whose directions make the angle θ with one another (Fig. 6.2). An alternative symbolic representation of the product $AB \cos \theta$ of one of these and the component in its direction of the other is $\mathbf{A} \cdot \mathbf{B}$. Thus

$$\mathbf{A} \cdot \mathbf{B} = AB \cos \theta;$$

the left side is simply a convenient notation for what is written on the right. The product of two scalars (one of them a component of a vector) or, alternatively, of three scalars, A, B, and $\cos \theta$, must itself be a scalar, so the quantity $\mathbf{A} \cdot \mathbf{B}$ is known as the scalar product of the two vectors **A** and **B**. An alternative name is the dot product.

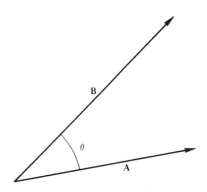

FIGURE 6.2

Using this notation, we could render Eq. (6.1) as

$$W = \mathbf{s} \cdot \mathbf{F},$$

or

$$W = \mathbf{F} \cdot \mathbf{s};$$

the work performed is the dot product of the force and the displacement of its point of application.

Energy

In physics, energy is defined as the **capacity for doing work.** Thus a man drawing water from a well has, initially, a capacity for doing work, and in the process of its performance this capacity is reduced because of the occurrence of certain chemical changes; some part of the man is "consumed," combining with the oxygen he breathes to form carbon dioxide and water, which are exhaled or otherwise eliminated. Similarly, in a steam locomotive hauling a train, coal is consumed chemically by combining with oxygen, generating heat, which is necessary for "raising steam" in the boiler. If it is an electric locomotive, we have, in many cases, the same generation of heat from burning coal, but this time it is in a distant power station, and this heat is then responsible, via a steam engine, for actuating an electric generator, which feeds "electric power" into the line. We may correspondingly speak of chemical energy, in the as-yet-unconsumed tissues of the man, or in the coal, in conjunction with oxygen. Also, as we shall see later, we may speak of heat energy in the locomotive or power station, or electrical energy, because they all have, directly or indirectly, a capacity for doing work, and the energy may be measured by the work that is capable of being performed as an accompaniment of their going out of existence. Also we have, evidently, the possibility of the conversion of one form of energy into another, for example, chemical energy into heat, or heat into electrical energy.

Work and energy necessarily have the same dimensions. From Eq. (6.1) it is evident that these are ML^2T^{-2}.

The forms of energy we have been discussing are only a few of several met with in nature. In due course we shall discuss these, and some others, further. In the present chapter we shall confine ourselves mainly to a consideration of a form of energy not yet mentioned, mechanical energy, and we shall see whether, in the realm of mechanics at least, the concept of energy is a useful one.

MECHANICAL ENERGY Mechanical energy may be subdivided into three main forms: **gravitational potential energy,**

elastic potential energy, and **kinetic energy.** We shall now briefly discuss these in turn.

Gravitational Potential Energy Let us again consider the case where water is drawn from a well. One possible agency for doing this would be that depicted in Fig. 6.3, a mass equal to that of the bucket of water being suspended from a rope attached to both of them which passes over a pulley. To simplify the discussion, let us idealize the situation by supposing that the mass of the rope may be neglected and that no energy is dissipated because of flexure of the rope or friction in the pulley bearings. Then if the bucket has to be raised a height h, the mass supplying the counterweight will fall through the same height in raising it, doing an amount of work mgh in the process, where m is the mass of the bucket and its contained water. When the bucket has been raised the capacity for doing work of the suspended mass has accordingly been reduced by the amount mgh. An alternative way of saying this is that its gravitational potential energy was greater by mgh before the bucket was raised than after.

Of course, the same considerations apply in reverse to the bucket of water. Before being raised its gravitational potential

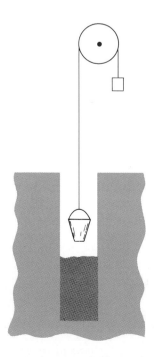

FIGURE 6.3 System for drawing water from a well.

energy is less by the amount mgh than after. In the process of its being raised there is merely a transfer of energy from one mass to the other.

Gravitational potential energy is to be regarded as the energy associated with the *position* of a body and not depending on its motion. Quite generally, if a body of mass m is at a height h above some agreed standard or "zero" level, its gravitational potential energy relative to that level is mgh.

Elastic Potential Energy In our discussion of the oscillating mass on a spring we saw that the force F exerted on the spring in stretching it is proportional to the elongation s,* so that we may write

$$F = ks,$$

where k is a constant characteristic of the spring. Hence a plot of F against s will be a straight line passing through the origin, of slope k, as shown in Fig. 6.4. Using this plot, let us consider how much work is done on the spring by the external agency when the extension is increased from a certain value s_1 to a slightly greater value s_2, the increase corresponding to the width of the strip shaded in the figure. During this increase the force is practically constant, being represented by the height of the strip. The product of the force and distance moved, or work done on the spring, is therefore represented by the area of the strip. But the whole triangular area shown in the fig-

---

* The symbols F and s are here used in place of the variables F_l and l of the previous discussion. We no longer have any use for the subscript in F_l. And in the present discussion it will be convenient to reserve the symbol l for the final extension of the spring and to use another symbol, s, to denote the smaller extensions passed through en route.

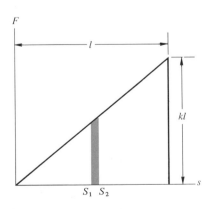

FIGURE 6.4 Relation between tension and extension of a spring.

ure may be thought of as consisting of a series of such strips, each of which represents the work done for the corresponding increment of elongation. The total work done on the spring in stretching it to its final elongation l is therefore represented by the area of the triangle, of base l and height kl; that is, it is $\frac{1}{2}kl^2$.

If we now consider the reverse operation of relaxation of the spring, it is evident that during each small change in length an amount of work represented by the area of the corresponding strip is done by the spring, and thus during relaxation the spring does a total amount of work on the external system equal to $\frac{1}{2}kl^2$. This, therefore, is the energy E initially stored in the stretched spring:

$$E = \tfrac{1}{2}kl^2. \tag{6.3}$$

The above derivation of the expression for E is really, in effect, a graphical integration. A much shorter procedure would have been to use the corresponding algebraic integration. Thus

$$E = \int_0^l ks\,ds = \tfrac{1}{2}kl^2.$$

However, the longer derivation is given here because of the detailed physical picture it provides of the performance of work accompanying stretching or relaxation.

Obviously, the same expression for the energy would be obtained if the spring were compressed instead of stretched.

Kinetic Energy Let a particle of mass m be moving with an initial velocity v_0, and let a force F be exerted on it by some external system in a direction opposing its motion, so that it will eventually be brought to rest. Then, while it is being brought to rest, it must exert on the external system a force of reaction F' equal to $-F$; and, if the particle comes to rest after traveling a total distance s, the corresponding work performed by it on the external system will be $F' \cdot s$. This must be the original capacity of the particle for doing work because of its motion; in other words, it is the original kinetic energy of the particle.

While the particle is coming to rest its acceleration is F/m. If we substitute this for a in Eq. (2.4) and give the final velocity v in this equation the value 0, we obtain for s the value $-mv_0^2/2F$ or $+mv_0^2/2F'$. Hence the product $F' \cdot s$ is $\frac{1}{2}mv_0^2$.

A somewhat more general derivation of this result, which makes use of the calculus, and in which F is not assumed to be constant, is the following. As before, F/m is the acceleration of the particle, but this is now not necessarily constant. Again writing s for displacement, we have for the work W performed by the particle in coming to rest,

$$W = \int F' \, ds = - \int F \, ds = -m \int a \, ds = -m \int \frac{dv}{dt} \, ds$$

$$= -m \int \frac{ds}{dt} \, dv = -m \int_{v_0}^{0} v \, dv = \tfrac{1}{2} m v_0^2.$$

This result having been obtained by one or other of the above methods, we may now, for convenience, drop the subscript from v_0 and write for the kinetic energy E_k of the particle corresponding to any value of its velocity v the expression

$$E_k = \tfrac{1}{2} m v^2. \tag{6.4}$$

Like the definition of force

$$F = ma$$

used in its derivation, the expression for E_k has only limited validity; it applies only for particle velocities that are small compared with the velocity of light. A completely general expression for E_k in terms of the rest mass* m_0 of the particle, its velocity v, and c, the velocity of light, has been derived by Einstein in his theory of relativity.

New Way of Regarding Force

Let us consider a particle on which a single force, or, if there is more than one force, a resultant force F is exerted. Then any increment δE_k in its kinetic energy can be expressed in terms of this force and the corresponding increment δs in the component of its displacement in the direction of the force by the equation

$$\delta E_k = F \, \delta s.$$

Writing this in the alternative form

$$F = \frac{\delta E_k}{\delta s},$$

and then proceeding to the limit of smallness of the variables con-

---

* The concept of "rest mass" was referred to briefly in Chapter 4, under the heading "Momentum." Mass is a quantity that increases with the velocity of the body concerned, at first imperceptibly slowly, then at an increasingly rapid rate as the velocity of light is approached. Accordingly, the rest mass of a body may be defined as its mass at velocities that are small compared with the velocity of light, or, more strictly, as the asymptotic limit approached by the mass as the velocity of the body approaches zero.

sidered, we have

$$F = \frac{dE_k}{ds}.$$

Thus we see that **force is the distance rate of increase of kinetic energy.** This concept of force is alternative to our earlier definition of this quantity as mass times acceleration, or time rate of change of momentum.

The reader should note carefully the wording of the last three sentences. In these the expressions appear in order: "the variables considered," "distance rate," and "time rate." In proceeding from

$$\delta E_k = F \,\delta s$$

to

$$F = \frac{dE_k}{ds},$$

we were concerned with only one independent variable, the position of the particle, as given by s. The time occupied by any change in s is irrelevant to the discussion, and so this need not be considered. Instead of having to indicate this in words, we might have conveyed the fact that we were disregarding any independent variable or variables other than s simply by writing the last equation in the form

$$F = \frac{\partial E_k}{\partial s}. \tag{6.5}$$

The right side of this equation, which is known as the **partial derivative** of E_k with respect to s, means, by convention, "the distance rate of increase of E_k, no other independent variable having any relevance to the matter under consideration." Alternatively, it could mean, although actually it does not in this case, "the rate of increase of E_k with s, all other independent variables being constant, or 'frozen.'" Which meaning is to be attached to the expression is usually indicated sufficiently clearly by the context of the mathematical discussion. On the same principle, instead of having to state, in words, that dp/dt is to be understood as the time rate of change of momentum, with the implication that we are not concerned with any associated variation of momentum with position, we might more conveniently have written the expression in partial-derivative form, $\partial p/\partial t$, so conveying precisely the same thing in mathematical symbolism.

Like

$$F = \frac{\partial p}{\partial t},$$

the alternative definition of force,

$$F = \frac{\partial E_k}{\partial s},$$

has the advantage over

$$F = ma$$

of applying quite generally, not merely in the limiting case where the velocity of the particle is small compared with the velocity of light.

In our development of the idea of work and the related quantity energy, we made use of the concept of force, this being thought of as the more fundamental quantity. However, future generations of physicists may well prefer to regard energy as the more fundamental, with force a derived quantity, corresponding to Eq. (6.5).

Conservation of Energy

During the oscillation of the bob of a simple pendulum there is a rhythmic variation of its height, and thus of its potential energy, between a minimum when the bob is at the midpoint of its swing and a maximum when it is at either extremity. Concurrently, the kinetic energy is a maximum at the midpoint of swing and zero at the extremities, when the bob is momentarily at rest. We shall now investigate these energy changes in more detail.

We may imagine the arc described by the bob as divided into a large number of short, virtually straight, sections. In effect, these will constitute a series of inclined planes, the angle of inclination to the horizontal varying from one plane to the next. Let us consider, in particular, one of these, indicated in Fig. 6.5. This will be inclined at the same angle ϕ to the horizontal as the string makes with the vertical.

While the bob is passing from one end of the section, P, to the other, Q, the component of its weight mg in the direction of motion is $-mg \sin \phi$, and the corresponding acceleration is $-g \sin \phi$. Substituting this for a in Eq. (2.4) and writing v_Q for v, v_P for v_0,

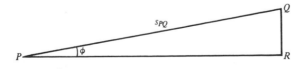

FIGURE 6.5 Inclined-plane equivalent of short section of arc described by pendulum bob.

and s_{PQ} for s, we have

$$v_Q{}^2 = v_P{}^2 - 2g \sin \phi \cdot s_{PQ}.$$

If, now, we write the heights of the points P and Q above an agreed level as h_P and h_Q, respectively, we may substitute $(h_Q - h_P)/s_{PQ}$ for $\sin \phi$, so obtaining the equation

$$v_Q{}^2 = v_P{}^2 - 2g(h_Q - h_P).$$

Finally, multiplying throughout by $\frac{1}{2}m$, and rearranging, we have

$$\tfrac{1}{2}mv_P{}^2 + mgh_P = \tfrac{1}{2}mv_Q{}^2 + mgh_Q.$$

We see from this result that, although there is a continual interchange, backward and forward, between potential and kinetic energy during the oscillation, the sum of the two remains constant throughout.

It may readily be shown that we have a corresponding state of affairs in the case of a mass executing vertical oscillations at the end of a supporting spring. Here the sum of the kinetic energy, gravitational potential energy, and elastic potential energy remains constant.

In the absence of losses such as that due to air resistance, which gradually cause the oscillations to die down, we could regard the simple pendulum and a mass oscillating on a spring as purely mechanical systems, and the total mechanical energy in these, the sum of the potential and kinetic energy, would then be constant throughout their oscillations. In such ideal systems we should, in other words, have **conservation of energy.**

Actually neither of these systems *is* purely mechanical. Their mechanical energy is gradually dissipated, because of air resistance and other factors. In all such systems there may be shown to be a certain heating effect associated with the loss of mechanical energy. But heat is also a form of energy, for heat may be made to do work, as for example in a steam engine. May it be that when, in the systems considered, the heat generated is taken into account, the total energy would still be found to be conserved, even though mechanical energy is gradually lost? This question will be dealt with in Chapter 15.

Let us now inquire more closely into what really happens when one system "does work" on another. What, for example, happens when two equal masses are suspended from the ends of a light string passing over a frictionless pulley (Fig. 6.6) and one mass

FIGURE 6.6

moves down while the other moves up, the former doing work via the string on the latter? Here we merely have a transfer of potential energy from one mass to another, the latter gaining what the former loses. Next let us consider the case of a body falling freely under gravity. Here we may say that the earth, by the mechanism of its gravitational attraction, does work on the falling body, exerting a force over a distance. But from the point of view of energy, we may equally well regard it as a conversion of potential energy into kinetic energy. Finally, what happens when a horse tows a barge along a canal? Here chemical energy derived from the horse's intake of food is converted finally to heat generated in the barge and the water of the canal on account of the resistance the latter offers to the motion of the barge. In all these cases we speak of work being done merely as a certain mechanical aspect of what is in reality either a transfer of energy from one body or system to another, or a conversion of energy from one form to another, or both. It is important that we should be quite clear in our minds concerning this; otherwise we shall be in a perpetual muddle and apt to fall into the error of regarding work as a form of energy, which it is not. Energy is not converted into work or work into energy, but energy is converted from one form to another or transferred from one system to another, mechanical work in some cases being performed as a feature of the conversion or transfer and enabling the magnitude of the energy conversion or transfer to be measured.

149

It is interesting to reflect, in this connection, that we can, in the nature of things, never become aware of the existence of energy except when it is being converted or transferred.

CONSERVATIVE FORCES It will now be useful to consider more generally ideal systems in which only mechanical energy items are involved, that is, in which there are no sliding frictional or viscous forces ("dissipative" forces) to generate heat.

As the first example let us take the case of a body falling freely under gravity in a vacuum. If the height of this body, of mass m, above an agreed "zero" level is h, then, as we have seen, its potential energy relative to this level is mgh. When at the height h, let the body be moving with a velocity v_0 whose downwardly directed vertical and horizontal components are v_{01} and v_{02}, respectively. Then, on reaching the zero level its downwardly directed vertical velocity component v_1 will be such that

$$v_1{}^2 = v_{01}{}^2 + 2gh,$$

while the horizontal component will still be v_{02}. Hence the resultant velocity v at the zero level must be given by the equation

$$v^2 = v_1{}^2 + v_{02}{}^2 = v_{01}{}^2 + 2gh + v_{02}{}^2.$$

But

$$v_{01}{}^2 + v_{02}{}^2 = v_0{}^2,$$

so that we have

$$v^2 = v_0{}^2 + 2gh,$$

and hence, multiplying throughout by $\frac{1}{2}m$ and rearranging,

$$\tfrac{1}{2}mv^2 - \tfrac{1}{2}mv_0{}^2 = mgh.$$

Thus we see that the gain in kinetic energy is just equal to the loss of potential energy; the total mechanical energy, kinetic plus gravitational potential, is constant.

Actually the system just discussed is a two-body system, consisting of the relatively enormously massive earth and the "falling" body. We have been able to confine our attention to a single body and treat it as falling in a constant gravitational field only by neglecting both the acceleration of the earth toward it and also the variation of g with height above the earth's surface. It is not difficult to generalize the discussion, considering any two-body system, in which there is no restriction with regard to either the ratio of masses of the two bodies or any variation of their distance apart. The result of this more general analysis would be the same—that under ideal conditions (absence of dissipative forces) the sum of the kinetic and gravitational potential energies of the two bodies must be conserved.

Further generalization to an ideal system of many bodies with mutual gravitational attraction conforming to Newton's law between the members of each pair would be mathematically more difficult. If the analysis were carried out for such a many-body system it would show that here, too, total mechanical energy must be conserved.

In both the two-body system and the many-body system we are concerned with "central" forces, forces that act between pairs of bodies along the line joining them. Also these forces are, for each pair of bodies, a unique function of the distance between them; in the cases we have considered, where the forces are gravitational, their variation with distance conforms to an inverse-square law.

In other purely mechanical systems in which we have established energy conservation we had a corresponding state of affairs, but with a different law of variation of force with distance between the bodies or between one body and a point having a fixed position relative to another. In the case of a simple pendulum oscillating with small amplitude, the force on the bob, directed toward the midpoint of swing, was found to be proportional to the displacement from this point. With a pendulum having a finite arc of swing, the restoring force, although no longer strictly proportional to the displacement, is still a unique function of it. And, effectively straight-line motion being replaced by motion along the arc of a circle, the force on the bob may still be regarded as quasi-central. Finally, in the case of the oscillating mass suspended by a spring we had two central forces acting simultaneously: one, the practically constant gravitational force, directed toward the center of the earth, and the other, the elastic force, directed toward the point corresponding to zero elongation of the spring.

The basic feature of all these systems is that for any given pair of bodies or pair of fixed points on bodies the mutual force between them is a unique function of the distance separating them and is directed along the line joining them. And in all these cases energy is conserved. It may be shown quite generally that in all systems of exclusively central forces conforming to a unique distance law energy must necessarily be conserved. Thus the law of conservation of mechanical energy under conditions where only such forces operate is put on a firmer basis than merely by showing it to apply in a few special cases. Forces of this kind are appropriately known as **conservative forces.**

151

Units of Work and Energy

The unit of work and of energy in the CGS system is the **erg;** it is equal to the work performed when the point of application of a force of 1 dyn is moved a distance of 1 cm in the direction of the force. In the MKSA system the unit is the **joule** (J), being equal to the work performed when the point of application of a force of 1 N is moved in the direction of the force through 1 m. A joule is equal to 10^5 (the number of dynes in a newton) times 10^2 (the number of centimeters in a meter) ergs, that is, 10^7 ergs. The corresponding FPS unit (seldom used) is the foot-poundal, whose name is self-explanatory.

The erg, the joule, and the foot-poundal are true scientific units, being perfectly definite. In addition to these, for certain engineering purposes where extreme accuracy is not required, gravitational units are often used. The CGS unit is the **gram-centimeter,** corresponding to the exertion of 1 gram weight through 1 cm, equal to g ergs (g in cm/sec²); the MKSA gravitational unit is the **kilogram-meter,** equal to g joules (g in m/sec²); and the FPS gravitational unit is the **foot-pound,** equal to g foot-poundals (g in ft/sec²). Curiously enough, common usage with regard to the order of naming the force and the distance is not self-consistent, the force coming first in the two metric systems, whereas in the British system the distance comes first. No significance is to be attached to this.

Power

Power is defined as the **rate of doing work,** or the **rate of energy conversion or transfer.** The CGS unit of power is the erg per second, but because of its extreme smallness this unit is rarely used. The unit in the MKSA system is the **watt** (W), this being defined as the rate of energy conversion of 1 J/sec. A watt is equal to 10^7 ergs/sec. A larger unit, widely used in engineering, is the **kilowatt** (kW), defined as 1000 W. The gravitational FPS unit is the **foot-pound per second,** and a larger unit commonly used in engineering is **horsepower** (hp), defined as 550 foot-pounds/sec. Taking g as 9.80 m/sec², 1 hp is equal to 746 W or 0.746 kW. The dimensions of power are ML^2T^{-3}.

Problems

6.1   A box loaded with bricks, of mass 100 kg, is to be pushed up an incline of 20°. If the coefficient of sliding friction between the box and the inclined surface is 0.3, how much work must be performed in pushing the box 10 m up this surface?

6.2　Ten books, each 6 cm thick and weighing 2 kg, are lying horizontally on a table. How much work is required to construct from these a stack 60 cm high?

6.3　A uniform metal rod 1 m long and of mass 1 kg hangs vertically from a pivot at one end. How much work has to be performed in displacing it through an angle of 60° from its equilibrium orientation?

6.4　If a body of mass 2 kg is raised from a height of 995 km above the surface of the earth to 1005 km, approximately by how much does this increase its energy?

6.5　The combined mass of a man, a platform on which he sits, and a movable pulley from which the platform is suspended (see the figure) is 75 kg. By pulling on the free end of the rope as shown, he raises his height above the ground by 1 m. What force must he exert downward on the rope, and how far does the free end of the rope move down? How much work has the man performed? Neglecting friction in the pulleys, the weight of the rope, and air resistance, check that the work performed by the man is the same as the increase in gravitational potential energy of the system.

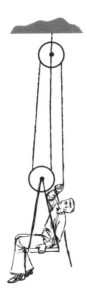

Figure for problem 6.5.

6.6　The natural (unextended) length of a helical spring, whose mass is negligible, is 1 m. This spring is attached at one end to a fixed support, and a 100-g mass is suspended from the other end. This causes the spring to stretch by 5 cm. The spring is now extended 5 cm extra by pulling downward on the mass. Calculate the work done on the system (spring plus suspended mass) during this movement. If the mass were pushed

153

upward 5 cm, instead of being pulled downward, would the work done on the system be the same or different? Explain your answer. The mass is now set into vertical oscillation. Calculate its period.

6.7   A car of mass 1000 kg is traveling at a speed of 80 km/hr. Calculate its kinetic energy expressed in gravitational units, kilogram-meters.

6.8   A block of mass 1 kg resting on a horizontal surface of smooth ice has a steady horizontal force of 2 kg applied to it for 10 sec. How far does it move during this time, and what are its final velocity and kinetic energy? Check that this kinetic energy is the same as the work done on the block. Consider also the cases where (a) twice the force is applied for the same time, and (b) the same force is applied for twice the time. How, in each of the three cases, does the rate of performance of work (power) vary with time? (Neglect friction and air resistance.)

6.9   The coefficient of limiting static friction between the tires of a car and the road is μ. Neglecting rotational dynamical effects on the wheels and in the engine, use considerations of kinetic energy and work to show that on a horizontal road the minimum stopping distance for an initial velocity v is $v^2/2\mu g$.

6.10   A simple pendulum 1 m long has a bob of mass 2 kg. How much work is required to displace the pendulum from its equilibrium state to a horizontal orientation? If the pendulum swings from the horizontal orientation, what will be the velocity, acceleration, and kinetic energy of the bob at the instant when it passes through the lowest position of its path?

6.11   A stick of length l, to the ends of which masses m and $2m$ are attached, is free to rotate with its length perpendicular to a horizontal axis of constraint through its center. The stick is held horizontal and then released. Neglecting the mass of the stick, friction, and air resistance, use the principle of conservation of energy to calculate the angular velocity with which the stick swings through its vertical orientation.

6.12   A sled is released on a track consisting of hardened snow sloping down the side of a hill. By considering what happens along a short section of the track, show that the sum of the potential and kinetic energies of the sled is constant. (Neglect friction and air resistance.)

6.13   The components of a two-body system mutually attracting each other gravitationally have masses m_1 and m_2, are a distance r apart, and describe circular orbits about a common center. Where is this center in relation to the two bodies? Are the angular velocities of the bodies in their orbits the same or different, and what are these? How much work would have to be performed in increasing the distance between the bodies from r to infinity, leaving both bodies at rest?

6.14   Show that for an oscillating mass on a spring the sum of the kinetic energy, gravitational potential energy, and elastic potential energy is constant.

6.15   A mass of 10 g hanging from the end of a spring oscillates up and down with an amplitude of 8 cm and makes two complete oscillations in one second. Calculate (a) the acceleration of the mass at the extremities of its motion, (b) the velocity of the mass at the center of the motion, (c) the kinetic energy at the center of the motion and hence the

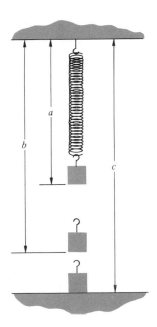

Figure for problem 6.16.

energy at any part of the motion, and (d) how much the spring is stretched by hanging the mass on it when the mass is stationary.

6.16   The lower surface of a block of mass m, suspended from a helical spring, is a distance a below the top fixed end of the spring when held so that the extension of the spring is zero (see the figure). When the block is released and at rest this distance is increased to b. On the basis of this information, and using the principle of conservation of energy, calculate the velocity with which the block will strike the floor, a distance c below the point of attachment of the fixed end of the spring, when the block is released from rest at the height where the spring has zero extension. ($c - b < b - a$.)

6.17   A gun of mass 4 kg fires a bullet of mass 20 g. The bullet leaves the gun with a velocity of 300 m/sec. Find the velocity of recoil of the gun, its kinetic energy, and the kinetic energy of the bullet. How much energy has been used up in firing the gun, and what was the original form of this energy?

6.18   The nucleus of a plutonium atom explodes into two fragments, one a cerium nucleus, of mass 135 units, and the other a krypton nucleus, of mass 105 units. Calculate the ratio of the initial kinetic energy of the cerium nucleus to that of the krypton nucleus.

6.19   A molecular hydrogen ion (H_2^+), of mass m, and traveling with velocity v, makes a direct hit against a helium atom at rest, of mass $2m$, so that the latter, after the collision, moves directly forward. Assuming the collision to be "elastic," that is, one in which none of the original

155

kinetic energy is converted to energy of any other form, find the velocities of the two particles after the collision and the direction of motion of the hydrogen ion. Also compare the final kinetic energies of the two particles.

If the initial kinetic energy of the hydrogen ion were just equal to the energy required to ionize the helium atom, that is, to detach an electron from it, leaving both the separated electron and the He^+ ion at rest, might it be possible for such ionization to result from the collision? Give reasons for your answer.

6.20 A particle of mass m_1 traveling with velocity v_1 makes an elastic collision (a collision in which no kinetic energy is lost) with a stationary particle of mass m_2. After the collision the first particle moves in the original direction with velocity v_1' (this may be either positive or negative) and the second particle with velocity v_2. Show that

$$v_1' = \frac{m_1 - m_2}{m_1 + m_2} v_1$$

and

$$v_2 = \frac{2m_1}{m_1 + m_2} v_1.$$

Find an expression for the proportional loss of kinetic energy suffered by the first particle in consequence of the collision, and apply your result to the cases where (a) the first particle is an electron making an elastic collision with a gas molecule ($m_1 \ll m_2$), (b) the first particle is a positive ion and the second a neutral molecule of the same kind of gas ($m_1 \doteq m_2$), and (c) the first particle is a helium ion and the second a hydrogen molecule ($m_1 \doteq 2m_2$).

6.21 A bullet is fired horizontally from a rifle into a block of wood of mass 1 kg suspended from a support 3 m above its center, and the block is thereby set swinging with an amplitude of 40 cm. Given that the mass of the bullet is 5 g, find its velocity before impact. Compare the kinetic energies before and after impact, and comment on what you find.

6.22 A particle of mass m_1 traveling with velocity v_1 collides with a stationary particle of mass m_2, after which the first particle has a velocity v_1' in the original direction and the velocity of the target particle in this direction is v_2. Denoting the original kinetic energy by E_{ko} and the total final kinetic energy by E_{kf}, show that, for a given required energy conversion from kinetic energy to some other form (for example, energy of excitation or of ionization in the case of molecular particles), E_{ko} is least when $v_1' = v_2$. Express $E_{ko}/(E_{ko} - E_{kf})$ in terms of m_1 and m_2 for this case.

SECTION IV

RIGID-BODY MECHANICS

PRINCIPLES OF ROTATIONAL DYNAMICS

Having developed the concepts of force, mass, and energy in our study of particle dynamics, we are now ready to consider the mechanics of bodies having extension in space, these being regarded as systems of particles. In this chapter and in Chapter 8 we shall confine our attention to **rigid bodies,** extended bodies whose configurations are not affected by the application of forces to them. Actually the concept of a rigid body is an idealization; such bodies do not exist in nature. Nevertheless, it is permissible to make this idealization, because the deformations produced in ordinary solids by such forces as those with which we shall be concerned may be regarded as negligibly small.

In the present chapter we shall consider the basic principles of rotational dynamics. In Chapter 8 we shall apply these principles in a discussion of two special cases of rotational motion: rotational oscillations and gyrosopic motion.

Torque

The agency to which we ascribe changes in the angular velocity of a rotating body about an axis is known as the **torque** acting on the body. This is the rotational analogue of the linear quantity force considered in Chapter 4. There are two independent approaches to the quantitative formulation of torque, one experimental, the other theoretical.

LAW OF MOMENTS In the experimental approach we try to find a quantitative measure of the importance, or "moment," of a force for producing rotation, or tending to produce rotation, of a body about some well-defined axis of constraint. We do this by balancing against each other the moments of two forces tending to produce rotation about this axis in opposite directions. Working with forces whose lines of action, while perpendicular to the axis of constraint, do not pass through it, what we then find is that the moment L of a force is proportional to the product of the force F and the perpendicular distance d from the axis of constraint to the

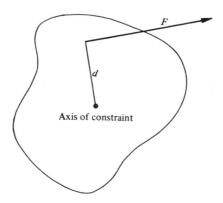

FIGURE 7.1   Moment L of a force inferred from experiments in statics is $F \cdot d$.

line of action of the force (Fig. 7.1), or, in certain units,

$$L = F \cdot d. \tag{7.1}$$

Instead of speaking of the moment of a force, we can, instead, refer to the torque (turning tendency) produced by it; the two terms are synonymous.

So far, so good. However, the use of Eq. (7.1) is not possible in all situations where a torque is exerted, nor is the manner in which it has been established completely satisfying.

In the first place, not all torques are ascribable to well-defined forces. What, for example, are these forces, and what are their lines of action, in the case of a shaft transmitting torque from a ship's engine to the propeller?

Secondly, we began by defining torque as the agency to which we ascribe *changes in the state of rotation* of a body. Yet the experiment by which Eq. (7.1) has been established was one in statics; there was no rotation, or change in state of rotation. Although we could use Eq. (7.1) without hesitation when dealing with any problem in statics, our concern in this chapter is more with dynamics than with statics.

We may extricate ourselves from these difficulties by considering the whole question of torque and the moment of a force theoretically. We shall find that, with torque properly defined quantitatively, the "law of moments" expressed by Eq. (7.1) may be derived theoretically, without any appeal to experiment. Accordingly, the term "law of moments" is, strictly speaking, a misnomer, because the use of the word "law" implies an experimental basis.

GENERAL DEFINITION OF TORQUE Torque, a rotational quantity, is best defined quantitatively as the complete rotational analogue of the linear quantity force. As we saw in Chapter 6, force may be defined as the distance rate of increase of translational kinetic energy, or, in symbols, by the equation

$$F = \frac{\partial E_k}{\partial s}.$$

Correspondingly, we may define torque as the angle rate of increase of rotational kinetic energy, that is, now writing E_k for *rotational* kinetic energy, by the equation

$$L = \frac{\partial E_k}{\partial \theta}, \tag{7.2}$$

angle being the rotational analogue of distance.

Before we can use this definition, we must first find an expression for the kinetic energy of a rotating body. For simplicity, let us suppose that the body is rotating about a fixed axis. The body may be thought of as a system of particles, and any one of these, selected at random, will have a kinetic energy $\frac{1}{2}mv^2$, where m is the mass of the particle and v its velocity. This will be the contribution of the particle in question to the total kinetic energy. If r is the distance of the particle from the axis of rotation and ω is the angular velocity, then v is equal to $r\omega$, and so the kinetic energy of the particle may be written $\frac{1}{2}mr^2\omega^2$. But ω is the same for all the particles; only m and r vary from one particle to another. Hence, if the masses of the separate particles constituting the body are m_1, m_2, m_3, . . . , and their distances from the axis are r_1, r_2, r_3, . . . , respectively, we may write for the total kinetic energy

$$E_k = \tfrac{1}{2}\omega^2(m_1r_1{}^2 + m_2r_2{}^2 + m_3r_3{}^2 + \cdots),$$

or

$$E_k = \tfrac{1}{2}\omega^2 \sum mr^2, \tag{7.3}$$

the symbol Σ being the conventional abbreviation for "the sum of all terms of the form . . . ."

If we compare this equation with Eq. (6.4),

$$E_k = \tfrac{1}{2}mv^2,$$

we observe that there is a close correspondence between the right sides of these two equations. The angular velocity ω of the rotating body is, of course, the rotational analogue of the linear velocity v of the particle. And the analogue of the mass m of the particle is $\Sigma \, mr^2$, this being a definite quantity characteristic of the rotating

161

body for the axis in question. This quantity is known as the corresponding **moment of inertia** of the body and is denoted by the symbol I:

$$I = \sum mr^2. \tag{7.4}$$

The name "moment of inertia" aptly describes the physical meaning of this quantity—the importance of the distribution of mass (inertia) among the constituent particles of the body as regards rotation about the axis in question. With I thus defined, we may rewrite Eq. (7.3) in the more concise form

$$E_k = \tfrac{1}{2}I\omega^2. \tag{7.5}$$

The above derivation of the expression for kinetic energy depends on the assumption that the capacity for doing work of a rigid rotating body is the same as if we could break it up into its constituent particles and allow each of these, separately, to do work on an external system in being brought to rest. Although this assumption appears entirely reasonable, it is, nevertheless, still an assumption. Its ultimate justification is that it works. On the basis of this assumption we may account, quantitatively, for a great variety of phenomena, and no known phenomenon of nature is inconsistent with it.

We shall now make use of our formula for rotational kinetic energy in developing more immediately useful quantitative expressions for torque than that given by the defining equation (7.2).

If in this equation we substitute for E_k its value according to Eq. (7.5), we find that

$$L = \frac{\partial E_k}{\partial \theta} = \frac{\partial}{\partial \theta}\left(\tfrac{1}{2}I\omega^2\right) = I\omega\frac{\partial \omega}{\partial \theta} = I\frac{\partial \theta}{\partial t}\frac{\partial \omega}{\partial \theta} = I\frac{\partial \omega}{\partial t},$$

or, writing $\dot{\omega}$ as a convenient abbreviation for $\partial\omega/\partial t$,

$$L = I\dot{\omega}. \tag{7.6}$$

Note the complete analogy between this equation and the equation

$$F = ma = m\dot{v}$$

in particle mechanics.

An alternative definition for force is, as we saw in Chapter 4,

$$F = \frac{\partial p}{\partial t},$$

where p is the momentum of the particle acted upon, mv. Correspondingly, if we rewrite Eq. (7.6) in the form

$$L = I \frac{\partial \omega}{\partial t} = \frac{\partial}{\partial t}(I\omega),$$

and we define the product $I\omega$ as the **angular momentum** P of the rotating body, that is, we define P by the equation

$$P = I\omega, \tag{7.7}$$

we obtain the analogous equation for torque,

$$L = \frac{\partial P}{\partial t}. \tag{7.8}$$

Thus we see that torque may be regarded either as the angle rate of change of kinetic energy, as expressed by Eq. (7.2), or as the time rate of change of angular momentum.

We may now obtain a further useful relation, expressing power in terms of torque and angular velocity. Writing Eq. (7.2) in the form

$$dE_k = L \, d\theta,$$

and dividing both sides by dt and at the same time changing to partial-derivative notation, we have

$$\frac{\partial E_k}{\partial t} = L \frac{\partial \theta}{\partial t} = L\omega.$$

But $\partial E_k / \partial t$, the rate at which the energy of the body increases, is necessarily the same as the rate at which work is done on it, that is, the power. Hence for this we have

$$\text{Power} = L\omega. \tag{7.9}$$

THEORETICAL DERIVATION OF THE "LAW OF MOMENTS"
Let us again consider a body with a fixed axis of constraint, and let a force applied to this body be resolved in three directions mutually at right angles, these being (1) parallel to the axis about which the body is free to rotate, (2) radial, and (3) perpendicular both to the axis and to the radius through the point of application. Of these three only the last is in the direction of motion of the point of application and does work on the rotating body. The other two, being at right angles to the displacement of the point of application, do not have the performance of any work associated with them.

Let the point of application of the force be at the distance r from the axis, and let component (3) of the force be F'. Then during the short time δt while the body rotates through the small angle $\delta\theta$

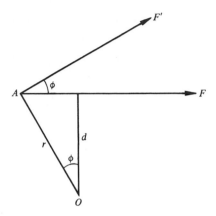

FIGURE 7.2 Moment L of a force derived theoretically is $F' \cdot r$ or $F \cdot d$.

in the sense corresponding to the motion of the point of application of the force,* the distance this point moves will be $r \cdot \delta\theta$, and hence the work that the force does on the body is $F' \cdot r \cdot \delta\theta$. This must be the amount by which the kinetic energy of the body is increased:

$$\delta E_k = F' \cdot r \cdot \delta\theta.$$

Hence, dividing, both sides of this equation by $\delta\theta$, proceeding to infinitesimals, and applying our definition of torque as expressed by Eq. (7.2), we have

$$L = \frac{\partial E_k}{\partial \theta} = F' \cdot r. \tag{7.10}$$

In Fig. 7.2 let the axis of rotation, perpendicular to the plane of the diagram, pass through O, and let A be the point of application of the force. Let F be the total component of the force in the plane perpendicular to the axis, that is, items (2) and (3) above combined, and let d be the perpendicular dropped from O on the line of action of F. Then, because F' is equal to $F \cos \phi$ and d is equal to $r \cos \phi$, where ϕ is the angle shown in the figure, we have

$$L = F' \cdot r = F \cos \phi \cdot r = F \cdot r \cos \phi = F \cdot d. \tag{7.11}$$

Thus the moment of a force is equal to the component of the force in a plane perpendicular to the axis multiplied by the perpendicular distance between the axis and the line of action of the force. This is the "law of moments" previously referred to, derived theoretically.

---

* This is, of course, to be taken algebraically; $\delta\theta$ may have either sign.

COUPLE In the foregoing discussions we have supposed the body to be so constrained that the only motion possible is rotation about a certain fixed axis. To limit the motion in this way it is obviously necessary for the constraining system to exert certain forces on the body, and if we knew what these forces were we could think of the body as free, and acted upon by the appropriate *system* of forces, including the constraining forces.

Let us consider from this point of view a particularly simple case, that in which there is a symmetrical distribution of mass about the axis of constraint and where a single external force, perpendicular to this axis but not passing through it, acts on the body. Under the action of this force the body will be set in rotation, but because for each element of mass there is a corresponding equal element on the other side of the axis moving with equal but opposite velocity, the total momentum of the body, $\Sigma \ mv$, does not undergo any change. This means that all the forces acting on the constituent particles of the body, when compounded vectorially, must have the resultant zero. Hence, if we may assume that "internal" forces exerted between the particles cancel out in pairs (see the section "Interaction between Bodies," Chapter 4), the externally applied force and the constraining force must also be equal and opposite. Such a system of two equal and opposite forces not in the same straight line is known as a **couple**. We may, then, regard the body as free and acted upon by this couple, which gives it an angular acceleration about its axis of symmetry, at right angles to the plane defined by the two forces.*

In the case under consideration the two forces are not symmetrically disposed about the axis of rotation; one passes through this axis, the other does not. This might be thought to constitute a difficulty. It may, however, be shown that the **moment of a couple,** as it is called, that is, the resultant of the moments of the two separate forces, is the same about all axes perpendicular to the plane of these forces. The forces could therefore, if required, be replaced by an equivalent pair of forces that *are* symmetrically disposed about the axis. This we shall now proceed to prove.

Let the two forces represented in Fig. 7.3, each of magnitude F, both be in the plane of the paper, and let O and O' be the traces of two axes at right angles to this plane. Then the resultant moment of the two forces about the axis through O is $d_1F + d_2F$, $= (d_1 + d_2)F$, $= d \cdot F$, where d is the perpendicular distance between the lines of

---

* It seems fair to assume that the plane defined by the two forces must be at right angles to the axis of symmetry; otherwise, the body would surely not be set rotating about this particular axis.

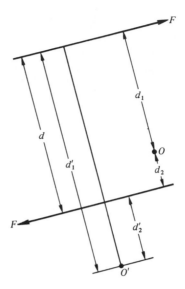

FIGURE 7.3   Moment of a couple.

action of the forces, the so-called **arm of the couple**. Here the two separate moments are added because they both have the same sense. Now let us consider the resultant moment about the axis through O'. The separate moments are $d_1'F$ clockwise and $d_2'F$ anticlockwise. These are opposite in sense, so we must on this occasion subtract the lesser from the greater, obtaining $(d_1' - d_2')F$, $= d \cdot F$, clockwise, as before. Thus the moment of a couple is the same about all axes perpendicular to the plane defined by the forces, its magnitude being given by the product of one of the forces and the arm of the couple.

Rotational Quantities as Vectors

Let us now consider whether some of the angular quantities we have dealt with conform to the geometrical construction for composition considered in Chapter 2, and so qualify to be included in the group of quantities known as vectors. The angular quantities so far discussed are angle (θ), angular velocity (ω), angular acceleration ($\dot{\omega}$), torque (L), angular momentum (P), and moment of inertia (I).

ANGLE   An angle is the rotational analogue of the linear quantity distance, and, as with distance, no idea of direction is included in this concept. An angle is therefore simply a scalar quantity.

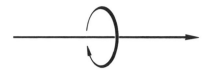

FIGURE 7.4 Representation of an angular displacement by a directed straight line.

ANGULAR DISPLACEMENT This quantity is closely related to an angle but, in contrast to the latter, *is* thought of as having a direction, and could, therefore, conceivably be a vector. At first sight it may not appear that the direction of a rotation is quite the same thing as is that of a linear directed quantity, such as a linear displacement. It is, nevertheless, possible to represent a rotation by a directed straight line, or an arrow, just as we can a linear displacement. By convention the straight line is given the direction of the axis about which the rotation occurs, and its "sense" bears the same relation to the rotation as the axial movement bears to the rotation in a right-handed screw (see Fig. 7.4). On the basis of this convention the magnitude and direction (including sense) of a rotation may be completely specified by the length and direction of a straight line.

Let us now see whether angular displacement conforms to the triangle rule for composition. Thus in Fig. 7.5 let *PQ* and *QR* each represent a rotation through a right angle about an axis having the corresponding direction, these two axes being taken as at right angles to one another. If angular displacement is a vector, an object should have the same final orientation whether (1) it is first rotated

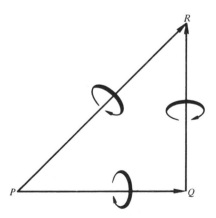

FIGURE 7.5 Test of whether finite angular displacements combine vectorially.

through a right angle about an axis parallel to PQ and then through another right angle about an axis parallel to QR, (2) these two rotations are carried out in reverse order, or (3) it is given a simple rotation about an axis parallel to PR of $\sqrt{2}$ right angles. But if we try this, say with a book, we shall find that in fact the final orientation is different in each of the three cases. It must therefore be concluded that angular displacements are not vectors; at least *finite* angular displacements are not.

ANGULAR VELOCITY Although, as we have just seen, finite angular displacements are not vectors, it may be shown that infinitesimally small angular displacements *are*, and from this it follows— one has merely to divide each such displacement by the time it occupies, dt—that angular velocities are also vectors. We shall now prove that angular velocities are vectors by a different method.

Let a rigid body be rotating simultaneously about two axes OA and OB intersecting in the point O, as indicated in Fig. 7.6, and let the distances OA and OB be proportional to the corresponding angular velocities ω_a and ω_b, respectively; that is, let

$$\omega_a = k \cdot OA, \qquad \omega_b = k \cdot OB,$$

where k is a constant. Then if the vector rule for composition applies for angular velocities, the resultant angular velocity ω should be equal to $k \cdot OC$, where OC is the diagonal of the parallelogram $OACB$, and OC should be the axis of the resultant rotation.

Let us investigate whether this is so, in two stages.

In the first place, if OC is the axis of the resultant rotation, the point C should be stationary. Let us see if it is. Let perpendiculars CM and CN be dropped on the prolongations of OA and OB, respectively. Then, owing to the rotation about the axis OA, the velocity of C will be $\omega_a \cdot CM$ in a direction perpendicular to the plane of the

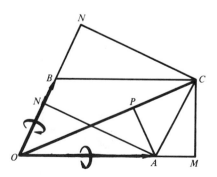

FIGURE 7.6 Vector parallelogram for composition of angular velocities.

diagram and out of the paper, and, owing to the rotation about OB, it will be $\omega_b \cdot CN$ into the paper. The resultant velocity of C will thus be $k(OA \cdot CM - OB \cdot CN)$ out of the paper. But each product within the parentheses is equal to the area of the parallelogram. Hence the resultant velocity of C is zero; C, as well as O, is stationary. In other words, OC must be the axis of the resultant rotation.

Now let us see what is the magnitude of the angular velocity ω about this axis. Let us take it as clockwise, looking in the direction from O to C. (This does not beg the question; an anticlockwise rotation would merely make ω negative.) To find ω, let us find an expression for the velocity of some point, say A, (1) in terms of the component angular velocities ω_a and ω_b, and (2) in terms of the resultant angular velocity ω. These linear velocities must necessarily be equal.

Obviously A, being on the axis of one of the component rotations, cannot have any linear velocity due to this rotation. Its linear velocity is due entirely to the rotation about the other axis and must be equal to $\omega_b \cdot AN'$, where N' is the foot of the perpendicular dropped from A on OB. The direction of this will be into the paper. And in terms of the resultant angular velocity ω about OC the velocity of A must be $\omega \cdot AP$, where P is the foot of the perpendicular dropped from A on OC. The direction of this must also be into the paper. Equating these two expressions for the velocity of A, we thus have

$$\omega \cdot AP = \omega_b \cdot AN'.$$

Hence

$$\omega = \frac{\omega_b \cdot AN'}{AP}$$

$$= \frac{k \cdot OB \cdot AN'}{AP}$$

$$= k \cdot \frac{\text{area of parallelogram}}{AP}$$

$$= k \cdot OC.$$

We see that the resultant angular velocity ω is not only about the axis OC, but is represented in magnitude by the length of OC according to the same proportionality factor k as applies to the representation of the component angular velocities ω_a and ω_b by OA and OB, respectively. Hence we must conclude that angular velocity is a vector.

ANGULAR ACCELERATION That angular acceleration is a vector follows from the fact that angular velocity is a vector in

169

exactly the same manner as for the corresponding linear quantities (see Chapter 2).

TORQUE In Chapter 4 we saw that by multiplying each side of a vector triangle of linear accelerations by the mass of the particle accelerated (a scalar quantity), we should obtain a triangle whose sides are proportional to the component and resultant mass $\times$ accelerations, and from this it was inferred that forces may be compounded by the vector-triangle method, that is, that forces are vectors.

We might suppose that if we applied the corresponding argument to the analogous rotational quantities torque would be proved to be a vector. But there is a difficulty here. In the case of the linear quantities there was only one scalar quantity involved, the mass of the particle. However, the corresponding rotational quantity, moment of inertia, is in general different for different axes, and consequently it is not possible to prove that torque is a vector quite as simply as it is to prove that force is a vector.

At least some measure of satisfaction in this matter is afforded by the following discussion. Let us consider the special case of a spherical body, having a spherically symmetrical distribution of mass. The moment of inertia of such a body must be the same about all axes through the center, and any applied "pure" torque—that is, a torque due to a system of forces which, if added vectorially, would have zero resultant—must produce rotation about an axis through the center corresponding in direction to that of the torque; rotation about any axis not through the center would imply a change in $\Sigma\, mv$ and therefore a resultant force. In the case of such a body, not only must the angular acceleration be about the axis indicated, but it must be proportional to the torque with the same proportionality factor, $1/I$, irrespective of the direction of the torque, I being the same for all central axes. For this body we may construct a triangle, such as that in Fig. 7.7, to represent two simultaneously applied torques and their resultant, this triangle being similar and

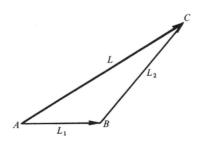

FIGURE 7.7 Vector triangle for composition of torques.

similarly orientated to the vector triangle representing the angular accelerations produced by these torques and the resultant of these angular accelerations. *As far as this body is concerned*, therefore, we may treat torque as a vector.

Now, just as force may be regarded either as the dp/dt produced in a body to which it is applied or as a property of the agency that applies it, for example tension in a spring, correspondingly with torque. It may be regarded either as the dP/dt produced in a body to which it is applied or as a property of a system of forces, for example a couple, or such as may be imagined to be associated with a twist in a wire. But a system of forces must be thought of as having an existence independent of the body to which it is applied; it cannot be affected by a change in the nature of this body. Accordingly, a combination of systems of forces corresponding to L_1 and L_2 in Fig. 7.7 must give a system of forces corresponding to L. This means that not only is Fig. 7.7 a vector triangle of torques for the special case where the body to which the torques are applied has a spherically symmetrical distribution of mass, but it must be expected to remain so when this body is replaced by another. Thus it appears that, quite generally, torque is a vector.

ANGULAR MOMENTUM If we may assume that torque is a vector, it follows at once that infinitesimal increments of angular momentum are vector quantities. For

$$L = \frac{\partial P}{\partial t},$$

so the vector triangle whose sides are L_1, L_2, and L, as in Fig. 7.7, may at once be converted into a corresponding similar triangle having sides dP_1, dP_2, and dP, respectively, simply by multiplying L_1, L_2, and L by dt. Thus infinitesimal increments of angular momentum are vectors. It does not necessarily follow from this that finite angular momenta are vectors. However, they may well be, and in the theory of gyroscopic action to be developed later it is assumed that a finite angular momentum can be compounded vectorially with an infinitesimal change in this quantity. The success of this theory suggests that finite angular momenta as well as infinitesimal increments are, indeed, vectors.

MOMENT OF INERTIA Although moment of inertia, besides having magnitude, also has different values for different axes, we cannot ascribe to it a sense; its value is the same for both senses of rotation about a given axis. There obviously cannot, therefore, be any question of moment of inertia conforming to the triangle law of composition applicable to vectors; moment of inertia is not a vector.

171

Forces on an Extended Body

We shall now consider certain important properties of an extended body that will enable us to perform calculations concerning its behavior without having on each occasion to take detailed account of the positions and masses of all the separate particles of which it is composed.

In dealing with an extended body it is convenient to think of it as consisting of a large number of separate, although interconnected, particles, and to differentiate between "external" and "internal" forces. By external forces is to be understood those that are exerted on the body in question by external agencies, and by internal forces we mean those which the constituent particles exert on one another. Now in all such systems of discrete particles that we are able to study in detail—an outstanding example is the solar system—there is overwhelming evidence that all the internal forces are exerted between pairs of particles, that they act along the line joining the members of each such pair, and that they are equal and opposite. Therefore, it seems reasonable to assume that this is universally true, and in what follows we shall make this assumption.

Let us for simplicity consider only the components of forces and accelerations in a particular direction—say the x components. These quantities will then be subject to the rules of simple non-vector algebra. Let the x components of the external forces exerted on the separate particles, of masses m_1, m_2, . . . , m_n, be F_1, F_2, . . . , F_n, while those of the internal forces exerted on the same particles are f_1, f_2, . . . , f_n, respectively. Then if a_1, a_2, . . . , a_n denote the corresponding components of acceleration of these particles, we must have

$$m_1 a_1 = F_1 + f_1,$$
$$m_2 a_2 = F_2 + f_2,$$
$$\cdot$$
$$\cdot$$
$$\cdot$$
$$m_n a_n = F_n + f_n,$$

whence

$$\sum ma = \sum F + \sum f.$$

Now

$$f_1 = f_{21} + f_{31} + \cdots + f_{n1},$$
$$f_2 = f_{12} + f_{32} + \cdots + f_{n2},$$
$$\cdot$$
$$\cdot$$
$$\cdot$$
$$f_n = f_{1n} + f_{2n} + \cdots + f_{(n-1)n},$$

where f_{21} is the x component of the force the second particle exerts on the first, and so on. And, as we at once see, for every item on the right sides of these equations in which the subscripts are in a certain order there is another in which the same subscripts are in the reverse order. But according to our assumption, $f_{12} = -f_{21}$, $f_{13} = -f_{31}$, and so on. Hence

$$\sum f = 0,$$

and therefore

$$\sum ma = \sum F.$$

To emphasize that we are dealing only with the x components of accelerations and forces we may rewrite this equation

$$\sum ma_x = \sum F_x. \tag{7.12}$$

Obviously the two corresponding equations for the y and z directions,

$$\sum ma_y = \sum F_y \tag{7.13}$$

and

$$\sum ma_z = \sum F_z, \tag{7.14}$$

must also apply.

In dealing with a single particle, we regarded that single force as the resultant of a system of separate applied forces that would produce in it the same acceleration as do the separate forces acting simultaneously. Correspondingly, we may regard that single force (if there is one) as the resultant of a system of forces applied to an extended body that would produce the same effects, as regards both translation and rotation, as does the system. As translational effects we obviously have $\sum ma_x$, $\sum ma_y$, and $\sum ma_z$, and if these are to be the same in the two cases, the resultant force must (as for a particle) be the vector sum of the separate forces. Thus the magnitude and direction of the resultant force are both determined. And its line of action must be such that the moment of the resultant force about any axis is the algebraic sum of the moments of the individual forces. Only in the case of a pair of forces constituting a couple, or of a number of such pairs, is there no single resultant force.

Center of Mass

Dynamical problems involving rigid bodies may be greatly simplified if we make use of the special properties of a point charac-

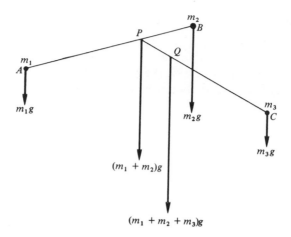

FIGURE 7.8 Single-force equivalent of the weights of a system of particles.

teristic of any such body known as the center of mass. Let us therefore now see how this point is defined and investigate its special properties.

We begin by considering the special case where the applied forces are the weights of the constituent particles.

Let us in the first instance consider two particles only of the system, those labeled A and B in Fig. 7.8, of masses m_1 and m_2, respectively. Then, as may readily be shown, the single force $(m_1 + m_2)g$ has a moment about any axis which is the same as the algebraic sum of the moments of the weights $m_1 g$ and $m_2 g$ if this force passes through the point P that divides the straight line joining the particles in the inverse ratio of the weights and therefore of the masses. Hence this force through P is the resultant of the weights of the two particles.

Having found this point P, let us now join it by a straight line to a third particle C, of mass m_3. The resultant of the force $(m_1 + m_2)g$ passing through P and the weight of the third particle must similarly pass through the point Q on PC such that

$$\frac{QP}{QC} = \frac{m_3 g}{(m_1 + m_2)g} = \frac{m_3}{m_1 + m_2}.$$

Next we might extend the procedure to include a fourth particle, then a fifth, and so on, until finally all the particles of the body have been dealt with. The final point through which the resultant of all the forces passes is known as the **center of gravity** of the body.

From the manner of obtaining it, it is obviously a perfectly definite point, at least for the particular order in which the particles have been taken. It will now be shown that the point found is independent of this order.

Let the n particles of the body, of masses $m_1, m_2, \ldots, m_n$, have x coordinates, relative to some selected Cartesian frame of reference, of $x_1, x_2, \ldots, x_n$, respectively. Then the x coordinate of P, which we shall denote by x_{12}, must be such that

$$\frac{x_{12} - x_1}{x_2 - x_{12}} = \frac{m_2}{m_1},$$

or

$$x_{12} = \frac{m_1 x_1 + m_2 x_2}{m_1 + m_2}. \tag{7.15}$$

Next, denoting the x coordinate of Q by x_{123}, we have

$$\frac{x_{123} - x_{12}}{x_3 - x_{123}} = \frac{m_3}{m_1 + m_2}. \tag{7.16}$$

Hence, eliminating x_{12} between (7.15) and (716.), and simplifying, we find that

$$x_{123} = \frac{m_1 x_1 + m_2 x_2 + m_3 x_3}{m_1 + m_2 + m_3}, \tag{7.17}$$

which is of a form similar to Eq. (7.15). Continuing in this manner until finally all the particles have been dealt with, we eventually obtain for the x coordinate $\bar{x}$ of the center of gravity of the whole system the expression

$$\frac{m_1 x_1 + m_2 x_2 + \cdots + m_n x_n}{m_1 + m_2 + \cdots + m_n};$$

that is,

$$\bar{x} = \frac{\sum mx}{\sum m}. \tag{7.18}$$

From the form of this expression it is evident that the value of $\bar{x}$ is independent of the order in which the points are taken.

Obviously, by considering the other horizontal components in similar fashion, we obtain the corresponding expression for the y coordinate of the center of gravity

$$\bar{y} = \frac{\sum my}{\sum m}. \tag{7.19}$$

Finally, by rotating both the Cartesian frame of axes and the body together through a right angle, about either the x or the y axis, and correspondingly making all the weights parallel to the y or to the z axis, we have the conditions for obtaining the third equation of this form,

$$\bar{z} = \frac{\sum mz}{\sum m}. \tag{7.20}$$

These three equations uniquely define the position of the center of gravity.

In the continuous case the center of gravity is found by reference to the continuous extension of these sums, so that

$$\bar{\mathbf{r}} = \frac{\int \mathbf{r}\rho \, dV}{\int \rho \, dV}, \tag{7.21}$$

where ρ is the density of the material.

It will be observed that g does not appear in the final expressions for $\bar{x}$, $\bar{y}$, and $\bar{z}$, or for $\bar{\mathbf{r}}$. The point they define has a significance in mechanics quite apart from gravitational action, and is accordingly known as the **center of mass** of the system.

For a system of particles situated in a uniform gravitational field "center of gravity" and "center of mass" are interchangeable terms, but this is no longer so if the system is not in a gravitational field or if the field is not uniform. In such a case the idea of a center of gravity loses its meaning. But even under these conditions the center of mass still remains a perfectly definite point, which is of considerable importance in the dynamical theory of the system.

It should be noted that the center of mass of any systems of particles, not merely rigid bodies, may be found. For example, we may speak of the center of mass of the earth–moon system, of a binary star, of the solar system, or of our galaxy.

For the determination of the center of mass of an extended solid body it is, of course, necessary to use Eqs. (7.18), (7.19), and (7.20) in their equivalent integral forms. Let us consider, for example, the case of a cone whose axis is perpendicular to its base, as shown in Fig. 7.9. From symmetry we know that the center of mass must be somewhere on the axis. Our problem, therefore, reduces to finding where it is on this axis.

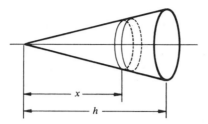

FIGURE 7.9 Construction for calculating the position of the center of mass of a homogenous conical body.

For integration purposes, we consider the cone, of height h, whose material is of density ρ, to consist of a number of thin slices, at various distances x from the apex and of thickness dx, as indicated in the figure. Then, because the area of such a slice is proportional to the square of its distance from the apex, by the proportionality factor k, let us say, its volume will be $kx^2\,dx$, and its mass $\rho kx^2\,dx$. Correspondingly, its mass times distance from the apex will be $\rho kx^2\,dx$. Accordingly, for the distance x of the center of mass of the cone from its apex, we have

$$x = \frac{\int_0^h \rho kx^3\,dx}{\int_0^h \rho kx^2\,dx} = \frac{\int_0^h x^3\,dx}{\int_0^h x^2\,dx} = \frac{3h}{4};$$

that is, the distance of the center of mass from the base is $h/4$.

From Eqs. (7.18) to (7.20), or the equivalent equations in integral form applying to continua, we may now find expressions for the velocity and acceleration of the center of mass. From

$$\bar{x} = \frac{\sum mx}{\sum m}$$

we obtain by two successive differentiations expressions for the x components of the velocity and acceleration of the center of mass,

$$\bar{v}_x = \frac{d\bar{x}}{dt} = \frac{\sum mv_x}{\sum m} \tag{7.22}$$

and

$$\bar{a}_x = \frac{d\bar{v}_x}{dt} = \frac{\sum ma_x}{\sum m} = \frac{\sum F_x}{\sum m} \tag{7.23}$$

177

Corresponding expressions may be obtained for the y and z components.

From Eq. (7.23) and the corresponding equations for the y and z components of the acceleration we see that the motion of the center of mass is the same as if the whole mass of the body or system were concentrated there and all forces were applied at this point. Hence, if the system also rotates, it must be about an axis through the center of mass.

Although the motion of the center of mass is independent of the lines of action of the external forces, the latter do determine the rotational response of the system. In Fig. 7.10 let the force 1 be applied to a rigid body whose center of mass is at C. Let us imagine two additional forces, 2 and 3, to be added, equal and parallel to 1, but oppositely directed, acting through C. Because these are equal and opposite, and act along the same line, they obviously cannot have any effect on the motion of the body. They do, however, serve the purpose of making clear how the line of action of the force actually applied affects the body rotationally; for 1 and 2 together constitute a couple, whereas 3 will affect only the translational motion of the center of mass.

We could carry out the same procedure for each of any number of applied forces. All the forces compounded vectorially would then give a single resultant force through the center of mass, determining (in conjunction with the total mass of the body) the acceleration of this point. Similarly, all the couples could be compounded vectorially to give a single resultant torque acting on the body.

It should be noted particularly that the moment of each couple is equal to the moment of the corresponding externally applied force about the axis perpendicular to the force that passes *through the center of mass* but *not about any other axis having this direction.*

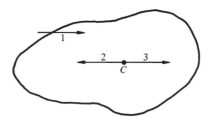

FIGURE 7.10 Equivalence of any force applied to a body to a corresponding force through the center of mass plus a couple.

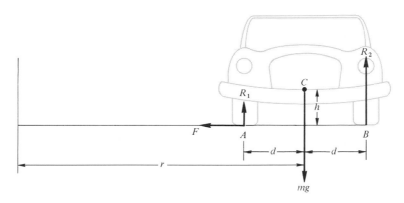

FIGURE 7.11 System of forces exerted on a car traveling along a curved horizontal road.

As an example of the translational and rotational effects of a system of forces, let us consider a car traveling along a curved horizontal road, neglecting gyroscopic effects associated with the rotating wheels.*

In Fig. 7.11 let r denote the radius of the curve in which the car, of mass m, is traveling with angular velocity ω, and let the inner and outer wheels make contact with the ground at A and B, where the vertical components of the forces the road exerts on the car are R_1 and R_2, respectively. Let the sum of the horizontal (frictional) forces acting at A and B be F. Let the height of the center of mass, C, above the ground be h, and let its distance from the planes of the wheels on either side be d.

The car has no acceleration vertically, so

$$R_1 + R_2 = mg. \tag{7.24}$$

Horizontally, the car is accelerated toward the center of the curve in which it is traveling, and for the associated centripetal force we must have

$$F = m\omega^2 r.$$

Finally, if the car does not overturn, there will be no angular acceleration about the horizontal longitudinal axis through C, so that the algebraic sum of the moments of all the forces about this axis must be zero,

$$(R_2 - R_1)d = Fh.$$

---

* The nature of these effects, and how to allow for them, will be considered later. In general, they are, quantitatively, of minor importance.

Eliminating F between these last two equations, we have

$$(R_2 - R_1)d = m\omega^2 rh, \tag{7.25}$$

and combining this with Eq. (7.24) we obtain for R_1 and R_2 the values

$$R_1 = \tfrac{1}{2}m\left(g - \omega^2 r\,\frac{h}{d}\right)$$

and

$$R_2 = \tfrac{1}{2}m\left(g + \omega^2 r\,\frac{h}{d}\right).$$

These equations are based on the assumption that the car does travel around the curve, that is, does not skid, and that it does not overturn. In order that it shall not skid, F must be equal to $m\omega^2 r$. If, however, the maximum frictional force the road is capable of exerting on the car is less than this, the requisite centripetal force cannot be applied, and the car will skid. In order that it shall not overturn, R_1 must be a force that the road is physically capable of exerting; that is, it must be positive. If, however, the values of the parameters are such as to give a negative value of R_1, then, because this is not physically realizable, the car must overturn. In order that the car shall not overturn, it is necessary for $\omega^2 rh/d$ to be less than g.

Values of Moment of Inertia

For homogeneous bodies of a number of simple geometric forms the moments of inertia may readily be calculated by straightforward integration.

Thus for a thin uniform rod of mass $M*$ and of length $2a$ the moment of inertia about a central axis perpendicular to its length is $Ma^2/3$. For a disk, again of mass M, and of radius r, the moment of inertia about an axis through its center and perpendicular to its plane is $\tfrac{1}{2}Mr^2$. And for a sphere the corresponding expression for the moment of inertia about a central axis is found to be $\tfrac{2}{5}Mr^2$. These three cases are indicated in parts (a), (b), and (c), respectively, of Fig. 7.12.

From the first two of these the moments of inertia about corresponding axes for a thin rectangle and for a cylinder immediately follow. A uniform rectangle may be supposed to be derived from

---

* In this section the mass of an extended body will be denoted by the capital letter M, the symbol m being reserved for the mass of one of its constituent particles or of some portion of the body.

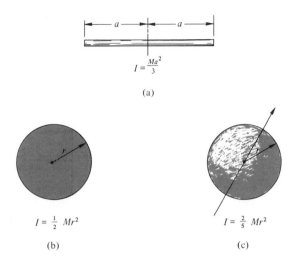

$$I = \frac{Ma^2}{3}$$

(a)

$$I = \frac{1}{2} Mr^2$$

(b)

$$I = \frac{2}{5} Mr^2$$

(c)

FIGURE 7.12   Moments of inertia about central axes of a thin rod, a disk, and a sphere.

a thin rod by smearing it along the direction of the axis of rotation previously taken, and if it is smeared a distance $2b$, a rectangle whose sides have semilengths a and b results (see Fig. 7.13a). In such smearing out, each original element of mass would be converted into a line parallel to the axis, which, because every part of it is at the same distance from the axis as before, makes the same contribution to the total moment of inertia. Hence the total moment of inertia of a uniform rectangle of sides $2a$ and $2b$ about a central axis in its plane parallel to the sides of length $2b$ is $Ma^2/3$, and that about the central axis parallel to the sides of length $2a$ is (by mathematical correspondence) $Mb^2/3$. Similarly, a cylinder may be imagined to be derived from a disk by extension along the direction of its axis, and by corresponding reasoning we see that the moment of inertia of such a cylinder, of radius r, having any length, and of mass M, about this axis is $\frac{1}{2}Mr^2$ (Fig. 7.13b).

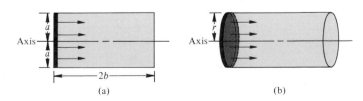

(a)

(b)

FIGURE 7.13   (a) A rod smeared out to form a rectangular plate.
(b) Extension of a disk to form a cylinder.

181

What, now, is the moment of inertia of a thin rectangle about a central axis *perpendicular to* its plane, or of a thin disk about a central axis *in* its plane? The latter may be found by direct integration, which yields the value $\frac{1}{4}Mr^2$, but it is interesting to obtain this result by a different method, the principle of which is applicable also to the evaluation of the moment of inertia of a rectangle about a central axis perpendicular to its plane.

Let us consider any two-dimensional body, with mutually perpendicular axes OX and OY drawn in its plane, as indicated in Fig. 7.14. Then if the moments of inertia about these axes are I_x and I_y, respectively, it may be shown that the moment of inertia I about the axis perpendicular to both of these and passing through their point of intersection is given by the simple relation

$$I = I_x + I_y.$$

For, if we take the axes as Cartesian axes of reference, the contributions of an element (particle) of mass m, and of coordinates x and y, to I_x, I_y, and I are my^2, mx^2, and mr^2, respectively, where r is the distance of the element from O. And r^2 is equal to $x^2 + y^2$, so we have

$$I = \sum mr^2 = \sum m(y^2 + x^2) = \sum my^2 + \sum mx^2$$
$$= I_x + I_y. \tag{7.26}$$

Let us now apply this result to the problem of finding the moment of inertia of a disk about a central axis in its plane. Let this be denoted by I_0, while I represents the moment of inertia about the central axis perpendicular to the plane of the disk. Then if we construct two mutually perpendicular axes in the plane of the disk intersecting at its center, the moments of inertia about

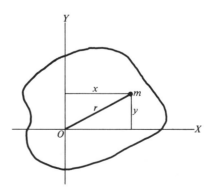

FIGURE 7.14

these correspond to I_x and I_y in the foregoing discussion, and because, as we see from considerations of symmetry, I_x and I_y must in this case be equal, we have

$$I = 2I_0.$$

But, as we have seen, I is equal to $\frac{1}{2}Mr^2$. Hence the value of I_0 must be $\frac{1}{4}Mr^2$. Similarly, by application of Eq. (7.26) to the case of the thin uniform rectangle previously considered, we find that the moment of inertia about the central axis perpendicular to its plane is $M(a^2 + b^2)/3$.

THEOREM OF PARALLEL AXES If we know the moment of inertia of any body about an axis through the center of mass, we may at once find the moment of inertia about any other, parallel, axis by applying the **theorem of parallel axes,** or **Steiner's theorem,** according to which

$$I = I_0 + Md^2,$$

where I_0 is the moment of inertia about the axis through the center of mass, I is the moment of inertia about the parallel axis, d is the distance between the two axes, and M is the mass of the body.

Let C in Fig. 7.15 denote the center of mass, and let the axis through this about which the moment of inertia is known be perpendicular to the plane of the paper. Let the parallel axis, at the distance d from C, pass through O, and let P be the projection on the plane of the diagram of the location of one of the particles of the body, of mass m.

Then

$$I = \sum mr^2$$

and

$$I_0 = \sum ms^2,$$

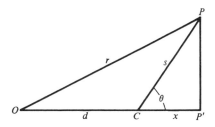

FIGURE 7.15 Construction for proving the theorem of parallel axes.   **183**

where r and s denote the distances of P from O and C, respectively. But r, s, and d are connected by the equation

$$r^2 = d^2 + s^2 + 2ds \cos \theta,$$

where θ is the angle CP makes with the prolongation of OC. Hence

$$\sum mr^2 = \sum md^2 + \sum ms^2 + 2\sum mds \cos \theta$$
$$= d^2 \sum m + \sum ms^2 + 2d \sum ms \cos \theta,$$

or

$$I = Md^2 + I_0 + 2d \sum mx,$$

where x is written for CP', the projection of CP on OC, this being reckoned positive if P' is to the right of C and negative if it is to the left. But C is the center of mass, so $\sum mx$ must, by the definition of this point, vanish. Hence, finally,

$$I = I_0 + Md^2. \tag{7.27}$$

As an interesting application of this theorem, let us consider the evaluation of the moment of inertia of a cylindrical body about a central axis perpendicular to the axis of the cylinder.

Let us imagine the cylinder, of semilength l and radius r, to be divided into a large number of thin slices such as that indicated in Fig. 7.16, and let us consider the contribution of one of these, of mass m and distance d from the center of the cylinder, to the total moment of inertia. The moment of inertia of such a slice about a central axis in its own plane is, as we have already noted, $mr^2/4$. By the theorem of parallel axes, the moment of inertia of this same slice about the parallel axis through the center of the cylinder is $(mr^2/4) + md^2$; similarly with all the other slices. We see that the moment of inertia of the whole cylinder is the sum of the moments of inertia of two bodies, each of equal mass, the first being obtained by squashing the cylinder longitudinally into a thin disk at its center, while the second is obtained by shrinking the cylinder

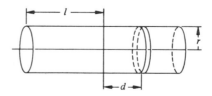

FIGURE 7.16 Construction for calculating the moment of inertia of a homogeneous cylindrical body about a central axis perpendicular to the axis of the cylinder.

radially into a thin rod located along its axis. The moments of inertia of these are $Mr^2/4$ and $Mr^2/3$, respectively. Hence the required expression for the moment of inertia of the cylinder about the axis in question is

$$I = M\left(\frac{r^2}{4} + \frac{l^2}{3}\right).$$

RADIUS OF GYRATION It is sometimes convenient to express the moment of inertia of a body about an axis through its center of mass in terms of its total mass M and a quantity k having the dimensions of a length known as its **radius of gyration.** This is defined by the equation

$$I = Mk^2. \tag{7.28}$$

For a homogeneous sphere of radius r the radius of gyration is $\sqrt{\frac{2}{5}}\, r$; for a uniform disk of radius r and an axis perpendicular to its plane k has the value $r/\sqrt{2}$, while for the same disk and an axis in its plane the value of k is $r/2$; and for a thin rod of semilength a and a perpendicular axis k is $a/\sqrt{3}$.

Statics

We may now consider as a limiting case of dynamics that in which a body is not acted upon by either a resultant force or a resultant torque, so that if the body is initially at rest it remains at rest. The body is then said to be **in equilibrium** under the action of the forces applied to it. The study of equilibrium conditions is known as **statics.**

If the body can move only by rotating about a fixed axis, the rotational condition of absence of torque is satisfied if the algebraic sum of the moments of all the forces about this axis is zero. In view of our definition of torque as the angle rate of change of rotational kinetic energy, and of our theoretical derivation of the "law of moments," in both of which we imagined the body to be restricted in its motion to rotation about a fixed axis, this is self-evident. In all cases where there is such an axis of constraint, or an axis that behaves as such, it is not only convenient, but perfectly legitimate, to take moments about this axis when applying the rotational condition for equilibrium. This is, for example, the usual procedure in investigating the equilibrium of a beam balance; moments are taken about a horizontal axis through the central knife-edge.

The possibility of taking moments about any axis of constraint may be felt to be in conflict with the statement made earlier—that

185

the torque due to any force is the moment of that force about an axis through the center of mass, but not, in general, about any other axis.

Actually there is no conflict; this is a special case. Where there is a fixed axis of constraint, moments may be taken about this axis as an alternative to taking them about an axis through the center of mass. It may be wondered why, then, in the example of the car traveling around a curve, the line of contact of the outer wheels with the ground could not be taken as an axis of constraint. It is easily seen that if this line is so regarded and moments of *real* forces (not fictitious ones) are taken about this, we do not obtain the correct expression for the torque. The reason for this is that *with the car traveling around the curve* the line of contact of the outer wheels with the road in no sense constitutes a fixed axis of constraint.

The following example will serve to illustrate the kind of procedure generally adopted when dealing with problems in statics. Let a uniform horizontal beam, of length l and mass m, be supported at one end, A, where it is in contact with a wall, in the manner indicated in Fig. 7.17. The other end, B, is supported by a rope whose direction makes an angle θ with that of the beam, and a mass m' is hung from the end B. It is required to find (1) the magnitude and direction of the force R exerted by the wall and support on the end A, and (2) the tension T of the rope.

Denoting by ϕ the angle R makes with the horizontal, we have, resolving horizontally,

$$R \cos \phi = T \cos \theta;$$

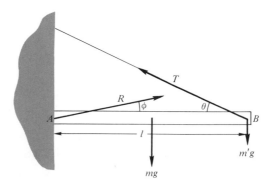

FIGURE 7.17 A typical problem in statics.

and, resolving vertically,

$$R \sin \phi + T \sin \theta = (m + m')g.$$

Then, taking moments about a horizontal axis through A, we find that

$$T \cdot l \sin \theta = mg \cdot \tfrac{l}{2} + m'g \cdot l,$$

or

$$T \sin \theta = \left(\frac{m}{2} + m'\right) g.$$

By means of these three equations we can solve for the three unknowns, T, R, and ϕ.

Problems

7.1 Find the torque exerted by a shaft which transmits 10 kW at 600 rpm.

7.2 Find the center of mass of a uniform thin wire bent in the form of a semicircle.

7.3 Find the center of mass of one quadrant of a uniform thin circular plate.

7.4 Find the center of mass of a thin, uniform, circular sheet of metal of radius r out of which a square of side $r/2$ has been cut in such a way that one corner of the square is at the center of the circle.

7.5 A narrow uniform pole 10 m high, initially resting on the ground in a vertical position, tilts about an axis through its base and falls on its side. What is the velocity of the top end of the pole as it strikes the ground?

7.6 Show that the moment of inertia of a uniform thin rod about an axis through one end normal to the length of the rod is $\frac{1}{3}ml^2$, where m is the mass of the rod and l its length.

7.7 Calculate the moment of inertia of an annular cylinder (hollow cylinder) of mass m and internal and external radii r_1 and r_2, respectively, about the cylinder axis.

7.8 Find the moment of inertia of a thin circular hoop (a) about the axis through the center perpendicular to its plane, (b) about a diametral axis, and (c) about a tangential axis in its plane.

7.9 Show that the moment of inertia of a homogeneous spherical body of radius r and mass m about an axis through its center is $\frac{2}{5}mr^2$.

7.10 A homogeneous metal sphere rolls from rest down an incline of 1 in 10. What will be its speed after it has covered a distance of 10 m? How long will it take to travel this distance? What must be the minimum value of the coefficient of limiting static friction to prevent slipping?

7.11 A homogeneous metal sphere on an inclined plane rolls down without slipping when the angle θ of inclination of the plane is less than θ_0 but slips when θ is greater than θ_0. Express the coefficient of limiting static friction μ as a function of θ_0.

7.12   A solid sphere, a solid cylinder, a thin hollow cylinder, and a hoop, each having the same radius, start from rest at the top of an incline, and roll (without slipping) to the bottom. In what order will they arrive there? How, if at all, will the motions of the centers of these bodies depend on their radii?

7.13   The inner and outer radii of a homogeneous hollow cylinder are r_1 and r_2, respectively. What is the radius of gyration r_g of the cylinder for rotation about its axis?

7.14   The moment of inertia of a flywheel, initially at rest, is 2 kg-m^2, and its radius is 60 cm. A tangential force of 60 kg weight is applied to the rim for 12 sec. How many revolutions does it make in this time, and what are the final values of the angular velocity and kinetic energy?

7.15   An extended body is mounted frictionlessly on a horizontal axle in such a way that the axis of rotation passes through the center of mass of the body. A string is wrapped tightly around the axle and supports a mass m. This mass is observed to fall with acceleration a. Calculate (a) the moment of inertia of the rotating system, and (b) the tension in the string.

7.16   A light belt of negligible thickness passes over and engages frictionally with a wheel of diameter 15 cm which is frictionlessly mounted on a horizontal axle, and masses of 2 kg and 4 kg are attached to its ends. It is observed that, starting from rest, the 4-kg mass falls 1 m in 1.5 sec. Find (a) the velocity of the masses at this time, (b) the tensions of the belt on the two sides of the wheel, and (c) the moment of inertia of the rotating system.

7.17   Around a homogeneous solid cylinder near its two ends light cords are wound, and the free ends of these are attached to the ceiling of a room so that with the cords taut the axis of the cylinder is horizontal (see the figure). Neglecting the radii of the cords relative to the radius of the cylinder, taking the cords between the cylinder and the ceiling to be essentially vertical, and assuming that there is no slip between the cords and

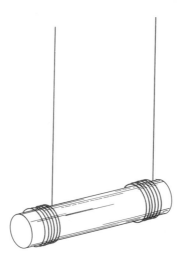

Figure for problem 7.17.

the cylinder, find the linear acceleration of the axis of the cylinder due to the earth's gravitational pull when it is released. Find also the tensions in the cords.

7.18 A cotton reel rests on a rough horizontal surface, and the thread is pulled horizontally with a force F in the manner indicated in the figure. Show that, in the absence of slip between the rim of the reel and the surface, the thread will wind up. By equating the power which the force applies to the reel to the rate of increase of the total kinetic energy, obtain a formula connecting the acceleration a of the center of the reel, the moment of inertia I, the mass m, the inside radius r_1, the outside radius r_2, and F.

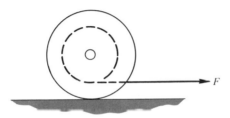

Figure for problem 7.18.

7.19 A small puck of mass m lies on a frictionless horizontal plane (see the figure). In (a) the puck is connected to one end of an inelastic weightless cord. The other end is wrapped around a narrow vertical peg rigidly fastened to the plane. In (b) the other end of the string passes through a small hole in the plane and can be pulled from below the plane. In both cases the puck is started with an initial angular velocity ω_0 and with its center of mass describing a path of initial radius of curvature r_0. In (a) the string shortens as it wraps around the narrow peg. In (b) the string shortens because it is slowly pulled through the small hole. Calculate the linear and angular speed of the puck when the "radius" has shortened to half its initial value. Compare linear and angular momenta and kinetic energies in both cases. What is conserved in each case, and why?

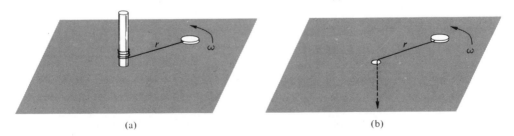

(a)   (b)

Figure for problem 7.19.

7.20 A homogeneous sphere of radius r is projected along a rough horizontal table with velocity v and angular velocity $6v/r$ in the direction opposite to that required for rolling. Find an expression for the time taken for the sphere to return to its starting point, and the velocity with which it does so, the coefficient of kinetic friction being μ.

7.21 A light horizontal rod is mounted symmetrically on a light vertical axle, and two small rings, each of mass m, are engaged with the rod, along which they can slide frictionlessly, one on either side of the axle. Light strings under tension keep the rings at equal distances r_1 from the axis of rotation while the system rotates freely (without friction and without the application of any external torque) with angular velocity ω_1. The lengths of the strings between the bodies and the axle are now progressively increased until the radius of the circle in which the rings rotate is r_2. (a) Neglecting the moment of inertia of the rod and axle, calculate the new angular velocity ω_2. (b) Show that the kinetic energy of the rotating system is reduced by $m\omega_1^2 r_1^2(1 - r_1^2/r_2^2)$. (c) Show that this is precisely the amount of work done by the masses on the constraining system as these move outward.

7.22 A space satellite spinning at 5 rps and having a radius of gyration of 0.5 m pushes out two metal tapes from opposite sides at right angles to its axis of rotation, the tapes then being in the same straight line passing through the axis. Each tape finally extends to 20 m from the axis and has a mass one thousandth of that of the satellite. How fast does the satellite now rotate? (Tapes may be assumed to have contributed zero moment of inertia before extension and to have uniform mass per unit length.)

7.23 Supposing the density of the earth, of radius 6371 km, to be uniform, calculate by how much the radius would have to be reduced by shrinkage for the length of the day to be decreased by 1 sec.

7.24 If a flywheel in the form of a disk of radius 50 cm and mass 700 kg rotates at 300 rpm and loses 1 percent of its angular velocity in 0.1 sec, what is (a) the power delivered, and (b) the mean value of the torque exerted, during this time, by the flywheel?

7.25 A uniform thin metal rod of mass m and length l lies on the surface of a horizontal sheet of ice, and a horizontal force F is applied to one end in a direction perpendicular to the long dimension of the rod. Assuming friction between the rod and the ice to be negligible, find the initial acceleration of the center of the rod and the initial angular acceleration of the rod.

7.26 At what height above a table must a cue, held horizontally, hit a billiard ball of radius r so that it rolls, without sliding, from the start? (Neglect friction.)

7.27 A thin uniform rod of length l describes a vertical circle about an axis through one end. What must be its angular velocity when the other (free) end is at its lowest point in order that the rod shall be momentarily at rest when this end is vertically above the axis of rotation?

7.28 The beam of a balance weighs 200 g, and its center of mass is 2 mm below the central knife-edge. Given that the arms of the balance are each 15 cm long, calculate the sensitivity (angular deflection per unit excess mass on one pan) of the balance.

7.29 A uniform circular table of diameter 1.5 m is supported on three vertical legs equally spaced around a circle of radius 0.6 m. The mass of the table (including legs) is 20 kg. What is the least vertical force applied at a point on the edge of the table (a) upward, and (b) downward, that will cause the table to tilt?

7.30   A uniform plank of mass 50 kg rests horizontally on the curved surface of a fixed cylindrical drum of radius 1 m, the axis of the cylinder being horizontal and normal to that of the plank. How far can a man of mass 80 kg walk along the plank from its center before it slips, if the coefficient of friction between the plank and the drum is 0.4?

7.31   A uniform ladder of mass m is held at an angle θ to the ground by a rope attached to it at its upper end which makes an angle of 90° with the ladder, and a man of mass $4m$ climbs up the ladder. How does (a) the vertical component and (b) the horizontal component of the force exerted by the ground on the lower end of the ladder vary with the position of the man on the ladder? When has the ratio of the latter to the former component its maximum value? For what angle θ has this maximum its greatest value?

7.32   Consider the moon to execute a circular orbit with center at the center of the earth. Disregard the effect of the sun and consider only the contribution of the moon to the tidal action on the earth. Use the fact that the daily rotation of the earth on its axis, being much faster than the monthly revolution of the moon around the earth, results in tidal "friction" tending to lengthen the day. Make other appropriate simplifying assumptions and show that at the present time the distance from earth to moon is increasing.

ROTATIONAL OSCILLATIONS AND GYROSCOPIC ACTION

This chapter is divided into two parts, each dealing with an important application of the general principles of rotational dynamics developed in Chapter 7. In Part I we shall consider rotational oscillations of simple harmonic type. The timing of such oscillations in the case of a system having a known moment of inertia is a convenient and accurate means for evaluating the relevant "control," the restoring torque per unit angular displacement from the equilibrium configuration. In Part II we shall discuss the interesting phenomenon known as "gyroscopic action," in which a torque applied to a spinning body causes the axis of spin to rotate about an axis perpendicular to itself.

PART I
ROTATIONAL OSCILLATIONS

General Theory

A rigid body will execute rotational oscillations of simple harmonic type if the control is such that a rotation of the body through an angle θ from its equilibrium orientation gives rise to a restoring torque L' (Fig. 8.1) which is proportional to this angle, that is, if

$$L' = \kappa\theta,$$

where κ is a constant, or

$$L = -\kappa\theta, \tag{8.1}$$

where L, equal to $-L'$, is the torque acting in the same direction as θ is measured. As we saw in Chapter 7, the torque applied to a body and the resulting angular acceleration are related by the equation

$$L = I\dot{\omega}.$$

If, therefore, we make the corresponding substitution for L in

FIGURE 8.1   Twist θ of a body suspended by a wire and corresponding restoring torque L'.

Eq. (8.1) we obtain the relation between the angular acceleration and θ,

$$\dot{\omega} = -\frac{\kappa}{I}\,\theta,$$

or

$$\frac{d^2\theta}{dt^2} = -\frac{\kappa}{I}\,\theta.$$

This is of the form

$$\frac{d^2s}{dt^2} = -ks$$

discussed under the heading "Simple Harmonic Disturbances" in Chapter 2. It was there shown that the period of such a disturbance is $2\pi\sqrt{1/k}$. If in this we substitute for k its value κ/I, or $|L/\theta|/I$, we obtain for the period of the rotational oscillations under consideration the formula

$$T = 2\pi\sqrt{\frac{I}{|L/\theta|}}. \tag{8.2}$$

This is the rotational analogue of Eq. (5.6),

$$T = 2\pi\sqrt{\frac{m}{|F/x|}}.$$

Rigid–Body Pendulum or Compound Pendulum

Let a rigid body be supported along the line of a knife-edge engaging with a horizontal plate, as indicated in Fig. 8.2, and let it execute oscillations about this knife-edge as axis. Let the distance from the axis of oscillation to the center of mass of the body be l, the moment of inertia about a parallel axis through the center of mass I_0, and the mass of the body m. Then by the theorem of parallel axes, the moment of inertia I about the axis of oscillation is given by the formula

$$I = I_0 + ml^2. \tag{8.3}$$

When the angular displacement of the pendulum about its rest position is θ, the moment of its weight about the knife-edge is $mgl \sin \theta$. This is the restoring torque; the torque in the direction taken as positive is $-mgl \sin \theta$. For small amplitudes $\sin \theta$ may be taken as sufficiently nearly equal to θ itself, so that

$$L = -mgl\theta,$$

or

$$\left| \frac{L}{\theta} \right| = mgl. \tag{8.4}$$

Substituting in Eq. (8.2) the expressions for I and $|L/\theta|$ given by Eqs. (8.3) and (8.4), respectively, we obtain the formula for the period,

$$T = 2\pi \sqrt{\frac{I_0 + ml^2}{mgl}}. \tag{8.5}$$

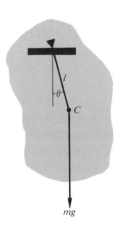

FIGURE 8.2 Rigid-body pendulum.

An important application of the rigid-body pendulum is the precision pendulum method of determining g, using Kater's reversible pendulum, which was referred to in Chapter 5. The theory of the rigid-body pendulum may also be applied in finding an expression for the period of oscillation of a beam balance in terms of its sensitivity and the effective moment of inertia of the oscillating system. The latter may be shown to be the moment of inertia of an idealized system in which the scale pans and load are imagined to be shrunk to massive points located at the end knife-edges.

Determination of the Gravitational Constant G

We have seen that there is a force of attraction between every pair of particles, given by the equation

$$F = G \frac{m_1 m_2}{r^2},$$

where m_1 and m_2 are the masses of the particles, r is their distance apart, and G is a universal constant.

Among the precision methods of determining G, one of the simplest, although by no means the earliest, was direct measurement of the force exerted on a sphere suspended on one arm of a sensitive beam balance by a larger fixed sphere placed immediately below it. This method was used by Poynting and others between 1878 and 1903.*

Antedating this work by nearly a century was the classical determination carried out by Henry Cavendish in 1797–1798, using a torsion balance. This is a device in which torque applied to a wire or fiber is measured by the angle through which it is twisted, torque being proportional to twist. Cavendish's apparatus was greatly improved in design by Boys, who, between 1889 and 1895, succeeded in measuring G to a higher degree of accuracy than either Poynting or Cavendish.

The principle of the Cavendish–Boys method is shown schematically in Fig. 8.3. Two equal spheres are attached to the ends of a light frame which is suspended by a torsion fiber. On one side of each of these is placed a larger sphere, one at A and the other at B, so that the forces of attraction between adjacent pairs of spheres constitute a couple, twisting the suspended system through an

---

* A full account of this and all earlier determinations of G is given in Poynting's book *The Mean Density of the Earth*, Charles Griffin & Co. Ltd., London, 1894.

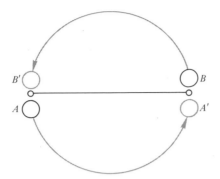

FIGURE 8.3   Principle of static part of Cavendish–Boys experiment for determining G.

angle. The attracting spheres are then placed in the positions A' and B', so producing a twist in the opposite direction. The angle turned through by the suspended system resulting from the change of the positions of the attracting spheres is measured, and then it is a simple matter to calculate the forces between adjacent pairs of spheres in terms of the arm of the couple and the torque per unit twist of the suspension fiber, and so, finally, to calculate the value of G.

The torque per unit twist may be determined from the period of oscillation of the suspended system and its moment of inertia, using the formula

$$T = 2\pi \sqrt{\frac{I}{|L/\theta|}},$$

or

$$\left| \frac{L}{\theta} \right| = \frac{4\pi^2 I}{T^2}.$$

This is the broad principle of the method. In practice, cross attractions must be allowed for, suitable precautions have to be taken against disturbances by air currents, and the apparatus must be designed, dimensionally and otherwise, in such a way as to give a high degree of accuracy.

The most recent determination of G, which is probably the most reliable of all, is one made in 1942 by Heyl and Chrzanowski, by a modification of the Cavendish–Boys method. The result they obtained was

$$G = 6.673 \pm 0.003 \times 10^{-8} \text{ dyn-cm}^2/g^2$$
$$= 6.673 \pm 0.003 \times 10^{-11} \text{ N-m}^2/\text{kg}^2.$$

197

MASS OF THE EARTH   Knowing the values of both g_0 and G, it is a simple matter to calculate the mass of the earth. The force which the earth, of mass m, exerts on a body at its surface, of mass μ, is

$$F = G\,\frac{m\mu}{r^2},$$

where r denotes the radius of the earth. The force per unit mass of the attracted body, g_0, is therefore Gm/r^2. Hence

$$m = \frac{g_0 r^2}{G}.$$

Substituting 9.82 m/sec² for g_0, 6.731 × 10⁶ m for r, and 6.673 × 10⁻¹¹ N-m^2/kg² for G, we then find the value of m in kilograms,

$$m = 5.97 \times 10^{24}\text{ kg.}$$

PART II
GYROSCOPIC ACTION

In our study of particle dynamics we dealt with changes of motion and the associated forces both (1) where the motion is confined to a straight line, and (2) where the particle moves with uniform speed in a circular path. Although, in considering rotational quantities, we have discussed the analogue of the first case, we have not yet dealt with that of the second case. We shall now proceed to an investigation of this case.

General Theory

Let a body be spinning with constant angular velocity ω about an axis for which its moment of inertia is I, and let this axis of spin rotate about an axis perpendicular to itself with angular velocity Ω. Such rotation of the axis of spin, known as **precession,** will obviously not occur spontaneously; it must be brought about by an applied torque. The occurrence of precession in response to this torque is known as **gyroscopic action.** We shall now investigate how this torque, L, is related to I, ω, and Ω.

Let OA and OB in Fig. 8.4 represent the angular momentum, of magnitude $I\omega$, of the spinning body at times t and $t + \delta t$, respectively, the angle between them, $\delta\theta$, being equal to $\Omega\,\delta t$. With $\delta\theta$ infinitesimally small, the arc AB becomes a side of the vector tri-

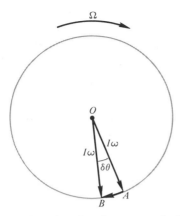

FIGURE 8.4 Construction for development of theory of gyroscopic action.

angle OAB, and represents the change in angular momentum, δP, to the same scale as OA and OB represent $I\omega$. But, by geometry, AB is equal to $OA \cdot \delta\theta$. Hence

$$\delta P = I\omega \cdot \delta\theta$$
$$= I\omega\Omega \, \delta t,$$

which, on dividing both sides by δt and proceeding to infinitesimals for δP and δt, becomes

$$\frac{dP}{dt} = I\omega\Omega. \tag{8.6}$$

We have seen that the time rate of change of angular momentum is one of two alternative quantitative expressions for the torque, L, applied to a body. In our discussion dealing with this we considered only the case of a body having a fixed axis of constraint. There is, however, no reason why our definition should be limited to this case. Let us therefore now agree to define torque as the time rate of change of angular momentum quite generally, that is, vectorially where necessary, so that our previous discussion then refers simply to a special case—that where all the vectors concerned have the same direction. The quantity defined in this wider sense still has the same dimensions and is surely of the same physical nature. We may, therefore, confidently expect any mathematical reasoning based on the widened definition and applied vectorially still to be valid. Assuming this, and substituting L for dP/dt in Eq. (8.6), we obtain the relation

$$L = I\omega\Omega. \tag{8.7}$$

It should be noted that the precession occurs about an axis which is perpendicular both to the axis of spin and to that of the applied torque, and that its sense is such as tends to convert the existing spin into one agreeing in direction with the applied torque.

There is a close analogy between gyroscopic action and uniform circular motion of a particle. Thus in the preceding paragraph we have merely to substitute the words "rotation" for "precession," "direction of motion" for "axis of spin," "force" for "torque," and "velocity" for "spin," to express the situation relating to uniform circular motion. We may also note the correspondence between Fig. 8.4 and the right-hand part of Fig. 2.4, and between Eqs. (8.7) and (5.1) in the form

$$F = mv\omega.$$

An instructive experiment illustrating gyroscopic action is the following. A wheel is mounted on an axle AB, which is suspended by a string attached to one end as indicated in Fig. 8.5. We see that the weight mg of the suspended system and the vertical component of the tension of the string constitute a couple, whose axis is horizontal and at right angles to the axis of spin. In response to this couple the system precesses about a vertical axis, the sense of the precession being such that, if continued through a right angle, it would bring the spin into agreement with the applied torque. Actually, the axis of the applied torque itself rotates as the system precesses, so we may think of the precession as constantly "chasing" the applied torque. The faster the spin, the slower is the precession,

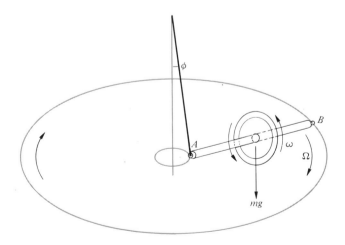

FIGURE 8.5   Wheel spinning about a horizontal axle supported by a string attached to one end.

because with a given applied torque L the product of ω and Ω must, according to Eq. (8.7), always be the same. Also if, with a given angular velocity of spin, a mass is hung on the end B of the axle, thereby increasing the applied torque, the rate of precession will be increased in proportion.

It will be observed that while the end B describes a large circle, the end A by which the axle is suspended describes a smaller circle, the string correspondingly tracing out a cone. Let the semiangle of this cone be denoted by ϕ. This must be such that the horizontal component of the tension T of the string constitutes the necessary centripetal force $m\Omega^2 r$, where r is the radius of the circle in which the center of mass of the system rotates. Thus, because

$$T \sin \phi = m\Omega^2 r$$

and

$$T \cos \phi = mg,$$

we must have

$$\tan \phi = \frac{\Omega^2 r}{g}. \tag{8.8}$$

As the angular velocity of spin ω is increased, Ω must decrease correspondingly, in accordance with Eq. (8.7); and, as we see from Eq. (8.8), this would of itself cause ϕ to decrease. But with this decrease in ϕ must be associated a decrease in r. A fortiori, therefore, ϕ must decrease with increasing angular velocity of spin.

APPLICATION OF TORQUE ABOUT AXIS OF PRECESSION
We have seen that a change in magnitude, but not in direction, of the applied torque simply brings about a corresponding change in the rate of precession. What if, now, instead of changing the magnitude of the existing torque, we applied an additional torque about the axis of precession, as if to hurry or retard the precession? We should find that just as the original applied torque did not cause the axis of spin to rotate about the axis of this torque, so the secondary applied torque does not actually hurry or retard the precession. Instead, it gives rise to a secondary precession which is related to the spin and the secondary torque according to exactly the same rule as we have already noted for the primary torque and the resulting precession. Applying this rule to the system of Fig. 8.5 we readily see that "hurrying the precession" gives rise to a secondary precession about the axis of the original applied torque, but in the opposite sense, so that the point B rises. "Retarding the precession," on the other hand, results in a secondary precession agreeing in direction and sense with the original applied torque, the point B correspondingly going down.

201

An alternative line of reasoning for predicting the effects of these secondary torques would be in terms of their vectorial combination with the primary torque.

Applications and Consequences

Gyrosopic action has a large number of important applications in modern technology and also numerous interesting consequences. Of these we shall here consider only three: the gyrocompass, the top, and the precession of the equinoxes.

GYROCOMPASS   A gyrocompass is essentially a wheel (gyroscope) of large moment of inertia which is set spinning at high speed with its axis in a horizontal direction and constrained to remain horizontal, or almost so, although free to revolve about a vertical axis. The full theory is rather complex, but the underlying idea can readily be understood by considering the special case where the compass is used at the equator.

The wheel is mounted in a housing which floats in a liquid. Accordingly, the axis of spin is free to revolve in a horizontal plane but is prevented from departing appreciably from this plane; any tendency to dip is immediately resisted by a torque about a horizontal axis perpendicular to the axis of spin. This torque gives rise to a precession that eventually brings the axis of spin into the north–south orientation.

Let Fig. 8.6 represent a view of the earth from a point outside it in the prolongation of the axis, with the North Pole nearest the observer. The rotation of the earth viewed from such a point will be anticlockwise. Let us suppose that the axis of the spinning gyroscope is initially pointing east–west, and let the arrow on the right

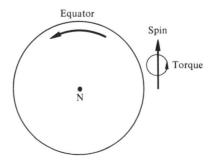

FIGURE 8.6   Diagrammatic representation of gyrocompass on the equator.

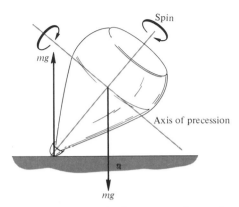

FIGURE 8.7   Forces exerted on spinning top, and associated precession.

be a vectorial representation of the angular momentum of the spin. As the earth rotates, the western end of the compass will tend to dip too far down in the liquid on which it floats and the eastern end not far enough. This will give rise to a torque as indicated, which will cause the compass to precess about a vertical axis. The sense of this precession will be such as to tend to bring the axis of spin into parallelism with the axis of the earth, the direction of the spin agreeing with that of the earth's rotation.

TOP   A spinning top leaning over at an angle to the vertical is acted upon by a system of forces that causes it to precess in the manner indicated in Fig. 8.7.

For a given torque, the faster the spin, the slower the precession, and vice versa. At the beginning of its career, when the top is spinning very fast, the motion of that part of the peg which is in contact with the ground gives rise to a frictional force acting on the top at right angles to the plane of the diagram and toward the observer. This force accelerates the top as a whole in this direction until it attains such a velocity that the frictional force is balanced by air resistance. The slowly precessing top therefore describes a circular path of large radius. At the same time, the frictional force exerts a moment about the axis of precession (passing through the center of mass) in the sense of hurrying the precession. As we have seen, such a secondary applied torque gives rise to a secondary precession opposing the primary torque. The tilt of the top therefore decreases, the circle in which it moves becomes smaller, and after a time it stands practically upright on a single spot. Eventually, however, when ω has become quite small, any tilt of the axis of spin tends to increase. This is because Ω is now so large

203

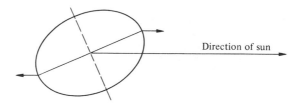

Direction of sun

FIGURE 8.8 Gravitational torque exerted by sun on spheroidal earth.

relative to ω that the precession predominates over the spin in determining the direction of relative motion at the point of contact of the peg with the ground, the direction of the frictional force is reversed, and the secondary precession is now in agreement with the primary torque.

PRECESSION OF THE EQUINOXES We have seen that the earth is ellipsoidal in shape, having an equatorial diameter (12,756.8 km) somewhat in excess of the polar diameter (12,711.8 km). Also, the plane of the earth's equator is inclined at a substantial angle (23°27′4.04″) to the plane of its orbit (plane of the "ecliptic"). Owing to the combination of these two circumstances, not only does the sun exert on the earth as a whole the centripetal force that keeps it in its orbit, but, in addition, its gravitational action on the equatorial bulge gives a torque, and this causes the earth's axis to precess.

In Fig. 8.8 the torque exerted on the equatorial bulge—greatly exaggerated—is indicated by the arrows. The exertion of this torque may readily be explained in terms of the variation with position of the sun's gravitational field. This, varying with distance from the sun according to the inverse-square law, is greater on the nearer side of the equatorial bulge than on the remote side. We may imagine the actual field to be replaced by a uniform field, of magnitude equal to its mean value,* and, superimposed on this, a varying field, the latter being directed toward the sun on the near side of the bulge and away from the sun on the remote side. The uniform component of the field gives a resultant centripetal force through the center of the earth, and the varying component produces the torque.

Although the torque is always about an axis parallel to the tangent to the earth's orbit at the position of the earth at midsummer and at midwinter and always has the same sense about

---

* This will approximate very closely to the field existing at the center of the earth.

this axis, it is not constant in magnitude; it is greatest at midsummer and at midwinter, when the direction of the sun makes its greatest angle with the equatorial plane, and zero at the times of the equinoxes. Correspondingly, the precession is not steady, but has two maxima per year. This unsteady precession is known as "nutation" (nodding). Actually the sun is not the only agency exerting a torque on the earth; the moon also contributes, and the nutation due to the combined actions of the sun and moon is correspondingly rather complex. The resulting motion of the earth's axis is thus similar to that of a somewhat jerky top, although on an enormously longer time scale; a complete period of the precession of the earth's axis occupies about 25,800 years.

An interesting consequence of this precession is the quite appreciable variation in the stars visible in the Northern and Southern Hemispheres, with extremes 12,900 years apart. Another is that in a complete period there is one more summer and winter at any particular place than there are revolutions of the earth in its orbit about the sun.

Problems

8.1   A homogeneous spherical ball of radius r is released from rest near the bottom of a hollow spherical shell of larger radius R and describes periodic motion as it rolls back and forth inside without slipping. Show that the oscillation is simple harmonic, and obtain an expression for its period in the case where $r \ll R$.

8.2   Find the period of a simple pendulum having a bob whose radius r, although not negligible, is small compared with the length l of the pendulum. Use the algebra of small quantities to find the correction factor to be applied to the simple formula previously obtained. (Consider the simple pendulum having a bob of finite dimensions as a rigid-body pendulum.) What is the correction factor if l and r have values of 30 cm and 1 cm, respectively?

8.3   A circular hoop of radius r is suspended on a horizontal nail. Calculate the period of small oscillations of the hoop (a) when these oscillations occur in its own plane, and (b) when they occur in a direction at right angles to this.

8.4   A movable knife-edge system can be clamped to a long thin rod of length l anywhere along its length, providing for oscillation of the rod about an axis perpendicular to and intersecting its own axis. Neglecting the mass of the knife-edge system, find the distance x of the axis of oscillation from one end of the rod for which the period of oscillation has its minimum value.

8.5   A thin uniform rod of length l oscillates with small amplitude about an axis through one end which is perpendicular to the axis of the rod. Find the period of this oscillation. What would be the length of a simple pendulum that has this period? Is there a point on the rod at which a body of small dimensions but significant mass could be clamped to it without affecting the period of oscillation? If so, determine the position of this point.

8.6 In dealing with a beam balance as an oscillating system the scale pans plus contents may be idealized as massive points located at the end knife-edges. What is the theoretical justification for this idealization?

8.7 A beam balance has a pointer 25 cm long moving over a scale whose divisions are each of length 1.2 mm, and the sensitivity of the balance is 3.8 scale divisions per centigram. The distance between the scale-pan knife-edges is 12 cm. Given that the period of oscillation of the balance is 15 sec, find the effective moment of inertia of the suspended system. If each scale pan were loaded with a 100-g mass, what would now be the period of oscillation?

8.8 A thin circular hoop of mass 1.57 kg and radius 18.2 cm is suspended from a vertical wire with its plane vertical and executes rotational oscillations about the wire as axis. Given that the period of oscillation is 7.28 sec, find the torque that would be necessary to twist the hoop through one complete rotation from rest.

8.9 A frame supported by a torsion fiber oscillates with a period of 5 sec. When a thin horizontal rod of length 12 cm and mass 20 g is attached to the frame with its center of mass in the prolongation of the fiber the period is 8 sec. Find (a) the moment of inertia of the frame, and (b) the torque per unit twist of the suspension.

8.10 In an experiment of the Cavendish–Boys type for determining G, the suspended spheres were each of mass 47.5 g and diameter 2.04 cm, and the attracting spheres were each of mass 6.37 kg. The horizontal distance between the centers of the suspended spheres was 6.04 cm, and the distance between the center of each suspended sphere and that of the adjacent attracting sphere in the static part of the experiment was 6.39 cm. When the attracting masses were moved from one pair of attracting positions to the other the suspended system turned through an angle of 0.0152 rad, and when they were removed altogether the period of oscillation was 5 min, 2.3 sec. Neglecting cross attractions and the mass of the frame, find the value of G.

8.11 The propeller of an aircraft and the rotating system to which it is attached have a total moment of inertia of 50 kg-m² and rotate at 100 rps. What is the mean torque exerted by the engine on its supports as a consequence of the aircraft turning through 90° in 5 sec?

8.12 A bicycle wheel of diameter 70 cm is mounted on an axle, set spinning at 250 rpm, and supported by a string tied to the axle 30 cm from the center of the wheel. How fast will the wheel precess with the axle horizontal? What would have to be the length of the string for the center of the wheel to describe a circle of radius 35 cm? (Neglect the mass of the axle and assume the whole mass of the wheel to be concentrated in the rim.)

8.13 The spinning component of a gyrocompass at the equator is cylindrical in form, has a mass of 1 kg and a diameter of 10 cm, and rotates at 80 rps. The frame in which it spins also has a mass of 1 kg, and the compass is supported from a point 10 cm higher than its center of gravity.

The axis lies in the E–W meridian but is slightly tilted, the line joining the center of gravity to the point of support making an angle of 0.6° with the vertical. Find the direction of the axis 10 seconds later.

8.14   The moment of inertia of a top about its axis of spin is 300 g-cm² and its center of mass is 3 cm from the point on which it spins. If it is spinning at 30 rps and its axis is inclined at an angle of 25° to the vertical, (a) What is the angular velocity of precession about the momentary axis of precession? (b) With what angular velocity does the point of contact of the peg with the ground describe its circular path? (c) Is the direction of motion of the bottom of the peg in its circle the same as, or opposite to, the direction of spin?

8.15   A homogeneous body in the form of an oblate ellipsoid of revolution of given eccentricity spins about its axis of symmetry with angular velocity ω as it describes a circular orbit about a gravitationally attracting body of relatively much larger mass. The orbit, of radius r, is large compared with the linear dimensions of the orbiting body, and the angle between the axis of spin of the orbiting body and the plane of the orbit is θ. Show that the mean rate of precession Ω of the orbiting body depends only on ω, r, and θ, being independent of the size of the body.

8.16   Given that the mean distances of the sun and moon from the earth are 1.495×10^8 km and 3.84×10^5 km, respectively, and that the mass of the sun is $2.72_5 \times 10^7$ times the mass of the moon, find the ratio of the torques (due to the earth's equatorial bulge) exerted by these bodies on the earth for equal angles between the earth's axis and the line joining the center of the earth to that of the attracting body. Does your answer also apply to the relative tide-raising effects of the sun and moon? Explain.

SECTION V

CONTINUUM MECHANICS

9

HYDROSTATICS AND HYDRODYNAMICS

The subject of study in this chapter will be the basic mechanical properties and behavior of fluids—liquids and gases. It is to be understood that what we are referring to here is what we might call the "bulk" properties of fluids, not surface properties; surface phenomena in liquids will be considered separately, in Chapter 12.

We shall divide the material of the present chapter into two parts: Part I will deal with hydrostatics, the study of fluids at rest, and Part II with hydrodynamics, the study of fluids in motion.

PART I

HYDROSTATICS

Density

Density, defined as the mass per unit volume of a material, may be expressed in grams per cubic centimeter, kilograms per cubic meter, or pounds per cubic foot, according as we are using the CGS, MKSA, or FPS system of units. It should be noted that the density of a material in the MKSA system is numerically 1000 times that in the CGS. In the FPS system the density is numerically 62.5 times that of the same material in CGS units.

In most of the indirect methods of measuring density what we actually do is to find the **relative density,** or **specific gravity,** defined as the ratio of the density of the material in question to that of a standard liquid, usually water. Owing to thermal expansion, the density of water, as of other materials, varies with temperature. It is therefore necessary to specify the temperature at which the water is regarded as standard. Water has the peculiarity of having its minimum volume, or maximum density, at 3.98°C. It is therefore convenient to compare the density of other materials with that of water at this temperature, regarded as a standard liquid.* At this temperature the density of water is 0.999973 g/cm³. For all ordinary purposes, where we are not concerned with extreme accuracy, we may adopt the value 1.0000 g/cm³. Thus if we take water at 3.98°C as our standard liquid and work in the CGS system, the numerical values of the absolute density and the relative density of any given material may for almost all practical purposes be taken as identical.

If in an experimental determination of relative density the temperature of the water used is not sufficiently close to 3.98°C, it may be necessary to make allowance for the variation of its density with temperature. Whether it is worthwhile doing this will depend on the temperature and the accuracy of the determination in question. Table 9.1 gives values of the density of water at various temperatures between 0° and 100°C.

Table 9.2 lists examples of densities of ordinary solids and liquids at 20°C.

The densities of gases are very sensitive to their temperatures and pressures. The nature of the variation will be studied later. Meanwhile we may note that dry air at 20°C and at a pressure of 1 atm has a density of 0.0012 g/cm³.

From the known masses and volumes of the earth and sun we may at once calculate the mean densities of these bodies. The mean density of the earth is found to be 5.51 g/cm³; that of the sun is 1.42 g/cm³.

TABLE 9.1   Density of Water

Temp. (°C)	Density (g/cm³)	Temp. (°C)	Density (g/cm³)
0	0.99984	50	0.98805
3.98	0.99997	60	0.98322
10	0.99970	70	0.97779
20	0.99820	80	0.97181
30	0.99565	90	0.96532
40	0.99222	100	0.95836

* Strictly speaking, it is also necessary to specify the pressure, because all material is compressible. The pressure taken as standard is 1 standard atmosphere. However, in practice, we may generally neglect the effects of pressure variations, because the density of water changes by only 0.005 percent per atmosphere of pressure change.

TABLE 9.2 Representative Densities

Material	Density (g/cm³)
Lithium	0.534
Ethyl alcohol	0.79
Water	1.000*
Aluminum	2.70
Iron	7.8
Copper	8.9
Lead	11.3
Mercury	13.55
Gold	19.3
Platinum	21.4

* The density of ice at 0°C is 0.917 g/cm³.

The sun is a fairly representative star, and its mean density is, as we have seen, of the same order of magnitude as the densities with which we are familiar on the earth. There is, however, a class of stars known as "white dwarfs," whose densities are, by these standards, quite fantastically high. An example of these is the faint companion of the bright star Sirius; this has a mean density of about 60,000 g/cm³. Other white dwarfs have comparable densities. It is believed that the white dwarfs are representatives of a late stage of stellar evolution, and that the sun, in common with most other stars, is destined eventually to pass through this stage.

Of a much higher order of density than even white-dwarf stars is nuclear matter, constituting atomic nuclei. It has been estimated that the density of nuclear matter is of the order of 10^{14} g/cm³. At the extreme other end of the scale is intergalactic space, which is believed to be within a factor of perhaps 100 or 1000 of 10^{-32} g/cm³.

Pressure

Pressure is the force per unit area exerted normally on a surface. The CGS unit of pressure is the dyn/cm², and the MKSA unit is the N/m². The N/m² is equal to 10 dyn/cm².

On occasion, pressures are given in gravitational units; thus we sometimes speak of atmospheric pressure being about 15 lb (weight) per square inch, or about 1030 g (weight) per square centimeter. This is, however, satisfactory only when we are not concerned with high accuracy, because these gravitational units vary in magnitude from one locality to another, with variations in g. They are also not suitable in theoretical discussions.

In a fluid at rest there can be no question of the exertion of a force on a surface in any direction other than at right angles to the surface, whether this is the surface of a solid with which the fluid is in contact, or a boundary of some part, or "element," of the fluid itself which we may wish to consider. That this is so follows at once from considerations of symmetry; any tangential component of the force would represent a bias in an arbitrary direction along the surface.

PRESSURE, ORIENTATION OF SURFACE, AND DEPTH   At a given point in a fluid the pressure must be independent of the orientation of the surface against which it is exerted. Also, the pressure varies in a definite manner with depth. These are conclusions that may be drawn from considerations of the equilibrium of a thin cylindrical portion of fluid, such as is indicated in Fig. 9.1.

Let the axis of the cylinder be inclined at an angle θ with the horizontal, and, while one end surface—say the upper—is at right angles to the axis, let the normal to the other end surface make an angle ϕ with the axis. We shall now investigate whether the pressure, p', exerted on the lower end surface varies with ϕ, and how it is related to p, the pressure on the upper end surface.

Let us consider the components in the axial direction of all forces exerted on the cylinder. The pressure forces exerted on the sides are everywhere at right angles to the axis, so the only forces with axial components will be those exerted on the two end surfaces and the weight. The force due to the pressure on the upper end

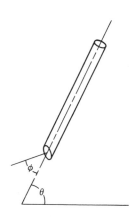

FIGURE 9.1   System for investigating variation of pressure with depth in a fluid.

surface is pA, where A is the area of cross section of the cylinder, and is in the axial direction. The area of the lower end surface is $A/\cos \phi$, and the force due to the pressure on this surface is $p'A/\cos \phi$. The axial component of this force is $(p'A/\cos \phi) \cos \phi$, that is, simply $p'A$. The weight of the portion of fluid under consideration is ρAlg, where ρ is the density and l is the mean length of the cylinder, and the component of this in the axial direction is $\rho Alg \sin \theta$, that is, ρAgh, where h $(= l \sin \theta)$ is the difference in height between the two ends. And, because for equilibrium the algebraic sum of the components of the forces in any direction must be zero, we must have

$$p'A = pA + \rho Agh,$$

or

$$p' = p + \rho gh. \tag{9.1}$$

This gives the variation of pressure with depth. It also shows that the pressure at any point must be the same for all orientations of surface, because the expression for p' is independent of ϕ.

Equation (9.1) applies as between any two points in a homogeneous body of fluid even in cases where the straight line joining the points is not everywhere within the fluid. For example, let a liquid of density ρ be contained in a vessel of the form depicted in Fig. 9.2, and let any two points 1 and 2 within it be joined by a series of rectilinear steps, as shown. Then an equation of type (9.1) will give the difference of pressure between the two ends of each such step, and by adding all these equations algebraically, increments of pressure and of depth all being given their appropriate signs, we must obtain the equation

$$p_2 - p_1 = \rho gh,$$

relating the pressures p_1 and p_2 at the points 1 and 2, respectively, and the depth h of 2 below 1.

FIGURE 9.2

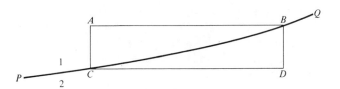

FIGURE 9.3 Imagined nonhorizontal boundary between two fluids at rest.

PASCAL'S PRINCIPLE From the foregoing, a principle first established experimentally by Blaise Pascal (1623–1662), and hence known as **Pascal's principle,** follows at once. According to this an increment of pressure applied at any point within a confined body of fluid is transmitted unattenuated to all parts of the fluid. Only if this is so can the difference in pressure at two points be a unique function of the difference in height between them, as we have shown it to be, and therefore independent of the pressure increment applied.

A direct application of Pascal's principle is the hydraulic or Bramah press, examples of which are certain forms of wool or paper presses, dentists' and barbers' chairs, hydraulic car jacks, and so on.

FREE SURFACE OF LIQUID From Eq. (9.1) it follows that the surface of separation between a liquid and a gas, or between two liquids that do not mix, must be horizontal.* For, in Fig. 9.3, let PQ represent the boundary between an upper fluid (liquid or gas) 1, of density ρ_1, and a lower fluid (liquid) 2, of density ρ_2. Let B and C be two points on the surface PQ, and let A be a point in the upper fluid vertically above C and on the same horizontal level as B, while D is a point in the lower fluid vertically below B and on the same horizontal level as C. Let AC and BD be denoted by h. Then if p_1 is the pressure (necessarily the same) at A and B, and p_2 that at C and D, we have both

$$p_2 - p_1 = \rho_1 g h$$

and

$$p_2 - p_1 = \rho_2 g h,$$

whence

$$\rho_1 h = \rho_2 h,$$

and with ρ_1 and ρ_2 unequal, this last equation can be satisfied only if h is zero, that is, if BC is horizontal.

---

* We are here neglecting surface-tension effects.

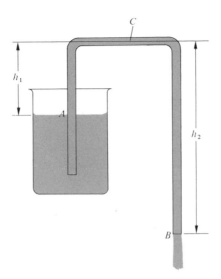

FIGURE 9.4 Principle of the siphon.

SIPHON A siphon in action is shown in Fig. 9.4. It is started by sucking liquid into the tube until it has reached a point on the outlet side below the level of A. The liquid thereafter flows out spontaneously.

The theory of the siphon is as follows: If we applied Eq. (9.1) for pressure differences as for a liquid at rest, we should find for the pressure at the highest point of the tube, C, the value $\rho g h_1$ below atmospheric. However, at B, where the liquid emerges into the atmosphere, the pressure must be atmospheric, as well as at A, and so (it would seem) the pressure at C must be $\rho g h_2$ below atmospheric. But obviously the pressure at C cannot have two different values simultaneously. The fallacy in the above reasoning is that we have applied to a liquid in motion equations that are valid only for a fluid at rest. In flowing through the tube the liquid encounters a certain viscous resistance (see Part II), which increases with the rate of flow. In consequence of this the pressure difference between A and C must be greater than $\rho g h_1$; otherwise, the flow could not be maintained. And that between B and C must be less than $\rho g h_2$; that is, the pressure at C must be less than it would be for a liquid at rest. Even at A the pressure *inside the tube* is less than atmospheric. The rate of flow adjusts itself to a value such that the resistance correction, applied along the whole length of the tube, gives zero pressure difference between the surface of the liquid in the reservoir (*outside* the tube) and B, the point of discharge.

MEASUREMENT OF PRESSURE We shall now consider the instrumentation of pressure measurement and incidentally take

217

note of certain important values of pressure, particularly those relating to the atmosphere.

A device for measuring pressure is known as a **pressure gauge,** or **manometer.**

U-Tube Manometer The unique relation between pressure difference and difference in height in a fluid at rest is turned to practical account in the **U-tube manometer,** depicted in Fig. 9.5. If the pressure difference to be determined is that between two bodies of gas, whose density is small enough to be neglected, and these are connected to the two limbs of a U tube containing a liquid of density ρ, then this difference of pressure, $p_2 - p_1$, is given by the relation

$$p_2 - p_1 = \rho g h,$$

where h is the difference in level between the free surfaces of the liquid in the two limbs.

The liquid most commonly employed is mercury, because of its low vapor pressure, and consequently generally negligible contamination of the gases whose pressures are being measured, and also because of its high density, permitting the use of manometers of convenient dimensions.

Pressures, or pressure differences, are often referred to in terms of the "head" h of liquid used in its measurement; for example, we might speak of a pressure, or pressure difference, of 20 mm of mercury, instead of the corresponding quantity expressed in absolute units, $2 \times 13.6 \times 980$ dyn/cm^2, or $0.02 \times 1.36 \times 10^4 \times 9.8$ N/m^2. This is very convenient, and we easily get into the way of thinking of pressures in millimeters of mercury; for example, we may say that the vapor pressure of mercury at room temperature (referred to above) is about 2×10^{-3} mm of mercury. It must, however, be remembered that in assigning numerical values to items in a theoretical formula expressed, as it usually is, in absolute units,

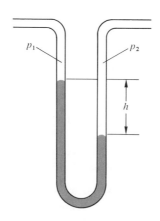

FIGURE 9.5   Open U-tube manometer.

it is always necessary to express any pressure occurring in such a formula in dyn/cm², or N/m², as the case may be.

A further consideration is that not only does g vary with locality, as already noted, but also ρ varies with temperature; thus whereas at 0°C the density of mercury is 13.595 g/cm³, at 20°C it is only 13.546 g/cm³, the difference amounting to nearly 0.4 percent. In all cases where we are concerned with high accuracy it is necessary either to state, in addition to h, the values of g and ρ (or the temperature) that apply, or to avoid the difficulty by expressing the pressure directly in absolute units.

The manometer shown in Fig. 9.6 measures only a *difference* of pressure; it does not measure either p_1 or p_2 by itself. If, however, the pressure on one side of a manometer is reduced to zero, then the head of liquid observed measures the whole pressure on the other side. In the closed U-tube manometer depicted in Fig. 9.6, the space above the mercury in the closed end has been evacuated of all gas other than mercury vapor, whose pressure (about 2×10^{-3} mm of mercury) may in the present connection be regarded as negligible. The pressure of the gas in contact with the mercury on the other side then corresponds to a head h of mercury, where h is the difference in height between the mercury surfaces in the two limbs; that is, the absolute value of the pressure is $\rho g h$.

Mercury Barometer An interesting special case of a closed U-tube manometer is that in which the open limb is simply exposed to the atmosphere, as indicated in Fig. 9.7(a); we then have a form of **barometer,** a manometer for measuring atmospheric pressure. Barometers of this form are sometimes known as J-tube barometers.

The type of mercury barometer shown in Fig. 9.7(b) may be thought of as derived from part (a) by changing the J tube to a

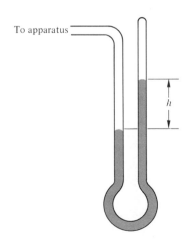

FIGURE 9.6 Closed vacuum U-tube manometer.

system of two coaxial tubes, the outer of these being simply a dish or bowl. The vacuum above the mercury in the tube of this barometer is normally established by first filling the tube completely with mercury, and then, while temporarily closing the open end with a finger, inverting it over a bowl of mercury. This type of barometer was first developed by the Italian physicist and mathematician, Evangelista Torricelli (1608–1647), and the vacuum above the mercury is usually called, after him, a "Torricellian vacuum."

ATMOSPHERIC PRESSURE For the specification of certain physical quantities it is convenient to refer them to a particular pressure, and perhaps also temperature, at which they are measured, this pressure, and, where required, temperature, being regarded as standard. A standard of pressure frequently used is the "standard atmosphere," defined as the pressure corresponding to a head of 760 mm of mercury at 0°C ($\rho = 13.595$ g/cm³) at sea level and latitude 45° ($g = 980.616$ cm/sec²). Expressed in absolute units, this pressure is $13.595 \times 980.616 \times 76$ dyn/cm², that is, 1.01323×10^6 dyn/cm², or 1.01323×10^5 N/m². The corresponding temperature taken as standard is usually 0°C. Thus "standard temperature and pressure" (STP) is understood as meaning a temperature of 0°C and a pressure of 1 standard atmosphere. We might say, for example, that the density of dry air at STP is 0.001293 g/cm³.

The pressure of the air at any given altitude is readily accounted for in terms of the weight of superincumbent air above a given horizontal surface. Let us consider the equilibrium of a vertical column of air of cross-sectional area A such as is depicted in Fig. 9.8, the column extending to the "top of the atmosphere," that is, to a sufficient height for the pressure at the top to be regarded as negli-

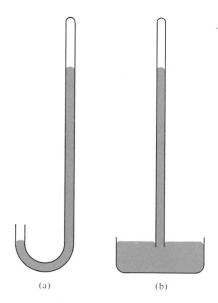

(a)   (b)

FIGURE 9.7   Two forms of mercury barometer.

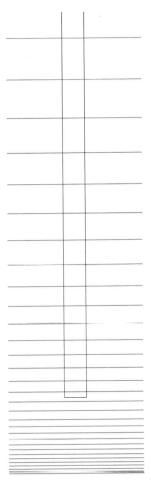

FIGURE 9.8 Column of air to whose weight pressure in the atmosphere is attributable.

gible.* Let the pressure at the bottom of this column be p. Then the upward force exerted on it must be pA, and for the air in the column to be in equilibrium this must be equal to its weight W. Thus $p = W/A$.

For a given cross-sectional area A, W decreases with increasing altitude of the bottom of the column, so p must decrease correspondingly. But the density decreases with diminishing pressure, and hence the *rate* at which W, and therefore also p, decreases with increasing altitude must progressively decrease. Some idea of this is given by Table 9.3.

---

*It is known that at an altitude of 100 km the pressure is of the order of only 10^{-3} mm of mercury. If, therefore, we do not wish to specify pressures more accurately than to the nearest 10^{-2} mm, say, we may regard 10^{-3} mm as negligibly small, and correspondingly we may think of about 100 km as the effective height of the atmosphere.

TABLE 9.3 Pressures at Various Altitudes

Altitude (m)	Pressure (mm Hg)
0	760
1000	674
2000	596
3000	526
4000	463
5000	405
6000	354
7000	308
8000	267
9000	230
10,000	198

Archimedes' Principle

It is a matter of everyday observation that bodies wholly or partially immersed in a fluid of appreciable density experience an upthrust, so that their apparent weight—actual weight minus upthrust—is less than when not so immersed. It becomes of interest, then, to investigate the magnitude of this upthrust.

The upthrust is evidently the vector resultant of the forces exerted on all the various elements of surface of the body due to the pressure of the surrounding fluid. Actually to compound all these forces in the case of a body of irregular shape would be an exceedingly complicated and tiresome procedure. Fortunately, however, there is a much simpler alternative, suggested by the consideration that pressure in a fluid is a unique function of depth, being independent of the nature of the surface on which the pressure acts. Thus if we removed the body and replaced it or that portion of it

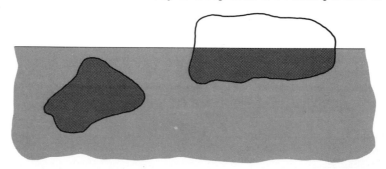

FIGURE 9.9 Systems used for theoretical derivation of Archimedes' principle.

that was immersed (shaded in Fig. 9.9) by a body of the fluid having exactly the same configuration, the same upthrust would be exerted on this portion of fluid as was formerly exerted on the body. But this portion of fluid would be in equilibrium—it would simply constitute a part of the general body of fluid at rest. The upthrust must therefore just balance the weight of this portion of fluid. In other words, **the upthrust exerted on a body wholly or partially immersed in a fluid is equal to the weight of fluid displaced.** This is Archimedes' principle.

It should be noted that Archimedes' principle is not a law; it is not a summary of experience, although it may readily be verified experimentally. It is simply a logical deduction from certain general considerations concerning the exertion of pressure.

Although the applications of Archimedes' principle with which we are most familiar are those in which the fluid concerned is a liquid, gases are also, on occasion, capable of exerting quite appreciable upthrusts. An obvious example of this is the upward force exerted by the surrounding air on a balloon. A less obvious case, frequently lost sight of, is the role played by Archimedean upthrusts in ordinary weighing. As we shall now see, neglect of these is capable of debasing the accuracy of determinations of mass quite appreciably.

BUOYANCY CORRECTION IN WEIGHING In weighing, Archimedean upthrusts must be exerted both on the standard masses used and on the body being weighed. The standard masses are usually of brass, having a density of about 8.4 g/cm³. Fractional masses (fractions of a gram) are sometimes of aluminum (density 2.7 g/cm³) and sometimes of platinum (density 21.4 g/cm³), but these usually constitute only a small fraction of the total standard masses used, so the error incurred in assuming that the mean density of the standard masses is simply that of brass will generally be negligible.

Let the mean density of the standard masses be denoted by D, that of the body being weighed by ρ, and that of the air by δ.* Then if m is the mass of the body and m_1 that of the standard masses, the resultant forces exerted on the masses on the two sides of the balance will be equal when

$$mg - \frac{m}{\rho}\delta g = m_1 g - \frac{m_1}{D}\delta g,$$

---

* This is of the order of 0.0012 g/cm³ but varies with temperature, pressure, and moisture content.

m/ρ and m_1/D being the volumes of air displaced by the body being weighed and the counterbalancing standard masses, respectively, or

$$m\left(1 - \frac{\delta}{\rho}\right) = m_1\left(1 - \frac{\delta}{D}\right). \tag{9.2}$$

If ordinary standard masses are used, δ/D will be very small compared with 1, and in most cases the same will be true of δ/ρ. We may then simplify Eq. (9.2) by using the algebra of small quantities, so obtaining the expression for the mass to be determined,

$$m = m_1\left(1 - \frac{\delta}{D} + \frac{\delta}{\rho}\right). \tag{9.3}$$

From Eq. (9.3) we see that if we are weighing a light material, whose density is of the order of 1 g/cm³, using standard masses of brass, the true mass m will exceed the apparent mass m_1 by something of the order of 0.1 percent. If, on the other hand, the body being weighed has a very high density, say in the neighborhood of 20 g/cm³, m will be less than m_1 by roughly 0.01 percent. In either case the buoyancy correction will be important if we are concerned in a precision determination of mass. Only if it so happens that ρ and D are approximately equal will it be permissible to neglect this correction in precision work.

The buoyancy correction is quite commonly ignored, owing simply to failure to appreciate its importance. Thus the result of a painstaking quantitative chemical analysis could be seriously vitiated by failure to apply the correction. Unfortunately, many of the data given in standard tables of physical constants are probably infected by such inaccuracies.

Pressure–Volume Relation for Gases

This may readily be investigated by means of the apparatus depicted in Fig. 9.10. The diagram is self-explanatory. A quantity of the gas under investigation is imprisoned in the closed tube and subjected to various pressures by raising and lowering the adjacent open tube. This pressure p is $p_A + \rho g(h' - h)$, where p_A is atmospheric pressure (ascertained by reading a barometer), ρ is the density of mercury, and h and h' are heights above some arbitrary "zero" level. If the closed tube is of uniform cross-sectional area A, the volume V of the imprisoned gas is Al, that is, is proportional to l, so that l may be taken as a measure of the volume. Thus by reading the values of l corresponding to a series of values of $h' - h$

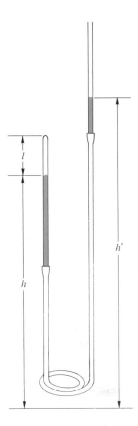

FIGURE 9.10 System used for experimental investigation of pressure-volume relation for gases.

we may obtain the corresponding volumes and pressures, and we may try plotting the results in various ways in an attempt to obtain a straight-line plot, which is easily interpreted algebraically. If, for example, we plotted pV against p or against V we should obtain a horizontal straight line. Or if we plotted V against $1/p$, or p against $1/V$, we should obtain a straight line passing through the origin. Interpreting either of these plots algebraically, we should conclude that the pressure–volume relation is

$$pV = \text{const.} \qquad (9.4)$$

The first to carry out such an investigation was Robert Boyle, in 1660, and Eq. (9.4) is known, after him, as **Boyle's law.**

It is to be understood that this investigation should be carried out *at constant temperature*, and that Eq. (9.4) applies only for constant-temperature conditions. To emphasize this, we may say that this equation represents the isothermal pressure–volume relation for a gas.

Strictly speaking, Boyle's law is only an approximate law, not an exact law. For gases at ordinary or low densities, however, Eq. (9.4) is a very close approximation indeed to the truth. Thus for air at ordinary temperatures and pressures the change in volume is found to be within 0.2 percent of what it should be for a given change in pressure according to Eq. (9.4). For all gases the approximation of this equation to the truth improves as the pressure, and with it the density of the gas, is decreased.

It would be difficult to imagine any simpler pressure–volume relation than is expressed by Eq. (9.4), which means that if we increase the pressure by any given factor we decrease the volume by the same factor. The extreme simplicity of this law is surely significant; it suggests that the gaseous form of matter must be of corresponding simplicity.

Bulk Modulus of Elasticity

Not only gases, but also liquids, and even solids, suffer a change in volume in response to a pressure change. A measure of the pressure increase needed to produce a given relative decrease of volume is the so-called **bulk modulus of elasticity** K of the material. This is defined by the equation

$$K = -\frac{dp}{dV/V} = -V\frac{dp}{dV},\tag{9.5}$$

the minus sign being required to make K positive, because dp/dV is essentially negative.

The law of variation of volume with pressure being known, the corresponding value of the bulk modulus is readily found. In this connection it is important to note that the pressure–volume relation depends on the thermal conditions under which the volume changes occur. There are two important special cases. One is that in which the volume change occurs **isothermally,** that is, at constant temperature. The corresponding elastic modulus is known as the **isothermal bulk modulus** of the material. It may be shown that to keep the temperature of a sample constant during an expansion, heat must be supplied to it from an outside source, while in a compression heat must be extracted from it. If this heat interchange with the surroundings is prevented, the volume change is said to occur **adiabatically,** and the corresponding modulus of elasticity is known as the **adiabatic bulk modulus.**

Although the isothermal and adiabatic bulk moduli of gases differ appreciably from one another, the difference in the case of liquids and solids is only slight.

ISOTHERMAL BULK MODULUS OF A GAS As we have just seen, the pressure–volume relation for a gas at constant temperature is

$$pV = \text{const.} \qquad (9.4)$$

The isothermal bulk modulus may be found by differentiating this with respect to V and making the appropriate substitution in Eq. (9.5). We have

$$V \frac{dp}{dV} + p = 0,$$

and so

$$K = -V \frac{dp}{dV} = p; \qquad (9.6)$$

that is, the bulk modulus of elasticity of a gas is simply the prevailing pressure. At standard atmospheric pressure, for example, the isothermal bulk modulus of any gas would, to the accuracy with which Boyle's law applies, be equal to 1.013×10^6 dyn/cm², or 1.013×10^5 N/m².

Another useful physical concept is that of **compressibility,** defined as the reciprocal of the bulk modulus of elasticity. For example, the isothermal compressibility of a gas at standard atmospheric pressure would be 0.987×10^{-6} dyn⁻¹-cm², or 0.987×10^{-5} N⁻¹-m².

The adiabatic bulk modulus of a gas will be considered in Chapter 17.

BULK MODULI OF LIQUIDS The bulk moduli of liquids are of a higher order of magnitude than those of gases, and correspondingly the compressibilities are of a lower order of magnitude. For example, water at 15°C has an isothermal bulk modulus of 2.04×10^{10} dyn/cm² and a compressibility of 4.89×10^{-11} dyn⁻¹-cm². The corresponding figures for mercury are 2.62×10^{11} dyn/cm² and 3.82×10^{-12} dyn⁻¹-cm².

The bulk moduli of solids will be considered in Chapter 10.

PART II
HYDRODYNAMICS

In hydrodynamical theory we shall be concerned with two factors additional to those that play a role in hydrostatics. These are inertial effects and a property of fluids known as viscosity. In preparation for the quantitative study of these it will be useful to note the main characteristics of fluid flow that have been observed.

Types of Flow

There are two types of flow of a fluid, orderly and disorderly. It is characteristic of the former that if an indicator such as a fine stream of colored fluid is injected at any point in such a manner as to disturb the flow as little as possible, the indicator, partaking of the general motion of the fluid, will move downstream in a definite line, showing that there is a unique, steady direction of flow at each subsequent point reached. Such a line is known as a **streamline,** and the orderly flow of a fluid, known as "streamline" flow, may be represented by a system of streamlines. In the disorderly, or **turbulent,** type of flow, on the other hand, there are no definite lines of flow—the fluid moves in irregular paths downstream, a characteristic feature of which is the formation of eddies. Broadly, fluid flow is of streamline type at relatively low rates of flow, becoming turbulent at high velocities. It is not proposed to pursue the discussion of turbulent flow in this book.

Unless the velocity is uniform throughout the body of a fluid having streamline flow, there must be a change of velocity with distance as we traverse it, either along the streamlines or transverse to them, or both together. In the first case there is said to be a **longitudinal velocity gradient,** defined quantitatively as the rate of change of velocity with distance along the streamlines. This condition is characterized by a convergence or divergence of the streamlines along the direction of flow. In the second case we have a **transverse velocity gradient,** correspondingly defined as the rate of change of velocity with distance in a direction at right angles to the streamlines.

Transverse Velocity Gradient

228 A transverse velocity gradient always exists when a fluid is in contact with, and flows parallel to, the surface of a solid. An example is the flow of air parallel to the ground, which we call a wind. Another is the flow of water in a pipe or in a stream.

The simplest case to consider is the one-dimensional one such as we should have if, for example, a steady wind blows from the north throughout a considerable region, and within this region the wind velocity is everywhere a unique function of height above the ground; apart from its variation with height, it does not vary with position in either the north–south or the east–west direction. Let us imagine an ideal case of this kind, where the ground is everywhere horizontal and where there are no protuberances such as trees or bushes to create local eddies; and let us suppose that the wind velocity v is measured at different heights d above the ground. Such measurements would show that within a considerable height range v is simply proportional to d, as is indicated in Fig. 9.11. In other words, the transverse velocity gradient, v/d, is everywhere the same.

Note that the proportionality of v to d means that where d is zero, so is v; there is no actual *sliding* of the air over the ground.

We have a corresponding state of affairs where the fluid is a liquid, as in the vicinity of the bed of a stream. It is not difficult to devise a suitable experiment in which one-dimensional conditions are simulated and in which velocities are measured by observing small suspended particles. Such an experiment would show that in the steady flow of liquids, as of gases, the velocity v is proportional to d, the distance from the solid surface.

As v varies continuously with d, it is misleading to refer to "layers" of fluid of finite thickness sliding over one another; there is no sliding anywhere. We may nevertheless think of an infinite number of layers each of infinitesimal thickness moving relative to one another in this manner. In this sense the flow of a fluid in which there is a transverse velocity gradient is sometimes referred to as "laminar flow."

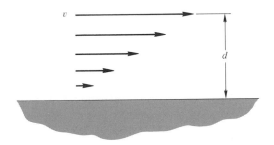

FIGURE 9.11   Laminar flow in a fluid.

Longitudinal Velocity Gradient

This occurs in regions where the streamlines either converge or diverge along the direction of flow. This they obviously must do in a pipe of variable section. Here, however, we cannot have *only* a longitudinal velocity gradient, unaccompanied by a transverse one. This complication can be avoided by considering, not the flow in the whole pipe, but that in an infinitely narrow axial "tube of flow." A tube of flow may be defined as an imaginary tube bounded by streamlines. A narrow tube of flow, although obviously not one of infinitesimal cross section, is indicated by the light curves of Fig. 9.12, the heavy lines representing the walls of the pipe. From considerations of symmetry it is evident that there can be no transverse velocity gradient on the axis of a circular pipe along which a fluid is flowing. Hence an infinitely narrow axial tube of flow is a system within which only one velocity gradient can occur, that is, a longitudinal gradient. An important hydrodynamical theorem applying to such a tube of flow, Bernoulli's theorem, will be considered later in this chapter.

Viscosity

If a solid body is drawn through a fluid it is found to be acted on by a force resisting its motion. This force is not primarily due to the bulk of the body; a resistive force is likewise exerted on a thin plate moved in a direction parallel to itself, or on a thin-walled hollow cylinder drawn through the fluid in the direction of its axis. The property of the fluid giving rise to this effect is known as its **viscosity.**

The resistive force per unit area exerted on either surface of the thin plate or on the internal or external surface of the thin-walled hollow cylinder is found to be independent of the nature of these surfaces, at least as long as the surfaces are reasonably smooth, so as not to create eddies at asperities. For a given fluid and not-

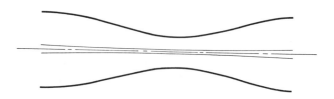

FIGURE 9.12   Narrow axial tube of flow in a pipe of variable section.

too-rough surfaces this force per unit area, F/A, is determined solely by the transverse velocity gradient in the region of the fluid immediately adjacent to the surface, known as the **boundary layer.** It is, indeed, found to be simply proportional to the transverse velocity gradient in this region, that is,

$$\frac{F}{A} = \eta \frac{v}{d}, \tag{9.7}$$

where the proportionality factor η is a property of the fluid. This quantity, defined as the ratio of F/A to v/d, is the viscosity of the fluid. Its dimensions are $ML^{-1}T^{-1}$. In honor of Poiseuille, a nineteenth-century French physician who was a pioneer investigator in this field, the CGS unit of viscosity is called the **poise** (P). The MKSA unit, the **decapoise,** is equal to 10 P.

TERMINAL VELOCITY An interesting phenomenon involving viscosity is the fall of a body through a fluid, for example a raindrop or a snowflake falling through air. As the velocity of such a body increases in response to gravity, it encounters a progressively increasing viscous resistance, or "drag," until finally this resistance attains a value equal to the downward force acting on the body. From now on the body is not subjected to any resultant force, so its velocity ceases to increase. The final velocity attained is known as the **terminal velocity.** Other things being equal, the smaller the object, the smaller is the terminal velocity. This is because the ratio of surface area to volume is greatest for small bodies. Because for a given body viscous drag is proportional to its surface area, while for given densities of the body and fluid the weight of the body minus the Archimedean upthrust exerted on it is proportional to its volume, equilibrium between the viscous drag (exerted upward) and the resultant downward force is reached at lower velocities for small bodies than for large ones. Thus the terminal velocity of a small creature such as an ant falling through air is quite small; consequently, an ant does not sustain any damage on striking the ground, no matter through what height it has fallen.

The British mathematical physicist Sir George Stokes (1819–1903) investigated this phenomenon quantitatively for the special case where the falling body is spherical, obtaining the formula

$$F = 6\pi\eta rv \tag{9.8}$$

for the viscous drag experienced by a body of radius r traveling with a velocity v through a medium of viscosity η. This velocity v will be the *terminal* velocity of a falling body if the sum of the viscous drag F and the Archimedean upthrust $\frac{4}{3}\pi r^3\delta$ just balances the weight of the body $\frac{4}{3}\pi r^3\rho$, ρ and δ being the density of the body and that

231

of the surrounding medium, respectively. Equating these two quantities, therefore, we obtain the expression for the terminal velocity,

$$v = \frac{2g}{9\eta} r^2 (\rho - \delta). \tag{9.9}$$

This equation, known as **Stokes's formula,** applies not only to falling bodies, but also in cases where δ is greater than ρ, for example where the body is a spherical bubble of gas rising through a liquid. Here the formula gives a negative value for v, the terminal velocity downward. This is as it should be, because, of course, in such cases the body travels upward through the fluid.

FLOW THROUGH A PIPE We have seen that a fluid cannot actually slide along a surface of a solid. For example, in the case of a pipe through which a fluid is flowing there can be no sliding along the containing walls; infinitely close to these walls the velocity must be zero. We have also seen that along the axis of such a pipe the transverse velocity *gradient* must be zero. Hence the velocity must presumably vary in some way from zero at the tube walls to a maximum along the axis, for example as indicated by the arrows in Fig. 9.13.

Let us now consider in particular the case of a long horizontal uniform pipe in which the volumetric rate of flow is constant with time, using Eq. (9.7) to investigate the flow quantitatively. Not only does this equation give the tangential force per unit area exerted on the surface of a solid body over which a fluid is moving; it also applies to any boundary we may wish to consider within the fluid, separating two portions of it, such that the boundary is everywhere parallel to the streamlines. In the course of an investigation of the flow of blood along arteries, veins, and capillaries, Poiseuille applied the equation to cylindrical surfaces coaxial with a circular pipe of internal radius r conveying a fluid of viscosity

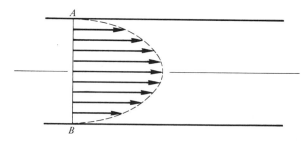

FIGURE 9.13 Variation of velocity of flow with distance from the axis of a pipe.

η, and, by a simple process of integration, obtained the equation

$$q = \frac{\pi r^4}{8\eta} \cdot \frac{p_1 - p_2}{l} \qquad (9.10)$$

for the volumetric rate of flow q in terms of the distance rate of fall of pressure in the direction of flow, $(p_1 - p_2)/l$. This is known as **Poiseuille's formula.**

An incidental item of information that comes out of the analysis is that if we now imagine Fig. 9.13 to apply to the case considered, the envelope of all the arrowheads, indicated in section by the dashed curve, is a paraboloid of revolution. Particles of the fluid occupying the plane AB perpendicular to the axis at a given time will at some later instant of time all be on this paraboloid, and the volume included between the plane AB and the paraboloid divided by the relevant time interval will be the volumetric rate of flow. This, divided by the internal cross-sectional area of the tube, is the mean velocity of flow. This comes out to just one half of the axial velocity.

It is important to realize that the envelope is paraboloidal only in the special case where the tube is long and uniform and the volumetric rate of flow is constant. It is no longer paraboloidal in regions where the pipe is constricted or where it widens out.

MEASUREMENT OF VISCOSITY The viscosity of a fluid may be determined in several ways. A convenient method that is available for most ordinary liquids and gases is to pass the fluid through a long horizontal tube provided with two side arms some distance apart for connection to a manometer, as indicated in Fig. 9.14, and

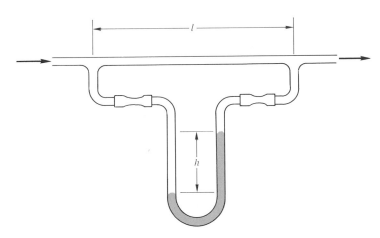

FIGURE 9.14 Measurement of pressure gradient associated with flow of fluid along a uniform horizontal pipe.

to measure both the pressure difference between the two sides of the manometer and the volumetric rate of flow. Thus if a U-tube manometer is used, the bottom of which contains mercury, of density ρ, and the remainder the fluid whose viscosity is being determined, of density ρ', then if h is the difference in height between the mercury surfaces in the two arms, the pressure difference $p_1 - p_2$ is evidently equal to $(\rho - \rho')gh$. Substituting this in Poiseuille's formula, we then have

$$\eta = \frac{\pi r^4}{8q} \cdot \frac{(\rho - \rho')gh}{l},$$

where the symbols have their usual significance. If l, r, and h are expressed in cm, g in cm/sec², ρ and ρ' in g/cm³, and q in cm³/sec, then η will be in poises. If, on the other hand, the quantities are expressed in MKSA units, the value of η will be obtained in decapoises.

Viscosities of Various Fluids   A few typical examples of values of viscosity for a range of materials are given in Table 9.4.

We see that in general the viscosity of a liquid decreases rapidly as the temperature rises.

The enormous range covered is also noteworthy. For example, at room temperature pitch has a viscosity about 10^{12} times that of water. What water will do in 1 second, therefore, pitch under similar conditions will take about 10^{12} seconds (30,000 years) to do. A

TABLE 9.4   Viscosities of Various Fluids

Fluid	Viscosity (P)
Dry air at 20°C	1.81×10^{-4}
Water	
at 0°C	0.0179
at 20°C	0.01006
at 100°C	0.00284
Glycerine	
at 0°C	46
at 20°C	8.5
at 30°C	3.5
Pitch	
at 0°C	5.1×10^{11}
at 15°C	1.3×10^{10}
Lime-soda glass	
at 575°C	1.1×10^{12}
at 710°C	4×10^{10}

good impression of the high viscosity of pitch may be formed by filling with this material a large funnel with a wide short stem and observing from time to time the progress of its exit. If this is done, it will be found that large drops form, and fall, qualitatively in much the same manner as water would do if falling from a capillary tube, but on an enormously longer time scale. The time interval between the formation of successive drops will depend on temperature and dimensions, but in most climates it will be of the order of years.

Glass is not usually thought of as a liquid, but a liquid it is—an exceedingly viscous one—even at ordinary temperatures. A true solid, such as a metal below its melting point, or crystalline quartz, diamond, or rock salt, has its constituent atoms arranged in a definite geometrical pattern, with which is associated a corresponding crystal structure. We know from x-ray diffraction experiments that this is not true of glass; this material has no crystal structure. Again, if a true solid is deformed to a moderate extent by a system of forces applied to it, it will return to its original configuration when the application of the forces is discontinued. If this is done with glass, however, it will not return fully to its original shape when the forces are removed. For example, if a long piece of glass tubing, selected for its straightness, is hung from its two ends, and reexamined after a year, or a few years, it will be found to have a distinct curvature. This means that it has begun to flow under the forces applied to it.

LUBRICATION   The technological processes in which viscosity plays a major role are many and varied. We shall confine ourselves here to noting only one of these—lubrication.

The function of a lubricant is to "wet" the engaging surfaces (see Chapter 12), flow between them, and keep them separated from one another. Because the surfaces do not come into contact, there is, strictly speaking, no friction; the tangential force (that is, force along the surface) is determined simply by the viscosity of the film of lubricant, its thickness, and the relative velocity of the surfaces. And the thickness of the film of lubricant is, in turn, determined by its viscosity and running conditions. In any given case there is an optimum viscosity for which the tangential force is a minimum.

THEORY OF VISCOSITY   Not only is viscosity important technologically; its existence also raises interesting questions of a theoretical nature. Just why does one portion of a fluid exert a tangential force on an adjacent portion across a surface of separation parallel to the streamlines? And why, for a given area of this surface, is the force proportional to the transverse velocity gradient? Again,

235

why, for liquids, does the viscosity decrease so rapidly with rising temperature?

The exertion of viscous forces must, basically, somehow be due to molecular interactions or exchanges across the surface to which the forces are tangential. Accordingly, any fundamental theory of viscosity must be based on a quantitative discussion of such interactions or exchanges. Although it would take us too far to develop a theory of viscosity for liquids, we may note here that a very satisfactory and fairly simple molecular theory of viscous forces has been worked out for gases, under the general heading of the kinetic theory of gases. This will be considered in Chapter 15.

Bernoulli's Theorem

We have seen that a transverse velocity gradient in a fluid necessarily gives rise to the exertion of a tangential force on an appropriately oriented surface. It may also be shown that with a longitudinal velocity gradient there must be associated a variation along the streamlines of normal force per unit area, that is, of pressure. This was first investigated quantitatively by the Swiss mathematician and scientist Daniel Bernoulli (1700–1782).

Let us consider this problem more specifically as it applies to the flow of a fluid of negligible compressibility (a liquid) along a pipe of variable cross section. As a sample of the fluid moves along the pipe, it exchanges both momentum and energy with other samples, and one might accordingly search for a way of basing a quantitative discussion on either of these exchanges. Actually, it is not at all clear how this might be done in the case of the momentum exchange. On the other hand, it is a simple matter to set down an equation based on energy considerations. Let us therefore do this, basing our discussion on the idealization that we are concerned here only with mechanical forms of energy, that is, with gravitational potential and kinetic energies.

The procedure will be to equate the net work performed on the sample selected to the energy gained by it. This can be done most simply if we take as our sample one that moves along a narrow axial tube of flow, such as is shown initially between the imaginary transverse boundaries P and Q and at a subsequent instant of time between P' and Q' in Fig. 9.15.

Under the steady-flow conditions we are here considering, the state of the liquid between P' and Q is the same at the initial and final times, so this common portion of liquid need not be taken into

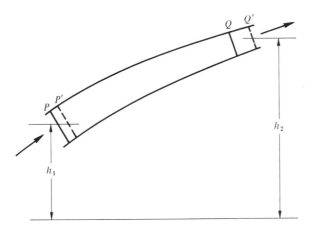

FIGURE 9.15 Section of narrow tube of flow discussed in derivation of Bernoulli's theorem.

account in any calculation of energy gain; this gain is simply the amount by which the energy of the liquid between Q and Q' exceeds that between P and P'.

Let the mean height above a standard reference level of the portion of liquid between P and P' be h_1 and its velocity v_1, and let the mean height of the liquid between Q and Q' be h_2 and its velocity v_2. Let the density of the liquid be ρ, and let the volume between P and P' (or between Q and Q') be V. Then the potential energy of the liquid between Q and Q' exceeds that between P and P' by $\rho V g(h_2 - h_1)$, and the corresponding gain in kinetic energy is $\frac{1}{2}\rho V (v_2{}^2 - v_1{}^2)$. The total energy gain is the sum of these two quantities,

$$\rho V[g(h_2 - h_1) + \tfrac{1}{2}(v_2{}^2 - v_1{}^2)].$$

We now have to find an expression for the net work done on the sample as it moves from between P and Q to between P' and Q'. As it moves, work is done *on* the sample by the liquid behind and *by* the sample on the liquid in front. The net work performed on the sample is therefore the former item minus the latter.

Let the pressure between P and P' be p_1, let the area of the boundary in this region be A_1, and let the distance the boundary moves in the time interval considered be d_1. Then the force exerted on the boundary is $p_1 A_1$, and the work done on the sample by the liquid behind is $p_1 A_1 d_1$, that is, $p_1 V$. Similarly, the work done by the sample on the liquid in front is $p_2 V$, where p_2 is the pressure in the region between Q and Q'. The net work done on the sample is, therefore, $(p_1 - p_2)V$.

237

We now, finally, equate this net work performed on the sample to the total energy gain. Doing this, and canceling out the common factor V, we have

$$p_1 - p_2 = \rho g(h_2 - h_1) + \tfrac{1}{2}\rho(v_2{}^2 - v_1{}^2),$$

or

$$p_1 + \rho g h_1 + \tfrac{1}{2}\rho v_1{}^2 = p_2 + \rho g h_2 + \tfrac{1}{2}\rho v_2{}^2;$$

that is,

$$p + \rho g h + \tfrac{1}{2}\rho v^2 = \text{const.,} \qquad (9.11)$$

where p, h, and v denote the pressure, height, and velocity, respectively, at any part of the tube of flow. This is Bernoulli's theorem.

We may note, incidentally, that p is the pressure not only within the narrow axial tube of flow considered but also at all points throughout the corresponding transverse section of the pipe, if variations in height over this section can be neglected. If this were not so, there would be a radial component of flow of the liquid, actually a radial acceleration, in response to the radial pressure gradient.

It must be remembered that in deriving Bernoulli's equation we have made the idealization that there is no conversion of mechanical energy to any other form, for example, to heat. This is equivalent to neglecting viscous drag on the boundaries of the tube of flow considered. Actually this is not strictly permissible, and accordingly Eq. (9.11) is an approximation only, although, for liquids of low viscosity, a good one. An amended discussion in which proper account is taken of the effects of viscosity would show that the magnitude of the expression on the left side of Eq. (9.11) must fall with increasing distance along the direction of flow.

The following special cases are of particular interest. In considering these we shall make the simplifying assumption that the viscosity of the liquid is sufficiently small for Eq. (9.11) to be taken as valid:

(1) *Velocity constant (constant cross section of pipe):* Here Bernoulli's equation becomes simplified to

$$p + \rho g h = \text{const.,}$$

as for a liquid at rest.

(2) *Pipe horizontal (h constant):* In this case we have the simplification of Bernoulli's equation,

$$p + \tfrac{1}{2}\rho v^2 = \text{const.}$$

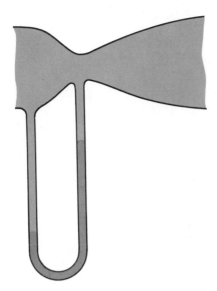

FIGURE 9.16 Venturi meter.

VENTURI METER This is an important application of the case of a horizontal pipe of variable cross section. Because the velocity is greatest where the cross section is least, the pressure must decrease at constrictions and increase wherever the tube widens out. Let us consider a tube of variable section such as is depicted in Fig. 9.16, through which a liquid, for example water, is flowing. If some form of differential manometer is connected between a relatively wide part of this tube and a constriction, this will register a lower pressure at the constriction than in the wider region, the difference in pressure varying with the volumetric rate of flow. This is the principle of the **Venturi meter,** used for measuring volumetric flow rates.

VELOCITY OF EXIT FROM CONTAINER An interesting application of Bernoulli's principle is the calculation of the velocity of exit of a liquid through an orifice in the containing vessel in terms of the depth of the orifice below the free surface. In Fig. 9.17 let us consider the regions 1 and 2 inside and outside the orifice, respectively. At 1 the liquid will be practically stationary, and the pressure p_1 will be equal to $p_A + \rho g h$, where p_A is atmospheric pressure, ρ the density of the liquid, and h the depth below the free surface. At 2 the pressure is simply p_A. Applying Bernoulli's equation, we have

$$p_A + \rho g h = p_A + \tfrac{1}{2}\rho v^2,$$

or

$$v^2 = 2gh,$$

239

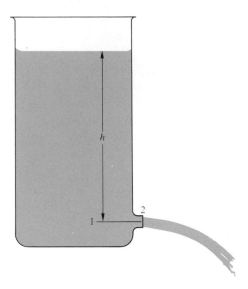

FIGURE 9.17 Flow of a fluid from orifice in a container.

where v denotes the velocity of exit along the axis. We see that this velocity is the same as would be acquired by a body falling freely from rest through the height h. It is interesting to note that this result might also have been obtained directly from a consideration of the law of conservation of energy.

As we have noted, the velocity cannot actually be uniform throughout a section perpendicular to the direction of flow. Consequently, the *mean* velocity of exit from the orifice shown in Fig. 9.17 must be less than $\sqrt{2gh}$, although if the viscosity is sufficiently small and the exit pipe is short, it will not be much less. If, in this case, the exit pipe is bent vertically upward, the jet issuing from it will attain nearly the height of the free surface of the liquid in the tank.

Problems

9.1 Calculate the weight of helium in a balloon of volume 100 m³, given that under the prevailing conditions of temperature and pressure the mean density of this gas within the balloon is 1.66×10^{-4} g/cm³.

9.2 A rectangular opening in a reservoir wall is closed by a vertical door 1 m high and 1.5 m wide, and the top edge of the door is 6 m below the surface of the water. Find (a) the resultant force exerted by the water on the door, and (b) the line of action of this force.

9.3   The wall of a reservoir is vertical inside and sloping outside, with a plane outside surface. It is 10 m high, 2 m thick at the top, and 8 m thick at the base. The material of which it is constructed has a density of 2.3 g/cm³. If the water is about to overflow, what are (a) the resultant force per meter length of the wall which the water exerts on it, (b) the overturning moment of this force, and (c) the moment of the weight of the wall, both moments being taken about the outer edge of the base?

9.4   A thin-walled glass sphere of diameter 15 cm is evacuated of air. Find the compressive force to which the glass is subjected per centimeter length on any great circle.

9.5   A bottle without a stopper is placed on one pan of a sensitive beam balance, and standard masses are added to the other pan so that the system is just balanced. A small insect of mass m and mean density ρ flies into the bottle and hovers but does not land. What change, if any, will have to be made to restore the balance?

9.6   Show that in a body of fluid which has an upward acceleration a, which may be either positive or negative, the rate of increase of pressure with depth, dp/dh, is $\rho(g + a)$.

9.7   A helium-filled balloon is attached by a string to the floor of a truck at rest on a horizontal road. The truck is sealed so that no drafts can enter from the outside. Describe the behavior of the balloon as the truck begins to accelerate forward along the road.

9.8   Discuss the problem of trying to light a cigarette with a match in a space satellite orbiting the earth.

9.9   It has been proposed to encapsulate a man (suitably equipped for breathing) in a water-filled cubicle as a device for permitting the man to survive much higher than normal accelerations during lift-off or reentry of space vehicles. Discuss.

9.10   A truck carrying a water tank along a straight horizontal road moves forward with uniform acceleration a and continues doing so long enough for the free surface of the water to settle down to a steady angle of inclination to the horizontal. What is this angle?

9.11   A body of liquid of density ρ, contained in a cylindrical vessel and having a free surface, rotates with constant angular velocity ω about a vertical axis (see the figure). Considering a small cylindrical element of the liquid having a radial axis between r and $r + \delta r$, show that the condition that the surrounding liquid exerts on it the requisite centripetal force is that

$$\delta p = \rho \omega^2 r \, \delta r,$$

where δp is the excess of pressure on the outer end of the cylindrical element over that on the inner end, and so

$$\frac{dp}{dr} = \rho \omega^2 r.$$

Show that to satisfy this condition the free surface of the liquid must be of paraboloidal form conforming to the equation, referred to the intersection

241

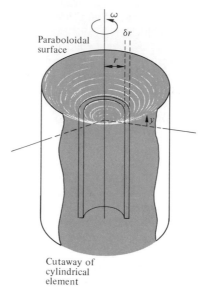

Figure for problem 9.11.

of the free surface with the axis of rotation as origin,

$$y = \frac{\omega^2 r^2}{2g}$$

for an axial cross section.

9.12   A tank of water is placed on one scale pan of a beam balance, and standard masses are added to the other to bring the scale into balance. A brass cylinder held by a string is lowered into the water without touching the sides or bottom of the tank. Is the balance disturbed? Explain.

9.13   A rectangular pontoon floating in water is required to carry a load of 6000 kg, and the depression due to this load must not exceed 15 cm. Find the horizontal area of the pontoon.

9.14   A block of wood floats in water with one half of its volume submerged, and in oil the proportion of its volume submerged is 0.8. Find the densities of the wood and the oil.

9.15   A rectangular block of wood of density 0.7 g/cm³, 100 cm² in area, and 10 cm deep floats in water. What mass placed on its upper surface will submerge it to a depth of 9 cm?

9.16   Given that the densities of ice and of seawater are 0.917 and 1.028 g/cm³, respectively, calculate what fraction of an iceberg is seen above the surface of the ocean.

9.17   A 400-kg wooden log is floating with 0.20 of its volume above water level. Given that the density of iron is 7.8 g/cm³, find the least mass of an iron object attached to the underside of the log that will sink it.

9.18   One end of a narrow glass tube is closed and the other end joined to a wider tube. By manipulating the amount of air trapped in this

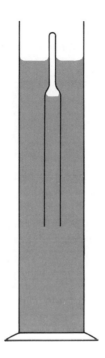

Figure for problem 9.18.

system it may be made to float in water with a portion of the narrow section protruding above the surface (see the figure). Investigate the conditions under which the equilibrium of the system when so floating will be stable; that is, under which the system can oscillate up and down with small amplitude.

Show that if the composite tube is pushed down far enough it will not float up again when released, but will sink further.

9.19 A thin rod is attached to a point on the wall of a vessel containing water in such a way that it is free to rotate about a horizontal axis through the point of attachment (see the figure). Find the specific gravity of the material of the rod, given that, when the rod is in equilibrium, a fraction $1/n$ of its length is above the surface of the water.

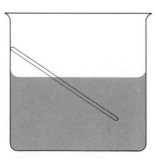

Figure for problem 9.19.

9.20 The densities of hydrogen, helium, and air are 0.084, 0.166, and 1.203 kg/m³, respectively. What volume is displaced by a hydrogen-filled balloon which has a total lift of 1000 kg? What would be the lift if helium were used instead of hydrogen?

9.21 Explain why a balloon inflated with hydrogen or helium has a definite "ceiling," or height to which it will rise, while a submarine, if it begins to sink and makes no adjustments, will go to the bottom of the ocean.

9.22 A submerged submarine of mass 10⁶ kg (including ballast) passes from the North Sea into the Baltic. Given that the densities of sea-water in the North Sea and in the Baltic are 1028 and 1007 kg/m³, respectively, calculate what mass of water should be discharged from the ballast tanks.

9.23 An aluminum cylinder has an apparent weight in air of 47.324 g. When immersed in a liquid it has an apparent weight of 32.191 g. What is the density of the liquid?

$$\text{Density of standard masses} = 8.75 \text{ g/cm}^3,$$
$$\text{Density of aluminum} = 2.70 \text{ g/cm}^3,$$
$$\text{Density of air} = 1.29 \times 10^{-3} \text{ g/cm}^3.$$

Discuss the accuracy required in each of the above values. (Use mathematics of small quantities.)

9.24 Lead shot is poured into a glass tube with a plane closed end perpendicular to its axis until it almost sinks in water. The cross-sectional area of the vessel containing the water is much larger than the external cross-sectional area of the tube. How much work must be done to lift the tube and its contents just clear of the water surface?

9.25 A narrow cylindrical object with its center of mass close to one end can float in a liquid with its axis vertical and half its length, l, submerged. This object is held with its axis vertical and its upper surface at a height h above the surface of a quantity of this liquid contained in a large tank, and then released. Under what circumstances, if any, is its subsequent motion simple harmonic? (All necessary simplifying assumptions may be made, including neglect of viscosity.) If simple harmonic motion can occur, derive expressions for its frequency and amplitude.

9.26 A solid metal cylinder 20 cm long floats upright in mercury, of density 13.6 g/cm³. When the cylinder is displaced downward and allowed to oscillate vertically, it executes simple harmonic motion with a period of 0.42 sec. What is the density of the metal?

9.27 A beaker contains water with a block of wood floating in it, and with the beaker at rest the wood floats half submerged. State, with explanation, what fraction of the wood will be submerged if the beaker and contents (a) are carried upward in an elevator with a uniform speed of 10 m/sec, (b) are carried upward with an acceleration of 5 m/sec², (c) are carried downward with an acceleration of 5 m/sec², (d) fall freely under gravity, (e) are contained within a space vehicle orbiting the earth at a distance from its center of 7000 km.

9.28 A faulty mercury barometer containing air above the mercury registers 72 cm when the pressure of the atmosphere is 76 cm Hg. If the top of the tube is 100 cm above the mercury in the bowl, what is the true value of the atmospheric pressure the barometer registers as 68 cm, the temperature being the same?

9.29　A vessel A of capacity 4 liters contains a perfect gas at atmospheric pressure. It can be connected via a valve and tube of negligible volume to a vessel B of capacity $\frac{4}{9}$ liter. The valve is initially closed and B is evacuated. The valve is then opened long enough for the pressures in A and B to become equal, and then after it is closed B is again evacuated. How many times must this process be repeated in order to reduce the pressure in A to 10^{-4} atm? (It may be assumed that the temperature of the gas remains constant.)

9.30　How many times as long will it take for a given quantity of glycerine, of density 1.3 g/cm³, to drain out of a container through a tube at 20°C as for the same quantity of water, using the same arrangement?

9.31　Find the terminal velocities of (a) cloud particles (water droplets) of 0.05-mm radius, (b) raindrops of 0.2-mm radius, (c) raindrops of 1-mm radius, and (d) spherical hailstones of 6-mm radius, the viscosity of the air being 1.8×10^{-4} P. (Density of ice = 0.917 g/cm³.)

9.32　With what terminal velocity will an air bubble of diameter 1 mm rise in water at 20°C?

9.33　Emery powder, comprising particles of various sizes, is stirred up in a vessel filled to a height of 10 cm with water. Assuming the particles to be spherical, calculate the sizes of the largest particles that will remain in suspension after (a) 1 hr, and (b) 24 hr. (Density of emery = 4.0 g/cm³; viscosity of water = 0.010 P.)

9.34　Apply the condition for zero rate of change of total momentum of a cylindrical portion of fluid flowing in a long, uniform tube—that the forward force due to the difference between the pressures at the two ends is just balanced by the viscous drag on the sides. Hence, by integration, and using the boundary condition of zero velocity at the internal surface of the tube, obtain an expression for the velocity v as a function of the distance from the axis. Finally, by means of an integration based on this, derive Poiseuille's formula. Show that in this case the curved surface indicated by the dashed line in Fig. 9.13 is a paraboloid, and that the mean velocity of flow within the tube is one half that along the axis.

9.35　Water is flowing with a mean velocity of 2 m/sec along a horizontal pipe of internal diameter 1 cm, the outlet of which is open to the air. Connected to the tube 5 m from the outlet is a long vertical tube open to the air at the top. To what height will water rise up this tube? (Viscosity of water at the prevailing temperature = 0.011 P.)

9.36　A disk 20 cm in diameter is mounted midway between two parallel fixed plates and rotates at 300 rpm when the whole system is immersed in oil. If the distance between the disk and each plate is 2 mm and a power of 100 W is required to maintain the rotation, what is the viscosity of the oil?

9.37　A Venturi meter in a water-supply system has diameters at the junctions of the manometer tubes of 30 cm and 10 cm, and the pressure difference observed is 10 cm Hg (density = 13.6 g/cm³). Neglecting viscosity, calculate the volumetric rate of flow of the water.

9.38 Water emerges into the atmosphere from the end of a pipe of internal diameter 6 mm at the rate of 100 cm³/sec. Just behind the exit end of the pipe is a constriction. What minimum diameter may the constriction have if the pressure at the constriction is not to fall to zero? (Under conditions where the pressure in a pipe approaches zero, vaporization and the release of dissolved gas occurs, producing bubbling. This is known as "cavitation.")

9.39 Kerosene, of density 0.83 g/cm³, is contained in a vessel under a total internal pressure of 3 atm. Neglecting viscosity, calculate the velocity with which the kerosene will escape from a circular hole in the side of the vessel.

9.40 Two meters below the surface of water in a tank, water squirts out through a circular hole at the rate of 1 cm³/sec. Assuming the resistive effect of viscosity to be negligible, calculate the radius of the hole.

9.41 A rainwater tank is mounted on stumps so that its base is at a height H above the horizontal ground. The tank is filled to a depth h. A hole is punched in the side of the tank a distance x below the surface. Find the value of x such that the range of the emerging stream, assumed not to be retarded significantly by viscous effects, shall be a maximum. Discuss the result in the particular case where $H = 3$ m and $h = 2$ m.

9.42 A circular hole of area a is punched in the bottom of a tank whose cross-sectional area is A. Assuming the resistive effect of viscosity to be negligible, show that the time required for the water level to fall from h_1 to h_2 is given by the equation

$$ t = \frac{\sqrt{2}\,A}{a\,\sqrt{g}}\,(h_1{}^{1/2} - h_2{}^{1/2}). $$

9.43 Water at 15°C escapes from a container through a horizontal tube 10 m long and of internal diameter 1 cm, joined to the container 2 m below the free surface of the water. Given that at 15°C the viscosity of water is 0.0114 P, and neglecting the Bernoulli drop of pressure at the entrance of the tube, calculate the rate at which the water escapes. Investigate whether the neglect of the Bernoulli effect is justified.

9.44 Water at 15°C (viscosity as in Problem 9.43) is siphoned from a tank through a 4-m length of hose of 1.3 cm internal diameter, the exit end of the hose being 1 m below the surface of the water in the tank. Calculate the rate of flow of the water (a) neglecting viscosity, and (b) neglecting the Bernoulli effect. Is the rate of flow controlled predominantly by viscosity or by the Bernoulli effect in this case? Will the actual rate of flow be less than your answer (a), between (a) and (b), or greater than (b)?

9.45 It was desired to pump water from A to B through the pipeline shown. The pump was chosen to provide sufficient pressure to overcome the static head h plus extra to overcome viscosity and provide kinetic energy. The pipe was initially empty. Before water emerged from the upper end the pump motor burned out in consequence of the necessary pressure being much higher than had been expected. Can you explain? The explana-

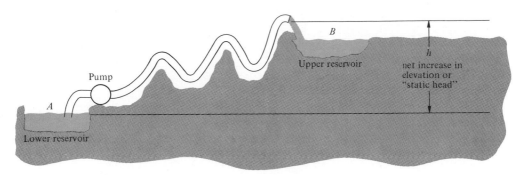

Figure for problem 9.45.

tion involves the fact that the water did not completely fill the pipe as it spilled over the top of each bend so that in the top of each bend some air was trapped.

10

DEFORMATIONS OF SOLIDS

In our discussion of the mechanics of "rigid" bodies in Chapter 7 we dealt with situations in which the deformations produced in solids by forces applied to them played no significant role. However, truly rigid bodies do not exist in nature, and the changes of form or volume that actually occur are not only of considerable theoretical interest but are often also of great practical importance.

We shall divide the subject matter of this chapter into three parts. In Part I we shall consider **elastic deformations.** These, like the volume changes suffered by fluids, which we discussed in Chapter 9, are reversible; that is, on reestablishment of the original pressure or on removal of the deforming force system the status quo is completely restored; the body returns to its original volume and/or form. But not all deformations in solids are reversible; under the action of sufficiently strong deforming forces a solid may suffer a deformation some of which persists after removal of the force system. In such a case the residual deformation is classified as **plastic,** and it is to a consideration of plastic deformations that Part II of this chapter will be devoted. In Part III we shall discuss the subject of **friction.** As we shall see, the inclusion of this topic in a chapter on deformations in solids is appropriate in view of the important role which these deformations, particularly plastic deformations, play in the processes of friction.

PART I
ELASTIC DEFORMATIONS

Three Elastic Moduli

There are three moduli of elasticity figuring in the theory of elastic deformations: the bulk modulus, Young's modulus, and the shear modulus or rigidity. We shall consider these in turn.

BULK MODULUS OF ELASTICITY This has already been defined in relation to fluids in Chapter 9. For solids, too, the bulk modulus K is defined in terms of the volume V of the specimen and

TABLE 10.1 Bulk Moduli and Compressibilities

Material	Bulk modulus (dyn/cm²)	Compressibility (dyn⁻¹-cm²)
Lead	5.0×10^{11}	2.0×10^{-12}
Aluminum	7.46×10^{11}	1.34×10^{-12}
Copper	1.31×10^{12}	7.6×10^{-13}
Steel	1.81×10^{12}	5.5×10^{-13}
Graphite	3.3×10^{11}	3.0×10^{-12}
Diamond	2×10^{12}	5×10^{-13}

the rate at which this decreases with increase of hydrostatic pressure p to which the body is subjected, by the equation

$$K = -V \frac{dp}{dV}.$$

Isotropic solids suffer a change in volume, but do not alter their geometrical form, when the hydrostatic pressure exerted on their external surfaces is changed.

Most solids have bulk moduli which are significantly greater than those of most liquids, and the compressibilities are correspondingly smaller. The bulk moduli and compressibilities of a selection of solids are given in Table 10.1.

YOUNG'S MODULUS If a solid rod is subjected to tension or compression in the direction of its axis, it becomes elongated or shortened, and if the tensile or compressive force per unit cross-sectional area does not exceed a certain limit, known as the **elastic limit** of the material, then on its being relaxed the rod returns to its original length. Let tensile forces F less than those corresponding to the elastic limit be applied along the axis of a rod of length L and cross-sectional area A, as indicated in Fig. 10.1, and in consequence let the length be increased by the amount l. Then the tensile force per unit area, F/A, is known as the tensile **stress** applied to the specimen, and the proportional deformation, l/L, is called the **strain.** And the ratio of stress to strain is defined as **Young's modulus E:**

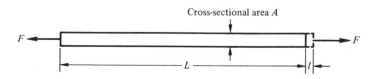

FIGURE 10.1 Stretching of a rod under tension.

$$E = \frac{F/A}{l/L} = \frac{FL}{Al}.$$

This equation is to be taken algebraically; if the "tensile" force F is negative, l is also negative. The same value of Young's modulus is found for both tensile and compressive stresses.

Perhaps the material whose elastic properties are most familiar to us is india rubber. If for this material we plot stress against strain, we obtain a curve of the form shown in Fig. 10.2. We see that here the ratio of stress to strain is a variable quantity, its value depending on the magnitude of the stress applied. However, for this material, as for all others, the tangent to the stress–strain curve at the origin has a finite slope, and because a sufficiently short section of any curve may be regarded as a straight line, we may take the stress as proportional to strain within the corresponding limits of these quantities. Young's modulus is then given by the slope of the tangent at the origin.

With some materials it is impossible to apply sufficient stress for a nonlinear stress–strain relationship to be obtained; the elastic limit occurs at a value of the stress for which the strain is still quite small, and *up to this limit* no curvature of the stress–strain plot is detectable; stress is proportional to strain. The statement of this proportionality is known as **Hooke's law,** after Robert Hooke (1635–1703), who was the first to observe it. There are, however, good reasons for believing that if the elastic limit did not occur so early, the plot obtained would definitely be curved, as it is with india rubber, although not necessarily in the same way.

RIGIDITY A tensile stress applied to a body not only increases its length but, at the same time, decreases its thickness; thus it alters the shape of the body. The increase in length is pro-

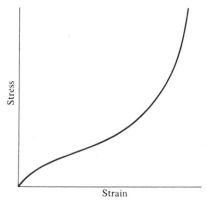

FIGURE 10.2 Relation between stress and strain for a specimen of india rubber.

portionately greater than is the decrease in cross-sectional area, so there is a resultant increase in volume. The volume increase is exactly one third of that which would be produced by the application of a negative pressure of magnitude equal to that of the tensile stress, that is, of three such tensile stresses mutually at right angles. It appears, then, that Young's modulus is a hybrid kind of elastic modulus; we should expect it to be related in some way to the bulk modulus, on the one hand, and, on the other, to another modulus of elasticity now to be considered, the so-called **rigidity,** which is concerned with changes of shape only, unaccompanied by volume changes.

Let us consider a block of material such as is depicted in Fig. 10.3, to which equal tensile and compressive stresses are applied at right angles. We should hardly expect these simultaneously applied stresses to produce any change in dimensions in the direction at right angles to them, and in the absence of such change there should be no change in volume; for the proportional increase in length in the direction of the tension must be equal to the proportional decrease in that of the compression. Presumably, then, this stress system would produce a change in shape only. Let us now examine both the stress system and the change in shape more closely.

With the system of normal stresses indicated in Fig. 10.3 there are necessarily associated **tangential stresses** exerted on surfaces orientated at 45° to these. That this is so may be seen by considering the equilibrium of a triangular prismatic section of the block in the region of the top left corner, as shown in Fig. 10.4. Let the normal stress (force per unit area) on each of the faces *AB* and

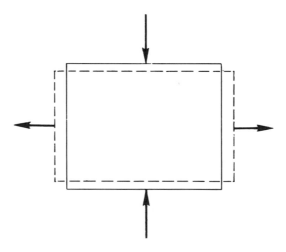

FIGURE 10.3 Effect of application of equal tensile and compressive stresses in directions at right angles.

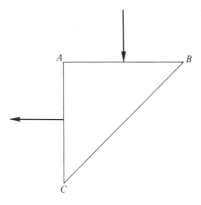

FIGURE 10.4

AC be P, and let the area of each of these faces be A. Then there must be a force PA acting downward on the face AB and an equal force PA acting from right to left on the face AC. In order that the section under consideration shall be in equilibrium there must therefore be a tangential force of magnitude $\sqrt{2}PA$ acting on the hypotenusal face directed from C toward B. But the area of this face is $\sqrt{2}A$. Hence the tangential force per unit area acting on this face, the "tangential stress," is P; that is, it is of the same magnitude as the normal stresses acting on the faces AB and AC.

By similarly considering the equilibrium conditions for other prismatic sections, it may readily be shown that a portion of the block in the form of a rectangular parallelepiped bounded by faces orientated at 45° to those of the block is subjected to a system of tangential stresses each of magnitude P, and to corresponding tangential forces proportional to the areas of the faces of the parallelepiped, as indicated in Fig. 10.5. These tangential stresses are known as **shearing stresses.**

Similar reasoning may be used to show that if a block is subjected to a system of shearing stresses as in Fig. 10.6, then a portion of it in the form of a rectangular parallelepiped with faces at 45° to those of the block must have normal stresses of the same magnitude acting on these faces, the corresponding forces being proportional to the areas, as indicated.

Let us now consider the nature of the deformation produced by a system of shearing stresses. Let the block represented in Fig. 10.7 be deformed by shearing stresses as shown by the arrows. The deformation will obviously be of the kind indicated, giving an elongation of dimensions in the direction at 45° to the faces corresponding to the tensile stress, and a contraction in the direction at

253

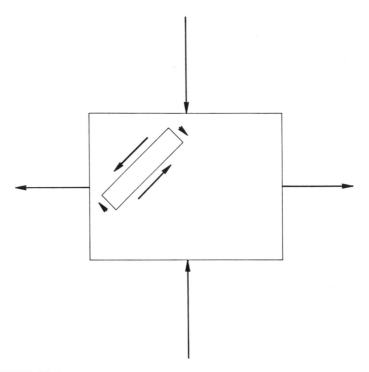

FIGURE 10.5

45° corresponding to the compressive stress. With this kind of
deformation it is found convenient to define the proportional
deformation, or **strain,** as the quotient of l, the distance moved by
one face relative to the opposite face, and L, the distance between
these two faces. For $l \ll L$, this is obviously the same as the angle θ
indicated in the figure.

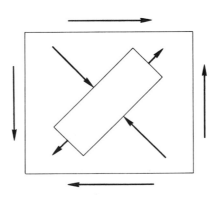

FIGURE 10.6

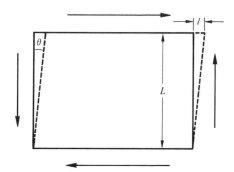

FIGURE 10.7   Effect of application of shearing stress system.

The **shear modulus,** or **rigidity,** of the material is now defined as the quotient of stress and strain. Thus if F is the tangential force exerted on a side of area A, and the resultant strain is θ, the rigidity G is defined by the equation

$$G = \frac{F/A}{l/L} = \frac{F/A}{\theta}.$$

Hooke's law and the existence of an elastic limit apply to shearing as well as to tensile and compressive stresses and strains.

Relation among the Elastic Moduli

It may be shown that the three elastic moduli are related by the equation

$$\frac{1}{E} = \frac{1}{3G} + \frac{1}{9K}.$$

In all cases, the strain is the quotient of two lengths and is therefore dimensionless. Accordingly, the dimensions of the elastic moduli are the same as those of stress, which, in all cases, is defined

TABLE 10.2   Elastic Moduli (in units of 10^{11} dyn/cm²)

Material	E	G	K
Copper	12.3	4.55	13.1
Brass	9.7–10.2	3.3–3.5	10.65 (approx.)
Steel	20–21	7.9–8.9	18.1 (approx.)
Silica	5.18	3.2	1.4

as a force per unit area. Thus the dimensions of the elastic moduli are $ML^{-1}T^{-2}$.

In Table 10.2 the values of the three elastic moduli are given for a selection of materials in CGS units. The corresponding values in MKSA units are smaller by a factor of 10.

PART II
PLASTIC DEFORMATIONS

When a solid body is subjected to a stress exceeding its elastic limit, it will not return to its original configuration when the stress is released. Beyond the elastic limit some materials fracture. Others, notably metals, suffer considerable further, nonelastic, deformation while fully retaining their structural soundness. This nonelastic further change of form is known as **plastic deformation,** or **plastic flow.** On removal of the stress from a body that has passed its elastic limit there is only a partial restoration of its original form, and the residual deformation is known as a **permanent set.**

Plastic Flow and Work Hardening

The features of the behavior of a metal under stress just referred to are represented graphically in Fig. 10.8. Between O and

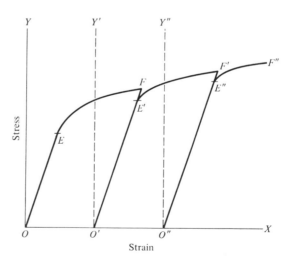

FIGURE 10.8   Typical variation of stress with strain for a metal stressed beyond the elastic limit.

E the stress–strain diagram is linear ("Hookean") and reversible. E represents the elastic limit of the material. If the stress is increased beyond this limit, the strain increases more than in proportion to it, as represented by the curve EF. At this stage the material is flowing plastically.* If at some point along this curve, say at F, the stress is now progressively released, the strain decreases with it according to a new linear relationship, as represented by FO', until finally, at O', the stress is reduced to zero, and the residual strain, OO', represents the permanent set. If now the stress is once more built up, the stress–strain diagram is retraced from O' to some point E' on $O'F$, but beyond E' it once again becomes curved, this curvature corresponding to a further plastic flow. If this is arrested, say at F', and the stress is progressively decreased to zero and then increased again, we have a repetition of the former type of behavior, represented by the line $F'O''$ for diminishing stress and $O''E''$ merging at E'' into the curve $E''F''$ for subsequent stress increase—and so on, although, at ordinary temperatures, not without limit. A stage is eventually reached where no further plastic flow can occur; at this stage structural breakdown takes place, the material fracturing at a point of weakness.

The wire in its condition corresponding to the point O' may be regarded as a new material, and elongations suffered by it may be considered as a proportional increase of its new length. The line $O'E'F'$ then becomes a new stress–strain diagram referred to the axes $O'Y'$ and $O'X$, the part $O'E'$ representing the Hookean part of the characteristic. Similarly, $O''E''F''$ represents the elastic and plastic behavior of the wire after it has undergone further plastic deformation, relative to the axes $O''Y''$ and $O''X$. It is found that all the Hookean portions of the characteristics are parallel to one another, but that the elastic limits, at E', E'', . . . , occur at progressively higher values of stress than the original limit corresponding to the point E. In this respect, the material, far from suffering any incipient mechanical breakdown, has actually been improved mechanically; its elastic range has been increased. In the sense that plastic flow now takes place less easily, the material is "harder" than before. The hardening, occurring as it does in consequence of plastic deformation, or "working," of the material, is known as "work hardening."

It should be noted that the condition of the wire corresponding to the points O', O'', . . . is not different *in kind* from that at O,

---

* If it were not for the onset of plastic flow, the stress–strain curve would continue to be linear far beyond E, the correspondingly extended part of the curve still being short enough to be indistinguishable from a straight line.

for the wire as received from the manufacturer is already in a work-hardened condition; it has been very considerably deformed plastically in the process of wire drawing.

The fact that the successive elastic limits E', E'', . . . occur at lower stresses than those corresponding to the points F, F', . . . is believed to be due to a certain degree of recovery in the condition of the material that takes place while the stress is being released and reimposed, a recovery of the same kind as occurs in annealing. This will be considered later.

Structural Changes

All deformations, elastic as well as plastic, obviously represent a change in the relative positions of the constituent atoms, or groups of atoms, of the material. Let us, therefore, consider the general nature of the arrangement of atoms in a solid and speculate on the changes that might occur in this arrangement when the solid is deformed.

Some solids are obviously crystalline, for example, quartz, diamond, rock salt, and galena. Others, notably metals, do not at first sight appear to be crystalline. Nevertheless, by appropriate heat treatments it is possible to grow crystals of metals exhibiting characteristic crystal forms of the same kinds as are familiar to us in minerals. Moreover, a crystal of any particular metal has physical properties—density, melting point, electrical conductivity, and so on—which are identical with those of the same metal in a form that is not obviously crystalline. This suggests that fundamentally the two are the same.

The regular geometrical form of a crystal would be incomprehensible except as a manifestation of a corresponding regularity and orderliness in the arrangement of the constituent atoms. In 1912 von Laue, assuming x rays to be propagated as waves, conceived the idea of sending a narrow beam of x rays through a crystal and seeing whether this would be broken up into a regular pattern of diffracted rays* corresponding to the presumed regular arrangement of the atoms. Soon, two of his research students, Friedrich and Knipping, succeeded in demonstrating such diffraction, obtaining a regular pattern of spots on a photographic plate located beyond the crystal. Analysis of such spots, now known as "Laue spots," has yielded precise information concerning the arrangement of the atoms. We now know that in each crystal the atoms are arranged in

---

* The theory of diffraction will be dealt with in Chapter 24.

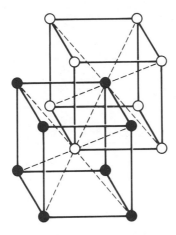

FIGURE 10.9 Arrangement of Cs+ and Cl− ions in cesium chloride.

a regular three-dimensional pattern characteristic of the material. Such an orderly array of the atoms is known as a "space lattice." Figure 10.9 shows the lattice arrangement in a crystal of cesium chlorides, the black and white circles representing the centers of cesium and chlorine ions. Such a lattice, in which there are three mutually perpendicular directions along which the general arrangement and the spacings of consecutive atoms are identical, is known as a "cubic lattice," and units of material that have this kind of regularity of atomic structure throughout are called "cubic crystals."

Although a sample of a material that has grown into a regular geometrical form always has a regular internal structure, the converse is not true; it is quite common for a mass of material whose external appearance in no way suggests a crystalline structure nevertheless to consist of units in each of which the atoms are arranged according to a regular pattern. This applies in particular to metals, which are, in general, not easily grown in the form of crystals exhibiting a crystalline external appearance. In what follows, the term "single crystal" is to be understood as referring to uninterrupted regularity of atomic arrangement throughout, not necessarily to any corresponding regularity of external form.

In describing the lattice type of a chemical compound such as an alkali halide we concern ourselves only with atoms of a single kind. The lattice of Fig. 10.9 is known as a *simple* cubic lattice. There are two other forms of cubic lattice. In one, a "unit cell," with an atom of a particular kind at each corner, has an extra atom of the same kind at the center. This is called a "body-centered cubic lattice." In the other, known as a "face-centered cubic lattice," there is an extra atom at the center of each of the six faces of the unit cell.

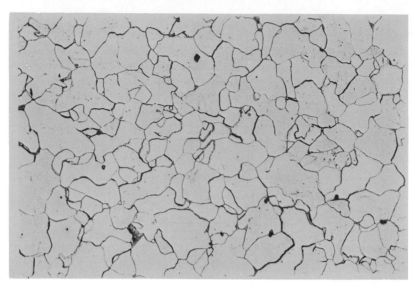

FIGURE 10.10   Etched surface of a metal, showing grain structure. (Courtesy of Mr. J. Waring, Department of Metallurgy, University of Queensland.)

Metals, in which of course there is only one kind of atom, all have either body-centered or face-centered cubic lattices; none are simple cubic.

If an ordinary specimen of a metal is etched and examined under a microscope, it is seen to be made up of innumerable interlocking "grains," separated from one another by irregularly shaped boundaries. Figure 10.10 is a photograph of such an etched surface. The metal thus resembles a three-dimensional jigsaw puzzle. From x-ray diffraction evidence we know that within each particular grain there is a continuous space lattice; thus each grain is a crystal. The principal lattice directions are differently orientated in adjacent grains. The grain boundaries are therefore simply regions of misfit between identical lattices that may meet at any angle.

What happens when a metal is deformed? A priori we might think of two kinds of distortions: distortion of the lattice within each separate grain, and relative movement between the grains at their boundaries. Obviously the latter cannot occur without the former, so there must in any case be lattice distortion, whether accompanied by movement along the grain boundaries or not.

Let us now see what may be inferred from experiment, confining our attention, for simplicity, to the case of cubic crystal lattices, which are isotropic with regard to their elastic properties.

We find that single large grains or crystals have elastic moduli that are effectively identical with those of the corresponding fine-grained material. From this it is clear that elastic deformation must be solely a matter of lattice distortion. Furthermore, it is found that large single grains or crystals are more easily plastically deformed than is the fine-grained material. This means that plastic deformation, far from being helped, is to some extent inhibited by the presence of grain boundaries; that is, relative movement along these cannot be a contributory factor in plastic deformation. Quite apart from these considerations, we can hardly conceive of any movement between irregularly shaped grain boundaries that would not lead to the occurrence of empty spaces within the material, and of such there is no experimental evidence, except, occasionally, at high temperatures. Thus there is ordinarily no relative movement along the grain boundaries in either elastic or plastic deformation.

We shall now consider how a deformation within the elastic range calls into being a state of stress within the material tending to restore it to its original configuration. Let parts (a) and (b) of Fig. 10.11 represent planes of atoms in a normal and in a distorted lattice, respectively. In both cases adjacent atoms must exert strong forces on one another. The configuration (a) is one of perfect symmetry with regard to three directions mutually at right angles, and we should expect this to correspond to either maximum or minimum energy of the system. If the energy were a maximum, the system would be unstable, which it obviously is not. Hence the energy must be a minimum, and with any departure from this state, as in (b), there must be associated a tendency for the system to return to state (a), just as a small displacement of the bob of a simple pendulum from its rest position brings into being a corresponding restoring force. This "tendency" for the system (b) to revert to (a) is, of course, manifested as the corresponding state of stress of the material.

Let us now compare the arrangements of the atoms represented

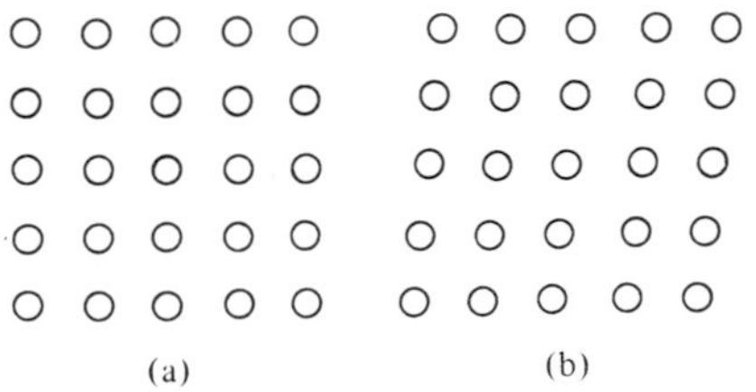

FIGURE 10.11 Diagrammatic representations of normal and distorted lattices.

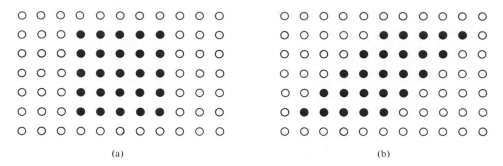

(a)   (b)

FIGURE 10.12 Imagined extreme case of distortion in which normal arrangement of atoms is restored.

by black circles* in parts (a) and (b) of Fig. 10.12, the latter being imagined to arise from the former by a shift of each horizontal plane of atoms relative to the one below by just one atomic spacing. State (b) may be regarded either as a more extreme lattice distortion than (b) in Fig. 10.11 or as no distortion at all, for obviously the state of symmetry in the three mutually perpendicular directions characteristic of the normal lattice has been restored, and there can consequently no longer be any stress within the material. Hence a deformation carried to this point—supposing it to be physically realizable in the manner imagined—would be a *plastic* deformation. Although actual plastic deformations do not, in fact, occur in quite this manner, the above discussion has nevertheless led to the expectation of an essential feature of a plastic deformation—relative slip between two adjacent atomic planes.

Such a deformation as was visualized in Fig. 10.12 would necessitate the simultaneous movement of whole planes of atoms relative to adjacent planes. It is interesting to note that as far as the halfway stage such movement must be elastic, although the stress–strain relationship would certainly not be Hookean. With increasing strain the stress would at first increase, but must presently decrease again, becoming zero at the halfway stage, where the system would be in unstable equilibrium. The maximum stress would occur at perhaps something like the quarter-way stage, where the strain would be far larger than any elastic strain ever observed in metals. Correspondingly, the stress would be enormous. Actually plastic flow always sets in much earlier than this, at relatively low stress, and from this we must infer that whole planes of atoms do

---

* The positions of some of the surrounding atoms (not all of them the same in the two diagrams) are also indicated, but it is on the atoms represented by the black circles that we are particularly concentrating our attention in the present discussion.

not, in fact, move simultaneously.

LATTICE IMPERFECTIONS The only possible alternative to the simultaneous movement of whole planes of atoms would appear to be a succession of jumps of single atoms or groups of atoms. Although such a process would not be possible in a perfect crystal lattice, it is not difficult to see how it might occur in crystals containing certain kinds of localized imperfections, or "dislocations." One type of dislocation believed to play an important role in plastic deformations is indicated in section in Fig. 10.13. This is known as an "edge dislocation." In the upper plane indicated in the figure the atoms are too closely spaced, but below this plane the lattice is normal. If, now, each atom in the upper plane in the region where the spacing is too close moves forward as indicated by the arrows, the dislocation as a whole moves forward by one atom spacing. If such forward motion is repeated, each *atom* in the upper part of the crystal moves forward a total distance of one atom spacing as the dislocation sweeps past it.

The process just considered could not produce a slip greater than one atom spacing, which is much less than such slips as are actually found to occur. This difficulty and others may be resolved by invoking, in addition to an edge dislocation, another kind of lattice imperfection known as a "screw dislocation." An edge and a screw dislocation, acting in conjunction, could bring about such relatively large slips as are actually observed.

There are strong reasons for believing that all crystals must contain dislocations of various kinds. An elaborate theory of these has been developed, according to which different kinds of dislocation may give rise to movement within crystals in different ways. Moreover, in certain circumstances, dislocations can react with one another, sometimes annihilating each other, sometimes becoming trapped in particular regions of the lattice, and sometimes giving birth to new dislocations.

Such slips as are observed always occur by the gliding of one plane of atoms over another, and lattice planes along which slip takes place are known as "slip planes." It is found that slips gen-

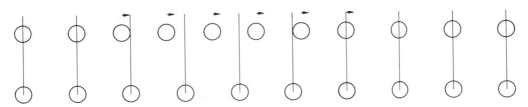

FIGURE 10.13 Edge dislocation in a line of atoms.

263

FIGURE 10.14 Surface of plastically deformed metal, showing slip lines
(Reprinted from G. R. Wilms and W. A. Wood, *Journal
of the Institute of Metals*, vol. 75, no. 8, p. 698.)

erally occur only between planes in which the atoms are most closely
packed; thus in each grain the slip planes can have only certain
definite directions.

The spatial distribution of dislocations is a matter of statistics.
We therefore cannot expect that dislocations favorable to slipping
will occur in every plane of atoms having one of the requisite orientations. Also, where there are such dislocations, they will presumably
not all enable slip to take place with equal ease. At stresses that
exceed the elastic limit by only a small amount, only those slip
planes come into operation along which slip can occur relatively
easily, while at higher stresses further slip occurs along planes containing less favorable dislocations.

The locations of the slip planes may be observed in a rod of
metal plastically deformed by stretching. If the surface of the metal
is highly polished before being stretched, it becomes dull as it is
pulled out, and under a microscope a multitude of fine slip lines
appears (Fig. 10.14).* From their spacing and the measured distortion of the metal it may be inferred that when slip occurs, a block
having a thickness of several thousand planes slides over the block
below by a distance of some hundreds of atoms.

---

* The curvature of some of the lines in this figure is not at variance with
the statement that slip occurs along planes; it is due merely to the curvature
of the exposed surface of the grains.

The theory of dislocations and of the role they play in the deformation of metals has now been developed to a point where the main phenomena can be satisfactorily accounted for. It would take us too far to enter into a detailed discussion of this theory, so we shall merely note a few of the more important general conclusions reached.

Not all lattice imperfections favor slipping; in addition to those that do, there are other kinds that tend to inhibit it. Among the latter are not only certain dislocations of the material under consideration but also the lattice disturbances due to foreign atoms. Thus impure metals are in general much harder—less plastically deformable—than are pure metals.

A familiar example of the role played by impurities in inhibiting movements of atoms in a lattice is the characteristic hardness of steel. Owing to the presence of atoms of such impurities as carbon and nitrogen within the lattice, steel has a much higher elastic limit than pure iron. A system such as this is known as a "solid solution." The solubility of impurities increases rapidly with temperature, and if very hot iron containing these impurities is suddenly "quenched," for example by being plunged into water, only a relatively small proportion of the impurity that has become incorporated in the lattice has time to go out of solution; the quenched material retains far more of it in the dissolved state than the equilibrium value at ordinary temperatures. This renders it very hard and brittle. Actually there is more involved here than the mere inhibition of slipping in the original crystals; the impurities also tend to prevent a change of lattice structure from that characteristic of high temperatures to another type normally found at lower temperatures, the latter being more favorable to plastic deformation than the former. Both these processes depend on rearrangements of atoms, so we may think of the presence of the impurity as tending to inhibit atomic movements generally.

The brittleness of the quenched material can be reduced, with retention of sufficient hardness, by subsequently subjecting it to a carefully controlled heat treatment, or "tempering." This brings some of the dissolved material out of solution and into the form of inclusions of Fe_3C (cementite) and iron nitride, and at the same time allows the iron to recrystallize into the form characteristic of lower temperatures.

Perfectly regular crystal lattices, devoid of dislocations, are not formed in nature except in very special and unusual circumstances. In order to account for the properties of pure, soft metals it is necessary to postulate the existence of about 10^8 dislocations

265

per square centimeter in each plane of closest packing. When slip occurs, those dislocations on which slipping depends migrate to regions where they are eventually trapped, after which they can no longer function. At the same time new dislocations, of a kind that tend to inhibit slipping, are brought into being, and the material becomes progressively harder. In a heavily work-hardened metal there are some 10,000 times as many dislocations as there are in the unworked material.

Dislocations can migrate, and the lattice readjust itself, by virtue of thermal agitation of the atoms about their mean positions, and some such readjustment may take place even at room temperature. This is responsible for the "reloading" line in a stress–strain diagram such as Fig. 10.8 not being exactly the same as the "unloading" line obtained after exceeding an elastic limit. To obtain large effects it is generally necessary to heat the metal for a sufficient length of time at a higher temperature. This is known as "annealing." In the process of annealing the extra dislocations brought into being by slipping gradually disappear, the material reverting to its former soft condition. Effective annealing temperatures vary considerably from one metal to another; in general, the higher the melting point of a metal, the higher the temperature to which it has to be raised to anneal it. An interesting case is that of lead, which, after having been work-hardened, reverts to its former soft condition after being left for a few hours at room temperature; it requires no special heating to anneal it.

Plastic deformation is of the utmost importance in industrial metal working. Bending, wire drawing, rolling, panel beating, pressing, spinning, and riveting are all processes depending on plastic deformation. In some cases these operations are performed on the cold metal, which becomes correspondingly work-hardened. Where, however, the desired deformation is too large to be carried out cold without risk of fracturing the material, it is done in stages, with intermediate annealings, or, alternatively, the metal may be deformed hot, as in forging.

PART III
FRICTION

We have several times had occasion to refer to frictional forces. We shall now study these more closely.

We classify as forces of friction those "passive" forces that tend always to prevent or retard motion, irrespective of the direction in which this motion takes place, or tends to do so.

Friction is of two main kinds: (1) **static,** and (2) **kinetic** or **sliding** friction. In static friction the two engaging surfaces do not rub against each other; their engagement somehow prevents relative motion. In kinetic friction, on the other hand, rubbing does occur, and this is invariably accompanied by the generation of heat.

Static friction is, on the whole, useful, and it is difficult to imagine what life would be like without it. The possibility of making string, rope, and fabrics of all kinds depends on static friction, the strength and durability of the materials depending on the friction between the fibers. And, but for friction, knots would come undone—indeed, it would not even be possible to tie them in the first place! Also the action of belts, pulleys, and friction drives in machinery depends on static friction. Finally, but for static friction we should not be able to walk, or to start, stop, or change the direction of wheeled vehicles.

Kinetic friction, on the other hand, is generally a nuisance, this being, for example, responsible for the "dissipation" of energy (conversion of other forms into heat) in machinery and for its eventual wearing out. Among the few items on the credit side are the action of the clutch of an automobile (during starting) and the action of brakes.

Main Phenomena

Both static and kinetic friction may be demonstrated very simply by means of the apparatus depicted in Fig. 10.15. A force F is applied horizontally, for example by means of masses on a scale pan suspended from the end of a string passing over a pulley, to a block resting on a horizontal board, the force being gradually increased until the block begins to move. While the block is stationary it obviously has no resultant force acting on it; that is, it is in equilibrium under the action of the forces applied to it. A

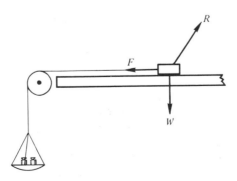

FIGURE 10.15 System for investigating frictional phenomena.

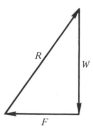

FIGURE 10.16

vector diagram of these forces will therefore be a closed figure. Two of these forces are the weight of the block, W, acting vertically downward, and the horizontal applied force, F, acting from right to left. There is also a third force, R, a force of reaction, applied to the block by the board. The closed vector triangle representing these forces is shown in Fig. 10.16.

The force of reaction R may be resolved vertically and horizontally, the vertical component obviously being equal in magnitude, although opposite in direction, to the weight W, while the horizontal component is equal to F. The latter, acting from left to right, is the force of friction—static friction.

As the applied force F is gradually increased, the force of static friction increases correspondingly, up to a certain limit, beyond which the block begins to move. At this limit the board is said to exert its limiting static friction on the block. When, on increasing F beyond this limit, the block begins to move, it is observed to do so with acceleration, showing that the (now kinetic) force of friction is less than the limiting static friction. The magnitude of the kinetic friction can be found by trial, adjusting F to such a value that, when the motion of the block is started, for example with the hand, it continues without acceleration.

If the block is a rectangular parallelepiped, with the three pairs of opposite sides of unequal area, then, provided that the surfaces are all similar, it will be found that both limiting static friction and kinetic friction are independent of which face rests on the board.

If the block is loaded, for example by placing masses on top of it, it will be found that both limiting static and kinetic friction are proportional to the total weight of the block plus load. The ratio of the tangential to the normal force is known as the **coefficient of friction** and is usually represented by the symbol μ.

Laws of Friction

Our earliest records of the experimental study of friction go back to Leonardo da Vinci (1452–1519), who found that frictional force is proportional to load and is independent of the area of the engaging surfaces. These laws were rediscovered by Amontons in 1699 and verified by Coulomb in 1781, who was the first to distinguish clearly between static and kinetic friction. The work of these investigators has led to the formulation of the following generalizations, or "laws," of friction:

1. Limiting static friction is greater than kinetic friction.
2. Kinetic friction is independent of the speed of rubbing, if moderate.
3. Frictional force is proportional to the normal force between the surfaces.
4. Frictional force is independent of the area of the surfaces in contact.

Although these so-called laws of friction, still commonly quoted quite uncritically, are useful for the purposes of a preliminary orientation in the subject and for rough engineering applications. a little experimentation and reflection will show that they really cannot be accepted without considerable qualification.

In the first place, even on the assumption that the apparent area of contact is the true area, these statements only have approximate validity. This is only another way of saying that they do not take all the relevant factors into account. Thus careful observation will show that neither the coefficient of limiting static friction nor the coefficient of kinetic friction can be precisely determined for a given pair of surfaces, these quantities varying in magnitude somewhat from one experiment to the next, despite superficial similarity of conditions. Accordingly, only approximate values can be assigned to them. As long as this is well understood, however, these approximate quantities may nevertheless be made to serve a useful purpose in dealing with many engineering problems.

As an extreme case of the variation of μ for a given pair of surfaces, we may note the case of glass engaged with, or sliding on, glass. If the surfaces are carefully prepared "optical flats," with departures from the plane nowhere amounting to more than a few millionths of an inch, and if these, after being cleaned, are put together, they may "seize," and it will then be very difficult indeed to separate them again, even when the pressure normal to the surfaces is given a large *negative* value. In such a case the whole idea of a coefficient of friction loses its meaning. Incidentally, the

269

liability to seize increases with time of contact, and if optically worked surfaces such as we have considered are engaged with one another it is advisable to slide or pull them apart not too long afterward if one wishes to avoid running undue risk of seizure.

A similar time effect is observed to some extent in other cases also; limiting static friction tends to increase with preceding contact time. It is also found that limiting static friction for a given normal force is increased by a larger normal force having been applied previously with the same contact.

Sliding friction is approximately independent of the speed of rubbing only between rather narrow limits; at high speeds the friction usually decreases appreciably.

Undoubtedly the "law" that is open to the most serious criticism is the fourth one, according to which the frictional force is independent of contact area. Although *apparently* this statement gives a very satisfactory account of what is observed experimentally, a little reflection will show that we have very little a priori knowledge indeed of what constitutes the the true area of contact in any given instance. For example, in the case of a block of the form of a rectangular parallelepiped resting on a flat board, how can we be sure that the true area of contact is any less when the block is, say, on end, than when it is "on the flat"? On reflection it soon becomes evident that we can have no such assurance. For, from a molecular point of view, all surfaces we are able to prepare are exceedingly rough; they are very mountainous country indeed, and it is impossible for true contact (whatever this might mean) to be established except at a very few points, that is, over a very minute fraction of the apparent area of engagement. As Bowden has graphically described the situation, a molecule in one of the hollows endeavoring to establish contact with the upper surface would be something like a mouse on the floor of a barn trying to touch the roof with its whiskers.

The nature of surfaces being what it is, the number of separate areas of true contact corresponding to a small normal force between two surfaces cannot be less than three, on the same principle that a chair must have at least three legs. Actually any legs beyond three on a chair are redundant, or would be if the chair, and also the floor, were completely rigid. In fact, however, no solids are truly rigid; they all "give" to some extent when subjected to a system of forces. Thus we have all had experience of a four-legged chair only three of whose legs will ordinarily touch the floor simultaneously, but which will rest on all four legs when a heavy person sits on it. Similarly, we should expect that when a sufficient normal force is applied between two surfaces there might be more areas of true contact than the minimum of three. Owing to deformation under pressure, we should also expect the contact areas themselves to grow with increasing normal force, in much the same way as may be observed when two rubber balls are pressed together.

The possible increase in number of the contact areas with increasing normal force and the probable simultaneous growth in size of each such area both suggest that the total true contact area must surely increase as the normal force increases. We do not at this stage know quantitatively just how the former varies with the latter. The simplest law of variation would be one of proportionality, and on this basis it is quite possible that the third and fourth alleged laws of friction should be replaced by the following statements:

a. The true area of contact increases statistically in proportion to the normal force, irrespective of the apparent area.

b. The frictional force is proportional to the true contact area.

These statements do not contradict the third law of friction previously listed; indeed, taken together, they lead directly to it. However, with the proposed replacement of 3 and 4 by a and b, law 3 does become redundant. On the other hand, the fourth so-called law is now seen to be quite wrong.

Fortunately, if the engaging surfaces are metals, it is possible to form an estimate of the true contact area by carrying out certain electrical measurements. With a suitable electric circuit (which it is not proposed to discuss here) the strength of the current carried across the apparent contact will be a measure of the true contact area. According to Bowden, the results of such measurements show that

The real area of contact is indeed very small; it varies with the load, but for flat steel surfaces it may be less than one ten-thousandth of the apparent area. The experiments . . . also . . . show that the real area of contact is almost independent of the size of the surfaces. It is also very little influenced by the shape and degree of roughness of the surfaces: it depends mainly on the load which is applied to them, and is, in fact, directly proportional to the load.*

It will be seen that these results are in complete accord with conjectures a and b and show that these do, in fact, correspond to the true state of affairs.

A third heading under which friction is sometimes classified, in addition to static and kinetic friction, is **rolling friction,** such as we have when a ball is rolled, and reference is sometimes made to

* F. P. Bowden, *Science News*, No. 4, (1947), p. 139.

a "coefficient of rolling friction." The mechanism of rolling friction will be discussed, along with those of static and kinetic friction, presently. In the meantime we may note, as an item of interest, that the coefficient of rolling friction is in general very small in comparison with the other two coefficients. For example, whereas the coefficient of friction for steel sliding on steel is of the order of 0.2, the corresponding quantity for a steel ball rolling on steel is only about 0.002. It is, of course, because of the extreme smallness of rolling friction that ball and roller bearings are so extensively used in machinery.

Apropos of ball and roller bearings, it is interesting to reflect that these represent the same kind of advance over a wheel mounted on an ordinary axle as a simple wheeled vehicle is an advance on a sled, which is simply dragged over the ground. The idea of a wheel mounted on ball bearings is, so to speak, that of a wheel raised to the second degree.

Modern Theory of Friction

We have seen that the true area of contact between the engaging surfaces is only a minute fraction of the apparent area. From this it follows that the pressure at the actual contacts must be enormous. Bowden's investigations have shown that it is, in fact, sufficient to cause even such a hard metal as steel to flow plastically; the summits of the undulations are crushed down, the contact areas thereby being increased until the pressure is just at the limit of what would cause the metal to flow further. This pressure is so intense that it may bring about actual welding at the contacts, small metallic junctions thus being produced.

Whether or not such localized welding occurs, there must always be a considerable degree of interlocking of the surfaces, which, as we have seen, are molecularly rough. This interlocking must be an important item—indeed, perhaps the main item—in static friction. And, if the tangential stress is increased to the point where sliding occurs, the roughness of the surfaces must give rise to a ploughing of some of the protuberances through others, with consequent surface damage in the form of localized plastic deformation and abrasion.

ROLE OF SURFACE CONTAMINATIONS According to what we have been saying, one would expect the roughness of the surfaces to be the main factor determining the coefficient of friction in any given case. Somewhat surprisingly, this is not so; the role played by roughness is a relatively minor one. Of far greater importance is the state of cleanliness of the surfaces. According to Bowden and Tabor,* even a single molecular layer of grease from the atmosphere or from the fingers may produce a considerable change in friction. Grease may even migrate from the back to the front surface of a block, like a two-dimensional gas or liquid. Extraordinary precautions must be taken if such layers of grease, "adsorbed" on surfaces, are to be avoided or, if present, removed. If thoroughly clean surfaces are required, it is necessary to manipulate them in a high vacuum.

It is not generally realized that metal surfaces exposed to the atmosphere are automatically contaminated, either by an invisibly thin film of oxide or at least by an adsorbed layer of oxygen or some other gas, for example water vapor or carbon dioxide. All ordinarily observed frictional effects are conditioned by such contaminations.

Bowden and Tabor describe an experiment in which an attempt was made to produce, and then engage with one another, clean iron surfaces, under good vacuum conditions. They point out that this attempt is unlikely to have been completely successful; almost certainly some oxygen was adsorbed from the residual gas or a minute amount of oxide formed. Nevertheless, it was found that under these relatively clean conditions the surfaces adhered strongly—presumably became welded together—wherever they touched, and a very strong force was needed to pull them apart.

According to Bowden and Tabor, *some* formation of such junctions occurs even under ordinary conditions of exposure to the atmosphere. This would presumably be due to localized destruction of the oxide films at asperities under the influence of the enormous pressures exerted. Therefore, to the resistance that must be ascribed to ploughing must be added the item of shearing or pulling apart of momentarily formed welds. It would appear that under good vacuum conditions the latter effect makes a much more important contribution to the total friction than the former.

Even under only moderately good vacua, and without special heat treatment of metal surfaces to remove contaminations, frictional effects are observed to be much larger than in air. This must be due to the evaporation of a significant proportion of the adsorbed gas, ordinarily functioning as a species of lubricant.

---

* F. P. Bowden and D. Tabor, *Friction and Lubrication*, Methuen & Co. Ltd., London, 1956.

EFFECTS OF LOCALIZED HEATING Associated with the extreme smallness of the areas of contact is a highly localized frictional heating when sliding occurs. Evidence for this has been obtained by connecting dissimilar metals sliding over one another to a suitable form of voltmeter and, with the momentarily engaging areas of metal functioning as a thermojunction (see Chapter 28), observing the electromotive force generated. Such observations indicate that exceedingly high temperatures are developed; even though the general temperature of the metals is hardly affected by the sliding, the regions in actual contact are quite commonly raised to temperatures of the order of 500° to 1000°C, or even more.

These local hot spots play an important role in a number of processes involving rubbing, for example polishing, the seizing of metal surfaces in contact with one another, the frictional welding of plastics, and the initiation of chemical reactions brought about by friction and by impact. Thus it has been found impossible to initiate the explosion of nitroglycerine unless metals are employed whose melting points exceed 480°C, because the melting point sets the upper limit to the temperature reached.

It is of particular interest to note the role played by melting points in the process of polishing. This appears to depend primarily on the melting, by localized frictional heating, of the material being polished. The experiment has been tried of rubbing the alloy known as Wood's metal (m.p. 72°C) with a block of camphor (m.p. 178°C). Although Wood's metal is much harder than camphor, it is the Wood's metal, not the camphor, that is polished by such rubbing, this being due, apparently, to its lower melting temperature. On the other hand, camphor is found not to polish any materials having a higher melting point than itself, such as tin (m.p. 232°C), white metal (m.p. 233°C), lead (m.p. 327°C), or zinc (m.p. 419°C).

All the polishers in Table 10.3 have been found to be effective on the materials higher up on the list (which, with one exception, is in the order of increasing melting points) but not on any lower down.

Similar results have been obtained for the polishing of non-metals. Thus calcite (m.p. 1333°C) is not polished by cuprous oxide (m.p. 1233°C) but is readily polished by zinc oxide (m.p. 1850°C).

It will be seen that there is a slight anomaly in the table with regard to zinc and oxamide. There are also a few others, in which high-melting but rather soft materials, for example gold, can be polished by plastic flow rather than by melting. Such processes are, however, always much slower than the polishing produced by a high-melting polisher.

TABLE 10.3 Selective Polishers

Metal	M.p. (°C)	Polisher	M.p. (°C)
Wood's metal	72		
		Camphor	178
Tin	232		
White metal	233		
Lead	327		
Zinc	419		
		Oxamide	417
Speculum metal	745		
		Lead oxide	888
Nickel	1083		
Molybdenum	2630		
		Chromic oxide⎫ Stannic oxide⎰	Very high

LUBRICATION Lubrication, insofar as it prevents actual contact between surfaces, eliminates sliding friction. On the other hand, it introduces a new item of resistance to movement of its own, the resistance due to the viscosity of the lubricant. Nevertheless, the overall effect of lubrication with a suitable oil or grease is to reduce resistance to motion substantially.

ROLLING FRICTION It was long believed that when a ball rolls along a groove there must be some element of interfacial slip, sliding friction thus contributing appreciably to the resistance experienced. If this were so, rolling friction should be reduced by lubrication. However, experiments by Tabor have shown that this is not the case; rolling resistance is almost unaffected by the presence of lubricants. It is now believed that rolling friction arises primarily from losses associated with what is known as "elastic hysteresis." By this is meant the failure of an elastically deformed body to regain its original configuration immediately on release of the stress. This necessarily involves a loss of mechanical energy, a conversion of mechanical energy to heat.

Problems

10.1 To what pressure, in atmospheres, would a specimen of lead have to be subjected to increase its density by 0.1 percent? What would be the corresponding elastic potential energy of the lead per unit volume? (Assume Hooke's law to apply throughout the relevant pressure range.)

10.2 A steel elevator cable has an elastic limit of 3000 kg weight per cm². Calculate the maximum upward acceleration that can be given to a 1000-kg elevator by a supporting cable system of total cross-sectional area 3 cm² if the stress is not to exceed 0.2 of the elastic limit.

10.3   Calculate how much work must be performed to stretch 10 m of steel wire, of diameter 1.53 mm, by 5 mm, Young's modulus being 2.0×10^{11} N/m².

10.4   A steel wire 40 m long has a cross-sectional area of 0.5 mm². Young's modulus for steel is 2.0×10^{11} N/m². (a) Compute the force constant (tension per unit extension) of the wire. (b) If a 10-kg load is suspended from the wire and pulled down a short distance below its equilibrium position, what will be the frequency of vibration when the load is released?

10.5   A brass wire of length 2.5 m and a steel wire of length 1.5 m, each of which has a diameter of 1 mm, are connected end to end to form a composite wire of total length 4 m. Given that the values of Young's modulus for brass and steel are 10^{11} N/m² and 2×10^{11} N/m², respectively, calculate (a) the tensile force required to increase the length of the composite wire by 2 mm, and (b) the increases in length of the brass and steel sections separately.

10.6   A wire of length $2l$ is laid horizontally on a bench and the position of the midpoint is accurately marked. One end is then lifted so that the wire hangs vertically and stretches elastically under its own weight. Show that a small element of wire of length δx a distance x from the point of support is stretched by an amount $(\rho g/E)(l - x)\,\delta x$, where ρ and E represent the density and Young's modulus, respectively, of the wire. Hence find the distance of the mark from the new midpoint of the wire. (Neglect variations of cross-sectional area and density with tension.)

10.7   A thin uniform rod of length l and density ρ rotates about a vertical axis passing through one end with an angular velocity ω which is sufficient to cause the rod to describe a cone having a semiangle of almost 90°; that is, the rod is practically horizontal. Find how the tensile stress in the rod is distributed along its length, and obtain an expression for its elongation in terms of l, ρ, ω, and Young's modulus E. (Neglect variations of cross-sectional area and density with tension.)

10.8   (a) A beam of rectangular cross section has sides of widths b and h and is bent into a circular arc of mean radius r about an axis perpendicular to the length of the beam and parallel to the sides of width b. The planes of the end surfaces of the beam are perpendicular to the adjacent edges, not only before bending but also after. This means that the bending must stretch the outer portions of the beam while it compresses the inner portions, the change in length of the "fibers" in both portions being proportional to their distance from the so-called neutral surface, the surface parallel to the sides of width b and midway between them. Show that the torque L required to bend the beam is given by the formula

$$L = \frac{Ebh^3}{12r}.$$

(b) If a bearer or a joist used in the building of a house is put "on the flat," instead of on edge with the sides of the greater width h vertical, in what ratio will its flexure be increased under a given load? (Express your answer in terms of b and h.)

10.9   A mass m is suspended from the free end of a horizontal cantilever of rectangular cross section and of length l, whose mass is negligible. A rectangular Cartesian coordinate system is chosen with x axis horizontal and y axis vertically downward, the cantilever being in the xy plane. On the assumption that the bending is small, so that d^2y/dx^2 may be identified with $1/r$, where r is the radius of curvature of the beam, write down the differential equation for the locus of the neutral surface, neglecting shearing effects. Solve this equation, and so find the deflection of the end of the beam due to bending only.

10.10   The cantilever of Problem 10.9 has a breadth b, a height h, and a length l. Calculate the elastic energy stored in it when its free end is pushed down a distance d.

10.11   What are the dimensions of (a) torque and (b) rigidity? Assuming that the torque L required to produce a twist θ in a wire of length l is proportional to θ/l (Hooke's law) and also that it depends on the radius a of the wire and the rigidity G of the material, these quantities entering as powers, show by the method of dimensions that

$$L \propto Ga^4 \frac{\theta}{l}.$$

10.12   A ring suspended with its plane vertical from a vertical steel wire oscillates rotationally about the wire as axis with the period T. Find the period of oscillation of a geometrically similar ring of the same material having twice the linear dimensions suspended from a steel wire of (a) twice the linear dimensions, and (b) half the linear dimensions.

10.13   A hollow cylinder of length l has an internal radius r and an external radius $r + \delta r$, δr being small compared with r (see the figure). Show that when this is twisted through an angle θ the wall of the cylinder

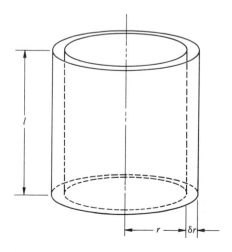

Figure for problem 10.13.

has a shear strain of $r\theta/l$, and hence show that the torque required to produce the twist is $(2\pi G\theta/l)r^3\,\delta r$.

10.14 Using the result quoted in Problem 10.13, show, by integration, that the torque L required to produce a twist θ in a *solid* cylinder of length l and radius a is given by the formula

$$L = \frac{\frac{1}{2}\pi Ga^4\theta}{l}.$$

10.15 Find the torque required to produce a twist of one complete rotation in a tungsten wire of length 50 cm and diameter 0.2 mm, given that the rigidity of tungsten is 1.7×10^{11} N/m². If a horizontal uniform thin rod of mass 100 g and length 20 cm is suspended at its center from this wire, what will be the period of torsional oscillation of the wire?

10.16 Calculate the energy stored in the tungsten wire of Problem 10.15 when given a twist of one complete rotation. By what factor is this energy increased by doubling the twist?

10.17 The steel driveshaft of a car is 1.5 m long and has a diameter of 5 cm. Through what angle is one end of the shaft twisted with respect to the other when it is transmitting a power of 20 kW at 30 rps?

10.18 Obtain a formula for the relevant coefficient of friction between a body and a plane in terms of the angle the latter makes with the horizontal when the body is (a) on the point of sliding, or (b) moves down the plane with uniform velocity after having been started. Apply this formula to find the coefficients of limiting static and kinetic friction in the case where the angles referred to above are (a) 20° and (b) 15°.

10.19 Find the acceleration of the body referred to in Problem 10.18 when the plane is tilted at an angle of 30° to the horizontal.

10.20 A steel block of height 5 cm is drawn over a horizontal steel plate by a horizontally applied force. Assuming that pressure welding occurs over all the contact areas and that the entire resistance to sliding is due to the shearing of these welds, find, in order of magnitude, the fraction of the area of the lower face of the block which is at any moment in true contact with the plate. Take the shearing strength as 3500 kg weight/cm² and the coefficient of friction as 0.2. The density of steel is 7.8 g/cm³. Do the assumptions listed give an upper or a lower limit for the true contact area?

11

WAVES IN A CONTINUUM

Our treatment of waves in Chapter 3 was a mathematical *description* of wave motion rather than a complete theory; it dealt only with the kinematical aspects of such motion. In the present chapter we shall supplement this by developing the full dynamical theory of the propagation of small-amplitude transverse waves along strings and longitudinal waves in fluids. We shall then consider a number of acoustic phenomena associated with the propagation of waves in fluids.

Wave Propagation along a Stretched String

The simplest kind of wave propagation to consider dynamically is the propagation of plane-polarized waves along a stretched string. The procedure we shall follow will be to find the resultant force exerted transversely on a short "element" of the string and equate this to the product of the mass of the element and its acceleration. This will give us a differential equation whose integral is the wave equation. From the latter it is then a simple matter to obtain an expression for the velocity of propagation of the waves.

Let the curved line in Fig. 11.1 represent part of the string, which is in tension, the tensile force being F, and let the mass per unit length of this string be μ. We shall suppose that the inclination of the string to its undisturbed direction, which we shall call the x direction, is everywhere small, so that, to a sufficiently close approximation, we may identify an element δl of its length with the corresponding δx.

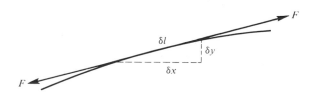

FIGURE 11.1   Short section of stretched string along which a wave disturbance is traveling.

The element δl is acted on by the tangential force F at each end, but if the element is curved these forces will not have the same direction, and so there will be a resultant force acting on the element. Because the inclination of the string to the x direction is everywhere small, we may consider the x components of the two end forces to balance each other. However, the y components will not be equal. We shall now obtain an expression for the resultant force, which will be in the y direction.

The y component of the force exerted on the left-hand end of the element is $-F \sin \theta$, where θ is the angle the string makes with the x direction at this point; or, to a sufficiently close approximation, $-F(\partial y/\partial x)_1$,* where $(\partial y/\partial x)_1$ denotes the value of the slope $\partial y/\partial x$ at the left-hand end of the element. Similarly, the y component of the force exerted on the element at its right-hand end is $F(\partial y/\partial x)_2$ where $(\partial y/\partial x)_2$ denotes the slope of the string at this end. The resultant force, obtained by adding these, is, therefore,

$$ F\left[\left(\frac{\partial y}{\partial x}\right)_2 - \left(\frac{\partial y}{\partial x}\right)_1\right]. $$

But $(\partial y/\partial x)_2$ exceeds $(\partial y/\partial x)_1$ by the rate at which the slope increases with x along the interval considered, multiplied by δx, that is, by $(\partial^2 y/\partial x^2)\, \delta x$. Hence the resultant force is $F(\partial^2 y/\partial x^2)\, \delta x$.

The acceleration $\partial^2 y/\partial t^2$ of the element is the quotient of the resultant force acting on it and its mass. We have just obtained an expression for the former, and the latter is obviously $\mu\, \delta x$. Thus

$$ \frac{\partial^2 y}{\partial t^2} = \frac{F(\partial^2 y/\partial x^2)\, \delta x}{\mu\, \delta x}, $$

or

$$ \frac{\partial^2 y}{\partial t^2} = \frac{F}{\mu}\frac{\partial^2 y}{\partial x^2}. \tag{11.1} $$

The most general solution of this differential equation is

$$ y = f_1\left(x - \sqrt{\frac{F}{\mu}}\,t\right) + f_2\left(x + \sqrt{\frac{F}{\mu}}\,t\right), \tag{11.2} $$

where the symbols f_1 and f_2 represent any two continuous functions of the variables $x - \sqrt{F/\mu}\cdot t$ and $x + \sqrt{F/\mu}\cdot t$, respectively. The reader may check that this is indeed the case by performing the

---

* Note that $\partial y/\partial x$ stands for the rate of increase of y with x at a particular instant of time ($t = $ const.), and $\partial y/\partial t$ stands for the rate at which y increases with time for a particular position along the string ($x = $ const.); correspondingly for $\partial^2 y/\partial x^2$ and $\partial^2 y/\partial t^2$.

corresponding successive partial differentiations with respect to t, x being treated as a constant, and to x, t being treated as a constant. The first term represents a wave of some form or other, corresponding to f_1, traveling with form unchanged in the positive x direction, while the second represents one, of form corresponding to f_2, traveling in the negative x direction.

To obtain the velocity v of either of these waves, let us consider one of the terms, say the first. For the displacement y to be the same at the time $t + \delta t$ and at the position $x + \delta x$ as it is at the time t and the position x, we must have

$$(x + \delta x) - \sqrt{\frac{F}{\mu}}\,(t + \delta t) = x - \sqrt{\frac{F}{\mu}}\,t,$$

or

$$\delta x = \sqrt{\frac{F}{\mu}}\,\delta t.$$

Hence, the velocity of propagation of a wave disturbance being, by definition, the rate at which the location of a particular disturbance (for example, displacement y) moves along, or $\delta x/\delta t$, we have for this

$$v = \sqrt{\frac{F}{\mu}}. \tag{11.3}$$

We may note that special cases of the most general solution, represented symbolically by Eq. (11.2), are the sine waves traveling in the positive and negative x directions,

$$y = A \sin \frac{2\pi}{\lambda}\left(x - \sqrt{\frac{F}{\mu}}\,t\right)$$

and

$$y = A \sin \frac{2\pi}{\lambda}\left(x + \sqrt{\frac{F}{\mu}}\,t\right),$$

respectively, the constants of integration A (amplitude) and λ (wavelength) having any values we please.

Wave Propagation in a Fluid

The simplest case to consider is that of plane waves. The procedure leading to the wave equation is similar to that followed for waves traveling along a stretched string. Here, as in that case, we shall begin by finding an expression for the resultant force exerted

281

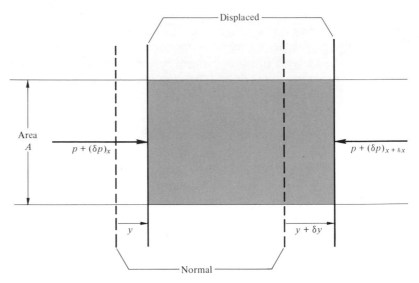

FIGURE 11.2 Displaced and stressed portion of a fluid through which longitudinal waves are propagated.

on an element of the medium and equating this to the product of the mass of the element and its acceleration.

Figure 11.2 shows an element of the fluid in which the waves are being propagated. The fluid is of density ρ, and the element under consideration has a cross-sectional area A, and, *in its undisturbed condition*, lies between parallel planes whose distances from a given plane of reference are x and $x + \delta x$. At a particular instant let the left-hand boundary of this element be at the distance y to the right of its undisturbed position, while the right-hand boundary is displaced a distance $y + \delta y$ to the right.

Let the pressure of the undisturbed fluid be p, and at the moment under consideration let the pressure on the left-hand boundary of our sample be $p + (\delta p)_x$ and that on the right-hand boundary $p + (\delta p)_{x+\delta x}$. Then the resultant force exerted on the sample in the direction from left to right (the positive x direction) is

$$-A[(\delta p)_{x+\delta x} - (\delta p)_x].$$

Considering δp as a function of the variables x and t, we may regard the quantity in brackets as minus the product of the rate at which δp increases with x at the given instant and δx. Thus the expression for the resultant force becomes

$$-A \frac{\partial}{\partial x} (\delta p) \, \delta x.$$

We may express δp in terms of the relative compression of the sample and the bulk modulus of elasticity K. Thus

$$K = \frac{\text{stress}}{\text{strain}} = \frac{\delta p}{-\partial y/\partial x},$$

or

$$\delta p = -K \frac{\partial y}{\partial x}.$$

Substituting this in the expression for the resultant force, we obtain for the latter $AK(\partial^2 y/\partial x^2)\ \delta x$.

The mass of our sample of fluid is $\rho A\ \delta x$, and its acceleration is $\partial^2 y/\partial t^2$. Equating the resultant force with the product of the mass and acceleration, and eliminating the factor A which is common to the two sides, we obtain the differential equation

$$K \frac{\partial^2 y}{\partial x^2} = \rho \frac{\partial^2 y}{\partial t^2},$$

or

$$\frac{\partial^2 y}{\partial t^2} = \frac{K}{\rho} \frac{\partial^2 y}{\partial x^2}. \tag{11.4}$$

This equation is of the same form as Eq. (11.1), obtained in our discussion of wave propagation along a stretched string, with K/ρ replacing F/μ. It therefore has the corresponding general solution

$$y = f_1\left(x - \sqrt{\frac{K}{\rho}}\, t\right) + f_2\left(x + \sqrt{\frac{K}{\rho}}\, t\right), \tag{11.5}$$

the first and second terms representing forward- and backward-traveling wave disturbances, respectively. And the corresponding expression for the velocity of either of these is

$$v = \sqrt{\frac{K}{\rho}}. \tag{11.6}$$

Velocity of Waves in a Gas

For K in Eq. (11.6) we must take the adiabatic bulk modulus, for the sequence of compressions and rarefactions (Fig. 11.3) occurring in any given sample of the medium is far too rapid to allow heat exchange between it and the surroundings. It may be shown (see Chapter 17) that the adiabatic bulk modulus for a gas is γp, where γ is a dimensionless quantity characteristic of the gas. Its magnitude for monatomic gases is about 1.66 and for diatomic gases about 1.4. The precise nature of this quantity, and a procedure for

283

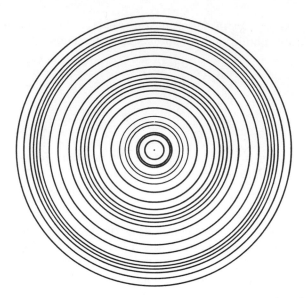

FIGURE 11.3 Compressions and rarefactions in a train of spherical longitudinal waves traveling out from a source. The compressions and rarefactions are indicated by small and large spacings, respectively, between circles.

measuring it, will be considered in Chapters 15 and 17, respectively. Substitution of γp for K in Eq. (11.6) gives us

$$v = \sqrt{\frac{\gamma p}{\rho}}. \tag{11.7}$$

It is useful to express the velocity in a gas in an alternative form, in terms of the molecular weight M of the gas, the universal gas constant R, and the absolute temperature T, and γ. It will be shown in Chapter 15 that for a sample of gas of mass m and volume V the quantities p, V, m, and T are related by the equation

$$pV = \frac{m}{M} RT.$$

Thus for p/ρ in Eq. (11.7) we have

$$\frac{p}{\rho} = \frac{pV}{m} = \frac{RT}{M};$$

and so, making the corresponding substitution in Eq. (11.7), we obtain for the velocity the expression

$$v = \sqrt{\frac{\gamma RT}{M}}. \tag{11.8}$$

As is implied by Eq. (11.7) and is brought out more obviously in Eq. (11.8), the velocity in a particular gas is independent of the pressure; it depends only on the temperature.

It will now be interesting to use one of these equations to calculate the velocity in a specific case. Thus let us consider the case of dry air at 0°C. It will be convenient to base the calculation on Eq. (11.7), using the value for the density of dry air at STP, 1.293 kg/m³. Standard atmospheric pressure is 1.0132×10^5 N/m², and, as we shall see in Chapter 15, the most reliable value of γ for dry air is 1.403. Substituting these values in Eq. (11.7), we obtain for v the value of 331.5 m/sec, with an uncertainty of about 0.1 m/sec. As we see from Eq. (11.8), the theoretical velocity at any other temperature may be calculated on the basis of the proportionality of v to $\sqrt{T}$.

The presence of water vapor in air should be to increase the velocity somewhat, because water vapor, although decreasing γ slightly, more than compensates for this by increasing p/ρ. Thus in air at 20°C and at a pressure of 1 atm whose humidity is 100 percent, the velocity calculated from Eq. (11.7) is greater than that in dry air at the same temperature by about 0.35 percent.

Identification with Sound

Waves in a material medium may be identified with sound from the following evidence. In the first place, sound needs a material medium for its propagation. Also, sources of sound are, characteristically, vibrating bodies. These two facts, considered together, can surely only mean that sound consists of elastic waves traveling through the medium. If the medium is a fluid, for example air or water, the waves must be longitudinal, for a fluid has only bulk elasticity; it has no rigidity, such as would be required for the propagation of transverse waves. Accordingly, in a fluid sound must be propagated as a series of compressions and rarefactions. As we shall now see, this inference is confirmed, in the case of air, by the close agreement between experimentally determined velocities of sound in this medium and the velocities of pressure waves at the corresponding temperatures predicted theoretically. Similar agreement is found for other media.

EXPERIMENTAL DETERMINATIONS OF THE VELOCITY OF SOUND IN AIR The first recorded experimental method for determining the velocity of sound in air, employed by Mersenne and Gassendi in the seventeenth century, was to measure the interval between seeing the flash of a distant gun and hearing the report.

This method has been used, with progressive improvements in technique, by a number of subsequent workers. Among the later determinations the work of Regnault, carried out in 1864, deserves special mention; Regnault, for the first time, used electrical recording of both the time of firing the gun and that of the reception of the sound. One of the most accurate determinations by the gun-firing method was carried out in 1917 and 1918, during World War I, by E. Esclagnon, who obtained a corrected value for the velocity in dry air at 0°C (assuming the theoretical variation with temperature and humidity as for pressure waves) of 330.9 m/sec. This is in very good agreement with the theoretical value for pressure waves, 331.5 m/sec. Note that the assumption underlying the correction does not beg the question: A calculation of the velocity of pressure waves under the prevailing conditions of temperature and humidity would have yielded a value in equally good agreement with the measured value of the velocity of sound.

It was pointed out by T. C. Hebb in 1905 that measurements of the time sound takes to travel over a considerable distance are impaired by variations of temperature, wind, and humidity along the line traveled. To avoid this difficulty, Hebb undertook a precision measurement of the wavelength in free air of the pressure waves—manifested as sound—emitted from a source of accurately known frequency. He employed two paraboloidal mirrors facing each other whose distance apart could be varied, generating standing waves between these, and observed electrically the sound intensity at the two foci. By measuring the distances separating the mirrors corresponding to a series of maxima and minima, Hebb was able to determine the wavelength to an accuracy of 1 part in 1000. The value he obtained for the velocity of sound in dry air at 0°C was 331.3 m/sec. On repeating his experiments in 1919 he obtained the value 331.4_4 m/sec. These values are in excellent agreement with the theoretical value of the velocity of pressure waves in dry air at 0°C, 331.5 m/sec, and this agreement is to be regarded as very satisfactory confirmation of the theory of these pressure waves. The fact that the existence of these waves was manifested as sound shows that these are the same. At the same time the slight discrepancy between Esclagnon's measured value of the velocity of sound and the theoretical velocity of pressure waves is accounted for along the lines suggested by Hebb.

Velocities in Liquids and Solids

As is well known, sound travels in liquids and solids as well as in gases.

A rough idea of the velocity of sound in a liquid may be arrived at theoretically from the fact that most ordinary liquids have densities of the order of 1000 times that of air at ordinary temperatures and at sea level, whereas their bulk moduli are of the order of 10,000 times as great. Hence the quotient of bulk modulus and density is something like 10 times and the square root of this roughly 3 times that of air. These considerations lead us to expect velocities of sound in these liquids of the order of 1000 m/sec. And velocities of this order we do, in fact, find experimentally.

Solids, unlike fluids, have rigidity as well as a bulk modulus, and can therefore carry transverse as well as longitudinal waves. Longitudinal waves traveling in an extended solid, such as a solid region of the earth's interior, are known as "bulk waves," and transverse waves are called "shear waves." These, depending as they do on different elastic moduli, or combinations of moduli, travel with different velocities, bulk waves being faster, usually by a factor between 1.5 and 2. Seismic waves may be of either kind, but only bulk waves can be transmitted through the liquid regions of the earth. Analysis of seismic waves, originating from the same source, and emerging at different points on the surface of the earth after traversing different paths through it, has furnished valuable information concerning the constitution of the earth's interior.

If a solid is in the form of a rod, free to expand and contract laterally in response to axial compression or extension, a third type of wave may travel along it which depends on Young's modulus E, its velocity being given by the formula

$$v = \sqrt{\frac{E}{\rho}}.$$

Such waves are known as "rod waves." The velocity of these is intermediate between those of bulk and shear waves in the same material but is usually much closer to the former than to the latter.

A simple and reasonably accurate method for determining the velocity of rod waves is one devised by Kundt. In this, standing waves are produced both in a solid rod and in air, and the wavelengths are measured. Then, because the frequency is the same for both, the ratio of velocities is simply that of the two wavelengths.

Kundt's apparatus is shown in Fig. 11.4. A wide glass tube

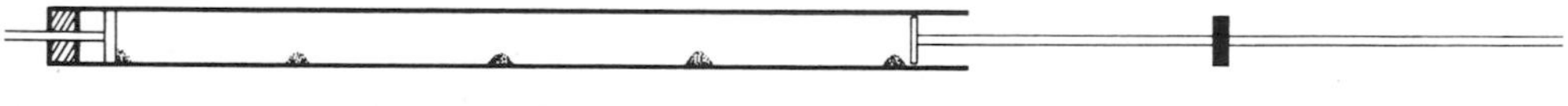

FIGURE 11.4 Kundt's apparatus.

is provided with an adjustable piston at one end, while into the other is inserted a light loose-fitting piston attached to the end of a long rod clamped at its center. The rod is set into vibration by pulling a wet cloth or a piece of chamois leather, held in the hand, along it. In this way a standing wave is set up in the rod with a node at the center and an antinode at each end, accompanied by the emission of a high-pitched note. Lycopodium powder is sprinkled along the bottom of the glass tube, inside, and the adjustable piston moved into one of the positions for which the lycopodium powder becomes violently agitated at intervals along the tube and tends to collect in small heaps or ridges at intermediate points, where the motion of the air is least. These points of least disturbance are the nodes of a standing wave established in the air inside the tube. The distance between two nodes near the ends is measured and the number of node-to-node intervals between them counted. The former divided by the latter gives the semiwavelength of the sound in air, and the length of the rod, being the distance between two adjacent anti-nodes, is the semiwavelength in the rod. The ratio of the wavelengths in the rod and in air is now multiplied by the known velocity of sound in air at the prevailing temperature, to give the velocity of the rod waves.

The velocities of sound in a number of media, including also a few gases, are listed in Table 11.1.

TABLE 11.1 Velocities of Sound

Medium	Velocity (m/sec)[a]
Carbon dioxide (20°C)	270
Air (20°C)	343.4
Helium (20°C)	1005
Hydrogen (20°C)	1332
Water (20°C)	1484
Ethyl alcohol (20°C)	1177
Turpentine (25°C)	1225
Mercury (20°C)	1451
Glass (crown)	5260–6120
Brass	3130–3450*
Steel (tool)	5150*
Wood, oak (with grain)	4100*
Wood, pine (with grain)	3600*
Rubber (natural)	1600
Concrete	4250–5250
Granite	5400

[a] Figures given for solids refer to bulk waves, except those marked with an asterisk, which refer to rod waves.

Effects of Wind and Temperature Gradients on Audibility

We are all familiar with the effect of wind on audibility. If a wind is blowing toward us from a distant source of sound, we hear it substantially better than if there is no wind, whereas if the wind is blowing in the opposite direction, we hear it only relatively faintly, if at all.

To the layman the explanation appears quite simple. In the former case the wind "helps the sound along"; the air distance the sound has to travel from source to observer is less than the actual distance, and it is this, he believes, that makes the sound louder. On the other hand, if the wind blows in the unfavorable direction, the air distance is increased, making the sound correspondingly fainter.

Let us now look into this quantitatively. Even a moderate wind, of velocity, say, 30 miles/hr, has quite a marked effect on audibility. What relative decrease or increase in air distance will such a wind produce?

The velocity of sound at 15°C is about 760 miles/hr. Compared with this a wind velocity of 30 miles/hr is very small indeed—only about $\frac{1}{25}$ of the velocity of sound—and correspondingly such a wind will decrease or increase the effective distance between source and observer (according as it blows toward the observer or away from him) by only about 4 percent. This is totally inadequate to account for the effect observed—is, in fact, insufficient to be detected by the unaided ear.

We are apt to forget that, owing to viscosity, there must be a transverse gradient of wind velocity, the velocity increasing with height above the ground. It is this velocity gradient that is primarily responsible for the effect of wind on audibility. In still air at a uniform temperature the condensations and rarefactions would spread out from the source as a series of spheres, as indicated in Fig. 11.5(a). But where there is a wind gradient, the tops of these

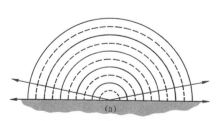
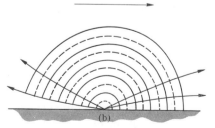

FIGURE 11.5   Effect of wind on acoustic wave surfaces and associated sound rays.

wave surfaces are carried forward more than the lower parts, resulting in a distortion of the wave surfaces of the kind shown in (b). Correspondingly, the sound "rays" will be curved in the manner indicated. Such bending of the sound rays is known as **refraction.** An observer on the ground who is downwind from the source will always receive one of these rays and so will hear the sound well. But if he is upwind and more than a certain distance away, the sound will pass over his head. An interesting corollary is that an observer at a sufficient height above the ground is able to hear just as well upwind as downwind.

Audibility upwind may similarly be improved by sufficiently increasing the height of the source, while the observer remains on the ground. The observer will then receive sound rays that left the source with a sufficient downward component of velocity not to pass over his head, despite their having been bent upward in their journey against the wind.

Not only a transverse velocity gradient of a wind, but also a temperature gradient has important effects on audibility. This is due to the fact that the velocity of sound decreases with falling temperature. Where there is the normal decrease of temperature with increasing altitude, the parts of the wave surfaces near the ground will travel faster than will those higher up, and so all rays will be bent upward. An observer near the ground and at a sufficient distance away from a source which is also near the ground will therefore not hear the sound, the rays passing over his head; or he will hear it only faintly, receiving merely a secondary disturbance spreading out laterally from these rays. As in the case of an unfavorable wind, so also here, a moderate elevation of either source or observer, or both, above the ground improves audibility wonderfully.

The temperature gradient close to the surface of the earth varies considerably with the time of day or night and with weather conditions, and the audibility of sound varies correspondingly. The usual kind of gradient, which is unfavorable to audibility, is particularly marked in the middle of a hot sunny day. The ground then becomes heated strongly by the absorption of radiation from the sun and passes on some of its heat to the adjacent layers of air, while the upper layers remain relatively cool. Audibility is then particularly poor. On the other hand, in the evening following such a day, the earth cools quickly if the sky is clear, and the adjacent layers of air become cooled correspondingly. Under these conditions there is often a reversed temperature gradient near the earth's surface, the temperature increasing with height. Sound rays are then bent downward, and audibility is abnormally good.

ZONES OF AUDIBILITY In the case of sounds of considerable intensity, for example those due to gunfire or other explosions, it is now well established that the audibility does not simply fade away with increasing distance from the source, but that beyond an inner zone, where this actually is observed to happen, there is an outer zone where the sound is again heard quite clearly. Between these two "zones of audibility" there is a "zone of silence," of considerable extent. Usually the inner zone extends out to some 50 km from the source, beyond which the zone of silence begins. Then, between 100 and 200 km from the source, the second zone of audibility begins, and this has a width of the same order as the preceding zone of silence.

In the inner zone, the time at which the sound arrives is that which we should expect from the known velocity of sound. But in the outer zone it arrives substantially later than would be expected on this basis, and this suggests that it has not come by the most direct route. And because the only possible indirect route is one via the upper atmosphere, it is here that we must look for the explanation of the phenomenon.

We have already seen that the normal temperature gradient near the earth's surface bends sound rays upward. For them to be bent down again to reach the surface in the outer zone of audibility it would be necessary to have a reversed temperature gradient higher up, or, alternatively, a progressive change in the composition of the atmosphere of a kind that would favor this refraction. The second alternative can be ruled out at once, in view of reliable estimates that the sound waves must be turned downward at a height of the order of 40 km, where the atmosphere is known to have substantially the same composition as at the earth's surface.

We are left, then, with the first alternative, a reversed temperature gradient. We now know from independent evidence that this is indeed a feature of the upper atmosphere. As we shall see in Chapter 17, the temperature ceases to fall with increasing height beyond about 11 km, remaining constant at about 56°C from this height up to at least 30 km. This is known from measurements made with sounding balloons, which have been made to rise to about 30 km but which are difficult to send much higher. Fortunately, temperature measurements made with sounding balloons cannot only be verified, but extended far beyond balloon range, by exploding grenades carried up by rockets and released at various heights. The location of each grenade at the moment of explosion is found by simultaneous photographic recording of the flash at different stations and calculation from the records. Also, the time that elapses

291

between seeing the flash and hearing the report is observed. When this has been done for a number of heights, it is a simple matter to calculate, from differences, the time it takes for sound to travel from one explosion point to another, and hence the velocity of sound in this region. Then, substitution for v, γ, R, and the effective mean M in Eq. (11.8) gives us T. In this way it has been found that at some 30 to 40 km above the ground the constant-temperature region gives place to a region in which the temperature gradient is positive, the temperature rising, with approximately 10 km of extra height, from about $-56°C$ to something like $+25°C$. This is quantitatively adequate to explain the second zone of audibility.

Audibility in the second zone is not symmetrical about the source; in summer, audibility is strongest in the western part of this zone, whereas in winter it is strongest in the eastern part. This is presumably due to winds blowing at high altitudes, these modifying the audibility pattern that would be produced by the temperature gradient alone.

Reflection

Sound waves are reflected whenever they encounter a discontinuity of the medium, for example when they strike a wall or the surface of a sheet of water.

In Chapter 3, where we considered the reflection of waves traveling along a stretched string, we saw that the absence of displacement at the fixed end means that there must be a reversal of phase accompanying the reflection that occurs here. By corresponding reasoning we see at once that this must be true of sound waves also; the phase of the displacement is necessarily reversed when sound waves are reflected from a surface that does not itself vibrate with an amplitude comparable with that of the waves in air or other gas in which the sound is traveling. It is important to take account of this phase reversal when considering the production of standing waves within a tube; thus phase reversal accompanies reflection at the closed end of an organ pipe.

Incidentally, it should be noted that it is only the displacement, not the pressure, whose phase is reversed on reflection. Thus a condensation is reflected as a condensation and a rarefaction as a rarefaction.

Sound is reflected by all objects, large and small. If the object is small, it reflects sounds diffusely, that is, in all directions—it may, in fact, be said to *scatter* the sound. If it is large, however, the reflection is, in the main, regular; it conforms to the same definite geo-

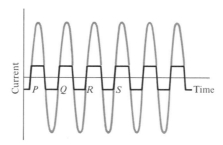

FIGURE 11.6   "Square-wave form" of electric current produced by sound waves in a device for measuring frequency.

metrical laws as apply in optics. Of course, "large" and "small" are relative terms, and it is found that an object which is large enough to reflect sound rays of high pitch and correspondingly short wavelengths in definite directions may reflect sounds of low pitch diffusely. It appears that if the wavelength of a sound is short relative to the dimensions of any particular object, this sound will be reflected regularly, whereas if the wavelength is relatively long the reflection will be diffuse.

Pitch and Frequency

It is explained in more elementary textbooks how the fundamental frequency of a note may be determined with the aid of a siren, or with a sonometer, and how by such means the relation between pitch and frequency has been established. Matching the note in question with that produced by the siren or the sonometer is involved in both these methods.

An equally convenient electrical method of measuring frequency, in which the subjective feature of pitch discrimination is eliminated, is the following. The sound in question is received by a microphone, which converts the mechanical oscillations of the sound waves into oscillations of electric current; that is, it produces alternating current of the same frequency. The latter is then amplified, and in each cycle the currents in both directions are prevented from exceeding a certain magnitude,* giving what is called in the jargon used in this field a "square-wave form"— actually we are not dealing with waves but with a variation of current with time. This variation is such as is indicated by the solid line in the graph of Fig. 11.6. It is then arranged, by the

---

* If this is greater than the amplitude of any of the harmonics, there will be no complication because of these.

appropriate electrical circuitry, that each sudden change of current in a particular sense, such as occurs at the points P, Q, R, S, $\ldots$, gives a trigger action, causing a current pulse of definite known magnitude, that is, a definite total quantity of electricity, to flow in another circuit. Then the total quantity of electrical charge that passes a given point in this circuit in a given time interval must be proportional to the number of pulses; that is, to the frequency. Accordingly, the registration of a current-measuring instrument through which the current is made to pass indicates the frequency on a known scale.

This is an example of an interesting trend in modern technology. Electrical techniques are probably the most highly developed and versatile of any field, and measurements made with their aid are correspondingly convenient and accurate. For this reason, electrical methods are coming into increasingly widespread use for the measurement, not only of electrical, but also of nonelectrical quantities.

Measurements of the frequencies of notes of various pitches have shown two things. One is that, except for a very minor effect of loudness on pitch to be considered presently, the pitch of a note is determined uniquely by its frequency. For example, the note of frequency 440 Hz has been standardized internationally as the tuning A of an orchestra. The other result of these measurements has been to show that, at least to a close approximation, any given musical interval corresponds to a definite *ratio* of frequencies. For example, a rise in pitch by an octave corresponds to a doubling of the frequency. Other intervals, too, are given, each by a definite frequency ratio; thus, for a fourth the ratio is 4/3, for a major third it is 5/4, and for a major tone it is 9/8.

Accurate work in relating pitch to frequency has shown that although frequency is the predominant factor determining the sensation of pitch, loudness also has some slight effect. Up to a frequency of about 2000 Hz, the pitch is found to fall perceptibly with increasing loudness. Above this frequency, pitch and loudness increase and decrease together. This effect of loudness on pitch is, however, very small; except where sounds of relatively high intensity are involved, it may be neglected.

For certain purposes it is useful to have handy, portable, reliable standards of frequency, for example for tuning musical instruments. The best known of these are tuning forks. A tuning fork consists of a bar of steel bent into a narrow U and provided at its base with a stem, for convenience of holding. When the fork is struck, the free ends of the U are set into vibration, alternately approaching and receding from one another, the frequency of the vibration depending on the dimensions of the fork and on the density and elastic properties of the steel. The backward and forward bending (change in curvature) of the arms that occurs during vibration is strongest at the base of the U and becomes progressively less toward the free ends; consequently, the center of mass of the U is farthest from the point of attachment of the stem when the free ends are closest together, and it is nearest to this point when they are farthest apart. The center of mass of the *whole system* tends to be stationary; this means that the stem must vibrate up and down, with the same frequency as the arms of the U vibrate laterally.

Owing to the smallness of the area of the relevant surfaces of its prongs, a vibrating tuning fork itself radiates very little sound into the surrounding air. However, if its stem is held against a table, this is set into vibration, and, because its area is relatively large, it functions as an efficient sounding board, the sound now being heard quite clearly.

The frequency of a tuning fork varies slightly with temperature; hence when using it for precision work it is necessary to allow for any variations from the temperature for which its frequency was determined.

Resonance

If the pitch of a sonometer is adjusted to equality with that of a tuning fork and the sonometer wire is plucked while the stem of the tuning fork is held against the sounding board, the fork will be set into oscillation. Conversely, if the fork is struck and held against the sounding board, this will cause the sonometer wire to vibrate. These are examples of the phenomenon known as **resonance,** the response of a system having a natural frequency of oscillation to a succession of impulses having the same frequency.

Actually, some response to periodic impulses occurs even if the frequency of these is not the same as any one of the set of natural frequencies with which the body can vibrate. We have already noted an extreme example of this in the forced oscillations set up in a table against which the stem of a vibrating tuning fork is held. Here the natural frequencies are totally different. Such forced oscillations are best considered as a rather extreme case of the response of a system having one natural frequency to stimulation at another frequency. If the two frequencies agree fairly closely, the response is usually referred to as resonance, whereas if they are quite different it is generally called a forced oscillation. In fact, one merges into the other; there is no sharp dividing line between resonance and forced oscillation.

A good way to study resonance qualitatively is to use as the vibrating system a partially enclosed body of air. A few examples of partial enclosures whose contained air is capable of oscillating and, under suitable stimulation, does so with quite definite frequencies, are shown in Fig. 11.7. An ordinary milk bottle belongs to type (a) in this figure; its contained air may be made to vibrate, emitting a note of characteristic frequency, by blowing into its mouth in an appropriate manner. This note may be varied by pouring different quantities of water into the bottle; the more water it contains and so the smaller the volume of the remaining air, the higher will be the pitch of the note. If by adjusting the amount of water in the bottle this pitch is tuned by ear to that of a tuning fork, then vibration of the air in the bottle may be stimulated by holding the vibrating fork over its mouth instead of by blowing into it; the note heard will be much louder when the tuning fork is held immediately above the mouth of the bottle than when it is some distance away.

This phenomenon raises an interesting question. We have, it would seem, a fixed rate of energy output from the fork, regardless of whether the resonating system is present or absent. Yet in the case where the resonator is present, the rate of arrival of acoustic energy at the ear is greatly increased. How can this be?

There appears to be only one possible answer: The acoustic power output from the fork must be greater with the resonator present than in its absence. A prong of an isolated fork vibrates in still air and, in doing so, must create eddies, dissipating its energy without generating much sound. But when this prong is placed over the mouth of the resonator, there is a rhythmic movement of the

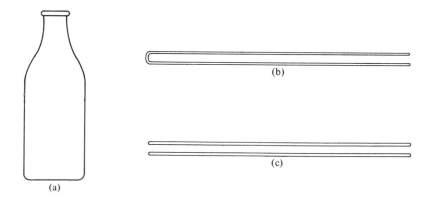

(b)

(c)

(a)

FIGURE 11.7   Different kinds of acoustic resonators.

air surrounding it produced by the vibration of the body of air in the resonator, and this modifies the forces between the prong and the surrounding air. The fork and the vibrating body of air are now acoustically "coupled." As a mathematical analysis of the situation would show, the fork now does more work in generating acoustic energy than in the absence of the resonator.

If, after the quantity of water in the bottle has been adjusted for resonance with the tuning fork, water is added or withdrawn, so changing the natural frequency of the contained air, it will be found that, provided the change has been sufficiently small, the air can still be set in vibration by the fork, producing an increase in the loudness of the sound heard. However, this increase will not be as great as when the natural frequencies have been more closely matched.

Changing the amount of water in a bottle is a rather clumsy way of varying the natural frequency. A more convenient system with which to study the response of a body of air to stimulation at or near its natural frequency is one usually discussed in more elementary texts—a long vertical glass tube whose lower end dips into water, so that the length of the air column above the water may be varied at will. This corresponds to the closed pipe shown in Fig. 11.7(b). At a series of emergent lengths of tube—differing from one another in steps of half-wavelengths—this air column gives maximum resonant response when a vibrating tuning fork is held over the open end. If the length of air column is increased or decreased slightly, above or below one of these values, *some* response is still observed, although this is less than the maximum.

It may be shown mathematically that if the periodic impulses to which a system is exposed are maintained at constant amplitude, the oscillation set up in the system will have the same frequency as that of the impulses, and its amplitude will approach asymptotically a limiting value depending on the strength of the impulses, the damping (resistance to motion), and the difference between the frequency of the impulses and the natural frequency. For impulses of a given strength, the less the damping, the greater will be the amplitude of the oscillation produced.

In Fig. 11.8 the amplitude of the oscillation finally attained is plotted against the frequency f of the impulses for three different degrees of damping, the damping being least for curve 1 and greatest for curve 3. In all cases the maximum occurs for a value of f equal to the natural frequency f_0, and it is, of course, highest where the damping is least. But also, the less the damping, the greater is the relative rate at which the amplitude falls off with the differ-

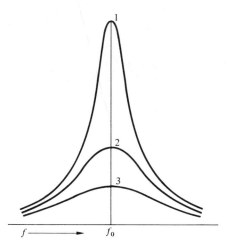

FIGURE 11.8 Sharpness of resonance as determined by degree of damping.

ence between f and f_0. Where the damping is light we have "sharp resonance," whereas if it is heavy the resonance is said to be "flat."

Quality

The quality of a note is found to be determined by the overtones sounded simultaneously with the fundamental and the amplitudes of these relative to that of the fundamental.

These factors determining quality are most conveniently ascertained by analyzing what is known as the **wave form** of the disturbance. By the wave form of a sound is meant a plot of displacement against time. In the case of most musical sounds such a plot repeats itself periodically, with the frequency of the fundamental*; hence all that is necessary is to ascertain the wave form for one period. This gives us complete information concerning overtones, for, as the French mathematician Fourier showed in 1822, "any single-valued periodic function of time can be expressed as a summation of a simple harmonic series having frequencies which are simple multiples of that of the given function." This statement is known as **Fourier's theorem.**

---

* This is not true in cases where not all the overtones are harmonics, that is, where the frequencies are not all simple multiples of that of the fundamental, as in the sound produced by a horn. In such cases the analysis of the sound is less simple than that described here.

The wave form may be ascertained by allowing the sound to impinge on the diaphragm of a special nonresonant microphone that converts mechanical displacements into variations of electric current and then, by means of suitable electrical equipment, to display these on the screen of a cathode-ray tube.

Fourier showed how to analyze mathematically a periodic wave form into its constituent simple harmonic components, with their appropriate relative amplitudes and phases. But this procedure tends to be rather laborious, so the analysis is now usually carried out by means of so-called "harmonic analyzers," machines that have been devised for this purpose. A periodic curve and its simple harmonic components are shown in Fig. 11.9.

Such analysis of wave forms has its counterpart in synthesis. A set of sine curves having frequencies proportional to the natural numbers 1, 2, 3, . . . , and any amplitudes (including zero) may be compounded graphically into a composite wave form, and the corresponding sound may then be produced by mechanical–electrical means, or the synthesis may be carried out in some other way. In principle, this is like compounding a corresponding set of simple harmonic oscillations, for the time variation of the disturbance in any such oscillation may be represented by a sine curve. It is not necessary actually to use such curves in the process of composition; thus, if the simple harmonic disturbances are electrical in nature, they may be combined purely electrically into a composite periodic electrical disturbance. This, in fact, is what is done in the electric organ, the resulting periodic electrical disturbance being finally converted into mechanical oscillations of the diaphragm of a loudspeaker and thence into sound.

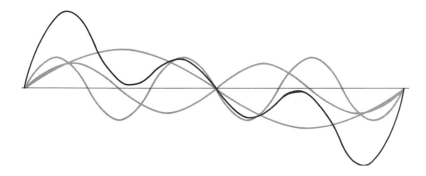

FIGURE 11.9 Fourier components of complex wave disturbance.

For given frequencies and amplitudes of the component sine curves the wave forms vary to a marked degree with the phase relations between the components at, say, the moment when the fundamental disturbance is passing through zero. In Fig. 11.9 all the component disturbances are shown in phase at the extreme left-hand end of the figure, all passing through zero simultaneously and changing in the same direction. But this need not necessarily be the case, and it is easily seen that a change in the phase relations at this point must give rise to a corresponding change in the wave form. This raises the question whether, perhaps, the quality of the corresponding sound is affected also. From experiments designed to answer this question it appears that, in fact, the sound is independent of the phase relations between the harmonics; it depends only on which harmonics are present and on their relative amplitudes.

Sound analysis reveals that the difference between musical sounds and noises is essentially one of regularity of wave form. With musical sounds the wave form, however complex, repeats itself at regular intervals corresponding to the period of the fundamental. The wave form of a noise, on the other hand, has no set regularity of pattern, although it may still possess a certain recognizable "character." Actually there is no definite borderline between musical sounds and noises; some sounds used in music do not strictly conform to the definition of musical sounds given above, in that at least some of the overtones are anharmonic, and so the wave form does not repeat itself *in detail* from one period to the next.

It is important to distinguish between the wave form we are discussing here and the form of a transverse standing wave in a stretched string or of a longitudinal standing wave in an organ pipe. The former is a plot of displacement (or, alternatively, of pressure) at a given point in the path of the sound against *time*, whereas the latter is a plot of displacement (or pressure) at a given instant against *distance*. Although the former is something quite definite for any musical sound heard, the latter changes from one moment to the next.

Problems

11.1   A wire of length 120 cm and mass 56 g is stretched by a load of 15 kg. Calculate the frequency of the fundamental vibration and the velocity of the corresponding transverse waves in the wire.

11.2   A piano wire is tensioned until it emits the same frequency as a 440-Hz tuning fork. Its tension is then increased until 2 beats per second are heard. What is the fractional amount by which the tension of the wire has been increased? (Neglect changes in the dimensions of the wire with tension.)

11.3 A long uniform rope of length l and total mass m is suspended vertically from one end. At a certain instant it is struck lightly at the top end, producing a transverse impulse, which is propagated down the rope. At the same instant a body is released from rest and falls freely from the top of the rope. How far from the bottom does the body pass the impulse? What mass M should be hung from the bottom of the rope so that both reach the bottom together?

11.4 Two strings of masses μ_1 and μ_2 per unit length are joined (for example, knotted) at $x = 0$ and subjected to a tension F. Waves traveling along the first string are partially reflected and partially transmitted at the junction. Assuming that the frequencies of the waves in the two strings are the same and that both strings have the same slope near $x = 0$, show that the ratio R of the amplitudes of the reflected and incident wave trains is given by the equation

$$R = \frac{v_1 - v_2}{v_1 + v_2},$$

where v_1 and v_2 are the velocities of propagation in the first and second strings, respectively. Hence determine the conditions at the junction which produce reflection with a change of phase.

11.5 A sinusoidal wave train of wavelength λ and amplitude A travels along a string of mass μ per unit length having tension F. Determine the energy of the wave train per unit length of the string.

11.6 The time interval between the emission of a pulse of sound and the arrival of its echo from a distant rock wall on a day when the temperature of the air is 20°C is 2.4 sec. How far away is the wall?

11.7 Calculate the rate of increase of the velocity of sound with temperature for air at 20°C.

11.8 A long tube of internal diameter 16 mm is provided with a plunger at one end, and the combination is used as a closed organ pipe of adjustable length in an atmosphere of argon at 20°C and at standard atmospheric pressure. When a tuning fork having a frequency of 512 Hz is held near one end, it is found that the minimum length of the pipe at which it resonates is 15.1 cm, and that it again resonates when its length is 46.2 cm. The fact that the second length is not exactly three times the first is attributable to an end effect, giving an effective length of the pipe in excess of its actual length by a certain fraction f of its diameter. (a) Given, in addition to the above information, that at standard atmospheric pressure and 20°C the density of argon is 1.662 kg/m³, calculate the value of γ for argon. (b) What is the value of the correction factor f for an open end of an organ pipe?

11.9 The atmosphere breathed by divers subjected to high pressures for long periods of time is often a mixture of helium and oxygen rather than nitrogen and oxygen because nitrogen has undesirable effects at high pressure. Divers breathing this artificial atmosphere have some difficulty understanding each other because of a peculiar high-pitched quality imparted to their voices by this atmosphere. Discuss this change in voice pitch.

11.10 As explained in the text, temperatures in the upper atmosphere may be determined by carrying grenades up to different heights in rockets, exploding them, and measuring the time intervals between observation of the flash and reception of the report at ground level. In two cases the heights were 11 and 12 km, and in the latter case the time interval between sighting of the flash and reception of the report was 3.38 sec longer than in the former. What was the mean temperature of the air between the 11-km and 12-km levels? Assume that the ground observation point is directly below the balloon.

11.11 A lightning discharge, which is heard as thunder, occupies only a very small fraction of a second; yet the thunder is heard as a prolonged rumbling. Bearing in mind the great length, and also the crookedness, of the discharge path, can you account for both the long duration of the thunder and its rumbling nature? Could you, from the duration of the thunder in any given case, arrive at an order-of-magnitude estimate of the length of the spark?

11.12 If the pressure amplitude in a train of sound waves is increased by a factor of 4, by what factor is the intensity of the sound (energy flux per unit transverse area) increased?

11.13 Show that the intensity I of sinusoidal sound waves in a gas is given in terms of its pressure amplitude p_0, the velocity v of the waves, and the density ρ of the gas by the equation

$$I = \frac{p_0{}^2}{2\rho v,}$$

and that in terms of the corresponding displacement amplitude y_0 and frequency f the corresponding equation is

$$I = 2\pi^2 \rho v y_0{}^2 f^2.$$

11.14 A man walks on a hard pavement midway between two walls which produce multiple reflection of the sounds of his footsteps. Each time one of his heels strikes the ground he hears a note which he recognizes as 2 octaves below the tuning A of an orchestra, the latter having a frequency of 440 Hz. Given that the temperature is 15°C, calculate the distance between the two walls.

11.15 The velocity of sound in water at 20°C has been measured as 1450 m/sec. What is the adiabatic bulk modulus of elasticity of water at this temperature?

11.16 Stationary waves are set up in a Kundt's tube by the longitudinal vibration of a steel rod 1 m long clamped at the center, and heaps of powder are formed within the tube at intervals of 7.0 cm. What is the ratio of the velocity of rod waves in the steel to that of sound in air at the prevailing temperature?

11.17 Large waves on the surface of deep water travel with a velocity which depends in a simple fashion on g and on their wavelength λ. Given that when $\lambda = 31.5$ m the velocity is 7.00 m/sec, find the relationship among v, g, and λ.

For ripples $v = \sqrt{2\pi} \cdot f(\sigma, \lambda, \rho)$, where σ = surface tension,* ρ = density, and f stands for a function of the variables indicated. Find $f(\sigma, \lambda, \rho)$.

For waves of intermediate size

$$v = \left(\frac{\lambda g}{2\pi} + \frac{2\pi\sigma}{\lambda\rho} \right)^{1/2}.$$

Obtain an expression for the wavelength for which the velocity is a minimum.

11.18   Explain why the sound produced by a plucked string must contain overtones. Comment on the harmonics that will be represented when the string is plucked in the middle.

---

* As will be explained in Chapter 12, this is a property of the surface of a liquid of the nature of a transverse force per unit length exerted in opposite directions on the two sides of any line in the surface.

12

LIQUID SURFACES

Having dealt with the mechanics of particles and of a one-dimensional continuum (wave propagation along a stretched string) and three-dimensional continua, we shall now consider a species of two-dimensional continuum, the surface of a liquid, with particular emphasis on a property of such a surface known as surface tension. Surface tension is important both because of its many technical applications and because of its involvement in other fields of physics.

In this chapter we shall first consider the physical reasons for the existence of a surface tension in liquids and deal qualitatively with the relation between surface tension and surface energy per unit area. Next we shall discuss generally some of the main features of the behavior of liquid surfaces involving surface tension, such as the formation of drops and bubbles, the pressure difference between the inside and outside of these, and the wetting and non-wetting of solids by liquids. Finally, we shall pass in review some of the more important consequences and technical applications of surface tension.

With a basic understanding of surface-tension effects established in the present chapter, we shall be equipped to undertake a theoretical discussion of the role played by these in certain fields of physics not usually classified as mechanical. For example, as we shall see in Chapter 16, the vapor pressure of a liquid not only varies with temperature, but also, because of surface tension, is affected to quite an important degree by the curvature of the liquid surface if the radius of curvature is small. This variation of vapor pressure with curvature plays a key role in the boiling of a liquid and in its condensation from vapor, and is responsible for the formation of raindrops from smaller droplets in clouds under certain conditions. Again, as we shall see in Chapter 20, an electrostatic charge residing on a surface gives the effect of a negative pressure. In the case where the charged surface is that of a liquid this is equivalent to a reduction of its surface tension; and it may be shown that this, in turn, means a reduction of the vapor pressure of a small charged droplet below that of an uncharged droplet of the same size. An important application of this electrical effect is the Wilson cloud chamber, an instrument in which the tracks of ionizing particles traveling through a gas are rendered visible. The ionization chamber has proved a most valuable and versatile instrument in research in the fields of nuclear and particle physics, whose importance in the development of our knowledge in these fields it would be difficult to exaggerate.

Condition of Surface Molecules

To understand the behavior and the spatial distribution of the molecules in the region of a liquid surface it is necessary to inquire more closely into the nature of intermolecular forces than we have done hitherto. Only when we have formed a clear picture of what we mean by a surface shall we be in a position to consider a physical explanation of the surface phenomena to be dealt with in this chapter.

In our discussion of the kinetic theory of gases in Chapter 15 we shall consider collisions of molecules with one another and with the walls of the containing vessel. The fact that colliding molecules rebound means that when their distance of separation falls below a certain value there must be a strong force of mutual repulsion between them. The existence of these repulsive forces must also be inferred from the fact that liquids and solids have a high bulk modulus of elasticity, that is, a low compressibility.

In addition to inferring the existence of repulsive forces at short distances, we must also postulate a mutual attraction between molecules separated by slightly greater distances. There are several lines of evidence leading to this conclusion. Thus without such attraction it would be difficult to understand how a solid rod could sustain a tensile stress. Again, these forces are implied by the existence of a latent heat of evaporation (see Chapter 14), that is, work needed to transfer molecules from a solid or a liquid to its vapor. From such evidence it must be inferred not only that there are attractive as well as repulsive forces between molecules, but that the attractive forces are of relatively longer range, so that for distances of separation appreciably greater than those between nearest neighbors in the interior the resultant forces are attractive.

Within a crystal lattice the atoms are in a state of equilibrium, in the sense that when occupying their normal sites they are not acted upon by any resultant forces, their neighbors being symmetrically disposed about them. In the interior of a liquid, also, a molecule cannot *on the average* be acted on by any resultant force tending to drive it in a particular direction, although at any moment it generally will experience such a force because its neighbors are somewhat asymmetrically disposed about it. In either case we may idealize the situation by supposing that none of the molecules in the interior are acted upon by any resultant force.

Let us now consider a molecule in the surface of a liquid, in the first instance idealizing the liquid as an assemblage of molecules having a constant population density throughout its whole volume, right up to the surface, where it changes abruptly to the much smaller value characteristic of the vapor or gas beyond. Such a surface molecule is acted on not only by its immediate neighbors but also by somewhat more distant ones, and the forces exerted on it by these other molecules have a resultant directed toward the interior of the liquid. A molecule immediately underneath the surface, up to distances of perhaps one or two molecular diameters, might also be subjected to an unbalanced force system, with an inwardly directed resultant, although this must be weaker than in the case of a surface molecule. The resultant forces acting on molecules in the interior (practically zero), just under the surface, at the surface, and just beyond the surface (a vapor molecule) are indicated in Fig. 12.1.

We should naturally expect that in consequence of the forces exerted on them, the surface and near-surface molecules must move from the surface region into the interior of the liquid. This they undoubtedly do, but then their places must immediately be taken by other molecules; for the liquid must continue to have a surface. There is thus a constant interchange between surface molecules and molecules in the interior of the liquid.

The mechanism of this interchange becomes obvious when it is considered that—as we shall see in Chapter 16—a feature of all molecular aggregates at temperatures above the absolute zero is a continuous, restless motion of the molecules in directions distributed

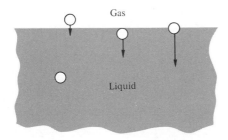

FIGURE 12.1 Forces exerted on molecules at and near the surface of a liquid.

at random and continually changing. Not only are the directions of motion randomly distributed, but there is also a statistical distribution of velocities about a mean. At any particular moment a small proportion of the molecules below the surface will have velocities well in excess of the mean, and such of these as also happen to be moving toward the surface are likely presently to reach it, or even to travel slightly beyond it, before finally falling back into the interior, after which other molecules will make similar excursions.

This picture obliges us to modify our ideas of the nature of a liquid surface. There can after all be no sudden discontinuity between a liquid and a vapor or gas, for different molecules must make different excursions outward, varying with the momenta with which they approach the surface region from within the liquid. Hence, statistically, the population density of molecules must fall off gradually, within a certain region, from that in the interior of the liquid to that well outside. This transition *region* is therefore what constitutes the "surface" of the liquid; it is not a surface in the geometrical sense, but a surface region.

Surface Energy and Surface Tension

Because molecules from the interior of a liquid can reach the surface region ("surface," for short) only against a retarding force, they must, on arrival there, possess a greater potential energy than before. Not only do they have a greater *potential* energy than the molecules inside, but their *total* energy is greater than the average for the interior of the liquid, because they have been derived selectively from the region below the surface on the basis of their relatively high kinetic energies. Hence, the greater the surface area of a given quantity of liquid at a given temperature,* the greater must be its total energy. Accordingly, we are led to the concept of "surface energy"—a mean energy per molecule possessed by surface molecules in excess of that characteristic of the molecules in the interior.

This concept of surface energy suggests an explanation of the formation of liquid drops, based on an analogy with the simple pendulum. Just as the equilibrium position of the bob of a simple pendulum is that in which its energy is a minimum, so also with a

---

* As is indicated in Chapters 15 and 16, the temperature may be taken as determined by the mean kinetic energy of the molecules in the *interior* of the liquid.

liquid; the configuration toward which it tends is that in which its surface area, and so its energy, is a minimum. The configuration that has the smallest surface area for a given volume is that of a sphere, and so it is that freely falling liquid drops are spherical, as are bubbles of gas within a liquid.*

If the surface of a liquid is to be increased without drawing upon the energy of the liquid itself for the extra energy required, it will obviously be necessary to transfer energy to it from some external source, for example by doing work on it. Actually it may be shown that to increase the area at constant temperature, or "isothermally," it is necessary not only to do work on the liquid but also to transfer a certain amount of heat to it.† The mechanical item is, however, quantitatively the more important of the two.

An interesting way to show, experimentally, that work must be performed in order to increase the area of a liquid surface is shown in Fig. 12.2. A wire is bent to form three sides of a rectangle and suspended from one arm of a balance, as shown, the ends dipping into a soap solution contained in a dish. After the system has been counterpoised, the wire frame is pushed down until the horizontal portion at the top touches the soap solution and is then released. A thin soap film is now drawn up by the frame as it ascends, and it is found that extra standard masses have to be added to the counterpoising scale pan to restore the balance with the frame in its former position. The extra masses are of a higher order of magnitude than the mass of the soap film, which is very thin—usually 10^{-3} mm or less in thickness.‡ Thus if, for example, the soap film is 5 cm wide, 5 cm high, and 10^{-4} cm thick, its volume is only 2.5×10^{-3} cm^3 and its mass would be about 2.5×10^{-3} g. The extra mass required in the counterpoising scale pan would, however, be about $\frac{1}{4}$ g, that is, about 100 times the mass of the soap film. Thus the film must exert a force downward on the horizontal part of the frame greatly in excess of its own weight, and if the frame moves upward, increasing the total surface area of the film, a corresponding amount of work is performed.

The transverse force per unit length along a surface is known as the **surface tension** of the liquid. Thus if the force F is exerted

---

* Falling liquid drops are truly spherical only if they fall in a vacuum. However, even in a resistive medium, the distortion produced by viscosity effects is usually very small indeed; to a close approximation the drops are still spherical. The same is also true of bubbles of gas rising through a liquid, provided they are not too large.

† If the supply of heat is withheld, the temperature falls.

‡ The thickness of a transparent film is readily measurable by an optical interference method.

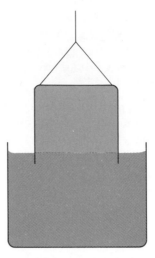

FIGURE 12.2 Simple device for demonstrating surface tension.

along the surface at right angles to a length l, the surface tension σ is defined by the equation

$$\sigma = \frac{F}{l}. \tag{12.1}$$

For example, in the experiment just considered, if W is the weight of the extra standard masses necessary to restore the balance with the soap film within the frame, and the width of the frame is d, then, because the film has two surfaces (front and back), and the total distance l across which the surface tension is exerted is therefore $2d$, we should have

$$\sigma = \frac{W}{2d}.$$

The value of σ for a soap solution is about 25 dyn/cm.

While surface tension is a transverse force per unit length exerted *along the surface*, a "tangential" force, the basic reason for the existence of a surface tension is, as we have seen, the *inwardly directed* force experienced by surface molecules. Even though we can think of no direct mechanism to account for the exertion of a tangential force per unit length, we are obliged to postulate the existence of such a force in order to account for the work per unit area associated with an increase in the surface. We are reminded of the alternative definition of force as the space rate of change of energy. We have an analogous situation here, and it would appear that, here at least, energy is a more fundamental concept than force.

Drops and Bubbles

Let us now discuss certain quantitative relationships concerning drops and bubbles; we shall see later that these have important applications.

As our first example, let us consider a drop falling freely under gravity, for example a drop of oil or mercury falling in a vacuum. Because each portion of such a drop must have the same acceleration, g, the resultant of all forces acting on it must be simply its own weight; that is, such forces as may act on it other than its weight must cancel each other out.

It will now be interesting to see what forces are exerted on the portion of the drop indicated by shading in Fig. 12.3, in addition to its weight. Let us consider a short element of length l of the circumference of the circle which divides the shaded portion from the remainder of the drop and whose radius is $r \sin \theta$. There must be a force σl acting transversely to this element and tangentially to the surface, and the component of this parallel to the center line shown is $\sigma l \sin \theta$. The sum of all such components around the circumference of the circle is $\sigma \sin \theta \cdot 2\pi r \sin \theta$, that is, $2\pi r \sigma \sin^2 \theta$, acting in the upward direction in the figure. But because the resultant of all forces exerted on this portion of the drop, apart from its weight, is zero, there must be an equal force acting downward. The only possible force that can act downward is one that would be due to a pressure p within the drop, and the condition it must satisfy is that $\pi r^2 \sin^2 \theta \cdot p$ shall be equal to $2\pi r \sigma \sin^2 \theta$, that is, that

$$p = \frac{2\sigma}{r}. \tag{12.2}$$

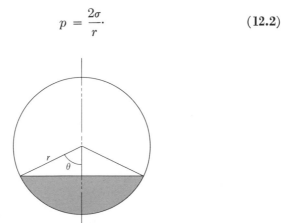

FIGURE 12.3 Portion of liquid drop (shaded) in equilibrium under the action of surface-tension and pressure forces.

Hence the liquid inside the drop must be subjected to a pressure of this magnitude.

Next let us consider a small drop falling through a resistive medium, which has attained its terminal velocity such that the upward force due to viscosity just balances its weight. The tiny droplets constituting a cloud or mist come into this category. Because the volume, and so the weight, decreases much faster with diminishing radius than does the total surface-tension force for a given value of θ (see Fig. 12.3), the weight and the opposing upward force due to viscosity both become relatively negligible items for a very small drop and so need not be taken into consideration. We may therefore regard any portion of the drop, for example again the shaded portion indicated in Fig. 12.3, as in equilibrium under the action of pressure and surface-tension forces alone. If the pressure of the atmosphere through which the drop is falling is P, the upward force on the horizontal projected area of the curved surface is $\pi r^2 \sin^2 \theta \cdot P$; the upward force due to surface tension is, as before, $2\pi r\sigma \sin^2 \theta$; and for equilibrium we must have a pressure $P + p$ inside such that, again,

$$p = \frac{2\sigma}{r}.\qquad\qquad (12.2)$$

We see that in both cases there must be a pressure inside the drop exceeding that outside by $2\sigma/r$.

Corresponding considerations apply to a small bubble of gas within a liquid, with the Archimedean upthrust replacing weight, this becoming a relatively negligible item if the bubble is sufficiently small. In considering the equilibrium of a portion of gas it may be felt to be hardly justifiable to invoke the same surface-tension forces as figured previously in the discussion of the equilibrium of a portion of liquid. However, this difficulty may be avoided by including with the portion of the gas bubble in question a thin layer of the liquid adjacent to it, as indicated in Fig. 12.4. The remainder of the reasoning is then exactly the same as that just discussed for a small drop, and leads to the corresponding conclusion—that the pressure within the bubble exceeds that outside by $2\sigma/r$.

It may be shown that in cases where the surface is not spherical a corresponding equation gives the difference of pressure between its two sides. Characteristic of *any* surface are two radii of curvature: the radii of maximum and minimum curvature, occurring in planes perpendicular both to the surface and to one another. Let these radii be r_1 and r_2, and let them be reckoned positive or negative according as the centers of curvature are on the same side

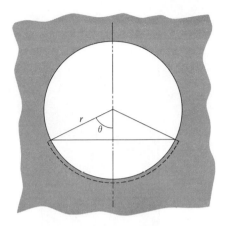

FIGURE 12.4 Portion of liquid adjacent to bubble in equilibrium under the action of surface-tension and pressure forces.

of the surface as the observer or on the opposite side. Then it may be shown that the pressure on the side of the observer exceeds that on the opposite side by p, where

$$p = \sigma \left(\frac{1}{r_1} + \frac{1}{r_2} \right).$$ (12.3)

Obviously the equation

$$p = \frac{2\sigma}{r}$$ (12.2)

applying to a spherical surface merely represents the special case where r_1 and r_2 are equal.

A spherical soap bubble may be regarded as a gas bubble within a very slightly larger liquid drop, and so the excess-pressure item $2\sigma/r$ arises twice, once for the inner surface and once for the outer. Accordingly, the pressure inside a soap bubble must exceed that outside by $4\sigma/r$. This result may also readily be obtained directly, by considering the equilibrium of an appropriate portion of the soap film.

Capillarity

We should naturally expect solids, as well as liquids, to have a surface tension, even though its effects are not so readily observable, owing to the resistance of solids to change of form. If such a surface tension exists, it might at least be expected to manifest itself in some way at the regions of contact between solids and

313

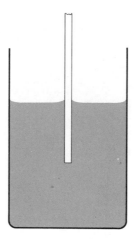

FIGURE 12.5 Section through surface of water in contact with glass.

liquids. We do, in fact, observe variations in the behavior of solids in contact with a given liquid, with regard to the angle at which the liquid meets the solid. For example, water makes contact with clean glass in such a way that the two surfaces are tangential to one another, in the manner indicated in Fig. 12.5, and is correspondingly said to "wet" the glass. On the other hand, there are certain other surfaces that water does not meet tangentially, that is, does not wet; for example fingernails and toenails, and feathers.

A well-known example of nonwetting is mercury in contact with glass, as in the case where a globule of mercury rests on a horizontal glass plate. This assumes the form indicated in Fig. 12.6, meeting the glass at a definite angle characteristic of the two materials. The angle is conveniently specified as that shown in the figure, α, which for mercury and glass is obtuse and has a value of about 140°. This angle is known as the **angle of contact.**

The existence of a definite angle of contact necessarily leads to a rise or fall of a liquid in a capillary tube, except only in such cases where the angle of contact happens to be 90°. If the angle is acute, the liquid rises, while if it is obtuse it falls. This follows at once from the relation we have found between the curvature of a

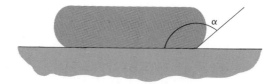

FIGURE 12.6 Globule of mercury resting on a glass plate.

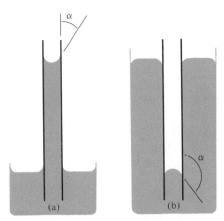

FIGURE 12.7 Rise (*a*) or fall (*b*) of a liquid in a narrow tube inserted in a body of the liquid, according to whether or not the angle of contact is acute or obtuse.

surface and the pressure difference between its two sides. This phenomenon is known as **capillarity.**

If the tube is sufficiently narrow, we may neglect the height range occupied by the curved meniscus in comparison with the total vertical distance through which the liquid rises or falls. The pressure under the surface may then be regarded as everywhere the same, and with a constant pressure difference between the two sides of the surface the curvature must be the same over the whole meniscus; that is, the meniscus must be spherical. We then have the state of affairs depicted in Fig. 12.7. In part (*a*) the pressure above the surface of the liquid is atmospheric, while immediately below the surface it is $\rho g h$ below atmospheric. But if the internal radius of the tube is r, the radius of the meniscus is, as we see from Fig. 12.8, $r \sec \alpha$. Hence Eq. (12.2) becomes, in the present case,

$$\rho g h = \frac{2\sigma}{r \sec \alpha},$$

or

$$h = \frac{2\sigma \cos \alpha}{\rho g r}. \tag{12.4}$$

Similarly, in Fig. 12.7(b), the pressure above the meniscus is atmospheric, while just below the surface it exceeds atmospheric pressure by $\rho g h'$, and Eq. (12.2) becomes

$$h' = \frac{2\sigma \cos \theta}{\rho g r}.$$

315

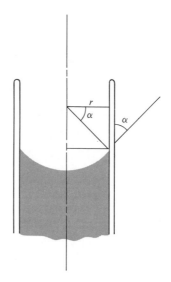

FIGURE 12.8

We may, if we wish, express both results algebraically by the single equation (12.4), where h is the height (negative where the meniscus is depressed) to which the liquid *rises* in the capillary.

Measurement of Surface Tension

Of the various methods available for measuring the surface tension of a liquid, the bubbling method developed by Jaeger is probably the best, and we shall accordingly consider only this method here.

The apparatus is shown in Fig. 12.9. A drawn-down tube with a clean-cut end is inserted to a measured depth h in the liquid and air bubbled through, the air pressure necessary for this being measured with a manometer. The reading of the manometer minus the

FIGURE 12.9   Jaeger's method of measuring surface tension.

FIGURE 12.10 Stages in the formation of a bubble.

pressure (above atmospheric) corresponding to the depth of immersion of the tube in the liquid then gives us the amount δp by which the maximum pressure inside the bubbles exceeds that outside. This pressure must be provided to sustain continuous bubbling, and its value is related to the minimum radius of curvature r and the surface tension as in Eq. (12.2), with δp substituted for p. As will be seen from Fig. 12.10, which shows a bubble at successive stages of its growth, the bubble has its minimum radius of curvature when it is hemispherical, and its radius is then the same as the internal radius of the tube at the orifice. The latter may be measured with a traveling microscope. Both δp and r being known, the value of the surface tension may then be calculated from the equation

$$\sigma = \frac{r\,\delta p}{2}.$$

Some typical values of surface tension are given in Table 12.1.

TABLE 12.1 Surface Tensions of Various Liquids

Liquid	Temp. (°C)	Surface tension (dyn/cm)
Water	0	75.6
	20	72.8
	100	61.5
Ethyl alcohol	0	23.5
	20	21.7
	70	17.3
Olive oil	20	33.5
Saturated solution of sodium oleate (soap solution)	20	25.0
Glass	Working range	400–500
Lead (in contact with hydrogen)	327	444
Mercury	20	485
Oxygen	−193	15.7

Stability of Liquid Films

It is a matter of everyday observation that certain liquids, for example soap solution, seawater, and beer, froth readily, air bubbles that are formed on the surfaces of these remaining for a considerable time before they burst. On the other hand, such air bubbles as may form on the surface of water, alcohol, or mercury are ephemeral in the extreme. What is it that determines which of these types of behavior a liquid will exhibit?

If a list were drawn up of all the liquids that froth readily, it would be observed that they are all impure. This strongly suggests that the feature which renders thin films of these liquids relatively stable is their composite nature.

Before embarking on a discussion of the special properties of composite liquids, let us consider what is implied by the statement that a film is stable. It means that if, for any reason, it begins to stretch and so becomes relatively thin locally, a system of forces is set up that drives liquid in toward the thinner part of the film from neighboring thicker regions, so tending to restore the uniformity of the film. Such a situation could arise in either of two ways, represented by parts (a) and (b) of Fig. 12.11. In part (a) the concavity of the two sides of the film associated with the local thinning gives a system of forces that evidently tends to restore uniformity of the film. The situation represented by part (a) must arise in all cases where there is local thinning, and corresponds to the tendency of the total surface area to become a minimum. However, not all liquid films are found in practice to be stable, and from this we must infer that the forces due to the very slight differences of curvature that actually arise are too weak to counteract such disruptive influences as air convection currents, and so on, to which films are normally exposed. We must therefore look for an alternative, much more powerful, mechanism, making for stability. Such a mechanism is indicated in Fig. 12.11(b), the forces arising from a

(a)

FIGURE 12.11   Possible mechanisms that might account for the stability of a liquid film.

surface-tension gradient brought about by the local thinning. The forces would be as shown if the surface tension were greatest where the film is thinnest.

Such a variation of surface tension could hardly occur in a pure liquid; certainly it could not in one whose temperature is uniform, because for each pure liquid the surface tension is a unique function of temperature. The lack of stability of films of pure liquids would thus be accounted for.

Let us now consider the equilibrium between a dissolved material, for example soap, in the surface and in the body of a liquid. If we regard the surface as in a state intermediate between those of liquid and vapor, and thus different from that of the bulk liquid, we should hardly expect that under equilibrium conditions the ratio of soap to water molecules in the surface would be the same as in the body of the liquid. If it is different, and there is a higher concentration of soap at the surface than in the interior of the film, then any stretching, increasing as it does the amount of surface, would necessitate a redistribution of soap as between surface and interior for the restoration of equilibrium. It is useful to think of the interior and the surface as competing for soap molecules. We have supposed that for equilibrium the surface needs proportionately more soap molecules than does the interior. Stretching a film and so producing more surface would then (so to speak) increase the demand, but with only the same number of soap molecules to go around. Thus the total available supply of these molecules would be more thinly spread; there would be relative impoverishment of both surface and interior. Owing to the fall in the concentration of soap molecules in the surface, the surface tension is reduced below the value characteristic of pure water by a smaller amount than before; that is, it is increased beyond its former value. And, if the stretching is uneven, the resulting surface tension must be greatest where the film has been thinned most, as indicated in Fig. 12.11(b). Thus we have the necessary condition for stability.

Applications and Consequences of Surface Tension

The role played by surface tension in everyday life and in technology is not generally appreciated. Its importance is indicated by the following examples.

GLASS WORKING Either consciously or unconsciously a glass worker has almost continually to take account of surface tension, and the success of his work often depends fundamentally on its

319

FIGURE 12.12 Section through leg of insect supported on surface of water.

operation. For example, in the blowing of a bulb, it is because of the action of surface-tension forces that the bulb acquires its spherical shape. And in the making of a joint the smoothness of the glass and the uniformity of wall thickness at the junction between the two components is due to the action of these same forces. Surface tension also plays a key role in several other aspects of glass working.

CAPILLARITY Capillarity is essentially a matter of wetting or nonwetting. Its importance will perhaps be sufficiently indicated by the following three examples: the spreading of moisture through soils, the rising of oil up the wicks of oil lamps, and the spreading of oil over the relevant surfaces in lubrication.

INSECTS WALKING ON WATER The legs of insects that are not wet by water may be supported on the surface of the water after the manner shown in Fig. 12.12. In the region where the water makes contact with a leg, the surface-tension forces have an upward component, and the resultant of these forces exerted on all the legs plus the product of the pressure difference between the upper and lower surfaces and the relevant projected area is equal to the weight of the insect.

On the same principle, boats that float on water may be constructed of wire mesh coated with wax so that the wire is not wet by the water. For the water to be able to squeeze through the holes in the mesh, the curvature of its surface would have to be greater than can be brought about by the available pressure difference

FIGURE 12.13 Model boat constructed of non-wettable wire mesh floating on water.

between the two sides of the mesh. The actual situation where the boat is supported on the water is shown in Fig. 12.13.

RAINPROOF FABRICS Fabrics which, while letting air through freely, are rainproof function in the same manner as the sieve boat just discussed. The cloth is made nonwettable by impregnating it with resin, by a process that leaves a thin film of resin on each fiber. Raindrops falling on a garment made of cloth so treated are, in the absence of the requisite pressure, unable to penetrate the spaces between the fibers, and simply roll off.

FLOTATION OF MINERALS This is an industrial process by which particles of different kinds of material are differentiated and sorted out according as they are or are not wet by a liquid in which they are immersed. A liquid is chosen which, in addition to wetting one kind of particle and not the other, froths readily. The rock with its contained ore is crushed, and the finely divided material then introduced into a large bath containing the appropriate liquid. This is agitated and caused to froth copiously by blowing air through it. If a particle that is wetted encounters an air bubble, it will, as Fig. 12.14(a) shows, be unable to enter it; if it did begin to do so it would immediately be ejected by surface-tension forces. Such particles will therefore tend to collect at the bottom.* On the other hand, a particle that is not wetted, on encountering an air bubble, will necessarily be drawn into it, as indicated in Fig. 12.14(b), and if the particle is small enough, the Archimedean upthrust on the bubble and its imprisoned particle will exceed the weight of the latter, so that they will rise to the top. By drawing off the froth the nonwettable particles may be separated from the wettable.

In practice, the liquid is always water containing a suitable additive, known as a "frother," which causes agitation of the mixture to produce a reasonably stable foam. Other materials, known as "collectors," "activators," or "depressants," may also be added to impart the necessary properties to the particles of the mixture. Thus a collector is a material that is held to the surface, or "adsorbed," probably as a monomolecular layer, on the mineral to be floated, and which prevents it from being wetted. A depressant, on the other hand, causes to be wetted the type of particle that it is not desired to float. An activator, found necessary in some cases, is a substance that reacts with the surface molecules of the mineral in such a way as to enable the collector to be adsorbed.

---

* It is assumed that both kinds of particles are denser than the liquid, so that Archimedean upthrusts are not sufficient to bring either to the surface.

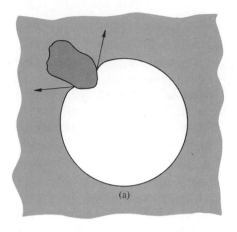

(a)

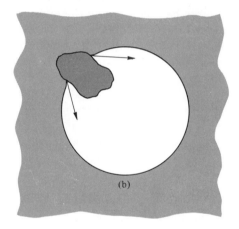

(b)

FIGURE 12.14   (a) Particle wet by liquid in flotation tank unable to
enter air bubble.
(b) Unwettable particle being drawn into bubble.

Flotation is the process almost universally used for the concentration of the valuable materials in low-grade ores, which are
first crushed to a powder and then treated as described above.

Problems

12.1   Falling drops of different liquids formed under different conditions have various sizes. In all cases the release of the drops sets them
into vibration with respect to their horizontal plane of symmetry. Find by

the method of dimensions how the period of vibration depends on the radius of the drop, its density, and the surface tension of the liquid.

12.2   How much work must be done to divide a droplet of mercury of diameter 4 mm into two equal smaller droplets at 20°C? ($\sigma_{Hg} = 485$ dyn/cm.)

12.3   A liquid at atmospheric pressure whose surface tension is 50 dyn/cm is heated until bubbles of vapor begin to form and grow within it. Find, in atmospheres, the pressure within a bubble necessary to prevent it from collapsing when its diameter is (a) 250 Å, (b) 1000 Å, and (c) 4000 Å. (1 Å = 10^{-10} m. Neglect variation of surface tension with temperature.)

12.4   Chips of ceramic material may be added to water in a test tube being heated by a bunsen flame in order to keep the contents of the tube from being expelled by the sudden creation of a single large bubble of vapor. Explain.

12.5   In an experiment for finding the surface tension of a liquid by the bubbling method, the exit end of the air-supply tube has an internal diameter of 0.78 mm and is 3.2 cm below the surface of the liquid. Given that the density of the liquid is 0.79 g/cm³ and that to sustain bubbling the pressure of the air has to be maintained above atmospheric by an amount which reads 38 mm on a water manometer, calculate the surface tension of the liquid.

12.6   A closed glass vessel which is wet by water has a small circular pinhole of diameter 0.04 mm at the bottom. It contains air at atmospheric pressure and 25°C, and it is immersed in a tank of water, also at 25°C, with the pinhole 30 cm below the surface. Find (a) whether water will leak into the vessel, and (b) the approximate temperature to which the system must be raised for air to bubble out of the vessel. (Pressure of air $\propto 273 + \Theta$, where Θ is the Celsius temperature.)

12.7   A rectangular dish is made from wire gauze, each mesh of which has internal dimensions of 1 mm square, and the surface of the wire is treated with a material such that it is not wet by water. Approximately to what depth may the bottom of the dish be depressed below the free surface before water begins to flow in through the meshes?

12.8   Two coaxial circles of wire, each of diameter 5 cm, are held some distance apart with a cylindrical soap bubble between them and with bubbles on the outer sides which are sections of spheres (see the figure). Calculate the radius of curvature of the ends.

Figure for problem 12.8.

12.9   How high will water, which wets glass, rise in a vertical glass tube of internal diameter 0.5 mm?

12.10   A fine glass capillary tube of the form shown in section in the figure dips into water. How far will the water rise up the tube? Explain your answer.

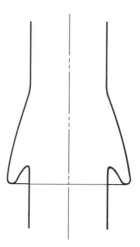

Figure for problem 12.10.

12.11   A perpetual-motion device consists of a fine glass capillary tube dipping into water and bent over at the top. Water rises up the tube by capillary action, and on reaching the open end pours out onto a water-wheel. Would such a device work? If not, why not?

12.12   Bearing in mind the explanation of the origin of surface tension, discuss the following statement: "At a glass–water interface, the surface tension of water is negative."

12.13   Calculate how high a liquid of specific gravity 1.1 will rise in a vertical tube of diameter 0.6 mm, given that the surface tension of this liquid is 0.04 N/m and the angle of contact is 14°.

12.14   What correction should be applied to the readings of a mercury barometer whose tube has an internal diameter of 4 mm?

12.15   How does surface tension affect the operation of a hydrometer? In general, would you expect the effect to be more serious with a variable-immersion than with a fixed-immersion hydrometer, or vice versa?

12.16   Two vertical glass plates are held parallel to one another and 1 mm apart in a basin of water so that they protrude some distance above the water surface. To what height will the water rise between them, and with what force per unit length will the plates be drawn toward one another? Explain the mechanism whereby this force is exerted on the plates.

12.17   A ring of glass is cut from a tube of 8.4 cm internal diameter and 8.8 cm external diameter. This ring, with its lower edge horizontal, is suspended from one arm of a balance so that the lower edge is in contact with water in a dish, and is in the same plane as the free horizontal surface of the water remote from the ring. It is found that an additional mass of

3.97 g must be placed on the other scale pan to compensate for the pull of surface tension on the ring. Calculate the surface tension of the water. If the edge of the ring were (a) 1 mm higher or (b) 1 mm lower than the free horizontal surface of the water, how would this affect the extra mass necessary to achieve a balance?

12.18 The air entering the lungs passes through subdividing passages ending finally in several hundred million small sacs called alveolar sacs. Consider the sacs to be roughly spherical and to have a radius of about 10^{-1} mm when inflated. If the sacs were lined with ordinary water, what pressure would be required to keep them inflated, assuming the pressure acted only against the surface tension of the water? What difficulty arises when you try to explain the breathing process on this basis if it is remembered that the sacs are not all of the same size? Nature has solved this problem in a remarkable way. The surface tension of the fluid lining the sacs is *not constant*. As the sac size decreases the surface tension also decreases, ranging from about 50 dyn/cm down to about 5 dyn/cm.

12.19 In certain beverages containing alcohol a thin film may be observed to climb some distance up the edge of the glass. The top edge of the film swells as liquid is "pumped" into it and droplets or "wine tears" are formed which drop back down the side of the glass. The explanation rests on the fact that alcohol reduces the surface tension of water and that alcohol evaporates more quickly than water from the large surface area of a thin film. Complete the explanation qualitatively.

SECTION VI

HEAT

13

TEMPERATURE AND ITS MEASUREMENT

In the subject of heat, one of the most basic concepts is that of temperature; virtually all quantitative work in this branch of physics involves temperature measurement in one way or another. We begin by defining temperature as precisely and scientifically as we know how to at this stage, hoping that we may later be able to establish a link with mechanical or other concepts in physics. The nature of this link, in which temperature is given a statistical meaning in the mechanics of large assemblages of particles, will appear in Chapter 15, in which we shall consider the kinetic theory of gases.

Definition of Temperature

Temperature is usually defined, in the first instance, in terms of a selected property of a selected body or material. If P is such a property, the definining equation for the temperature Θ becomes

$$\Theta = f(P),$$

where f is an agreed function. Concerning this, the simplest and most obvious choice would be a linear relationship, according to which

$$\Theta' - \Theta \propto P' - P, \tag{13.1}$$

where $\Theta' - \Theta$ and $P' - P$ denote the changes in temperature and property, respectively. If, with this as our defining relation, we further decide on a proportionality factor and a reference point, we then have a corresponding **scale of temperature.**

However, before we can proceed to a detailed discussion of scales of temperature, it is necessary to consider a serious limitation of *any* temperature scale defined as above—that it applies only to the particular body or system whose property P is used in the definition. Thus let us suppose that P denotes the length of a rod of brass. Then, according to our definition, measurements of this length will only give us information concerning the temperature of this brass rod; we shall have no basis for determining the temperature of any

329

other body, for example a beaker of water. It would be no escape from this difficulty to insert the brass rod in the water and then to measure its length; this would still only enable us to find the temperature of the rod. We cannot simply assert that the temperatures of the rod and the water when in contact are the same; indeed with the definition as it stands, and in the absence of a separate definition of the temperature of the water, such a statement would be meaningless.

We can extricate ourselves from this situation, where every body is its own thermometer, only by invoking the laws of thermal equilibrium and redefining temperature. Let us suppose that two bodies, A and B, are brought into intimate contact with one another, and that we observe, say, their dimensions as indications of their temperatures. Then it will be found that after a sufficient lapse of time these dimensions remain constant. When this state of affairs has been attained, A and B are said to be in **thermal equilibrium** with one another. Now let a third body, Θ, sufficiently small not to be capable of disturbing either A or B appreciably, be brought into thermal equilibrium first with A and then with B, some property P of Θ being used as an indication of its temperature. Then the final value of P will be found to be the same on each occasion.

If now the temperatures of A and B are defined, not, as hitherto, in terms of their own properties, but of the registrations (the values of the property P) of Θ when in thermal equilibrium with them, the results of our experiments mean that when two bodies (A and B) are in thermal equilibrium with one another they are at the same temperature. This law is quite general, and must therefore apply to Θ also; that is, if Θ is in thermal equilibrium with a body, then Θ must have the same temperature as this body. The body Θ in terms of whose selected property temperatures are defined is conveniently designated a **thermometer.** We may now proceed to a detailed discussion of scales of temperature.

We may represent our defining relation (13.1) by any straight-line graph, such as one of those drawn in Fig. 13.1, in which the temperature axis is numbered to an arbitrary scale of **degrees.** Then each straight line we choose to construct will define a corresponding scale of temperature. Each is characterized by its own particular slope, corresponding to the proportionality factor in (13.1), and its zero, or the value of the property P to which zero temperature is assigned.

For the temperature scales now in use, both the proportionality factor and the zero were originally made to depend on two **fixed points,** the temperatures assigned to particular values of the

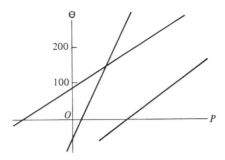

FIGURE 13.1   Graphical representation of temperatures Θ on different
scales defined as varying linearly with a selected property
P of some body or material.

property. In 1954 the two most important scales, the Celsius* and
the Kelvin scales, were redefined, but the numerical changes consequent on the changes in definition are so small that for the purposes
of the present book they may be disregarded. Let us, therefore, for
simplicity, now consider the original definitions. These, besides
being in practice almost exactly equivalent to those by which they
have now been replaced, will provide the necessary background for
appreciation of the latter, which we shall discuss later in this chapter
and in Chapter 16.

Let us, for example, suppose that P represents the length of
the stem of an ordinary mercury-in-glass thermometer occupied
by mercury, and let P_F and P_B denote the values of this length at
the "freezing point" and "boiling point" (at standard atmospheric
pressure) of water, respectively. On the Fahrenheit scale the temperature assigned to the freezing point is 32° and that assigned
to the boiling point is 212°. On the Celsius scale the corresponding
temperatures are 0° and 100°, respectively, and in the "absolute"
or Kelvin scale the values of the temperatures are 273.15° and
373.15°, respectively. The lines correspondingly labeled in Fig. 13.2
represent these scales. It is readily seen that if F, C, and K denote
the numerical values of the temperature on the Fahrenheit, Celsius,
and Kelvin scales, respectively, at a particular thermal condition,
for example, the melting point of a given material, then

$$C = K - 273.15 = \tfrac{5}{9}(F - 32).$$

---

*In 1948 the Ninth General Conference on Weights and Measures
recommended that the name "centigrade," until that time in common use in
English-speaking countries for this scale, should be abandoned and "Celsius"
used instead.

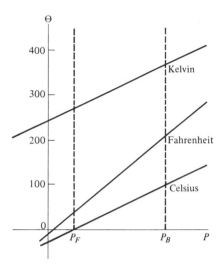

FIGURE 13.2 Fahrenheit, Celsius, and Kelvin temperatures represented graphically in terms of a property P on which their definitions are based.

The derivation of the Kelvin from the Celsius scale has a particular significance of its own, to be discussed later, and is of considerable importance. On the other hand, the question of the relative merits of the Celsius and Fahrenheit scales is of no scientific interest whatsoever. From a practical point of view it is obviously undesirable to use them both. Actually, the Fahrenheit scale is rapidly falling into disuse in scientific literature. We shall, therefore, not concern ourselves with this scale henceforward.

If, in the defining relation (13.1) for temperature, we insert the values of the proportionality factor and zero appropriate to the Celsius scale, then, as is at once evident from Fig. 13.3, we obtain the equation

$$\frac{\Theta - 0}{P - P_0} = \frac{100}{P_{100} - P_0},$$

or

$$\Theta = \frac{P - P_0}{P_{100} - P_0}, \tag{13.2}$$

defining the Celsius temperature in terms of the value of the property P, the symbols P_0 and P_{100} denoting the values of the property at the freezing point (0°C) and boiling point (100°C) of water, respectively.

We have so far used the designation "Celsius scale" in the general meaning intended by its originator, Celsius, based on a

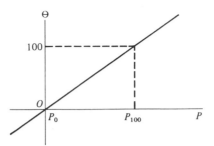

FIGURE 13.3 Celsius temperature ϴ represented graphically in terms of a property P with special reference to the values of P at the freezing and boiling points of water.

somewhat arbitrary choice of the property P, for example the apparent volume of a quantity of mercury in a glass container. It is important to realize that with the scale so defined there are as many Celsius scales, all different, as there are usable properties of different materials or systems on which to base the defining equation (13.2). All these scales must, of course, by definition, agree with one another at the two fixed points 0° and 100°, but in general they will not do so elsewhere.

To see this, it will be helpful to consider an example. Let us consider in the first instance the Celsius scale as defined by Eq. (13.2), where P denotes the length of the stem of a mercury-in-glass thermometer occupied by mercury. It is this length we are concerned with, so let us now write l instead of P, and let the temperature scale defined by this thermometer be denoted by Θ_{Hg}. On the basis of *this* temperature scale, l, *by definition*, increases uniformly with temperature, being expressed by the equation

$$ l = l_0(1 + \epsilon \Theta_{Hg}), $$

where ϵ is written for $(l_{100} - l_0)/100l_0$.

Similarly, let a Celsius temperature Θ_R be based on the electrical resistance R of a given coil of metal wire. This "resistance temperature" will be defined by another equation of the form of (13.2),

$$ \Theta_R = 100\,\frac{R - R_0}{R_{100} - R_0}, $$

and R may accordingly be expressed in terms of Θ_R by the equation

$$ R = R_0(1 + \gamma \Theta_R), $$

where γ is written for $(R_{100} - R_0)/100R_0$. R is, by definition of Θ_R, a linear function of Θ_R. But is R also a linear function of Θ_{Hg}, and l of

333

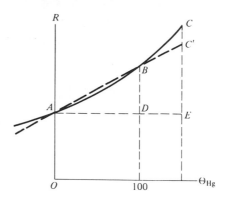

FIGURE 13.4   Variation of the resistance R of a metal wire with a Celsius temperature Θ_{Hg} as registered by mercury-in-glass thermometer.

Θ_R? Are Θ_{Hg} and Θ_R the same elsewhere than at the freezing and boiling points of water?

To answer this question, let the mercury-in-glass thermometer and the coil of wire both be immersed in, and so be in thermal equilibrium with, a suitable liquid, which is gradually heated; and let R be determined at a series of values of Θ_{Hg}, the results being represented graphically as in Fig. 13.4. It would be found that this graph is not a straight line, and from this it follows that Θ_R does not in general agree with Θ_{Hg}. For, by definition,

$$\Theta_R = 100 \, \frac{R - R_0}{R_{100} - R_0}.$$

Thus the value of Θ_R corresponding to the point C would be $100(CE/BD)$. And at this same point the value of Θ_{Hg} is $100(AE/AD)$, and this is the same as $100(C'E/BD)$, where C' is on the straight line passing through A and B. Because C and C' are different points, it follows that Θ_R and Θ_{Hg} are not the same.

What are we to do in this situation? Which particular Celsius scale, of all those available, shall we decide to regard as *the* Celsius scale, the standard scale?

Standard Celsius Scale

Let us suppose all possible thermometers to be in thermal equilibrium with one another at a series of temperatures as indicated by one of them, and let us observe how well they agree or how badly they disagree with one another in their readings. In such a comparison we should find that there are certain obvious "outsiders," whose readings disagree violently with those of the majority. For example, one such would be a water thermometer, constructed similarly to the mercury-in-glass thermometer but with water replacing mercury. These outsiders we should naturally discard. On examining the remainder, we should discover that all gas thermometers in which the property P used is either the volume at constant pressure or the pressure at constant volume agree among one another much more closely than, in general, do any of these with the others, or the others with one another; gas thermometers are in a class of their own. We therefore also discard all nongas thermometers.

Our final retention of only gas thermometers as candidates for the standard on which to base a temperature scale is supported by the consideration that gases, conforming as they do to such a simple p–V relationship as Boyle's law, would appear to be a correspondingly simple form of matter

We are confronted with one remaining difficulty: The scales given by different gas thermometers are not in *exact* agreement with one another. Perhaps this has some connection with the fact that no real gas exhibits an elastic behavior strictly in accordance with the ideal law formulated by Boyle. However, for all gases it is found that deviations from Boyle's law decrease asymptotically toward zero with increasing rarefaction. Correspondingly, it is found that the temperature scales based on gases tend asymptotically toward complete agreement as the degree of rarefaction increases. An infinitely rarefied gas would, according to all indications, behave precisely in accordance with Boyle's law, and such a gas is conveniently designated as **perfect** or **ideal.** As our choice of standard thermometer, then, let us finally decide, in principle, on an ideal-gas thermometer, a gas thermometer with an infinitely rarefied filling of *any* kind of gas.

In practice this solution of our problems is obviously inapplicable, for an infinitely rarefied gas is a vacuum. We may, however, find what would be the scale based on the behavior of an ideal gas by a process of extrapolation, using observations made with a real gas. The defining equations for temperature based on variations of volume at constant pressure and of pressure at constant volume sor an ideal gas, in which the property P of Eq. (13.2) is the volume V and the pressure p, respectively, of the gas, are really only special cases of the more general equation

$$\theta = 100 \, \frac{pV - (pV)_0}{(pV)_{100} - (pV)_0}. \tag{13.3}$$

335

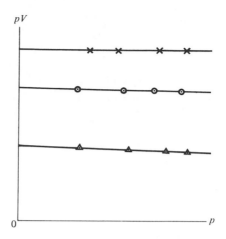

FIGURE 13.5 Berthelot's method of determining the products pV for a specimen of an ideal gas at various temperatures by graphical extrapolation of the corresponding products obtained at different pressures with a real gas.

This follows at once from the fact that for an ideal gas at a given temperature the product of pressure and volume is constant (Boyle's law). The plot of pV against p at any constant temperature for such a gas would be a horizontal straight line. For a real gas it is not quite horizontal but is indistinguishable from a straight line, except at rather high pressures. Hence we may extrapolate the values of pV observed at various temperatures with a real gas to zero pressure, in the manner indicated in Fig. 13.5, so obtaining the corresponding ideal-gas values characteristic of these temperatures. This is known as Berthelot's method, after Berthelot, who first used it. In practice it is more convenient to find the temperature scale defined by the reading of a constant-pressure or constant-volume thermometer having a real-gas filling and then to apply corrections to this scale based on the observed deviations of the gas from Boyle's law.*

Reassurance concerning the soundness of our choice of the ideal-gas scale as standard is furnished by the fact that the graduations of this scale may be shown theoretically to be identical with those of a scale based on the so-called "second law of thermodynamics," this being the Kelvin scale previously referred to. The second law of thermodynamics is one of the most firmly established

---

* For a discussion of these corrections, see, for example, M. N. Saha and B. N. Srivistava, *A Treatise on Heat*, The Indian Press Ltd., Allahabad, India, 1958, pp. 11–15.

of all scientific laws, and the Kelvin scale may be defined independently of the particular properties of any specific material.

Gas Thermometers

The gas in a gas thermometer has of necessity to be held in a container, or "bulb," and there must be a liquid boundary to the gas imprisoned—in practice this liquid is always mercury. This mercury must not be exposed to the temperature to be measured; if it were, it would freeze at low temperatures, and at higher temperatures its vapor pressure would become an intolerable complication to temperature measurement. This means that in practice a gas thermometer must consist of three main parts, the bulb, a connecting tube, and a mercury manometer, as shown in Fig. 13.6, which refers more specifically to a constant-volume gas thermometer, parts (a) and (b) corresponding to relatively low and high temperatures, respectively, of the bulb. Here the volumes

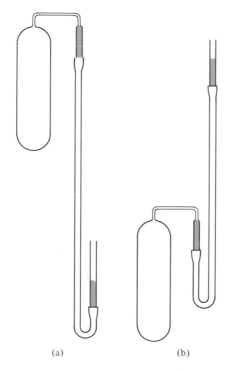

(a)   (b)

FIGURE 13.6 Constant-volume gas thermometer (a) at a relatively low temperature and (b) at a relatively high temperature.

occupied by the gas in the connecting tube and in the manometer tube above the mercury constitute "dead space"—an undesirable, but unavoidable, feature of the instrument. This dead space is always made as small as practicable relative to the volume of the bulb. For example, in an instrument used by Harker and Chappuis in 1900 for the purpose of establishing the ideal-gas scale, the bulb had a volume of 1100 cm³, but the total dead space amounted to only 0.5 cm³.

However small the dead space is made, it is never negligible, and in accurate work corrections have to be applied on account of it. Another complication is the thermal expansion and contraction of the bulb, which also has to be allowed for. Fortunately, although the application of all the necessary corrections is exceedingly tedious, the accuracy of the final result obtained is not appreciably affected by these. Just how all the corrections are applied is explained in specialized books—it would take us too far to discuss these corrections here.

As already mentioned, Fig. 13.6 refers to a constant-volume thermometer. An alternative form of gas thermometer is a constant-pressure instrument. In this, expansion of the gas can only be taken up by a lowering of the mercury in the manometer tube; and if high temperatures are to be measured this tube should be wide, to accommodate more easily the gas that has been driven out of the bulb. The situations at the lowest temperature of the range to be covered and at a relatively high temperature will then be as shown in the parts (a) and (b), respectively, of Fig. 13.7. It is easily seen that at high temperatures the greater proportion of the gas may well be in the manometer tube. This would at first sight appear to be an undesirable feature of the instrument. However, if the manometer tube is kept at a constant temperature throughout, this

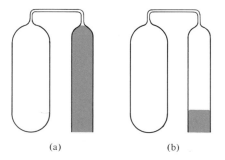

(a)   (b)

FIGURE 13.7   Constant-pressure gas thermometer at two temperatures.

being the same as that which the bulb has when the manometer tube is occupied entirely by mercury, as shown in Fig. 13.7(a), this is not really so. Under these conditions it is not difficult to calculate, from the volume of the bulb and the volume of gas in the manometer tube, what the total volume of the gas would be if it were all at the temperature of the bulb.

With a constant-pressure thermometer, as with the constant-volume instrument, the thermal expansion of the bulb and the finite volume of the connecting tube enter as complicating factors, for which corrections have to be applied.

In an ingenious form of the constant-pressure thermometer designed by Callendar, known as the "compensated gas thermometer," this instrument is capable of giving temperature determinations whose accuracy is in no way inferior to those obtained with a constant-volume thermometer.

The temperature Θ found with a gas thermometer will not be strictly that defined by the behavior of an ideal gas, because, in the interests of accuracy of measurement, these thermometers are always filled with a gas having an appreciable concentration. In any given case the corrections to be applied to convert the scale obtained to the ideal-gas scale may be calculated from the observed deviations from Boyle's law, in the manner already referred to. It is found that with certain selected gas fillings the corrections are in most cases so small that they are hardly worth applying.

The gas whose behavior most nearly approaches that of a perfect gas is helium, and this is now the standard gas used at all temperatures below a few hundred degrees Celsius. At higher temperatures helium diffuses through the bulb walls at an appreciable rate, irrespective of the material of which the bulb is made. For this reason thermometers intended for use at high temperatures are filled, instead, with nitrogen.

The best bulb material tried so far for high-temperature work is a platinum–rhodium alloy, containing about 20 percent rhodium. When suitable precautions are taken, such a bulb will not stretch plastically or change its shape as long as a temperature of about 1600°C is not exceeded. This is, at the present time, the upper limit of practicable gas thermometry.

The fall in the pressure of the gas in a constant-volume thermometer that accompanies a reduction in temperature gives a useful downward extension of the range over which temperatures can be measured; for, the lower the pressure, the lower the temperature at

339

which the gas will liquefy. However, an ordinary U-tube manometer is not a suitable instrument for measuring low pressures accurately. For accurate gas thermometry at very low temperatures it becomes necessary to employ special means to measure the pressure; for example, a McLeod gauge can be used. Unfortunately this increases the effective dead space, by the volume of the bulb of the McLeod, but this can be allowed for. In practice it is not possible to push the lower limit of gas thermometry below about $-272°C$. This limit may be reached by using a helium thermometer whose filling has a pressure of about 100 mm Hg at 0°C.

Kelvin Scale of Temperature

The study of the behavior of an ideal gas and of the measurement of temperature based on this behavior leads naturally to the introduction of a new temperature scale.

Our definition of temperature in terms of the properties of an ideal gas implies that there is a lower limit beyond which a temperature cannot be reduced. This limit corresponds to zero volume of the gas kept at constant pressure or zero pressure at constant volume.

A property of a material known as its coefficient of volume expansion, or expansivity, is defined as the increase in the volume of a sample of the material per degree rise in temperature divided by V_0, the volume at 0°C. By our definition of the standard Celsius scale of temperature, this coefficient, β, is for a perfect gas independent of temperature, and so at a temperature θ the volume V must be equal to $V_0 + V_0\beta\theta$, that is, to $V_0(1 + \beta\theta)$:

$$V = V_0(1 + \beta\theta).$$

Much painstaking experimental work extending over many years has been devoted to determining the value of β in this equation for an ideal gas. At present the value considered most reliable is 0.0036610 per degree Celsius. Correspondingly, by virtue of the validity of Boyle's law, the temperature coefficient of pressure variation at constant volume is also 0.0036610 per degree Celsius; this is also the value of β' in the constant-volume equation

$$p = p_0(1 + \beta'\theta);$$

and β and β' are both equal to η in the equation

$$pV = (pV)_0(1 + \eta\theta).$$

According to these equations, zero volume* or zero pressure (as the case may be) would both occur at the temperature $\theta = -1/\eta$, which, for the value of $\eta (= \beta)$ cited, is $-273.15°C$.

From the theoretical point of view it would appear desirable to have a temperature scale whose zero corresponds to the lowest value a temperature can possibly have. Such a scale is the so-called **absolute** or **Kelvin** scale, whose graduations are of the same magnitude as those of the Celsius scale but whose zero is defined as occurring $1/\eta$ degrees farther down. Thus with the value of η quoted above, the absolute zero of temperature, $0°K$, corresponds to $-273.15°C$.

Historically, Lord Kelvin, the originator of the temperature scale now known by his name, did not define this scale in terms of the behavior of an ideal gas, but based his definition on the second law of thermodynamics (see Chapter 18). However, it may be shown that the two definitions are necessarily equivalent, and for the purposes of introducing the new scale the present definition has the advantage of greater simplicity.

At a meeting of the International Committee of Weights and Measures held in 1954 it was decided to redefine the Kelvin scale, and with it the Celsius scale, using the triple point of H_2O as a single fixed point. The triple point is the temperature at which ice, water, and water vapor are all in equilibrium with one another. This is $0.0098°$ above the ice point. It was decided that in the new definition of the Kelvin scale the temperature assigned to the triple point should be $273.16°K$ exactly, while on the Celsius scale this temperature should be called $0.01°C$ exactly, the numerical values of temperatures on the Celsius scale always being $273.15°$ higher than on the Kelvin scale. All Kelvin temperatures would now be defined by the equation

$$T = 273.16 \frac{pV}{(pV)_{tr}},$$

where $(pV)_{tr}$ is the value of the product pV for the supposed sample of perfect gas at the triple point.

According to these new definitions the ice point on the Celsius scale becomes $0.0002°C$ and, as a simple calculation would show, the steam point becomes $99.996°C$. We see that these are sufficiently close to $0°$ and $100°C$, respectively, to justify the use of the latter figures for most purposes.

---

*Obviously the attainment of zero volume is an ideal situation that would be realizable only with an ideal gas. Any real gas would liquefy at a sufficiently low temperature, and its molecules must always occupy a finite volume.

Calibration of Other Than Gas Thermometers

Although gas thermometers are our best means of establishing an acceptable temperature scale, they are unfortunately both troublesome to use and extremely cumbersome. For these reasons they are rarely employed in practice as working instruments; other, much more convenient, thermometers are available for this purpose. The only use to which gas thermometers are put is in the calibration of the latter.

In performing such a calibration it would be possible, in principle, to place the two thermometers side by side in a suitable bath or uniform-temperature enclosure and to read them simultaneously while the temperature is taken through a series of values. In practice, however, the following procedure is found more convenient. The temperatures at a number of fixed points are first determined as accurately as possible by means of gas thermometers. These fixed points are usually melting points or boiling points of pure materials at standard atmospheric pressure. These known temperatures are then reproduced for the calibration of other, nongas, thermometers. A selection of fixed points suitable for calibration purposes is listed in Table 13.1.

TABLE 13.1 Fixed Points of Temperature Scale

Material	Phenomenon	Temperature (°C)
Hydrogen	Boiling	−252.78
Neon	Boiling	−245.99
Nitrogen	Boiling	−195.80
Oxygen	Boiling	−183.00
Carbon dioxide	Subliming	−78.51
Mercury	Melting	−38.87
Ice	Melting	0.00
Water	Boiling	100.00
Tin	Melting	231.85
Cadmium	Melting	320.9
Lead	Melting	327.4
Zinc	Melting	419.45
Sulfur	Boiling	444.6
Antimony	Melting	630.5
Aluminum	Melting	660.1
Potassium chloride	Melting	770.3
Silver	Melting	960.5
Gold	Melting	1063.0
Copper	Melting	1083.0
Palladium	Melting	1555

Practical Thermometers

The principal kinds of thermometer normally used in the laboratory are platinum resistance thermometers, thermocouples, mercury-in-glass and other liquid-in-glass thermometers, vapor-pressure thermometers, semiconductor and carbon thermometers, paramagnetic salts, and radiation pyrometers. For a full discussion of these the reader must be referred to specialist books on heat. We shall here confine ourselves to a brief review.

PLATINUM RESISTANCE THERMOMETERS A platinum resistance thermometer, in addition to being a working instrument that lends itself well to accurate thermometry within a certain range, is also reproducible to a high degree of precision. It may accordingly be made to serve as a "standard substandard" instrument against which other types of thermometer may be calibrated.

The property that serves as a measure of temperature is the electrical resistance of a fine coil of platinum wire. The difficulty that current leads to and from this coil have to be provided, along which the temperature necessarily varies, is avoided by the provision of an identical pair of "dummy" leads adjacent to the working leads and suitable circuitry that automatically subtracts the resistance of the dummy leads from that of the working coil and leads. The purest platinum available is used and the variation of resistance with temperature determined. Other "pure" specimens will give slightly different resistance–temperature curves, but because these are always of the same general character it is sufficient to measure the resistance at just a few standard points. This enables the necessary correction to be ascertained as a function of resistance over the whole range of temperatures in which this thermometer is used. This extends from −200°C to about +1000°C.

THERMOCOUPLES In 1821 it was observed by Seebeck that if the ends of a metal wire A (Fig. 13.8) are joined to a wire of a different kind of metal B, and the junctions P and Q are maintained at different temperatures, an electromotive force (emf) is developed, so that on joining the free ends of the wires B to a galvanometer G the latter registers a current. This phenomenon is known as the **Seebeck effect,** or the **thermoelectric effect.** The emf depends on the kinds of metal used and the temperatures of the two junctions. The phenomenon accordingly lends itself to the measurement of temperature. The system of wires BAB is known as a thermocouple, and P and Q are known as **thermojunctions.**

The galvanometer coil will not in general be of the same kind of metal as either A or B, so there will usually be at least three

343

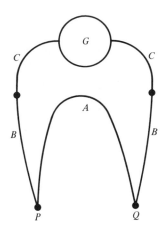

FIGURE 13.8 Thermocouple thermometer.

kinds of metal involved in the circuit. However, as long as the junctions of B with a third metal C (for example, copper) are both at the same temperature, the nature of the metal C has no effect on the emf.

In thermometry based on the Seebeck effect it is usual to keep one of the junctions at the definite temperature of 0°C by having it in thermal contact with melting ice. The other junction is exposed to the temperature to be measured. Although the use of a galvanometer, preferably in the form of a millivoltmeter, suffices for rough measurements, a potentiometer must be used in precision work.

By means of a few suitably selected thermocouples it is possible to measure, with fair accuracy, temperatures ranging from about −250°C to 1700° or 1750°C.*

LIQUID-IN-GLASS THERMOMETERS The most commonly used thermometers of this type are mercury-in-glass thermometers, which are too well known to need detailed description. In the usual form of "mercury" thermometer there is a vacuum in that part of the stem not occupied by mercury (Fig. 13.9), but if such a thermometer were used above about 100°C, distillation of mercury into the upper part of the stem would give trouble. To "blanket" distillation in thermometers designed for use at relatively high temperatures it is usual to fill the capillary above the mercury column with nitrogen at a suitable pressure.

---

* How temperatures beyond the range of gas thermometry are determined for use in the calibration of these instruments will appear in the section "Radiation Pyrometers."

It is not generally realized that mercury thermometers are in no sense precision instruments. In the first place, it is not considered worth the trouble to calibrate them, either directly or indirectly, against a gas thermometer. Instead, the interval between the 0°C and 100°C marks, obtained by immersion in melting ice and in steam at atmospheric pressure, respectively, is simply divided into 100 equal parts, and the graduation is then extended above and below this range. The consequent deviations from the standard scale are attributable mainly to the slight nonuniformity in the expansion of mercury, but the expansion characteristics of the glass used also have some effect. At the lower limit of these thermometers, −38.9°C, the reading is usually too low by about 0.4°, and at 200°C it is between 0.1° and 1.1° too high. At higher temperatures the discrepancies are more serious; for example at 500°C these thermometers read between 20° and 30° too high. A number of other defects of these instruments severely limit their use for precision

FIGURE 13.9 Liquid-in-glass thermometer.

work. Among these are difficulties associated with the "emergent stem," nonuniformity of capillary bore, a tendency of the mercury to stick in the capillary, resulting in jerky movement, sensitivity to pressure, and progressive changes in the volume of the bulb occurring both for a considerable period after manufacture and also after prolonged use at a high temperature.

Offsetting these defects are certain undoubted merits of mercury thermometers: They are of small size and small heat capacity, convenient, and direct reading. These properties recommend them for many purposes where high accuracy is not required.

Alcohol is also commonly employed as a filling for thermometers intended for use at ordinary temperatures, alcohol being cheaper than mercury. It is usual to color the alcohol to render it more easily visible.

If it is desired to use a liquid-in-glass thermometer at temperatures below the freezing point of mercury it must be filled with a suitable low-freezing-point liquid. Thus a filling material such as ethyl alcohol (freezing point $-114.9°C$) or n-pentane (f.p. $-131.5°C$) may be used to give a useful downward extension of range.

These organic liquids have much higher expansivities than mercury and therefore do not require such fine-bore capillaries. Their expansions are, however, less uniform than is that of mercury.

VAPOR-PRESSURE THERMOMETERS   We have seen that at very low temperatures ordinary gas thermometry is very troublesome and that below about $1°K$ it ceases to be practicable. Fortunately, reliable temperature determinations can be made at low temperatures much more conveniently by measuring the vapor pressures of certain liquids (see Chapter 16). One of these liquids is hydrogen. By thermodynamical reasoning incorporating the results of certain measurements, it is possible to establish a reliable theoretical relation between the pressure of the vapor in equilibrium with a liquid (its vapor pressure) and the absolute temperature. In the case of hydrogen the temperature may be expressed in terms of the measured vapor pressure in the range from the critical point for this element (Chapter 16), $33°K$, where the vapor pressure is about 9700 mm Hg, down to the triple point, $13.8°K$, where the pressure is about 53 mm. (Of course, the temperature–pressure relation in this range could also be obtained empirically, using a conventional gas thermometer.) The hydrogen vapor-pressure thermometer is much more convenient as a working instrument between these temperatures than gas thermometer.

SEMICONDUCTOR AND CARBON THERMOMETERS Certain materials, such as silicon, lead sulfide, germanium, and so on, although they look like metals, are not actually true metals; they are classified as "semiconductors." Semiconductors have the property that their electrical conductivity increases extremely rapidly with temperature, according to a law of the same general nature as applies to the vapor pressure of a liquid. A few of these lend themselves well to low-temperature thermometry based on this property. An example is germanium containing a trace of indium or arsenic as an impurity. This has been found to give reproducible temperature readings to within 0.001 degree between 1° and 5°K. Of course, the germanium thermometer has to be calibrated against an absolute thermometer.

Carbon, also, exhibits a useful variation of conductivity with temperature at low temperatures. It may be used as a convenient secondary thermometer giving readings reliable to within 0.1 degree between 1° and 20°K.

THERMOMETRY BELOW 1°K It is possible to extend absolute thermometry below 1°K right down to the lowest temperatures attainable by making use of certain paramagnetic salts (see Chapter 31) as thermometers. The magnetic susceptibility of such a salt varies with temperature according to a theoretically known law.

RADIATION PYROMETERS Thermometers designed for use at very high temperatures are often referred to as **pyrometers.** Of special importance in this field are the instruments which make use of the radiation emitted by hot bodies.

With rising temperature a body first becomes "red hot," then appears yellow, and at still higher temperatures becomes "white hot." Accompanying these changes in the quality of the radiation there is a rapid increase in its intensity. These phenomena suggest that it might be possible to measure high temperatures by determinations of the quality and/or intensity of the radiation emitted. Actually, a quantitative experimental investigation of the phenomena of radiation at temperatures up to the limit of the range of gas thermometry, in conjunction with theoretical reasoning involving the second law of thermodynamics (see Chapter 18) and one of the basic theories of modern physics known as the quantum theory, has resulted in the formulation of certain laws of radiation which may now be regarded as among the best founded of all the laws of physics. As we shall now see, it is possible, by making use of these laws, to extend the range of practicable thermometry up to the highest temperatures; there is no upper limit. This has been done with two different types of instrument to be described below:

347

the total-radiation pyrometer and the optical pyrometer. The discussion of these involves some elementary physics in the fields of optics and electricity not yet considered in this book. Readers who are not familiar with this material from their earlier studies may postpone the reading of the remainder of this chapter to a later stage.

BLACK-BODY RADIATION Different bodies at the same temperature radiate with different total intensities and with different spectral distributions of intensity. It is therefore not possible to find any unique law connecting the intensity or quality of radiation emitted from the exposed surface of a body with its temperature. However, it may be shown that the radiation within a uniform-temperature enclosure in equilibrium with the body containing this enclosure or cavity is a unique function of the temperature; it is completely independent of the material constituting the body. This "cavity radiation" may be observed by allowing a sample of it to escape through a small window, as indicated in Fig. 13.10; if the window is sufficiently small relative to the dimensions of the cavity, its effect in disturbing the intensity and quality of the radiation is negligible. For any given temperature, cavity radiation may be shown to be identical with the radiation emitted by a perfectly absorbing, or black, body. Hence this radiation is also known as "black-body radiation."

Total Intensity Like a gas or a vapor, cavity radiation may be shown theoretically to exert a pressure on its containing walls; and the quantitative predictions of the theory may be, and have been, verified experimentally. Accordingly, it would in principle be possible to construct a heat engine in which the "working substance" is the radiation contained within a cylinder, pushing a piston out as it expands; in this way work would be performed, just

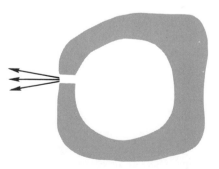

FIGURE 13.10 Black-body radiation issuing through a small window in a uniform-temperature enclosure.

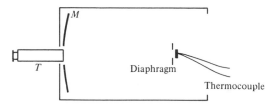

FIGURE 13.11   Total-radiation pyrometer.

as in the case of steam expanding against a moving piston in a steam engine. By applying thermodynamical reasoning to an engine utilizing radiation as its working substance, it may be shown that the total energy E of cavity radiation escaping through a window per unit area per unit time, or the power radiated per unit area from a black body, is related to the absolute temperature T by the equation

$$E = \sigma T^4,$$

where σ is a constant. The law stating that E is proportional to the fourth power of the temperature is known as the **Stefan–Boltzmann law,** and the proportionality constant σ is known as the **Stefan–Boltzmann constant.** Its value may readily be found experimentally, from measurements within the range of gas thermometry. In this way, and also theoretically from the quantum theory of black-body radiation (see below), it has been established that

$$\sigma = 5.6687 \times 10^{-5} \text{ erg/sec/cm}^2/\text{deg}^4$$
$$= 5.6687 \times 10^{-8} \text{ W/m}^2/\text{deg}^4.$$

Total-Radiation Pyrometer   This is an instrument for measuring high temperatures based on the Stefan–Boltzmann law, developed by Féry, and shown diagrammatically in Fig. 13.11. Radiation emerging from a small aperture in a uniform-temperature enclosure is focused by a concave mirror M on a diaphragm, provided with a central hole. It is arranged that the image of the aperture, observed through the telescope T, more than covers this hole. The radiation passing through the hole falls on a blackened metal receiver to which is attached a thermocouple connected to a sensitive voltmeter, whose registrations give the temperature of the source.

To avoid, as far as possible, selectivity of reflection by the mirror, this is usually made of polished nickel. This metal approximates well to the requirement of nonselectivity.

The instrument is, in practice, calibrated against a gas thermometer by using a number of fixed points established with the latter, for example standard melting and boiling points, within

349

the limits of practical gas thermometry. Within these limits, any selectivity of reflection by the mirror, absorption by the air or by the lenses of the telescope, and so on, are clearly of no consequence, because the calibration makes no direct use of the Stefan–Boltzmann law. However, these become serious complicating factors when the instrument is used for measuring temperatures beyond the range of gas thermometry. Such use constitutes one of the most important applications of the instrument. We shall now consider this.

Let T be a temperature above the upper limit of gas thermometry which it is desired to measure. This may be determined with the aid of a rotating disk, one side of which is interposed in the path of the radiation incident on the pyrometer. Narrow windows in the form of sectors are cut in this disk (see Fig. 13.12), which is made to rotate at high speed. The area of the rotating disk on which the radiation directed toward the pyrometer is incident is indicated by the dashed circle. With this device the mean energy of the radiation intercepted by the instrument per unit time ("energy flux") may be reduced to any desired fraction f (for example, $\frac{1}{100}$) of what would be received in the absence of the disk. Let f be adjusted to such a value that the reading of the voltmeter corresponds to some temperature T_0 within the range in which the instrument has been calibrated. Then T may be calculated in terms of T_0 and f as follows.

Let the energy flux incident on the instrument, determining its reading when the temperature of the source is T_0, be Φ_0, and let the flux which would reach the instrument in the absence of the disk when the temperature is T be Φ. Then, since

$$\Phi \propto T^4,$$

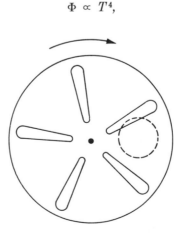

FIGURE 13.12 Rotating disk used for extension of pyrometer calibrations to higher temperatures.

we must have

$$\frac{T}{T_0} = \left(\frac{\Phi}{\Phi_0}\right)^{1/4}$$

But with the rotating disk interposed, the reading when the temperature is T is the same as it is without the disk when the temperature is T_0. Hence the mean flux in the former case is the same as the steady flux in the latter; that is,

$$f\Phi = \Phi_0.$$

Substituting $f\Phi$ for Φ_0 in the expression for the ratio of temperatures, we obtain the relation

$$T = \frac{T_0}{f^{1/4}}.$$

The temperature T having been determined in this way, the rotating disk may now be removed and, with the radiation from the source passing freely to the instrument, the reading observed. This will be the calibration point for the temperature T. Proceeding in this manner for a whole series of temperatures above the limit of gas thermometry, the calibration may be extended upward as far as is desired.

In practice some selective absorption is unavoidable. Also the receiver does not absorb radiation equally efficiently at all wavelengths. Corresponding corrections must therefore be applied.

PLANCK'S RADIATION FORMULA The energy δE radiated by a black body at any given temperature per unit area per unit time within the narrow wavelength interval $\delta\lambda$ between λ and $\lambda + \delta\lambda$ is obviously proportional to $\delta\lambda$; that is, denoting the proportionality factor by the symbol E_λ, we may write

$$\delta E = E_\lambda \, \delta\lambda,$$

and E_λ, known as the **emissive power** of the black body, is a function of both λ and T. According to a quantum theory of radiation developed by Planck in 1900,

$$E_\lambda = \frac{c_1 \lambda^{-5}}{e^{c_2/\lambda T} - 1},$$

in which

$$c_1 = 2\pi c^2 h$$

and

$$c_2 = \frac{ch}{k},$$

351

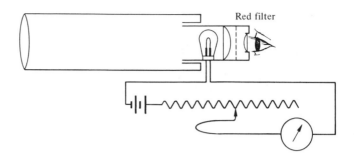

FIGURE 13.13 Optical pyrometer.

where c is the velocity of light, k is Boltzmann's constant, or the gas constant per molecule (see Chapters 15 and 22), and h is a quantity of fundamental importance in quantum physics, known as Planck's constant, to be introduced in Chapter 23. Numerically,

$$c_1 = 3.7407 \times 10^{-16} \text{ W-m}^2$$

and

$$c_2 = 1.4386 \text{ m-deg.}$$

Clearly, in all cases where the product λT is sufficiently small compared with c_2, making $e^{c_2/\lambda T}$ correspondingly large, the -1 in the denominator of the expression for E_λ may, with negligible error, be omitted, the formula then reducing to

$$E_\lambda = c_1 \lambda^{-5} e^{-c_2/\lambda T}.$$

This is Wien's formula, guessed at by Wien before the development of Planck's theory and found to accord well with experimental measurements in the visible range of wavelengths and at all temperatures that can be established by ordinary means in the laboratory.

Optical Pyrometer This is an instrument designed for measuring high temperatures whose use is based on Wien's radiation formula. It employs as nearly as possible radiation of a single wavelength—in practice a narrow range of wavelengths centered on an effective mean wavelength near the red end of the visible spectrum. Red is selected for the radiation used because this is the first visible radiation to be emitted at an appreciable rate by a body as its temperature is increased from a relatively low value.

The instrument is shown diagrammatically in Fig. 13.13. It consists essentially of a telescope in the focal plane of whose eyepiece the filament of a small lamp is located, and in which a monochromatic red filter is placed somewhere in the light path beyond the lamp. The current through the lamp filament can be adjusted by a rheostat. The circuit also includes an ammeter, whose scale may be calibrated to read directly in degrees, in the manner now to be considered.

The aperture in the uniform-temperature enclosure whose temperature is to be measured is first focused on the focal plane of the eyepiece. The current through the lamp filament is then adjusted for disappearance of the filament against its background, that is, so that it is neither seen dark against a relatively bright background (current too small) nor bright against a relatively dark background (current too large). The corresponding temperature, measured by some independent means, is then marked on the dial of the ammeter. This is done for a series of temperatures below the upper limit of gas thermometry, and so the corresponding portion of the calibration is obtained. In none of this is Wien's formula involved. It does become involved, however, in extending the calibration to higher temperatures. Here the same rotating-disk procedure is used as we have already considered in connection with Féry's total-radiation pyrometer. It is a simple matter, using Wien's formula, to calculate any temperature T above the range of gas thermometry in terms of the lower temperature registration T_0 corresponding to disappearance of the filament which is obtained when the rotating disk is interposed in the path of the radiation, the fraction f of this radiation which passes through the sectors cut in the disk, and the mean effective wavelength λ transmitted through the monochromatic filter. In practice, because no filter is truly monochromatic, allowance has to be made for the slight shift of the mean effective wavelength transmitted which occurs when the temperature is increased.

The optical pyrometer may also be used for finding the temperature of a nonblack radiator, a "gray" body, if the relevant emissive properties of a body made of the material in question have previously been determined. This may be done, for example, by drilling a deep hole in a block of this material, and then, holding this at a series of temperatures, at each temperature (1) finding the true temperature T of the body by adjusting for disappearance of the filament against the image of the mouth of the cavity, and (2) noting the lower temperature T_b corresponding to the calibration of the instrument when the filament is made to disappear against the image of the outer exposed surface of the body. T_b is known as the "brightness temperature" of the surface in question for the mean effective wavelength in which the filter transmits. The relation between T and T_b having been found in this way for a considerable range of temperatures, the true temperature of any body having this kind of surface may at once be ascertained from its measured brightness temperature.

Problems

13.1   At what temperature do a Celsius and a Fahrenheit thermometer have identical readings?

13.2   A platinum wire has a resistance of 12.102 Ω* at the ice point, 16.775 Ω at the steam point, and 31.781 Ω at the boiling point of sulfur. Find the constants a and b in the formula

$$R = R_0(1 + a\Theta + b\Theta^2)$$

giving the resistance R of any specimen of this material in terms of its resistance R_0 at 0°C and the Celsius temperature Θ on the standard scale.

13.3   A copper–constantan thermocouple, one junction of which is kept in melting ice, gives a voltmeter reading of 4.23 mV† when the other junction is at the boiling point of water. What is the thermocouple Celsius temperature of this other junction when the voltmeter reading is (a) 2.92 mV, (b) 10.65 mV, and (c) −4.82 mV?

13.4   The following table gives the resistance R of a tungsten wire as determined at various temperatures Θ registered by a mercury-in-glass thermometer:

Θ (°C)	0	50	100	150	200	250	300
R (Ω)	47.8	56.6	65.6	75.2	85.1	95.6	106.3

Plot these results, and, using your graph, find (a) the Celsius temperature on the tungsten-resistance scale corresponding to 180°C on the mercury-in-glass scale, and (b) the Celsius temperature on the mercury-in-glass scale corresponding to 270°C on the tungsten-resistance scale.

13.5   A chromel–alumel thermocouple, one junction of which is immersed in melting ice, gives the following voltmeter readings V on a millivoltmeter when the other junction is at temperatures Θ on the standard gas scale:

Θ (°C)	V (mV)	Θ (°C)	V (mV)
0	0	700	29.1
100	4.1	800	33.3
200	8.1	900	37.4
300	12.2	1000	41.3
400	16.4	1100	45.1
500	20.6	1200	48.8
600	24.9	1300	52.4

What is the Celsius temperature on the thermocouple scale corresponding to the standard temperature 875°C, and what is the standard Kelvin temperature T when the Celsius temperature on the thermocouple scale is 1050°C?

* This is the conventional symbol for the ohm, the unit of resistance.
† V stands for volts, the unit of emf, and 1 mV = 10^{-3} V.

13.6   When the bulb of a constant-volume gas thermometer is immersed in melting ice, the level of the mercury in the free manometer arm is 36.4 cm below the standard level in the "fixed" arm communicating with the bulb. When the bulb is now immersed in a bath at a different temperature it is found that, in order to restore the level of the mercury in the fixed arm, the level in the free arm must be raised to 45.3 cm above this. Given that an adjacent mercury barometer reads 75.8 cm, calculate the temperature of the bath (a) on the Kelvin scale, and (b) on the Celsius scale. (Neglect dead space and other small corrections.)

13.7   The gas imprisoned in a constant-pressure thermometer just occupies the whole of the bulb, of volume V_0, when this is immersed in melting ice. When the bulb is raised to a higher temperature, the gas, at the same pressure, expands, through a connecting tube of negligible volume, into the wide-bore manometer tube maintained at 0°C whose volume above the mercury is V. What is the temperature of the bulb? (Neglect small corrections.)

13.8   Two glass vessels, of volume 4 liters and 1 liter, respectively, are connected by a tube of negligible volume. Initially, they are both at 20°C and contain dry air at a pressure of 76 cm Hg. The larger vessel is now raised to a temperature of Θ°C while the smaller remains at 20°C. Given that the final pressure is 91.7 cm Hg, find Θ. Assume that the coefficient of expansion of glass is negligible, and treat air as a perfect gas.

13.9   If temperature is involved in a theoretical relationship, do you consider it immaterial whether this is expressed in the Celsius or the Kelvin scale, or is one of these scales the more likely to be relevant theoretically? Give reasons for your answer.

13.10   The vapor pressures of alcohol at a series of temperatures are as follows:

Θ (°C)	−20	0	20	40	60	80
p (mm Hg)	3.34	12.73	44.0	133.4	350	812

Find a method of plotting the relationship between the vapor pressure and temperature which gives a good approximation to a straight-line graph, and hence obtain an algebraical formula expressing vapor pressure as a function of temperature.

13.11   When the orientation of a mercury thermometer at a given temperature is changed from horizontal to vertical, the reading is depressed. One reason for this is the stretching of the bulb consequent on the increase in pressure, but there must also be some compression of the mercury itself. Given that the proportional increase in volume of mercury per degree rise in temperature is 1.8×10^{-4} and that the bulk modulus is 2.6×10^{11} dyn/cm², find whether this latter effect is of any importance in practical mercury thermometry.

13.12   Radiation from a cavity in a body at a high temperature is viewed in an optical pyrometer whose filter transmits a narrow band of wavelengths having an effective mean at 6670 Å. Between the source of radiation and the pyrometer is a disk rotating at high speed in which

355

sectors are cut such that one-hundredth of the light incident on it passes through the pyrometer. With the disk interposed the current that causes the filament to disappear against its background is the same as when viewing black-body radiation at 1550°K without the disk. Using Wien's formula, and neglecting any variation with temperature of the effective mean wavelength for transmission through the filter, calculate the temperature of the body.

13.13   The 6670-Å wavelength quoted in Problem 13.12 is an effective mean for the two temperatures considered. For what reason do you think the mean effective wavelength transmitted through the filter for a particular temperature of the cavity might vary slightly with this temperature, necessitating a correction on this account?

13.14   An experimenter wishes to determine the temperature of a body from the measured emissive power E_λ at 6670 Å for the radiation emitted from a cavity in the body. Find the percentage error in his temperature determination which the use of Wien's formula instead of Planck's introduces in the two cases where the true temperature is (a) 2000°K and (b) 3000°K.

14

HEAT AND THERMAL TRANSPORT

In this chapter a selected group of thermal effects will be considered phenomenologically, this being a necessary preliminary to their theoretical analysis either in subsequent chapters or in more advanced studies which the student will undertake at a later stage.

Thermal Expansion

Let the **coefficient of linear expansion,** or **linear expansivity,** of a body be represented by the symbol α and the **coefficient of volume expansion,** or **volume expansivity,** by β. Then, these quantities being defined by the equations

$$\alpha = \frac{1}{l_0} \frac{dl}{d\Theta}$$

and

$$\beta = \frac{1}{V_0} \frac{dV}{d\Theta},$$

where l_0 and V_0 are the length and volume, respectively, at 0°C, it is easily seen that in the case of an isotropic solid, which has both kinds of expansivity, with α independent of direction, we must have, to a close approximation,

$$\beta = 3\alpha.$$

Let us consider a cube of side l and volume V. Then, differentiating the equation

$$V = l^3$$

with respect to Θ, we have

$$\frac{dV}{d\Theta} = 3l^2 \frac{dl}{d\Theta},$$

and so, dividing both sides by $V_0 \ (= l_0{}^3)$,

$$\frac{1}{V_0} \frac{dV}{d\Theta} = 3 \left(\frac{l}{l_0}\right)^2 \frac{1}{l_0} \frac{dl}{d\Theta};$$

that is,

$$\beta = 3 \left(\frac{l}{l_0}\right)^2 \alpha,$$

or, since in practice l/l_0 is always very close to 1,

$$\beta = 3\alpha,$$

to the corresponding degree of approximation.

The thermal expansivities of most materials vary to a greater or lesser extent with temperature. Thus in Fig. 14.1 the linear expansivities of aluminum, copper, and an alloy of iron and nickel (64 percent Fe, 36 percent Ni) known as Invar are plotted against temperature. In view of such temperature variations, it is not in general sufficient to quote a figure for the thermal expansivity of a material without at the same time specifying the temperature at which this figure applies. When this is not done it is generally to be understood that the figure is a mean value for a relatively restricted range of "ordinary" (for example, room) temperatures, or for some special temperature range under discussion which is of particular interest.

Approximate values of the linear expansivities of a number of materials are given in Table 14.1.

Many applications and consequences of thermal expansion are discussed in more elementary textbooks. These will not be considered here.

Heat as a Physical Quantity

Several phenomena, some involving a change of temperature, others not, suggest that under certain circumstances something passes from one body to another, something we call **heat.** Examples are (1) the melting of a solid, for example ice; (2) the vaporization of a liquid, for example the boiling of water; and (3) a rise in the

TABLE 14.1 Values of $10^6\alpha$ at Ordinary or Slightly Elevated Temperatures for a Selection of Materials (per °C)

Aluminum	25	Stainless Invar	
Steel	11	(36% Fe, 54% Co, 10% Cr)	<0.1
Brass	19	Lime-soda and lead glasses	8–10
Copper	17	Pyrex glass	3.3
Nickel	13	Fused silica	0.5
Platinum	9	Sulfur	70
Tungsten	4.5	Concrete	10
Invar			
(64% Fe, 36% Ni)	0.9		

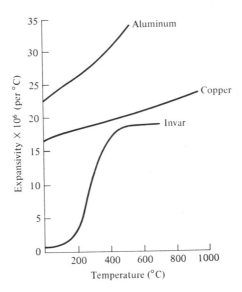

FIGURE 14.1 Temperature variation of expansivities of aluminum, copper, and Invar.

temperature of a body without a change of state. With these phenomena in mind we could tentatively define the quantity of heat transferred as proportional to the change produced, for example the amount of solid melted, the amount of liquid vaporized, or the temperature change produced in a particular body.

Now, changes of state, such as we have in the melting of a solid or the boiling of a liquid at a fixed pressure, are found to occur at a constant temperature; the temperature does not change until the whole of the solid has been converted to a liquid or the whole of the liquid has been vaporized. For this reason the heat transferred is known as "latent" (hidden) heat. It appears, then, that a definition of heat in terms of the quantity of material converted from one form to another in a change of state must surely be sound in principle; thus we can hardly imagine the process of melting the second gram of ice to have any features not possessed by that of melting the first. On the other hand, such a definition is rather limited in its usefulness; it is applicable only in cases where the heat to be measured is transferred to or from the particular material, for example ice or water, under the conditions specified (pressure, and so on), producing a change of state.

The other alternative, of defining heat in terms of the temperature change produced in a standard material, is more generally useful, although its acceptability is not immediately obvious. Actually, such a definition has been adopted in practice, and the

359

unit of heat, known as the **calorie,** has been defined as the quantity
of heat required to raise the temperature of 1 g of water by 1 degree
Celsius. The question now arises whether calories at all levels of
temperature are the same, as is implied by this definition of quantity of heat, and, if they are not the same, how closely they approximate one another.

One way in which this question could be decided would be to
bring a given mass of water at a series of initial temperatures into
thermal contact with ice at 0°C and determine the quantity of ice
melted on each occasion while the water, giving up heat to the ice,
cools down to 0°C. If calories at all temperature levels were equal,
the quantity of ice melted would be proportional to the fall in
temperature of the water. If not, it could readily be calculated from
the precise manner in which the former varies with the latter how
one calorie compares with another. Fortunately it is possible to
carry out such a series of determinations with considerable accuracy,
using an ingenious "ice calorimeter" devised by Bunsen, which will
now be described. This almost completely avoids heat exchanges
with the surroundings and correspondingly reduces the allowance
that has to be made for them.

Bunsen's ice calorimeter is shown in Fig. 14.2. The material
to be tested is introduced into the tube T, which is sealed into the
bulb B and has a shell of ice, some 5 to 10 mm thick, frozen to its
exterior. The remainder of the space between the two tubes is
occupied by water in thermal equilibrium with the ice, except that
the lower part of the bulb, together with a J tube fused onto it at

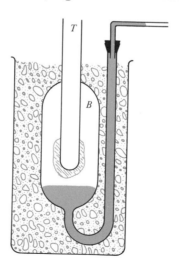

    FIGURE 14.2   Ice calorimeter.

TABLE 14.2 Specific Heats of Water at Various Temperatures

Temp. (°C)	Specific heat (15°-cal/g/deg)	Temp. (°C)	Specific heat (15°-cal/g/deg)
0	1.0076	50	0.9988
10	1.0015	60	0.9997
15	1.0000	70	1.0009
20	0.9997	80	1.0025
30	0.9982	90	1.0046
40	0.9983	100	1.0072

the bottom, contains mercury. The upper end of the J tube is engaged with a narrow-bore tube bent at right angles in such a manner as to allow axial adjustment at the junction. The horizontal part of the narrow-bore tube contains the end of the thread of mercury and is graduated. The shell of ice is formed around the tube T initially by circulating chilled alcohol through it internally or by pouring some ether into it and promoting its rapid evaporation (with extraction of latent heat of vaporization) by bubbling air through it. Tube T is then emptied, and the whole apparatus is immersed in melting snow or in a mixture of ice and water.

The apparatus is used mainly for finding the specific heats of liquids and solids, but it can also be used for comparing calories at different temperature levels, by the procedure already indicated. In this, the quantity of ice melted on each occasion is proportional to the distance the end of the thread of mercury in the horizontal tube moves back, this movement being due to the reduction in volume of the H_2O as it turns from ice to water.

The quantity of heat per gram per degree of temperature increase required to raise the temperature of water may conveniently be designated the specific heat of water at the temperature level in question. Experimentally determined specific heats of water at various temperatures, expressed in 15-degree calories* per gram per degree, are given in Table 14.2. We see that calories at different levels are not equal, and, correspondingly, that unless the temperature level to which it is referred is specified, the calorie is not to be regarded as a truly scientific unit of heat. Nevertheless, the calorie is sufficiently nearly constant with temperature for its use without such specification to be justified in certain cases. Just as it is often convenient to employ a mercury thermometer for the

---

* The 15-degree calorie is defined as the quantity of heat required to raise the temperature of 1 g of water from 14.5° to 15.5°C.

measurement of temperature or to determine a force in gravitational units (variations in g being neglected), so also in heat measurements requiring only moderate accuracy it is customary to use the calorie as the unit of quantity.

The calorie as we have defined it is often referred to more specifically as the "gram-calorie," or "small calorie," to distinguish it from the calorie of the dieticians, which is 1000 times as large. To avoid confusion, the latter is sometimes called the "kilogram-calorie," or "kilocalorie." It is also sometimes written with a capital, as "Calorie." In physical literature the term calorie always means the gram-calorie.

Specific and Latent Heats

Both of these are important properties of material, which have considerable theoretical interest. In this chapter we shall simply define them and note their values for a selection of materials. The specific heats of gases will be discussed theoretically at some length in Chapter 15. A more detailed discussion of these, and also a theoretical analysis of the specific heats of solids, is usually included in introductory textbooks on modern physics.* Latent heats will be discussed further in Chapters 16 and 18.

SPECIFIC HEAT The specific heat of a material is defined as the quantity of heat per unit mass per unit temperature increase that has to be imparted to it to raise its temperature. Using the symbol c for specific heat and Q for heat imparted to a sample of mass m, the definition in mathematical form is

$$c = \frac{1}{m}\frac{dQ}{d\Theta}.$$

It is important to realize that the specific heat is not an absolute constant for each material; its value depends on the conditions under which it is measured. In general, the specific heat not only increases with increasing temperature, particularly at low temperatures, but it also varies with pressure and volume conditions. Two cases of particular interest are the specific heat at constant pressure, c_p, and that at constant volume, c_v. Although there is a substantial difference between these for gases, the difference between c_p and c_v for solids and liquids is relatively small and can usually be neglected. In the standard experimental methods, normally studied in a laboratory course, it is usually c_p that is measured rather than c_v.

---

* See, for example, Chapter 6 of Volume III of this book.

TABLE 14.3 Specific Heats

Material	Specific heat (cal/g/deg)	Material	Specific heat (cal/g/deg)
Aluminum	0.22	Sodium	0.29
Copper	0.093	Sulfur	0.18
Diamond	0.14	Tin	0.055
Graphite	0.19	Tungsten	0.034
Gold	0.030	Zinc	0.093
Ice, −20 to 0°C	0.53	Alcohol (ethyl),	
Iodine	0.054	25°C	0.58
Iron	0.11	Glycerine, 0°C	0.54
Lead	0.031	Mercury, 20°C	0.033
Nickel	0.11	Air, atmospheric	
Platinum	0.032	pressure	0.24
Silicon	0.18	Steam, atmospheric	
Silver	0.056	pressure, 100–200°C	0.48

The specific heats at constant pressure of a number of materials are listed in Table 14.3. The values given are those at slightly elevated temperatures, for example about 50°C, unless otherwise stated. For most materials the temperature variation of specific heat is not important at ordinary or higher temperatures.

LATENT HEAT In its more specific and quantitative meaning the latent heat of a material is defined as the quantity of heat per unit mass required to change the material from a solid to a liquid, or from a liquid to a vapor, without change of temperature. According as it refers to fusion (melting) or vaporization it is called the latent heat of fusion or the latent heat of vaporization. The term "latent heat" is also used in a more general sense as the heat transferred to or extracted from *any* (unspecified) mass of material to bring about the relevant change of state. It is always clear from the context which meaning is intended.

The latent heats of fusion of a number of materials are given in Table 14.4.

For any particular material the latent heat of vaporization varies somewhat with temperature, increasing in value as the temperature is reduced. For example, whereas the latent heat of vaporization of water is 539.5 cal/g at 100°C, at 0°C it is about 596.5 cal/g.

The latent heats of vaporization of a number of materials at their atmospheric-pressure boiling points are given in Table 14.5.

363

TABLE 14.4 Latent Heats of Fusion

Material	Melting point (°C)	Latent heat (cal/g)
Aluminum	658	76.8
Ammonia	−75	108
Benzene	5.4	30.6
Ice	0	79.6
Lead	327	5.47
Mercury	−39	2.82
Naphthalene	80	35
Platinum	1750	27.2
Potassium	62	15.7
Silver	961	21.
Sulfur	115	9.37

Dulong and Petit's Law

A survey of the specific heats of various materials leads to a generalization that is highly significant theoretically.

Examination of the specific heats of the solid elements listed in Table 14.3 shows that, with few exceptions, the specific heat is approximately inversely proportional to the atomic weight. How well this inverse proportionality holds may be seen from Fig. 14.3, in which the specific heats of most of the solid elements in Table 14.3 are plotted against the reciprocals of their atomic weights A. From left to right the points are for Pb, I, Ag, Cu, Ni, S, Si, Al, Na, and C. In the case of the last named of these, the upper point refers to graphite and the lower to diamond. We see that there is a good approximation to proportionality for most of the elements plotted. A notable exception is carbon in its two crystalline forms. Others are beryllium and boron. We shall consider these exceptions presently. For the great majority of the solid elements, which conform

TABLE 14.5 Latent Heats of Vaporization

Material	Boiling point (0°C)	Latent heat (cal/g)
Alcohol	78.4	205.7
Mercury	358	68
Nitrogen	−195.8	50
Sulfur dioxide	−10.8	96
Water	100	539.5

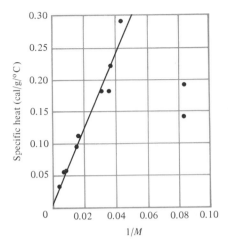

FIGURE 14.3   Relation between specific heats of elements and atomic weights.

reasonably well to the law of inverse proportionality between specific heat and atomic weight, the mean value of the product of the two has the numerical value of about 6. This product, representing the heat capacity of 1 gram-atom, that is, the heat required to raise the temperature of 1 gram-atom by 1 degree, is known as the **atomic heat** of the element.

The fact that the atomic heats of the great majority of solid elements are approximately the same was first noted by Dulong and Petit in 1819, and is known, accordingly, as **Dulong and Petit's law.**

The existence of exceptions to this law is at first sight rather puzzling. We must keep in mind, however, that we have somewhat arbitrarily chosen as the values of the specific heats to be considered those corresponding to about 50°C. Can it be that the exceptional elements would be brought into line if we proceeded on some other basis with regard to temperature?

Although in all cases the specific heat is found to be practically temperature-independent above a certain level of temperature, it falls off with reduction of temperature below this level, at first slowly, then rapidly, and finally, at absolute zero, it vanishes altogether. In Fig. 14.4 the atomic heats of silver, silicon, and diamond are shown as a function of temperature. We see that while at 50°C the curve for silver, which conforms well to the law, has become nearly horizontal, that for silicon, which conforms less well, is still rising fairly rapidly with increasing temperature. And for diamond

365

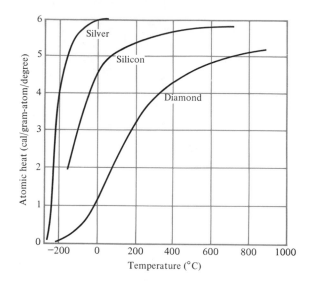

FIGURE 14.4 Variation of atomic heats of elements with temperature.

the value of the atomic heat at 50°C is only a small fraction of the final value; this is not attained until the temperature is well above 1000°C. The curves for other elements are similar in character. It so happens that at 50°C and beyond, most of them are practically horizontal. But even the exceptional ones that are still rising strongly at 50°C eventually flatten out at higher temperatures at about the same level. If instead of arbitrarily taking the 50° values, we took the limiting values approached asymptotically with increasing temperature, we should find that there are no longer any exceptions; at sufficiently high temperatures the atomic heats of all solid elements are approximately the same. We shall be in a better position to formulate a theory of this when we have made a closer study of the thermal behavior of gases.

Heat Transfer

The processes by which heat is transferred from one place to another, or from one body to another, are those known as **conduction, convection,** and **radiation.**

CONDUCTION   Let us (Fig. 14.5) consider two parallel planes within a medium such that everywhere in one plane the temperature is Θ, while throughout the other it is $\Theta + \delta\Theta$, and let Cartesian coordinates be so chosen that these planes are perpendicular to the x axis. Let the x coordinate of the first plane be x and that of the second

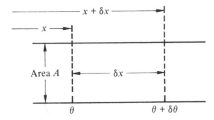

FIGURE 14.5

$x + \delta x$. Then it is found that through any area A normal to the x axis heat is conducted in the negative x direction at a rate that is proportional both to A and to the temperature gradient, which, at the limit of smallness of δx, may be written $d\Theta/dx$. Thus the rate of heat flow in the *positive* x direction, dQ/dt, is given by an equation of the form

$$\frac{dQ}{dt} = -kA\,\frac{d\Theta}{dt},$$

where k is a positive constant. This constant is known as the **thermal conductivity** of the medium; it is the rate of heat flow per unit area per unit temperature gradient. Accordingly, it may be expressed in cal/sec/cm/deg. In view of the fact that heat is generally measured in calories, the cal/sec/cm/deg is a convenient practical unit. In theoretical discussions a better unit to employ is the erg/sec/cm/deg or the J/sec/m/deg, according as we are using the CGS or the MKSA system of units.

Thermal conductivity does not have a unique value for any given material but varies somewhat with temperature. In the case of metals it has a negative temperature coefficient, decreasing with rising temperature and assuming larger values as the temperature is lowered.

The values of thermal conductivity at ordinary room temperatures (except in the case of ice) for a selection of materials are given in Table 14.6.

As we shall see later, heat is the energy of molecular motion or, in the case of solids, the energy of the vibrating atoms situated at the lattice points. It would appear, therefore, that conduction of heat must occur by the handing on of such energy by the more energetic molecules or atoms to their less energetic neighbors. On this basis we should rather expect the thermal conductivities of all materials of a given kind of structure, for example all solids, to be of about the same order of magnitude. However, inspection of

367

TABLE 14.6   Thermal Conductivities (in cal/sec/cm/deg)

Silver	1.00	Cork	0.0001
Copper	0.92	Quartz	
Aluminum	0.57	Parallel to	
Magnesium	0.36	axis	0.022
Iron	0.18	Perpendicular	
Lead	0.083	to axis	0.013
Glass (lime-soda)	0.0016	Ice	0.0053
Brick	0.0014	Mica	0.0014
Wood		Sulfur	0.00055
(ordinary kinds)	0.00033–0.00041	Water	0.0014
Asbestos (paper)	0.00025	Hydrogen	0.0004
Diatomaceous powder	0.00017	Air	0.00006

Table 14.6 shows that this is by no means the case. Whereas solid metals have conductivities between about 0.1 and 1 cal/sec/cm/deg, the range of conductivities of nonmetallic crystalline solids is well below this, by a factor of about 100. What can be the reason for this?

The property that particularly distinguishes metals from nonmetals is their electrical conductivity; electricity is conveyed through metals by electrically charged particles known as electrons, which are free to move in response to an electric field. Insulating solids, on the other hand, normally have no free electrons. Can it be that the same free electrons which are responsible for the high electrical conductivity of metals also play the predominant role in the conduction of heat in these materials?

Not only is this suggested by the fact that both thermal and electrical conductivity are much higher for metals than for nonmetallic crystalline solids, but the suggestion is strongly supported by a fairly detailed correspondence between the two conductivities for the various metals. This will be seen on comparing Table 14.7, giving the electrical conductivities of a number of metals at 20°C,

TABLE 14.7   Electrical Conductivities of Metals (in $\Omega^{-1}/m \times 10^7$)

Silver	6.3
Copper	5.9
Aluminum	4.1
Magnesium	2.4
Iron	1.0
Lead	0.49

with Table 14.6. Thus, both thermally and electrically, silver has the highest conductivity of all metals, closely followed by copper, and throughout the list of metals selected the order of decreasing conductivities is the same in both cases. Indeed, there is even a rough approximation to constancy of the ratio of the two conductivities.* From this evidence, the inference seems inescapable that in the conduction of heat through metals the free electrons do, indeed, play the predominant role, the direct transfer of energy between vibrating atoms being of relatively minor importance.

In addition to the transfer of heat by the vibrations of atoms in crystal lattices and, in metals, by the motions of free electrons, a third possibility should be considered. It is well known that hot bodies emit energy in the form of electromagnetic radiation and that this may be absorbed by other bodies and reconverted into heat. There is no valid reason for assuming that the production of such radiation is confined to the *surfaces* of hot bodies; indeed, it would be extraordinary if it were. Instead, we must surely think of this as a volume effect, radiation being produced throughout the whole body. Also it is known that radiation of certain wavelength ranges is transmitted through any given nonmetallic material, the transmitted radiation being absorbed more or less strongly as it travels. In view of these considerations it would appear that an item in thermal conduction through nonmetallic bodies might well be the volume emission, transmission, and absorption of radiation.

CONVECTION In addition to the more familiar cases of convection considered in most elementary textbooks, it should be noted that examples of very large-scale convection currents occurring in nature are those we observe as ocean currents and winds. Atmospheric convection currents are of particular interest because of the role they play in determining temperatures at various altitudes. This subject will be discussed in Chapter 17.

RADIATION If we hold a hand some inches below an electric lamp or a hot iron, we experience a sensation of warmth; yet no heat can have been conveyed to the hand either by conduction or by convection. Similarly, if we expose ourselves to direct sunlight we are warmed, even thought no heat (in the sense ordinarily understood) can be conveyed to us from the sun across the vacuum of interplanetary space. What actually happens is that energy is emitted by the hot body (the lamp filament, the hot iron, or the sun) in the form of electromagnetic waves, covering a wide range

---

* This was first noticed by Wiedemann and Franz in 1853, and the statement of the approximate constancy of the ratio of the two conductivities is known, after them, as the Wiedemann–Franz law.

of wavelengths. The propagation of these waves we call **radiation.**
When such radiation encounters another body, some of it is absorbed, that is, converted from radiation into some other kind of
energy, and usually the latter takes the form of heat. Thus we
have, first, the process of emission from the hot body, that is, the
conversion of some of its heat energy to radiation. This radiation
is then transmitted across the intervening space to another body.
Finally, it is absorbed by the latter, the energy of the radiation
being reconverted to heat.

Owing to the circumstance that radiation is emitted from hot
bodies, it is often called "heat radiation," and one frequently finds
in physical literature such expressions as "radiation of heat,"
"radiant heat," and "heat rays." Although the use of such expressions may well have appeared justified 100 years ago, before the
nature of the radiation was fully understood, their continued use
at the present time is most unfortunate. The false distinction between "heat radiation" and "light" to which this usage has given
rise, with the implication that these are different in nature, has
been responsible for endless confusion in this field of physics. Thus
there is a widespread but quite erroneous impression that "heat
radiation" is confined to the longer wavelengths of the electromagnetic spectrum, and this idea persists in spite of the obvious
fact that, for example, the filament of an incandescent lamp emits
not only visible light but even some ultraviolet radiation, and that
these, no less than the radiation of longer wavelengths, are absorbed
by bodies on which they impinge and so are converted into heat.

On the score that all this radiation is propagated in the form
of electromagnetic waves, it might well be contended that the
radiative transfer of energy from a hot body to another body
should not, strictly speaking, be considered under the heading of
heat transfer. This contention would be based on the argument that
it is not heat, as such, that is transferred, but energy in the form of
electromagnetic waves which happens to have been derived from
heat—energy which, normally, is destined to be reconverted to heat
on absorption. However, it is difficult to draw a sharp line between
radiative and nonradiative energy transfers in which heat is lost by
one body and gained by another. As we have already noted, radiation may well play a significant role in the process of thermal conduction. Also, although it may not be quantitatively important, the
energy present within a hot body in the form of radiation and in
equilibrium with it is, at least in principle, one of the items contributing to its total energy, which we call heat. In view of these
considerations and also in the interests of convenience, we therefore
include radiative energy transfers in which the radiation arises

from and is reconverted into heat under the general heading of heat transfer.

14.1   Calculate the increase in length of 100 m of copper wire in a power line when its temperature increases from $-20°$ to $+50°$C.

14.2   A circular hole in an aluminum plate has a diameter of 1.000 cm at 20°C. What will be its diameter when the temperature of the plate is raised to 100°C?

14.3   The density of gold is 19.30 g/cm³ at 20°C, and its coefficient of linear expansion is 1.43×10^{-4} deg⁻¹. What is the density at 80°C?

14.4   A barometer reading taken at the equator on a day when the temperature is 25°C is 758.3 mm Hg, the graduated scale of the instrument being of brass. Find the corrected reading in terms of the head of mercury at 0°C corresponding to "standard" g ($= 980.62$ cm/sec²), given that the scale graduations are correct at 15°C, and that

$$g \text{ at equator} = 978.03 \text{ cm/sec}^2,$$
$$\alpha \text{ for brass} = 1.9 \times 10^{-5} \text{ deg}^{-1},$$
$$\beta \text{ for mercury} = 1.8 \times 10^{-4} \text{ deg}^{-1}.$$

14.5   A clock with a brass pendulum ($\alpha = 1.9 \times 10^{-5}$ deg⁻¹) runs correctly at 20°C. How much will it gain or lose per day at 10°C?

14.6   A watch with a brass balance wheel ($\alpha = 1.9 \times 10^{-5}$ deg⁻¹) and steel hair spring runs correctly at 20°C. Find how much it will gain or lose per day at 10°C, given that Young's modulus for steel, to which the elastic control associated with bending is proportional, varies with temperature Θ according to the equation

$$E = E_0(1 - \gamma\Theta),$$

the value of γ being 2.4×10^{-4} deg⁻¹.

14.7   Given that for steel $\alpha = 1.1 \times 10^{-5}$ deg⁻¹ and $E = 2.1 \times 10^{12}$ dyn/cm², and the elastic limit under tension occurs at 2000 kg weight/cm², calculate the drop in temperature from that at which railroad lines are laid and welded together end to end that can be accommodated elastically.

14.8   A copper calorimeter (a vessel having the form of a can or beaker) of mass 130 g contains 100 g of water at 15°C. After an additional 200 g of water at 30°C has been poured in, and the contents stirred, the temperature is found to be 25°C. What is the specific heat of copper?

14.9   The temperature of 200 g of alcohol contained in a copper calorimeter of mass 400 g is 36°C. Into this is poured 1000 g of copper turnings at 0°C, and after the contents of the calorimeter have been stirred the temperature is found to be 22.4°C. What is the specific heat of alcohol?

14.10   A quantity of lead shot of mass 400 g is heated in a steam jacket to 100°C and then dropped into 300 g of water at 15°C contained in an aluminum calorimeter of mass 23 g. The temperature after stirring is found to be 18.3°C. What is the specific heat of lead?

14.11 The time t is observed for a 100-g copper calorimeter containing 500 g of water to cool under controlled conditions from a temperature of Θ_1 to Θ_2. The water is then replaced by the same volume of another liquid, of mass 400 g, and it is found that the time taken for this to cool under the same conditions from Θ_1 to Θ_2 is $0.52t$. Assuming the rate of heat loss to be a unique function of temperature for the conditions applying, calculate the specific heat of this liquid.

14.12 A copper calorimeter of mass 274 g contains 305 g of water at 31.6°C, and into this 52.1 g of ice at 0°C is poured. After all the ice has melted, while the contents of the calorimeter are stirred, the temperature is found to be 26.4°C. What is the latent heat of fusion of ice?

14.13 Twelve grams of turpentine at 15°C are poured into an ice calorimeter, and this causes the end of the mercury thread in the capillary tube to recede 12.9 cm. Given that the internal diameter of this tube is 1 mm, that the densities at 0°C of ice and water are 0.917 g/cm³ and 1.000 g/cm³, respectively, and that the latent heat of fusion of ice is 79.6 cal/g, calculate the specific heat of turpentine.

14.14 How many calories of heat are needed to convert 40 g of ice at −20°C to steam at 200°C at atmospheric pressure?

14.15 If 1000 g of steam at 200°C is mixed with 20 g of water at 30°C, what will be the final state? (Assume that heat exchange with the surroundings is negligible.)

14.16 A quantity of hot iron, of mass 300 g, is dropped into a copper calorimeter of mass 150 g containing 220 g of water at 20°C. This causes the water to boil, with the loss of 5 g of steam. What was the original temperature of the iron?

14.17 Ten grams of steam at 150°C and 60 g of ice at −20°C are added to 200 g of water at 20°C contained in a copper calorimeter of mass 100 g, and the contents are stirred. Assuming heat exchange with the surroundings to be negligible, calculate the final temperature.

14.18 An evaporating pan has a heating area of 1.5 m² and a thickness of 2 mm. The outer heating surface is maintained at a temperature of 180°C, and the pan contains water boiling at 100°C. Given that the thermal conductivity of the material of the pan is 0.13 cal/sec/cm/deg, calculate the quantity of water evaporating per minute.

14.19 A compound slab consists of two uniform sheets of material in intimate thermal contact. Show that when heat is conducted through this slab the temperature gradients in the two components are inversely proportional to their thermal conductivities.

14.20 A layer of ice of thickness x has formed on the surface of a lake. Assuming the temperature of the upper surface to remain constant at −5°C, find an expression for the increase of thickness occurring in a short time δt. Hence find the time required for the thickness to increase from 5 cm to 10 cm. (Treat the specific heat of ice as negligible compared with its latent heat of fusion.)

14.21 Imagine a homogeneous sphere whose surface is maintained at a constant temperature and at whose center heat is generated at a constant rate dQ/dt, this being conducted radially outward. Show that, if the thermal conductivity k is independent of temperature, the temperatures

Θ_1 and Θ_2 at distances r_1 and r_2 from the center conform to the equation

$$\Theta_1 - \Theta_2 = \frac{dQ/dt}{4\pi k}\left(\frac{1}{r_1} - \frac{1}{r_2}\right).$$

14.22   The inner surface of a cylindrical pipe is kept at an effectively constant temperature by a liquid at this temperature flowing through it, and the outer surface is kept at an effectively constant lower temperature by a bath in which it is immersed. Show that if the thermal conductivity k of the material of the pipe is constant, the temperatures Θ_1 and Θ_2 at points within the pipe at distances r_1 and r_2, respectively, from the axis are related to the rate of radial heat flow per unit length dq/dt by the equation

$$\Theta_1 - \Theta_2 = \frac{dq/dt}{2\pi k}\ln\frac{r_2}{r_1}.$$

14.23   A steel pipe of inner and outer radii 1.5 cm and 2.0 cm, respectively, conveys steam at approximately atmospheric pressure at the rate of 2 g/sec, and at a certain point thermocouples attached to the inside and outside of the pipe register temperatures of 120° and 115°C, respectively. Given that the thermal conductivity of the steel is 0.11 cal/sec/cm/deg, calculate the decrease of temperature of the steam per unit length of pipe at this point.

15

THE KINETIC THEORY OF GASES

Up to now we have dealt with heat as a mysterious something that passes from one body to another. To be sure, reference has been made to it as the energy of molecular motion, but no evidence has so far been adduced in support of this description. It is now desirable to try to advance beyond this stage, by establishing a theory of heat in mechanical or other terms, so integrating heat with other branches of physics. As a first step in this direction, let us consider in some detail the thermal behavior of gases. The choice of gases in preference to other forms of matter is indicated by the apparent simplicity of the gaseous state, referred to in Chapter 13; we may infer this from the simple pressure–volume relation applying to gases.

Equation of State for a Gas

We have considered two proportionality relations applying to an ideal gas: Boyle's law, for a gas at constant temperature, and also the relation expressing the definition of temperature, for example in terms of volume at constant pressure. It will now be useful to see if we can combine these, together with other relevant relations, into a single "equation of state" for such a gas.

In addition to the two proportionality relations just referred to, there is one that is rather self-evident—sometimes referred to as Guericke's law—to the effect that at constant pressure and temperature the volume of a given kind of gas is proportional to its mass. Finally, there is Avogadro's law, according to which, for a given mass, pressure, and temperature, the volume of a gas is inversely proportional to its molecular weight M.

Altogether, then, we have the following four relations:

Guericke's law:

$$V \propto m, \qquad p, \Theta, \text{ and kind of gas constant,}$$

Boyle's law:

$$V \propto \frac{1}{p}, \qquad m, \Theta, \text{ and kind of gas constant,}$$

Definition of temperature:

$$V \propto 1 + \eta\Theta,^* \qquad m, p, \text{ and kind of gas constant,}$$

Avogadro's law:

$$V \propto \frac{1}{M}, \qquad m, p, \text{ and } \Theta \text{ constant.}$$

To these should be added the statement, which we may refer to as Charles's law, that η has the same value for all gases.

In view of Charles's law, it is obvious that all four of the above relations are merely special cases of a single, comprehensive relation,

$$V \propto \frac{m}{M} \frac{1 + \eta\Theta}{p},$$

or

$$pV \propto \frac{m}{M} (1 + \eta\Theta).$$

Since we are concerned only with proportionality, we may divide the right side by the constant η, obtaining the more useful relation

$$pV \propto \frac{m}{M} \left(\frac{1}{\eta} + \Theta \right),$$

which is the same as the equation

$$pV = \frac{m}{M} RT, \qquad\qquad (15.1)$$

where T denotes temperature on the Kelvin scale and R is a universal constant, the so-called **gas constant.** This equation is known as the **equation of state** for a gas.

CHARLES'S LAW   Most textbooks give this name to what we have referred to as the definition of temperature—that, at constant pressure,

$$V = V_0(1 + \eta\Theta),$$

$\eta(= \beta)$ being a constant that has the same value for all gases.

This could properly be called a law only if temperature on our standard scale could be, and had been, reliably determined other-

---

* The symbol η is here written for the coefficient rather than β, to which it is equal, in view of the fact that the definition of temperature on the gas scale has been given impartially in terms of the variation of either V or p, or, more generally, in terms of the variation of the product pV.

wise than on the basis of the behavior of a gas. Now the graduations of an ideal-gas thermometer may, in fact, be shown theoretically to be identical with those of a scale defined by Kelvin (1824–1907) without reference to the particular properties of any specific material. However, Charles (1746–1823) certainly did not determine his temperatures by the procedure corresponding to Kelvin's definition. Actually it would not be feasible to do so, even today, because in practice determinations of temperature on the standard (ideal-gas or Kelvin) scale having any pretense to accuracy can only be based, ultimately, on the behavior of a gas. Little is known about the details of Charles's work, but because mercury thermometers were current in his time, he presumably used one of these. Gay-Lussac (1778–1850), who repeated the measurements of Charles, succeeded in achieving substantially greater accuracy, and he certainly did use a mercury thermometer. Gay-Lussac's investigation led him to the conclusion that "in general all gases, by equal degrees of heat, under the same conditions, expand proportionately just alike."

Charles and Gay-Lussac both regarded temperature as being given by a mercury thermometer. But, for reasons set out in our general discussion of temperature scales, we no longer think of the mercury thermometer as in any sense a fundamental instrument, and Charles's so-called law must therefore be revised accordingly. From the present-day point of view, the mercury thermometer used by Charles and by Gay-Lussac can legitimately be regarded only as a means for establishing *sameness* of temperature on two different occasions, that is, of repeating a temperature without being able to assign a numerical value to it. The main result of Charles's and Gay-Lussac's work may be expressed mathematically by the statement that at constant pressure

$$\frac{V_1'}{V_1} = \frac{V_2'}{V_2}$$

for a given (although numerically unknown) pair of temperatures, where the subscripts refer to two different gases. This is equivalent to the statement that η, which for any sufficiently rarefied gas is, according to the way we now define the standard temperature scale, a constant *by definition*, is also the same for all gases. It is to *this* statement—that η is independent of the nature of the gas—that the name "Charles's law" is correctly given, *not* to the statement that η is a constant. An incidental further result of Charles's and Gay-Lussac's work is that there is agreement (now known to be approximate only) between the scales of a gas and a mercury-in-glass thermometer in each of which the fundamental interval

377

(between the freezing point and the boiling point of water) is divided into the same number of equal parts.

VALUE OF THE GAS CONSTANT The usual statement of Avogadro's law in its quantitative form is that at STP 1 gram-molecule ("mole") of any kind of gas occupies 2.2414×10^4 cm^3. Now, standard atmospheric pressure is 760 mm Hg, or 1.01325×10^6 dyn/cm^2. "Standard" temperature (0°C) is 273.15°K. And m for 1 gram-molecule is M. Hence substituting these values in Eq. (15.1) we have

$$1.01325 \times 10^6 \times 2.2414 \times 10^4 = 273.15R,$$

whence

$$R = 8.314 \times 10^7 \text{ ergs*/gram-molecule/deg,}$$

or

$$R = 8.314 \times 10^3 \text{ J*/kilogram-molecule/deg.}$$

Molecular Picture of a Gas

Whether rightly or wrongly, we naturally think of a molecule as having some more or less definite size, some definite occupancy of space. On the assumption that this is so, it is possible, by comparing the density of a gas with that of the corresponding liquid or solid, to form a reasonable estimate of the mean distance between adjacent molecules in a gas relative to their effective linear dimensions.

Let us, for example, consider the gas nitrogen. At STP the density of nitrogen is 1.25×10^{-3} g/cm^3. *Liquid* nitrogen, on the other hand, has a density of 0.79 g/cm^3, that is, 630 times that of gaseous nitrogen at STP. This can only mean that in the latter the mean distance between centers of neighboring molecules is $(630)^{1/3}$, or 8.5 times what it is in the former; molecules in the gas are 8.5 times farther apart than in the liquid. Now, from the relative difficulty of compressing liquids and solids we may infer that any such spaces as there may be between adjacent molecules in these are very small in comparison with the effective dimensions of the molecules themselves; to a first approximation we may regard the molecules as touching one another. Accordingly, it appears that the mean distance between the centers of adjacent molecules in nitrogen at STP is some 8.5 times the effective diameter of a single molecule.

Similar considerations apply to other gases; quite generally we may say that at ordinary temperatures and pressures inter-

---

* It will be recalled, for example from our discussion of Bernoulli's principle in Chapter 9, that the dimensions of pV are those of work or energy.

molecular distances in gases are of the order of magnitude of 10 molecular diameters.

From the observed phenomenon of gaseous diffusion we know that the molecules must be in a state of continual movement. We need to postulate such movement also to account for the fact that a gas will occupy the whole of the space made available to it. And obviously there is no reason why there should be any preferred direction of the molecular velocities. Accordingly, we have the following picture of a gas: It consists of an assemblage of molecules moving in all sorts of directions and separated from one another by distances that are large compared with the dimensions of the individual molecules.

It is not proposed to invoke prematurely the detailed information concerning molecular sizes and structures gained from advanced physical studies. In the absence of precise knowledge concerning these properties of molecules they are often visualized as hard elastic spheres, like billiard balls. This picture, which has already been implied by the use of the term "effective diameter," is often useful, even though we know it to be a gross oversimplification. In justification of such an idealization, we may reflect that there is no preferred direction in space applying to a large assemblage of molecules; such an assemblage has "spherical symmetry" of properties. This encourages us to postulate spheres of some definite size—which also do not have any property relating to a preferred direction—as the *equivalent* of actual molecules, from a statistical point of view. We have, then, an idealized picture of a gas such as is shown in Fig. 15.1, the arrows representing the instantaneous velocities of the individual molecules.

We know from such phenomena as elasticity and surface tension that molecules in close proximity exert forces on one another, and we must suppose that in a gas the centers of mass of molecules must from time to time come sufficiently close to one another to call into play strong mutual repulsive forces. We may aptly describe such an occurrence as a "collision" between the molecules. By their nature, such collisions must in general lead to rather abrupt changes in the directions of travel and in the speeds of the molecules concerned, although, exceptionally ("glancing" collisions), these changes will only be slight. Small changes in direction and speed may also result from close approaches between centers of molecules, which, although not close enough to bring repulsive forces into play, are yet sufficiently close to give rise to slight changes in the directions and speeds of travel resulting from a mutual attraction. In a first attempt to set up a quantitative kinetic theory it is usual

379

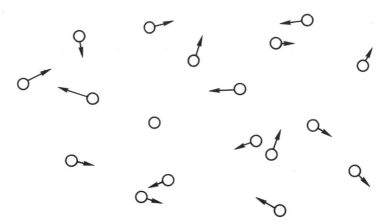

FIGURE 15.1 Diagrammatic representation of instantaneous velocities of molecules of a gas.

to assume that the effects of close approaches are negligibly small. It is also usual to consider molecules as having sufficiently definite effective sizes and forms to justify drawing a sharp distinction between a hit (collision) between two molecules and a miss. As we shall see, a kinetic theory based on these assumptions is quite impressively successful.

It will be observed that the arrows drawn in Fig. 15.1 are not all of equal length, implying that the molecular velocities are not all the same. That there must in fact be a distribution of velocities may be seen as follows. Let us suppose that we begin with an assemblage of molecules moving in random directions but all with the same velocity, and let us consider a particular pair of colliding molecules, labeled 1 and 2 in Fig. 15.2. Then if the collision occurs as shown, the impulse given to molecule 1 will produce an increase

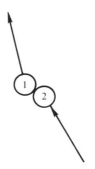

FIGURE 15.2 Collision between two gas molecules.

in its velocity (which incidentally will have a new direction), while molecule 2 will suffer a decrease in its velocity. Quite generally, collisions will cause some molecules to be speeded up and others to be slowed down, so that eventually the velocities of the molecules must settle down to a certain statistically determined distribution.

Molecules of a gas confined within a vessel must make collisions with the containing walls as well as with one another. If we treat as unimportant the forces exerted between molecules during a close approach, we must, to be consistent, also neglect the forces that the molecules exert on the containing walls before and after collisions with these; we must assume that only during collisions with the walls are (repulsive) forces exerted on these. This means that the pressure which the gas exerts on the containing walls must be attributed simply to the continual bombardment of these by the gas molecules.

The collisions that the molecules make with one another and with the walls of the containing vessel cannot change the total kinetic energy of the molecules, for with any such change there would be a corresponding change in the momenta of the molecules bombarding the walls and so in the pressure exerted by the gas at constant volume. But no such spontaneous change is observed; the pressure remains constant with time. Hence statistically we may consider all collisions to be "perfectly elastic," that is, energy-conserving. This does not necessarily mean that in individual cases some molecular kinetic energy is not converted into energy of some other form, for example energy of molecular "excitation" of some sort, or radiation. But it does mean that for every such event there must be a compensating event occurring somewhere else or at some other time in which the same energy conversion takes place in reverse. And, because we are discussing the molecular behavior of the gas *statistically*, we may ignore all such energy conversions; **the gas behaves statistically as if all collisions were perfectly elastic.** This perfect elasticity of collisions is usually listed as one of the *assumptions* of the kinetic theory of gases, along with certain other features of molecular behavior. Actually, as we see, it is not an assumption at all; the statistical equivalence of perfect elasticity is an inference drawn from the observed behavior of a gas.

The mean distance a molecule travels between successive collisions with other molecules is known as the **mean free path.** Obviously, the mean free path must be of a higher order of magnitude than the mean distance separating neighboring molecules. It is even larger compared with the dimensions of an individual molecule.

381

Behavior of a Gas in Terms of Molecules

Having established the equation of state for a gas, based on its observed macroscopic behavior, and considered the careers of its constituent molecules, we are now ready for our main task, that of endeavoring to account for the various mechanical and thermal properties of gases in terms of molecules. Among these properties are the pressure a gas exerts on its containing walls, its thermal expansion at constant pressure, its thermal conductivity, its viscosity, its diffusion characteristics, and its specific heats c_v and c_p.

EXERTION OF PRESSURE Our picture of the exertion of pressure is of innumerable punches delivered by the bombarding molecules of the gas at randomly distributed points on the inner surface of the container. This being so, it is clear that to have any prospect of success in deriving a formula for the pressure in terms of molecules we must approach our problem statistically. But in addressing ourselves to this task we are at once confronted with what appear to be formidable difficulties. Thus the molecules strike the surface with velocities distributed according to an as-yet-unknown law; they impinge at all sorts of angles to the normal; and, because the surface is molecularly rough, there can be no question of equality of the angles of incidence and reflection. Again, the bombarding molecules have complicated careers in the gas, making innumerable collisions with other molecules and rebounding in new directions.

In such a situation, rather than be daunted by all the difficulties, it often pays, in the first instance, to pretend that they do not exist. We first solve the problem for an extremely simplified and idealized system, and then consider how the difficulties might be dealt with, one by one. This is what we shall now do.

Let us consider an idealized gas (specified below) contained within a cubical box of side l, three of whose edges meeting at a corner define the x, y, and z axes of a Cartesian coordinate system, as indicated in Fig. 15.3, and let us investigate the pressure exerted on one of the sides of this box, say the one shaded, the side $x = l$.

Let there be N molecules in the box, each of mass μ but of zero molecular volume, so that the complication of intermolecular collisions does not arise. And let us imagine that the inner surfaces of the box, instead of being molecularly rough, are smooth, so that each molecule incident at a certain angle to the normal will rebound, making the same angle with the normal on the other side.

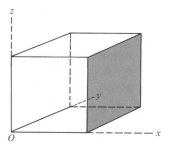

FIGURE 15.3

To evaluate the force exerted internally on the side $x = l$, which, divided by l^2, will give the pressure, we shall have to make use of the definition of force as time rate of change of momentum; thus we shall have to obtain an expression for the total change in molecular momentum occurring at this side per unit time. This will be the force which this side exerts on the gas, and, because action and reaction are equal and opposite, it will also be the force that the gas exerts on the side of the box.

We shall begin by calculating the contribution of a single randomly selected molecule to this force. Let this molecule have velocity components whose magnitudes are $|v_x|$, $|v_y|$, and $|v_z|$. Owing to the assumed equality of angles of incidence and reflection at each collision with a side, the molecule will continue to have velocity components of these magnitudes indefinitely. At each impact with the side $x = l$ the x component of momentum of the molecule will be changed from μv_x to $-\mu v_x$; the other components will not be affected. The change in momentum per impact with this side will therefore be $2\mu|v_x|$, and this will be in the negative x direction. And the period that elapses between one impact with this side and the next is the double distance $2l$ divided by the relevant velocity component $|v_x|$; that is, it is $2l/|v_x|$. Accordingly, the number of impacts per second of this molecule with the side in question is $|v_x|/2l$. This number, multiplied by the change of momentum per impact, $2\mu|v_x|$, is therefore the contribution of the molecule to the force on the side $x = l$. This is v_x^2/l.

We have now merely to add all such quantities as this for the N molecules in the box to obtain the total force. This is

$$\frac{\mu}{l} \sum_{r=1}^{N} v_{rx}^2,$$

the separate x components of velocity being labeled v_{1x}, v_{2x}, . . . , v_{Nx}. Hence, dividing this by l^2, the area of the side, we have for the pressure

$$p = \frac{\mu}{l^3} \sum_{r=1}^{N} v_{rx}^2 = \frac{\mu}{V} \sum_{r=1}^{N} v_{rx}^2,$$

where V is the volume of the box.

For a large assemblage of molecules there is no preferred direction in space, so we must have, on average,

$$\sum_{r=1}^{N} v_{rx}^2 = \sum_{r=1}^{N} v_{ry}^2 = \sum_{r=1}^{N} v_{rz}^2,$$

and so the expression for the pressure may also be written

$$p = \frac{\mu}{3V} \left(\sum_{r=1}^{N} v_{rx}^2 + \sum_{r=1}^{N} v_{ry}^2 + \sum_{r=1}^{N} v_{rz}^2 \right),$$

and because for each molecule

$$v_x^2 + v_y^2 + v_z^2 = v^2,$$

where v is the velocity of the molecule, this reduces to

$$p = \frac{\mu}{3V} \sum_{r=1}^{N} v_r^2.$$

Let us now, finally, multiply both numerator and denominator of this expression by N, the total number of molecules, so obtaining

$$p = \frac{N\mu}{3V} \frac{\displaystyle\sum_{r=1}^{N} v_r^2}{N}.$$

The second factor on the right of this equation is obviously the mean value of v^2 for all the molecules, which we may write as $\overline{v^2}$, and $N\mu$ is the total mass m of the gas within the box. Also m/V is the density ρ of the gas. Accordingly, the expression for the pressure may be written

$$p = \tfrac{1}{3}\rho\overline{v^2}. \tag{15.2}$$

We shall now consider the various complications listed earlier.

Our ignorance of the precise law of distribution of velocities among the gas molecules has not, after all, mattered; Eq. (15.2) does not depend in any way on this law.

Statistically, the inequality of the angles of incidence and reflection on the inner surface of the containing vessel is not of any consequence either. This is because for every change from, say, v_x to v_x' in the x component of the velocity of a molecule resulting from its reflection from one of the sides of the cube parallel to the x axis, there must, on average, be a compensating change in the x component of the velocity of another molecule from v_x' to v_x resulting from a reflection at or near the same value of x. There will be similar compensation for any changes in $|v_x|$ occurring in reflections at the surfaces $x = 0$ and $x = l$. This must be so, for otherwise no unique distribution of velocities among the molecules could be maintained. In a gas of high molecular concentration the two mutually compensating changes may be taken to occur not only practically at the same value of x but also practically simultaneously, and the second molecule may be considered as taking over the role of the first, and vice versa. Clearly, this cannot affect the pressure exerted on any face of the cube. In a rarefied gas the compensating change may take place either earlier or later, with the "earlier" balancing the "later" statistically. It is not difficult to see that in this case, also, the pressure cannot be affected.

Similar considerations apply to changes in v_x (or one of the other velocity components) resulting from intermolecular collisions; essentially the same argument applies. However, it should be noted that in the arguments relating to collisions of molecules either with walls or with other molecules it is assumed that the times occupied by these collisions are short compared with the periods between successive collisions. This assumption, corresponding closely to what is observed in the behavior of billiard balls, appears eminently reasonable.

Finally, there is the matter of the finite sizes of the molecules to be considered, a complication not listed with the others. As we saw earlier, in all except very highly compressed gases, the mean free path is large compared with the effective diameter of a molecule. Accordingly, the correction required in the paths traveled by the centers of molecules as a result of their finite sizes is generally negligible.

MEANING OF TEMPERATURE Let us now consider the physical significance of Eq. (15.2). Writing m/V for ρ in this, and multiplying both sides by V, we have

$$pV = \tfrac{1}{3}m\overline{v^2}. \tag{15.3}$$

The appearance of the product of a mass and the square of a velocity in this equation suggests translational kinetic energy. The expres-

385

sion for the total translational kinetic energy E_{tr} of the molecules of our sample of gas is

$$E_{tr} = \tfrac{1}{2}\mu \sum_{r=1}^{N} v_r{}^2,$$

which, clearly, is the same as

$$E_{tr} = \tfrac{1}{2}N\mu \frac{\sum\limits_{r=1}^{N} v_r{}^2}{N} = \tfrac{1}{2}m\overline{v^2}. \tag{15.4}$$

Thus eliminating $m\overline{v^2}$ between Eqs. (15.3) and (15.4), we have

$$pV = \tfrac{2}{3}E_{tr}. \tag{15.5}$$

Let us now compare Eq. (15.5) with the equation of state for a gas,

$$pV = \frac{m}{M} RT. \tag{15.1}$$

Equating the right sides of these, we have

$$E_{tr} = \frac{3}{2}\frac{m}{M} RT, \tag{15.6}$$

whence we see that the sum of the translational kinetic energies of all the molecules in a gram-molecule ($m = M$) is $\tfrac{3}{2}RT$. If N is Avogadro's number, the number of molecules in a gram-molecule, the mean translational kinetic energy of a single molecule is $\tfrac{3}{2}kT$, where k is written for R/N. This constant k is known as **Boltzmann's constant.**

It is now clear that we can attach a meaning to the temperature of an ideal gas in mechanical terms: It is simply something proportional (by the factor $\tfrac{2}{3}k$) to the mean translational kinetic energy of the molecules.

VAN DER WAALS'S EQUATION In addition to occupying finite volumes, molecules must also exert forces of attraction on one another; otherwise there would be no tendency for gases at low temperatures to condense into liquids or solids. For both these reasons the simple kinetic theory of gases that we have been considering can give no better than a good first approximation to the behavior of real gases. In 1879 J. D. van der Waals therefore attempted to take account of the finite volumes and mutual interactions of the molecules and, without having to make any detailed assumptions concerning the latter, showed that a second approximation to the behavior of real gases should be given by the equation (for 1 mole)

$$\left(p + \frac{a}{V^2}\right)(V - b) = RT,$$

where a and b are constants characteristic of the gas. The constant a arises from the mutual interactions of the molecules, while b is $4\sqrt{2}$ times the effective volume occupied by all the molecules. It is found that if a and b are given appropriate values, van der Waals's equation does indeed represent the behavior of actual gases very closely.

For a mass m of gas the equation would assume the more complex form

$$\left[p + \left(\frac{m}{M}\right)^2 \frac{a}{V^2}\right]\left(V - \frac{m}{M}b\right) = \frac{m}{M}RT.$$

Note that when V is large, this equation reduces to the simple form of the ideal-gas equation (15.1).

MOLECULAR VELOCITIES The equation

$$p = \tfrac{1}{3}\rho\overline{v^2}, \tag{15.2}$$

which may generally be taken to apply with sufficient accuracy to real gases, gives us information concerning the molecular velocities. For example, on substituting known values for p and ρ for the case of nitrogen at STP,

$$p = 1.013 \times 10^6 \text{ dyn/cm}^2$$

and

$$\rho = 1.25 \times 10^{-3} \text{ g/cm}^3,$$

we have for this case

$$\overline{v^2} = \frac{3p}{\rho} = \frac{3 \times 1.013 \times 10^6}{1.25 \times 10^{-3}}$$
$$= 2.431 \times 10^9 \text{ cm}^2/\text{sec}^2,$$

whence

$$\sqrt{\overline{v^2}} = 4.93 \times 10^4 \text{ cm/sec}$$
$$= 493 \text{ m/sec}.$$

The square root of the mean value of the square ("root-mean-square," or rms, value) of the velocity is not quite the same thing as the mean velocity $\bar{v}$. It may be shown by statistical reasoning that the two are related by the equation

$$\bar{v} = \sqrt{\frac{8}{3\pi}}\ \sqrt{\overline{v^2}} = 0.921\sqrt{\overline{v^2}},$$

387

whence, for the case just considered,

$$\bar{v} = 0.921 \times 493 = 454 \text{ m/sec.}$$

In view of the fact just noted, that E_{tr} is uniquely determined by the temperature, $\bar{v}$ must have the same value for a given temperature irrespective of the pressure.

It is interesting to compare the mean velocity of the molecules, $\bar{v}$, with the velocity of propagation of sound waves, v_s. Eliminating p/ρ between the equation

$$\bar{v} = \sqrt{\frac{8}{3\pi}}\sqrt{\bar{v^2}} = \sqrt{\frac{8}{3\pi}}\sqrt{\frac{3p}{\rho}} = \sqrt{\frac{8p}{\pi\rho}}$$

and the expression for the velocity of sound in a gas

$$v_s = \sqrt{\frac{\gamma p}{\rho}},$$

we have

$$\frac{v_s}{\bar{v}} = \sqrt{\frac{\pi\gamma}{8}}.$$

As γ has its highest value for monatomic gases, 1.66 (see Table 15.1), and the corresponding value of $v_s/\bar{v}$ is 0.81, we see that v_s must always be less than $\bar{v}$, although it is of the same order of magnitude. Obviously it would not be physically possible for pressure waves to travel faster than do the individual molecules. For nitrogen, whose γ is 1.405, v_s is equal to $0.76\bar{v}$.

The mean molecular velocities are of the same general order for most gases at ordinary temperatures—a few hundred meters per second. It is of interest to compare with this the "velocity of escape" from the moon. The evaluation of this velocity was set as a problem in Chapter 5; it is 2367 m/sec. This is of the order of five times the mean molecular velocity of most gases. However, even though the velocity of escape is so much greater than the mean molecular velocity, it is still not nearly large enough to enable the moon to retain an atmosphere. This is because of the considerable spread of velocities about the mean; the velocity distribution is such that quite an appreciable proportion of the molecules of a lunar atmosphere would have velocities exceeding the velocity of escape. It would therefore not be long before the whole of such an atmosphere would "leak" away.

We can obtain a more generally useful and convenient expression than Eq. (15.2) from which to calculate the rms velocity by making use of the equation of state. Thus, eliminating pV between Eqs. (15.2) and (15.1), we obtain the formula

$$\overline{v^2} = \frac{3RT}{M}. \tag{15.7}$$

From this we see directly something that was implicit in the earlier formula—that for a given gas $\overline{v^2}$ is independent of the pressure, depending only on the temperature. This formula also shows that at a given temperature the mean molecular velocity is inversely proportional to the square root of the molecular weight. The rate at which a gas will diffuse through a porous partition or at which one gas will diffuse through another must clearly be proportional to the product of the number of molecules per unit volume and the mean molecular velocity. Accordingly, kinetic theory gives us the result that, other things being equal, the rate at which a gas diffuses is proportional to the square root of its molecular weight. This was first found to be the case experimentally by Graham, and the corresponding law is known, after him, as **Graham's law of diffusion.**

MEAN FREE PATH We have noted the fact that the mean free path of gas molecules at ordinary pressures and temperatures is of a higher order of magnitude than the linear dimensions of individual molecules. We shall now consider this quantitatively.

Any given molecule, such as A in Fig. 15.4, will, by virtue of its own motion and the motions of other molecules, collide with the latter from time to time. In view of the statistical nature of the spatial distribution of the other molecules at all times, it seems reasonable to ignore the motions of these, imagining them to be "frozen" in fixed positions. We shall not expect this idealization, which will greatly simplify our discussion, to introduce any serious error.

In each unit of time the molecule A will collide with certain other molecules. Which will these be? Let us idealize all the molecules as spheres of diameter d, and let us imagine the molecule A,

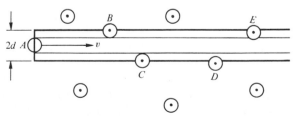

FIGURE 15.4 Idealized diagram showing collision possibilities for a selected gas molecule.

traveling in a straight line with velocity v as indicated, to carry along with it a concentric disk of diameter $2d$ whose plane is perpendicular to the direction of motion. Then this disk will sweep out a cylinder whose length increases at the velocity v. As is evident from the figure, the molecule A will make grazing collisions with other molecules whose centers lie on the periphery of this cylinder, such as B and C. It will not collide with any molecules such as D whose centers lie outside this cylinder, but it will collide with all whose centers lie within it, such as E. Of course, in any collision that is not grazing, the direction of motion of the molecule A will be changed. However, this does not affect the situation *statistically;* for statistical purposes we may, if we wish, imagine the molecule to continue along its original path despite any collision that it may have.

It will now be evident that the number of collisions molecule A makes per unit time is equal to the number of other molecules whose centers lie within the volume $\pi d^2 v$ which the imaginary disk sweeps out per unit time. If n is the number of molecules per unit volume, then the number of collisions A makes per unit time must be $\pi d^2 v n$. And during this time A has covered the total distance v, so the mean distance traveled between collisions must be $v/\pi d^2 v n$, that is, $1/\pi d^2 n$. A more rigorous analysis gives for this distance the value $1/\sqrt{2}\,\pi d^2 n$. Thus, for the mean free path λ we have

$$\lambda = \frac{1}{\sqrt{2}\pi d^2 n}.\qquad(15.8)$$

Interest also attaches to the **collision frequency** f, the number of collisions any given molecule makes with other molecules per second. This is given by the equation

$$f = \sqrt{2}\,\pi n d^2 \bar{v},\qquad(15.9)$$

where $\bar{v}$ is the mean molecular velocity.

Let us now see if we can arrive at some estimate of d in a particular case. In Chapter 22 we shall see that Avogadro's number, the number of molecules in a gram-molecule, is $6.02_5 \times 10^{23}$. We have noted that liquid nitrogen, in which the molecules are almost in contact with one another, has a density of 0.79 g/cm^3, that is, that 1 cm^3 has a mass of 0.79 g. This is $0.79/28$, or 0.028, of a gram-molecule, and therefore contains $0.028 \times 6.02_5 \times 10^{23} = 1.7 \times 10^{22}$ molecules. The volume expressed in cubic centimeters occupied by each molecule must be the reciprocal of this, 5.9×10^{-23} cm^3. This is the same as 59 Å^3. On the assumption of actual contact of the molecules with one another in liquid nitrogen and a "cubic" pattern of their arrangement, this would mean that the cube root of this, 3.9 Å, is the diameter of a nitrogen molecule. Actually, neither of these assumptions can be upheld, and we should expect the true effective diameter to be somewhat less than this. From other evidence it is believed that the effective diameter of a nitrogen molecule is close to 3.5 Å.

A knowledge of Avogadro's number also enables us to calculate the number of molecules per unit volume of a gas at any temperature and pressure; for example, at STP it is $6.02_5 \times 10^{23}$ divided by 2.24×10^4, that is, 2.76×10^{19} per cm^3. Substituting this for n and 3.5×10^{-8} cm for d in Eq. (15.8), and in addition 4.54×10^4 cm/sec for $\bar{v}$ in Eq. (15.9), we obtain for the values of λ and f in nitrogen at STP the values

$$\lambda = 6.7 \times 10^{-6} \text{ cm} = 670 \text{ Å}$$

and

$$f = 6.8 \times 10^9 \text{ sec}^{-1}.$$

It is interesting to note that for this gas at STP the mean free path is nearly 200 times the effective molecular diameter.

The mean free paths, in centimeters, of a number of gases at 20°C and 10^{-3} mm Hg pressure are as follows:

N$_2$	O$_2$	Air	A	H$_2$	He	CO$_2$
4.9	5.3	5.0	5.2	9.2	16.0	3.3

It will be seen that these are of the order of magnitude of the linear dimensions of small-caliber vacuum apparatus. At substantially lower pressures than 10^{-3} mm, most of the collisions made by the molecules are with the walls of the container and only relatively few with one another.

THERMAL CONDUCTIVITY Thermal conduction in a gas is basically a transport phenomenon, a transport of molecules having a higher mean energy across a plane in one direction and of molecules having a lower mean energy in the opposite direction.

A simple picture of this energy transport is as follows. Let the temperature gradient be in the z direction, and let us, as in the derivation of the pressure exerted by a gas, imagine one sixth of the molecules in any given volume to travel in the positive z direction and one sixth in the negative z direction. Now consider three planes, A, B, and C, each of unit area, perpendicular to the z direction and separated from one another by a distance equal to the mean free path λ, as indicated in Fig. 15.5. Then if the temperature at the plane B is T, those at the planes A and C will be $T - [\lambda(dT/dz)]$ and $T + [\lambda(dT/dz)]$, respectively. And if the molecular

391

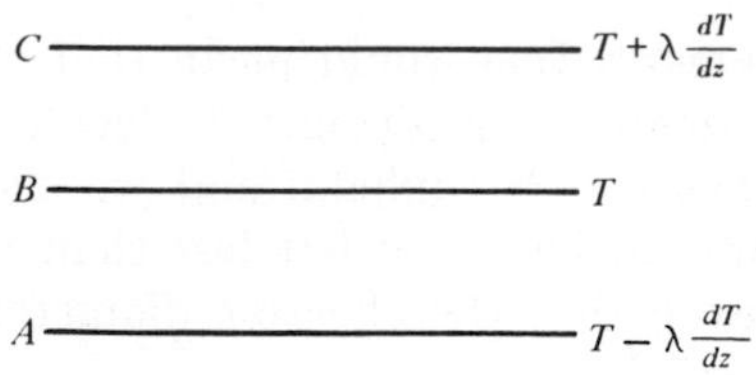

FIGURE 15.5

velocity characteristic of the temperature T is v (idealized as the same for each molecule), the mass of gas passing per unit time in each direction across the plane B will be $\rho v/6$. Let the specific heat of the gas at constant volume be c_v, and let us suppose that, by virtue of having made their last collisions at the plane A, the molecules traveling upward across the plane B all have the energy characteristic of the temperature $T - [\lambda(dT/dz)]$, while all those traveling downward across B have the energy characteristic of the temperature $T + [\lambda(dT/dz)]$ prevailing at the plane C. Then the rate of energy transport across B from below will be

$$\frac{\rho v c_v}{6}\left(T - \lambda\frac{dT}{dz}\right)$$

and that from above will be

$$\frac{\rho v c_v}{6}\left(T + \lambda\frac{dT}{dz}\right).$$

Hence the net rate of energy transport downward is

$$\frac{\rho v c_v \lambda}{3}\frac{dT}{dz}.$$

This divided by the temperature gradient dT/dz is therefore the thermal conductivity k:

$$k = \tfrac{1}{3}\rho v c_v \lambda.$$

A more rigorous analysis, in which account is taken of the distribution of velocities among the molecules, gives the coefficient $\frac{1}{2}$ instead of $\frac{1}{3}$; it also shows that the relevant velocity to use is the mean velocity $\bar{v}$. Thus

$$k = \tfrac{1}{2}\rho\bar{v}c_v\lambda. \tag{15.10}$$

Note that, because λ is inversely proportional to ρ, this formula implies that k at a given temperature should be independent of the pressure of the gas. This somewhat surprising prediction is well substantiated by experiment; it is found that the thermal conductivity is indeed independent of pressure except at very high pressures, where the mean free path is no longer large compared with molecular dimensions.

We may note also that since, as we shall see presently, $\bar{v}$ is proportional to $\sqrt{T}$, the theory predicts that k, which is proportional to $\bar{v}$, should also be proportional to $\sqrt{T}$. Experimentally, thermal conductivities of gases are found to increase with temperature somewhat faster than this.

Let us now work out the theoretical value of k for nitrogen at 0°C according to Eq. (15.10). The density of nitrogen at STP is 1.25×10^{-3} g/cm³, $\bar{v}$ at 0°C is 4.54×10^4 cm/sec, the measured value of c_v is 0.176 cal/g/deg, and the value we have calculated for λ at STP is 6.7×10^{-6} cm. Making the corresponding substitutions in the formula we find for k at 0°C the theoretical value of 3.4×10^{-5} cal/sec/cm/deg. The measured value agrees with this in order of magnitude, but is somewhat larger; it is 5.6×10^{-5} cal/sec/cm/deg.

The agreement between theory and experiment is, as we see, reasonable. It is probably as good as we could expect, in view of the crudeness of the billiard-ball model we have used and of the theory itself at some points.

VISCOSITY This, also, can be treated theoretically as a transport phenomenon. Again we consider three planes of unit area, A, B, and C, perpendicular to the z direction and separated from one another by the mean free path λ, as in Fig. 15.6. We suppose that the transverse (drift) velocity superimposed on the thermal velocities of the molecules is u at the plane B, $u - [\lambda(du/dz)]$ at the plane A, and $u + [\lambda(du/dz)]$ at the plane C. Then, as before, the mass of gas passing per unit time in each direction across the plane B is $\rho v/6$, and, as we readily see, the net rate of transfer of momentum across the plane B is $\frac{1}{3}\rho v\lambda(du/dz)$. This must be the force exerted from right to left by the gas above B on that below, and so, dividing this by the transverse velocity gradient du/dz, we obtain the theo-

FIGURE 15.6

393

retical expression for the viscosity. As in the case of thermal conductivity, a more rigorous analysis replaces the coefficient $\frac{1}{3}$ by $\frac{1}{2}$, and again the relevant thermal velocity of the molecules is $\bar{v}$, so that we have

$$\eta = \tfrac{1}{2}\rho\bar{v}\lambda. \qquad (15.11)$$

We see that η, like k, is independent of pressure and that it should be proportional to $\sqrt{T}$. Here again, the increase with temperature is, in fact, a little faster than this.

Substitution of the numerical values of ρ and λ for nitrogen at STP and of $\bar{v}$ for this gas at 0°C gives 1.9×10^{-4} P for the theoretical value of η. The measured value of η for nitrogen at 0°C is 1.66×10^{-4} P.

MECHANICAL EQUIVALENT OF HEAT   Let us consider, at the outset, a gas consisting of very simple molecules, particles that can neither rotate nor vibrate; such particles would have kinetic energies of translation only.

We have seen that when the temperature of such a gas is raised, the kinetic energy of its molecules increases by $\frac{3}{2}R$ per mole per degree. If the temperature increase occurs at constant volume, that is, under conditions where no external work is performed, this, according to our picture of a gas, is *all* that happens. But, as we know, if the temperature of a gas kept at constant volume is to be increased, heat must be imparted to it. These two things, the increase in the molecular kinetic energy of the gas and the conveyance of heat to it, can be reconciled only if we suppose that they are identical, that is, that the heat conveyed to the gas *is* the increase of energy that it acquires. On this assumption the calorie and the joule must simply be different units of the same kind of quantity, energy, and the question thus arises of how they are related quantitatively.

The heat capacity per gram-molecule, or "molecular heat," at constant volume is the same for all the monatomic gases; it is 2.98 cal/gram-molecule/deg. This must be equal to $\frac{3}{2}R$, and, as we have seen, R has the value of 8.314 J/gram-molecule/deg. Hence it follows that 1 cal must be equal to $\frac{3}{2}(8.314/2.98)$, or 4.18 J.

This is correct only if the whole of the energy of the molecules of a monatomic gas is indeed translational. Fortunately, it is possible to obtain an independent evaluation of the calorie. If this is found to agree with that just made, the agreement will constitute evidence of the correctness of our assumption.

We have seen that, according to the kinetic theory, the total translational kinetic energy of the molecules of an ideal gas is a function of temperature only; it is independent of the volume occupied. In cases where, in addition to moving as a whole, the molecules also rotate and/or vibrate, we should rather expect the ratio of the additional forms of energy to the translational energy to be constant with respect to volume at any particular temperature. If this is so, the *total* energy, also, would be independent of volume at any particular temperature.

This expectation was borne out in an experiment carried out by James Joule (1818–1889), in which air, whose behavior approximates that of an ideal gas, was allowed to expand without the performance of any external work. The apparatus is indicated diagrammatically in Fig. 15.7. A and B are two steel cylinders connected by a pipe in which there is a tap. The whole is immersed in a vessel of water. With the tap closed, Joule filled one cylinder, A, with air at a pressure of 20 atm, while the other was evacuated. Then, after noting the temperature of the water with a sensitive thermometer, he opened the tap, allowing part of the air in A to flow into B. On now stirring the water and again observing the temperature, Joule could detect no change. No heat was conveyed to or from the gas and no work was done on or by it, so the total energy of its molecules must have remained unchanged. Thus the total molecular energy at the temperature at which the experiment was carried out was shown—at least to a good first approximation— to be independent of volume, that is, to depend on the temperature only.

Let us now consider the heating of a gas, not at constant volume but at constant pressure. From Joule's experiment we must

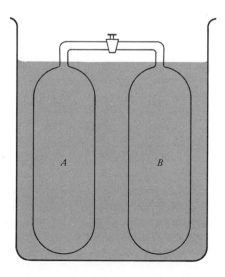

FIGURE 15.7 Joule's experiment on the unresisted expansion of a gas. **395**

infer that whether the gas is allowed to expand or not the increase of molecular energy is the same; this depends only on the rise in temperature. If the gas expands, a corresponding amount of work is performed by it, equal to the product of the pressure and the increase in volume. At constant pressure, therefore, the heat supplied to the gas has not merely to increase the molecular energy; some of it also has to be available for the performance of the external work.

If the initial volume per gram-molecule of the gas at the temperature T is V, and the final volume at the (higher) temperature T' is V', then

$$pV = RT$$

and

$$pV' = RT',$$

whence

$$p(V' - V) = R(T' - T).$$

We see that the mechanical work performed by the expanding gas per degree rise in temperature, $p(V' - V)/(T' - T)$, is R. This is, therefore, the amount by which the molecular heat at constant pressure, C_p, exceeds that at constant volume, C_v:

$$C_p = C_v + R. \tag{15.12}$$

The molecular heats of a number of gases at constant pressure and at constant volume, together with the differences and ratios of these, are given in Table 15.1. It will be seen that $C_p - C_v$ is very nearly the same for all these gases. Taking the mean value for the gases of simpler molecular constitution, from helium to oxygen, and substituting this and the value we have found for R, we have

$$1.99 \text{ cal} = 8.314 \text{ J},$$

or

$$1 \text{ cal} = 4.18 \text{ J}.$$

TABLE 15.1   Molecular Heats of Gases (cal/mole/deg) at 20°C and Atmospheric Pressure

Gas	C_p	C_v	$C_p - C_v$	C_p/C_v
He	4.97	2.98	1.99	1.66
A	4.97	2.98	1.99	1.66
H_2	6.86	4.87	1.99	1.408
N_2	6.92	4.93	1.99	1.405
O_2	7.04	5.04	2.00	1.396
N_2O	8.85	6.81	2.04	1.30
C_2H_6	11.47	9.40	2.07	1.22

The agreement between this and the previous value obtained for the calorie constitutes indirect evidence of the correctness of our assumption that monatomic molecules have energies of translation only.

From Eq. (15.6) it follows that for a monatomic gas $C_v = \frac{3}{2}R$ and $C_p = C_v + R = \frac{5}{2}R$, so the ratio of the two molecular (and also specific) heats must be

$$\gamma = \frac{C_p}{C_v} = \tfrac{5}{3},$$

and we see from our table that the experimental values of γ for helium and argon are in close agreement with this.

EQUIPARTITION OF ENERGY Examination of Table 15.1 shows that simple ratios occur not only between items in rows but also between those in the columns. Thus the values of C_v for the three diatomic gases are approximately the same, and their mean, 4.95, is almost exactly $\frac{5}{3}$ of 2.98, the value for the monatomic gases helium and argon. In the light of the foregoing discussion this can only mean that for every three units of translational kinetic energy possessed by the molecules of a diatomic gas, two additional units are associated with other modes of motion—rotation, and/or vibration. If we may assume simply the former of these and imagine a molecule of a diatomic gas as a kind of dumbbell, the occurrence of the ratio 3:2 between translation and rotation suggests a correspondence with the numbers of "degrees of freedom" for these two modes of motion. For, a body can move *as a whole* in three mutually perpendicular, and therefore independent, directions, but a system with masses concentrated at the ends can *rotate* only about two mutually perpendicular axes, these being at right angles to its own axis. Thus a point, to which we have idealized a helium or an argon atom (molecule), has only three degrees of freedom, these being translational. However, a dumbbell, representing a diatomic molecule, has five, three translational and two rotational.

It would appear, then, that statistically the energy is equally divided between the possible modes of motion, or degrees of freedom. The mean energy corresponding to each degree of freedom is $\frac{1}{2}RT$ per gram-molecule, or $\frac{1}{2}kT$ per molecule.

Accordingly, if the number of degrees of freedom of the molecules of a gas is n, C_v is equal to $n(R/2)$, and so by virtue of Eq. (15.12) we have

$$\gamma = \frac{C_p}{C_v} = \frac{C_v + R}{C_v} = \frac{(R/2)(n + 2)}{(R/2)n} = \frac{n + 2}{n}.$$

397

We have already noted how, in the case of the monatomic gases, for which n is equal to 3, γ has the value $\frac{5}{3}$. For diatomic gases, n is equal to 5, and so γ should be $\frac{7}{5}$. As we see from the table, the experimental values of γ for hydrogen, nitrogen, and oxygen are very close to this.

We have found how the energy is distributed among the different modes of motion from an examination of experimental data, but the **theorem of equipartition of energy,** as it is called, may also be deduced theoretically. This was achieved by James Clerk Maxwell (1831–1879). The mathematical analysis is complicated, but physically it depends simply on an application of the ordinary laws of mechanics to the collisions of gas molecules. The theorem states that the mean energy corresponding to each degree of freedom is the same for all the degrees of freedom of a system in equilibrium.

Problems

15.1   What is the volume of 10 g of nitrogen at 15°C and 60 cm pressure?

15.2   What mass of air (assumed dry) is contained in a room of dimensions 12 m × 10 m × 3 m at standard atmospheric pressure and 20°C?

15.3   An oxygen cylinder has a volumetric capacity of 50 liters, and oxygen is forced into it until its weight has increased by 2 kg, the temperature being 20°C. What is the increase in the pressure of the oxygen?

15.4   How much external work is done when 10 g of nitrogen is heated from 15° to 45°C at constant pressure?

15.5   Two vessels, A and B, of volumes V and $2V$, respectively, are connected by a narrow tube of negligible volume. After both vessels have been filled with air at 760 mm pressure and 25°C, A is immersed in a mixture of ice and water, and B is immersed in steam at 100°C. Neglecting thermal expansion of the containers, calculate the final equilibrium pressure.

15.6   A vessel A of capacity 4 liters contains a perfect gas at atmospheric pressure. It can be connected via a valve and tube of negligible volume to a vessel B of capacity $\frac{4}{9}$ liter. The valve is initially closed and B is evacuated. The valve is then opened long enough for the pressures in A and B to become equal, and then, after it is closed, B is again evacuated. Assuming that there are no complications such as leakage, release of adsorbed gas from walls, and so on, calculate the number of times this process must be repeated to reduce the pressure in A by a factor of 10^4. (Assume that the temperature of the gas remains constant throughout.)

15.7   A narrow glass tube of uniform internal cross-sectional area A extends vertically upward from a large vessel of volume V containing air at atmospheric pressure p_0. A steel ball bearing of mass m which just fits in the tube is placed in its upper end and released, so that as it falls it compresses the enclosed air. Show that, if volume changes occur isothermally, the pressure inside the vessel when the ball has fallen a distance x will be approximately $p_0(1 + Ax/V)$. Hence show that the equation of motion of the ball in the tube is

$$m \frac{d^2x}{dt^2} = mg - \frac{p_0 A^2}{V} x,$$

and so find the minimum length of tube required if the ball is to execute simple harmonic oscillations, and deduce an expression for the period. (Assume that there is no leakage of air past the ball and no friction between the ball and the tube.)

15.8   An isothermal atmosphere at temperature T exerts a pressure p_0 at ground level. By considering the equilibrium of an element of such an atmosphere under a gravitational field g, and integrating, show that the pressure at a height h is of the form $p = p_0 e^{-h/H}$. Find H in this expression.

15.9   A vessel containing a gas at a constant temperature T rotates about a fixed axis with angular velocity ω. Show that when the gas rotates with the same angular velocity as the vessel in which it is contained, there is a radial pressure gradient such that

$$\frac{dp}{dr} = \rho \omega^2 r,$$

where ρ is the density of the gas at the distance r from the axis. Hence show that

$$\frac{dp}{p} = \frac{M\omega^2}{RT} r \, dr,$$

and by integrating this find how p varies with r.

15.10   A box of volume $2V$ is divided into two equal parts by a thin partition. One part is evacuated and the other contains a gas. If the partition is porous and is equivalent to a wall with a large number of very small holes of total cross-sectional area A, show that the time required for the pressure difference between the two parts to fall to half its original value is given by the equation

$$t = \frac{3V \ln 2}{Av},$$

where v is the molecular velocity. (It may be assumed that all the molecules have the same speed, and that one-third move at right angles to the partition while the remainder move parallel to its surface.)

15.11   In the "box" theory of the exertion of pressure considered in the text, no account was taken of the action of the earth's gravitational field on the gas molecules. Now take this into account by considering the career of a molecule projected vertically upward with velocity v from the bottom face of such a box, and find by how much the time-averaged force exerted by a molecule on the top face falls short of that exerted on the bottom face. Compare the value found with the average force exerted on either face on account of bombardment at temperature T and show that the former is negligible compared with the latter at all ordinary temperatures and with boxes of ordinary dimensions.

399

15.12 Van der Waals's equation may be regarded as an approximate equation of state applying not only to the gaseous phase of a substance but also to the condensed phases. Regarding it in this way, express this equation as a cubic in V, and, by making use of the fact that at the critical point it must have three equal roots and hence be of the form

$$(V - V_c)^3 = 0,$$

show that the critical volume, pressure, and temperature should be given by the equations

$$V_c = 3b,$$

$$p_c = \frac{a}{27b^2},$$

and

$$T_c = \frac{8a}{27Rb},$$

respectively.

15.13 Calculate the root-mean-square value of the gas-kinetic velocity of hydrogen atoms (not molecules) at 1500°K. Would a hydrogen atom in the upper atmosphere traveling vertically upward with this velocity qualify for escape from the earth's gravitational field? (Assume that it starts at an altitude of 600 km.)

15.14 Calculate the mean velocities of molecules of H_2 and UF_6 (molecular weight 352) at 20°C.

15.15 A radio tube of volume 30 cm³ contains residual gas whose pressure at 300°K is 10^{-8} mm Hg. How many molecules are left in the tube?

15.16 What is the mean translational kinetic energy of a molecule of gas at 20°C? What is the mean translational kinetic energy corresponding to the velocity component in a particular direction in space?

15.17 Calculate the total translational kinetic energy of all the molecules of air contained in a volume of 1 liter at standard atmospheric pressure and 20°C.

15.18 Calculate the pressures that (a) 10 g of oxygen and (b) 1 g of helium would separately exert on the walls of a 10-liter flask in which these gases are contained at 20°C. What would be the mean free paths of the molecules in the two cases? What would you expect concerning the mean free path in the *mixture* of gases if both were contained together in the flask at the temperature specified? How would this compare in order of magnitude with the effective molecular diameters? Would the mean free path be the same or different for the two kinds of molecule? What would you expect to be the pressure exerted by the mixture of gases?

THERMAL BEHAVIOR OF MATERIALS

In Chapter 15 we considered a number of aspects of the behavior of gases in terms of molecules and developed a successful quantitative theory of gases on this basis. With this background we are now in a position to extend the scope of our discussions to include thermal phenomena other than those relating exclusively to gases. This will be our main concern in the present chapter.

Dulong and Petit's Law

From our analysis of the molecular heats of gases we inferred that the mean energy corresponding to each degree of freedom of the molecules is the same. In this discussion we considered the three translational degrees of freedom of all molecules and the two additional rotational degrees of freedom of diatomic molecules, and it was implied that by a degree of freedom we meant simply a possible mode of motion. However, although a degree of freedom is indeed the same as a mode of motion in the case of molecules having linear and rotational motion, it is not the same in all cases. Strictly, the number of degrees of freedom of a body is defined as the number of independent squared terms entering into the expression for the energy of the body.

Maxwell's equipartition theorem gives us a partial explanation of the law of Dulong and Petit concerning the atomic heats of the solid elements referred to in Chapter 14. First, we may imagine such a solid to be in thermal equilibrium with a gas. Then, according to the equipartition law, the mean energy per atom of the solid associated with each degree of freedom will be the same as that per gas molecule corresponding to each degree of freedom, $\frac{1}{2}kT$. Now, the atoms in a solid cannot normally move away from their lattice sites, neither can they rotate; they can merely vibrate about their mean positions. In each dimension the energy of an atom may be expressed as the sum of two squared terms, one representing kinetic energy and the other elastic potential energy. And because there are three independent directions in which the atom can vibrate, it has a total of six degrees of vibrational freedom. Hence the mean energy per atom must be $6 \times \frac{1}{2}kT$, or $3kT$, and the energy per gram-atom is $3RT$. But R expressed in calories per gram-atom per degree has the numerical value 1.99, which, for present purposes, we may take as 2. Thus the atomic heat, $3R$, is about 6 cal/gram-atom/deg. This is in agreement with experimental data for sufficiently high temperatures.

The failure of Dulong and Petit's law at lower temperatures still requires explanation. It cannot be explained by classical mechanics. Apparently the classical laws of mechanics do not apply universally; here they do not apply at low temperatures. It would appear that only at higher temperatures are all the possible modes of motion fully excited. This would also explain the apparent absence of vibration in diatomic gas molecules at ordinary temperatures. In fact, according to the quantum theory, first postulated by Max Planck (1858–1947) in 1900 and since developed into one of the most important theoretical structures in the whole scheme of physics, full excitation of all possible modes of motion cannot occur until the temperature is sufficiently high.

Changes of State

We shall now consider the processes of melting (fusion) and evaporation from the point of view of atomic or molecular behavior.

MELTING We saw in Chapter 10 how it has been learned from the Laue spots produced by x-ray diffraction that in all true solids the atoms are arranged in a regular three-dimensional pattern, or space lattice. Theoretical analysis shows that the proportion of the energy diffracted away from the incident direction to form the Laue spots must be at its maximum when the arrangement of the atoms is completely regular, and that with any loss in regularity the spots must become less intense, although without suffering any deterioration in definition. Precisely such an effect was observed by Debye, in 1913, with diamond and a crystal of KCl. The Laue spots were photographed with the diamond at room temperature and at 1200°C and with the KCl at room temperature (20°C; Fig. 16.1) and at 400°C. In both cases Debye found the intensity of the spots to be substantially less at the higher temperature. The reduction in the regularity of the lattice which this reveals can, presumably, only be due to the vibrations of the atoms about their mean positions occurring with substantially greater amplitude at the higher temperatures than at room temperature. The observed specific heats of solids constitute corroborative evidence of increasing energy of the vibrations of the lattice points with rising temperature.

FIGURE 16.1 Laue spot X-ray photograph of potassium chloride at 20°C taken with Mo radiation. (Courtesy of S. K. Tan and S. Raman, Laboratory for Crystallographic Research, Rensselaer Polytechnic Institute, Troy, New York.)

At each temperature the amplitudes of the atomic vibrations must presumably be distributed about a mean, and atoms possessing exceptionally high energies can no longer be held by their neighbors and consequently break away, either singly or in molecular groups. They do not, however, have sufficient energy to escape altogether from the surface of the solid, and, owing to continual energy exchanges between adjacent atoms, those that have broken away from surface lattice sites will presently lose their extra energies and will become reattached at other sites on the surface. Although a small minority of the atoms always have enough energy to become detached from the lattice, below a certain temperature the solid does not yet melt. As the temperature rises, the rate of detachment increases and that of reattachment decreases until, at a certain temperature, these two rates just balance one another. This is the melting point of the material, where solid and liquid can coexist in contact with one another, in a state of dynamic equilibrium. In the liquid, the atoms or molecules ("particles") wander about in complicated paths, exchange energies with one another, and, as in the solid, have a distribution of energies characteristic of the temperature. The least energetic of the particles of the liquid, on coming into contact with the surface of the solid, are captured by this; they become reattached to the lattice. The slightest—almost imperceptible—rise of temperature above the melting point will increase

403

the rate of detachment and decrease that at which atoms become reattached, so giving a net change of solid to liquid (melting), while, on the other hand, a minute fall in temperature below the melting point will bring about a net change from liquid to solid (freezing).

It is selectively the particles of highest energy that break away from the surface of the solid in melting, so the mean energy of the lattice per unit mass is thereby decreased unless energy to compensate for the loss is fed in from outside. Only by the supply of such energy (heat) can the temperature of the solid be maintained. The heat so supplied is the latent heat of fusion.

The melting points of all materials vary to some slight extent with pressure. For example, the melting point of ice is reduced by 0.0075°C for each pressure increase of 1 atm; the pressure would have to be increased by 133 atm to depress the melting point by 1°C.

The phenomenon of the melting of ice under extreme pressure and its resolidification, or "regelation" (Fig. 16.2), on release of the pressure, is well illustrated in the following experiment. A wire, heavily weighted at both ends, is hung over a block of ice at its atmospheric-pressure melting point. The high pressure on the ice immediately under the wire causes it to melt, and so the wire is pulled downward. As the water flows up past the wire, it immediately freezes again, because it is no longer under high pressure.

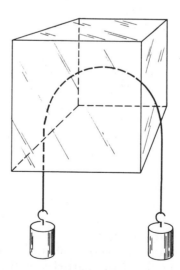

FIGURE 16.2   Weighted wire, exerting pressure against a block of ice at 0°C, "cuts" right through it without severing one half from the other.

Thus, the wire gradually cuts right through the block of ice but without severing it; at the end the block is still intact.

The depression of the melting point of ice under pressure is also responsible for the flow of glaciers, for the clogging of snow underfoot, and, in part, for the lubrication of ice skates by the melting of some of the ice immediately underneath.* In all these cases, regelation occurs on release of the pressure.

In the melting of ice under pressure and the subsequent regelation the latent heat of fusion plays an important role. To melt, the ice needs latent heat, and this it takes from itself, so becoming colder and more difficult to melt. The process is correspondingly slowed down, the supply of heat from the surrounding parts of the ice depending on thermal conduction. In the subsequent regelation this latent heat is set free again, restoring the temperature of the ice to its original value.

Although the melting point of ice is reduced by an increase of pressure, most materials exhibit the opposite behavior, an increase of melting point with pressure. Which way the melting point varies with pressure depends on whether the solid occupies a greater or lesser volume than does the liquid. If, as in the case of ice and water, the volume of the solid is the greater, the melting point is decreased by pressure. On the other hand, if the volume of the solid is less than that of the liquid, the melting point rises with increase of pressure. That it must be so may be shown by an argument based on the second law of thermodynamics. This will be discussed in Chapter 18.

FREEZING MIXTURES Heat is needed not only to melt a solid, but also, in general, to dissolve it; for, in dissolving, no less than in melting, it passes into the liquid state. Hence an important reduction in temperature may be brought about by merely dissolving a solid in a liquid, the heat required for solution being taken from the materials themselves. Simultaneously with the passage of the solid into solution, the freezing point of the solvent is very substantially reduced. Consequently, the final temperature produced can easily be well below the freezing point of the pure solvent. In some cases, where water is the solvent, the effect is enhanced by using ice instead of water, so occasioning two solids to pass into the form of a liquid. Table 16.1 lists the ingredients of some representative "freezing mixtures" and the corresponding approximate values of the temperatures they produce.

---

* Frictional heating also contributes to melting, this being particularly important at relatively low temperatures.

TABLE 16.1   Freezing Mixtures

Parts by weight	Temp. (°C)
1 of NH_4NO_3, 1 of water	-15
8 of Na_2SO_4, 5 of water	-17
2 of snow or crushed ice, 1 of NaCl	-18
3 of snow, 4 of cryst. $CaCl_2$	-48

VAPORIZATION   As we shall see later, there is for every gas a critical temperature above which it cannot be condensed into a liquid or solid, while below this temperature it can—by the application of sufficient pressure. A gas below this critical temperature is often referred to, more specifically, as a "vapor." But apart from the possibility or otherwise of condensation, there is no essential difference in behavior between a vapor and a gas. The same kinetic theory applies to each, and in this sense a vapor *is*, to all intents and purposes, a gas.

There is an interesting connection between evaporation and surface tension. In our study of that subject we saw that there is no actual geometrical discontinuity at the surface of a liquid, but that the molecular population density changes continuously over a finite distance from that in the bulk liquid to that characteristic of the vapor outside. We saw also that this transition region is populated by molecules the vast majority of which are on an outward excursion but destined finally to fall back into the liquid, these molecules possessing higher than average energies, although not enough to enable them to escape. However, some molecules, far less numerous than those in the transition region which will eventually return, approach the surface from inside with sufficient outwardly directed momenta to carry them right through the region of retarding forces at the surface and into the vapor beyond. Actually, not every molecule that qualifies for escape in this way will necessarily do so, because it may collide with other molecules on the way and be knocked back. But statistically these collisions are less important than might be supposed, for reasons analogous to those considered in the discussion of the kinetic theory of gases.

The selective loss of the most energetic molecules of a liquid in evaporation must result in a decrease in the mean molecular energy and therefore in the temperature of the liquid that remains. If the temperature is to be maintained, heat must be supplied to the liquid in proportion to the loss by evaporation and sufficient to compensate for the fall in mean energy. This is the latent heat of vaporization.

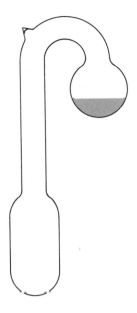

FIGURE 16.3 Cryophorus.

Similar considerations apply to the evaporation of a solid, although here it is only the atoms or molecules in the surface that escape.

The cooling effect of evaporation from a liquid may be demonstrated rather spectacularly with the aid of a device known as a "cryophorus," shown in Fig. 16.3. This consists of a glass vessel of the form indicated, containing water and water vapor, but otherwise evacuated. With all the water in the upper bulb, the lower one is immersed in a freezing mixture or in liquid air. By virtue of the great reduction of vapor pressure within it, this bulb functions as a pump, which rapidly withdraws the water vapor that has evaporated from the surface of the water in the upper bulb and condenses it. In a few minutes the water freezes.

Vapor Pressure

Let us now consider the equilibrium condition where the rate of evaporation of molecules from the surface of a liquid or solid is just balanced by the rate of condensation on this surface of molecules derived from the vapor. For simplicity, let us suppose that a liquid is contained in a closed vessel and that the space above the liquid has been evacuated of air. Then some of the molecules of the liquid will escape and occupy this evacuated space, making multiple collisions with one another and with the walls of the containing vessel, until, eventually, they impinge on the surface of the liquid from which they originally evaporated. The impact on this surface of each molecule that is not reflected marks the end of its career in the vapor; it is drawn back into the liquid. Such molecules as are reflected (if any) will on some subsequent occasion be reabsorbed by the liquid. But, as molecules return to the liquid, others are leaving it, and very rapidly a state of dynamical equilibrium is established in which the rates of escape and return balance one another. This is the condition known as **saturation.** The higher the temperature, the greater the rate of escape per unit area, and, correspondingly, in equilibrium, the greater the rate of return. But the rate of return, in conjunction with the mean momentum with which the molecules approach the surface, is a measure of the pressure of the vapor. Accordingly, this pressure also increases with temperature. The pressure of the vapor in equilibrium with the liquid is known as the **saturated vapor pressure,** or simply the **vapor pressure,** of the liquid at the temperature in question. Similar reasoning applies to the equilibrium between a solid and its vapor.

Vapor pressures may be measured in a variety of ways. Thus a little of the liquid or solid in question may be floated up into the Torricellian vacuum above the column of mercury in a barometer, enough being introduced for some of the liquid or solid to remain as such at the end. The resulting decrease in the height of the mercury in the barometer tube is then a measure of the vapor pressure at the prevailing temperature. Alternatively a U-tube mercury manometer, open or closed as required, may be attached to an evacuated vessel containing a fragile "ampoule" with some of the liquid or solid and its vapor, but no air, inside. The ampoule is then broken by shaking, and the resulting pressure of the vapor filling the vessel measured. If the vapor pressure is very small, some other, more sensitive, form of manometer may have to be used.

An impression of the rapidity with which vapor pressures increase with temperature may be gained from the series of values in Table 16.2, which have been obtained for various materials.

In all cases it is found that if the logarithm of the pressure is plotted against the reciprocal of the absolute temperature, a very close approach indeed to a straight line is obtained. This will be seen, for example, in Fig. 16.4, where the data for ice, water, and mercury have been plotted in this manner.

TABLE 16.2 Variation of Vapor Pressure with Temperature

Material	Temperature (°C)	Vapor pressure (mm Hg)
Ice	−50	0.030
	−40	0.096
	−30	0.288
	−20	0.784
	−10	1.963
	0	4.58
Water	0	4.58
	20	17.51
	40	55.13
	60	149.2
	80	355.1
	100	760
	140	2709
	180	7514
	220	17,380
	260	35,760
	300	67,620
	360	141,870
Iodine	0	0.03
	15	0.131
	30	0.469
	55	3.08
	85	20
	117	100
	137	200
	160.9	400
	185.3	760
Nitrogen	−210.7	86
	−205.4	200
	−200.8	400
	−195.8	760
	−190.2	1386
	−182.2	2916
Mercury	0	1.6×10^{-4}
	50	1.22×10^{-2}
	100	0.276
	150	2.88
	200	17.81
	300	248.6
	400	1566

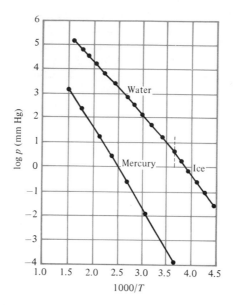

FIGURE 16.4   Temperature variation of vapor pressures of ice, water, and mercury.

An algebraical representation of a straight-line graph would be

$$\log p = a - \frac{b}{T},$$

where a and b are constants. By an argument based on the second law of thermodynamics it may be shown that, theoretically, the relation should be of the form

$$\log p = A - \frac{B}{T} + C \log T, \tag{16.1}$$

where A, B, and C are constants. Actually C is always so small (of the order of magnitude of 1 or 2) that the variation of p with T is dominated by the exponential factor. It is for this reason that the plot of $\log p$ against $1/T$ approximates to linearity. An alternative form of Eq. (16.1) is

$$p = \alpha T^n e^{-\beta/T}, \tag{16.2}$$

where α, n, and β are constants and e is the base of natural logarithms.

A relation of type (16.2), in which p may be replaced by some other quantity, is characteristic of a whole group of phenomena, evaporation being merely one member of the group. Others are thermionic emission, diffusion through a solid, the conductivities of both solid electrolytes and electronic semiconductors, and the speeds of chemical reactions. It is thus strongly suggested that all these things have, in their essential mechanism, some basic feature in common. It is now known that this is indeed the case, the common feature being a so-called "activation energy." All the processes listed depend on energy jumps of one kind or another. In evaporation there is a jump in the potential energy of a molecule when it passes through the surface; in diffusing through a solid a foreign atom jumps from one position in the lattice, through a state of higher energy, to a neighboring one; and so on. It is the minimum energy required for such a jump that is known as the activation energy. This is given by the constant B in Eq. (16.1), to which it is proportional.

Triple Point

From Table 16.2 it will be seen that the vapor pressure is the same for ice at 0°C as for water at 0°C. But this is so only because the vapor pressures are not given with sufficient accuracy to show a difference. Although the difference between the two vapor pressures is exceedingly small and it would be extremely difficult to measure it experimentally, it may be shown theoretically that there must *be* a difference. There is only one temperature at which the two vapor pressures are the same, and at this temperature ice and water, besides being in equilibrium with their vapor, are also in equilibrium with one another. This temperature is known as the **triple point** for the substance H_2O. As we saw in Chapter 13, the triple point for H_2O is 0.0098°C above the freezing point of water at atmospheric pressure.

The difference in temperature between the atmospheric-pressure freezing point of water and the triple point can be calculated from the information given earlier in this chapter concerning the rate of variation of the freezing point with pressure. This is a fall of 0.0075°C for each pressure increase of 1 atm and a corresponding rate of rise for a decrease of pressure. Hence for a pressure decrease from 760 mm to 4.6 mm (the approximate vapor pressure at the triple point) the freezing point must rise by $[(760 - 4.6)/760] \times 0.0075°$, that is, by 0.0075°C.

There is an apparent discrepancy between this result and the accepted amount, 0.0098°C, by which the triple point is higher than the ice point. The latter is based on measurements for which an accuracy to within 0.0001° can be claimed. The explanation would

411

appear to be that the ice point is not defined as the temperature where *pure* ice and water are in equilibrium with one another at a pressure of 1 standard atmosphere. In practice the standard atmosphere is always applied by exposing the ice and water to *air* at this pressure. Thus the ice point is the temperature at which ice and air-saturated water are in equilibrium at a pressure of 1 atm. Because the presence of any dissolved material depresses the freezing point, this makes the temperature difference between the ice point as defined and the triple point somewhat greater than the value calculated above.

The existence of a triple point is not peculiar to H_2O; all substances have their characteristic triple points. For example, the triple point of carbon dioxide is $-56.6°C$. At this temperature, solid and liquid carbon dioxide, besides being in equilibrium with one another, are also both in equilibrium with carbon dioxide vapor at the same pressure, this being 5.11 atm.

Unsaturated and Supersaturated Vapors

A vapor isolated from its condensed phase (liquid or solid) may have its pressure either reduced below or increased above that corresponding to saturation. This could, for example, be done by expanding or compressing it at constant temperature. In the former case it is said to become **unsaturated,** in the latter, **supersaturated.** The vapor when out of contact with its condensed phase behaves just like any other gas, as long as it remains a vapor. A supersaturated vapor will not condense unless either (1) suitable nuclei are present, or (2) the degree of supersaturation exceeds a certain limit, beyond which spontaneous condensation will occur.

PARTIAL PRESSURES   Let us suppose that into a vessel containing only a liquid and its vapor a quantity of some other gas is introduced, and let us consider how the presence of this gas will affect the careers of the vapor molecules. We shall assume that the volumes of the molecules themselves, of both kinds, are negligibly small compared with the volume occupied by the gas–vapor mixture; also that the forces which the molecules exert on one another may be disregarded.

A proportion of the vapor molecules escaping from the liquid, and also of those rebounding from the containing walls, will now collide with gas molecules and be knocked back. But, for every such event there will, on the average, be a compensating event in which a vapor molecule about to reenter the liquid or collide with a wall is prevented by a collision with a gas molecule from so doing. Statistically, therefore, collisions between vapor and gas molecules have no effect either on the concentration of vapor molecules or on the contribution these make to the bombardment of the walls, that is, to the pressure. Likewise, the presence of the vapor can have no effect on the contribution the gas molecules make to the pressure. The separate contributions of the vapor and the gas to the total pressure are known as their **partial pressures,** each of which is, as we have just seen, the same as it would be if no other gas or vapor were present. We could easily extend the reasoning to the case of a mixture of several gases that have no chemical action on one another. The mutual independence of the partial pressures, which was first established experimentally by Dalton, is known as **Dalton's law of partial pressures.**

Vapor Pressure and Curvature of Surface

In our discussion of the retarding forces to which a molecule is subjected in passing through the surface of a liquid into the vapor, the geometrical form of this surface was not specified. It is quite conceivable that these forces, and with them the vapor pressure, might vary with the curvature of the surface. We shall now look into this question.

Let us consider the system shown in Fig. 16.5. We suppose that within a closed vessel, which has been evacuated of air, a

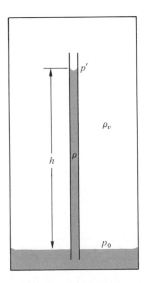

FIGURE 16.5

413

liquid is in equilibrium with its vapor, part of the liquid surface being plane, while another part—the meniscus in the capillary tube at a height h above it—is curved. We wish to investigate how the pressure p' of the vapor in contact with the curved surface is related to the vapor pressure p_0 characteristic of the plane surface.

Considering the density ρ_v of the vapor as a variable and writing y for the height above the plane surface, we have for the increment of pressure δp in the vapor corresponding to a small increment δy in y,

$$\delta p = -\rho_v g \, \delta y.$$

By applying the equation of state for a gas (assumed sufficiently nearly ideal) we may substitute for ρ_v the expression Mp/RT, so obtaining the equation

$$RT \frac{\delta p}{p} = -Mg \, \delta y,$$

which, on making δy and δp infinitesimal, and integrating, gives us

$$RT \ln \left(\frac{p_0}{p'} \right) = Mgh.* \tag{16.3}$$

We may get rid of the unwanted gh in this equation by making use of the expression for the excess of pressure on the concave side of a curved liquid surface over that on the convex side necessary to maintain it in equilibrium, which we established in Chapter 12. Now neglecting ρ_v in comparison with the density ρ of the liquid, we have for the excess of pressure on the concave side of the meniscus over that on the convex side simply ρgh, and because this pressure difference must be equal to $2\sigma/r$, where σ is the surface tension of the liquid and r the radius of curvature of the meniscus, we may substitute $2\sigma/r\rho$ for gh in Eq. (16.3), so obtaining

$$RT \ln \left(\frac{p_0}{p'} \right) = \frac{2\sigma}{r} \frac{M}{\rho} . \tag{16.4}$$

We may find the vapor pressure p'' for a convex surface by corresponding reasoning. Applying this to the situation of Fig. 16.6, which is self-explanatory, we obtain the equation

$$RT \ln \left(\frac{p''}{p_0} \right) = \frac{2\sigma}{r} \frac{M}{\rho}. \tag{16.5}$$

---

* Here we adopt the convention of writing "ln" to denote the natural logarithm, or $\log_e$.

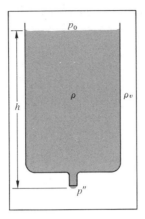

FIGURE 16.6

From these equations we may readily evaluate p_0/p' or p''/p_0 in any given case. For bubbles in water or water droplets at 300°K ($\sigma = 73$ dyn/cm) we obtain the following values:

r (mm)	10^{-3}	10^{-4}	10^{-5}	10^{-6}
p_0/p' or p''/p_0	1.001	1.011	1.114	2.95

A water droplet of radius 10^{-5} mm would contain 140,000 molecules, while in one of radius 10^{-6} mm there would be only 140 molecules.

Equation (16.5) is valid only for electrically neutral droplets. An electric charge residing on the surface of a droplet has an important effect on the pressure within it.

Equations (16.4) and (16.5) may be taken as applying not only where there is only the liquid and its vapor present but also where p_0, p', and p'' are partial pressures of vapor in another gas.

From the foregoing discussion it follows that excess water vapor can begin to separate out from moist air or other gas when it is cooled below the dew point only if, at the very beginning of condensation, the surface curvature of the water is not too high. This condition is satisfied at the surfaces of relatively large objects exposed to the vapor which are wetted by water; on such surfaces dew is deposited as soon as the temperature falls below the dew point. Condensation can also occur, in the form of mist, or fog, or cloud, on suitable nuclei held in suspension in the gas. These must be of sufficient size; the curvature of their surfaces, and so of the

415

surface of water condensing on them, must be such that p'' does not differ appreciably from p_0. Condensation can also occur on smaller nuclei, but only after the requisite degree of supersaturation has been reached. Electrically charged molecules, or groups of molecules, known as "ions," can also function as nuclei for condensation; thus, at a value of the ratio of actual to saturation vapor pressure of about 4.2 water will begin to condense on negative ions, and at a ratio of about 5.7 condensation will also take place on positive ions.

Even in the complete absence of nuclei, condensation can still occur. Such spontaneous condensation requires a supersaturation ratio in the neighborhood of 8. To this, according to Eq. (16.5), would correspond a value of r of the order of only 10^{-7} mm, and the volume of a sphere of this radius would contain less than a single molecule. Obviously, the whole theory underlying Eq. (16.5) must break down for sizes of droplets which no longer contain a large number of molecules; it becomes meaningless to speak of surface tension when we are dealing with assemblages of only a few molecules, not to mention less than a single one! Possibly, at high degrees of supersaturation, spontaneous condensation takes place on fortuitous concourses of molecules functioning as nuclei.

Atmospheric Precipitation

This occurs in three main forms: (1) snow, (2) hail, and (3) rain. Much of the rain that falls over the surface of the globe originates as ice and melts on the way down. This we may conveniently refer to as "cold rain." But by no means all rain is of this kind; a large proportion comes from clouds no part of which is below the freezing point of water. This, which we may call "warm rain," is particularly prevalent in maritime regions and is initiated by a different process. We shall discuss the formation of snow, hail, cold rain, and warm rain under separate headings below.

GENERAL CONSIDERATIONS The lower part of the atmosphere, in which weather phenomena occur, is known as the **troposphere** (Fig. 16.7, not to scale). This varies in height with latitude and with the season of the year, the mean height varying between about 8 km (26,000 ft) at the poles and about 16 to 17 km (52,000 to 56,000 ft) at the equator.

Throughout the whole of the troposphere, suitable condensation nuclei are present in sufficient concentration to prevent the occurrence anywhere of any appreciable degree of supersaturation. When in any region the temperature falls below the dew point,

FIGURE 16.7 The troposphere (not to scale). The depth of the troposphere varies from 8 km to 12 km in high and middle latitudes, and from about 16 km to 17 km at the equator.

the excess moisture condenses out on at least some of these nuclei. Important among the nuclei are minute particles of dust, of both terrestrial and meteoric origin, and particles of salt, derived from sea spray. In their nature and in size, these nuclei cover a considerable range, and they are not all equally effective as centers for condensation. Actually some of them, consisting of or contaminated with hygroscopic salts, have a certain amount of water condensed on them even at ordinary, nonsaturated humidities. When the temperature falls below the dew point, water vapor condenses selectively on the most favorable nuclei only, and the concentration of water droplets in the resulting cloud does not usually much exceed 100 per cm^3. As condensation proceeds further, no fresh droplets are produced; the existing droplets merely grow in size. In practically all circumstances the share of vapor received by each droplet is sufficient for it to reach a radius of about 10^{-3} cm.

The rate at which droplets of such minute size fall through the air is negligibly small; a simple calculation based on Stokes's formula (see Chapter 9) shows that for a radius of 10^{-3} cm, the terminal velocity is only of the order of 1 cm/sec. With such a slow rate of fall it would take many hours for the particles to fall from clouds of usual heights through still air to the ground. Actually, the air in clouds is normally far from motionless; it is part of a huge convection current, and the velocities involved are of a higher order of magnitude than the terminal velocities of the droplets. Consequently, in an up-current the droplets move upward with the air containing them. If the movement of the air is downward, toward regions of higher temperature, the entire cloud usually disperses by evaporation; under practically no circumstances can there be any question of these tiny droplets falling as rain. For effective precipitation it is necessary for much larger droplets to be formed from the original cloud particles. And, because each cloud has only a finite lifetime, the process must not be too slow. We shall now study the various growth processes that lead to precipitation.

SNOW Even at temperatures far below the freezing point, atmospheric water vapor does not usually condense as ice, and water droplets formed above the freezing point and carried up into colder regions do not normally freeze, even at temperatures considerably below 0°C. Water at temperatures below the freezing point is said to be **supercooled.** Supercooling is quite common, and degrees of supercooling of as much as 35°C have been observed. Although this condition of supercooling may continue almost indefinitely, it is not, strictly speaking, a stable condition; and if supercooled water comes into contact with a particle of ice or with some other solid of sufficiently similar lattice structure it immediately freezes, with liberation of latent heat. Mechanical shock also appears to be effective in initiating freezing. Supercooled cloud or mist is particularly dangerous to aircraft flying through it, this being the cause of icing.

Among the condensation nuclei present in the atmosphere is an exceedingly small proportion that are favorable to the formation of ice crystals. These are known as "ice nuclei." Their occurrence is so rare that even in clouds with temperatures as low as $-20°C$ the proportion of ice crystals to supercooled droplets is often only about 1 in 10,000 and sometimes even as low as 1 in 1,000,000.

The clue to the mechanism of the formation of snow is contained in the discontinuity of slope in the vapor-pressure plot for ice and water in Fig. 16.2. If, in this figure, the line for the vapor pressure of water is produced beyond the freezing point to regions of lower temperature, it will be found to lie above the vapor-pressure line for ice. This means that supercooled water has a higher vapor pressure than has ice at the same temperature. Consequently, in a cloud containing both water and ice particles, vapor must distill over from the water to the ice, the latter particles growing in size at the expense of the former. Given sufficient time before the eventual dispersal of the cloud as vapor (something of the order of an hour) and a suitable concentration of ice nuclei, the ice particles will grow to the size of snow crystals and then fall at an appreciable speed. This will not happen if the ice nuclei are too numerous, for then the share of water vapor each receives, and so the ultimate size of the crystals, will be too small. On the other hand, if the ice nuclei are too few in number, the total precipitation will be trifling. If conditions are right, the formation of the falling ice crystals will be followed by the aggregation of these into snowflakes as they fall through the cloud and collide with one another in their fluttering or tumbling motions.

HAIL A process for the conversion of supercooled water to ice, alternative to distillation, is contact of a supercooled droplet with an existing ice particle. Such contact immediately converts the water into ice, with the release of latent heat. This is the process responsible for the production of hail.

An essential condition for hail formation is a strong updraft of air in which the cloud is carried rapidly into regions of progressively lower temperatures. Under such conditions vigorous condensation occurs on all particles alike, water as well as ice. However, owing to the difference in vapor pressures, the ice particles receive a greater proportion of vapor and grow more rapidly than the water droplets. In consequence of this, they presently begin to fall through the cloud more rapidly than do the water droplets, and the latter are collected in the process, freezing immediately on contact. The associated release of latent heat accelerates the updraft, and at the same time the progressive growth of the hailstones causes them to fall ever faster, and so the sweeping process and rate of growth are accelerated. In a very dense cloud this growth is almost explosive, and hailstones begin to fall out of the cloud within about 10 minutes. Large hailstones have fall speeds of the order of 15 m/sec and can be formed only in clouds having exceptionally strong updrafts, sufficient to keep them in the cloud for several minutes.

Figure 16.8(a) indicates the layered structure of a hailstone and (b) shows a typical trajectory.

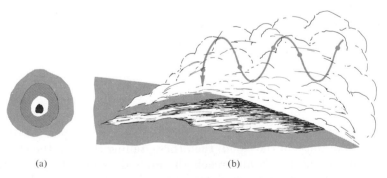

(a) (b)

FIGURE 16.8 (a) Layered structure of a typical hailstone.
(b) Hail formation. The curve shows a typical trajectory of a hailstone.

COLD RAIN If hail, formed in the manner just described, melts on the way down, it becomes cold rain. Alternatively, cold rain may result from the melting of falling snowflakes.

Potential rain clouds do not by any means always give rain. This may, for example, be due to the lifetime of the cloud not being long enough for the rain-producing processes to become effective, or the ice nuclei may not be sufficiently numerous. In such circumstances the cloud may sometimes be induced to give rain by artificially "seeding" it with some material that will initiate the freezing of the water droplets. One seeding agency that has been found effective is "dry ice"—solid carbon dioxide—dropped from an aircraft into the top of the cloud. Another, which is of particular interest, is silver iodide. This has a lattice structure of the same kind as that of ice, and the lattice spacing, also, is almost the same in the two cases. It appears that the correspondence is sufficiently close for the supercooled water, if at a temperature of $-6°C$ or lower, to be "fooled" into freezing onto the silver iodide, just as it would onto ice; the closely similar lattice evidently produces the right kind of disturbance to upset the unstable condition of the water droplets. Unlike dry ice, the relatively coarse particles of which fall readily even through rapidly upward-moving air, silver iodide is usually released from aircraft flying just above the base of a potentially rain-producing cumulus cloud, where there is a strong updraft which carries it in; or, in the case of stratiform cloud, it is released within the cloud, at about the $-6°C$ level.

WARM RAIN It is believed that in rain clouds above $0°C$ the raindrops grow by the same sweeping process that produces hail. This is initiated by the settling through the cloud of a few droplets that are rather larger than the average. These relatively large droplets could have been formed by the chance coalescence of smaller ones, or they may have begun earlier, being formed on particularly efficient condensation nuclei. It has been calculated that, whatever may be the origin of the relatively large droplets, if these are only twice as big as the average they could in favorable circumstances grow by the coalescence mechanism into raindrops in about 20 minutes.

Warm clouds, as well as cold ones, may be seeded artificially. If water is sprayed into the top of the cloud in droplets, rather larger than those of the cloud, then these, falling through the cloud, will grow by accretion and reach the ground as useful rain. Estimates of the number of droplets sprayed into the cloud and those reaching the ground as rain indicate that—as we should expect—the two numbers are equal. However, the raindrops have something like 100 times the diameter of the droplets in the spray; thus about 1 million times more water comes down as rain than is sprayed into the cloud.

Boiling

The phenomenon of boiling is in one respect analogous to that of precipitation from a vapor; unless suitable nuclei are present, boiling will not begin at the temperature where a plane surface of the liquid would be in equilibrium with its vapor. In the absence of nuclei, bubbles have to begin at a very small size, where the equilibrium vapor pressure is considerably reduced, and therefore a very much higher temperature has to be produced locally within the liquid for the pressure of the vapor within the bubble to equal that of the liquid outside. But mere equality of the two pressures is not sufficient; to give mechanical equilibrium the pressure inside the bubble must exceed that outside by $2\sigma/r$. Actually, of the two effects, this is quantitatively much the more important. To produce this excess pressure the temperature has to be raised to a still higher value. The equilibrium is, of course, essentially unstable, and, as the bubble grows, not only does the pressure of the vapor within it increase but also $2\sigma/r$ decreases; hence the growth is explosive. This is the cause of the "bumping" so often experienced when a liquid is boiled. This phenomenon tends to be particularly troublesome where conditions are not favorable to convection, as when a liquid is being boiled in a test tube, and where there is little or no dissolved air present.

Under conditions where convection is entirely prevented and nuclei are excluded, the temperature of water exposed to atmospheric pressure may be raised far above 100°C before vaporization within the body of the water occurs. For example, in an experiment where a drop of water was suspended in oil, it was found not to vaporize until the temperature reached 178°C, when it exploded.

Even when suitable nuclei, such as broken glass, sand, metal wire, and so on, are present and boiling occurs smoothly, the formation of bubbles of vapor below the surface is dependent on the temperature there being greater than at the surface, because the pressure increases with depth. This condition will be satisfied if the heating agency is applied below the surface and heat transfer by convection and conduction is sufficiently slow. On the other hand, if heat exchange within the liquid is rapid, it will tend to vaporize from its surface only, without the formation of any bubbles below the surface.

421

Complete Isothermals

We have noted the rapidity with which the pressure of a vapor in contact with a plane surface of a liquid increases with temperature. With this is associated a correspondingly rapid rate of increase with temperature of the density of the saturated vapor. If we assume the saturated vapor to conform sufficiently closely to the equation of state for an ideal gas and substitute $\rho RT/M$ for p in Eq. (16.2), we obtain for the variation of density with temperature an equation of the same general form,

$$\rho = \frac{M}{R}\,\alpha T^{n-1}e^{-\beta/T}.$$

At the same time as the density of the vapor is increasing, that of the liquid must fall, owing to thermal expansion, and so the question arises whether, with continued increase in temperature, the two densities ultimately converge, and what the state of the material then is, liquid or gas.

The answer to these questions is provided by a study of a series of isothermals, such as is shown for carbon dioxide in Fig. 16.9. Following one of the lower-temperature members of this family of curves from right to left, for example the lowest one shown, we have first a curved portion AB representing the unsaturated vapor, whose form approximates to that characteristic of an ideal gas. At the point B saturation is reached, and from B to C the plot represents the vapor in equilibrium with its liquid, and is therefore horizontal. At C the whole of the vapor has been condensed to liquid, and from this point the curve rises almost vertically, owing to the very small compressibility of the liquid.

If we now examine this series of isothermals, we see that the horizontal portion becomes progressively shorter with increasing temperature, finally becoming a mere point of inflection. This confirms our expectation with regard to the convergence of densities. The temperature at which the horizontal portion finally vanishes is known as the **critical temperature** of the material, Θ_c. Beyond this there is no condensation of vapor to liquid; above the critical temperature it is impossible to liquefy a gas by compression. The pressure at the point of inflection is called the **critical pressure,** p_c.

Critical temperatures, pressures, and densities for a number of gases are given in Table 16.3, together with the vapor pressures at 20°C of those materials whose critical temperatures lie above this value, and also the atmospheric-pressure boiling points.

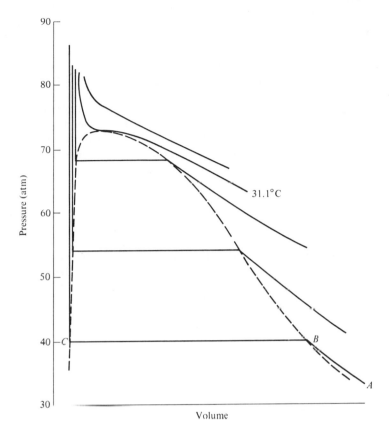

FIGURE 16.9   Pressure-volume isothermals for carbon dioxide.

TABLE 16.3   Critical and Other Data

Substance	θ_c (°C)	p_c (atm)	p at 20°C (mm Hg)	B.p. (°C)	ρ_c (g/cm³)
H_2O	374	218	17.5	100	0.33
SO_2	155	79	2462	−10.8	0.52
NH_3	130	115	6388	−33.5	0.235
CO_2	31.1	73	44,717	(−78.5)*	0.46
O_2	−118	50		−183.0	0.43
N_2	−146	33		−193.9	0.33
H_2	−239.9	12.8		−252.8	0.033
He	−268	2.26		−268.8	0.069

* This, being the temperature at which *solid* carbon dioxide is in equilibrium with its vapor at atmospheric pressure, is not actually a boiling point as are the other figures in this column.

Problems

16.1 One liter of oxygen is collected over water, the pressure in the collecting vessel being 756 mm Hg and the temperature 25°C. Calculate the mass of the oxygen.

16.2 What mass of water will be condensed in a vessel of volume 1 m³ if it is filled with saturated air at 30°C and then cooled to −50°C?

16.3 What is the ratio of the density of air saturated with water vapor at 20°C and 760 mm pressure to that of dry air at the same temperature and pressure?

16.4 Air at 20°C of relative humidity 40 percent (that is, containing water vapor whose partial pressure is 40 percent of saturation pressure) is contained in a volume of 50 liters. This air is compressed isothermally to 25 liters. What will now be the relative humidity?

16.5 The isothermal compression of the air referred to in Problem 16.4 is continued beyond 25 liters down to 10 liters, and during this further compression water condenses. What is the mass of the condensed water?

If the original air of relative humidity 40 percent occupying a volume of 50 liters was at a pressure of 760 mm Hg, what is the final pressure when the volume is 10 liters?

16.6 Show that, according to the simplified version of the kinetic theory discussed in this book, the mass rate r at which molecules escape from a liquid per unit area of surface would be given by the formula

$$r = \sqrt{\frac{p\rho}{12}},$$

where p is the vapor pressure and ρ is the density of the vapor.* Apply this to the case of water at 20°C, the vapor pressure at this temperature being 17.5 mm Hg.

16.7 Find, by trial, a method of plotting the vapor pressure of mercury as a function of temperature, given in Table 16.2, in such a way as to obtain a straight-line graph. Using this graph, find what would be the vapor pressure (a) at the freezing point of mercury, −39°C, and (b) at 800°C. Express the relationship you have found graphically in the form of an algebraic equation.

16.8 Using the vapor-pressure data given in Table 16.2 and assuming the value of 50 dyn/cm for surface tension, estimate the temperature at which a bubble of steam of radius 1000 Å would be in (unstable) equilibrium with water contained in a vessel open to the atmosphere.

16.9 One may estimate very roughly the heat of vaporization of a liquid by regarding the liquid as a group of tightly packed cubic molecules. The process of vaporization "frees" the cubes so that the amount of surface area increases from the relatively insignificant area of the initial liquid volume to the large final surface area of the molecular cubes. The heat of vaporization is then estimated to be the energy required to make this extra surface. Perform this calculation, using the surface tension of water, the appropriate molecular data, and Avogadro's number.

* The true formula is $r = \frac{3}{2}\sqrt{p\rho/12}$.

17

THE FIRST LAW OF THERMODYNAMICS

In Chapter 15 we saw that the heat content* of a gas may be identified, at least to a close approximation, with the kinetic energy of its constituent molecules, and in Chapter 16 we saw that we could apparently equate the heat content of a solid with the energy of vibration of its atoms. Reasoning from the particular to the general, it seems that we may accordingly infer that heat is simply the total molecular or atomic mechanical energy in these systems.

This was not the way in which heat was identified as energy historically, nor does the line of reasoning followed give us complete assurance that such identification applies exactly and in all cases. Even in the cases considered, the reasoning is persuasive rather than rigorous. Thus the success of van der Waals's equation shows that the mutual forces between the molecules of a gas are not quite negligible, and that accordingly there is an energy item—potential energy—which we should really have taken into account in our discussion of the molecular heats of gases, and whose neglect is presumably responsible for the fact that the molecular heats at constant volume or at constant pressure of comparable gases, for example diatomic gases, are not exactly the same. Similarly, the atomic heats of solids, even at high temperatures, differ slightly from one another. Also, there are other energy items which might, and in many cases certainly do, make a contribution to the total, but which we have not allowed for in our discussion of the molecular heats of gases and the atomic heats of solids. Among these are the energies of excitation, ionization, and dissociation of gas molecules, and the energy of electromagnetic radiation within the material of, and in equilibrium with, hot bodies. The question to which we must now find an answer is this: If all possible forms of energy— mechanical, electrical, radiation, and so on—within a hot body were taken into account, would there be an exact correspondence between the sum total of all these energies and what we measure as its heat content?

---

* The question of the desirability of using the expression "heat content" will be discussed later in this chapter. Meanwhile there can be no serious objection to its use in the present context, where we do not contemplate the performance of any work by the system under consideration.

An answer to this question which is intellectually completely satisfying has been obtained, not by considering atoms or molecules at all, nor by any attempt at itemization of the energy within bodies, but by quite general reasoning based on certain historical experiments carried out by Joule, Hirn, and others during the nineteenth century. The answer was, as the reader will no doubt have guessed, in the affirmative. The process of finding it, and a discussion of its consequences, will constitute the subject matter of this chapter.

Heat and Conservation of Energy

The first kind of energy conversion involving heat to be investigated quantitatively was that from mechanical energy to heat. This investigation was carried out by Joule. In the course of a long series of experiments beginning in the year 1840 Joule dissipated mechanical energy in a variety of ways to produce heat, and measured the ratio between the work performed and the heat produced. Among the methods employed were the churning of water, the churning of mercury, and forcing water through a narrow pipe. In all cases the ratios found were—within the limits of experimental uncertainty—the same, between 4.1 and 4.2 J of work being required to generate 1 cal of heat. Figure 17.1 shows Joule's churn; the energy

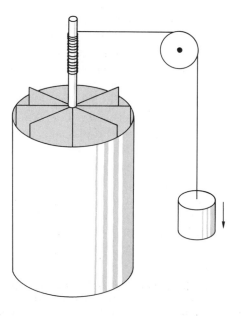

FIGURE 17.1 Joule's churn.

added to the liquid equals the loss in potential energy of the mass as it falls.

In 1862 Hirn, an Alsatian engineer, investigated the converse of this process, determining the ratio of work performed to heat used up by a steam engine. He measured the heat supplied to and lost by the engine, as well as the work performed by it, and in two determinations found for the ratio of work performed to heat that went out of existence the values of 4.05 and 4.12 J/cal. In view of the great experimental difficulties of this investigation, the agreement of these results with Joule's must be regarded as eminently satisfactory.

A weighted mean of more recent and more accurate determinations of the ratio of mechanical energy expended to heat produced, to which the name **mechanical equivalent of heat,** already used in Chapter 15, has been given, is 4.186 J per 15-degree calorie. It will be noted, incidentally, that this is in very satisfactory agreement with the value calculated in Chapter 15 in our discussion of the kinetic theory of gases.

The statement of the interconvertibility of heat and mechanical energy according to a constant conversion ratio depending on the units in which these quantities are expressed is known as the **first law of thermodynamics.** This is a specialized statement of a more general law of classical physics, embracing within its scope all possible forms of energy, known as the **law of conservation of energy.** Although, as Einstein has shown, this law must now be combined with the classical law of the conservation of mass in the single relativistic law of conservation of mass–energy, we need not concern ourselves with this in most situations of everyday life, where there is no detectable change in total mass. The experimental basis for the law of conservation of energy not only includes that for the first law of thermodynamics but also embraces corresponding evidence in other fields of classical physics. In addition to mechanical energy and heat we also have electrical energy, chemical energy, radiation energy, and so on. All of these are interconvertible according to mutually consistent conversion ratios, depending on the units in which these energies are expressed.

The realization, in the mid-nineteenth century, of the existence of this fundamental quantity, energy, in various forms, and of its total conservation, marked the beginning of a new era in physics. The introduction of this new concept and the explanation of phenomena in terms of it have brought order out of the earlier chaos of empiricism. Several investigators played their parts in the birth

427

of the idea. Foremost among these, in addition to Joule, were Robert Mayer (1814–1878) and Hermann von Helmholtz (1821–1894). All three came to substantially the same conclusions at about the same time and independently of one another.

HEAT AS A "FORM OF ENERGY"   Among the various forms of energy which, as we have just seen, are interconvertible, heat is peculiar in that, strictly, it does not exist in its own right as do, for example, mechanical energy, electrical energy, and so on. As we have noted in Chapters 15 and 16 and again in the introduction to the present chapter, it commonly consists of a mixture of different kinds of energy, for example molecular kinetic and potential energies, electrical energy, radiation energy, and so on, in statistical equilibrium with one another; and the proportions in which these items contribute to the mixture differ on different occasions and in different systems.

We see, then, that "heat content" and "heat energy" are merely alternative names for the internal energy of the body under consideration. This internal energy is in general made up of several items, so it does not really make sense to say that heat is "a form of energy"; it would be better to say that it is a collection of several different forms. However, to avoid circumlocution it will be convenient to continue to use the statement "Heat is a form of energy" on the understanding that this is merely an abbreviation for "Heat is a collection of a number of kinds of energy."

In this connection it should be noted that the various forms of matter known to us possess considerable energy even at the absolute zero of temperature. This is known as "zero-point energy" and is not included in what we refer to as heat content in the present context.

CONTINUOUS-FLOW CALORIMETER   In certain of the calorimetric experiments usually included in introductory laboratory courses in physics, specific heats are determined by a procedure commonly referred to as the "method of mixtures." In these experiments, allowance usually has to be made for heat loss to, or gain from, the surroundings, and for the heat capacity of the calorimeter. The uncertainty attaching to these corrections appreciably reduces the accuracy of the determination. A way of escape from these difficulties lies in the use of the continuous-flow method. In this the experimental uncertainty arising from heat exchange with the surroundings is much smaller than in the usual method of mixtures. Also, as with the ice calorimeter, the heat capacity of the apparatus does not enter into the calculation.

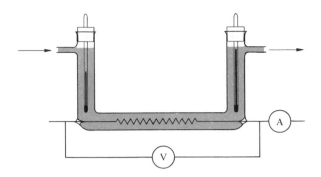

FIGURE 17.2   Continuous-flow calorimeter.

A continuous-flow calorimeter is shown in Fig. 17.2. A horizontal glass tube has vertical extensions attached at both ends, and inlet and outlet tubes are joined to these near the top as shown, while thermometers* indicate the temperatures of the liquid entering and leaving the calorimeter. Running along the axis of the horizontal tube is an electric heater, consisting of a coil of suitable resistance wire, and a steady current is passed through this while liquid flows through the apparatus at a steady rate. The current strength is measured with an ammeter A and the potential difference between the ends of the heater with a voltmeter V. (For highest accuracy this should be a potentiometer.) Then, as will be shown in Volume II, if I is the current in amperes and V the potential difference in volts, the rate of conversion of electrical energy to heat, expressed in watts, is IV. The mass m of liquid which passes through the apparatus in a measured time t is ascertained by collecting this liquid in a beaker and weighing. Then, on the assumption that no heat is lost to the surroundings and that the mechanical energy converted to heat in the calorimeter on account of viscosity is negligible, the specific heat c of the liquid must be given by the equation

$$c = \frac{IVt}{m(\Theta_2 - \Theta_1)},$$

where Θ_1 and Θ_2 are the temperatures of the liquid at the inlet and outlet ends, respectively, c being expressed in joules per degree per unit mass.

The time taken for a particular sample of liquid to flow from the inlet to the outlet end is much shorter than the time usually

---

* For highly accurate work the mercury thermometers shown in the diagram would be replaced by thermometers of a more sensitive and reliable type.

occupied by a specific-heat determination by the method of mixtures, and the heat-loss correction is correspondingly less important. Provision of suitable thermal insulation—not indicated in the diagram—further reduces this item.

The rate of conversion of mechanical energy to heat because of viscosity may be estimated in any given case, at least roughly, using Poiseuille's equation, and the corresponding correction applied, if significant. However, in most cases it will be negligibly small, and it may always be made so by suitable design and use of the apparatus, except in the case of liquids of very high viscosity.

This method may be employed, as an alternative to the use of an ice calorimeter, to find the relation between calories at different temperature levels. Thus by passing water through the calorimeter, always beginning with the same initial temperature Θ_1 and heating it to different final temperatures Θ_2, the quantities of heat necessary to raise the temperature of a given mass of water by different amounts may be ascertained in mechanical units (joules), and hence, from differences, calories at different temperature levels compared. Obviously, the information yielded by such experiments is not merely the *ratios* of calories at different temperature levels but their *values* in mechanical units. Thanks to the great accuracy of electrical measurements, this method is actually superior to mechanical methods (such as stirring water) for finding the mechanical equivalent of heat.

Heat and Internal Energy

Our identification hitherto of the heat content of a body with its internal energy raises the question of whether it is useful or desirable to have two different terms signifying the same thing.

Actually, as we have seen, these terms do not signify *quite* the same thing; the heat content of a body is not the whole of its internal energy, but only that part of it which is subject to change, that is, the non-zero-point part of it. Thus, in the case of a perfect gas, it includes only the translational and rotational kinetic energies of the molecules, plus, as we now know, the vibrational energies which these acquire at very high temperatures. But it does not include the normal orbital energies of the electrons or the energies residing within the nuclei of the atoms.* This suggests that it might be

---

* The structures of atoms and molecules and their energies are discussed in most introductory texts on modern physics, for example in Volume III of this book.

reasonable to use the two terms with different meanings, internal energy for the sum of all the energy items, including the zero-point energy, and heat only for that part of the internal energy which is subject to change in thermal processes.

There is, however, an important objection even to this identification of heat with the non-zero-point part of the internal energy. This is at once evident when we consider the mathematical equation commonly used to express the first law of thermodynamics,

$$\delta U = \delta Q - \delta W, \tag{17.1}$$

in which δU denotes the increase in the internal energy of a system, δQ the heat supplied to it, and δW the external work done by the system.

This equation shows that while U is by definition the internal energy of the system, Q cannot represent its heat content as hitherto visualized, either with or without the zero-point-energy item; for if it did, δU and δQ must always be equal, and so δW would necessarily have to be zero; that is, without δW always equal to zero the equation could not possibly be valid. At the same time it is difficult to think of heat other than in terms of internal energy.

A way out of this difficulty is the following. We define δQ *not* as the amount by which the internal energy of the system under consideration is increased, except only in the special case where no external work is performed ($\delta W = 0$), but as the amount by which the internal energy of some *other* body or system in thermal contact with it is *decreased*. With δQ defined in this way the physical meaning of the equation is that the gain in the internal energy of the system under consideration (δU) is equal to the loss of internal energy of some other system in communication with it (δQ), less the external work (δW) performed by the system.*

Joule–Kelvin Experiment

We have seen how Joule, in his experiment on the unresisted expansion of air, was unable to detect any change in temperature accompanying the expansion. At the time Joule carried out this work, he realized that the considerable heat capacity of his apparatus must render this type of experiment rather insensitive, and

---

* For simplicity, we may imagine that the other system, whose internal energy is decreased by δQ, is not involved in any performance of work; otherwise, such work would have to be included in δW.

shortly afterward he began, in collaboration with Lord Kelvin, to put the matter to a much more searching test by means of a continuous-flow experiment, in which heat-capacity limitations to sensitivity would not arise. In this experiment the pressure of the gas under investigation was reduced by a large amount by passing it through a porous plug of high resistance, and the temperature of the gas was measured on both sides of the plug. No external work was performed by the gas in passing through the plug, so the expansion was, in this sense, "unresisted"—at least within the plug itself.

Without entering into a full discussion of the theory of this experiment, we may note that Joule and Kelvin found by means of it that in general the unresisted expansion of a gas is, after all, accompanied by a temperature change. From the existence of the van der Waals cohesive (attractive) forces between the molecules, we should expect some of the kinetic energy of the molecules to be converted by the expansion into potential energy. The corresponding loss of kinetic energy should be registered as a fall of temperature; the expansion of any gas, if truly unresisted, should be accompanied by cooling.

Under the conditions of the porous-plug experiment, however, the temperature does not necessarily always fall; in certain circumstances it may rise. This is because there is in general a residual energy interchange with the surroundings, in the form of work done on or by the gas, and if work is done on the gas this will tend to make the temperature rise. In Fig. 17.3 let us consider a sample of gas initially between A and A' whose left-hand and right-hand boundaries travel, in a given time, to B and B', respectively. Then if V_1 and V_2 are the volumes of gas between A and B and between A' and B', respectively, there will be a net amount of work done on the gas equal to $p_1V_1 - p_2V_2$. If the gas were ideal and the temperature were the same on both sides, this would be zero. Actually, neither of these conditions is strictly fulfilled. Thus, whereas the cohesive forces of the molecules tend to bring about a reduction of temperature, the finite volume occupied by the molecules, which is responsible for the quantity b in van der Waals's equation, tends to make p_2V_2 less, relative to p_1V_1, than it other-

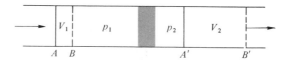

FIGURE 17.3   Joule-Kelvin experiment on the expansion of a gas through a porous plug.

wise would be. Which of the two opposing tendencies is the greater depends on the gas and the experimental conditions. For example, Joule and Kelvin found that whereas for each atmosphere of pressure reduction at ordinary temperatures air is cooled by 0.21°C and carbon dioxide by 1.00°C, hydrogen is heated by 0.04°C.

Characteristic of each gas there is a definite temperature, known as the **inversion temperature,** such that when a gas is expanded as in the porous-plug experiment, above this temperature it is heated, while expansion at a lower temperature results in cooling. For air the inversion temperature is somewhat below 100°C. For hydrogen it is about −80°C.

Isothermal Volume Change of a Perfect Gas

As we saw in Chapter 15, the unresisted expansion of a perfect gas does not give rise to any change in the temperature, and we might accordingly be inclined to call this an isothermal expansion. By convention, however, the term "isothermal" is not used to describe an unresisted expansion; it is reserved for a more specific meaning. It is applied only to the case where the expansion of the gas, besides taking place at a constant temperature, is *fully resisted*, that is, where the gas does the maximum possible amount of work in expanding, as it would do in pushing a piston along a cylinder. Correspondingly, an isothermal compression is one in which work is done on the gas to compress it, the gas meanwhile being kept at a constant temperature. During an expansion, heat has to be supplied from an external source to keep the temperature constant, while during a compression heat must be extracted.

Isothermal volume changes are of particular interest because the total internal energy of the gas—the total kinetic energy of the molecules—remains constant throughout. Consequently, in an isothermal expansion the whole of the heat supplied is utilized in doing work on an external system, while, if the volume change is a contraction, the whole of the work done on the gas produces heat in a body or bodies other than the gas itself.

For future reference it will be useful to calculate how much work is performed when a gas, initially occupying a volume V_1, expands at a constant temperature T to the final volume V_2. This is given by the equation

$$W_{12} = \int_{V_1}^{V_2} p \, dV;$$

433

and if in this we substitute for p its value according to the equation of state, $(m/M)(RT/V)$, we have

$$W_{12} = \frac{m}{M} RT \int_{V_1}^{V_2} \frac{dV}{V} = \frac{m}{M} RT \ln \left(\frac{V_2}{V_1} \right). \qquad (17.2)$$

Adiabatic Volume Change

When a gas undergoes a volume change in such a way that (1) there is no heat interchange with the surroundings, and (2) if the gas expands it performs the maximum amount of work on some external system in doing so, while, of course, contraction is effected by doing work on it, then the volume change is said to be **adiabatic.** Adiabatic volume changes are of considerable importance. Thus, together with isothermal changes, they are involved in the Carnot cycle, a cycle of operations performed by an ideal heat engine, to which many actual heat engines approximate. Again, the cooling that accompanies an adiabatic expansion is turned to account in one of the processes employed in the liquefaction of gases. Also, the temperature changes accompanying adiabatic volume changes play a basic role in certain atmospheric phenomena.

As no energy in the form of heat is communicated to or withdrawn from a gas undergoing an adiabatic volume change, the work it performs in expanding must be at the expense of its store of internal energy, while if it is compressed, and so work is done on it, its internal energy is correspondingly increased. Let the volume of the gas be increased by the small positive or negative amount δV at the pressure p. Then the work done is $p \, \delta V$, and this must be equal to the decrease of internal energy, $-(m/M)C_v \, \delta T$:

$$\frac{m}{M} C_v \, \delta T + p \, \delta V = 0.$$

On dividing throughout by δT and proceeding to the limit of smallness for δT, this becomes

$$\frac{m}{M} C_v + p \frac{dV}{dT} = 0. \qquad (17.3)$$

Now, from the equation of state for a gas, Eq. (15.1), we have

$$p \frac{dV}{dT} + V \frac{dp}{dT} = \frac{m}{M} R,$$

434   so that, eliminating m/M between the last two equations, we obtain

the equation

$$C_v \left(p \frac{dV}{dT} + V \frac{dp}{dT} \right) + Rp \frac{dV}{dT} = 0,$$

or, because R is equal to $C_p - C_v$,

$$C_v \left(p \frac{dV}{dT} + V \frac{dp}{dT} \right) + (C_p - C_v)p \frac{dV}{dT} = 0,$$

which simplifies to

$$V \frac{dp}{dT} + \gamma p \frac{dV}{dT} = 0. \tag{17.4}$$

On integrating this, we obtain the equation for an adiabatic volume change,

$$pV^\gamma = \text{const.} \tag{17.5}$$

From this equation we may readily calculate the change in temperature corresponding to any given adiabatic volume change. For, in addition to

$$p_2 V_2{}^\gamma = p_1 V_1{}^\gamma,$$

or

$$\frac{p_2 V_2}{p_1 V_1} = \left(\frac{V_1}{V_2} \right)^{\gamma-1}, \tag{17.6}$$

we have, because of our definition of temperature in terms of the product of the pressure and volume of a gas,

$$\frac{p_2 V_2}{p_1 V_1} = \frac{T_2}{T_1}. \tag{17.7}$$

Hence, equating the right sides of Eqs. (17.6) and (17.7), we find that

$$\frac{T_2}{T_1} = \left(\frac{V_1}{V_2} \right)^{\gamma-1}. \tag{17.8}$$

We may also calculate the temperature change in terms of the change in pressure. Writing

$$p_2 V_2{}^\gamma = p_1 V_1{}^\gamma$$

in the form

$$(p_2 V_2)^\gamma p_2{}^{1-\gamma} = (p_1 V_1)^\gamma p_1{}^{1-\gamma},$$

and remembering that pV is proportional to T, we have

$$\left(\frac{T_2}{T_1} \right)^\gamma = \left(\frac{p_2}{p_1} \right)^{\gamma-1},$$

or

$$\frac{T_2}{T_1} = \left(\frac{p_2}{p_1} \right)^{(\gamma-1)/\gamma}. \tag{17.9}$$

It is interesting to consider the reason for these temperature changes in terms of molecular motions. Thus, let us suppose a gas is contained in a thermally nonconducting cylinder and is allowed to expand by pushing a piston outward. On the average, a molecule will rebound from the piston after impact with the same speed *relative to the piston* as that with which it was incident. But the piston is moving outward, so the speed *relative to the cylinder* must be less after impact than before. Accordingly, the collisions of the molecules with the moving piston slows them down, correspondingly reducing the temperature. Similarly, if the gas is being compressed, the collisions with the piston will speed them up and so raise the temperature.

ADIABATIC BULK MODULUS OF A GAS In Chapter 11 it was stated without proof that the adiabatic bulk modulus K of a gas at a pressure p is γp, and this was then substituted in Eq. (11.4) to obtain the expression for the velocity v of sound in the gas,

$$v = \sqrt{\frac{\gamma p}{\rho}},$$

or, in terms of the temperature T of the gas and its molecular weight M,

$$v = \sqrt{\frac{\gamma RT}{M}}.$$

That the adiabatic bulk modulus is indeed γp follows at once from the definition of a bulk modulus,

$$K = -V\frac{dp}{dV}$$

and the expression for dp/dV corresponding to the equation

$$pV^{\gamma} = \text{const.} \tag{17.5}$$

for an adiabatic volume change. For, differentiating this equation with respect to V, we have

$$V^{\gamma}\frac{dp}{dV} + \gamma p V^{\gamma-1} = 0,$$

from which, on dividing by $V^{\gamma-1}$ and rearranging, we have for the bulk modulus, defined as above,

$$K = -V\frac{dp}{dV} = \gamma p.$$

EXPERIMENTAL DETERMINATION OF γ As we see from Eq. (17.5), the quantitative expression for an adiabatic volume change involves the ratio of specific heats, γ, and thus it becomes important that we should have some reliable means for determining this quantity experimentally.

The specific heats of gases at constant volume and at constant pressure can be measured directly, and their ratio then gives us γ. However, γ can be determined with rather greater accuracy by a method devised by Clément and Desormes, based on Eq. (17.5). A large flask, provided with a stopcock, is connected to a sensitive manometer, whose volume, together with that of the connecting tube, is negligibly small compared with the volume of the flask. The flask is first filled with the gas whose γ we wish to determine, at a pressure p_1 somewhat exceeding atmospheric, and the value of this is noted on the manometer. The stopcock is then quickly opened and closed again, the gas expanding adiabatically while the stopcock is open. Finally, when the gas has settled down again to room temperature, the corresponding pressure p_2 is read. These two pressure readings, in conjunction with the atmospheric pressure p_0, enable us to calculate the value of γ. For, if V_1 and V_2 are the volumes of a given mass of the gas before and after opening the stopcock, we have

$$p_1 V_1{}^\gamma = p_0 V_0{}^\gamma,$$

and so

$$\log p_1 - \log p_0 = \gamma(\log V_0 - \log V_1),$$

or

$$\gamma = \frac{\log p_1 - \log p_0}{\log V_1 - \log V_0}.$$

But since the initial and final temperatures are the same, we must have

$$p_1 V_1 = p_2 V_0,$$

that is,

$$\frac{V_0}{V_1} = \frac{p_1}{p_2},$$

so that

$$\log V_0 - \log V_1 = \log p_1 - \log p_2.$$

Hence, finally, we have

$$\gamma = \frac{\log p_1 - \log p_0}{\log p_1 - \log p_2}.$$

Unfortunately, this procedure is attended by serious experimental difficulties. In its exit from the flask the air gains a certain

437

amount of momentum, as a result of which it "overshoots itself," and the somewhat excessive outrush of air is followed by an inrush, and so on. There are successive expansions and contractions of an oscillatory nature, and the moment of closing the tap will not necessarily be that when the pressure inside the bulb is the same as that outside. On the other hand, it is not permissible to wait until the oscillations have died down before closing the tap, for by then an appreciable amount of heat will have been transferred to the gas by the bulb; it is essential that the opening and closing of the tap be carried out quickly.

Röntgen, in 1873, found these difficulties to be much reduced by the use of a wide stopcock. By this means he found a value of γ for dry air of 1.4053. Further refinements and modifications were introduced by Lummer and Pringsheim in 1898 and by Brinkworth in 1926. Lummer and Pringsheim's value of γ for dry air at temperatures between 4° and 16°C was 1.4025, and Brinkworth's for air at 17°C was 1.4032. Probably a good weighted mean of these results would be 1.403.

Variation of Temperature with Altitude in the Troposphere

In Chapter 14 reference was made to the huge convection currents existing in the troposphere, which we on the earth's surface experience as winds. As we shall now see, it is these convection currents that are responsible for the fall of temperature with increasing altitude.

As a preliminary, let us take note of the situation as it is observed. Table 17.1 gives typical values of the pressure and temperature of the air at a number of heights above sea level for average conditions in temperate latitudes. We see that, although the temperature falls at a substantial rate with increasing height up to an altitude of about 11 km, it remains constant thereafter, at least as far as 30 km. It is in the lower region, the troposphere, extending in these latitudes up to about 11 km, that all the important weather phenomena occur. Here there is considerable agitation, a complex of convection currents giving a continual interchange of air samples between various altitudes. Above this is the "stratosphere," a relatively calm region, in which there is little vertical movement of the air.

The variation of temperature with altitude in the troposphere is readily accounted for by considering what must happen to a sample of air taking part in a convection current. As this gains

TABLE 17.1 Atmospheric Pressures and Temperatures
at Various Altitudes

Altitude (km)	Pressure (mm Hg)	Temp. (°C)
0	760	15
1	674	8.5
2	596	2
3	526	−4.5
4	463	−11
5	405	−17.5
6	354	−24
7	308	−30.5
8	267	−37
9	230	−43.5
10	198	−50
11	170	−56.5
12	145	−56.5
13	127	−56.5
14	106	−56.5
15	90	−56.5
30	9	−56.5

altitude, in an up-current, it comes into regions of progressively lower pressure, and so it expands. This expansion is accompanied by the performance of work, for the air in our sample pushes against other air in expanding; thus the expansion is adiabatic, or at least approximately so. Consequently, the temperature falls. Conversely, when, later, the sample loses altitude in the downward-moving part of the convection current, it is compressed; work is done on it, and so the temperature rises.

If we idealize the situation by assuming that we are dealing with dry air and that its expansion and contraction are truly adiabatic, we must expect the temperature to fall with decreasing pressure according to Eq. (17.9). Substituting for γ in this equation its value for dry air, 1.403, we obtain the relation

$$\frac{T_2}{T_1} = \left(\frac{p_2}{p_1}\right)^{0.2872}.$$

Let us now see what temperature fall this should give us between sea level and 2000 m, adopting the values of pressure given in Table 17.1. We find that according to our equation T_2/T_1 should be 0.932, so that with $T_1 = 288°K$, T_2 should be $268°K$, or the Celsius temperature should be $-5°$. Thus the actual temperature fall is only about two thirds that predicted.

439

It will now be interesting to see how well the formula applies higher up in the atmosphere, say between 4000 and 8000 m. Substituting the values of pressure at these altitudes given in the table, we find that T_2/T_1 should be 0.854. With $T_1 = 262°$K, T_2 should be 224°K, to which corresponds a Celsius temperature of $-49°$. We see that here, again, the observed temperature fall is only about two thirds that predicted, although actually the ratio is a trifle higher on this occasion.

Some of the discrepancy between the observed and predicted rates of temperature fall with increasing height may be attributed to our predicted rates being based on the assumption that we are dealing with *dry* air. Actually the atmosphere is by no means dry; and condensation of moisture, occurring in the process of cooling, can release sufficient latent heat to reduce the temperature fall quite substantially.

Let us consider the case of air expanding from 4000 to 8000 m. For the purposes of our calculation we shall be concerned with the specific heat of air at constant pressure, for we are dealing with the discrepancy between the actual and calculated temperatures *at the final pressure reached,* that at 8000 m. For dry air this specific heat is 0.24 cal/g/deg. Hence, to warm 1 g of dry air by 12°C would require just under 3 cal. The actual moisture content of the air would not affect this appreciably, so let us assume that 3 cal is required. For the latent heat of vaporization of ice in the temperature region with which we are here concerned we may take the round figure of 700 cal/g. To account for the discrepancy between the observed and calculated rates of temperature fall, we should therefore require the cooling from $-11°$ to $-37°$C to be accompanied by the precipitation of $3 \div 700$, or 0.0043 g of ice in each gram of air. Let us compare with this the vapor contents of the air at these two temperatures under saturation conditions.

From the plot of the logarithm of the vapor pressure of ice against $1/T$ in Fig. 16.4 we can at once find the vapor pressures at the two temperatures in which we are interested. They are 1.79 mm at $-11°$C and 0.10 mm at $-37°$C. Now, taking the mean molecular weight of air as 28.8, 1.79 mm of water vapor in air at 463 mm pressure represents $(18/28.8)(1.79/463) = 2.42 \times 10^{-3}$ g of water vapor per gram of air. And 0.1 mm in air at 267 mm represents $(18/28.8)(0.1/267) = 2.3 \times 10^{-4}$ g of water vapor per gram of air. The difference is 0.00219 g. This is about half the amount we have calculated to be necessary to account for the difference between the observed and predicted rates of fall of temperature with increasing altitude. If the air were not saturated at the lower altitude, not even this much ice would be condensed. However, the relative humidity at these altitudes would generally be quite high, even if not 100 percent. Thus, we may conclude that the latent heat released by the condensation of water accompanying expansion accounts for an appreciable proportion of the discrepancy under consideration.

In our calculation we have considered the expansion of air in an up-current and the accompanying condensation. Generally, the condensed ice or water will remain as part of the body of air considered in the form of cloud, and later, when this air moves downward, it will evaporate again. When this happens, the calculation will apply (in reverse) also to the contraction that occurs in a downcurrent. On the occasions when a proportion of the ice or water condensed at the higher altitude falls as snow or rain, there would be correspondingly less water available for evaporation, and so the air would be at a higher temperature on arrival at the lower altitude.

How can we account for the part of the discrepancy between the calculated and observed rates of temperature fall that still remains? Surely this must be due to the expansions and contractions not being truly adiabatic. There might, for example, be a radiative interchange of energy—in effect a "heat exchange"— between different parts of the atmosphere. There must presumably also be some radiative energy interchange between the surface of the earth (whether land or water) and the atmosphere.

The rate of fall of temperature with increasing altitude is referred to by meteorologists as the "lapse rate," while the rate calculated on the assumption of truly adiabatic volume changes is known as the "adiabatic lapse rate."

Liquefaction of Gases

A number of methods are used to liquefy gases, varying with their physical properties.

LIQUEFACTION BY ISOTHERMAL COMPRESSION The easiest gases to liquefy are those whose critical temperatures are relatively high, for example carbon dioxide, sulfur dioxide, and ammonia. These can be liquefied simply by compression at ordinary temperatures, accompanied by the extraction of the heat equivalent of the work of compression as far as the saturation point and of the latent heat of condensation thereafter. For example, if carbon dioxide is compressed at 20°C, it will become saturated at a pressure of 59 atm, and further reduction of volume at this temperature will bring about condensation.

441

If the pressure on a liquid produced in this manner is released without heat being supplied to it from an external source, selective evaporation of its most energetic molecules will result in substantial cooling. On progressive reduction of the pressure of carbon dioxide vapor in contact with liquid carbon dioxide, the temperature of the latter will fall until, at $-57°C$, it will begin to freeze. At this temperature the pressure of the vapor in equilibrium with it will be 5.11 atm. With further evaporation the temperature and pressure will remain constant at these values until the whole of the liquid is frozen. If the pressure is now further reduced, further evaporation occurs, and the temperature of the solid will fall still lower, until, at 1 atm, it will attain the value of $-78.5°C$, the "temperature of dry ice." Thus, if a cylinder of carbon dioxide at room temperature, the greater part of whose contents is normally in the liquid form, is so orientated that the outlet is below the free surface of the liquid, and the valve is opened, a mixture of solid carbon dioxide and its vapor will be blown out, and the former may be collected in a cloth bag tied to the outlet pipe. This is by far the most convenient way to obtain dry ice for occasional laboratory use.

Cooling by evaporation is an essential feature of household and industrial refrigeration. A refrigerator may be regarded simply as a reversed heat engine. In the operation of a steam engine, steam is evaporated at a high temperature and pressure and subsequently condensed at a much lower temperature. More heat is supplied at the high temperature than is rejected at the lower temperature, the difference corresponding to the work performed by the engine. In a refrigerator, on the other hand, the working substance is allowed to evaporate at a low temperature, extracting the requisite latent heat from its surroundings, and is later condensed under considerable pressure and at a higher temperature, for example room temperature. At this temperature more heat is rejected than is supplied at the low temperature, the difference corresponding to the work done by the motor. Various working substances are used in refrigerators, for example, sulfur dioxide, carbon dioxide, and ammonia. Another substance extensively used in modern refrigerators is dichlorodifluoromethane, commercially known as Freon.

Gases such as oxygen, nitrogen, hydrogen, and helium, which have a very low critical temperature, must be cooled below this temperature before they can be liquefied. We shall consider two methods by which this can be done.

LINDE PROCESS This depends on the reduction in temperature accompanying the "unresisted" expansion of a gas below its inversion temperature. With large pressure reductions this cooling

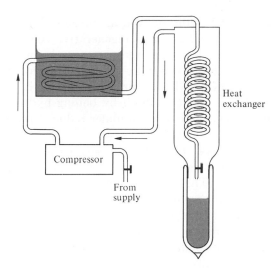

FIGURE 17.4 Liquefaction of a gas by Linde's method.

can be quite substantial. In our discussion of the porous-plug experiment of Joule and Kelvin we noted that air at ordinary temperatures is cooled by 0.21°C for each atmosphere of pressure reduction. At this rate the cooling produced by a reduction of pressure from 200 to 20 atm would be of the order of 40°C, and at lower temperatures it would be greater than this.

A schematic diagram of Linde's arrangement is shown in Fig. 17.4. Highly compressed gas at a pressure of the order of 200 atm has its heat of compression extracted by passage through a water-cooled tube and then enters the "heat exchanger," shown in the figure. The compressed gas passes down the coiled tube, and then, after expansion through a valve at the bottom down to a pressure of about 20 atm, passes up over the coil on its way back to the compressor. In so doing the expanded and cooled gas extracts heat from the oncoming gas within the coil, and this therefore arrives at the valve at a lower temperature than that of the gas which preceded it. This, in turn, expands through the valve, attaining a still lower temperature, and so on, until finally the gas leaving the valve is so cold that some of it condenses, the liquid being collected in a vacuum flask.

The yield of liquefied gas may be increased substantially by precooling the compressed gas before it enters the heat exchanger. This might be done, for example, by passing it through a tube cooled by dry ice (−78.5°C).

443

Hydrogen and helium, as well as oxygen and nitrogen, may be liquefied by Linde's process of regenerative cooling if they are precooled below their inversion temperatures before entering the heat exchanger. Hydrogen is usually precooled by passing it through a copper coil immersed in liquid air. Helium may be cooled below its inversion temperature ($-238°C$) by boiling hydrogen. In practice it is usual to boil the hydrogen under reduced pressure.

CLAUDE PROCESS   Obviously the cooling produced by expansion through a valve cannot be as great as that resulting from adiabatic expansion. The first air liquefier in which cooling was effected by making the compressed gas perform work was developed during the early years of the present century by Claude. The main difficulty to be overcome was that of finding a lubricant for the piston that would not solidify at the very low temperatures reached. The lubricant eventually used for the early stages of operation was petroleum ether. This was replaced by liquid air after liquefaction had begun. The cooling was arranged to be regenerative, as in the Linde process, and the work done by the air in expanding was made to help in driving the compressor. In more modern liquefiers the original Claude process has been replaced by a combination of this with the Linde process; the performance of external work cools the gas to below the inversion temperature, and the gas is then further expanded through a valve, where liquefaction is produced by the Joule–Kelvin effect.

Although theoretically the efficiency of Claude or Claude–Linde liquefiers must be greater than that of machines depending entirely on Joule–Kelvin cooling, the gain is actually only slight. Against this gain must be weighed the greater simplicity of the Linde process.

Several years ago, Kapitza succeeded in liquefying helium by a method in which it is no longer necessary first to produce liquid hydrogen for precooling the helium below its inversion temperature. This he did by using the Claude–Linde process. He avoided the lubrication difficulty by arranging for the piston to move in the cylinder without actually touching it, the clearance being made sufficiently small and the expansion stroke sufficiently rapid for the leakage of gas past the piston not to be serious.

In more recent times modified and improved helium liquefiers have been developed, one of the most widely used being the Collins liquefier. It would take us too far to embark on a discussion of these here.

Carnot's Cycle

In 1824 Sadi Carnot published an essay entitled *Reflexions sur la puissance motrice du feu,* in which he investigated the performance of an ideal heat engine such as can never be realized in practice. Its working substance was a perfect gas, and the working parts consisted of a piston moving without leakage and without friction within a cylinder. The cylinder and piston had zero heat capacity, and the system was imagined to be capable of manipulation in such a way that the gas could either be completely insulated thermally from its surroundings or put in perfect thermal contact either with a source of heat or with a "sink."

Remote though these conditions are from anything that could ever be established in practice, the impossibility of their realization constitutes no objection to their use in a theoretical investigation. It is one of the most interesting features of thermodynamical theory that it repeatedly invokes such idealized systems, yet the results obtained are not only of the profoundest theoretical significance but are also completely relevant to observations made in real life.

The cycle of operations imagined by Carnot, known as **Carnot's cycle,** could, in principle, be carried out not only with a perfect gas as working substance, but also with several alternative systems, for example water and steam, water and ice, a liquid film, and many others. The use of Carnot's cycle in this wider sense has been found exceedingly fruitful in the derivation of a number of important theoretical relationships. Some of these uses will be considered in Chapter 18. Also, a number of practical heat engines, among them the reciprocating steam engine, are modeled on Carnot's cycle, and the results of Carnot's investigation are of corresponding relevance to these. Here we shall confine ourselves to the consideration of the cycle in which the working substance is a perfect gas.

Carnot's cycle of operations, which is usefully represented graphically on a p–V diagram such as that shown in Fig. 17.5, is as follows. Beginning at a pressure p_1 and volume V_1, the gas is first allowed to expand isothermally at the temperature T_1 until its pressure has fallen to a lower value p_2 and its volume increased to the corresponding value V_2. During this expansion heat must be supplied to the gas from some source to keep its temperature constant. Let this quantity of heat be Q_1. Next, after isolating the gas from the source of heat, it is allowed to expand a further amount, this time adiabatically, to the still larger volume V_3, its pressure falling to p_3 and its temperature to T_2. At this lower temperature

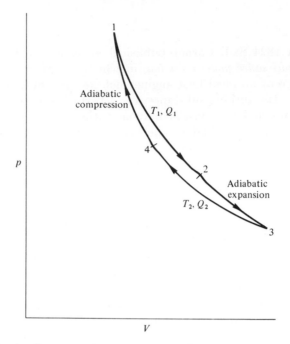

FIGURE 17.5   Carnot-cycle diagram for an ideal gas.

the gas is now compressed isothermally, the compression being stopped at a pressure p_4 and corresponding volume V_4 such that a further compression to the original volume V_1, carried out adiabatically, will restore the original pressure p_1. During the isothermal compression from the volume V_3 to V_4 the requisite quantity of heat Q_2 is withdrawn from the gas and communicated to a heat sink. After the final adiabatic compression to the original volume V_1 and pressure p_1 the cycle is complete; the gas is now restored to its original condition, as regards not only its pressure and volume but also its temperature, because temperature is a unique function of pressure and volume, actually being proportional to the product of these.

Let us now discuss this cycle quantitatively, concerning ourselves particularly with the net amount of work performed and the relation among Q_1, Q_2, T_1, and T_2.

The work performed during an isothermal volume change has already been investigated and is expressed by Eq. (17.2). Thus the work W_1 done by the gas during its isothermal expansion is $(m/M)RT_1 \ln (V_2/V_1)$, while the work W_2 done on it during its isothermal compression at the lower temperature T_2 is $(m/M)RT_2 \ln (V_3/V_4)$. We saw earlier in this chapter that in an isothermal expansion there is 100 percent conversion of the heat supplied, via the performance of external work, to some other form of energy, the internal energy of the gas itself remaining constant throughout. Hence we may write

$$Q_1 = W_1 = \frac{m}{M} RT_1 \ln \left(\frac{V_2}{V_1}\right). \qquad (17.10)$$

Correspondingly, for the quantity of heat rejected by the gas and the work done on it during the isothermal compression at T_2 we have

$$Q_2 = W_2 = \frac{m}{M} RT_2 \ln \left(\frac{V_3}{V_4}\right). \qquad (17.11)$$

Let us now investigate the ratios beteeen the volumes that appear in these expressions. We have

$$p_1V_1 = p_2V_2,$$
$$p_2V_2{}^\gamma = p_3V_3{}^\gamma,$$
$$p_3V_3 - p_4V_4,$$

and

$$p_4V_4{}^\gamma = p_1V_1{}^\gamma.$$

Equating the product of all the left sides of these equations to that of all the right sides, and dividing each product by the common factor $p_1p_2p_3p_4V_1V_2V_3V_4$, we find that

$$(V_2V_4)^{\gamma-1} = (V_3V_1)^{\gamma-1},$$

from which it follows that

$$\frac{V_2}{V_1} = \frac{V_3}{V_4}. \qquad (17.12)$$

We see that $(m/M)R \ln (V_2/V_1)$ and $(m/M)R \ln (V_3/V_4)$ in the expressions (17.10) and (17.11) for Q_1 and Q_2, respectively, are equal, and so

$$\frac{Q_1}{Q_2} = \frac{T_1}{T_2}. \qquad (17.13)$$

Equation (17.13) leads at once to an expression for the **thermodynamic efficiency,** η, of this ideal heat engine. The thermodynamic efficiency of a heat engine is defined as the ratio of the net work performed by it during a complete cycle to the heat supplied by the source. Obviously, the net work performed must be equal to the excess of heat supplied over heat rejected, $Q_1 - Q_2$, because it is necessarily the heat that goes out of existence that is effective in causing work to be performed. Hence

$$\eta = \frac{Q_1 - Q_2}{Q_1}. \qquad (17.14)$$

This equation must apply to *any* heat engine, whatever may be its nature. But in the case of the Carnot cycle at present under consideration the efficiency may also be expressed in terms of the two temperatures. For, from Eqs. (17.13) and (17.14), we have at once

$$\eta = \frac{Q_1 - Q_2}{Q_1} = \frac{T_1 - T_2}{T_1}. \qquad (17.15)$$

Problems

17.1 A certain waterfall is 300 m in height. If there were no complications, such as cooling by evaporation on the way down, by how much would the temperature of the water rise on reaching the bottom?

17.2 A car of mass 1000 kg is traveling along a horizontal road at 100 km/hr. How much heat, in calories, is developed in the brakes when it is brought to rest?

17.3 A lead bullet traveling at 150 m/sec strikes a target. Assuming that the heat derived from its kinetic energy is divided equally between the bullet and the target, calculate the rise in temperature of the bullet.

17.4 An electric heater passes 2 kW through a copper wall of area 100 cm² and thickness 1 mm. Find the difference in temperature between the two faces.

17.5 Imagine that the whole of the heat energy lost by 1 km³ of ocean water in cooling through 1°C could somehow be converted to mechanical energy which drives a 1-MW (10^6-watt) motor. For how long could the motor be operated? (Assume that seawater has the same specific heat as fresh water.)

17.6 What part of the latent heat of vaporization of water at 100°C is used to "push back" the surrounding atmosphere? Would this contribution be greater or less at a lower temperature, say at 90°C?

17.7 An electric heating coil is immersed in 1000 g of water contained in a thermos flask, and raises the temperature by 60 deg in 5 min. When the water is replaced by 800 g of another liquid, it is found that the same temperature rise occurs in 1.8 min. What is the approximate specific heat of the liquid? What additional factors must be taken into consideration if a more accurate determination of the specific heat is to be made by this method?

17.8 A liquid of density 0.873 g/cm³ flows through a continuous-flow calorimeter at the rate of 9.30 cm³/sec, and when steady-state conditions have been established with a 200-W electric heating coil switched on, the temperature of the liquid rises from 10.24°C at the inlet end to 24.60°C at the outlet. What is the specific heat of the liquid?

17.9 A mixture of hydrogen and oxygen contained in a rigid enclosure with heat- and sound-insulating walls is exploded by an electric spark. Apart from the energy of the spark, which is relatively small, what effect does this have on the internal energy of the system?

17.10 How much external work is performed by 10 g of nitrogen in the course of being heated at constant pressure from 15° to 45°C? How much is the internal energy of the gas increased, and what is the mechanical equivalent of the heat input?

17.11   A quantity of argon, of mass 20 g, and initially at 15°C and 760 mm pressure, is heated at constant pressure to twice its original volume. What is the final temperature? How much work has been performed by the gas in expanding, and by how much has its internal energy been increased? How much heat has had to be supplied to the gas?

17.12   In which kind of expansion of a gas, one at constant temperature or one at constant pressure, is the greater amount of work performed for a given expansion ratio? Which requires the greater heat input?

17.13   Calculate the work done by a gas in expanding isothermally from a volume of 1 liter at a pressure of 5 atm to 2 liters.

17.14   Assuming that in the system described in Problem 15.7 the volume changes of the enclosed air occur adiabatically, the ratio of the specific heats at constant pressure and at constant volume being γ, obtain an expression for the minimum length which this tube must have if the ball bearing is to execute simple harmonic motion, and find the period of this motion.

17.15   Show that if a molecule of mass m moving from left to right with velocity v is reflected normally from a surface moving from right to left with velocity u, where $u \ll v$, the gain in molecular kinetic energy is $2mvu$.

A thermally insulated cylinder of cross-sectional area A contains a monatomic gas at a temperature T. It is fitted with a piston which moves within the cylinder without leakage and without friction. Show that if the piston, initially at a distance l from the closed end of the cylinder, moves toward this end a small distance δx, the total increase in molecular kinetic energy is approximately $\frac{1}{3}Anmv^2\,\delta x$, where n is the number of molecules per unit volume, m the mass of each molecule, and v the molecular speed. Show that the mean kinetic energy per molecule increases by $\frac{1}{3}(mv^2/l)\,\delta x$. By making use of the fact that the temperature of a gas is proportional to the mean kinetic energy of its molecules, show that the compression results in a temperature rise δT given by the equation

$$\delta T = \frac{2T}{3l}\,\delta x.$$

Compare this result with that obtained by using the formula $pV^\gamma = \text{const.}$, with $\gamma = \frac{5}{3}$. (It may be assumed that all molecules have the same speed v, and that one third of the molecules are moving axially, while the remainder move in directions at right angles to the axis of the cylinder.)

17.16   One liter of air at 20°C and 760 mm pressure is compressed adiabatically to 200 cm.^{3} Find (a) the new pressure, (b) the new temperature, and (c) the work performed in compressing the gas.

17.17   A volume V of air at 0°C and 760 mm Hg has its pressure suddenly doubled. Find the new temperature and the new volume. How much work has been done on the gas in compressing it? By how much has its internal energy been increased?

17.18   Nitrogen at 25°C is to be expanded adiabatically until its pressure becomes 1 atm. What must be the initial pressure of the gas in order that its temperature after expansion shall be the boiling point of liquid nitrogen, $-195.8°C$?

17.19   Moist air at 25°C and 760 mm pressure is allowed to expand adiabatically to 665 mm. What is the final temperature? If the relative humidity was initially 40 percent, what is its value after the expansion? (Assume that $\gamma = 1.4$.)

17.20   A quantity of air is (a) expanded adiabatically until its volume is doubled and then heated at constant volume until its initial temperature is restored, or (b) expanded isothermally to twice its original volume. Compare the quantities of heat imparted to the gas in the two cases, and explain why, although the final state of the gas is the same in either case, the quantities of heat added are not the same.

17.21   In reducing the volume of a gas adiabatically, 10 J of work is done in compressing it. By how much has its internal energy been increased? If, instead, the same reduction in volume were to be effected isothermally, how much heat would have to be removed from it during the compression?

17.22   Two identical gases in identical cylinders provided with pistons are allowed to expand, the first isothermally and the second adiabatically. Assuming the cylinders to be infinitely long, so that the pistons may be pushed back any distance, show that the amount of work done in the isothermal expansion may become infinite, while in the adiabatic expansion it remains finite. For what physical reason would one expect this?

17.23   Plot a Carnot-cycle p–V diagram for 1 gram-molecule of an ideal gas for which $\gamma = 1.4$ between the temperatures 1000° and 300°K, with $p_1 = 2$ atm and $p_2 = 1$ atm (see Fig. 17.5). Find, graphically or otherwise, (a) the net work performed in this cycle, (b) the heat supplied to the gas during its isothermal expansion from p_1 to p_2, and (c) the heat rejected during the isothermal compression from p_3 to p_4.

17.24   An ideal-gas Carnot engine takes in 200 cal each cycle from the high-temperature reservoir at 200°C and rejects 130 cal to the sink. What is the temperature of the latter?

17.25   The thermodynamic efficiency of an ideal-gas Carnot engine which rejects heat to a sink at 15°C is 30 percent. What is the temperature of the source?

17.26   Two ideal-gas Carnot engines operate as a two-stage combination in the following manner: The first engine takes in a quantity Q_1 of heat per cycle at a temperature T_1 and rejects the quantity Q_2 at a lower temperature T_2. The second engine takes in the heat Q_2 at the temperature T_2 and rejects the quantity Q_3 at the still lower temperature T_3. Show that the efficiency of the combination is $(T_1 - T_3)/T_1$.

17.27   An ideal-gas Carnot engine is operated in reverse, functioning as an ideal refrigerator. It takes in a quantity of heat Q_2 per cycle from a heat reservoir at a temperature T_2 and rejects a larger quantity Q_1 of heat to a heat reservoir at a higher temperature T_1. Show that to operate this refrigerator a net amount of work $Q_2(T_1 - T_2)/T_2$ must be performed on it per cycle.

THE SECOND LAW OF THERMODYNAMICS

Ranking in importance with the first law of thermodynamics is another, the second law, which we shall consider in this chapter. Unlike the first law, which is concerned with the quantity of heat that is produced or goes out of existence in energy transformations involving heat, the second law deals with a preference found in nature for a certain *direction* of such energy transformations—from other forms of energy to heat.

The second law of thermodynamics differs in kind from any of the physical laws we have so far considered, in that it is a statement of an impossibility. Although one might expect little of value from a law of such negative character, there are in fact few laws in physics whose consequences are more far-reaching. A small selection of these will be considered toward the end of this chapter.

Degradation of Energy

If heat is one of the items in an energy conversion it is found that there is an asymmetry as regards the direction of the conversion. There is never any difficulty in converting any other form of energy completely into heat, but it is not possible to convert the whole of a given quantity of heat into any other form of energy without causing changes in other bodies. And, as heat is very commonly an item in energy transformations, there is a general tendency in nature for the availability of energy for conversion into other forms to decrease. There is no decrease in its total quantity, but it is becoming progressively less available for further transformations. It appears, then, that the universe, or at least that part or aspect of it which we have hitherto succeeded in investigating, is "running down"; it is approaching a condition sometimes referred to as "heat death."

As an example of this, let us consider the generation of electrical energy and its ultimate end. Let us suppose that this energy is produced by the combustion of coal in a steam engine, and let us consider all the energy transformations from the beginning. The first is the conversion of the chemical energy of the coal plus

451

atmospheric oxygen to heat. Some, but not all, of this is utilized, via the steam engine, in doing work on the electric generator, the remainder being rejected as heat to the condenser. In the operation of both the engine and the generator there is some unavoidable loss of energy by friction and the viscosity of the lubricant, with corresponding production of heat. However, most of the work done by the expanding steam is effective in producing electrical energy. A small proportion of the latter is dissipated as heat in the windings of the generator and in the transmission system, but the greater part is eventually used to drive machinery, to light lamps, to operate refrigerators, vacuum cleaners, electric irons, cookers, radiators, and so on. In almost all cases it is ultimately dissipated as heat residing in bodies at ordinary temperatures, and its availability is correspondingly lost.

There has been no lack of ingenuity or persistence in men's attempts to intervene in these natural processes and to reverse them, to convert the heat in the objects around us back to other forms of energy. However, by no means that have so far been devised has this been found possible without leaving changes in other bodies.

Statement of the Second Law

There are various possible formulations of the second law of thermodynamics. Perhaps as useful a formulation as any for present purposes is the following: **It is impossible to construct an engine that operates in a cycle and performs work by extracting heat from a body without returning any heat to another body at a lower temperature.**

There are two aspects of this statement that need amplification. The first is the reference to the performance of work. As we have seen, this means an energy conversion—in the present case from heat to some other form of energy, such as mechanical. The other point is the specification that the engine shall operate in a cycle, that is, in such a way that it is finally in the same condition as at the beginning. This ensures that, apart from the conversion of heat to some other form of energy that is envisaged, and the change in the recipient of the other form of energy, for example the lifting of a mass against gravity, there shall be no change left in any other body.

This proviso is important; without it, the statement simply would not be true. For example, if we placed a source of heat in thermal communication with a gas at the same temperature and we allowed the gas to expand isothermally, doing work by pushing a piston along a cylinder, heat would flow from the source to the gas during its expansion, maintaining its internal energy at a constant level, while the piston, exerting a force on some external system while it moves outward, performs work. At the end the internal energy of the gas would remain unchanged, but heat would have been lost by the source and the whole of this would have been utilized in the performance of work. But this would not violate the second law as stated above, because at the end the piston would be further along the cylinder than at the beginning, and this would constitute a change in the system used for the energy conversion; it would not have gone through a *cycle* of operations.

Having noted what, according to the second law, a heat engine working in a cycle cannot do, let us consider what it does in fact do. This is indicated schematically in Fig. 18.1(a), in which the circle represents the heat engine. In its cyclic operation this extracts heat Q_1 from a reservoir of heat at a certain temperature T_1, rejects a smaller amount of heat Q_2 to a reservoir at a lower temperature T_2, and converts the difference, $Q_1 - Q_2$, into some other form of energy via the performance of work W, equal to $Q_1 - Q_2$. What the statement of the second law in the form given above asserts is that Q_2, the heat rejected to the "cold" reservoir, is finite; it cannot be zero.

ALTERNATIVE STATEMENT OF THE SECOND LAW An alternative statement of the second law is the following: **It is impossible for any self-acting device, working in a cycle, to convey heat from a body at one temperature to another at a higher temperature.**

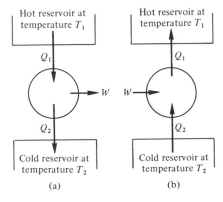

(a)   (b)

FIGURE 18.1 Reversible heat engine operating directly and in reverse.   **453**

The conveyance of heat from a body at a lower temperature to another at a higher temperature is precisely what a refrigerator does. But no refrigerator is self-acting; a refrigerator must have an external source of energy for its operation, so leaving changes in other bodies or systems, for example the consumption of coal in a distant power station. If the refrigerator is, like most household refrigerators, a mechanical one, its action may be represented schematically as in Fig. 18.1(b). For its operation, work W must be performed on its cyclically operating system, in response to which it will extract heat Q_2 from a cold reservoir and reject a greater amount of heat Q_1, equal to $Q_2 + W$, to a hot reservoir. We see that in its operation a refrigerator is really a heat engine running in reverse. What the second law in its alternative formulation asserts is that W must be finite, so precluding the possibility of the refrigerator being self-acting. Consequently, because of the first law, Q_1 must be greater than Q_2.

Although at first sight there appears to be no connection between the alternative statement of the second law and the original statement, the two statements are, in fact, completely equivalent. To see this, let us consider, in particular, the consequences of any violation of the law in its alternative formulation. If possible, let a quantity of heat Q be conveyed from a body at a temperature T_2 to another at a higher temperature T_1, without leaving changes in any other body. This having been done, the body at the higher temperature could be used as a heat source and that at the lower temperature as a sink, for an engine working in a Carnot cycle with a perfect gas as working substance. This engine would perform work equivalent to the fraction $(T_1 - T_2)/T_1$ of Q and would reject the remainder to the sink. The net result would be that the amount of heat $[(T_1 - T_2)/T_1]Q$ will have been extracted from the body at the lower temperature and converted into another form of energy in some recipient of energy, without leaving changes in any other body. But precisely this is what was alleged to be impossible in our original statement of the second law.

Conversely, it may be shown that if the original statement is not true, then the alternative statement is not true either. For, if possible, let heat be taken from a body and converted to mechanical energy in some system without leaving changes in any other body. This mechanical energy could then be reconverted to heat in a body at any higher temperature, for example by friction. But this is what has just been alleged, in the alternative statement of the second law, to be impossible.

Evidence for the Truth of the Second Law

If the second law were not valid, and accordingly some means could be devised for converting the heat residing in all the objects surrounding us to useful forms of energy, all our power problems would be at an end. It would no longer be necessary to mine coal or oil, to harness hydraulic power, or set up nuclear ("atomic") power stations; all we would have to do to have unlimited power at our disposal would be to set in operation the kind of process whose possibility is denied in the second law. As we have already noted, an enormous amount of ingenuity has been brought to bear in attempts to devise a system that would carry out such a process, but all to no avail. This is what we may call the negative part of the evidence on which the second law is based.

There is also evidence of a positive nature. A large number of theoretical predictions can be made on the assumption that the law is valid. When put to experimental test, these predictions are always verified. This constitutes strong evidence that the assumption on which the predictions are based is valid.

Maxwell's Demon

Notwithstanding all the evidence just considered, it is possible at least to *imagine*, even if not to realize, a process that would violate the second law. Such a process was, in fact, imagined many years ago by Maxwell. This process requires the intelligent intervention of a hypothetical being—to whom the name "Maxwell's demon" has been given—in the processes of nature. It is imagined that this being is able to see the individual molecules of a gas, to anticipate their arrival at a miniature window in a partition separating two compartments of a box containing gas, and to open and shut the window without expenditure of energy at just such times as would bring about a violation of the second law.

Let one compartment, A, contain gas at a higher temperature than that in the other compartment, B. Now, as we noted in our discussion of the kinetic theory of gases, there is a considerable spread of velocities among the molecules of a gas at any temperature, so that, at any given instant, not only do some of the molecules of a relatively cold gas have quite high velocities, but also some of the molecules of a hot gas will be moving relatively slowly. Maxwell's demon is supposed to sit at the window between the two compartments and watch the molecules approaching it from either side. Each time a particularly slow molecule in compartment A approaches the window, the latter is opened to let it pass through

455

into B, and then immediately shut again. For each such passage of a slow molecule from A to B a relatively fast molecule in the cooler gas is allowed to pass through from B to A by appropriate opening and shutting of the window. By repeated manipulations of this kind the average energy of the molecules in A, and so the temperature of the gas in this compartment, will be increased, while the temperature of the relatively cool gas in B will be reduced still further. Thus, if we were able to avail ourselves of a Maxwell demon and could construct the appropriate apparatus for him to manipulate, the second law would be violated.

Macroworlds and Microworlds

Unfortunately, there is no obliging Maxwell demon at our disposal, nor is it possible to create one. The scale on which we are built is far too gross for us to be able to construct apparatus suitable for manipulation by a Maxwell demon, even if one existed. And, if we had such an apparatus, we ourselves would not be able to operate it in place of a Maxwell demon, because our biological processes of perception and volition involve enormous numbers of molecules; it is inconceivable that we should be able to operate on individuals among the molecules of gas. We belong to the macroworld, and it appears that full entry into and effectiveness within the microworld is forever denied us.

Within this microworld the second law of thermodynamics no longer applies, even without the intervention of a Maxwell demon. If we consider sufficiently small volumes within a gas, it is obvious from the kinetic theory that the average kinetic energy of the molecules within these cannot be strictly constant but must fluctuate about a mean, with the fluctuations occurring on a very small time scale. Between two such volumes, adjacent to each other, there must occur, spontaneously, just such interchanges of energy, and corresponding changes of temperature, as we imagined Maxwell's demon to bring about between much larger volumes of gas. In this connection it must be borne in mind that the smaller the volume of gas considered and so the number of molecules contained within it, the more the concept of temperature tends to lose its meaning. The concept of temperature is a statistical one, and in the limit, where we are contemplating only a single molecule, this concept finally loses its meaning altogether.

Another example of the nonapplicability of the second law to the microworld which we can see before our very eyes is that known as "Brownian motion." This is a phenomenon first noticed by the botanist Robert Brown, in 1827, while observing aqueous suspensions of fine inanimate spores under a high-power microscope. Brown saw these spores dancing about in the wildest fashion. This irregular movement may be observed microscopically in the case of any sufficiently small particles in suspension in a liquid or gas, such as colloidal particles in a liquid or smoke particles in a gas (see Fig. 18.2). These particles behave like giant molecules, and observation of their motions reveals that they have a mean translational kinetic energy equal to that of the molecules of the liquid or gas in which they are suspended. This is an interesting extension of the law of equipartition of energies considered in Chapter 15. The molecular explanation of this movement is fairly obvious. It is due simply to the fact that on any particular side of such a minute particle the rate of bombardment by the molecules varies with time according to a statistical law, so that during any sufficiently short time interval the force due to this molecular bombardment is likely to be greater on one side than on the other, and the particle will be impelled in the corresponding direction. The direction of the resultant force varies in a random manner from one moment to the next; hence the irregular movement.

This Brownian motion of a small suspended particle is of special interest when considered from the point of view of the second law; it represents a continual violation of this law within the microworld to which it belongs. For, every time the kinetic energy of the particle increases, this occurs at the expense of the kinetic energies of the molecules of the liquid or gas in which it is suspended, that is, of the heat content of this liquid or gas.

We see, then, that the second law of thermodynamics is, essentially, a statistical law, applying only to large assemblages of

FIGURE 18.2 Typical path traversed by a suspended particle in Brownian motion.

molecules. Viewed from this angle, the impossibility of converting heat to macroscopic mechanical energy without leaving external changes corresponds to the impossibility of converting chaos into order. Strictly speaking, it is *not impossible* for order to emerge momentarily from chaos, but—for large assemblages of particles— it is overwhelmingly improbable. And it is certainly something we cannot bring about at will.

From the above considerations we should expect the second law to be derivable from a statistical investigation of the behavior of large assemblages of particles, the ordinary laws of mechanics applying to the individual particles. This is found indeed to be the case; it is shown in more advanced studies that thermodynamics may be considered simply as a branch of statistical mechanics,* the study of the behavior of large assemblages of particles.

Reversible and Irreversible Processes

If, by some means or other, it is possible to restore completely the situation existing prior to the occurrence of a particular process, this process is said to be **reversible.** Here restoration of the status quo is to be understood as applying not merely to the body or system immediately under consideration but to all others as well. If such restoration is not possible, the process is classed as **irreversible.** Both reversible and irreversible processes are of considerable importance in thermodynamics.

As an example of a process which, at least under certain conditions, is reversible, let us consider the expansion of a gas in which work is performed while heat is supplied to the gas from a source in thermal communication with it. An obvious condition for reversibility of this process is that the piston moves along the cylinder not only without leakage but also without friction or viscous resistance in the lubricant. Although absence of leakage and of frictional or viscous resistance could, in practice, never be realized, even separately, let alone simultaneously, this idealization is nevertheless permissible in a thermodynamical discussion.

Another condition for reversibility is that the heat source shall not at any stage be at a higher temperature than the gas, for heat will not spontaneously flow "uphill," as it would have to do

---

* We are here referring exclusively to *classical* statistical mechanics, in which no quantum theory is invoked.

if the process were reversed. It may be wondered how, if there is no temperature difference between the source and the gas, heat can possibly flow from one to the other. Obviously, it cannot. However, there is no lower limit to the temperature difference that must exist between the two to produce a heat flow, and this could in principle be made infinitesimally small, in which case, incidentally, the time occupied by the expansion would become infinitely long. Every stage of the expansion would then represent an infinitely close approach to a condition of equilibrium. By reducing the temperature difference toward zero, the condition of true reversibility may be approached as closely as we please. Let us, then, agree along these lines to regard such an expansion, occurring when there is an infinitesimal temperature difference between the source and the gas, as a process which for the purposes of our present discussion may be accepted as reversible.

An important special case of this process is that in which the heat capacity of the source is zero, that is, in effect, that in which the heat converted to another form of energy is derived exclusively from the gas itself. In this case the expansion of the gas would be adiabatic. An equally important special case is that in which the heat capacity of the source is infinite. Here the expansion would be isothermal, and the whole of the heat converted would come from the external source.

Let us now consider two examples of irreversible processes, for example the production of heat by friction and the unresisted expansion of a gas. By invoking the second law of thermodynamics it may be shown that these are irreversible processes, according to our definition of reversibility. In discussions of this kind it is permissible to make any reasonable simplifying assumptions that suit our convenience, and also to bring to our aid any devices we wish, provided we finally leave these in the same condition in which they were at the beginning.

The production of heat by friction is an example of the conversion of mechanical energy, say, to heat, which raises the temperature of the bodies concerned above that of the surroundings. Let us assume that the heat capacity of the surroundings is infinite, and that we make use of a gas, initially at the temperature of the surroundings, which we take through a succession of Carnot cycles. Each one of these begins with an adiabatic compression that raises the temperature of the gas from that of the surroundings to that of the bodies concerned in the frictional production of heat. We make these bodies function as the heat source in each cycle, while the surroundings function as the sink. By making the quantity of heat

extracted from the source during each isothermal expansion sufficiently small, we may provide for virtual constancy of the corresponding temperature of the source, so that the expansion really *is*,
practically speaking, isothermal, and the cycle is correspondingly
a true Carnot cycle, as previously discussed. By employing a sufficiently large number of such Carnot cycles, each working under
ideal conditions between the temperature of the source as left at
the end of the previous cycle and that of the sink, we may reduce
the bodies in which the heat was produced frictionally to their
original temperature, so restoring the status quo as far as *these* are
concerned.

The whole procedure so far discussed is reversible, because
each stage of a Carnot cycle, and therefore each Carnot cycle as a
whole, is reversible. And the result of this procedure is, in addition
to the restoration of the bodies to their original temperature, the
conversion of some of the heat that was produced in them to another
form of energy—the original form, let us say—and the transfer of
the remainder to the surroundings. *Complete* restoration of the
initial conditions would now require, in addition, the conversion of
this heat, residing in the surroundings, to the original form of
energy. But precisely this is what the second law states to be impossible. Hence it must be concluded that the production of heat by
friction is a thermodynamically irreversible process.

Let us now consider the unresisted expansion of a gas, as
investigated by Joule. Let us suppose we have a heat reservoir at
our disposal, at the same temperature as the gas and of infinite
thermal capacity. We could begin by compressing the gas isothermally while it is in thermal communication with this reservoir, until
it is restored to its original volume. The gas has now been brought
by a reversible process—the isothermal compression—to its original
condition. To completely reverse the unresisted expansion, that is,
to restore *everything* to its original condition, it would now be necessary to extract the heat that has been added to the reservoir during
the compression of the gas and to convert this into the kind of
energy that was used to do work on the gas in compressing it. But
this, according to the second law, is impossible. It must be concluded, therefore, that the unresisted expansion of a gas is an
irreversible process.

Carnot's Principle

It will be recalled that in dealing with Carnot's cycle in
Chapter 17 we were concerned with the quantity $(Q_1 - Q_2)/Q_1$,

460

which we defined as the "thermodynamic efficiency" of the cycle. The use of this expression is analogous to the use of the term "efficiency" in mechanics, where it is applied to the ratio of the work done by a machine to that done on it. In neither case are we concerned with the efficiency of an *energy conversion*, which, by the law of conservation, is always 100 percent. Instead, in the case of the machine, we are interested in the ratio of the energy made available to an external system by the machine (the work done by it) to that made available to the machine (the work done on it). The difference between these energy items appears as unavailable heat in the bearings, and so on. Similarly, in the present case, the energy Q_1 is made available to the heat engine, but not all of this is utilized in the performance of work, the portion Q_2 of it being rejected to the sink as unavailable heat. It will be convenient, in what follows, generally to refer to the thermodynamic efficiency of a heat engine simply as its "efficiency."

By invoking the second law of thermodynamics it will now be shown that no heat engine that takes in heat at a temperature T_1, works in a cycle, and gives out heat at a lower temperature T_2, can be more efficient than a reversible engine working between the same temperatures; and, incidentally, it will also be shown that all reversible heat engines working between these limits have the efficiency $(T_1 - T_2)/T_1$. This theorem was first propounded and proved by Carnot, and is known, accordingly, as **Carnot's principle.**

Let a reversible engine A take in the quantity of heat Q_1 per cycle entirely at the temperature T_1 and reject heat Q_2 entirely at the lower temperature T_2, so that the work W it performs per cycle is $Q_1 - Q_2$ and its efficiency η is $(Q_1 - Q_2)/Q_1$. Let us compare with the performance of this engine that of some other, B, which takes in the heat Q_1' per cycle at T_1 and rejects Q_2' at T_2, so that the work W' it performs per cycle when operating between these temperature limits is $Q_1' - Q_2'$ and its efficiency η' is $(Q_1' - Q_2')/Q_1'$.

Reversibility of the engine A means that when the cycle of operations is executed in the reverse direction the engine takes in during each cycle the quantity Q_2 of heat at the lower temperature T_2 and rejects the quantity Q_1 at T_1, while the amount of work $Q_1 - Q_2$ is done *on* the engine. Apart from the associated changes in the heat contents of the source and sink and the energy changes in the system on or by which work is performed, as the case may be, no changes are left in any other body.

Let this engine A be driven in reverse by the other engine B. Then, for the two engines so coupled, B running directly and A in reverse, we have, for each complete cycle, the following "balance sheet":

Engine(s):	A	B	$A + B$
Source gain:	Q_1	$-Q_1'$	$Q_1 - Q_1'$
Sink gain:	$-Q_2$	Q_2'	$Q_2' - Q_2$
Work done:	$-(Q_1 - Q_2) = -W$	$Q_1' - Q_2' = W'$	$W' - W$

If possible, let the efficiency of B be greater than that of A, so that

$$\frac{Q_1' - Q_2'}{Q_1'} > \frac{Q_1 - Q_2}{Q_1}.$$

There are two alternative methods of testing whether this is possible, according to which statement of the second law we use.

1. Let the engines be adjusted, for example with regard to size or quantity of working substance used, so that the source neither gains nor loses heat; that is, $Q_1' = Q_1$. Then the net work per cycle performed by the combination, $W' - W$, which is equal to $(Q_1' - Q_2') - (Q_1 - Q_2)$, becomes simply $Q_2 - Q_2'$, and this as we see from the balance sheet is equal to the sink loss. According to the assumptions made—that $\eta' > \eta$ and $Q_1' = Q_1$—this quantity $Q_2 - Q_2'$ must be positive. This means that the combination performs work simply by extracting heat from the sink and converting it to some other form of energy. According to our first statement of the second law, however, this is impossible. Hence B cannot be more efficient than A.

2. Let the engines be so adjusted that the net work performed by the combination is zero; that is, $Q_1' - Q_2' = Q_1 - Q_2$. Then if η' is greater than η, Q_1' must be less than Q_1, or the heat gained by the source, $Q_1 - Q_1'$, must be positive. Correspondingly, the heat gained by the sink, $Q_2' - Q_2$, which, in this case must be the same as $Q_1' - Q_1$, is negative. That is, the combination, without any performance of work, simply takes in heat at the lower temperature T_2 and gives it out at the higher temperature T_1. But this is what is stated in our second version of the second law to be impossible. Hence, again, it must be concluded that B cannot be more efficient than A; **no engine, working between given temperature limits, can be more efficient than a reversible engine working between these limits.**

Now let us apply this result to the special case where B, as well as A, is reversible. We now have the situation that not only can B not be more efficient than A, but also, because B is reversible, A cannot be more efficient than B. But if neither can be more efficient than the other, then, necessarily, both must be equally efficient; **all reversible engines working between given temperature limits have the same efficiency.**

In Chapter 17 we obtained an expression for the efficiency of one reversible engine—one working in a Carnot cycle and using a perfect gas as its working substance. The efficiency of this engine for temperature limits T_1 and T_2 was $(T_1 - T_2)/T_1$. Hence we must conclude that this is also the efficiency of all reversible engines working between these limits.

Efficiencies of Practical Heat Engines

Strictly speaking, practical heat engines do not always take a sample of the working substance through a complete cycle. Let us, for example, consider the action of a reciprocating steam engine. Cold water is fed into the boiler, it is there heated and converted into steam, and, communication having been established between the boiler and the cylinder of the engine, some of the steam enters the latter, driving the piston forward through a portion of its working stroke. Communication with the boiler is then cut off, and the steam expands approximately adiabatically through the remainder of the stroke. Finally, during the return stroke, the steam is rejected to a condenser whose temperature is not usually quite as low as that of the water fed in. At this stage the condensed water is in most cases simply rejected, being replaced later by a corresponding amount at a lower temperature. Thermodynamically, however, this is equivalent to feeding the condensed water itself back into the boiler after it has been cooled further, outside the engine, so completing its cycle. This last stage of cooling would be irreversible.

There are several other irreversible features in the cycle of operations of actual heat engines. Let us again consider the reciprocating steam engine as an example. In this, as in all other engines, a certain amount of dissipation of energy to unavailable heat must occur in the various working parts because of viscous loss in the lubricant and friction. Also, there is always some slight leakage of steam past the piston—an item of unresisted expansion. A much more important item of unresisted expansion arises because of the impracticability of completing the expansion right down to the pressure of the condenser; unresisted expansion occurs on establishing communication between the expanded steam and the condenser at the end of the working stroke. Again, during the latter part of the working stroke, in which the steam expands approximately adiabatically, the temperature falls correspondingly, and so the

463

cylinder walls are cooled. Consequently, at the beginning of the next stroke, when more steam is admitted, this is cooled on coming in contact with the walls. This cooling is irreversible. Also, there is always some heat loss from the boiler, the cylinder, and the pipe connecting these. Finally, there is some loss of useful energy in the viscous flow of the steam in the supply pipe from the boiler to the cylinder and, during the return stroke, in the pipe connecting the cylinder to the condenser. These are the most obvious items of energy loss, but there are also other, less obvious ones. Their combined effect is to reduce the thermodynamic efficiency to a value far below that which a reversible heat engine working between the same temperature limits would have.

Actually, reciprocating steam engines are notoriously inefficient—so much so that they are now obsolescent. Their upper limit of thermodynamic efficiency is only of the order of 10 percent. These engines are rapidly being superseded by turbines, whose efficiencies are much higher. Even in these, losses are considerable, so that the efficiencies actually attained fall substantially below those of corresponding reversible heat engines. About the best performance of a steam turbine so far attained is that of a huge engine developing 500,000 hp, fed with steam at a pressure of about 170 atm superheated to 565°C (838°K), and having a condenser working at an effective temperature of about 26°C (about 300°K). This has a thermodynamic efficiency of about 35 percent. Although this figure is high, as thermodynamic efficiencies go, it is only slightly more than half that which a reversible heat engine working between the same temperature limits would have, this being about 64 percent.

The most efficient heat engines currently in use are highly supercharged diesel engines; with these, thermodynamic efficiencies of as much as 43 percent have been achieved. Practically all diesel engines have efficiencies higher than 30 percent. The efficiencies of automobile engines using gasoline as fuel are somewhat lower—of the order of 25 percent.

Thermodynamic Scale of Temperature

In Chapter 13 we considered the general principle of the definition of a temperature scale, temperature variations being defined as proportional to the variations of some selected property of a chosen material or system. It now appears that the definition of a temperature scale need not necessarily be based on any property of any particular material; it could, instead, be based on the efficiencies of reversible heat engines working between different temperature limits. For a given pair of temperature levels there is, according to Carnot's principle, a unique value of the efficiency, $(Q_1 - Q_2)/Q_1$, of a reversible heat engine taking in heat at the upper level and rejecting it at the lower, this value being independent of the particular reversible cycle of operations or of the working substance used. And, as $(Q_1 - Q_2)/Q_1$ is uniquely determined by these temperature levels, so must be the value of the ratio Q_1/Q_2. Accordingly, it becomes possible to define the relation between the two temperatures in terms of Q_1/Q_2.

This possibility was first recognized by Lord Kelvin, who also realized that the definition might be formulated in a number of different ways. If the upper and lower temperatures are denoted by Θ_1 and Θ_2, respectively, we might define Θ_1/Θ_2 as equal to Q_1/Q_2, or to $(Q_1/Q_2)^2$, or to $\log Q_1/\log Q_2$, and so on. Of all these possible ways of defining temperature, the one actually chosen by Kelvin on the grounds of usefulness was that corresponding to the equation

$$\frac{\Theta_1}{\Theta_2} = \frac{Q_1}{Q_2}. \qquad (18.1)$$

According to this equation, for a given Q_2, equal increments of Q_1 would define equal increments of temperature. The scale of temperature so defined is known as the **absolute thermodynamic scale.**

At the end of Chapter 17 we considered one particular reversible heat engine, and we saw that if T_1 and T_2 denote the absolute temperatures on the perfect-gas scale of the source and sink, respectively, then for this engine

$$\frac{Q_1}{Q_2} = \frac{T_1}{T_2}. \qquad (18.2)$$

But, according to Carnot's principle, this must be true not only for the particular engine considered but for all reversible heat engines working between the same temperature limits. Hence, comparing Eqs. (18.1) and (18.2), we see that Θ_1/Θ_2 may be identified with T_1/T_2. The absolute perfect-gas scale and the absolute thermodynamic scale are identical; and both, with their degrees adjusted to correspond in magnitude to those of the Celsius scale, are also known as the Kelvin scale.

Although the possibility of defining a temperature scale without reference to the physical behavior of any particular substance is of considerable theoretical interest, it must not be imagined that there is the remotest prospect of our being able to establish this scale in practice in terms of its definition, that is, of measuring various melting points, boiling points, and so on, by using reversible heat engines. We saw in our discussion of actual heat engines how inefficient even the best of them are compared with a reversible engine working between the same temperature limits. This being so, the only means we have at present of realizing the Kelvin scale in practice is that provided, either directly or indirectly, by gas thermometry.

Clausius–Clapeyron Equation

By applying Carnot's principle to an appropriate system it is possible to derive an equation expressing the rate of change of the vapor pressure of a liquid with temperature in terms of the temperature, the latent heat of vaporization at this temperature, and the corresponding difference in volume per unit mass between the vapor and the liquid. This equation, from which important consequences can be deduced, was first derived by Clapeyron, by a method which, although not quite rigorous, is simple and straightforward. This derivation will be given below. A rigorous proof of the same equation was subsequently obtained by Clausius.

Let us consider again a system of isothermals such as that shown in Fig. 16.9, with particular reference to the horizontal portions AB and CD of two neighboring isothermals corresponding to the temperatures T_1 and T_2, as indicated in Fig. 18.3. Let M be a point in CD such that an adiabatic compression of the corresponding liquid-vapor mixture will just condense the vapor in it and give a liquid at temperature T_1, as represented by point A. And let N be a point such that an adiabatic expansion of the vapor at temperature T_1, as indicated by point B, will give the corresponding liquid-vapor mixture at temperature T_2. Then the closed figure $ABNM$ will represent a Carnot cycle, which is obviously reversible and whose thermodynamic efficiency is therefore $(T_1 - T_2)/T_1$.

We now have to make use of the fact that this efficiency is the ratio of the work performed by the working substance during a cycle of operations to the heat supplied at the higher temperature. To do this we shall need an expression for the work performed. As a preliminary to finding this, it will be shown how the work performed during any complete cycle of operations may be calculated from the corresponding p–V diagram.

Let the closed curve shown in Fig. 18.4 be the p–V diagram for a certain cycle of operations, and let us consider the two work items associated with small volume changes between V_1 and V_2,

466

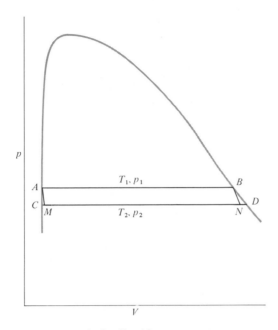

FIGURE 18.3 Carnot cycle for liquid-vapor system.

these volumes being represented by the two adjacent vertical lines. If the difference between these volumes is sufficiently small, the pressure changes occurring during the transitions from one to the other will be small enough to be neglected, and we may assign a particular value of the pressure to each transition. Let the pressure prevailing during the transition from V_1 to V_2 be p_{12}, while that corresponding to the transition from V_2 to V_1 is p_{21}. The work performed by the system during the former transition is $p_{12}(V_2 - V_1)$ and that performed on the system during the latter transition

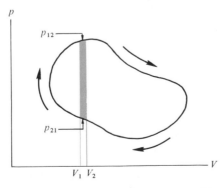

FIGURE 18.4 Work done during a cycle represented by area enclosed by p-V curve.

is $p_{21}(V_2 - V_1)$. The net work performed by the system during these two stages is therefore $(p_{12} - p_{21})(V_2 - V_1)$. This is represented by the shaded area of the corresponding strip in the figure. The same sort of reasoning will apply to each of the narrow strips into which the diagram can be divided. Hence the net work performed by the system during a complete cycle of operations is represented by the sum of the corresponding shaded portions of all these strips, that is, by the area enclosed by the curve.

Let us now apply this to the particular case of the Carnot cycle represented in Fig. 18.3. Here the work performed is represented by the area of $ABNM$. If the two temperatures are very close together, so that the distance between the horizontal lines AB and CD is correspondingly small, we need not distinguish between the lengths of AB and MN. Under these conditions we may, with negligible error, set the work performed during a cycle as equal to $(V_B - V_A)(p_1 - p_2)$, where the symbols have their obvious meanings. Accordingly, if Q_1 is the heat supplied at the temperature T_1, we have

$$\frac{(V_B - V_A)(p_1 - p_2)}{Q_1} = \frac{T_1 - T_2}{T_1}.$$

For simplicity, let us suppose that the sample of working substance to which Fig. 18.3 refers has unit mass. Then Q_1 is equal to L, the latent heat of vaporization at this level of temperature. Also, V_A is the specific volume (volume per unit mass) of the liquid and V_B that of the saturated vapor. Accordingly, it will be convenient to substitute the symbol V_l for V_A and V_v for V_B. Finally, let us write T instead of T_1 and dT and dp in place of $T_1 - T_2$ and $p_1 - p_2$, respectively; these substitutions are obviously justified mathematically when we proceed to the limit of smallness of these quantities. Thus our equation above becomes

$$\frac{(V_v - V_l)\, dp}{L} = \frac{dT}{T},$$

or

$$\frac{dp}{dT} = \frac{L}{(V_v - V_l)T}. \tag{18.3}$$

This is the Clausius–Clapeyron equation, usually referred to simply as **Clapeyron's equation.**

As a test of the validity of this equation, let us calculate the latent heat of vaporization of steam at 100°C. At this temperature the specific volume of steam is 1671 cm³/g and that of water is

1 cm³/g,* and so $V_v - V_l = 1670$ cm³. The rate of increase of vapor pressure with temperature at 100°C is 2.71 cm Hg per degree, or 2.71 × 13.6 × 980.6 dyn/cm²/deg. Hence, substituting these values in Clapeyron's equation, and also the value of T, 373.16°C, and dividing by 4.186 × 10⁷ to convert from ergs per gram to calories per gram, we obtain for L the value

$$\frac{373.16 \times 1670 \times 2.71 \times 13.6 \times 980.6}{4.186 \times 10^7},$$

or 538 cal/g. The value found by direct experiment is 539.5 cal/g. In view of the various experimental uncertainties involved in the comparison, the agreement between the two values must be regarded as eminently satisfactory.

Variation of Vapor Pressure with Temperature

If we treat V_l as negligibly small compared with V_v and assume for the latter the value it would have if the vapor were a perfect gas,

$$V_v = \frac{RT}{Mp},$$

then Clapeyron's equation takes the form

$$\frac{1}{p}\frac{dp}{dT} = \frac{ML}{RT^2}. \qquad (18.4)$$

Before we can integrate this, we must first consider how the latent heat L is made up. A part of this, which we may call the "external latent heat," is the work of expansion pV_v, which, being equal to RT/M, varies linearly with T. What remains we may refer to as the "internal latent heat" L_i. This is not necessarily a constant; it may well vary with temperature. In the absence of any precise theory of this quantity, it will be useful to consider the temperature variation of such experimental values of L as are available and of the corresponding values of L_i. Values of L for water measured at a series of temperatures are given in Table 18.1, together with the corresponding values of L_i, assumed to be equal to $L - RT/M$.

We see that L varies with temperature, not merely because of the term RT/M, but additionally because of a temperature variation

---

* The specific volume of steam is a little less than the value a perfect gas of molecular weight 18 would have, this being 1701 cm³.

TABLE 18.1   Variation of Latent Heat of Vaporization
of Water with Temperature

Temp. ($^\circ$K)	L (cal/g)	RT/M (cal/g)	L_i (cal/g)
273	597	30	567
293	585	32	553
313	574	34	540
333	563	37	526
353	551	39	512
373	539	41	498
393	525	43	482
413	511	45	466
433	497	48	449
453	482	50	432

of L_i. The values of both L and L_i are plotted against T in Fig. 18.5. Although these plots are not accurately linear, they are nearly so, and we may therefore write, as an acceptable approximation to the variation of L with T,

$$L = L_0 + cT,$$

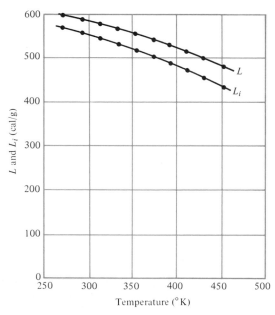

FIGURE 18.5   Temperature variation of total and internal latent heats of vaporization of water.

where L_0 and c are constants, the latter having a negative value in the present case. Substituting this in Eq. (18.4) we obtain

$$\frac{1}{p}\frac{dp}{dT} = \frac{ML_0}{RT^2} + c\frac{M}{RT},$$

or, writing n for cM/R,

$$\frac{1}{p}\frac{dp}{dT} = \frac{ML_0}{RT^2} + \frac{n}{T}. \tag{18.5}$$

The numerical value of c is -0.58, and the corresponding value of n is -5.2.

On integrating Eq. (18.5) we obtain the equation

$$\ln p - n \ln T = -\frac{ML_0}{RT} + \ln \alpha, \tag{18.6}$$

where $\ln \alpha$ is a constant of integration.

L_0 is the temperature-independent part of the latent heat per gram, so ML_0 is the corresponding quantity per gram-molecule and ML_0/N that per molecule, N being Avogadro's number. Hence, writing ML_0/R in the form $(ML_0/N)/(N/R)$, we see that this quantity is the same as λ/k, where λ is the temperature-independent part of the latent heat per molecule and k is Boltzmann's constant. Accordingly, we may write Eq. (18.6) in the form

$$\ln p - n \ln T = -\frac{\lambda}{kT} + \ln \alpha,$$

or

$$p = \alpha T^n e^{-\lambda/kT}. \tag{18.7}$$

This, the so-called **vapor-pressure equation,** corresponds to Eq. (16.2), the constant β in that equation corresponding to λ/k in our present discussion. As was indicated in Chapter 16, equations of this general type are of far-reaching importance in physics. In the vapor-pressure equation particular significance attaches to the quantity λ. This, or something approximating closely to it, is the activation energy of the evaporation process, the minimum energy a molecule in the liquid must have to qualify for escape.

Effect of Pressure on the Freezing Point

Clapeyron's equation applies not only to a liquid–vapor system but also to one consisting of a solid and a liquid. Thus, let us investigate the case of ice and water, imagining the system to be taken through a Carnot cycle. This may be done as follows. Ice, at a temperature T_1 and pressure p_1, is first melted, the corresponding heat supplied, Q_1, being equal to L per unit mass, where L is the latent heat of fusion of ice. The melting is actually not carried quite to completion, but enough ice is left so that on adiabatic compression this will just be converted to water, with a fall of temperature to T_2. If T_1 and T_2 are close enough together, the amount of ice that must be left at the end of the isothermal (melting) stage becomes negligibly small. At the lower temperature T_2 and the corresponding higher pressure p_2, the water is then frozen—again not quite completely. Finally, an adiabatic expansion just completes the freezing and restores the original pressure p_1 and temperature T_1. This cycle of operations is represented in the form of a p–V diagram in Fig. 18.6.

If, in this cycle, we use 1 g of H_2O and we write V_l and V_s for the specific volumes of the liquid and solid, respectively, we have for the work performed during the cycle $(V_s - V_l)(p_2 - p_1)$, and the heat supplied at the higher temperature, T_1, is L. Hence we have

$$\frac{(V_s - V_l)(p_2 - p_1)}{L} = \frac{T_1 - T_2}{T_1},$$

and if we proceed to the limit of smallness of $T_1 - T_2$ and $p_2 - p_1$, we may write this in the form

$$\frac{dp}{dT} = -\frac{L}{(V_s - V_l)T},$$

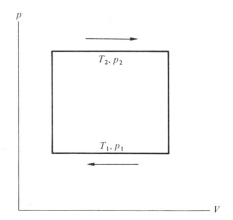

FIGURE 18.6   Carnot cycle for ice-water system.

which corresponds to Clapeyron's equation for a liquid–vapor system, or, alternatively,

$$\frac{dT}{dp} = - \frac{(V_s - V_l)T}{L}, \qquad (18.8)$$

the minus sign appearing on account of $p_1 - p_2$ being of opposite sign to $T_1 - T_2$.

Let us now, on the basis of this equation, investigate numerically the rate of change of the freezing point of water with pressure. The specific volumes of ice and water at 0°C are 1.0905 and 1.0002 cm³/g, respectively; thus $V_s - V_l$ is equal to 0.0903 cm³/g. And the latent heat of fusion of ice is 79.6 cal/g or 3.332×10^9 ergs/g. Hence, according to Eq. (18.8), the depression of the freezing point per atmosphere (1.0132×10^6 dyn/cm²) of pressure should be

$$\frac{1.0132 \times 10^6 \times 0.0903 \times 273.16}{3.332 \times 10^9},$$

or 0.00756 deg. This is in excellent agreement with the measured value of 0.0075 deg/atm.

It should be noted that dT/dp is negative only in such a case as that of ice and water, where the solid occupies a larger volume than the liquid. This is somewhat unusual; most liquids contract on freezing. In cases where V_l is greater than V_s, dT/dp must be positive; that is, the application of pressure raises the freezing point. This is evident from Eq. (18.8). In these cases the Carnot-cycle $p–V$ diagram would be as in Fig. 18.6; only the labels of the top and bottom horizontal lines would have to be interchanged, and the upper isothermal would correspond to melting and the lower to freezing.

Surface Tension and Surface Energy

In Chapter 12 it was intimated that the numerical values of surface tension and surface energy per unit area are not quite the same. It will now be shown on the basis of the second law of thermodynamics how these two quantities are related.

We could, at least in principle, have a heat engine in which the motive force is provided by surface tension. Let us, then, imagine such an engine, in which the working system is a liquid film and in which this film is taken through a Carnot cycle of operations.

473

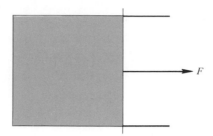

FIGURE 18.7   Mechanism of surface-tension heat engine.

Let us suppose a thin film of a liquid is formed between a rectangular wire frame and a wire that can slide frictionlessly along it in the manner indicated in Fig. 18.7. The area of this film may be changed either isothermally or adiabatically. As was explained in Chapter 12, when the film is stretched isothermally it is necessary to supply heat to it to keep its temperature constant. If, instead, it is stretched adiabatically, that is, this heat is withheld, the temperature must fall. Conversely, during an isothermal contraction it would be necessary to withdraw heat from the film, and an adiabatic contraction would give a rise in temperature.

We may now take the film through a Carnot cycle in the following manner, which is represented graphically as a σ–A diagram in Fig. 18.8, A denoting the total surface area of the two sides of the film. Beginning with the film as indicated by point M, at temperature T_2, it is allowed to contract isothermally, so doing work equal to $(A_M - A_N)\sigma_2$, where A_M and A_N are the initial and final areas and σ_2 is the surface tension at temperature T_2. During this stage the requisite amount of heat must be withdrawn from the film to keep the temperature constant, and at the end of it the condition is as represented by point N. Next, the film is allowed to contract a little further, this time adiabatically, thereby causing it to rise in temperature to T_1 and bringing it to the condition indicated by

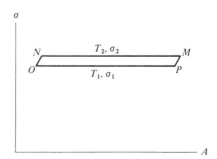

FIGURE 18.8   Carnot cycle for surface-tension heat engine.

point O. The film is now stretched isothermally at this temperature to such a stage (P) that a further small amount of stretching, carried out adiabatically, will cool the film to temperature T_2 and restore the original area, as indicated by point M. During this isothermal stretching the amount of work $(A_P - A_O)\sigma_1$ has to be done on the film, the symbols having their obvious meanings. During the whole cycle, the net work done by the film, apart from the items corresponding to the adiabatic stages, is $(A_M - A_N)\sigma_2 - (A_P - A_O)\sigma_1$. If the difference between the two temperatures is made sufficiently small, the work items associated with the adiabatic stages become correspondingly unimportant, and the value of $A_P - A_O$ approaches that of $A_M - A_N$. Hence, proceeding to the limit of smallness of $T_1 - T_2$ and $\sigma_2 - \sigma_1$, we may express the work performed during the cycle as $-(A_M - A_N)\, d\sigma^*$ and write our equation for the efficiency of this reversible heat engine in the form

$$- \frac{(A_M - A_N)\, d\sigma}{Q} - \frac{dT}{T},$$

where Q is the heat supplied during the isothermal expansion at the temperature T and $d\sigma/dT$ is the rate of increase of surface tension with temperature. This equation may be expressed somewhat more elegantly in the alternative form

$$q = -T \frac{d\sigma}{dT}, \tag{18.9}$$

where q denotes the heat supplied per unit increase in area.

During the stretching of the film its energy is obviously increased by the total energy supplied to it from outside. This is made up of two items: (1) the work done on the film, which necessarily represents an energy conversion or transfer, and (2) the heat supplied. As we saw in Chapter 12, the work per unit increase in area is numerically equal to the surface tension σ. And the corresponding amount of heat supplied is, as we have just seen, expressed by Eq. (18.9). Hence for the total surface energy per unit area, u, we have

$$u = \sigma - T \frac{d\sigma}{dT}. \tag{18.10}$$

As was explained in Chapter 12, the surface energy u per unit area is necessarily greater than σ. Hence, according to Eq. (18.10), $d\sigma/dT$ must be negative; and experimentally this is found indeed to be the case. The quantitative relation between u and σ for any given liquid is most conveniently found from the measured rate of varia-

---

* Here $d\sigma$ is intended to correspond to $\sigma_1 - \sigma_2$, not to $\sigma_2 - \sigma_1$.

tion of σ with T, using Eq. (18.10). Water in contact with air at 20°C, for example, has a surface tension of 72.75 dyn/cm, and $d\sigma/dT$ is −0.15 dyn/cm/deg. The corresponding surface energy per unit area at this temperature must therefore be 72.75 + (0.15 × 293), that is, 116.7 ergs/cm².

Problems

18.1   What power must the motor of a refrigerator have to convert 1000 kg of water at 20°C to ice at −10°C every 24 hr, with the refrigerator, having an efficiency of 70 percent of that of an ideal reversed heat engine, working between −15° and 40°C?

18.2   The latent heats of vaporization of mercury as found experimentally at a series of temperatures are given in the following table:

Temp. (°C)	0	60	120	180	240	300	360
Latent heat (cal/g)	73.43	72.93	72.45	71.98	71.52	71.06	70.64

Assuming mercury vapor to behave like a perfect gas, and also assuming a linear variation of latent heat with temperature approximated by the figures given in the table, derive a formula similar to Eq. (18.7) for the variation of the vapor pressure of mercury with temperature, finding the particular power of T in this formula, and also the numerical value of λ. Hence find the ratio of vapor pressures at 100° and 200°C, and compare this with the ratio obtained from Table 16.2.

18.3   Given that the melting point of lead is 327.4°C, the densities of solid and liquid lead at this temperature are 11.005 g/cm³ and 10.645 g/cm³, respectively, and the latent heat of fusion is 5.47 cal/g, find how the melting point is affected by an increase of pressure of 1 atm.

18.4   Show that while, on the one hand, above the temperature of maximum density of water (3.98°C), adiabatic and isothermal compressions of this liquid are accompanied by a rise in temperature and a release of heat, respectively, on the other hand, adiabatic and isothermal compressions carried out below the temperature of maximum density are accompanied by a fall in temperature and an absorption of heat, respectively.

18.5   The material of a wire has a positive coefficient of linear expansion. In principle, the wire, in a state of tension below the elastic limit, may be made to function as a heat engine, each cycle of which has four stages, one each of isothermal extension and contraction and one each of adiabatic extension and contraction. Describe this cycle in detail, and, basing your reasoning on Carnot's principle, show that (a) an adiabatic extension produces a drop in temperature, and (b) an isothermal extension is accompanied by an absorption of heat.

18.6   Find the radius of a drop of water such that its complete vaporization at 0°C involves neither cooling nor the supply of heat. (Assume surface tension and latent heat of vaporization to be independent of curvature of surface.)

18.7   If solar energy is to be used to power a heat engine there are two physical principles to contend with in obtaining optimum conversion of incident radiant energy to mechanical energy. One is the Carnot equation and the other is the radiation equation, which asserts that a hot body (the thermal reservoir of your heat engine in this case) will reradiate the incident energy at a rate proportional to the fourth power of the temperature. Discuss the interplay of these two effects. Actually the difficulties are partially alleviated by the fact that the reradiated energy is concentrated in a longer wavelength range which may be blocked by glass.

18.8   Imagine three very large heat reservoirs A, B, and C at temperatures $T_A > T_B > T_C$ (see the figure). (a) Suppose a Carnot engine

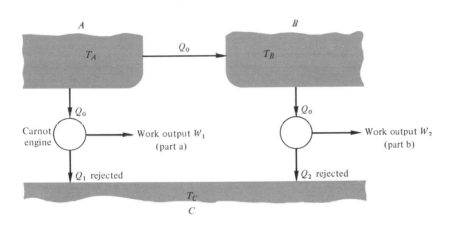

Figure for problem 18.8.

operates between A and C, withdrawing an amount of heat Q_0 from A. What is the fraction of this heat converted to mechanical (or other) energy? (b) Suppose that a conducting rod is connected between A and B and the same amount of heat Q_0 as in part (a) is conducted from A to B and then passed through a Carnot engine operating between B and C. What fraction of this heat Q_0 is now converted to another form of energy? (c) What is the difference between the energies converted in parts (a) and (b); that is, how much of the energy Q_0 became unavailable for conversion into another form in part (b) because it was first "degraded" to a lower temperature? (d) If we define the change in entropy S for the heat transfers in this problem to be $\delta S = \pm Q/T$, with the negative sign for heat removed and the positive for heat added, and where Q stands for the amount of heat added to or removed from a reservoir and T is the temperature of the reservoir, show that the loss in available energy between parts (a) and (b) may be written in terms of the entropy as $T_C \, \delta S$, where δS is the net increase in entropy incurred by conducting Q_0 from A to B before passing it into the Carnot engine.

SECTION VII

ELECTRIC CHARGES AND CURRENTS

19

BASIC PHENOMENA OF ELECTROSTATICS

An important question in physics concerns the nature of the forces that atoms and molecules exert on one another. A simple calculation shows that *gravitational* forces between atoms and molecules are totally inadequate to account for the observed phenomena of elasticity, surface tension, vapor pressure, and so on; they are many orders of magnitude too small. Even apart from the quantitative inadequacy of gravitational action, this could not be held responsible for the short-range mutual *repulsive* forces known to be exerted between these particles. Thus we need some other and much stronger force to account for atomic and molecular interactions. The well-known phenomena of attraction and repulsion between electric charges suggest that electrostatic forces might well supply this need. If interatomic and intermolecular forces are indeed electrical in nature, this raises the additional question of the distribution of charges within atoms and molecules. With this in mind as a fundamental requirement for a study of atomic structure to be undertaken later, and more immediately because of the great importance of electricity in its own right, we shall, in the present chapter, consider the basic phenomena of electrostatics, electricity at rest. Included in this study will be a discussion of electrostatic induction, in which the *motion* of electricity is involved. Notwithstanding this, electrostatic induction phenomena are traditionally included, together with the action of "electrostatic" generators to be considered in Chapter 21, under the general heading of electrostatics.

Electrification by Friction and by Induction

A large number of materials are known to us which, after having been rubbed against fur, silk, wool, and so on, will attract light bodies. This phenomenon may, for example, be observed very well with an ordinary Perspex ruler and a long strip of light paper. Let the paper be suspended vertically from one end by the experimenter's hand. Then if the ruler, after having been rubbed against a coat sleeve, is brought close to the paper, the latter will be attracted quite strongly to it, and will adhere to it, so that if the paper is now released by the hand holding it, the ruler alone will support its whole weight. Similar behavior may be observed with rubbed sealing wax, resin, ebonite, polystyrene, polylite, glass, and many other materials. However, no such effect is ordinarily observed when a stick of wood or a rod of metal is rubbed and brought close to a light body.

This behavior of rubbed Perspex, polystyrene, glass, and so on, suggests that after rubbing there is something on or in it that was not there before, and that it is to this "something" that the force of attraction exerted on the paper must be ascribed. To this hypothetical something we give the name **electricity,** and we refer to the body on or in which we suppose it to reside as **electrified, or electrically charged.**

There is an apparent asymmetry about this situation which we should not pass over lightly. It will be recalled that in our study of gravitation it was emphasized that gravitational attraction is always exerted mutually between two masses, and that to speak of an attract*ing* and an attract*ed* mass is merely to adopt one of two equally valid points of view. Here also, because action and reaction are equal and opposite, we may call either the polystyrene (say) or the paper the attracting body, the other being regarded as the attracted body. But one of these bodies (the polystyrene) is electrically charged, the other (the paper) apparently not. Thus it would appear that to the symmetry of the mutual exertion of force there does not correspond a symmetrical situation with regard to electrification.

This philosophical difficulty would be removed if it could be shown that while in the vicinity of the charged polystyrene rod the paper, too, acquires a charge. It is a simple matter to put this to the test, experimentally: The polystyrene is charged by rubbing, and the paper is held near it but not allowed to touch, so that there is not likely to be any *sharing* of charge by transfer. They may be kept apart by holding the polystyrene near a large sheet of glass or plastic while the paper is brought up close to it on the other side. Attraction will then be seen to occur through the glass or plastic just as it did previously through air. If the paper is now removed and immediately brought near another piece of paper, or to a table, or a finger, it will be found that attraction occurs between the two bodies, just as it did originally between the rubbed polystyrene and the paper. Electrification produced in a body, or part of a body, by bringing it near an already charged body is known as **electrification by induction,** and the charge so produced is called an **induced charge.**

The strip of paper may also be charged by rubbing, for example by pulling it quickly between a finger and thumb pressed against it (these must be dry), or by rubbing it with flannel or with silk.

Insulators and Conductors

The experiments described above are best performed on a dry day. Although polystyrene retains its charge quite well even in damp weather, paper is very sensitive to humidity, and under conditions of high humidity a charge induced on paper or produced on it by friction may not remain long enough to be observed.

The disappearance of a charge from an electrified body must presumably be due to leakage, either into the surrounding air or along the electrified body into the hand and thence away to the earth. Obviously leakage through air could not account for a *difference* in the rate of disappearance of charge from two electrified bodies; this difference must be due to a difference between the bodies themselves. Bodies such as polystyrene which, when held in one's hand in the manner indicated, retain their charge for a relatively long time are said to be good **insulators;** they do not readily conduct the charge away to other bodies with which they are in contact. On the other hand, a body on which a charge will remain only fleetingly under these conditions, or not at all, is known as a **conductor** of electricity. Thus paper, which would be classed as a reasonably good insulating material under conditions of low humidity, would rank as a conducting material in damp weather, when an appreciable amount of atmospheric moisture is absorbed by the fibers.

From the good insulating properties of polystyrene we must infer that the electricity produced by friction at the surface of this material could hardly be conducted through the body of the material to the inside. It would appear, then, that the charge resides where it is produced, *on* the electrified body, not *in* it. It will be shown in Chapter 20 that this is true also in the case of a charged conductor.

The fact that paper cannot be electrified by friction in damp weather, when it is not a sufficiently good insulator, suggests a possible reason for our failure to observe a charge on a metal or wooden rod after it has been rubbed; a charge might well be pro-

483

duced on this just as it is on polystyrene or sealing wax, but only to be conducted away to the hand, and thence to earth, as fast as it appears.

If this is indeed the explanation, it should be possible to prevent conduction to earth by providing the metal or wooden body with an insulating handle. If a metal sphere is mounted on a rod of polystyrene which is used as a handle, and the sphere is flicked with a piece of cloth, it, like rubbed polystyrene, Perspex, glass, and so on, may be observed to attract paper and other light bodies, showing that it has acquired a charge. It has now been established that *all* solid bodies, insulators and conductors alike, may be electrified by friction.

It is obviously impossible to make a real distinction between a rubbed and a rubbing body, and this suggests that not only do the polystyrene, metal sphere, and so on, acquire a charge, but so also must the cloth with which they have been rubbed. That this is indeed the case may easily be verified by mounting a piece of flannel on an insulating handle and rubbing another body, for example a rod of polystyrene, with this. It will now be found that the flannel, as well as the polystyrene, has become charged.

TRANSFER OF CHARGE From the fact that metals conduct electricity, it follows that if one of two metal bodies mounted on insulating handles is charged, for example by friction, then, on bringing these into contact some of the charge will be transferred to the second, hitherto uncharged body; the charge will now be shared between them. This may readily be verified experimentally, using the attractive force exerted on a paper strip as a criterion of electrification.

Classification of Charges

Our study of the exertion forces between charges, made with a body charged by friction and a paper strip on whose free end a charge is produced by induction, was useful only for the purposes of a preliminary orientation in this field; the possibilities of the method are too limited for a closer investigation. Thus, as will be evident later, the fact that an induced charge appears at all at the free end of the paper strip is due to the paper being sufficiently conducting for the charge to arrive there, and the possibility of observing the induced charge subsequently is due to the conductivity not being great enough for the charge to leak away before it can be observed.

A much more direct procedure is to produce charges by friction at the ends of two rods of good insulating material, where they can be relied upon to remain without appreciable leakage for a reasonable length of time, and to observe the mutual force between these. One rod may be suspended in a stirrup attached to a thread, in the manner indicated in Fig. 19.1, and the other held in the hand. The force the charged end of the latter exerts on the charged end of the suspended rod may then readily be observed.

If in such an experiment both rods are of the same material and their ends are similarly rubbed, it will be found in all cases that the force between the charged ends is one of repulsion. We have not yet considered the question of whether or not there might be more than one kind of charge, but in any case charges produced on the same kind of material in the same way must be presumed to be similar. It appears, then, that **like charges repel one another.**

Let us suppose that in one such experiment both rods are of polystyrene, one end of each having been rubbed with flannel. Now, with the same polystyrene rod suspended in the stirrup, let the rod held in the hand be, instead, ebonite, sealing wax, resin, and Perspex, all rubbed with flannel. In all these cases repulsion will be observed. On the principle that things that *do* the same *are* the same —similarity of behavior being the only criterion of sameness we can possibly have—it may be inferred that the charges produced on all these bodies are of the same kind. If it is felt that the observation of the mechanical action on rubbed polystyrene alone is not a sufficient test of sameness of behavior, other tests may be made, for example with suspended rods of rubbed ebonite, sealing wax,

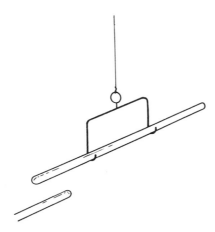

FIGURE 19.1 Simple device for demonstrating forces between electric charges.

and so on. If this is done, it will be found that the actions observed are in all cases mutually consistent. The early French investigator in this field, Charles du Fay (1698–1739), named electricity of this kind **resinous electricity.**

It is found experimentally that not all rubbed materials repel rubbed polystyrene, sealing wax, ebonite, and so on; some attract these. Thus glass* rubbed with silk, or polylite rubbed with flannel, is found to *attract* rubbed polystyrene, sealing wax, resin, and ebonite. Hence, because difference of behavior must mean difference in nature, the electricity produced on glass by rubbing it with silk or on polylite by rubbing it with flannel must be of a kind different from resinous electricity. Du Fay named electricity of this kind **vitreous electricity.**

It is found by such tests that all charges may be placed in one of two categories—two are necessary, and they are also sufficient. This division into two and only two categories is found to be completely satisfactory, in that all items are mutually consistent.

It is usually stated that there are two qualitative laws concerning the exertion of forces between charges. These are that (1) like charges repel one another, and (2) unlike charges attract. However, the second statement cannot be accepted as a law in the strict meaning of this term. In the performance of the experiments described above we do not know beforehand that the charges on, say, rubbed polystyrene and rubbed glass are different; hence the observation of the mutual attraction between these does not constitute a discovery that unlike charges attract. Actually, it is the other way around; difference in behavior, for example in the actions these exert on rubbed ebonite, is taken as the criterion of difference in kind; only after performing the experiment do we know that the two charges *are* different.

Attraction and repulsion are opposite effects, and this suggests that it might be useful, for the purposes of mathematical treatment, to label the charges giving rise to these effects in a particular case (for example, action on rubbed resin) positive and negative, instead of vitreous and resinous. This was first proposed by the eighteenth-century American investigator in this field, Benjamin Franklin (1706–1790). The choice of which label to attach to which kind of electricity was, of course, quite arbitrary—there was no argument

---

* Glass is a good insulator only under conditions of low humidity. However, a considerable reduction in the rate of leakage of charge that occurs in damp weather from the rubbed end of a glass rod may be effected by coating the other end, which is to serve as a handle, with shellac.

favoring either alternative over the other. As it happened, vitreous electricity was called positive and resinous negative.

Before we can discuss intelligently the usefulness or otherwise of these labels for mathematical purposes, we shall have to decide on what basis to call quantities of vitreous and resinous electricity equal in magnitude. The only reasonable definition of equality would appear to be equality of forces exerted on any given third charge held at a given distance from it. Let us suppose we have a long dumbbell, constructed of a thin insulating rod with metal spheres A and B mounted on the two ends, and that this is suspended horizontally on a torsion fiber, as indicated in Fig. 19.2. Let one of these spheres, B, be given a charge, and let another charged sphere, C, mounted on an insulating handle, be held in the same horizontal plane as A and B, and in such a position that the angle ABC is a right angle. Then the force that C exerts on B, whether one of attraction or of repulsion, will cause the suspension to twist, and the amount of twist for a given distance BC is noted. Let us suppose that the force is one of attraction. Then if another charge is substituted for C which repels B, but for the same distance away twists the suspension through the same angle (that is, exerts the same force on B), the two charges that have been held near B are defined as equal in magnitude, although different in kind.

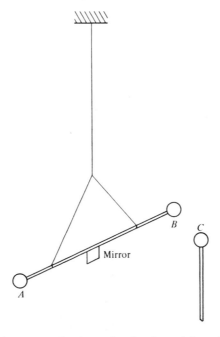

FIGURE 19.2 Apparatus for investigating law of force between charges.

Let us now investigate what happens when we charge a body by friction or by induction, considering first the case where the charge is produced frictionally. As we have already noted, a charge is produced on both bodies which rub against each other. Let a short sleeve of flannel be mounted on a long insulating handle, having an inside diameter such that it fits snugly over a long rod of polystyrene. A resinous charge may now be produced on the end of the polystyrene rod by engaging the flannel sleeve with it and twisting the polystyrene rod. Without withdrawing the rod, its presumably charged end and the flannel sleeve are placed together in position C of Fig. 19.2, for a test of the kind described above. No action is observed on B. Therefore, if, as we must presume, both the polystyrene and the flannel are charged, the charge on the flannel must be vitreous and equal in magnitude to the resinous charge on the polystyrene, the two forces exerted on B thus being equal and opposite, with zero resultant. That this is really the case can now be verified by withdrawing the polystyrene rod and testing the charge on its end and that on the flannel separately. The result obtained will be that the charge on the flannel is indeed vitreous and equal in magnitude to the resinous charge on the polystyrene.

Now let us produce electrification by induction, using a charge obtained frictionally on the end of a polystyrene rod as inducing charge. Two originally uncharged metal spheres, 1 and 2, are held in contact with one another a short distance from the charged end of the polystyrene rod as shown in Fig. 19.3, separated from one another while still near the polystyrene rod, and then removed from the neighborhood of the rod. On being tested, these spheres will now be found to have acquired induced charges, equal in magnitude but different in kind. The charge on 1 is vitreous, that is, different from that on the polystyrene rod, whereas that on 2 is

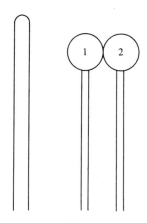

FIGURE 19.3 Separation of charges by induction.

resinous, that is, the same as that on the rod. If the spheres, far removed from the inducing charge, are now brought into contact with each other again, they will be completely discharged, reverting to their original condition.

These experiments and others indicate that all bodies ordinarily have within them equal quantities of vitreous and resinous electricity, which neutralize each other's external effects. It is further indicated that the process of charging, whatever its mechanism, is merely one of transfer of one or both kinds of charge (in opposite directions in the latter case) between two bodies or between parts of a single body. In this situation the mathematical labels suggested by Franklin are obviously much more convenient than du Fay's, and mathematically they are completely justified. They are, in fact, now universally employed, the terms "vitreous" and "resinous" having long since fallen into disuse.

Nature of the Conduction Process

As has been indicated above, there are three possible ways in which conduction of electricity might be supposed to occur: (1) Positive electricity might flow in one direction, negative electricity being immobile; (2) negative electricity might flow in the other direction, positive electricity being immobile; or (3) there might be a simultaneous flow of both kinds of electricity in opposite directions. A priori all three alternatives might be considered equally probable; each would account equally well for the observed mechanical effects, which are due simply to the excess of one kind of charge over the other in the various charged bodies.

Which of the possible alternatives actually applies will be investigated later for particular cases, for example that of conduction along metals. Meanwhile it will be convenient, in discussing electrostatic phenomena, to refer somewhat loosely to the movement of both positive and negative charges. It is clearly to be understood that this is done merely to avoid circumlocution, and that no reference to the actual physical process is intended. If, for example, we speak of a movement of positive charge in a certain direction, what we mean is either actual movement of positive electricity in this direction, or movement of negative electricity in the opposite direction, or, possibly, both of these together. In other words, we are speaking mathematically rather than physically.

Electroscope

An electroscope is a device designed to test for charge. In its usual form its construction is based on the observed mutual repulsion between like charges. Thus a long strip of metal foil, for example gold leaf or aluminum foil, folded at the center and mounted on an insulated metal support as indicated in Fig. 19.4, might be expected to serve as an indicator of charge. The idea would be, in the first instance, to try to transfer some of the charge from the body to be tested to the metal support. We should then expect this charge to spread over the conducting support and leaves, and the leaves, being similarly charged, would presumably repel each other and diverge.

Upon testing this idea we should find that it is, in fact, quite unnecessary to transfer any of the charge to the electroscope; mere approach of the charged body to the metal support causes the leaves to diverge. This is readily explained in terms of electrostatic induction, similar to that which was observed in the arrangement depicted in Fig. 19.3; a charge opposite to the inducing charge is induced on the metal support near which the inducing charge is held, and a like charge is induced on the leaves, which, being similarly charged, repel each other.

The instrument is found to be more sensitive if the top of the metal stem is provided with a disk, so presenting a larger area to the inducing charge, and its readings are more reproducible if the stem and leaves are surrounded by, although of course insulated from, an earthed metal casing. A window in this casing permits observation of the leaves. Finally, a single light flexible leaf adjacent to a stiff metal plate may replace the pair of leaves. This instrument is shown diagrammatically in Fig. 19.5, with the leaf diverged as when registering a charge.

An electroscope may be used not merely to detect *whether* a body brought near the disk is charged, but also to test the nature of such a charge. To do this it is necessary to give the electroscope an initial charge, for example by induction. Thus a rod of polystyrene, negatively charged by rubbing it with flannel, may be brought close enough to the disk to produce a moderate divergence

FIGURE 19.4 Primitive form of electroscope.

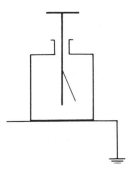

FIGURE 19.5   Electroscope.

of the leaf. With the polystyrene rod held in position, the disk is then momentarily touched with the finger; this causes the leaf to collapse. On now removing the charged rod, the leaf will be seen to diverge again, to the same extent as before. As we shall see later, the electroscope now has a positive charge, produced by induction. If, instead, a positively charged glass or polylite rod had been used as the inducing body, and the same procedure followed, the final charge on the electroscope would be negative. As may now easily be verified, if we approach a positively charged body to the disk of a positively charged electroscope, or a negatively charged body to the disk of a negatively charged electroscope, the leaf will diverge further. If, however, the charge on the body brought near the disk is of opposite sign to the charge on the electroscope, the leaf will fall. Actually the production of increased divergence is the only reliable test; for an uncharged conducting body brought near the disk will cause the leaf to fall, just as will the approach of a charge of opposite sign to that on the electroscope. The explanation of these phenomena will be evident after we have studied the next section.

Explanation of Electrostatic Induction

The experiment to which Fig. 19.3 refers suggests a simple qualitative explanation of electrostatic induction. When the spheres 1 and 2 are in contact with one another, they constitute what is in effect a single conducting body. This contains within itself equal quantities of positive and negative electricity. The negative charge on the polystyrene rod attracts the positive electricity toward itself and repels the negative electricity, causing sphere 1 to have excess positive and sphere 2 excess negative charge. These two kinds of excess charge in turn attract one another, and this mutual attraction sets a limit to the transfer of charge from one sphere to the other due to the negative charge on the polystyrene rod.

491

We may repeat this experiment in the somewhat different form indicated in Fig. 19.6. Here a long sausage-shaped metal body (a "conductor") on an insulating handle replaces the two spheres. The state of charge of any part of the conductor may now be tested with a proof plane. This consists of a small metal disk, for example a coin, mounted on an insulating handle. The disk is held flat against the part of the conductor to be tested, so becoming, in effect, part of its surface in this region. It is then removed, carrying a sample of the charge with it, and this is tested with an electroscope. This test will not only show the kind of charge in the region in question, or, as it is also called, the "polarity" of the charge, but it will also give some sort of impression of its surface density. Such a test, made in the case where the inducing charge is negative, will show the conductor to be charged as indicated in the figure. If, instead, the inducing charge were positive (for example if polylite rubbed with flannel were used), all the minuses in Fig. 19.6 would have to be changed to pluses, and vice versa, to show the situation we should then have.

Let us now suppose that with the rubbed polystyrene (say) still in position the conductor is touched momentarily with the finger. In this way the metal conductor, the body of the experimenter, and the whole earth on which the experimenter is standing momentarily become, in effect, a single conducting body, and negative electricity is able to flow away "to earth," while more positive electricity comes up from the earth to give a rather stronger positive charge than before. If, after the earth connection is broken, different parts of the surface of the metal conductor are again tested with the proof plane, it will be found that all parts of it are charged positively, but that the surface density is greatest at the end nearest the polystyrene rod.

In the light of these experiments it is a simple matter to explain the charging of an electroscope by induction.

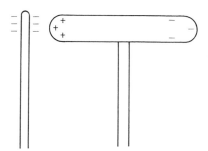

FIGURE 19.6 Production of opposite charges at ends of conductor by induction.

Coulomb's Law

The mutual force exerted between two charges must obviously depend on both the magnitudes of the two charges and the distance between them, and it might well also depend on the nature of the medium in which the charges are "immersed." Theoretically, the simplest kind of medium we can think of is a vacuum, and, fortunately from the point of view of convenience of experimenting, the forces exerted between charges in a vacuum are almost exactly the same as those found in air. If an electroscope is charged so that the leaf diverges, and the interior of the instrument is then evacuated, no change in the divergence will be apparent. We need not, therefore, at this stage distinguish between air and a vacuum. Experimenting in air, then, let us address ourselves to the task of ascertaining how the force between two charges varies with the magnitudes of these and the distance between them. The effect of substituting other media (glass, ebonite, oil, and so on) for air or a vacuum will be considered later.

As a necessary preliminary to our proposed investigation we must have some means of comparing given quantities of charge or, alternatively, of producing at will quantities having a known ratio to one another. Fortunately, by invoking considerations of symmetry, it is a simple matter to achieve the second of these alternatives. This may be done as follows.

In the first place it is convenient, even if not absolutely necessary, that we should be able to reproduce a given charge any number of times. The simplest way to do this is to charge a body by induction under conditions that can be repeated exactly. Thus let a polystyrene rod that has been charged by friction be mounted on a stand, and let a metal sphere mounted on an insulating handle be brought to within a definite, repeatable distance from it and, while there, momentarily earthed. Then, provided that these operations are always carried out in exactly the same manner and loss of charge from the polystyrene by leakage between one occasion and the next is negligible, it may be assumed that after having been earthed the sphere will always have the same charge.

After taking the sphere away from the inducing charge, we may now divide it into two, three, four or any number of exactly equal parts, by having available a number of other, geometrically identical spheres, all mounted on insulating handles. If the first sphere, which has been charged by induction, and a second sphere, geometrically similar but uncharged, are brought into contact, then, because of the symmetry of the situation, the charge must divide itself equally between the two spheres, and each will now

493

have just half the original charge. If, instead, the charged sphere is brought simultaneously into contact with two other spheres whose centers are at the corners of an equilateral triangle, the charge must divide itself into three equal parts. However, the use of only two spheres is experimentally the simpler procedure. By discharging one of the spheres after the first contact between two spheres and then again bringing them into contact, each sphere can now be made to have one quarter of the original charge; and by successive repetitions of this procedure the charge may be further subdivided, each time by a factor of 2.

By using a pair of such charged spheres in a Coulomb torsion balance* or other suitable device for measuring the mutual force between them, it is found that this force F may be expressed in terms of the two charges Q_1 and Q_2 and the distance r separating them by the equation

$$F = k \frac{Q_1 Q_2}{r^2}, \tag{19.1}$$

where k is a constant whose value, together with the units in which F and r are measured, defines the unit of charge.

If we adopt the convention that F denotes, more specifically, the force of *repulsion* between the charges, then Eq. (19.1) correctly takes account of the polarities of the charges Q_1 and Q_2. If they both have the same sign, we *do* have repulsion, whereas if they are of opposite sign, F is negative; and negative repulsion means attraction.

The law expressed by Eq. (19.1) was first established by Charles Coulomb (1736–1806) and is known, after him, as **Coulomb's law.** This is the fundamental law of electrostatics.

In the experiments leading to Coulomb's law the linear dimensions of the charged spheres should be small compared with the distance between them. It will be shown in Chapter 20 that, from the point of view of the exertion of force, a charge having a spherically symmetrical distribution about a center behaves like a point charge concentrated at this center. It might therefore be thought that, as long as any appreciable modification in charge distribution due to induction effects were avoided by keeping the charges small, it would be permissible to work with larger spheres and measure the distance between their centers. However, we are not at this point entitled to assume this property of a spherically symmetrically distributed charge.

---

* The essential parts of this instrument are shown in Fig. 19.2.

MKSA Unit of Charge

The "practical" unit of current, long used by electrical engineers and now adopted as the scientific unit, is the **ampere** (A). The unit of quantity of electricity is defined as that quantity which flows past a point in a conductor per second when it carries a current of 1 A. This unit of quantity is known as the **coulomb** (C).

The original definition of the ampere, and so of the coulomb, has its roots in a theoretically somewhat illogical system of electromagnetism, which has now, happily, been abandoned, or at least is rapidly on the way out. But this is not a good argument for abandoning the coulomb, and because the choice of the constant k in Eq. (19.1), which defines the unit of charge, must in principle be arbitrary, this choice has been made in such a way as to give the coulomb. This, with the ampere, is now so firmly established in electrical engineering that it would lead to endless confusion to attempt any charge.

For such a manipulation one would naturally expect to pay a price—an inelegant-looking numerical value of the constant k. This expectation is apparently borne out, for it is found that, to give the coulomb, the constant in Eq. (19.1) must have the numerical value

$$k = 8.99 \times 10^9,$$

where F is expressed in newtons and r in meters. If, on the other hand, F were expressed in dynes and r in centimeters, the numerical value of the constant would have to be 8.99×10^{18}.

In the electrical engineer's system, already referred to, the unit of potential difference, the volt (see later), is so chosen that the product of coulombs and volts gives joules. We are already familiar with the joule from our study of mechanics—it is the MKSA unit of energy. This circumstance decides our choice of the MKSA system in which to work when dealing with electricity rather than a CGSA system.

The inelegance of the numerical value of k appears less disturbing when it is realized that, numerically,

$$8.99 \times 10^9 = c^2 \times 10^{-7},$$

where c denotes the velocity of light in a vacuum, expressed in meters per second. This is no mere coincidence; it may be shown to be due to the fact that the propagation of light is an essentially electrical (or "electromagnetic") phenomenon. Thus we may write

Eq. (19.1) in the form

$$F = c^2 \times 10^{-7} \frac{Q_1 Q_2}{r^2}. \qquad (19.2)$$

It should be noted that although Eq. (19.2) is correct numerically, it does not balance dimensionally; However, let us not concern ourselves about this now; the dimensional aspect of the matter will be considered presently.

Perhaps we should feel completely happy if Eq. (19.2) did not contain the power-of-10 factor. But to demand this would be asking too much. It must not be forgotten that the old "absolute electromagnetic unit" of current, from which, by definition, the ampere differs by a factor of 10,* was based on the CGS system. It is, however, some comfort to reflect that the factor 10^{-7} is completely accounted for by the change from the CGS to the MKSA system in measuring force and distance, plus the factor of 10 involved in changing from the old absolute electromagnetic unit of current to the ampere. That this is so may easily be verified. We see, then, that the price we have had to pay for making the coulomb our unit of charge is actually a much smaller one than we might reasonably have expected, having regard to the extremely complicated and unnatural way in which the old electromagnetic unit of current was defined.

Even though Eq. (19.2) as written does not balance dimensionally, the appearance in it of the factor c^2 is highly significant theoretically. It draws our attention to the fact that the same property of the medium—a vacuum—which determines the velocity of light waves traveling through it also operates in determining the mutual force between charges.

In physics we must demand that every equation shall balance dimensionally. To achieve this balance in the case of Eq. (19.1), the constant k must be given dimensions as well as a numerical value. Thus we may write

$$k = 8.99 \times 10^9 \text{ N-m}^2/\text{C}^2,$$

or

$$k = 8.99 \times 10^9 \text{ kg-m}^3/\text{A}^2/\text{sec}^4.$$

We see that the constant k is a property of the medium whose numerical value depends on the units we employ for the measurement of mass, length, time, and electric current.

---

* The ampere was originally defined as one tenth of the absolute electromagnetic unit.

In the complete theoretical system of electrostatics it has been found more convenient that the proportionality factor expressing the property of the medium in Coulomb's law should appear in the denominator rather than in the numerator, and also that the factor 4π should be included. Purely in the interests of elegance of equations subsequently to be considered, therefore, Coulomb's equation for charges in a vacuum is written in the form

$$F = \frac{1}{4\pi\epsilon_0} \frac{Q_1 Q_2}{r^2}, \tag{19.3}$$

where ϵ_0, known as the **permittivity of free space** (a vacuum), has the value

$$\epsilon_0 = 8.85 \times 10^{-12} \text{ C}^2/\text{N}/\text{m}^2,$$

or

$$\epsilon_0 = 8.85 \times 10^{-12} \text{ A}^2\text{-sec}^4/\text{kg}/\text{m}^3.$$

Problems

19.1   The distance between the centers of two small spheres each of which carries a charge of 0.1 μC is 1 m. What force does each exert on the other?

19.2   Two small charged conducting spheres, each of mass 100 mg and carrying the same charge, are suspended from the same point by insulating threads of length 2 m. When in equilibrium the centers of the spheres are 10 cm apart. On the assumption that the diameters of the spheres are negligible compared with their distance apart, calculate the charge on each.

If the diameters of the spheres are *not* negligible, will the actual charge be greater or less than that calculated? Explain.

19.3   The mutual force of attraction between two unlike and unequal charges on a pair of identical small conducting spheres is measured when the spheres are separated by a certain distance. The spheres are now brought into contact, separated to their previous distance apart, and the mutual force between them, now of repulsion, again measured. If this force is of the same magnitude as before, what is the ratio of the magnitudes of the original charges?

19.4   Three identical charges Q are placed at the corners of an equilateral triangle of side l. What force will be exerted on each of these by the other two combined? Will the direction of this force depend on whether Q is positive or negative?

19.5   Three small spheres, each of mass m and each carrying the same charge Q, are suspended by light insulating threads of length l from the same point and come to rest at the corners of an equilateral triangle of side a. What is the magnitude of Q?

19.6   Equal charges q are placed at the corners of an equilateral triangle, and a charge Q of opposite sign to that of q is placed at the center of the triangle. What must be the ratio $|Q/q|$ for the resultant forces on all four charges to be zero?

19.7 Point charges Q, $-2Q$, and $3Q$ are situated at the corners A, B, and C, respectively, of an equilateral triangle of side l, and another point charge q, of the same sign as Q, is placed at the center of this triangle. What is the force acting on the charge q, and what is the angle between the direction of this force and the side AB of the triangle?

19.8 Charges $2Q$, $-Q$, and $-Q$ are situated at the corners A, B, and C, respectively, of an equilateral triangle of side l, and a charge q, of the same sign as Q, is placed on the line DA produced beyond A, where D is the midpoint of the side BC (see the figure). The distance between the charge q and the midpoint of the line AD is d. Show that when d is large compared with l, the force exerted on the charge q closely approximates $Mq/2\pi\epsilon_0 d^3$, where M is written for $\sqrt{3}\,lQ$, and that this force is in the direction of DA, away from the triangle.

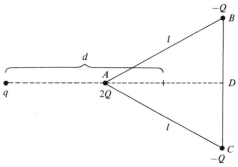

Figure for problem 19.8.

19.9 The same charges as in Problem 19.8 are situated at the corners A, B, and C of an equilateral triangle, but the charge q (of the same sign as Q) is on a line parallel to BC passing through the midpoint of the

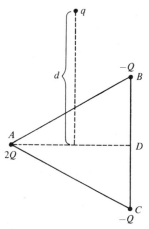

Figure for problem 19.9.

line AD (see the figure). Show that if the distance d of this charge from the midpoint of the line AD is large compared with l, the force exerted on it closely approximates $Mq/4\pi\epsilon_0 d^3$, and that the direction of this force is opposite to that exerted on the charge q in Problem 19.8.

19.10   A negative charge of magnitude $|Q|$ is placed midway between two fixed positive charges q and on the line joining them. Is the negative charge in stable or unstable equilibrium with respect to motion (a) along the line joining the charges q, and (b) at right angles to this line? For what ratio of q to Q would the resultant force on each charge be zero?

19.11   A particle of mass m which carries a charge q is situated in a gravitationally field-free region midway between two fixed point charges Q, the signs of q and Q being the same. The particle is constrained so that it is able to move only along the line joining the point charges Q. Find the frequency with which it will execute oscillations of small amplitude when it is disturbed.

19.12   Repeat Problem 19.11 for the case where q and Q have opposite signs and the particle is able to move only at right angles to the line joining the fixed charges.

19.13   A uniform thin metal ring of radius r is given a charge Q, and a particle of mass m which carries a charge q is placed at the center of the ring, Q and q being of opposite sign, and the particle being constrained so that it is able to move only along the axis of the ring. Assuming the system to be in a gravitationally field-free region, find the frequency with which the particle will execute oscillations of small amplitude when it is disturbed. Compare your result with that of Problem 19.12 and comment on what you find.

19.14   Basing your calculation exclusively on Coulomb's law, find how much work must be performed in bringing two point charges Q_1 and Q_2, initially an infinite distance apart, to within a distance r of one another.

20

ELECTROSTATIC FIELD THEORY

Having dealt with the basic phenomena of electrostatics, we shall now introduce the concept of an electrostatic field. As we shall see, the use of this concept leads to a great simplification in the mathematical discussion of situations where we are concerned with the electrostatic forces exerted on charged bodies. Although this is as far as we shall go in the present chapter, it should be appreciated that electrostatic fields have a significance far transcending their use as a mathematical tool in solving problems in electrostatics. This is so both practically and theoretically. As we shall see in Chapter 23, electron optics, with all its important applications in the development and design of such devices as radio tubes, x-ray tubes, cathode-ray tubes, and electron microscopes is essentially an aspect of electrostatic field theory. But in physics we are not concerned merely with practical applications, however useful these may be. On the theoretical side, perhaps the most exciting development involving electrostatic fields—in conjunction with magnetic fields—is the theory of electromagnetic radiation. In this theory the disturbances whose space and time variations constitute a wave propagation are electric and magnetic fields, at right angles to one another and to the direction of propagation, each of which depends on the time variation of the other.

Concept of Electric Field

By an electric field is meant a condition applying at a particular point in space determining the electrostatic force per unit charge that a small "test" charge located there would experience. If the magnitude of the test charge is q and this experiences a force $\mathbf{F}$ due to the presence of other charges, then the field $\mathbf{E}$ at the location of q is expressed by the equation

$$\mathbf{E} = \frac{\mathbf{F}}{q}. \tag{20.1}$$

Force is a vector and quantity of charge is a scalar, so the electric field, defined as the quotient of these, must be a vector, as indicated.

The specification that the test charge must be small may be taken as referring both to the spatial extension of the charge and to its magnitude. The former requirement is obvious, because we are concerned with the field at a point. The insistence on q being small in magnitude could be relaxed if the Coulomb forces exerted mutually between this and the charges regarded as the cause of the field did not have the effect of modifying the latter. In practice, however, if q were appreciable, such modification would usually occur, by the process of electrostatic induction. It is to emphasize the requirement that the test charge should be small that we use the lowercase letter q as the corresponding symbol.

Let us now consider the question of the polarity of the test charge. The force $\mathbf{F}$ may be regarded as the vector resultant of the separate Coulomb forces due to all the elements of charge to whose presence the field is attributed. Any charge distribution, whether over areas or within volumes, may always be regarded as made up of a large number of areal or volumetric "elements" of charge, and as long as these are small enough they may be treated mathematically as point charges and Coulomb's law applied to their interactions with the test charge. Let us suppose that with the test charge positive all the component Coulomb forces due to these elements of charge acting on it have been determined and are compounded vectorially to produce the resultant $\mathbf{F}$, and that this force is then divided by the positive quantity q to give the field $\mathbf{E}$. Now, instead, let the test charge be negative. This will cause the directions of all the component forces to be reversed, and consequently the direction of their resultant $\mathbf{F}$ must be reversed also. However, because q is now negative, the field $\mathbf{E}$ defined by the quotient $\mathbf{F}/q$ must have the same direction as before. Hence it is not necessary to specify the polarity of the test charge; the field defined by Eq. (20.1) is independent of this.

The unit of electric field is the **newton per coulomb.**

LINES OF FORCE   The term "electric field" is actually used in two senses. One of these is the specific quantity $\mathbf{E}$ at a particular point defined above. The other meaning is a more descriptive one, in which the reference is to a picture of how $\mathbf{E}$ varies in direction and magnitude from point to point throughout a considerable region. This dual use of the term, which is very convenient, will be adopted in the following pages; the context will always indicate which meaning is intended.

At least a qualitative picture of an electric field may be given in terms of **lines of force.** These are lines constructed within the region in which we are interested such that the tangent at any point P on a line gives the direction of the field at that point. The sense may be indicated by means of arrowheads. An example of an assemblage

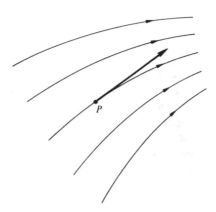

FIGURE 20.1 System of lines of force.

of lines of force representing a field pictorially is given in Fig. 20.1. It should be noted that since at any particular point the field can have only one direction, lines of force can neither meet at an angle nor cross. Also, in electrostatics there can be no component of the field tangential to a conducting surface, for if there were, electricity in the conductor would immediately be set in motion accordingly. Consequently, lines of force must always leave or approach the surface of a conductor in directions normal to the surface. Again, there can be no lines of force within the material of a conducting body. Finally, a line of force cannot both begin and end on the same conductor, for if it did it would be possible to obtain energy for nothing, simply by taking a positive charge along the line of force, from beginning to end, allowing it to do work on an external system, and then through the conductor, where there is no field, back to the beginning. After the charge has completed the round trip everything would be back in its original condition, but work would have been performed by the charge, and this would violate the law of conservation of energy.

Areal Density of Lines and Field Strength By definition, the direction of a line of force at any point gives us the corresponding *direction* of the field, as indicated by the arrow in Fig. 20.1. It is, however, perhaps not immediately obvious how we might obtain information concerning the variation of the *magnitude* of the field from one region to another from a pictorial representation of a system of lines. Apart from the directions of lines at various points, there is only one other feature of such a pictorial representation that seems likely to be of interest, and that is the degree of crowding of the lines. Might the field intensity be related to this in some way? It is surely an idea that seems worth pursuing. If we can find a

503

relation between the degree of crowding and the field intensity in one case, we should expect that this relation might well apply in all cases, even though it would not be easy to establish rigorously its general validity. Assuming, tentatively, that it does apply generally, we might then reasonably hope to obtain at least qualitative explanations of certain electrostatic phenomena relating to field intensity. Let us examine from this point of view the field due to an isolated charge Q located at a point in space. The force exerted on a test charge q situated at a distance r from this is, as we have seen, $qQ/4\pi\epsilon_0 r^2$. For the field E at the point where q is situated we therefore have

$$E = \frac{Q}{4\pi\epsilon_0 r^2}. \tag{20.2}$$

The lines of force due to a point charge obviously must be straight lines radiating out in all directions from the charge. As this is a spherically symmetrical situation, it is indicated that in our pictorial representation we should distribute the lines uniformly in space so that in section the field would appear as in Fig. 20.2.

Now let us construct a sphere of radius r about the charge Q, as indicated by the dashed circle, and let the total number of lines radiating out from this charge be N—the choice of this number would in the first instance be quite arbitrary. Then, because the area of a sphere of radius r is $4\pi r^2$, the number of lines crossing unit area at the distance r from the charge Q, the "areal density"

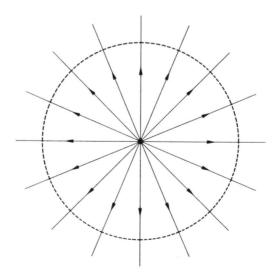

FIGURE 20.2   Spherically symmetrical system of lines of force radiating from isolated point charge.

of lines, must be equal to $N/4\pi r^2$. We see that both the field strength E and the areal density of the lines of force vary in the same manner with distance from the charge, conforming to an inverse-square law So, at least in this case and perhaps also quite generally, **the field is proportional to the areal density of lines of force,** where the area in question is taken in the plane to which the field is normal.

We have not yet considered how the number of lines radiating out from the charge should be related to the magnitude of the latter. An obvious choice would be a number *proportional* to Q. If such a number is chosen, then, as we see from Eq. (20.2), the factor by which E is proportional to the areal density of lines will be independent of Q.

Although, in principle, any ratio of N to Q would serve our purpose as well as any other, the choice that has in fact been made from considerations of mathematical convenience is to set N equal to Q/ϵ_0:

$$N = \frac{1}{\epsilon_0} Q.$$

This makes the areal density of lines not merely proportional to, but numerically *equal* to, the field strength E.

Q may have any value,* so proportionality of N to Q implies that N, also, can have any value, and only by the most unlikely chance will the N corresponding to an arbitrarily chosen value of Q be a whole number. The difficulty then arises of visualizing a fraction of a line. Any attempt at such visualization would patently be absurd; there can be no such thing as a fraction of a line. However, it must not be forgotten that lines of force are in any case merely a convenient mathematical abstraction; they have no real existence. There can therefore be no objection to an extension or generalization of this abstraction to include nonintegral values. This is only one of many mathematical abstractions, such as $\sqrt{-1}$, or n-dimensional space with n greater than 3, which, despite our inability to visualize them, are nevertheless exceedingly useful concepts.

Field Near a Pointed Conductor   As our first application of the idea of lines of force, let us consider a conductor of the form shown in profile in Fig. 20.3. Let lines be constructed that radiate out uniformly in all directions from a point such as P in this conductor, as indicated (dashed lines close to the conductor). At distances

---

* We are here ignoring the "graininess" of electric charge, that is, the fact that every charge is necessarily an integral multiple of the electronic charge e— see Chapter 22.

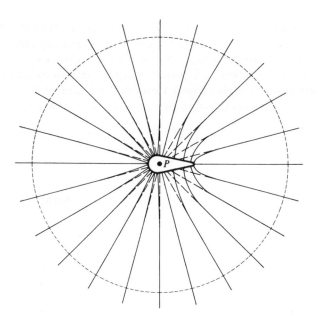

FIGURE 20.3 Lines of force radiating from pointed charged conductor.

from the conductor that are large compared with its linear dimensions, these lines may be taken to represent lines of force, for at such distances the charged conductor must give a field approximating that of a point charge. However, relatively close to the conductor the lines of force must depart from these lines, because the lines of force cannot meet the surface of a conductor other than normally to it. Hence the lines of force must be distributed in the manner indicated by the solid lines in the figure. These are closest together where they meet the conductor in the region of highest curvature, and it is therefore also here that the field is strongest.

This property of regions of high curvature—"points"—has a number of applications in the promotion of discharges between conductors through intervening gas. By the provision of points in appropriate places on conducting bodies correspondingly intense fields can be established. Beyond a certain limiting field strength in a gas, such as air, the insulation of the gas breaks down and charge flows freely. The mechanism by which this occurs will be discussed in Chapter 24. Meanwhile we may think of the flow of electricity following insulation breakdown as a spraying of charge from the point to a neighboring conductor or, alternatively (from a different point of view), as a collection of charge from the latter by the point.

This action of points is not only made use of in scientific apparatus of various kinds, but it also has important applications in everyday life. For example, charge sprayed from a pointed lightning conductor tends to neutralize the charges in overhead clouds and so reduces the danger of cloud-to-earth lightning strokes during thunderstorms. Also, arrays of points are commonly provided on the trailing edges of the wings of aircraft, promoting dissipation of electrostatic charges which they acquire in flight.

It should be noted that a high surface charge density is necessarily associated with the crowding of the lines of force emanating from or terminating on a region of high curvature of a charged conducting surface.

An alternative theoretical treatment of the existence of strong fields and high charge densities at points on charged conductors will be given later in this chapter under the heading "Equipotential Surfaces."

Electrostatic Induction in Terms of Lines of Force We have already noted that electrostatic induction may be explained qualitatively in terms of Coulomb forces. From his earlier studies the reader will presumably also be familiar with an alternative manner of explaining induction phenomena in terms of lines of force. These give a useful pictorial representation of what happens in various situations involving electrostatic induction.

Electrostatic Shielding Delicate electrical measurements are often adversely affected by induction due to stray charges. To guard against this, the entire apparatus may be surrounded by a metal enclosure. The screening is usually constructed of fine wire mesh, which, while providing adequate shielding, enables the experimenter to see the apparatus. The enclosure, with the apparatus inside it, may be regarded as electrically a little world of its own, which cannot be affected by any charges outside, for lines of force due to these charges, which may well terminate on the screen, cannot pass through it.

Further Consideration of Coulomb's Law

In all the experiments on which Coulomb based his law, the charges were held on metal spheres, and it was tacitly assumed that, despite this, their mechanical interactions would be the same as those between "pure" charges, not residing on any material. We now have to consider whether this assumption is justified, for each sphere "sees" only about one half of the other across the air or vacuum between them. How can the charges on the remote sides of the spheres, shielded as they are by intervening metal, exert forces on each other?

A rather more obvious case of shielding occurs in an electroscope. The leaves of this instrument constitute a rough approximation to an enclosure within which no field can exist, particularly if the leaves are wide and the divergence is small. We may well ask how, under these conditions, mutual repulsive forces can be exerted between the charges on the leaves, because each leaf must surely constitute an effective shield against the transmission of electric forces through it. Do the like charges, residing on the outside of the leaves, *really* repel each other, or is the divergence simply due to the attraction between the charge on each leaf and the charge of opposite sign on the adjacent part of the casing? Such attraction might be imagined to be exerted via the lines of force passing between these charges, as if these lines were in a state of tension (see Fig. 20.4).

However, we do not *necessarily* have to assume that the intervention of metal prevents the exertion of Coulomb forces between charges. In the electrostatic shielding of apparatus considered above, might not the provision of the shield protect the apparatus from the effects of external charges simply by enabling induced charges to appear on the shield whose Coulomb forces exerted on any charges inside exactly balance those of the external charges?

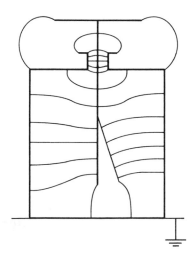

FIGURE 20.4 System of lines of force between components of charged electroscope.

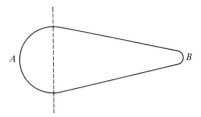

FIGURE 20.5

The question of whether, in fact, intervening metal prevents charges from exerting Coulomb forces on one another is answered by the following simple experiment. As our carriers of charge, let bodies of a form such as is shown in profile in Fig. 20.5 be used, and let the distance between two such charged bodies be large compared with their linear dimensions. Then the mutual force between them, as found, for example, with a torsion balance, will be the same whether the blunt ends A or the sharp ends B face one another. But in the former case the amounts of charge that "see" each other, without any metal intervening, are substantially smaller than in the latter. This is so not only because the amount of surface on the side A of the plane indicated by the dashed line is less than on the side B, but also because the charge *density* is less in regions of low than in those of relatively high curvature. It appears, then, that the intervention of metal has no effect on the exertion of Coulomb forces. We shall see later why this is so.

Gauss's Theorem

New developments in physics are often suggested by turning an equation back to front, inverting it, or otherwise manipulating it mathematically. As we shall now see, a simple manipulation of Eq. (20.2) leads to a very important theorem in electrostatics, formulated by Carl Friedrich Gauss (1777–1855) and known after him as Gauss's theorem.

Multiplying both sides of Eq. (20.2) by $4\pi r^2$ we have

$$E \cdot 4\pi r^2 = \frac{Q}{\epsilon_0}. \qquad (20.3)$$

The factor $4\pi r^2$ on the left side of this equation suggests a sphere of radius r centered on the point where the charge Q is located; $4\pi r^2$ would be the area of this sphere. But even with this sphere constructed, Eq. (20.3), as it stands, really tells us no more than Eq. (20.2), of which it is a rearrangement.

509

Perhaps the limited interest of Eq. (20.3) is due to the fact that it applies only to a particular kind of closed surface, a sphere, centered on a point charge. It would seem worthwhile to try to find a valid generalization of this equation. In the first instance we might see whether $\int E \, dA$, with the integration including every element of area of *any* closed surface, might be equal to $(1/\epsilon_0) \int dQ$, where the integration is carried out for all elements of charge enclosed by this surface.

It is at once clear that these two quantities cannot, in general, be equal. To see this, let us, for example, again consider the special case of a single point charge Q but with the enclosing surface being, not a true sphere centered on this, but only an approximation to the spherical form, in that it is crinkly instead of smooth. Obviously the crinkliness must substantially increase the surface area without much affecting the value of E at any part of the surface; hence the surface integral is substantially increased.

In making a second attempt at finding a valid generalization we might remember that in the situation to which Eq. (20.3) refers, the vector **E** is everywhere normal to the surface of the sphere and is directed from the inside to the outside of the enclosure. This feature was not retained in the first generalization attempted. So let us now try

$$\int E_n \, dA = \frac{1}{\epsilon_0} \int dQ, \tag{20.4}$$

where the symbol E_n denotes the outwardly directed component of the field normal to the element of area dA. Actually Eq. (20.4) *is* valid for all cases. We shall now prove that this is so.

We may regard any distribution of charges as made up of a number of point charges; in the case of a continuous charge distribution, this would be an infinite number of infinitesimal elements of charge. At any point under consideration each of these charges would make its own contribution to the field, and thus the components of all these items in any chosen direction, for example that of the normal to an element of surface, must be algebraically additive; the algebraic sum of all these components must be equal to the component of the resultant field in the same direction. Consequently, if it can be shown that Eq. (20.4) is valid for the case of a single point charge and any enclosing surface we wish to choose, it must follow that it is also true for any number of charges within the same closed surface.

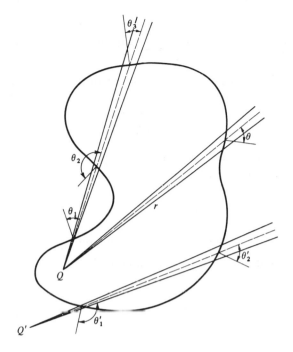

FIGURE 20.6 Point charges within and external to imaginary closed surface.

We shall first show that the equation

$$\int E_n \, dA = \frac{1}{\epsilon_0} Q,$$

which represents the special case of Eq. (20.4) for a single point charge Q situated within the closed surface, is valid. At the same time we shall show that the value of the integral is zero when the charge is outside the closed surface.

Let Q be a point charge located within the closed surface indicated in Fig. 20.6, and let Q' be another point charge outside. Let us consider cones with their apices at the charges Q and Q' defining a solid angle ω,* cones centered at Q cutting the surface once and three times, and one centered at Q' cutting it twice. The angles marked are those between the directions of the cones outward from the charges and the outwardly directed normals to the elements of area cut. Let the distances of these elements from the corresponding charges be r, r_1, r_2, r_3, r_1', and r_2', where r is the dis-

---

* The solid angle of a cone is defined as the ratio of the area of that portion of a sphere centered at the apex of the cone which is included within the cone to the square of the radius of the sphere.

tance marked and the subscripts and primes of the other radii correspond to those of the angles.

The field strength at each element of area is expressed by an equation of the form

$$E = \frac{1}{4\pi\epsilon_0} \frac{Q}{r^2},$$

in which E and r may have their appropriate labels in the form of subscripts and primes, and correspondingly

$$E_n = \frac{1}{4\pi\epsilon_0} \frac{Q}{r^2} \cos \theta, \tag{20.5}$$

where θ also has a label attached to it where appropriate.

For the element of area δA cut by a cone and within it we have either

$$\delta A = \frac{r^2 \, \delta\omega}{\cos \theta}$$

or

$$\delta A = - \frac{r^2 \, \delta\omega}{\cos \theta},$$

according as θ is acute or obtuse.

Multiplying Eq. (20.5) by one or other of the last two equations, we have either

$$E_n \, \delta A = \frac{Q}{4\pi\epsilon_0} \, \delta\omega$$

or

$$E_n \, \delta A = - \frac{Q}{4\pi\epsilon_0} \, \delta\omega,$$

again according as θ is acute or obtuse. For the cone centered at Q and cutting the surface once only, we have, θ being acute,

$$E_n \, \delta A = \frac{Q}{4\pi\epsilon_0} \, \delta\omega;$$

for the cone centered at Q and cutting the surface three times,

$$(E_n \, \delta A)_1 + (E_n \, \delta A)_2 + (E_n \, \delta A)_3 = \frac{Q}{4\pi\epsilon_0} \, \delta\omega,$$

the first and second (or the second and third) terms on the left canceling each other out; and for the cone centered at Q' and cutting the surface twice,

$$(E_n \, \delta A)_1' + (E_n \, \delta A)_2' = 0.$$

Obviously all cones emanating from any charge Q situated inside the closed surface must cut this surface an odd number of times, so that the algebraic sum of all such terms as $E_n \, \delta A$ is $Q \, \delta\omega / 4\pi\epsilon_0$. On the other hand, the number of times a cone emanating from a charge outside the surface is even, so the corresponding algebraic sum in this case is zero. Proceeding to the limit of smallness for $\delta\omega$ and summing over the whole surface, we see that whereas a charge outside makes no contribution to the surface integral, one inside contributes the amount $(Q/4\pi\epsilon_0) \int d\omega$; and, since $\int d\omega$ is 4π (the ratio of the area of a sphere to the square of the radius), this is equal to Q/ϵ_0. Thus

$$\int E_n \, dA = \frac{Q}{\epsilon_0} \quad \text{or} \quad 0,$$

according as the point charge Q is within the closed surface or outside.

It will now be shown, formally, how this result may be extended to cover the case of any charge distribution.

Let the individual charges enclosed by the surface be Q_1, Q_2, Q_3, . . . , Q_μ, while those outside the enclosure are Q_1', Q_2', Q_3', . . . , Q_ν', and let the components of the fields due to these along the outwardly directed normal to an element of surface be denoted by E_{1n}, E_{2n}, E_{3n}, . . . , $E_{\mu n}$, E_{1n}', E_{2n}', E_{3n}', . . . , $E_{\nu n}'$. Then since, as we have seen,

$$\int E_{1n} \, dA = \frac{1}{\epsilon_0} Q_1,$$

$$\int E_{2n} \, dA = \frac{1}{\epsilon_0} Q_2,$$

$$\cdot$$
$$\cdot$$
$$\cdot$$

$$\int E_{\mu n} \, dA = \frac{1}{\epsilon_0} Q_\mu,$$

and

$$\int E_{1n}' \, dA = 0,$$

$$\int E_{2n}' \, dA = 0,$$

$$\cdot$$
$$\cdot$$

$$\int E_{\nu n}' \, dA = 0,$$

513

we must have

$$\int (E_{1n} + E_{2n} + \cdots + E_{\mu n} + E_{1n}' + E_{2n}' + \cdots + E_{\nu n}') \, dA$$

$$= \frac{1}{\epsilon_0} (Q_1 + Q_2 + \cdots + Q_\mu);$$

and hence, since at each point on the closed surface

$$E_{1n} + E_{2n} + \cdots + E_{\mu n} + E_{1n}' + E_{2n}' + \cdots + E_{\nu n}' = E_n,$$

$$\int E_n \, dA = \frac{1}{\epsilon_0} \sum_{r=1}^{\mu} Q_r,$$

or, expressing the right side of this equation as an integral,

$$\int E_n \, dA = \frac{1}{\epsilon_0} \int dQ \qquad (20.4)$$

for all the elements of charge contained within the closed surface.

Nineteenth-century physicists, in their attempts to explain certain physical phenomena, felt that they would be going a long way in this direction if they could invent mechanical models. Accordingly, they showed great interest in working out hydrodynamical analogies of electrical, magnetic, and gravitational field phenomena. In the electrical case, positive charges were thought of as corresponding to sources in an infinite volume of incompressible fluid, where this fluid was continuously generated, while negative charges were supposed to correspond to "sinks," at which continuous annihilation of fluid occurred. In these two systems, one electrical and the other hydrodynamical, there was a close correspondence between electrostatic field strength and velocity in the fluid. The component of velocity normal to any element of area is equal to the volume crossing this area per unit time, or "flux," divided by the area, that is, to the normal component of the (areal) "flux density." The use of the terms flux and flux density was carried over into the analogous electrostatic system to signify the corresponding electrical quantities. Thus E was—and still is—called the **flux density** and $E_n \, \delta A$ the **flux** through the element of area δA. In these terms Eq. (20.4), which is the mathematical formulation of Gauss's theorem, may be expressed in words as follows: **The net flux emerging from any closed surface is equal to $1/\epsilon_0$ of the algebraic sum of all the charges contained within it.**

APPLICATIONS OF GAUSS'S THEOREM Gauss's theorem may be used to show where charges are, or are not, located, and also to calculate fields in a number of cases where direct calculation on the basis of Coulomb's law would be both difficult and

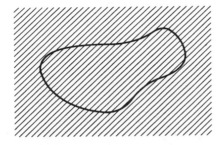

FIGURE 20.7 Gaussian surface entirely within a conducting material.

tedious, if not completely impracticable. We shall now consider a few such applications.

Location of Charge Let us imagine a closed surface (a "Gaussian surface") to be constructed entirely within the material of a conductor as indicated in Fig. 20.7, it being understood that this conductor does not convey any current. In any system in electrostatics there cannot be any field within the material of a conductor, for if there were, this field would produce a current; that is, we should not have an *electrostatic* situation, in which there is no movement of charge. Thus, because nowhere on the Gaussian surface under consideration is there a field, the surface integral in Eq. (20.4) must be zero, and with it the volume integral of charge; **there can be no charge within the material of a conductor.** Consequently, any charge the conductor may have must reside exclusively on its surface.

Hollow Conductor Let a Gaussian surface be constructed entirely within the material of the walls of a hollow conductor as indicated in Fig. 20.8. As in the previous case considered, the surface integral for this surface must be zero, and so **there can be no net charge within a hollow conductor.** In the absence of any insulated charge that has been introduced into the cavity through

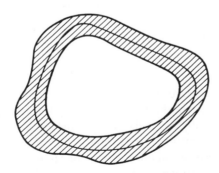

FIGURE 20.8 Gaussian surface entirely within material of hollow conductor.

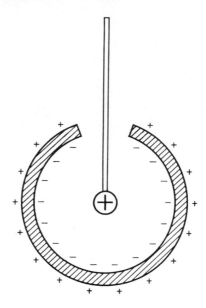

FIGURE 20.9 System of charges produced on internal and external surfaces of a hollow conductor by an insulated charge held inside the cavity.

a small opening and held there, as shown in Fig. 20.9, any charge held by the hollow conductor must therefore reside wholly on its outer surface. If an insulated charge is held in the cavity, an equal but opposite charge must be induced on the inner surface of the cavity; only in this way can the net charge within the Gaussian surface considered above be zero.

Let us now consider a Gaussian surface constructed outside the hollow conductor and enclosing this. Because the surface integral for this must correspond to (be equal to $1/\epsilon_0$ of) the net charge within it, and because the induced charge inside the hollow conductor is equal and opposite to the inducing charge, a like induced charge must reside on the outer surface of the hollow conductor of magnitude equal to the inducing charge. This must be additional to any charge the hollow conductor may have had initially.

If now the introduced conductor with its charge is allowed to touch the inside of the hollow conductor, the inducing and unlike induced charges will neutralize one another, and if the insulated conductor is then withdrawn through the opening it will take no charge with it. Its original charge may be thought of as having been conveyed in its entirety to the hollow conductor, for what was formerly the like and equal induced charge residing on the outer surface of the hollow conductor will still be there.

Obviously a whole succession of charges may be transferred completely to the hollow conductor in like manner, the insulated conductor with its charge being introduced through the opening each time, allowed to touch the hollow conductor, and then withdrawn.

If, following the introduction of the insulated charged conductor into the inside of an initially uncharged hollow conductor, the latter is momentarily earthed, this will allow the induced charge on the outer surface to escape while leaving unaffected the unlike induced charge inside. Subsequent withdrawal of the insulated charged conductor without making contact with the hollow conductor will leave the equal unlike induced charge on the hollow conductor, but to maintain the net charge within the hollow conductor at its required zero value the charge must now move to the outside.

The reader will recognize these procedures as Faraday's "ice-pail" experiments. We see that the various states of charge that Faraday established experimentally can be predicted theoretically simply by applying Gauss's theorem to the appropriate Gaussian surfaces.

Isolated Charged Sphere This is the first of three cases of geometrical symmetry that we shall consider here. Such cases lend themselves particularly well to mathematical analysis based on Gauss's theorem.

Let an isolated conducting sphere of radius a carry a charge Q, and let us imagine a Gaussian surface consisting of a larger sphere concentric with it, of radius r, as indicated in Fig. 20.10. The total flux Q/ϵ_0 must pass through this sphere, and from considerations of symmetry we must assume not only that it passes through radially but also that it is uniform over the surface. Consequently, the flux density, or field strength E, must be $Q/4\pi\epsilon_0 r^2$:

$$E = \frac{Q}{4\pi\epsilon_0 r^2}.$$

Comparing this with Eq. (20.2), we see that the field everywhere outside a uniformly charged sphere is the same as it would be if the whole of the charge were concentrated at its geometrical center.

We may note, parenthetically, that we might have treated gravitational attraction, which also conforms to an inverse-square law, along similar lines and come to the corresponding conclusion— that the gravitational field at any point outside a body with a spherically symmetrical distribution of mass is the same as if the whole mass of the body were concentrated at its geometrical center.

517

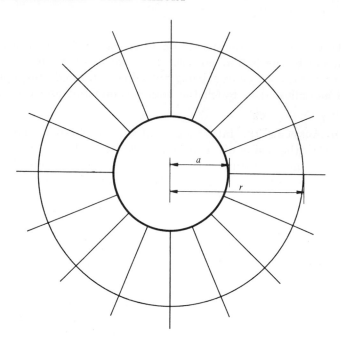

FIGURE 20.10 Spherical Gaussian surface external to and concentric with isolated charged sphere.

As a special case of some interest, let us consider the electrostatic field at a point which, although outside the charged sphere, is so close to it that its distance from the surface is negligible compared with the radius of the sphere. In this case r becomes, to all intents and purposes, a, so that the field strength is $Q/4\pi\epsilon_0 a^2$. Let this now be expressed in terms of the areal density of charge σ. This is $Q/4\pi a^2$. For the field just outside the surface of a uniformly charged sphere we have, therefore,

$$E = \frac{\sigma}{\epsilon_0}.\qquad(20.6)$$

Infinite Uniformly Charged Cylinder Let the charge per unit length of a long (ideally infinitely long) cylinder be λ, and let us investigate the field at some point outside, situated a distance r from the axis of the cylinder. For our Gaussian surface we take a cylinder of radius r and length l which is coaxial with the charged cylinder (Fig. 20.11). From considerations of symmetry we must assume the flux to be everywhere radial, so that none passes through the ends of the Gaussian surface; only the sides of this cylinder contribute to the surface integral. The charge contained within the Gaussian surface is λl, and so the total flux passing through it is $\lambda l/\epsilon_0$. The area of the sides is $2\pi rl$. Hence for the flux density, and so the field, we have

$$E = \frac{\lambda}{2\pi\epsilon_0 r}.\qquad(20.7)$$

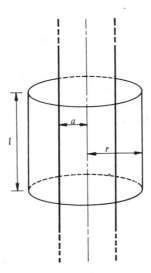

FIGURE 20.11   Cylindrical Gaussian surface of larger radius than and coaxial with infinite uniformly charged cylinder.

Again, as in the case of the charged sphere, if the radius of the charged cylinder is a, and r is made only infinitesimally larger than a, we have for the field just outside,

$$E = \frac{\lambda}{2\pi\epsilon_0 a} = \frac{2\pi a\sigma}{2\pi\epsilon_0 a} = \frac{\sigma}{\epsilon_0}.$$

Direct Investigation of Field Outside Charged Conducting Surface
The appropriate Gaussian surface to use here is a box-shaped surface of the kind indicated in Fig. 20.12, part of which is outside and the remainder inside the body of the conductor. Let the area of each of the sides parallel to the surface of the conductor be A. Then if the areal charge density is σ, the charge enclosed is σA and the flux passing out through the sides of the box is $\sigma A/\epsilon_0$. None of this passes through that part of the box which is within the material of the conductor, where the field is necessarily zero; the whole of it must pass out through the portion outside, in a direction normal to the surface. The flux density is therefore $\sigma A/\epsilon_0 A$, and so

$$E = \frac{\sigma}{\epsilon_0}, \tag{20.6}$$

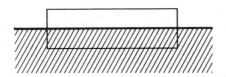

FIGURE 20.12   Gaussian surface of box form enclosing portion of surface of charged conductor.

519

in agreement with what we found for the limiting cases of a charged sphere and a long charged cylinder.

Implied in the derivation of Eq. (20.6) by the method just considered is an apparent paradox. The validity of Gauss's theorem, which we have used in this derivation, rests on the assumption that the presence of a conductor between a charge and a point does not affect the field at this point. Yet, although there is a field above the charged surface indicated in Fig. 20.12 in a vacuum, there is none below, within the conductor. How can this be?

The answer to this question lies in the fact that Fig. 20.12 represents only part of the charged conducting surface. For, whatever may be the geometrical form of the conductor, its surface must entirely enclose the conducting material, and in order to find the field within any part of this surface using Coulomb's law, we must take into account the entire surface and the charge distribution over this, and, in addition, any charges that may be present on other bodies.

TEST OF INVERSE-SQUARE LAW We have seen that the validity of Gauss's theorem rests on Coulomb's inverse-square law; if this law is correct, Eq. (20.4) necessarily follows. But if the mutual force between two charges varied with their distance apart according to any function of this distance other than the inverse square, Gauss's theorem would not be valid. Conversely, if the accuracy of Eq. (20.4) is established experimentally, this is equivalent to a verification of the inverse-square law.

As will readily be appreciated, the experimental observation that there is no field within an empty hollow charged conductor constitutes one verification of the validity of Gauss's theorem. As long ago as 1767—some twenty years before Coulomb carried out his torsion-balance experiments—Joseph Priestley realized that this observation constituted proof of the inverse-square law of force between charges. The absence of a field within a charged hollow conductor was subsequently confirmed in turn by Faraday, Cavendish, Maxwell, and Plimpton and Lawton. The experiments of Cavendish, Maxwell, and Plimpton and Lawton were in principle similar and gave successively higher accuracies in the establishment of the inverse-square law.

A simplified version of these experiments will now be discussed; actually this corresponds more closely to the experiment carried out by Maxwell than to Plimpton and Lawton's. The apparatus is shown diagrammatically in Fig. 20.13. It consists of two concentric hollow conducting spheres A and B, an electrometer E situated within the inner sphere, a steady high-voltage supply V, and a two-way switch S by means of which the outer sphere A can

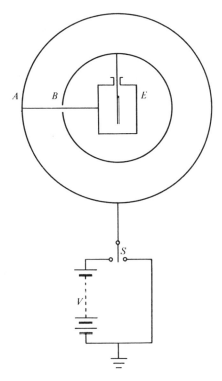

FIGURE 20.13 Principle of Plimpton-Lawton experiment.

be connected in turn either to earth or to the high-voltage supply.*
Plimpton and Lawton's inner conductor was not spherical, and in
place of the electrometer shown in Fig. 20.13 they used a galvanometer. And instead of switching alternately between earth potential and a steady high voltage, they applied to the outer sphere a low-
frequency alternating voltage whose frequency was tuned to agreement with the natural frequency of oscillation of the galvanometer.
By taking advantage of the resonance of the galvanometer in this
way they were able to achieve an extraordinarily high sensitivity.
But such features of Plimpton and Lawton's experiment are merely
modifications or refinements of those to which Fig. 20.13 relates; the
basic electrostatic principles of the method are not affected.

As will be seen from the figure, one electrode of the electrometer
is connected electrically to each sphere. Provision is made for tele-
scopic viewing of the registration of the electrometer through two
small holes (not shown), one in each sphere, the hole in the outer
sphere being closed electrically, although not optically, by spreading
a film of soap solution over it.

---

* As will be appreciated on reading the next section, dealing with the
concept of potential, connection of the high-voltage source to the outer sphere
conveys to it a substantial amount of charge.

If with a high voltage applied to sphere A any field were produced between the two spheres, this would have to be compensated by a field across the electrometer; otherwise it would be possible, by conveying a charge across from one sphere to the other, through the connecting wires, and back to its starting point across the electrometer, to have work performed for nothing. Actually, no registration is observed in the electrometer when high voltage is applied to A, showing that there is no field across it. There can therefore also be no field in the space between the two spheres. The zero reading of the electrometer also shows that there is no charge on the inner sphere; for any such charge could be established only by means of a charging of the electrometer functioning as a capacitor, and this would necessarily give a finite reading. Hence we have the situation that there is no charge on the inner sphere and no field in the space between it and the outer sphere. If, therefore, we imagine a closed surface between these spheres, enclosing the inner one, both sides of Eq. (20.4) relating to this surface are zero. For this case, therefore, Gauss's theorem applies.

At first sight this result may seem rather a negative one, but the whole point is that there is no field in this region, not only in the absence of charge within the enclosure, *but despite the existence of a very substantial charge on the outer sphere.*

We could not have zero field within the outer sphere on the basis of any but an inverse-square law of force between charges. This conclusion holds, irrespective of whether or not the spheres are accurately machined and concentric; indeed, the absence of registration of the electrometer would still prove the inverse-square law if the metal enclosures were of some entirely different form, for example if they were cubes. However, it is most easily seen that the absence of a field within the outer enclosure means an inverse-square law of force in the ideal case where this enclosure is an isolated sphere, uniformly charged over its surface area. Let us consider a point P within such a sphere, as indicated in Fig. 20.14. Let this point be the common apex of two narrow cones, generated by a straight line through this point which is rotated about a certain chord of the sphere functioning as the axis of the cones. Then, as is at once evident geometrically, the areas A_1 and A_2 on the surface of the sphere are proportional to the squares of the distances r_1 and r_2 of these areas from P. For the charges Q_1 and Q_2 on these areas, which are proportional to the areas, we therefore have

$$\frac{Q_1}{Q_2} = \frac{A_1}{A_2} = \frac{r_1^2}{r_2^2}.$$

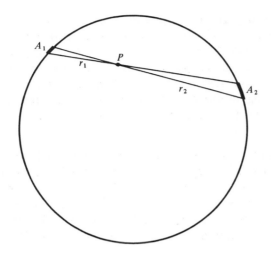

FIGURE 20.14

With an inverse-square law determining the oppositely directed fields E_1 and E_2 at P due to these charges, the ratio of the fields must be given by the equation

$$\frac{E_1}{E_2} = \frac{Q_1/r_1^2}{Q_2/r_2^2} = \frac{Q_1}{Q_2}\frac{r_2^2}{r_1^2} = \frac{r_1^2}{r_2^2}\frac{r_2^2}{r_1^2} = 1;$$

that is, there is no resultant field at P due to the charges on these areas. The whole area of the sphere may be regarded as consisting of such pairs of elements of area, so there can be no resultant field at P due to the whole charge on the sphere. But, with any other than an inverse-square law of force, there clearly must be a resultant field at all points within the sphere other than its geometrical center.

Maxwell estimated that n in the equation

$$F = k\,\frac{Q_1 Q_2}{r^n}$$

could not differ from 2 by more than 1 part in 40,000. Plimpton and Lawton, with their more accurate instrumentation, estimated the value of n to be 2 within about 1 in 10^9.

Potential

In electricity, as in mechanics, it is often useful to employ the concept of potential energy. However, in electricity it is generally convenient to reason not so much in terms of the potential energy possessed by any given charge but in terms of the potential energy per unit charge. This is known as the **potential** of the point where the charge is located, and, if some zero has been agreed upon, it is a definite property of this point, uniquely determined by the magnitudes and spatial distribution of all charges other than the test charge figuring in the definition. For the same reason as in our definition of field strength, the test charge is usually thought of as vanishingly small.

In principle, the choice of a zero is always arbitrary, although in practice a particular choice of zero often leads to greater simplicity of mathematical expression than would others. Obviously the choice of a zero cannot be involved in the *difference* of potential between two points—this is something perfectly definite. It is more specifically with difference of potential that we shall now concern ourselves.

The amount by which the potential energy of a test charge q at a point A in an electrostatic field exceeds that at another point B is, by the definition of potential energy, the work W_{AB} that the charge is capable of performing on an outside system in moving from A to B. Hence the **difference of potential,** or **potential difference** (PD), is expressed by the defining equation

$$V_A - V_B = \frac{W_{AB}}{q}, \tag{20.8}$$

where V_A and V_B denote the potentials at points A and B, respectively. It will be noted that in this equation it is not necessary to specify whether the test charge q is positive or negative, because with a change in the sign of q, W_{AB} changes sign also, and the quotient is therefore not affected.

The field that gives rise to this PD may be regarded as due to a system of point charges, and by thinking of it in these terms it may readily be shown that the PD between two given points is independent of the path taken by the test charge in moving from one to the other.

Let us consider first the simple case where the field is due to a single point charge Q located at O (Fig. 20.15). Let the test charge q move the short distance δl from P to P', the distance r from O thereby being increased by δr. Let the direction of PP' make with that of OP the angle θ. Then

$$\delta l = \frac{\delta r}{\cos \theta}.$$

The component of the Coulomb force in the direction of motion is $(Qq/4\pi\epsilon_0 r^2) \cos \theta$. Hence for the work δW performed by the charge

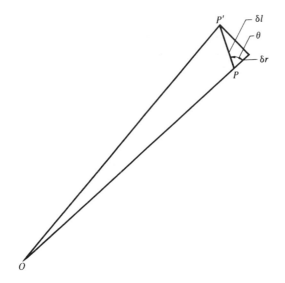

FIGURE 20.15

in moving from P to P' we have

$$\delta W = \frac{Qq}{4\pi\epsilon_0 r^2} \cos\theta \, \frac{\delta r}{\cos\theta} = \frac{Qq}{4\pi\epsilon_0} \frac{\delta r}{r^2}.$$

Any path traveled by the test charge from a point A to another point B may be thought of as a succession of infinitesimal elements of path such as that just considered (from P to P'), and hence in traveling from A, at the distance r_A from O, to B, at the distance r_B, the work W_{AB} performed is given by the equation

$$W_{AB} = \frac{Qq}{4\pi\epsilon_0} \int_{r_A}^{r_B} \frac{dr}{r^2} = \frac{Qq}{4\pi\epsilon_0} \left(\frac{1}{r_A} - \frac{1}{r_B} \right),$$

whence

$$V_A - V_B = \frac{W_{AB}}{q} = \frac{Q}{4\pi\epsilon_0} \left(\frac{1}{r_A} - \frac{1}{r_B} \right). \tag{20.9}$$

The independence of the path taken between the points by the test charge, previously referred to, is obvious from this equation.

If the field is due to a system of point charges (or elements of charge) instead of to a single one, we may think of the force on the test charge as made up vectorially of the corresponding items. Then, for each element of the path contemplated, all the work items will be algebraically additive, and so, finally, we obtain the expression for PD in this more general case,

$$V_A - V_B = \frac{\sum W_{AB}}{q} = \frac{1}{4\pi\epsilon_0} \left(\sum \frac{Q}{r_A} - \sum \frac{Q}{r_B} \right).$$

525

In the special case where B is at infinity and the potential at infinity is taken as the zero from which potentials are measured, this expression becomes simplified to

$$V = \frac{1}{4\pi\epsilon_0} \sum \frac{Q}{r},\qquad(20.10)$$

in which the subscript A has been dropped and V now stands for the potential at the point (previously the point A) in question. This potential may now be defined as the work per unit charge performed by the test charge in traveling from the point considered to infinity or, alternatively, performed *on* the test charge in bringing it *to* this point *from* infinity.

Equation (20.10), like Eq. (20.9), implies independence of the path taken.

In many situations it is mathematically a great convenience to be able to work in terms of potential, which is a scalar quantity, rather than to have to deal vectorially with fields.

The MKSA unit of potential or potential difference is that which corresponds to the measurement of the test charge in coulombs and of work or energy in joules. This unit is known, in honor of Alessandro Volta (1745–1827), a prominent pioneer investigator of electrical phenomena, as the volt (V). If we multiply the number of coulombs in the test charge by the PD between two points expressed in volts, we obtain the amount of work in joules performed on or by the test charge (as the case may be) when it is conveyed from one point to the other:

$$\text{coulombs} \times \text{volts} = \text{joules}.$$

Precision methods of measuring potential differences and also quantities of charge will be discussed later.

EQUIPOTENTIAL SURFACES An electrostatic field may be represented graphically either by lines of force or by the system of surfaces through which these lines of force everywhere pass normally. Because of the way these surfaces are defined, there can be no component of the field along any given such surface, so no work is done in moving a test charge along it, and hence the surface is an **equipotential surface;** all points on the surface have the same potential.

As an example, Fig. 20.16 shows a central section through the system of equipotential surfaces (solid lines) and the corresponding lines of force (dashed lines) in the region surrounding an isolated charged metal or other conducting sphere. In this case the equi-

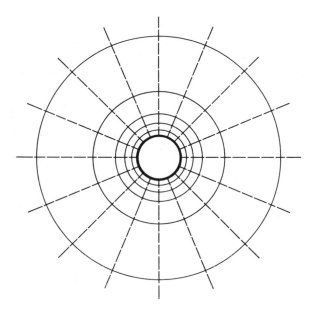

FIGURE 20.16 System of lines of force radiating from, and equipotential surfaces surrounding, isolated charged sphere.

potential surfaces are of course themselves spheres, all having the same center as the charged metal sphere. Another interesting case is depicted in Fig. 20.17. This shows a section through the system of equipotential surfaces due to a charged metal sphere in the vicinity of an extensive plane surface of a conductor, the section being one that passes through the center of the sphere and is per-

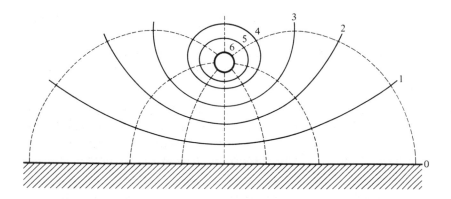

FIGURE 20.17 System of lines of force and equipotential surfaces in region of charged sphere and adjacent plane surface of conductor.

pendicular to the surface of the conductor. In this figure the potentials of the surfaces shown, relative to that of the plane conducting surface, are marked to an arbitrary scale—for example, the units *could* be volts; they would be for an appropriate relation between charge and distance of the sphere from the plane.

If, as in these figures, the equipotential surfaces drawn are those corresponding to a succession of equal steps in potential, then, from an inspection of the diagram, we may deduce how the strength of the field, as well as its direction, varies with position. Let us consider a movement of a test charge q by a short distance δs in the direction of the field, that is, at right angles to the equipotential surfaces. In this case, Eq. (20.8) may be written in the form

$$-\delta V = \frac{\bar{E}q\,\delta s}{q} = \bar{E}\,\delta s,$$

$\bar{E}$ denoting the mean value of the field along the path δs, or

$$\bar{E} = -\frac{\delta V}{\delta s}. \tag{20.11}$$

If we have equal steps δV between equipotentials, $\bar{E}$ between any two of them is inversely proportional to the corresponding δs; the field is everywhere inversely proportional to the distance between successive equipotential surfaces. This is the counterpart, in the representation of the field by equipotential surfaces, to the proportionality of field to areal density of lines.

It should be noted here that if we proceed to the limit of smallness of δs, Eq. (20.11) for the mean field becomes

$$E = -\frac{dV}{ds}; \tag{20.12}$$

the field is equal to the negative potential gradient. Accordingly, it is frequently expressed in such units as volts per meter, or volts per centimeter.

As an interesting application of Eq. (20.12), let us investigate how the field near the surface of an isolated charged conductor varies with the curvature of the surface, basing our reasoning on the associated system of equipotentials, instead of, as previously, on lines of force.

Consider a conductor of the form shown in profile in Fig. 20.18. The surface of the conductor is necessarily itself an equipotential surface. At a great distance from the conductor the field must approximate to that of a point charge, and accordingly the

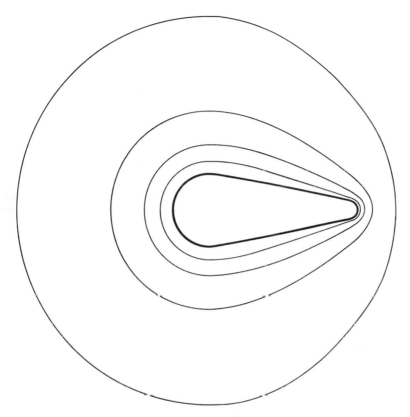

FIGURE 20.18 System of equipotentials surrounding charged pointed
conductor.

equipotentials must approximate to spheres. Hence, as the distance
from the conductor increases, there must be a gradual transition
in the form of successive equipotentials from that of the conductor
itself (at the surface of the conductor) to something approximating
very closely to a sphere. There is only one way in which this transition
can occur—as shown in the figure; the equipotentials near regions
of sharp curvature must be substantially closer together than else-
where. In the neighborhood of these regions, therefore, $|dV/ds|$, and
hence the field E, must be greater than in regions where the curvature is relatively small.

A section through a system of equipotential surfaces may be
regarded as analogous to lines of equal elevation—contour lines—
drawn on a map. Just as the steepest gradients are in directions at
right angles to the contour lines representing the topography of a
land surface, so, in the section under consideration, the electric
fields are directed at right angles to the equipotential lines. This

529

gravitational analogy is exceedingly useful, and is a powerful aid to the recognition of the properties of an electrostatic field. The representation of electrostatic fields by equipotential surfaces has led to important advances in the development of systems for guiding charged particles in an evacuated space and for focusing "beams" of such particles in various ways. This technical application of electrostatic field theory, known as "electron optics," will be studied in Chapter 23.

By bearing in mind that the surface of a conductor is necessarily an equipotential surface, it is often possible to guess with considerable accuracy the configurations of the equipotentials between such surfaces. It is largely this possibility that has made the representation of electrostatic fields by equipotential surfaces so useful.

Problems

20.1 Sketch qualitatively the system of lines of force in a plane in which two equal point positive charges are situated. To what does this system approximate in regions whose distance from the charges is large compared with the distance between these charges? What change would be produced in the field by substitution of negative for the positive charges?

20.2 Sketch qualitatively the system of lines of force in the plane of an equilateral triangle at the corners of which equal positive charges are situated.

20.3 Show by simple reasoning based on Coulomb's law that everywhere within a uniformly charged infinite hollow cylinder the field is zero.

20.4 Find by direct integration the field at a distance r from an infinitely long thin wire having a charge λ per unit length.

20.5 Charges $+Q$ and $-Q$ are situated on the x axis of a Cartesian coordinate system at the points $x = +a$ and $x = -a$, respectively. Show that the electric field intensity at the points $(x, 0)$ and $(0, y)$, where x and y are large compared with a, are given by the equations

$$E_x = \frac{M}{2\pi \epsilon_0 x^3}, \qquad E_y = 0,$$

and

$$E_x = -\frac{M}{4\pi \epsilon_0 y^3}, \qquad E_y = 0,$$

respectively, M being the dipole moment of the system of charges, defined as $2Qa$. By assuming that a dipole moment is a vector quantity, show that the equations of the lines of force in polar coordinates are of the form

$$r = A \sin^2 \theta,$$

where A is a constant. Hence sketch the shape of the field.

20.6   An electric dipole of moment M (see Problem 20.5) is aligned with an external field E. How much work would have to be performed to reverse the direction of the dipole, making it "antialigned" with the field?

20.7   Find the electrostatic field at a point on the axis of a circular wire loop that carries a total charge q. Use the expression found to determine the field intensity at a point (a) outside and (b) inside a uniformly charged hollow spherical metal shell that carries a total charge Q. Show how, considering gravitational instead of electric fields, the result obtained under (a) may be extended to show that any body having a spherically symmetrical distribution of mass produces a gravitational field outside it which is identical with that which would be produced if the whole mass of the body were concentrated at its geometrical center. What inference may be drawn from (b) concerning the gravitational field at a point within such a body?

20.8   Gauss's theorem may be used to show that the field due to a single isolated uniformly charged sphere at any outside point is the same as if the whole charge were concentrated at the geometrical center of the sphere. Show that it follows from this that the mutual force between two uniformly charged spheres is the same as if the charge on *each* sphere were concentrated at its geometrical center.

20.9   The maximum field that can be sustained by air at standard atmospheric pressure without insulation breakdown is 30,000 V/cm. What is the total charge on an isolated metal sphere of diameter 4 cm when the field at its surface has this value?

20.10   Write the equation for the gravitational analogue of Gauss's theorem, and show that if Gauss's theorem applies in electrostatics the analogous gravitational equation must also be valid. Hence show that at any point outside a body having a spherically symmetrical distribution of mass whose total mass is m, the gravitational field intensity is Gm/r^2, r being the distance of the point from the center of the body.

20.11   Use Gauss's theorem and the fact that an electrostatic field is conservative to prove that, in the absence of any charged body held within a hollow conductor, not only is the total charge on the inner surface of the conductor zero, but also no part of this surface can have any charge, regardless of the geometrical form of the conductor or of the charge residing on its outer surface.

20.12   Charges Q_1 and Q_2 reside on the surfaces of an isolated system of two concentric thin spherical metal shells of radii R_1 and R_2, respectively $(R_1 < R_2)$. How are the charges distributed between the inner and outer surfaces of these spheres? How does the field intensity vary with the distance r from the common center of the spheres (a) for $r < R_1$, (b) for $R_1 < r < R_2$, and (c) for $r > R_2$?

20.13   Two isolated infinite coaxial thin cylindrical metal shells of radii R_1 and R_2 $(R_1 < R_2)$ carry charges per unit length λ_1 and λ_2, respectively. How are the surface charge densities σ distributed between the inner and outer surfaces of these two cylinders? How does the field E vary with the distance r from the axis (a) for $r < R_1$, (b) for $R_1 < r < R_2$, and (c) for $r > R_2$?

20.14   Two infinite parallel metal plates have total charges γ_1 and γ_2 per unit area of plate. Find the separate areal charge densities σ_1, σ_2,

σ_3, and σ_4 on the four surfaces. Also find the field intensities both between the plates and in the region outside.

20.15   The electric field is everywhere radial and of constant magnitude E within a sphere of radius R. What must be the corresponding variation with distance r from the center of the sphere of (a) the volumetric charge density ρ, and (b) the potential V? Consider both $r < R$ and $r > R$.

20.16   Given that in a certain region the electric field intensity has Cartesian components $E_x = ax$, $E_y = ay$, and $E_z = 0$, where a is a constant, determine the shapes of the lines of force. Find the charge contained in a cylinder whose axis is along the z axis and whose radius and length are r and l, respectively.

20.17   At a certain point in an electrostatic field the potential has a gradient in a certain direction of 500 kV/m. What is the component in this direction of the force exerted on a positive charge of 1 μC?

20.18   Two isolated conducting spheres, one having a radius twice that of the other, are connected by a long thin wire and given a charge. How will the charge distribute itself between the two spheres? On which will the charge density be the greater?

20.19   Must a positively charged insulated conductor have a positive potential, the potential at infinity being taken as zero, or could its potential be negative? Explain.

20.20   Two raindrops of equal size having identical negative charges come into collision and coalesce to form a larger sphere. What effect does this have on (a) the surface density of charge, (b) the field at the surface, and (c) the potential relative to earth? Would you expect the electrical energy of the final single drop to be greater or less than the sum of the energies of the original drops? If there is a change, by what mechanism has this been brought about?

20.21   What are the values of the potential V as a function of r for the three cases referred to in Problem 20.12, the potential at infinity being taken as zero?

20.22   Using the expression found in Problem 20.7 for the field E at a point on the axis of a circular loop of wire which carries a charge q, show by integration that the potential at this point relative to zero at infinity is given by the equation

$$V = \frac{1}{4\pi\epsilon_0} \frac{q}{\sqrt{x^2 + a^2}},$$

where a is the radius of the loop and x is the distance of the point in question from its center. Show how this result follows immediately from Eq. (20.10), independently of any consideration of the field E, and check that E as found in Problem 20.7 is the same as $-dV/dx$.

20.23   Sketch the intersections of equipotential surfaces with the planes of the diagrams of Problems 20.1 and 20.2.

20.24   Two unequal point charges of opposite sign, of magnitudes $|Q_1|$ and $|Q_2|$, are placed a distance d apart. Show that the surface of zero potential, relative to infinity, is a sphere, and find its radius and the position of its center.

ELECTROSTATIC DEVICES

The devices to be considered in this chapter are electrostatic generators, electrostatic voltmeters, capacitors, and the attracted-disk electrometer. Certain theoretical topics involved in the operation of these devices will also be discussed.

Van de Graaff Generator

The action of the Van de Graaff generator depends essentially on (1) the fact that there can be no charge within a hollow conductor, and (2) the existence of an exceptionally strong field in the neighborhood of a charged metal point. Apart from its use of the action of points, the operation of a Van de Graaff generator may be regarded as corresponding to a continuously acting Faraday's ice-pail experiment, the experiment referred to being that in which a charged conductor on an insulating handle is introduced into the interior of an insulated hollow conductor and electrical contact with the latter then established.

Several devices have been developed in which the principle demonstrated by this experiment is used to build up large charges. Probably the first was Kelvin's "water dropper,"* in which a succession of charged drops of water fall through an insulated hollow conducting cylinder, each drop in turn making electrical contact with the cylinder while within it, so that the whole of its charge is transferred to the cylinder. Kelvin also suggested a modification of this device, in which the drops of water were replaced by metal segments attached to an insulating conveyor belt, and later such a machine was actually constructed by Righi. An important further step was taken in the early 1930s by R. J. Van de Graaff and collaborators working at the Massachusetts Institute of Technology, in which the metal segments were dispensed with, charge being simply sprayed onto the belt by a set of points well outside the hollow conductor and then collected by another set of points inside. They also considered a system in which the belt is

---

* See J. Gray, *Electrical Influence Machines*, Whittaker & Co., London, 1903.

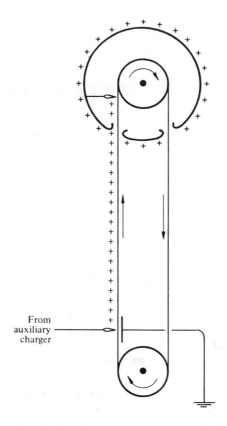

From
auxiliary
charger

FIGURE 21.1   Van de Graaff generator.

charged by friction before entering the hollow conductor, but found
the method of spraying the charge onto the belt by a set of points
to give the greater output.

   The Van de Graaff generator is shown schematically in its
simplest form in Fig. 21.1. An auxiliary charger such as a transformer–rectifier system (see Chapter 33) raises a set of charging
points to a high positive or negative potential relative to earth.
These points are arranged in a line occupying the width of the belt,
and an earthed plate is situated opposite these points on the other
side of the belt. The field produced between these is particularly
strong in the vicinity of the points, where a "corona discharge" is
produced, charge being sprayed from the points in the direction of
the plate. This charge is intercepted by, and retained on the surface
of, the belt, which carries it up into the interior of a hollow sphere,
usually called the "dome," which is mounted on an insulating
support. Here the charge is collected by a further set of points connected to the dome and at once conveyed to the outer surface

of the latter, where it remains. As more and more charge is collected the potential of the dome relative to the earth rises until the stage is reached where the rate of loss of charge from the dome, for example by corona discharge from asperities plus leakage along supports, just balances the rate of charging. In this arrangement the pulley and its supports would usually be connected electrically to the dome. (In the interest of simplicity this feature is not shown in the diagram.)

Various belt materials have been used in these machines; any material that insulates sufficiently well and also has suitable mechanical properties will serve. Among the materials that have been used successfully are silk, insulation paper, rubber, and nylon. The insulation of some belt materials deteriorates when exposed to a damp atmosphere, because of the absorption of moisture, and if such materials are used, precautions must be taken against exposure while the machine is in operation.

Van de Graaff and his collaborators soon realized that the output of the machine would be increased if provision could be made for the belt on its return journey from the dome to carry away a charge of opposite sign to that supplied by the charger. The simplest way of doing this is by means of the arrangement indicated in Fig. 21.2. Here some of the charge carried up by the belt is collected by points attached to the frame carrying the upper pulley. The remainder is collected by another set of points connected to the dome, and, because the pulley is charged with the same kind of electricity as

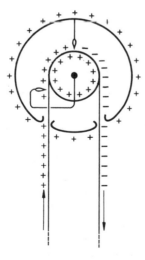

FIGURE 21.2 System for spraying charge of opposite polarity on belt before it leaves dome of generator.

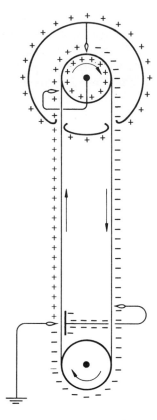

FIGURE 21.3 Self-exciting Van de Graaff generator.

that delivered by the belt, these points also spray charge of opposite polarity to the outer surface of the belt, and this is carried down by the belt on its return journey.

Obviously, what can be done in the region of the upper pulley within the dome can also be done near the lower pulley outside it; in the same way as some of the charge on the belt moving upward toward the upper pulley is made to induce a charge of opposite polarity which is sprayed onto the belt, to be carried down on the other side, so the latter charge, approaching the lower pulley, may be made to provide the requisite charge on the belt moving upward from this lower pulley. In other words, the provision of an auxiliary charger, such as is indicated in Fig. 21.1, is really unnecessary; the machine may be made self-exciting. Such a machine, in which no auxiliary charger is used, is shown in Fig. 21.3. The arrangement at the bottom, although the same in principle as that at the top, is modified in such a way as to avoid the necessity for insulating the lower (driving) pulley. Charge collected from the downward-traveling belt is communicated to a plate mounted behind the belt on the other side. This charged plate induces charge of opposite polarity which is sprayed onto the belt from points connected to earth as it travels upward toward the dome.

Of course, a machine of this kind has to be *started* in some way; it has to be "primed." Such priming often occurs spontaneously, for example through frictional charging of the inner surface of the belt in its engagement with the pulley, this causing a charge of opposite polarity to be sprayed onto the outer surface. Otherwise the machine may be primed by holding a cloth against the outer surface of the belt for a few seconds, so giving it an initial charge.

The potential relative to earth that is built up on the dome is limited by leakage along the supports and corona discharge, electricity ultimately being lost by these processes as fast as it is supplied by the action of the machine. The insulation of air at atmospheric pressure breaks down when the field reaches an intensity of about 30,000 V/cm, and leakage by corona discharge sets in when the field reaches this value either at projections or asperities on the dome or at similar projections on neighboring objects where strong fields are produced by induction. Thus a drastic reduction in the potential of the dome may sometimes be observed to occur in consequence merely of pointing a finger at it. Hence in the use of these machines it is important to avoid sharp curvatures on any part of the dome or on neighboring objects.

The field at which the insulation of the gas breaks down is proportional to the pressure—the reason for this will be appreciated when we consider the mechanism of the breakdown in Chapter 24.* Hence a substantial increase in dome potential may be achieved simply by mounting the machine within an enclosure containing air or other gas at a pressure of several atmospheres. In this way, working dome potentials of a few million volts may be established with machines of quite moderate dimensions. In the case of such "pressurized" machines it is usual for the enclosure in the region of the dome to be of metal, constructed in the form of a substantially larger sphere, or a hemispherical end of a cylinder, concentric with the dome and kept at earth potential.

---

* Although the proportionality between the field strength at which insulation breakdown occurs in a gas and the pressure holds strictly for all pressures here contemplated, this law ceases to be valid at very low pressures, where the mean free path of the gas molecules becomes comparable with the linear dimensions of the apparatus. Below a certain pressure the breakdown field strength increases with *decreasing* pressure.

537

For convenience, Van de Graaff generators are sometimes mounted with their axes horizontal instead of vertical.

CIRCULATION OF ELECTRICITY   It is important to realize that during the steady operation of a Van de Graaff or any other electric generator there is a continuous *circulation* of electricity; there is nowhere any accumulation of charge. Quite commonly the earth and the air surrounding the machine both play a part in this circulation. Let us, for example, consider how the electricity circulates in the system depicted in Fig. 21.1. Here positive charge comes up from the earth, is sprayed onto the upward-moving belt at the lower left-hand side, and is collected by the dome. When the breakdown field strength has been built up at the surface of the dome, this charge leaks away to the surrounding air and is carried by the field* back to earth, whence it once more supplies the points spraying the belt. Alternatively, we may think of negative electricity being collected from the belt by the same set of points and conveyed to earth and then traveling by leakage through the air to the dome, whence it is sprayed back onto the belt at the top.

Instead of passing through the air in the manner described, this circulation may be diverted through some apparatus served by the machine.

Thunderstorm Cell

This is a name given to a region of the atmosphere in which a thunderstorm is generated. During the period of its activity— usually something of the order of an hour—a thunderstorm cell is a gigantic electrostatic generator, of enormous power.

A thunderstorm cell may be created when air covering a large area becomes heated and humidified sufficiently to give rise to a strong convective updraft. In Chapter 16, where we considered the various mechanisms by which rain is produced, we saw that under certain conditions such an updraft may attain a very high velocity. When this happens, and other conditions are favorable, a thunderstorm will be generated.

---

* As will be appreciated later, for leakage to occur through a body of gas it is merely necessary for the field to exceed the insulation breakdown limit *somewhere* in the leakage path. Here collisions of electrons with neutral molecules produce copious ionization, and, by mechanisms to be discussed in Chapter 24, the supply of charged carriers becomes self-regenerating everywhere in the discharge path.

It is believed that there is more than one charging process operating in a thunderstorm, and a complete theory of thunderstorms has yet to be worked out. However, the basic features of the relevant mechanisms are now beginning to be understood; at least, by a process of elimination, those now considered possible have been reduced to a relatively small number. We shall now discuss briefly two alternatives for the main charging mechanism and a subsidiary mechanism which is believed to operate in conjunction with one or both of these.

Laboratory investigators have shown that when ice and water are in contact with one another an appreciable potential difference is established between them.* Which is positive and which negative depends on the nature of the impurities in the water, but with the impurities usually present in clouds the ice becomes negatively charged and the water positively. In the usual kind of precipitating cloud ice and water *are* in contact in the lower regions, below the $0°C$ level. Here the ice progressively melts on its way down, and if the hailstones are large enough, so that they acquire high terminal velocities relative to the rapidly upward-moving air through which they fall, some of the water adhering to them may be blown off and carried, as a positively charged mist, up into the higher regions of the cloud, leaving the hailstones with a negative charge. These would ultimately become negatively charged raindrops. We should thus have, as our final condition, a cloud whose upper regions carry a positive charge, while the base consists of negatively charged raindrops. This corresponds to what is, in fact, the normal situation in thunderclouds.

Recently another effect has been observed which may equally well be responsible for the main separation of charge observed in normal thunderclouds. When a supercooled water droplet freezes suddenly on impact with an ice crystal or a falling hailstone, it freezes first on the outside, forming a solid shell about an inner core of water; and when this, in turn, freezes, expanding as it does so, it bursts open the shell and sends out a shower of splinters. It has been found that these splinters carry away a positive charge, leaving the residue negatively charged. Whatever may be the mechanism of this separation of charge associated with splintering, the fact has definitely been established that it does, somehow, occur. The splinters,

---

* The process by which this occurs is not fully understood. It may perhaps be assumed to correspond broadly to the separation of charge which takes place in other cases where dissimilar bodies are brought into intimate contact, for example when polystyrene is rubbed with flannel or when a metal body is similarly rubbed.

which are carried by the updraft to higher levels, finally constitute the positively charged upper portion of the cloud, while the lower part consists of the negatively charged residues. Measurements that have been made of the amount of charge separation produced by the freezing of each supercooled droplet, and calculations based on these measurements, show that the effect is of the right order to account for the quantities of charge produced in a thundercloud.

Although one of these processes may be more efficient than the other, and so make the more important contribution to the main charge separation, they are not mutually exclusive; both processes may well operate simultaneously in thunderstorm cells, the former in the lower regions, below the 0°C level, and the latter above this level.

In most clouds it is found that in addition to the general negative charge of the lower portion of the cloud, there is also a local accumulation of positive charge in the region of turbulence at the base, associated with large raindrops. A typical cloud in which this lower positively charged region is present is represented in Fig. 21.4. This is believed to be due to the breaking up of drops, after they have become too large for stability, into drops differing greatly in size. When this happens the larger drops acquire a positive charge

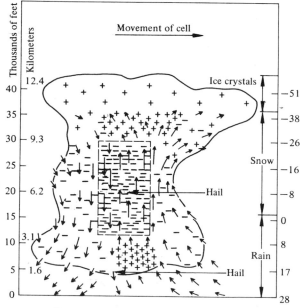

Arrows represent wind direction and velocity

FIGURE 21.4   Thunderstorm cell.

and the relatively small droplets become charged negatively. If—as would here be the case—the original large drop is already negative, a continuation of this process would tend to make it less negative and could eventually lead to its becoming positively charged. Presumably the process *can* continue, by virtue of the coalescence of larger drops within the turbulent region of the cloud, to again form drops that are too large for stability, so that these, like their predecessors, break up into smaller drops. The fine mist of very small droplets, carrying a negative charge, would soon be conveyed by air currents away from the region occupied by the large positively charged drops.

It has been estimated that an average lightning stroke discharges about 20 C and that the total charge generated by a thunderstorm cell during its active life is of the order of 3000 to 4000 C. The energy expended in an average lightning stroke is of the order of 10^{10} to 10^{11} J. All the thunderstorms occurring over the surface of the earth give mainly negatively charged rain, which, added to what is discharged in cloud-to-earth lightning strokes, charges the earth at an average rate of roughly 1800 A. Obviously such charging cannot continue indefinitely, and it is balanced by the flow, between the earth and the positively charged upper regions of the troposphere, of ions (see Chapter 24) created in the atmosphere by cosmic rays and radioactivity. This flow of ions—positive ions one way and negative the other—occurs in response to the field set up between the surface of the earth and positively charged clouds plus ionized regions of clear atmosphere that have been left behind by evaporated clouds. The field strength is about 1 V/cm.

Electrostatic Voltmeters

We have hitherto considered a gold-leaf electroscope* as an instrument for registering a charge, either on itself or residing on an object brought near it. As we shall now see, the deflection of the leaf or leaves is also determined uniquely by the potential difference between the leaf system and the casing. Consequently, the instrument may be calibrated to read this potential difference, for example in volts.

The system within the casing of an electroscope may be regarded as a little world of its own, the deflection of the leaf or leaves being determined simply by the charges on the leaf system

---

* The term "gold-leaf electroscope" is to be taken as including all instruments of this type, whether the leaf is actually of gold or of some other metal, for example aluminum.

and on the inner surface of the casing. As we shall see later, the force per unit area acting outward on any given portion of a leaf is determined uniquely by the surface density of charge, σ, residing on this portion, being, in fact, proportional to σ^2. Just outside the surface

$$E = \frac{\sigma}{\epsilon_0}, \qquad (20.6)$$

so the force per unit area is also determined by the value of E close to the surface. But the PD between the leaf system and the casing is, by definition, $\int E\, ds$ along any line of force from the leaf to the casing. Clearly, the greater this PD, the stronger must be the field along the whole of the path considered, including the region adjacent to the area of leaf in question, and the greater, therefore, is the force on this area. Thus the deflection must be determined uniquely by the PD, the two increasing and decreasing together.

It follows that if we connect the disk and casing of a calibrated electroscope electrically (by wires, for example) to any two conductors the potential difference between which we wish to measure, the PD may be found simply from the deflection. The deflection is most conveniently observed and measured, either on a screen onto which an optical image of the leaf system is projected or through a telescope provided with a graticule onto which the image is focused.

Any device, such as a calibrated electroscope, that enables us to measure potential differences is known as an **electrometer**. If it is an electrostatic device it may also be called an **electrostatic voltmeter.**

A form of electrometer that lends itself rather better to precise measurement than does the calibrated gold-leaf electroscope is one developed by Braun, shown in Fig. 21.5, whose indicator is a pivoted needle. The figure is self-explanatory.

By using electrostatic voltmeters of suitable design it is possible to measure potential differences ranging from a few hundred thousand volts down to 10^{-4} V. Among the most sensitive of these instruments is Kelvin's quadrant electrometer, whose construction and theory we shall not discuss here. The gold-leaf and Braun forms of electrometer usually measure potential differences up to 200 to 500 V.

Capacitors (Condensers)

A capacitor is a device consisting essentially of two metal electrodes separated by air or some other insulating medium. The electrodes are typically in the form of parallel plates or coaxial

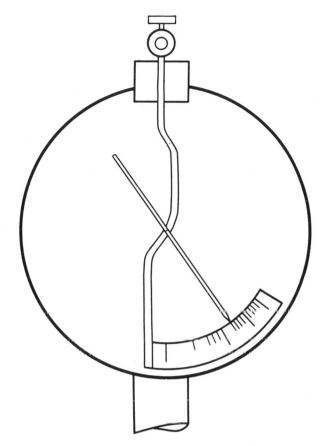

FIGURE 21.5 Braun's needle-type electrostatic voltmeter.

cylinders. When these are given charges of opposite polarity a field is set up in the intervening insulating medium. Capacitors are not only of considerable theoretical interest but also have important technical applications, particularly in alternating-current circuitry.

We shall begin our study by considering only air (or vacuum) capacitors. Later we shall discuss the effect of substituting some other insulating material for air.

The most important property of a capacitor is its **capacitance** (or **capacity**) C. This is defined as the ratio of the charge Q on one electrode* to the corresponding potential difference V between the electrodes:

$$C = \frac{Q}{V}. \tag{21.1}$$

---

* If the charge Q resides on one electrode, the other electrode has the equal and opposite charge $-Q$. In discussing capacitors it is in general convenient to disregard signs, simply referring to the magnitude of the charge on either electrode; that is, what we write as Q really means $|Q|$.

The MKSA unit of capacitance is the coulomb per volt, usually called the **farad** (F). Accordingly, in the defining equation (21.1) C is expressed in farads if Q is in coulombs and V in volts. The farad is much too large a unit for expressing conveniently the capacitances of actual capacitors, so these are usually expressed in microfarads (μF), nanofarads (nF), or picofarads (pF). These are of magnitudes 10^{-6} F, 10^{-9} F, and 10^{-12} F, respectively. Older names for 10^{-9} F and 10^{-12} F, which are now going out of use, are the milli-microfarad (mμF) and micromicrofarad ($\mu\mu$F), respectively. Still smaller units, which, however, one seldom has occasion to use, are the femtofarad (10^{-15} F) and the attofarad (10^{-18} F).

CONCENTRIC-SPHERE CAPACITOR A capacitor having any approximation to the concentric-sphere form is difficult to construct and is accordingly seldom used in practice. Nevertheless, some theoretical interest attaches to this capacitor, and for this reason we shall now consider it quantitatively.

Applying Gauss's theorem to any spherical surface between the metal spheres of this capacitor and concentric with them, we see at once that the field between these spheres is nowhere affected by the presence of the outer sphere. Thus the field in this region is the same as if the charged inner sphere were isolated in space. We have also seen that this field is identical with that which would be produced if the charge Q on the inner sphere were concentrated at its geometrical center. We may therefore obtain the potential difference between any two points A and B in the region between the metal spheres whose distances from the center are r_A and r_B, respectively, directly from Eq. (20.9),

$$V_A - V_B = \frac{Q}{4\pi\epsilon_0}\left(\frac{1}{r_A} - \frac{1}{r_B}\right).$$

A limiting case which particularly concerns us here is that in which r_A and r_B are the radii of the metal spheres themselves. The capacitance C of the capacitor is then given by the corresponding ratio of Q to $V_A - V_B$, $4\pi\epsilon_0/[(1/r_A) - (1/r_B)]$:

$$C = \frac{4\pi\epsilon_0}{(1/r_A) - (1/r_B)}. \tag{21.2}$$

A special case of some importance is that in which r_B is infinite. In practice this condition is approximated when a sphere is at a great distance from all other bodies—a distance of a higher order than its own linear dimensions. In this case we have, in effect, a capacitor one electrode of which is this sphere, while nearby portions of the earth and objects at earth potential constitute the other electrode. Under these conditions the expression for the capacitance becomes simplified to

$$C = 4\pi\epsilon_0 r, \qquad\qquad (21.3)$$

where r is the radius of the sphere.

As an interesting application of Eq. (21.3), let us calculate the maximum potential relative to earth which an effectively isolated sphere of radius r can sustain in air at atmospheric pressure without the field E at its surface causing insulation breakdown. This sphere might, for example, be the dome of a Van de Graaff generator.

The field E at the surface of the sphere is, as we have seen, given in terms of the charge density σ by the equation

$$E = \frac{\sigma}{\epsilon_0}.$$

Since

$$\sigma = \frac{Q}{4\pi r^2},$$

where Q is the total charge on the sphere and is the product of the capacitance C of the sphere and its potential V, we have, on substituting for C its value, $4\pi\epsilon_0 r$,

$$E = \frac{\sigma}{\epsilon_0} = \frac{Q}{4\pi\epsilon_0 r^2} = \frac{CV}{4\pi\epsilon_0 r^2} = \frac{4\pi\epsilon_0 r V}{4\pi\epsilon_0 r^2} = \frac{V}{r},$$

or

$$V = rE. \qquad\qquad (21.4)$$

It is found that the insulation breakdown field E in air is approximately 30,000 V/cm, or 3×10^6 V/m. Accordingly, with V expressed in volts and r in meters, we have, numerically,

$$V = 3 \times 10^6 r.$$

Thus, a Van de Graaff generator whose dome has a radius of 1 m should, ideally, be able to build up a dome potential of about 3 million volts in air at atmospheric pressure.

Another interesting case of a spherical capacitor is that in which the distance t between the surfaces of the spheres is small compared with the radius of either. Writing Eq. (21.2) in the form

$$C = \frac{4\pi\epsilon_0 r_A r_B}{r_B - r_A},$$

and substituting t for $r_B - r_A$ and r^2 for $r_A r_B$, we obtain for C the expression

$$C = \frac{4\pi\epsilon_0 r^2}{t}.$$

But $4\pi r^2$ is, to a close approximation, the area of each spherical electrode. Thus the capacitance per unit area is ϵ_0/t. A sufficiently small portion of this capacitor may be regarded as a parallel-plate capacitor. Hence, writing A for the area of such a portion, we obtain for the capacitance of a parallel-plate capacitor the expression

$$C = \frac{\epsilon_0 A}{t}. \tag{21.5}$$

PARALLEL-PLATE CAPACITOR Equation (21.5), for the capacitance of a parallel-plate capacitor, may be obtained more directly from Eq. (20.6) in conjunction with the definition of PD. If the density of the charge on the plates is σ and the area of the plates is A, then Q is equal to σA. And, according to Eq. (20.6), the field strength in the region between the plates is σ/ϵ_0. The corresponding potential difference V must be t times this, $\sigma t/\epsilon_0$. Dividing Q by V we obtain the expression (21.5) for the capacitance.

These derivations of Eq. (21.5) have been obtained on the basis of the tacit assumption that the field between the plates is uniform and is cut off abruptly beyond the edges. Actually this assumption is not justified; there must always be an edge effect, which will modify the capacitance. The edges of the plates represent regions of relatively sharp curvature, and there must accordingly be a somewhat increased charge density at and near the edges. There must also be some charge on the back surfaces of the plates near the edges. This means that if we write σ for the charge density in the central regions, Q must somewhat exceed σA. The capacitance must therefore be correspondingly greater than $\epsilon_0 A/t$.

COAXIAL-CYLINDER CAPACITOR In our discussion of applications of Gauss's theorem, we saw that if the charge per unit length of an infinitely long cylinder is λ, then at a distance r from the axis, r being greater than the radius of the cylinder, the field strength is $\lambda/2\pi\epsilon_0 r$. This will be the magnitude of the field regardless of whether the cylinder in question is at a great distance (compared with its radius) from all other bodies or is surrounded by a larger, coaxial, cylindrical metal surface. If in the latter case we denote the radii of the inner and outer cylinders by r_A and r_B, respectively, we have for the potential difference V between the cylinders the expression

$$V = \frac{\lambda}{2\pi\epsilon_0} \int_{r_A}^{r_B} \frac{dr}{r} = \frac{\lambda}{2\pi\epsilon_0} \ln\left(\frac{r_B}{r_A}\right). \tag{21.6}$$

If, instead of being infinitely long, the cylinder is of length l, which is large compared with $r_B - r_A$, then, neglecting end effects, we may write λl for Q, and so, dividing this by V, we obtain for C the expression

$$C = \frac{2\pi\epsilon_0 l}{\ln\ (r_B/r_A)}. \tag{21.7}$$

An interesting special case is that in which $r_B - r_A$ is small compared with either r_A or r_B, that is, where r_B/r_A is only slightly greater than 1. Let r_B/r_A be written in the form $1 + [(r_B - r_A)/r_A]$, or $1 + t/r$, where t is substituted for $r_B - r_A$ and r for r_A. Then the logarithm of this may be expanded according to the formula

$$\ln\ (1 + x) = \frac{x}{1} - \frac{x^2}{2} + \frac{x^3}{3} - \frac{x^4}{4} + \cdot\ \cdot\ \cdot\ ,$$

which applies to the case where x is less than 1. In the present case x, which corresponds to t/r, is very small indeed, and so the expansion need not be continued beyond the first term. We have, therefore, as a sufficiently close approximation,

$$\ln\left(\frac{r_B}{r_A}\right) = \ln\left(1 + \frac{t}{r}\right) = \frac{t}{r},$$

and accordingly Eq. (21.7) becomes simplified to

$$C = \frac{2\pi\epsilon_0 r l}{t},$$

which, on substituting A for $2\pi r l$, may be written

$$C = \frac{\epsilon_0 A}{t}. \tag{21.5}$$

Thus we have shown mathematically what is in any case obvious physically—that the cylindrical capacitor just discussed may be regarded as a "rolled-up" parallel-plate capacitor.

Dielectrics

An alternative name for an insulator is a **dielectric.**

If the space between the plates of a capacitor is occupied by a dielectric instead of by a vacuum, the capacitance is increased by a factor that is characteristic of the material of the dielectric. This factor, for which we shall use the symbol κ, is known as the **dielectric constant,** or **specific inductive capacity,** of the material. In all cases where a dielectric material replaces a vacuum between the plates of a capacitor, the expressions for capacitance given in Eqs. (21.2), (21.3), (21.5), and (21.7) should be multiplied by the factor κ.

MEASUREMENT OF DIELECTRIC CONSTANT A simple, although not very accurate, method of measuring the dielectric constant of a solid dielectric is the following. One plate of a parallel-plate air capacitor is connected to an electrostatic voltmeter and the other to earth, the former being charged so that the voltmeter gives a reading of suitable magnitude. A slab of the material in question, of a thickness such that it fits snugly between the plates of the capacitor, is then placed between them and the new voltmeter reading noted. The dielectric constant is then the ratio of the former potential difference to the latter.

The accuracy of this method is limited by two factors. One is the difficulty of matching the slab of dielectric exactly to the spacing between the plates of the capacitor. The other is the capacitance of the electrostatic voltmeter, which—as will be appreciated later—has the effect of reducing the ratio of potential differences below the true dielectric constant of the material.

In the method just considered, the charge on the plates of the capacitor is kept constant and readings are taken of the potential difference between them with and without the dielectric. An alternative is to work with a constant PD and measure the change in Q. This is the better alternative from the point of view of accuracy. Q is most conveniently measured by discharging the capacitor through a galvanometer and noting the ballistic throw (see Chapter 25). If the dielectric whose constant is to be measured is a liquid (for example, an oil), the capacitor may first be used as an air capacitor and then entirely immersed in a large vessel of the liquid; or better, if the capacitor is of suitable form, simply filled with liquid between the plates. In the more usual case where the dielectric is a solid, a good procedure would be to construct two geometrically identical capacitors, one with air and the other with the material in question as its dielectric.

In the above methods no distinction is made between air and a vacuum, and what is measured is, actually, only the ratio of the dielectric constant of the experimental material to that of air. However, in general, the error incurred in doing this is negligible, being less than the experimental error.

Other methods, which are more accurate than those considered above, are also available. A discussion of these would be beyond the scope of this book.

The dielectric constants of a number of materials are given in Table 21.1.

Dielectric Constant and Coulomb's Law   From the fact that the introduction of a slab of dielectric between the plates of a charged

TABLE 21.1 Dielectric Constants

Material	κ	Material	κ
Air (STP)	1.00059	Polystyrene	2.55
Castor oil	4.5	Glass	4–7.5
Ethyl alcohol	26	Porcelain	7
Water	80	Quartz ($\perp$ axis)	4.55
Paraffin wax	2.2	Quartz ($\parallel$ axis)	4.49
Rubber	2.4	Calcite ($\perp$ axis)	8.5
Ebonite	3.0	Calcite ($\parallel$ axis)	8.0
Shellac	2.9	Mica	7

parallel-plate capacitor reduces the potential difference between the plates by the factor κ, it must be inferred that the field strength in the region between the plates is reduced by this factor. The same inference may be drawn from the corresponding increase in capacitance of a capacitor of any other form which results from filling the whole interelectrode space with a dielectric.

It should be noted that this reduction in field strength is something we *infer* from the increase in capacitance; we do not observe it directly. Direct observation would generally be difficult, to say the least, and in the case of a solid dielectric it would be virtually impossible. Accordingly, it is misleading to present the equation

$$F = \frac{1}{4\pi\epsilon} \frac{Q_1 Q_2}{r^2}, \tag{21.8}$$

where ϵ is written for $\kappa\epsilon_0$, as "Coulomb's law" for the mutual force between point charges in any insulating medium, even though there is no reason to doubt that this equation correctly gives the force that *would* be observed under these conditions if such observation were practicable.

The quantity ϵ, defined as $\kappa\epsilon_0$, is known as the **permittivity** of the medium whose dielectric constant is κ.

It should be noted that while the concept of a permittivity ϵ characteristic of an insulating medium other than a vacuum is useful mathematically, it really has no basic physical significance. This is because, according to modern atomic theory, the constituent atoms of such a medium consist essentially of a system of positively and negatively charged particles whose occupancy of space relative to that of the atoms as a whole is utterly insignificant; practically the whole volume occupied by an atom is empty space, and the charged particles constituting it may be regarded as point charges within

549

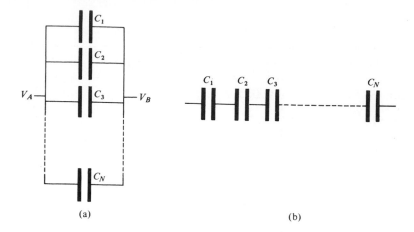

(a)   (b)

FIGURE 21.6 Capacitors (a) in parallel and (b) in series.

this space. The effect of applying a field is simply to distort the atoms, in the sense of displacing the positive charges slightly from their average zero-field sites in the direction of the field and the negative charges in the opposite direction. Thus the field obtaining within the dielectric is the resultant of the applied field and the field due to the displaced particles, all acting in a vacuum. In this sense there is only one permittivity in nature, ϵ_0.

Capacitors in Parallel and in Series

Arrays of capacitors in parallel and in series are indicated in Fig. 21.6(a) and (b), respectively. Let us consider the capacitances of these in turn.

CAPACITORS IN PARALLEL One electrode of each capacitor is connected to a common terminal, so all these electrodes are necessarily at the same potential, which we may denote by V_A. Similarly, all the other electrodes are at some other definite potential V_B. Consequently, there is the same potential difference V, equal to $V_A - V_B$, across the plates of each capacitor, and so, according to the defining equation (21.1) for capacitance, we must have for the charges on the separate capacitors

$$Q_1 = C_1 V,$$
$$Q_2 = C_2 V, \ldots,$$
$$Q_N = C_N V.$$

550   Hence for the total charge Q on all the plates connected to each common terminal we have

$$Q = Q_1 + Q_2 + \cdots + Q_N$$
$$= (C_1 + C_2 + \cdots + C_N)V.$$

We therefore have for the capacitance C of the whole system, defined as Q/V,

$$C = C_1 + C_2 + \cdots + C_N. \qquad \textbf{(21.9)}$$

CAPACITORS IN SERIES Each pair of connected plates in a system of capacitors arranged in series is uncharged initially. Each such pair must therefore still have zero total charge after the system has been charged, for no charge has been conveyed to, or abstracted from, the pair in the charging process. Thus if the charge $+Q$ resides on any one plate of such a pair, the charge $-Q$ resides on the other. Also the opposite charges on the plates of each capacitor must be equal in magnitude. We see then that the charge on each plate of each capacitor, throughout the whole series, must be of the same magnitude Q.

The potential differences established across the plates of the various capacitors of the series are given by the equations

$$V_1 = \frac{Q}{C_1}, \; V_2 = \frac{Q}{C_2}, \cdots, V_N = \frac{Q}{C_N}.$$

Also, the potential difference $V \; (= V_A - V_B)$ across the whole series is necessarily the sum of the separate potential differences:

$$V = V_1 + V_2 + \cdots + V_N.$$

Combining these last two equations, we have

$$V = Q\left(\frac{1}{C_1} + \frac{1}{C_2} + \cdots + \frac{1}{C_N}\right),$$

and so the capacitance C of the whole series, defined as Q/V, is given by the equation

$$C = \frac{1}{1/C_1 + 1/C_2 + \cdots + 1/C_N},$$

or

$$\frac{1}{C} = \frac{1}{C_1} + \frac{1}{C_2} + \cdots + \frac{1}{C_N}. \qquad \textbf{(21.10)}$$

Energy of a Charged Capacitor

We may regard the process of charging a capacitor as the conveyance of a succession of small elements of charge δQ from the negative plate to the positive, each against the momentarily existing

551

potential difference V between these. In principle this could be directly across the intervening dielectric, but in practice it usually is not. However, as we saw earlier, the corresponding work performed is independent of the path. Thus we may obtain an expression for the energy W^* of the charged capacitor by summing all the items $V\,\delta Q$ of work performed, beginning with the uncharged condition, when $V = 0$, and ending when the PD has its final value V_0, corresponding to which the charge on each plate is of magnitude Q_0. Making the elements of charge conveyed infinitesimal, we have for this energy

$$W = \int_0^{Q_0} V\,dQ.$$

To evaluate this we must make use of the fact that V is equal to Q/C. Making this substitution, we obtain for W the expression

$$W = \frac{1}{2}\frac{Q_0{}^2}{C},$$

which, by virtue of the relation among V, Q, and C (or among V_0, Q_0, and C) just referred to, may also be written in the alternative forms $\frac{1}{2}Q_0 V_0$ and $\frac{1}{2}CV_0{}^2$. Now dropping the subscripts, for which we have no further use, we may write for the energy of a capacitor of capacitance C having a charge of magnitude Q on each plate, the corresponding PD between the plates being V,

$$W = \frac{1}{2}QV = \frac{1}{2}\frac{Q^2}{C} = \frac{1}{2}CV^2. \tag{21.11}$$

Force on a Charged Surface

An expression for the force on a charged surface is most easily obtained by applying Eq. (21.11) to the special case of a parallel-plate capacitor the distance t between whose plates is small compared with the linear dimensions of these. By substituting the value obtained earlier for the capacitance C of such a capacitor,

$$C = \frac{\epsilon_0 A}{t}, \tag{21.5}$$

in the expression $\frac{1}{2}Q^2/C$ for the energy, we find that the latter can also be written in the form

$$W = \frac{1}{2}\frac{Q^2}{\epsilon_0 A}t. \tag{21.12}$$

---

* We use the symbol W for energy here, rather than E, to avoid confusion with electric field intensity.

Let us now suppose that, with Q constant, we increase the spacing between the plates from t_1 to t_2. From Eq. (21.12) we see that we must thereby increase the energy of the capacitor by

$$\frac{1}{2}\left(\frac{Q^2}{\epsilon_0 A}\right)(t_2 - t_1).$$

We do this by exerting a certain force F applied to one plate through the distance $t_2 - t_1$. Obviously, the corresponding work $F(t_2 - t_1)$ we perform must be the amount by which the energy of the capacitor is increased. Equating the two expressions, and canceling out the common factor $t_2 - t_1$, we find that

$$F = \frac{1}{2}\frac{Q^2}{\epsilon_0 A},$$

or

$$\frac{F}{A} = \frac{1}{2}\frac{Q^2}{\epsilon_0 A^2} = \frac{\sigma^2}{2\epsilon_0}, \qquad \textbf{(21.13)}$$

where σ is the surface density of the charge on either plate.

Because the charge density σ and the field intensity E between the plates are related by the equation

$$E = \frac{\sigma}{\epsilon_0},$$

we may substitute accordingly for σ in Eq. (21.12), obtaining the relation between force per unit area and field intensity,

$$\frac{F}{A} = \tfrac{1}{2}\epsilon_0 E^2. \qquad \textbf{(21.14)}$$

It may readily be shown that $\tfrac{1}{2}\epsilon_0 E^2$ is also the expression for the energy of the capacitor per unit volume of the space between the plates. This follows from the fact that the increase in energy brought about by increasing the spacing between the plates from t_1 to t_2 is $F(t_2 - t_1)$, where F is the force exerted on the plate moved. The associated volume increase is $A(t_2 - t_1)$. Hence the volume rate of energy increase is the former divided by the latter, F/A. Substituting for F/A its value according to Eq. (21.14) and writing V for volume (not PD this time), we have

$$\frac{W}{V} = \tfrac{1}{2}\epsilon_0 E^2. \qquad \textbf{(21.15)}$$

Substituting $\sigma\epsilon_0 E$ for σ^2 in Eq. (21.13), we see that a third alternative expression for the force per unit area is

$$\frac{F}{A} = \tfrac{1}{2}\sigma E.$$

On the principle that the force exerted on a charge is the product of the charge and the intensity of the field acting on it, we might perhaps have anticipated that F/A should be equal to σE, not half this. However, such an expectation rests on a misconception of the principle just referred to. In the first place, the field E between the plates is in part *due* to the charge in question, and we cannot expect this to act on the charge as if it had an independent existence. Also, apart from this objection, it should be noted that the charge is not "fully immersed" in the field; it is merely at one extremity of it. Hence there is no valid reason why the factor $\frac{1}{2}$ should occasion surprise.

Equation (21.13) may also be derived directly, using Coulomb's law. Let us consider the force exerted on a small area of one plate of a parallel-plate capacitor of infinite area on which the charge q resides. This is attracted by every element of charge on the opposite plate. From considerations of symmetry it is obvious that in calculating the resultant force exerted by these elements on the charge q we need consider only the normal components; the tangential components must all cancel each other out.

Let the distance between the plates of the capacitor, represented in Fig. 21.7, be t, and let us consider the annular area on the

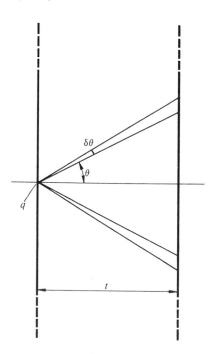

FIGURE 21.7

opposite plate between two conical surfaces constructed about the normal through the location of q as axis, which make with this axis the angles θ and $\theta + \delta\theta$, $\delta\theta$ being small. This area is easily seen to be $2\pi t^2 \tan\theta\,\delta\theta/\cos^2\theta$, or $2\pi t^2 \sin\theta\,\delta\theta/\cos^3\theta$. Each small element δA of this area, on which resides the charge $\sigma\,\delta A$, is at the distance $t/\cos\theta$ from the location of q, and therefore exerts on the charge q the force $q\sigma\,\delta A\,\cos^2\theta/4\pi\epsilon_0 t^2$, the normal component of which is

$$\frac{q\sigma\,\delta A\,\cos^3\theta}{4\pi\epsilon_0 t^2}.$$

The force δF on q due to the whole annular area is obtained by substituting $2\pi t^2 \sin\theta\,\delta\theta/\cos^3\theta$ for δA. In this way we find that

$$\delta F = \frac{q\sigma\,\sin\theta\,\delta\theta}{2\epsilon_0}.$$

The force F due to the whole charge on the opposite plate is now found by integration:

$$F = \frac{q\sigma}{2\epsilon_0}\int_0^{\pi/2}\sin\theta\,d\theta = \frac{q\sigma}{2\epsilon_0}.$$

From this we obtain at once the expression for the force per unit area,

$$\frac{F}{A} = \frac{\sigma^2}{2\epsilon_0}. \tag{21.13}$$

Clearly, if the charge resided not on one plate, but in the region between the two plates, we should have to integrate over the area of both plates, q being attracted by the charge on one plate and repelled by that on the other. The expression for F would therefore be doubled; that is, it would be $q\sigma/\epsilon_0$, giving for the field strength the value σ/ϵ_0, in harmony with Eq. (20.6), which we obtained by using Gauss's theorem.

Attracted-Disk Electrometer

We have seen that the force per unit area exerted on one plate of a parallel-plate capacitor is $\frac{1}{2}\epsilon_0 E^2$, where E is the field intensity between the plates. We may therefore express E in terms of F/A by the equation

$$E = \sqrt{\frac{2F}{\epsilon_0 A}}.$$

Accordingly, for V, the potential difference between the plates, we have

$$V = E \cdot t = \sqrt{\frac{2F}{\epsilon_0 A}} \cdot t. \tag{21.16}$$

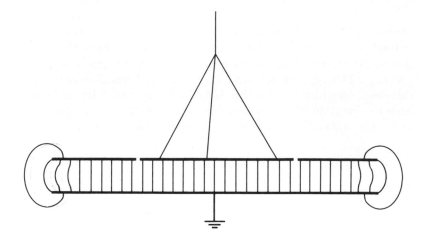

FIGURE 21.8   Attracted-disk electrometer.

This equation expresses in mathematical form the theory of the attracted-disk electrometer, which is essentially a parallel-plate capacitor so designed as to enable F/A to be measured to a high degree of precision, and so, t being known, enabling V to be calculated with corresponding accuracy.

The apparatus is represented schematically in Fig. 21.8. The central portion of the upper plate is in the form of a disk and is suspended from one arm of a balance. This is surrounded by, and separated by only a short gap from, a coaxial and coplanar annular plate. The plate, known as a "guard ring," is connected electrically, but not mechanically, to the suspended disk, and, like the lower plate, is secured in a fixed position. The function of the guard ring is to ensure that the field between the disk and the lower plate is uniform, edge effects existing practically only in the region of the outer edge of the guard ring. This is indicated by the system of lines of force shown in the figure. The balance is first adjusted, by suitable counterpoising, so that the disk is coplanar with the guard ring when both electrodes of the capacitor are at the same potential. The potential difference V is then applied and the requisite extra standard masses are added to restore the balance. Thus we have a measure of the electrostatic force of attraction, F, exerted by the lower plate on the disk. This, divided by the area of the circle which divides the gap between the disk and the guard ring into equal portions, gives the effective F/A. Hence, using Eq. (21.16), we have an absolute measure of the potential difference V applied between the plates.

The attracted-disk electrometer can obviously be used for calibrating a secondary instrument, such as a gold-leaf electroscope or any other form of voltmeter.

An incidental application of the attracted-disk electrometer which is of some importance is the absolute measurement of the charge Q residing on the disk. From Eq. (21.13) we have

$$\sigma = \sqrt{\frac{2\epsilon_0 F}{A}},$$

and hence

$$Q = \sigma A = \sqrt{2\epsilon_0 F A}. \tag{21.17}$$

This quantity may be discharged through any convenient secondary charge-measuring instrument (for example, a ballistic galvanometer (see Chapter 25)) and so used to calibrate the instrument.

Problems

21.1   A Van de Graaff generator has a spherical dome of diameter 2 m in the open atmosphere which is sufficiently far removed from surrounding objects for it to be regarded as virtually isolated in space. Given that the insulation breakdown field in air is 30,000 V/cm, calculate the upper limit of PD between the dome and earth that can be attained.

21.2   The dome of a Van de Graaff generator is maintained at a potential of 3 MV positive to earth. The belt conveys 500 μC/sec of positive charge to the dome and takes 400 μC/sec of negative charge away from it. What must be the power of the motor driving the generator, additional to that necessary to make up for frictional and viscous losses?

21.3   A parallel-plate air capacitor is charged to a PD of V and disconnected from the potential source. The separation between the plates is then increased to twice its former value. How, if at all, does this affect the PD across the terminals? (Neglect edge effects.)

21.4   An air capacitor consists of two coaxial metal tubes 30 cm long, the external diameter of the inner tube and the internal diameter of the outer tube being 9 cm and 10 cm, respectively. If the inner tube is given a charge of 5×10^{-8} C, what is the PD between it and the outer tube? Compare the capacitance of this capacitor with that of a parallel-plate capacitor whose plates each have an area equal to the mean of the cylindrical areas above and are separated by a distance of 5 mm. What are the effects of (a) increasing the spacing between the plates of the parallel-plate capacitor from 5 mm to 5 cm, and (b) inserting between the plates having the separation of 5 mm a slab of glass of this thickness, given that the dielectric constant of the glass is 6? (Neglect edge effects.)

21.5   If the insulator used in a parallel-plate capacitor whose plates are 0.1 mm apart has a dielectric constant of 5.2 and a dielectric strength (maximum field it is able to sustain without insulation breakdown) of 5×10^7 V/m, what is the maximum possible charge on either plate per unit area?

21.6   A Leyden jar is a capacitor having the form of a jar or bottle, usually of glass, provided inside and outside to a certain height with a coating of metal foil. Consider a Leyden jar that has a diameter of 10 cm, a depth of metal foil of 15 cm, and a glass thickness (sides and bottom) of 3 mm. Given that the glass has a dielectric constant of 6.2, find the capacitance of the jar.

21.7   A capacitor is constructed of a pile of 12 parallel sheets of metal foil, each of area 20 cm², separated from one another by 0.2-mm sheets of mica, with alternate metal sheets connected together. What is the capacitance?

21.8   How may virtually the whole of a given charge be communicated to the inner component of an initially uncharged insulated coaxial-cylinder capacitor? (Make any necessary simplifying assumptions.) What will be the electrical condition of the outer cylinder after this has been done? How will this be changed (a) by momentarily connecting the outer cylinder to earth, and (b) by then connecting the inner cylinder to earth? Will the potential differences (1) between the two cylinders, or (2) between the inner cylinder and earth, be affected by the operations (a) or (b)? If so, how?

21.9   Three capacitors have capacitances of 10, 20, and 60 μF. In how many ways can these be connected to give systems of different capacitances, and what are these? Show the connections in each case.

21.10   A 5-μF capacitor and a 20-μF capacitor are connected in series to points having a PD of 600 V. Find the charge on each capacitor and the PD across each. The charged capacitors are now isolated from the supply, disconnected from one another, and reconnected with the positive terminal to positive and negative to negative. Find the final charge on each and the final PD.

21.11   Three capacitors, of capacitances C_1, C_2, and C_3, are each independently charged to the same PD, V, and then connected as shown in the figure. What will be the charge and the voltage across each capacitor after the switch is closed?

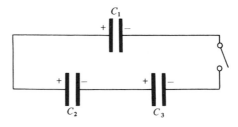

Figure for problem 21.11.

21.12   What is the capacitance of the network shown in the figure between the points A and B? Find the charges on the three capacitors farthest from A and B when the PD between these two points is V.

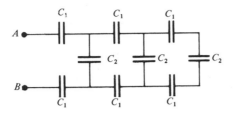

Figure for problem 21.12.

21.13 For flash pictures a photographer uses a capacitor of 30 μF and a charger that supplies 3000 V. Find the charge and energy expenditure in joules for each flash.

21.14 Three capacitors, of capacitances C_1, C_2, and C_3, are connected in series, and a total PD of V is established across them. What is the energy of the system? How is this energy distributed among the separate capacitors?

21.15 A parallel-plate capacitor has square plates 50 cm by 50 cm separated by a 0.5-mm glass slab of dielectric constant 6. Calculate the energy of the capacitor when a 200-V battery is connected across it. What change in the energy occurs if the glass is removed without touching the plates (a) while the battery remains connected across the capacitor, and (b) after the battery has been disconnected? By what processes are the energy changes brought about in the two cases? Is the PD between the plates affected by the withdrawal of the glass in the latter case? If so, what is the final PD?

21.16 A sheet of metal is introduced between the plates of a charged parallel-plate air capacitor without touching the plates and is held parallel to them (see the figure). Will the capacitance of the resulting system vary with the relative values of the gaps a and b shown in the figure? What effects on (a) the capacitance, (b) the PD between the capacitor plates, and (c) the electric potential energy will be produced by the insertion of the sheet of metal if its thickness is the fraction f of the separation between the plates of the capacitor? (Assume that, apart from thickness, the metal has the same dimensions as the capacitor plates, and neglect edge effects.)

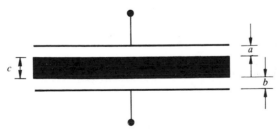

Figure for problem 21.16.

21.17 Repeat Problem 21.16 for the case where the sheet introduced between the capacitor plates is a dielectric of dielectric constant κ instead

559

of a metal. Compare the results obtained in the two cases, and comment on what you find.

21.18   In the figure the capacitors have the capacitances shown. What is the capacitance of the system between the points A and B? What is the potential difference between the points P and B when the charge on the 3-μF capacitor is 80 μC?

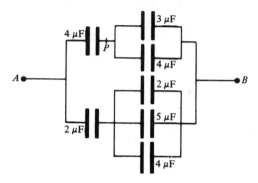

Figure for problem 21.18.

21.19   Apply the conditions that (a) during the charging process there is no accumulation of charge at any of the junctions P, Q, R, and S, and (b) every point must at all times have a definite potential, to work out the effective capacitance of the system shown in the figure between the points A and B.

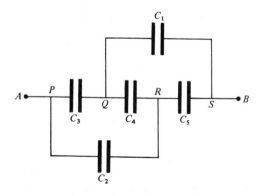

Figure for problem 21.19.

21.20   A Leyden jar is charged by connecting the inner coating to the dome of a demonstration model of a Van de Graaff generator, with the jar (a) resting on an earthed conductor, and (b) supported on a small table with polystyrene legs. In each case the jar is then discharged by a spark passing between the knob of the Leyden jar (connected to the inner coating) and an earthed metal sphere. In the former case the spark is not only of substantial length but is also very bright. In the latter case the spark, although of equal length, is relatively thin and feeble, dissipating so little energy that it may comfortably be taken through one's body. Explain. (*N.B.*: The length of spark passing between two electrodes is a measure of their PD.)

21.21   A concentric spherical capacitor has spheres of radii 7 cm and 10 cm and is filled with oil of dielectric constant 4 which does not withstand a potential gradient exceeding 120,000 V/cm. Calculate the maximum energy that can be stored in the capacitor.

21.22   A 20-μF capacitor is charged to a PD of 600 V and its terminals are then connected to those of an uncharged capacitor of capacitance 10 μF. Calculate (a) the final PD, and (b) the loss of electrical energy that occurs on making the connection. What becomes of this lost energy?

21.23   (a) Show by appropriate integration that the electric field intensity on the axis of a uniformly charged thin disk of radius a is

$$\frac{\gamma}{2\epsilon_0} [1 - x(a^2 + x^2)^{-1/2}],$$

where x is the distance from the disk and γ the total charge per unit area (both sides included) of the disk.

(b)   What is the limiting value of this at a point very close to the disk $(x \ll a)$?

(c)   Point out the relevance of the answer to (b) to the problem of finding the force per unit charge on one plate of a parallel-plate capacitor due to the combined Coulomb forces exerted by all the elements of charge on the opposite plate in the case where the plates extend to infinity in all directions in their planes, and the areal charge densities on the inner surfaces of the positive and negative plates are $+\sigma$ and $-\sigma$, respectively.

(d)   Hence, show that the force per unit area exerted on each plate in the direction toward the other is $\sigma^2/2\epsilon_0$.

(e) Show also that it may be inferred from the answer to (b) that at all points between the plates the field intensity is σ/ϵ_0.

21.24   Two large parallel flat metal plates are 8 mm apart, and the PD between them is 1000 V. Calculate (a) the areal density of charge on each plate, (b) the field intensity in the region between the plates, and (c) the force per unit area exerted on each plate.

21.25   A soap bubble of radius r suspended in air at a pressure p is given a charge Q. Will this cause it to expand or contract? Explain your answer.

Obtain an equation connecting p, Q, r, δr, and the surface tension T (use this symbol to distinguish it from surface charge density σ) of the soap solution. Assume that whatever change occurs does so isothermally.

21.26   A soap bubble in air at normal atmospheric pressure, of 10 cm diameter, is attached to the end of a thin insulating tube connected to an air reservoir of 1000 times its own volume. On giving the bubble an electric charge it is found to increase in size to 11 cm in diameter, the temperature of the air and atmospheric pressure remaining unchanged. Given that the surface tension of the soap solution is 0.025 N/m, find the charge on the bubble and the field intensity just outside it.

21.27 The attracted disk of an absolute electrometer has an effective diameter of 10 cm and is acted on by an electrostatic force of 0.1-g weight. What is the charge on the disk? If this disk is 5 mm above the lower plate, what is the PD between these?

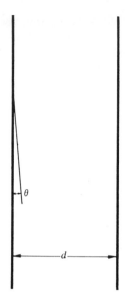

Figure for problem 21.28.

21.28 In the construction of an electrometer, a pair of parallel vertical metal plates are arranged so that their distance d apart is small compared with their linear dimensions (see the figure). To one plate a horizontal hinge is attached, on which is supported a wide metal leaf of mass μ per unit area on the side facing the other plate, and the PD to be measured is applied between the two plates. Assuming the deflection of the lower end of the leaf to be small compared with d, obtain an expression relating the angular deflection θ (assumed small) to the charge density σ, and hence to the PD applied. If μ is 1 mg/cm² and d is 5 cm, what is the PD required to produce a deflection of 0.05 rad?

BASIC ELECTRONICS I

In this chapter we shall first consider evidence from which the existence of certain fundamental negatively charged particles that we call electrons has been inferred, and it will be shown that an electric current in a metal consists simply of a flow of electrons within it. We shall then go on to discuss one of the main processes by which electrons may be extracted from metals, thermionic emission; and from a close study of the associated electron currents across a vacuum we shall (1) obtain important information concerning the variation of electrostatic potential with position in the region of metal surfaces, and (2) see how the ratio of charge to mass of an electron may be evaluated. Finally, we shall consider a classical series of experiments carried out by R. A. Millikan in the early years of the present century by which he was able to determine the electronic charge separately, and we shall review some of the important items of physical information obtainable from a knowledge of this charge.

These topics, with certain additional ones to be considered in Chapter 23, define what we may call "basic electronics." A more detailed study of the circuitry involved in the operation of electronic devices, often classified broadly as "electronics," cannot be undertaken at the present stage, but some of the more elementary features of this will be considered in Chapter 33 under the general heading of alternating-current theory.

Positive and Negative Electricity in Metals

We have so far left open the question of whether both kinds of electricity within a metal are mobile, responding to a field, or only one, and, in the latter case, which one.

The concept of electric charges within metals raises another question: Why are these charges normally confined to the metal in which they reside? What can it be that prevents them from leaving the metal, to pass out into the air or vacuum beyond its surface?

We might hope to gain some information concerning these questions by trying to bring about conditions that would favor the passage of electricity out of the metal. If we succeeded in extracting electricity from a metal and had it available for examination in a vacuum, we should then have the ideal conditions for investigating its nature.

FIELD EMISSION The first idea that might naturally occur to one would be to apply an exceptionally strong field to a metal surface, hoping that, in response to this, one or the other kind of charge might pass out beyond the surface.

As we saw in Chapter 20, the strongest field at right angles to the surface of a conductor exists at regions of high curvature, for example at points. It is on this that the spraying of charge from points, such as is used in the operation of a Van de Graaff generator or a Wimshurst machine,* depends. It is, however, important to realize that this is not the kind of effect we are now looking for, because it requires the presence of air or some other gas for its occurrence. Although the term used to describe this phenomenon, "the spraying of charge," is a convenient one, no charge originating within the metal actually does pass through the surface into the space beyond, other than in response to its bombardment by charged particles from outside. The issue should not be complicated by having a gas present. Only if we perform the experiment under clean conditions, with the metal situated in a vacuum, will it be possible to draw definite, unambiguous conclusions from any passage of electricity that may be observed to occur from the metal when exposed to a strong field.

Actually the emission of electricity from a metal into a vacuum does occur under the action of a sufficiently strong field, applied in the right direction. This effect, now generally known as **field emission,** has been known for many years, and has been the subject of considerable experimental and theoretical investigation. As we shall now see, important conclusions may be drawn from its main experimentally observed features.

The field strengths that have to be established at the surface of a metal for measurable currents to be observed are of the order of 10^8 V/cm. Such enormous fields can best be obtained by preparing an exceedingly fine point and then applying a high PD between this and an adjacent electrode. It will then be *at the point* that the field

---

* This is an electrostatic generator which, prior to the development of the Van de Graaff, was used extensively in electrostatic work. It is described in most of the older textbooks.

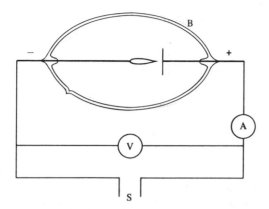

FIGURE 22.1 System for observation of field emission.

is really strong; elsewhere it will be orders of magnitude weaker. Under these experimental conditions, field emission is observable, but *only when the point is made the negative electrode*, never when it is made positive. In other words, only *negative* electricity can be extracted from a metal whose surface is exposed to a strong field.

Actually this "brute-force" procedure is not the only method available for extracting electricity from a metal; this may also be accomplished by less heroic measures. Some of these alternative methods will be considered presently. Meanwhile let us see what conclusions may be drawn from the main features of field emission.

MOBILITY OF CHARGES AND SURFACE POTENTIAL BARRIER

The experimental arrangement for observing field emission is shown schematically in Fig. 22.1. Two electrodes spaced apart from one another are mounted within an evacuated bulb B, one electrode terminating in a sharp point, the other not. These electrodes are connected to a suitable device S which generates and maintains a high PD between the electrodes; this might, for example, be a Van de Graaff generator or a Wimshurst machine. As has already been explained in Chapter 21, such electric machines in steady action maintain a continuous *circulation* of electricity, and in the present case this circulation occurs through the machine, to one electrode of the evacuated bulb, across the vacuum to the other electrode, and from there back to the machine. A voltmeter V is provided to measure the PD between the electrodes in the bulb, and a sensitive current-measuring instrument (ammeter A) is included in the circuit.

Mobility of Charges   In the operation of this apparatus we have a steady state; none of its components changes its condition with

565

time. Let us consider the implications of this with particular reference to the pointed electrode within the bulb. This is continuously
losing negative electricity, which passes across the vacuum to the
other electrode at a rate measured by the ammeter. But its condition
does not change with time, so this loss of negative electricity must be
compensated by a corresponding gain. Such a supply can occur only
along the wire leading to this electrode. Hence we must infer that
**the negative electricity within this metal wire must be
mobile.**

A priori, the current registered by the ammeter might be
assumed to be due to the flow of negative electricity through it in
one direction, or of positive in the other, or both together. If positive
electricity flowed through this instrument to the plate electrode
within the evacuated bulb, this must be emitted from the latter,
to be collected by the pointed electrode, for otherwise the plate
electrode could not remain in a steady state. Such collection of positive charge emitted from the plate must obviously occur regardless
of the form of the collecting electrode; it must be conditioned merely
by the existence of a field driving positive electricity toward the
latter. But experimentally it is found that if the left-hand electrode
is not made in the form of a sharp point, the ammeter registers no
current. Hence it must be inferred that no positive electricity is
emitted from the plate. The conclusion to which we are driven is,
then, that *only* negative electricity flows across the gap between the
two electrodes, and therefore only negative electricity flows along
the wires. That is, **only the negative electricity in a metal is
mobile; the positive is immobile.**

Surface Potential Barrier   The failure of negative electricity
to escape spontaneously from a metal could most naturally be
accounted for by supposing that it encounters inwardly directed
forces in the region of the surface. With such forces would necessarily be associated a PD between the inside and the outside, in the
sense that the potential outside is lower than that inside. Such a
PD is usually referred to as a **potential barrier.** It is analogous to
the gravitational potential barrier which prevents the escape of
water from a bowl by running up and over the sides. In the latter
case, let us consider a central section through the bowl, as depicted
in Fig. 22.2. This obviously is a graph, to some scale, of gravitational
potential energy per unit mass, or gravitational potential, of
water as a function of position along the surface of the bowl. The
analogy between this and the electric potential in the region of the
surface of a metal becomes more obvious if in the latter case we
consider *negative* potential, or potential energy per unit negative
charge, that is, minus conventional potential as defined in terms of a

FIGURE 22.2

positive charge. If we plot negative potential against position along a line at right angles to a metal surface, from inside to outside, according to our potential-barrier picture, we must obtain a graph such as that shown in Fig. 22.3. This resembles qualitatively the right-hand side of the section of the bowl containing water, shown in Fig. 22.2.

If this idea is right, the same potential barrier that normally prevents the escape of negative electricity must favor the escape of positive; positive charges must be acted on by *outwardly* directed forces in the region of the surface. That, despite this, positive charges do not spontaneously escape can only be due to their not being free to move; they must somehow be anchored to—perhaps even in some special way practically *constitute*—the metal atoms.

This argument cannot be applied in reverse; we cannot suppose that the failure of positive electricity to escape spontaneously is due to the conventional positive potential being higher outside than inside. For, if this were so, negative electricity, which field-emission experiments show to be mobile, would escape spontaneously, and this it does not do.

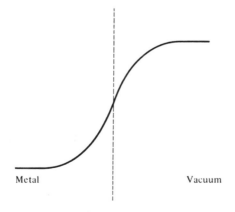

Metal   Vacuum

FIGURE 22.3 Potential barrier for electrons in the region of a metal surface.

Electrons

Electricity might equally well be continuous or occur in small ultimate units, or particles. We could make either assumption and see where it leads us. In making our choice we would naturally be influenced by the way the analogous question concerning matter has been resolved and by the fact that electricity is normally associated with matter. In view of these considerations, let us tentatively assume that electricity, as well as matter, occurs discontinuously, in the form of ultimate particles. We can then look for the evidence of appropriate experiments either to support this assumption or otherwise.

It is convenient to give a name to the assumed particles of negative electricity, which can move about within metals and be extracted from them. The name that has been chosen for such a particle is an **electron.** Whether or not all electrons are alike or are different from one another as are different kinds of atoms is a question we may leave open for the present.

We can account for the presence of the mobile electrons in a metal only by supposing that they have become detached from the originally electrically neutral atoms constituting the metal. And, because one atom is like another, and all atoms except those at or very near the surface have an identical environment, there can be no reason why only particular atoms should supply electrons, at least as far as atoms well below the surface are concerned; they must all be actual or potential suppliers of these. Whether, in fact, all atoms do simultaneously supply mobile electrons, or only a statistically determined proportion of them at any particular moment, is a question we need not consider at this stage. We shall also leave open the question of whether a neutral atom contains more than one electron, and, if so, whether some are more tightly bound than others, so that only one, or a few, are available for detachment.

Explanation of Surface Potential Barrier

The field in the region of the surface of a metal derives, in part, from the asymmetry of the electrostatic forces to which the surface atoms are subjected. An atom deep down in the metal has other atoms symmetrically disposed about it and is therefore not acted on by any resultant force. But the surface atoms have other atoms only underneath and beside them, none above. Consequently, these are distorted, in the sense that their electrons are, on the average, either farther in or farther out than the positive charges. We may idealize this situation by imagining the positive charges of the atoms and the electrons to be smeared out into uniform sheets of charge separated from one another by a short distance. Between these there must be an electrostatic field, just as there is between the plates of a charged capacitor. This field, which is due to the polarization of the surface atoms, we may conveniently call the "polarization field." The polarization field may be either electron-retarding or electron-accelerating outward, according to the polarity of the equivalent charged capacitor. To have an electron-retarding field and so a potential barrier, which is what we need, we have only to suppose that the electrons are, on the whole, farther out than the positive charges.

The polarization field is not the only one that contributes to the potential barrier. Let us consider the final stage of the journey of an electron caused by some means to leave the metal, where it has passed into the vacuum. If this were at rest it would induce a positive charge equal in magnitude to its own negative charge in the surface of the metal. It may be shown that the force of attraction between the electron and the charge it induces is the same as if the latter were concentrated at a point along the normal to the surface passing through the electron whose distance behind the surface is the same as that of the electron in front; that is, it is the same as if the induced charge were located at the optical image of the electron in the surface functioning as a plane mirror. For this reason the force of attraction toward the surface experienced by the electron is generally referred to as the "image force" and the equivalent field as the "image field."* Actually, the electron is not at rest; it is in motion outward, and the image field due to the induced charge might be thought to require some time to establish itself. However, certain experiments, whose discussion here would take us too far, have shown that beyond a few atomic diameters from the surface, electrons escaping from a metal are acted on by the same retarding force as they would be if they were at rest.

We cannot expect the image picture of the induced charge to apply in regions so close to the metal that the nearest atoms are only one or two atomic diameters away, for in these regions the electron is under the influence of the localized charges of the adjacent atoms. Even if this were not so, application of the image law would not be feasible because of the impossibility of specifying just where the surface is. The image picture only becomes valid beyond a cer-

---

* It is interesting to note that the image field does not exist except in the presence of the electron, which induces the image charge causing it.

tain distance from the metal, where the image field takes over from the polarization field, and extends, with diminishing magnitude, to relatively remote distances, where it asymptotically approaches zero. In practice we need not consider the image force beyond some 100 to 1000 Å from the surface; at distances greater than this its effects are negligible.

From this picture of a combined polarization and image field we can construct a probable form of potential barrier at the surface such as that shown in Fig. 22.3.

Thermionic Emission

The question now arises whether the escape of electrons from a metal might be brought about by some means other than the brute-force method of applying an enormously strong field. The problem seems to be that of imparting to the internal electrons enough energy to enable them to surmount the surface potential barrier. This must be energy of some form which is available for conversion to potential energy in the process of escape. It is difficult to think of any form such energy might have other than kinetic energy. This implies mass. Let us therefore assume, tentatively, that electrons, besides having a charge, also have mass. Expressed in terms of momentum the picture of escape we now have is of the outwardly directed momentum of an internal electron being sufficient to carry it through the retarding-force system it encounters in the region of the surface.

The normal method of imparting to the molecules of a liquid enough kinetic energy to enable them to escape is to heat the liquid. This suggests that by heating a metal in a vacuum some of the electrons within it might be given enough kinetic energy to surmount the potential barrier, so being "boiled out." And if the experiment is tried—at least with metals that do not vaporize readily or melt at too low a temperature—it works. The emission of electrons from a hot metal is known as **thermionic emission.** It is essentially an evaporation process, the evaporating particles being electrons.

The question is sometimes raised whether, as we have supposed, the electrons emitted thermally from a metal come from its interior, passing *through* the surface, or whether they come *from* the surface and are then replaced by other electrons coming from the interior. The answer is that both processes may occur side by

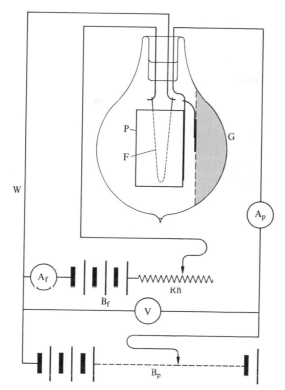

FIGURE 22.4 System for observation of thermionic emission.

side. But the important thing to note is that with a given potential barrier at the surface the same emission would be obtained either way; that is, whether the emission occurs directly from the interior or as a two-stage process. That this is so will be appreciated later by those readers who proceed to the study of statistical mechanics in Chapter 7 of Volume III.

A simple arrangement with which thermionic emission may be observed is indicated in Fig. 22.4. Within an evacuated bulb are mounted a fine metal filament F (for example, of hairpin form), which is the metal to be heated, and a surrounding "plate" P constructed of metal sheet in the form of a cylinder of circular or oval section. A good metal to use for the filament is tungsten, which has the highest melting point of all metals and does not evaporate at an appreciable rate until an exceedingly high temperature is reached. The plate may be made of nickel sheet, which, if predegassed in a vacuum or hydrogen furnace and again heated briefly before the tube is sealed off, does not subsequently give off

571

gas at an appreciable rate. A getter G,* dispersed at an appropriate stage during pumping, serves to maintain a good vacuum after the tube has been sealed off.

A tube such as this, which has, in effect, two electrodes, is known as a **diode.** Actually, if the source of electrons is a filament, as here, there are three electrodes, two for the filament and one for the plate. However, the two filament electrodes are regarded as one from the point of view of the thermionic functioning of the tube, in which we are concerned only with a hot source of electrons and an adjacent collector, the plate.

In the system shown in the figure, the requisite filament heating current is derived from a "filament battery" B_f and adjusted by means of the rheostat Rh, while another battery, the "plate battery" B_p, provides the necessary PD between the filament and plate, this PD being measured by the voltmeter V. By some suitable device (for example, a clip) provision may be made for any desired number of cells of the plate battery to be included in the plate circuit, giving correspondingly different values of the potential of the plate relative to that of the filament—the "plate voltage"— which we may designate by the symbol V_p. Strictly speaking, there is no unique PD between filament and plate, because there is necessarily a potential drop along the filament itself, from the positive to the negative end. However, this is usually small compared with the PD between any point on the filament and the plate, so it will simplify our discussion if we ignore the potential drop along the filament, considering simply the PD between, say, the negative end of the filament and the plate. Finally, a sensitive ammeter A_p measures the current I_p passing across the vacuum between the filament and the plate.

With this apparatus, let us suppose the filament to be held at a series of constant temperatures $T_1, T_2, \ldots, T_n$, at each one of which the plate current I_p is measured as a function of V_p.

If this is done, the battery being so manipulated that V_p is given negative as well as positive values, it is found that at no filament temperature is an anode current observable with the plate at a negative potential; only with V_p positive does

---

* This is a film of an electropositive metal (for example, barium) from a nickel-clad pellet, formed on the inner surface of the glass by distillation in a vacuum. Its purpose is to combine with, dissolve, or adsorb molecules or ions of residual gas and gas coming off from the glass or electrodes during the operation of the tube.

the ammeter A_p show a reading.* This is in harmony with what was observed in the case of field emission; whether the electricity is drawn out by a strong field or released thermally, only negative electricity can be extracted from a metal.

Before going on to examine quantitatively what happens when V_p is positive, we may note that essentially the same argument may be used here as in connection with the field-emission observations to show that only the negative electricity within metals is mobile. The filament circuit loses negative electricity at a steady rate, and this must be replenished by a flow of negative electricity along the wire marked W in Fig. 22.4. Also, if positive electricity were emitted from the plate this would presumably not be affected by the temperature of the filament,† but with the filament cold no current is registered by the ammeter A_p. Hence it may be inferred that *only* negative electricity flows across the gap, and correspondingly only negative electricity flows along the wire W or along the wire connected to the plate.

The same kind of argument may also be based on observations of photoelectric emission, to be discussed later.

Let us now suppose that I_p is plotted against V_p for each of a number of constant values T_1, T_2, T_3, T_4 of the filament temperature. Such a series of plots is shown in Fig. 22.5. We see that at each temperature the electron current increases with V_p, at first at an increasingly rapid rate, then less rapidly, until finally it ceases to increase any more, the curve becoming horizontal.‡ This last stage is known as **saturation** and must surely be interpreted as the condition where all the electrons emitted from the filament are drawn off and collected by the plate, none returning to the filament. In other words, the *saturated* plate current measures the whole of the electron emission from the filament, whereas unsaturated, or

---

* Actually, a very sensitive instrument does show a measurable current with V_p somewhat negative. This is because the electrons are emitted with finite velocities, and so a finite retarding voltage must be applied to the plate to prevent them from reaching this electrode. This is a second-order effect which, for simplicity, we shall disregard here.

† Actually there *could* conceivably be an effect conditioned by the light the plate receives from the filament when hot. However, experiments in which different kinds of light from an external source are directed onto the plate would show that, in fact, no irradiation of the plate with light releases positive electricity from it.

‡ Actually this is not strictly true; owing to an effect of the field on surface conditions at the filament, the current never quite ceases to increase with applied field. However, this increase is small, and may be ignored in the present discussion.

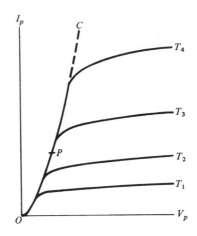

FIGURE 22.5 Current–voltage curves for thermionic diode.

"infrasaturation" currents, obtained at lower values of V_p, represent only a part of the emission, the remainder of the emitted electrons returning and being reabsorbed by the filament. Why they return we shall see later.

This behavior is analogous to that which we have discussed in connection with evaporating liquids; only it should be noted that the use of the term "saturation" is different in the two cases. Zero plate current corresponds to the case of a liquid in contact with saturated vapor, every evaporating molecule being replaced by one that impinges on the liquid from the vapor and is absorbed. Infrasaturation plate currents correspond to evaporation from a liquid in which a proportion of the evaporating molecules are withdrawn permanently from it, while the remainder return and are reabsorbed. Finally, saturated thermionic current corresponds to the case of an evaporating liquid—not easily realized experimentally—where no molecule that evaporates from the liquid ever returns. It is important to realize that there is no question of the *emission* from the filament being saturated or unsaturated; the rate at which electrons leave the filament is determined by the temperature alone. The plate *current*, on the other hand, corresponds to the difference between the rate of emission of electrons from the filament and the rate of return, that is, to the net or resultant rate of loss. Thus the emission is measured by the *saturated* plate current.

EMISSION AS A FUNCTION OF TEMPERATURE The tube shown in Fig. 22.4, while of relatively simple construction, is not quite suitable for precise quantitative measurements. This is because, owing to thermal conduction, the filament is not at the

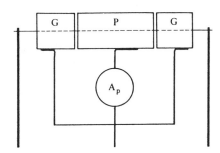

FIGURE 22.6 System for quantitative observation of variation of thermionic emission with temperature, and also of thermionic current with anode voltage at a given temperature.

same temperature throughout; it is cooled appreciably in the region of the ends where it is supported. This difficulty can be avoided by the use of guard rings, after the manner indicated schematically in Fig. 22.6, in which, for simplicity, the details of construction of the tube, including the bulb, are not shown. Here the filament is straight, and is kept so, despite thermal expansion, by a spring incorporated with one of the supports. The cylindrical plate is divided into three sections, separated by very small gaps. The ammeter A_p is connected only to the central section P, containing that portion of the filament which may be taken to be at a uniform temperature. The two end sections G function as guard rings. These collect electrons from the ends of the filament, which include the portions appreciably cooled by thermal conduction. All three sections are connected together beyond the ammeter and so are always at the same potential.

We may note here, parenthetically, that it is no longer usual to employ the term "plate" for the positive electrode in a thermionic tube. The usual name given to this electrode is the **anode.** Correspondingly, the negative electrode, from which electrons are emitted, is called the **cathode.**

For any given surface there is at each temperature a definite value of the emission i per unit area. In practice, however, it is not easy to determine this. Cathode surfaces are not smooth, and consequently what is obtained from the measured emission and the diameter and length of the filament is not i, the emission per unit true area of the cathode, but ρi, the emission per unit *apparent* area, where ρ, the "roughness factor," is the ratio of the effective true area to the apparent area. It is believed that effective roughness factors are usually somewhere in the neighborhood of 1.3.

575

TABLE 22.1   Emission from Tungsten at Various Temperatures

T (°K)	ρi (A/cm²)
1000	1.07×10^{-15}
1200	9.7×10^{-12}
1400	6.6×10^{-9}
1600	9.3×10^{-7}
1800	4.5×10^{-5}
2000	1.00×10^{-3}
2200	1.33×10^{-2}
2400	0.116
2600	0.716
2800	3.54

The thermionic emission from any emitting surface increases at an extremely rapid rate with rising temperature. This is shown in the case of tungsten, for example, in Table 22.1.

This kind of variation is reminiscent of the way vapor pressures vary with temperature, noted in Chapter 16, and the reason for this is not difficult to see: Thermionic emission is, essentially, a process of evaporation—of electrons—and the saturation current is a measure of the rate of evaporation. As we saw in Chapter 16, this is closely related to the corresponding vapor pressure.

That thermionic emission is, indeed, an evaporation phenomenon is shown very beautifully by a simple experiment which demonstrates that electrons, like the molecules of an evaporating liquid or solid, have a latent heat of evaporation. To show this we have merely to insert a switch in the plate circuit of a system such as that shown in Fig. 22.4. With the switch open, the filament is heated to a suitably high temperature and its brightness noted. If the switch is now closed, an appreciable dimming of the filament may be observed. When the switch is open, all the electrons emitted are reabsorbed, and so there is no net loss of electrons. But when the switch is closed, so connecting the plate to the battery B_p, the emitted electrons are withdrawn from the filament, with the corresponding loss by the latter of latent heat of evaporation.

This connection between thermionic emission and ordinary evaporation suggests that we plot the data of Table 22.1 in the same manner as in Chapter 16 we plotted vapor-pressure data, that is, that we plot log (ρi) against $1/T$. Such a plot is shown in Fig. 22.7. We see that it is a straight line, so that an empirical for-

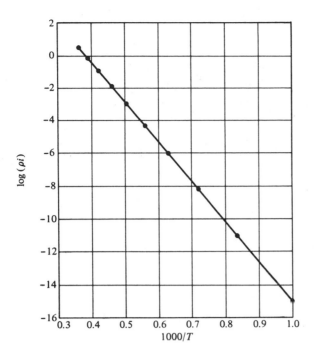

FIGURE 22.7

mula for the i–T relation is

$$i = a \cdot 10^{-b/T},$$

where a and b are constants, or

$$i = a \cdot e^{-b'/T},$$

where b' is another constant.

As we saw in Chapter 16 in our discussion of the algebraical representation of the variation of vapor pressure with temperature, we can represent experimental data, within the limits of the uncertainties attaching to their determinations, equally well by *any* equation of the general form

$$p = \alpha T^n e^{-\beta/T},$$

in which n is a not-too-large positive or negative number, for in any such expression the variation is dominated by the exponential factor—the power-of-T factor is relatively unimportant. With any suitable choice of n, an appropriate pair of values of α and β gives a satisfactory representation of the data.

In the quantitative theory of thermionic emission the conduction electrons within a metal are assumed to have thermal velocities distributed according to a statistical law about a mean, these velocities spreading to higher values with increasing temperature. All those sufficiently near the surface and having outwardly directed normal components of momentum large enough to carry them through the surface potential barrier and into the vacuum beyond are regarded as qualifying for escape. However, not all those so qualifying necessarily do escape; there may be some mechanism of internal reflection that sends a proportion of them back into the metal.

Although the surface potential barrier is not necessarily constant with temperature, various lines of evidence (whose discussion is beyond the scope of this book) indicate that any such variation of the height of the barrier as there may be is quite small. This small variation is taken into account in the theory that will now be considered.

The statistical velocity distribution to which the internal electrons are known to conform, when used in a theoretical discussion of the emission process according to the picture outlined above, leads to the formula

$$i = A T^2 e^{-B/T}, \tag{22.1}$$

where A is a constant whose value is affected by internal reflection (if any) of the electrons qualifying for escape and by the rate of variation with temperature of the surface potential barrier, while B is proportional to the extra energy that would have to be imparted to the most energetic electrons within the metal at a low temperature to enable them (if traveling in the right direction) to surmount the potential barrier. At low temperatures this extra energy may be identified with an energy step figuring in the full theory of thermionic emission known as the **work function, ϕ.**

Equation (22.1) is known as **Richardson's equation,** after O. W. Richardson, a pioneer investigator in the field of thermionic emission. From the form of this equation we see that log (i/T^2) varies linearly with $1/T$, and the corresponding plot is known as a "Richardson line." In practice, of course, it is generally only ρi, not i itself, that is determined experimentally, and so the Richardson line becomes a plot of log ($\rho i/T^2$) against $1/T$. The slope of this plot gives the value of the constant B for the emitter in question, while from the intercept on the log ($\rho i/T^2$) axis the value of ρA may be found.

Some values of ρA and B that have been obtained in this way for various emitters are given in Table 22.2. All but one of these emitters are clean metals, the exception being "thoriated tungsten,"

TABLE 22.2 Thermionic Data

	Emitter				
	W	Mo	Ta	Th	W–Th
ρA (A/cm²/deg²)	80	55	60	70	3
B (°C)	52,700	48,100	47,800	39,200	24,500

W–Th. This is tungsten covered by a single complete layer of thorium atoms. These are adsorbed on the underlying tungsten, being held to it by strong forces.

The case of thoriated tungsten is of particular interest. It will be observed that the value of A for this emitter is much lower than that for pure metals—by a factor of about 25. This is believed to be due to internal reflection of a large proportion of the electrons qualifying for escape. Despite this, a given emission is obtainable at a temperature some hundreds of degrees lower than from plain tungsten and also substantially lower than from bulk thorium. This is due to the low value of B. However, a price has to be paid for this relatively high emission efficiency. Beginning with tungsten containing about 1 percent of thoria, ThO_2, as an impurity, the filament is put through a certain heat treatment that results in chemical reduction of some of the thoria, accompanied by the evaporation of the tungsten oxide produced, followed by a layer of thorium atoms, diffusing outward, being "grown" on the surface of the tungsten. With the completion of this, the full thermionic activity characteristic of W–Th is developed. But this adsorbed thorium is highly vulnerable to the effects of residual gases. Unless the vacuum is particularly good, residual oxygen will itself become adsorbed on, and so contaminate, the thorium as it appears on the surface, "poisoning" the filament. And of course if air is admitted and the tube is then reevacuated, the whole procedure of growing a layer of thorium on the surface must be repeated. For this reason thoriated tungsten is only used for rather special purposes and in situations where a sufficiently good vacuum can be produced and maintained in a sealed-off vessel.

A particularly good thermionic emitter, which is used in most radio receiving tubes within which a good vacuum is maintained by a getter, is a so-called "oxide cathode." This consists of a metal such as nickel coated with a thin layer of a mixture of barium and strontium oxides in a fine state of subdivision. Like thoriated tungsten, oxide cathodes have to be taken through an appropriate heat treatment

579

to develop their full activity, which depends on the presence of some alkaline-earth metal in a state of solid solution in, and adsorbed as a monatomic layer on, the oxide. Also, like thoriated tungsten, oxide cathodes are vulnerable to poisoning by residual electronegative gases. The normal operating temperature of oxide cathodes in radio tubes is only about 1000°K—a dull red.

The low operating temperature of oxide emitters renders feasible a form of cathode known as an "indirectly heated cathode." This consists of a long, thin, hollow nickel body (for example, a cylinder) coated on the outside with alkaline-earth oxides. Within this cylinder a heater is provided, consisting of a tungsten or tungsten–molybdenum alloy wire coated with a refractory material which insulates it electrically from the cathode. Heat is transferred from the heater to the cathode by radiation. Indirectly heated cathodes not only have a large surface area but are also at the same potential throughout. This is an advantage in certain cases.

INFRASATURATION CURRENTS At anode voltages insufficient to give saturated thermionic currents, only a proportion of the emitted electrons, those emitted with sufficiently high velocities, get right across to the anode. The remainder, emitted with relatively low velocities, are turned back by a potential minimum in the interelectrode space and are reabsorbed by the cathode. The reason for this potential minimum is the existence of what is known as a negative **space charge** between the electrodes. This is due to all the electrons that are at any particular moment present in the interelectrode space, whether in transit straight across from cathode to anode or on an excursion toward the potential minimum and back to the cathode. An alternative, widely used name for an infrasaturation current is a "space-charge-limited current."

The space charge is sometimes described as a "cloud" of electrons in the interelectrode space. The use of this term is most misleading, suggesting, as it does, a stationary or almost stationary condition of the electrons, as if they hovered between the electrodes in the same way as a cloud of water droplets hovers in the atmosphere. Nothing could be further from the truth than such a picture. Far from hovering, all space-charge electrons, whether on an excursion toward the anode and destined to be reabsorbed by the cathode or in transit from the cathode across to the anode, have exceedingly short lifetimes in the interelectrode space, of the order of only a microsecond, 10^{-6} sec.

It is also sometimes suggested that when, with a large electron-accelerating field established between the electrodes, saturated electron current is drawn, there is no longer any space charge.

There certainly is—the electrons at any moment in transit from cathode to anode still constitute a space charge.

When the potential of the anode with respect to the cathode is sufficiently high for saturated current to be drawn, then, although, as we have just noted, there is still a space charge, there is no longer a potential minimum; the field is electron-accelerating right across. Consequently, all the emitted electrons, however small may be the initial velocities with which they leave the cathode, pass right across to the anode, which absorbs them.

The potential minimum that exists at the lower anode voltages at which the current is space-charge-limited really constitutes another potential barrier, additional to the one in the immediate region of the surface of the cathode, which the electrons have to surmount in order to escape permanently. Thus we may think of a total, composite potential barrier, having two stages or steps.

On referring to Fig. 22.5 it will be observed that the space-charge-limited curves, or "characteristics," at all temperatures up to the points where they bend over and approach saturation, all form part of a single curve, marked C in the figure. In view of the explanation of space-charge limitation in terms of an extra barrier this may seem strange. Let us consider the point P, regarded as a point on the curve characteristic of the temperature T_3. If the temperature is now raised to a substantially higher value T_4, will not this shift the velocity distribution toward higher velocities and so enable more electrons than before to surmount the composite barrier? Should not the point on the T_4 curve therefore be higher than P?

The number of electrons able to surmount *any given* potential barrier at T_4 must certainly be greater than at T_3. But is the potential barrier established at T_4 of the same height as at T_3? Obviously it cannot be—it must be higher. This is because more electrons now leave the filament. It can be shown that the raising of the height of the potential barrier almost exactly compensates for the increase in temperature, in the sense that the number of electrons which surmount the potential barrier and are collected by the anode in a given time is practically the same at both temperatures. Strictly speaking, the anode currents are not *exactly* the same, but the difference is negligibly small.

APPLICATIONS OF THERMIONIC EMISSION By far the most convenient and versatile method of producing free electrons in a vacuum is to liberate them from a metal thermionically. Such free electrons are required not only for physical research in many

581

fields but also in radio tubes (both transmitting and receiving), in x-ray tubes, in cathode-ray tubes, in electron microscopes, and in a large number of other commercial and industrial applications. Some of these applications will be discussed later in this and subsequent chapters.

Electronic Charge-to-Mass-Ratio

The space-charge-limited I_p–V_p relation has been calculated theoretically for certain geometrically simple configurations, such as, for example, that depicted in Fig. 22.6. As can easily be seen physically, the theoretical formula for any given case must involve the ratio of charge $e*$ to mass m of the electrons. For, the greater the charge associated with any given mass, the greater will be the force per unit mass in any given field, that is, the greater the acceleration will be. But the greater the acceleration, the shorter must be the time during which the electrons linger in the interelectrode space, and correspondingly the smaller will be their contribution to the space charge, and the larger, therefore, will be the current that passes.

For a thin straight filament stretched along the axis of a cylindrical anode of radius r, the theoretical formula for the space-charge-limited current per unit length is

$$\frac{I_p}{l} = \frac{8\sqrt{2}}{9} \, \pi\epsilon_0 \sqrt{\frac{e}{m}} \frac{V_p{}^{3/2}}{r}.$$

Applying this formula to experimentally determined space-charge-limited values of I_p as a function of V_p, we may calculate the value of e/m. This is 1.76×10^{11} C/kg.

Certain other methods are also available for evaluating the electronic charge-to-mass ratio. These all give the same result, within the limits of experimental uncertainty. The value at present accepted as most reliable, based on the most accurate determinations, is

$$\frac{e}{m} = 1.759 \times 10^{11} \text{ C/kg}$$

$$= 1.759 \times 10^8 \text{ C/g.}$$

---

* The symbol e is almost universally used for the electronic charge—the same symbol as that for the base of natural logarithms. There can be no real objection to this. Even though both quantities may on occasion appear in the same equation, the mathematical contexts are so different that there is never any danger of confusion.

The fact that the same value of e/m is found, regardless of the metal from which the electrons are derived or of the method used, is highly significant. It constitutes the strongest of evidence that **there is only one kind of electron.**

Millikan's Oil-Drop Experiment

Early attempts to measure the electronic charge were made by J. J. Thomson and others, and these yielded rough values. Later, a considerable refinement of the earlier methods, devised by R. A. Millikan, enabled him, in work carried out between about 1909 and 1917, to make the first really accurate determination. Millikan's method consisted essentially in observing the motions of tiny charged oil droplets in air (1) under the action of gravity alone, and (2) under the action of both gravity and an electric field simultaneously, and from these observations deducing the magnitude of the charge on each droplet. The highest common factor of all the charges found was then taken as the smallest step by which a charge could increase or decrease, that is, the electronic charge e.

Millikan's apparatus is represented schematically in Fig. 22.8. Two horizontal metal plates are arranged, one above the other, so as to constitute a parallel-plate air capacitor, and a small hole is provided in the upper plate to allow occasional oil droplets sprayed into a chamber within which the capacitor is housed to find their way into the intervening air space. A proportion of these droplets become charged in the process of their formation. This is often attributed to friction within the nozzle, but this idea surely cannot be accepted, because we cannot have friction between a fluid and a solid; moreover, the droplets constituting the spray presumably do not at any stage come into actual contact with the nozzle. Some

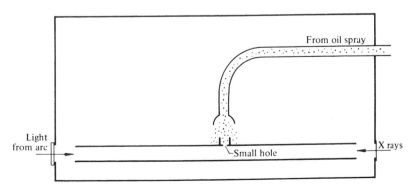

FIGURE 22.8 Apparatus used in Millikan's oil-drop experiment.

may, however, be blown off from a thin film of oil adhering to the nozzle walls, and so charged droplets may be produced in a manner analogous to the production of charged water droplets in thunderstorm cells by the blowing off of water from melting hailstones. There might also conceivably be a transfer of charge during the process of separation of a larger drop into two smaller ones within the nozzle. Whatever may be the mechanism, the important fact is that a proportion of the droplets do, somehow, acquire a charge.

The droplets that pass through the hole are illuminated by a strong beam of light directed between the two plates, and these may then be observed at right angles to the beam through a telescope, in whose field of view they appear as bright points of light, like stars. The vertical velocity of a droplet is determined by viewing it through the telescope and timing its rise or fall over a known distance.

The vertical motion of any droplet that carries a charge may be controlled by manipulating the PD, and so the field, between the plates. If a particular droplet, of density ρ, has a radius r and carries a charge q, and the field between the plates is E, the droplet will be acted upon by three forces—apart from a viscous resistance due to its motion, which will be considered presently. These are (1) its weight, $\frac{4}{3}\pi r^3 \rho g$; (2) the Archimedean upthrust of the surrounding air, $\frac{4}{3}\pi r^3 \delta g$, where δ is the density of the air; and (3) an electrostatic force Eq, whose direction depends on the sign of q and the direction of E.

We shall here be concerned with two special cases. One is that in which E is zero. In this case the resultant force is $\frac{4}{3}\pi r^3(\rho - \delta)g$ downward, and the terminal velocity v_1 with which the drop falls is, as we saw in Chapter 9, given by the equation

$$v_1 = \frac{2gr^2}{9}(\rho - \delta). \qquad (22.2)$$

The other special case is that in which the resultant force (apart from viscous resistance) is such as to impel the droplet upward. In this case the resultant force is $Eq - \frac{4}{3}\pi r^3(\rho - \delta)g$; and when the droplet has attained its terminal velocity v_2, this must be exactly balanced by the viscous resistive force, which, according to Stokes's law, is $6\pi r\eta v_2$. Thus

$$6\pi r\eta v_2 = Eq - \frac{4}{3}\pi r^3(\rho - \delta)g,$$

or

$$v_2 = \frac{Eq}{6\pi r\eta} - \frac{2gr^2}{9\eta}(\rho - \delta). \qquad (22.3)$$

We see that by adding Eqs. (22.2) and (22.3) we obtain the equation

$$v_1 + v_2 = \frac{Eq}{6\pi r \eta},$$

or

$$q = \frac{6\pi r \eta}{E}(v_1 + v_2). \tag{22.4}$$

The experimental procedure is as follows. A droplet is first selected whose motion responds to variations of PD applied between the plates, showing that it carries a charge. The plates are then connected electrically, making E zero, and the terminal velocity of the selected droplet is measured. This determination enables us to calculate the radius of the droplet, by substituting for all the other quantities in Eq. (22.2) and solving for r. Next, a field E is applied which causes this droplet to move upward, the new terminal velocity v_2 is measured, and q is calculated by substituting for all quantities on the right side of Eq. (22.4).

In principle this procedure could be followed with each of a number of selected charged droplets and the highest factor common to all the charges ascertained. In practice, however, Millikan found it more convenient to work with the same droplet throughout, causing this to take up a number of different charges. This he did by passing a beam of x rays between the plates of the capacitor. X rays have the effect of converting neutral molecules of air into charged particles known as gaseous ions. The nature of these will be discussed in Chapter 24. The ions formed in this way attach themselves from time to time to suspended oil droplets, altering their charge. Millikan kept a single droplet under observation for hours at a time, checking that the radius, as measured by v_1, remained constant, and observing a number of different values of v_2. From these the corresponding values of q were calculated.

Millikan and his co-workers, in the course of their investigations extending over several years, made observations with thousands of oil droplets. They found that not only the set of charges calculated from the observations made with any particular droplet, but *all* the charges, without exception, in whose determinations a large range of values of E were used, were integral multiples of a charge e, having the value 1.59×10^{-19} C. More recent measurements of the viscosity η have shown that the value used by Millikan was slightly in error. At the present time the value of the electronic charge regarded as most reliable, based on determinations by this and other methods, is

$$e = 1.6021 \times 10^{-19} \text{ C.}$$

Division of the electronic charge by the ratio of charge to mass gives us the mass of an electron. The value found for this is

$$m = 9.11 \times 10^{-31} \text{ kg}$$
$$= 9.11 \times 10^{-28} \text{ g.}$$

Information Deduced from Value of Electronic Charge

A number of important items of information may at once be deduced from a knowledge of the value of the electronic charge. We shall now consider these.

AVOGADRO'S NUMBER   From the laws of electrolysis it is deduced that each kind of ion of a monovalent element in electrolysis carries the same charge, each kind of ion of a divalent element carries twice this charge, and so on. It would be extraordinary if the charge carried by a monovalent ion were anything other than the electronic charge. Let us, then, tentatively assume that it is. It is found that to deposit 1 gram-atom of a monovalent element in electrolysis, 96,488 C* must be passed through the electrolyte. The corresponding number of electrons must be delivered by the cathode to the positive ions arriving there to change them into neutral atoms, and this number must also be delivered by the negative ions to the anode. To find this number, all we have to do is to divide 96,488 C by the electronic charge, 1.6021×10^{-19} C. The number so found is 6.0226×10^{23}. This is Avogadro's number, usually designated by the symbol N:

$$N = 6.0226 \times 10^{23} \text{ atoms per gram-atom.}$$

Although it is extremely unlikely that the assumption underlying this result—that we may identify the charge carried by a monovalent ion in electrolysis with the electronic charge—is wrong, it is nevertheless conceivable that it *could* be. It is also just conceivable that the smallest step by which charges can vary in the oil-drop experiment might not be identical with the charge carried by a thermionically emitted electron. In view of these possibilities, however remote, it is reassuring to note that Avogadro's number may also be determined from observations of suspended particles in an emulsion, on the basis of purely statistical considerations involving the law of equipartition of energies. This was done, many years ago, by J. Perrin; and later, using improved techniques, by

---

* It will be appreciated that this figure depends on the scale of values assigned to the atomic weights of the elements. For a discussion of this see Chapter 10 of Volume III.

Westgren, Svedberg, and others. In the most reliable of these determinations values of N were obtained agreeing within 2 to 3 percent with that calculated from the electronic charge. These methods do not lend themselves to high accuracy, and so a discrepancy of a few percent is of no significance. The agreement within the limits of experimental uncertainty must surely mean that both the assumptions on which our result is based are justified.

BOLTZMANN'S CONSTANT Avogadro's number being known, it is a simple matter to calculate the value of Boltzmann's constant k, the "gas constant per molecule." Thus

$$k = \frac{R}{N} = \frac{8.314}{6.0226 \times 10^{23}} = 1.3805 \times 10^{-23} \text{ J/deg.}$$

ELECTRON VOLT AS A UNIT OF ENERGY For many purposes the energy acquired by an electron in "falling" from rest through a potential difference of 1 V is a more convenient unit of energy than the joule. This unit is known as the **electron volt** (eV). The relation between the electron volt and the joule is obtained at once when it is remembered that the joule is a coulomb-volt. Thus, the electronic charge being 1.602×10^{-19} C, we must have

$$1 \text{ eV} = 1.602 \times 10^{-19} \text{ J.}$$

An important example of the use of this unit is in the theory of thermionic emission. We have considered Richardson's equation in the form

$$i = A T^2 e^{-B/T}, \tag{22.1}$$

in which, as we see at once, B must have the dimensions of a temperature. However, as we have noted, theory requires this quantity to be proportional to the low-temperature value of the work function, ϕ, the amount of energy by which the most energetic electrons in the cold metal fall short of the energy necessary to surmount the potential barrier. It appears desirable to substitute for B an expression that explicitly contains this quantity ϕ, and, because it is electrons with which we are concerned, to express this energy in electron volts.

As we have seen, Boltzmann's constant k has the dimensions of energy divided by temperature. Hence ϕ/k has the dimensions of temperature and may be identified with B. Thus we may substitute ϕ/kT for B/T. It is not only necessary that the numerator and denominator should both have the same dimensions, but it is also desirable that they should both be expressed in the same units. Thus if ϕ is expressed in electron volts, k should be expressed in electron volts per degree. We already have the value of k in joules

587

TABLE 22.3 Thermionic Data

	Emitter				
	W	Mo	Ta	Th	W–Th
ρA (A/cm²/deg²)	80	55	60	70	3
ϕ (eV)	4.54	4.15	4.12	3.38	2.63

per degree, 1.380×10^{-23} J/deg. For its value in electron volts per degree, therefore, we have

$$k = \frac{1.380 \times 10^{-23}}{1.602 \times 10^{-19}} = 8.61 \times 10^{-5} \text{ eV/deg}.$$

With ϕ expressed in electron volts, the numerical relation between B and ϕ is

$$B = \frac{1}{k}\phi = 1.161 \times 10^4 \phi,$$

or

$$\phi = \frac{B}{1.161} \times 10^{-4}. \tag{22.5}$$

We may now rewrite Richardson's equation in the more usual form,

$$i = AT^2 e^{-\phi/kT}. \tag{22.6}$$

Converting from B to ϕ according to Eq. (22.5), we now have for the thermionic constants of the emitters listed in Table 22.2 the values shown in Table 22.3.

ATOMIC MASSES Our knowledge of Avogadro's number enables us to calculate the masses of individual atoms. Oxygen, for example, has the atomic weight 16; that is, N atoms of oxygen, which constitute a gram-atom, have a combined mass of 16 g. Hence the mass of a single atom of oxygen is $16/N$, that is, 2.656×10^{-23} g.

Similarly, we may find the mass of a hydrogen atom, this being 1.673×10^{-24} g. It is interesting to note that the mass of an electron, 9.11×10^{-28} g, is only $\frac{1}{1837}$ that of a hydrogen atom.

SIZES OF ATOMS A final application of our knowledge of the electronic charge, and hence of Avogadro's number, concerns the sizes of individual atoms; we can form at least a rough estimate of these. We assume that in a solid the atoms are about as closely spaced from one another as they can be, that is, that neighboring atoms "touch" one another. We also have to assume some arrangement of these atoms in a lattice. The simplest assumption we can make is that of a simple cubic lattice, an arrangement in which the "unit cell" is a cube with an atom centered at each of the eight corners. X-ray diffraction (see Chapter 24) gives us precise information and shows that lattice structures are, in fact, usually less simple than this. However, the assumption of a simple cubic lattice, even when at fault, does not affect the order of magnitude of the result obtained. Let us, for example, consider the case of sodium. The density of sodium is 0.97 g/cm³, and the atomic weight is 23.0. Hence 1 cm³ contains 0.97/23.0 of a gram-atom, and so the number of atoms in 1 cm³ is this multiplied by N; that is, 1 cm³ contains

$$\frac{0.97}{23.0} \times 6.023 \times 10^{23} = 2.54 \times 10^{22} \text{ atoms.}$$

If d is the effective diameter of a single atom of sodium, the volume of a cube such that it will just contain 1 atom is d^3, and the number of these in 1 cm³ is therefore $1/d^3$. Numerically,

$$\frac{1}{d^3} = 2.54 \times 10^{22},$$

and so

$$d = 3.40 \times 10^{-8} \text{ cm}$$
$$= 3.40 \text{ Å};$$

that is, the effective diameter of a sodium atom is of the order of rather more than 3 Å.

From x-ray diffraction we know that the lattice structure of sodium is in fact not simple cubic but "body-centered cubic"; each unit cell has an atom not only at the eight corners of the cube but also at its center. It is not difficult to show that this necessitates a correction to the above calculation by a factor of $\sqrt{3}/2^{2/3}$, that is, of 1.092. Multiplication of our previous result by this gives us 3.72 Å for d.

Problems

22.1   A thin metal ring of radius a has a positive charge Q which is distributed uniformly around the ring. Show that an electron that is constrained to move along the axis of the ring can oscillate with a period T, where

$$T = \sqrt{\frac{16\pi^3 a^3 \epsilon_0 m}{|eQ|}},$$

e and m being the electronic charge and mass, respectively. (Assume that the amplitude of the oscillation is small compared with a.)

22.2   A beam of electrons of energy W passes between a pair of parallel deflector plates in a cathode-ray tube, of length l and separated by a distance d, the direction of the beam being along the length of the plates and parallel to them. Show that when the PD between the plates is V, the deflection s of the spot on the fluorescent screen, situated a distance L beyond the deflector plates, is given by the equation

$$s = \frac{e}{2} \frac{Vl}{Wd} \left(L + \frac{l}{2} \right).$$

Calculate s for the case where $W = 2000$ eV, $l = 4$ cm, $d = 1$ cm, and $L = 18$ cm.

22.3   Find the values of the constants ρA and ϕ for niobium from the following emissions I observed at a series of temperatures T from a central section 10 cm long of a niobium filament whose diameter is 0.3 mm:

T (°K)	I (A)
1441	6.35×10^{-7}
1504	2.83×10^{-6}
1563	9.77×10^{-6}
1618	2.94×10^{-5}
1678	8.28×10^{-5}
1732	2.07×10^{-4}
1786	5.27×10^{-4}
1832	1.051×10^{-3}
1883	2.23×10^{-3}
1937	4.60×10^{-3}
1984	8.89×10^{-3}

At what temperature would niobium give an emission of 100 mA/cm² of apparent surface? At what temperature would thoriated tungsten ($\rho A = 3.0$ amp/cm²/deg², $\phi = 2.63$ eV) give this emission?

22.4   The work function of tungsten is 4.54 eV. At 2400°K a tungsten filament of diameter 0.05 mm gives a thermionic emission of 1.84 mA per cm of length. When saturated thermionic current is drawn from this filament, how much of the power required per centimeter of its length to maintain its temperature is used up in providing the latent heat of evaporation of the emitted electrons?

22.5   (a)   A thermionic diode consists of parallel cathode and anode plates mounted in a vacuum, with the distance between these small compared with their linear dimensions. A space-charge-limited current of magnitude i per unit area flows from cathode to anode. Write (1) the equation for the electric field E in terms of the number n of electrons per unit volume (a function of the distance x from the cathode), (2) the equation for i in terms of n and the velocity v of the electrons, and (3) the equation of motion of the electrons in terms of E, the electronic charge e, and the electronic mass m.

(b) Assuming as a first approximation that at the surface of the cathode $E = 0$ and $v = 0$, find the relation between the time t after an electron has left the cathode and the distance x it has covered in its journey toward the anode.

(c) Hence show that

$$ i = \frac{4\sqrt{2}}{9}\,\epsilon_0\,\sqrt{\frac{e}{m}}\,\frac{V^{3/2}}{d^2}, $$

where d is the distance between cathode and anode and V the potential of the anode relative to the cathode.

22.6 The potential of the anode of a thermionic diode relative to the cathode falls at a uniform rate from $+100$ V to -100 V during 1 sec, then rises (again uniformly) back to $+100$ V in 0.1 sec. Assuming that at $+100$ V the electron current does not yet approach saturation, sketch a graph showing how this current varies with time.

22.7 The cathode of a thermionic diode is a long straight thin tungsten filament, which is stretched along the axis of three closely spaced cylindrical anodes each of internal diameter 2.54 cm, the two outer anodes serving as guard rings, while the electron current collected by the central anode, whose effective length is 7.62 cm, passes through a milliammeter. With this tube in operation, space-charge-limited currents I are observed at a series of potentials V of the anodes relative to the center of the filament as follows:

V (V)	I (mA)
35	17
47	28
55.5	36
68	48
75.2	54
89	72
108.5	97.7
129	130

Plot these results in an appropriate manner, and from your graph calculate the value of e/m.

22.8 An electron-accelerating potential difference of 200 V is applied between a plane thermionic cathode and a plane anode parallel to it and 4 mm distant from it. Neglecting the initial velocities with which the electrons are emitted from the cathode, calculate (a) the velocity with which they strike the anode, and (b) the transit time of an electron from cathode to anode. (Assume negligible space charge and hence uniform field between electrodes.)

22.9 A certain thermionic diode consists of a plane metal cathode and a parallel anode held at a positive potential V_0 relative to the cathode. The effect of the space charge is to produce a potential V in the region between the electrodes varying with the distance x from the cathode according to

the equation

$$V = V_0 \left(\frac{x}{d}\right)^{4/3},$$

where d is the distance between the electrodes. Assuming that the electrons are emitted from the cathode with negligible initial velocity, find their velocity in the interelectrode space as a function of x. Find also the transit time of an electron from cathode to anode. Calculate the transit time in the case where $d = 4$ mm and $V_0 = 200$ V. (Use the value of e/m given in the text.) Show that the total space charge Q in a cylinder of cross-sectional area A extending normally from cathode to anode is given by the equation

$$Q = -\frac{4}{3}\frac{\epsilon_0 V_0 A}{d}.$$

22.10   If, in Problem 22.8, the current flowing between the electrodes is 100 mA, what is the force that the stream of electrons impinging on the anode exerts on it?

22.11   In an oil-drop experiment of the type carried out by Millikan, the oil used had a density of 0.92 g/cm³ and the plates were spaced 6 mm apart. When the upper plate was made 807 V positive to the lower, one of the drops was observed to be stationary, but when the plates were connected this drop fell at the rate of 0.167 mm/sec. Given that the density and viscosity of the air were 0.0012 g/cm³ and 1.72×10^{-4} P, respectively, find the charge on the drop.

22.12   In an oil-drop experiment, charge magnitudes of 6.48×10^{-19} C, 11.55×10^{-19} C, 13.13×10^{-19} C, 21.34×10^{-19} C, and 27.50×10^{-19} C were found. What is the maximum value the electronic charge could have, to be consistent with these results? What is the probable error in this maximum value?

22.13   How many excess electrons must an isolated conducting sphere of 1-cm radius have for its potential to be 100 V below earth potential?

22.14   How many electrons per second pass a point in a wire along which a current of 1 A is flowing?

22.15   How many times greater is the acceleration of an electron in a field of 100 V/cm than that of a body at the earth's surface falling freely under gravity?

22.16   Express the mean translational kinetic energy of a molecule at 20°C corresponding to the velocity component in a particular direction in electron-volt units of energy.

22.17   From considerations of equilibrium between a thermionically emitting cathode and an "electron gas" in the vacuum outside, it may be shown that the velocity distribution among the electrons as they are emitted or reflected from the emitting surface is the same as that among the electrons approaching the surface from the electron gas. Also it may be shown that in the matter of energy distribution an electron gas in a vacuum is like a gas consisting of material molecules. Using this information, and assuming zero reflection, calculate the mean kinetic energy corresponding to the normal velocity components with which electrons are emitted from a cathode at 2000°K. (Express your answer in electron volts.)

BASIC ELECTRONICS II

Having dealt with thermionic emission and the fundamentals of the theory of electrons in Chapter 22, we shall, in the present chapter, discuss an interesting alternative process in which electrons are extracted from metals: photoelectric emission. The study of photoelectric emission is of particular value, not merely because of its many important applications in scientific research and in technology, but more especially because it gives a useful and simple introduction to what has become one of the cornerstones of the structure of modern physics, the quantum theory.

After the study of photoelectric emission and the introduction to the quantum theory to which it leads, the next main topic to be discussed in this chapter will be the generation of x rays, which may be regarded, broadly, as photoelectric emission in reverse. Consideration of the construction of x-ray tubes will then lead naturally to the subject of electron optics; and a study of the electron-optical systems of the thermionic triode and the modern cathode-ray tube will bring the chapter to a close.

Photoelectric Emission

In order to extract electrons from a metal otherwise than by the application of a strong electron-accelerating field to the metal surface (field emission) we need to impart to some of the electrons within the metal sufficient extra energy, over and above what they may already possess, to enable them to surmount the surface potential barrier. We have already considered one way of doing this, that of heating the metal to a high temperature. Let us now inquire whether certain alternative methods might not be available.

A priori, we might anticipate that almost any kind of energy made available at the surface of a metal could be converted into kinetic energy of the internal electrons, given a mechanism for such conversion. This consideration suggests that we try making various kinds of energy available, whether or not we have as yet any clear picture of a mechanism. One kind of energy that should be well worth trying is that of electromagnetic radiation impinging on the metal surface. Normally, when radiation is absorbed, it is converted into heat, or at least most of it is. However, some might well become available more specifically to the internal electrons, increasing the kinetic energy of these.

It has, in fact, been found that light incident on the surface of a metal *is* capable of releasing electrons from it. Historically, however, the experiments that led to the discovery of this effect, now known as **photoelectric emission,** were not inspired by such considerations as the above; indeed, they could not have been, for the background knowledge on which such reasoning could have been based did not exist at the time. Actually, the discovery was a by-product of investigations carried out by Heinrich Hertz, in 1887, on the generation and detection of what we now call radio waves, using a spark oscillator. Hertz observed that a spark would pass more easily between the electrodes of the oscillator when illuminated by light from another spark than when not so illuminated, and by following up this observation he was able to show that the effect was due to ultraviolet light falling on the negative electrode of the spark gap. Subsequently, Hallwachs and others investigated the phenomenon more closely, and it was shown that illumination of a metal by light in a suitable range of wavelengths causes negative electricity to be emitted from it. Electrons were not known at this time, and the recognition of the phenomenon as an emission of electrons did not come until much later, as a result of investigations carried out by Millikan.

THE MAIN PHENOMENA A simple experiment first carried out by Hallwachs, which demonstrates this effect, is the following. A sheet of freshly abraded zinc is bent into cylindrical form so that it may be stood up on one end, and is placed on the disk of an electroscope. A beam of light from a carbon arc provided with a silica condensing lens, or other light containing at least a proportion of ultraviolet radiation, is arranged so that it may be directed on to the zinc. Now, if the electroscope is negatively charged and the zinc plate is illuminated by the beam, the leaf will be seen to fall, showing that negative electricity is being lost. This we now know to be due to electrons—"photoelectrons," as they are sometimes called, after the means employed for releasing them—emitted from the zinc plate and following the lines of force between this and earth. If an obstacle is interposed to cut off the illumination, the fall of the leaf is at once arrested, to be continued only when the obstacle is removed.

If the electroscope is charged positively instead of negatively, no effect is observed; the electroscope retains its charge whether or not the zinc is illuminated. This does not mean that electrons are not emitted under the influence of the illumination, only that they are immediately driven back again onto the zinc by the electron-retarding field and are reabsorbed.

If the electroscope is charged negatively and a plate of silica is interposed in the path of the light, the leaf still falls, but somewhat more slowly, because of the absorption of some of the radiation by the silica. However, if a sheet of ordinary glass is interposed, the leaf at once stops falling; there is no longer any emission of electrons. The difference between silica and glass is that silica, while absorbing some of the ultraviolet radiation, transmits most of it (if the plate if not unduly thick), whereas a sheet of glass of ordinary thickness absorbs virtually the whole of the ultraviolet—at least that having wavelengths appreciably below the visible. This result is highly significant; it means that it is not the *intensity* of the light that determines whether or not photoelectric emission occurs, but its *quality*. The intensity merely determines the *rate* of emission, but if the radiation does not contain the right wavelengths no emission at all will occur, however intense the illumination may be.

If the zinc has not been reasonably freshly abraded (that is, has been allowed to become appreciably oxidized) or if some other metal is substituted for it, no effect is observed unless ultraviolet light of considerably shorter wavelengths is used.

PHOTOELECTRIC THRESHOLD The experiment described above is suitable only for a preliminary orientation; for a closer study of photoelectric emission it is necessary to protect the emitting metal—the cathode—from contamination by enclosing it in an evacuated bulb of a material transparent to radiation in the required wavelength range, or at least provided with a transparent window. An anode must also be provided, and this must be so disposed as not to obscure any substantial proportion of the radiation. Such a device is known as a **photoelectric cell,** a **photoelectric tube,** or a **phototube.** One design of phototube is shown, together with a circuit diagram, in Fig. 23.1. In this the cathode C is a portion of a hollow cylinder, while the anode A is a rod located along the axis of the cathode. B is a battery, and G is a galvanometer.

For a fully quantitative investigation of photoelectric emission it is necessary to ensure that the condition of the cathode is the same all over its surface, that is, that it is completely free of contamination, or, if contamination is present, that this is well defined and uniform, like the adsorbed layer of thorium on thoriated tungsten. The former alternative is the simpler of the two, but even to achieve this condition it is generally necessary to subject the cathode to

595

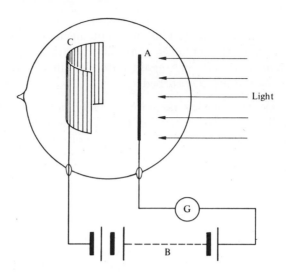

FIGURE 23.1   System for observation of photoelectric emission.

prolonged heat treatment in a vacuum.* The heat treatment distills off any contamination, while the vacuum, if sufficiently good, ensures against appreciable recontamination by residual gas during the course of the measurements to be made.

Let such a cathode, at room temperature, now be irradiated by "monochromatic" light (that is, by light of a single wavelength) or at least by light confined to a narrow range of wavelengths, while the anode is held at a positive potential with respect to the cathode, sufficient to draw off any electrons emitted. Let the wavelength in the first instance be relatively long. If it is sufficiently long, no emission will be registered by the galvanometer, however intense the radiation may be. Then let the experiment be repeated, using progressively shorter wavelengths. It will be found that at and below a certain threshold value λ_0 of the wavelength characteristic of the cathode a response *is* obtained. Figure 23.2 shows how the emission from potassium per unit area per unit intensity varies with the illuminating wavelength. The spectral response curves of other cathodes are qualitatively similar.

At all wavelengths below λ_0 the emission per unit area is proportional to the intensity of the light with which the cathode is irradiated.

---

* This could, for example, take the form of induction heating (see Chapter 34) or bombardment of the cathode with a stream of electrons from an auxiliary thermionic source.

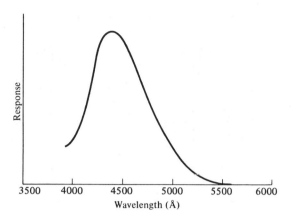

FIGURE 23.2 Variation of photoelectric emission from potassium with wavelength of exciting radiation of given intensity.

Quantum Theory

In casting about for an explanation of the existence of a threshold wavelength for each metal and for its variation from one metal to another, we should naturally seek to correlate it with the thermionic work function ϕ of the cold metal; for the basic idea on which we are working is that heating and irradiation are merely different methods of imparting to the internal electrons sufficient extra energy to enable them to escape. Let us, then, compare the values of ϕ and λ_0 for a number of metals, as listed in Table 23.1.

From an inspection of the table it will be seen that as ϕ increases λ_0 decreases, and it looks as if one might be inversely proportional to the other. To test this, $1/\lambda_0$ is plotted against ϕ, using the values of the table, in Fig. 23.3. We see that the plot is a straight line passing through the origin, showing that $1/\lambda_0$ is proportional

TABLE 23.1

Metal	ϕ (eV)	λ_0 (Å)
Cs	1.81	6.44×10^3
Ta	4.12	3.00×10^3
Mo	4.15	2.97×10^3
W	4.54	2.72×10^3
Fe	4.77	2.59×10^3
Pd	4.99	2.48×10^3
Re	5.1	2.48×10^3

597

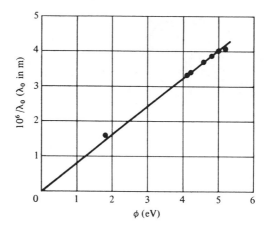

FIGURE 23.3 Plot of reciprocal of photoelectric threshold wavelength against thermionic work function for a number of metals.

to ϕ. This gives the possibility of determining ϕ photoelectrically, from the value of λ_0, using this proportionality. From the graph the relation is found to be

$$\frac{1}{\lambda_0} = 8.1 \times 10^5 \phi,$$

where λ_0 is measured in meters, or, since

$$f_0 \lambda_0 = c,$$

where f_0 is the threshold *frequency* and c is the velocity of light,

$$\phi = h f_0, \tag{23.1}$$

the value of the factor h being

$$h = \frac{1}{8.1 \times 10^5 c} = \frac{1}{8.1 \times 10^5 \times 3.00 \times 10^8} = 4.1 \times 10^{-15} \text{ eV-sec.}$$

The accuracy with which h can be determined in this way is severely limited by the experimental difficulties of both the thermionic and the photoelectric measurements, in particular the difficulty of obtaining a definite, reproducible condition of the cathode surface. This is aggravated by the circumstance that the thermionic and photoelectric measurements are normally made in different experiments, using different cathode samples. Fortunately, however, it is possible to avoid the necessity for making any thermionic measurements at all. This is done by using a single photoelectric cathode and artificially increasing its effective work function by accurately known amounts $\delta\phi$.

In our study of the action of a thermionic diode we saw that the electrons reaching the anode under space-charge-limited conditions have a composite potential barrier to surmount, consisting of two steps, one at the cathode surface, the other between this and the space-charge potential minimum. It is as if the current received by the anode were the full thermionic emission from a cathode of increased work function. In the corresponding case of a photocell the emission is usually much smaller, and in consequence space charge is unimportant. When the anode of a photocell is made in the form of a plate, parallel to a plane cathode, and a negative (that is, retarding) potential is applied to it, there is, over practically the whole voltage range, no space-charge potential minimum between the electrodes, despite the existence of a small space charge; the potential is lowest at the anode itself. With a retarding potential V applied to the anode, the total potential barrier that has to be surmounted by electrons reaching the anode is increased by V volts, and it is if the work function had been increased by the amount eV, where e is the electronic charge; that is, writing ϕ_{eff} for the effective work function, we have

$$\phi_{eff} = \phi + eV.$$

To each retarding potential applied to the anode there is found to correspond a threshold frequency f_0 for radiation falling on the cathode below which there is no perceptible anode current but beyond which a current is observable. This threshold frequency must be related to ϕ_{eff} in the same way as that observed normally (with the anode at a positive potential) is related to ϕ. In place of Eq. (23.1) we now have

$$\phi_{eff} = \phi + eV = hf_0, \qquad (23.2)$$

and so, if we vary the retarding potential V, we must have a corresponding variation of f_0, according to the equation

$$\frac{df_0}{dV} = \frac{e}{h};$$

that is, the slope of a plot of f_0 against V, such as is shown in Fig. 23.4, must be independent of the particular cathode used and must have the value e/h. Experimentally it is found that f_0–V plots made with different cathodes do indeed all have the same slope. Thus our theory is confirmed, and by multiplying the reciprocal of this slope as found with any cathode by e we may determine h. The value so found is 4.14×10^{-15} eV-sec. Accordingly, in absolute units, we have

$$h = 4.14 \times 10^{-15} \times 1.602 \times 10^{-19} = 6.63 \times 10^{-34} \text{ C-V-sec}$$
$$= 6.63 \times 10^{-34} \text{ J-sec.}$$

599

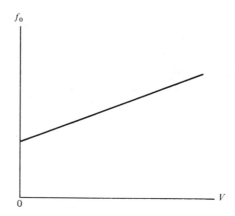

FIGURE 23.4 Variation of threshold frequency for a given photoelectric cathode with retarding potential applied to it.

Equation (23.1) and its equivalent Eq. (23.2) are of profound theoretical significance. They identify an energy jump with something proportional to the frequency of the radiation whose energy is made available for this jump. From radiation of an frequency less than f_0, however intense, an electron in the metal is unable to obtain sufficient energy—or, if it already possesses some energy, sufficient *extra* energy*—to enable it to surmount the potential barrier with which it is confronted. But such energy *is* available from radiation of higher frequencies, however weak. This, surely, can mean only one thing: We may imagine the energy of radiation to be, so to speak, tied up in bundles, or packets ("wave packets"), each of magnitude hf, where f is the frequency of the radiation. Only a whole packet can be transferred at a time; no more, no less. Thus we have a kind of atomicity of energy as well as of matter and of electricity. This discrete quantity of energy is known as an **energy quantum,** and the quantity h is known, after Max Planck (1858–1947), the originator of the quantum theory,† as **Planck's constant.** Radia-

---

* Actually, according to the modern theory of electrons in metals, the free electrons already have considerable kinetic energy inside the metal, even at 0°K.

† The introduction of the quantum theory here given is not the same as Planck's original development. Planck was led to the postulation of energy quanta by a study of the distribution of energy among the frequencies constituting full, or black-body, radiation. On the basis of this theory, developed by Planck in 1900, Einstein drew the logical consequences relating to photoelectric emission and predicted the linear relation between threshold frequency and retarding potential in a phototube, represented by Fig. 23.4. This prediction was subsequently verified by Millikan in 1916.

tion quanta, regarded in a certain sense as particles, are also called **photons.**

Quite apart from the existence of a photoelectric threshold, the photon aspect of radiation is definitely indicated by certain considerations of the intensity of radiation observed to excite an immediate photoelectric response. However feeble the illumination of a metal at a frequency capable of exciting photoelectric emission, electrons are found to be emitted immediately the surface is irradiated—there is no time lag. In the feeblest intensities giving observable photoelectric response the energy density is so low that, even assuming the effective area of collection of energy by an electron to be a circle with a diameter as large as a few atomic diameters (many orders of magnitude larger than an electron), it would, according to classical theory, take weeks for the electron to acquire sufficient energy for emission. The absence of any observable time lag, even in such a case, can only mean that, rather than being spread continuously over the radiation, the energy is bunched in concentrated packets (that is, photons) which immediately make the requisite energy available to the electrons they encounter.

There are several distinct ways of evaluating h, some of which are much more accurate than the photoelectric method. The value at present regarded as the most reliable is

$$h = 6.625 \times 10^{-34} \text{ J-sec.}$$

The quantum theory is not merely something necessary to account for the phenomena of photoelectric emission and, as the footnote on p. 600 indicates, radiation from hot bodies; it far transcends in scope and importance its application to these special fields. Accordingly, Eq. (23.1) must be regarded merely as a special case of the more general equation

$$E = hf, \tag{23.3}$$

where E is the energy of the quanta in electromagnetic radiation of frequency f available for conversion to *any* other form.

The availability of the energy of radiation in discrete quanta is something we cannot hope to "explain," in the sense of finding a mechanism for it in terms of our ordinary experience. Like the atoms of the chemical elements, or electrons, radiation quanta are entities that we can never experience directly, but whose existence we are obliged to infer from experimental evidence. It is solely because of the absence of direct experience of quantum processes in our everyday lives that the ideas of the quantum theory appear to us so revolutionary.

601

We can account qualitatively for the form of the tail of the photoelectric response curve shown in Fig. 23.2 as follows. Let the outwardly directed normal to the surface be taken as the positive x direction in a Cartesian coordinate system. Then the kinetic energy of an electron within the metal is $\frac{1}{2}m(v_x^2 + v_y^2 + v_z^2)$, where v_x, v_y, and v_z are the three velocity components. However, an electron can escape only if (1) v_x is positive, (2) the electron is sufficiently near the surface not to share its energy with other electrons or with atoms on the way out, and (3) $\frac{1}{2}mv_x^2$ exceeds the difference of potential between the inside and outside of the metal; the other two terms in the expression for the energy can obviously make no contribution toward enabling the electron to surmount the potential barrier. It seems reasonable to assume that the directions of travel of the internal electrons after they have absorbed radiation quanta would be distributed uniformly in all directions, or at least that there would be no strong bias, if, indeed any bias at all, in favor of the positive x direction. On this basis, the greater the frequency of the radiation absorbed, and so the resulting kinetic energy of the electrons, the greater must be the probability that the item $\frac{1}{2}mv_x^2$ exceeds the potential-energy barrier. In this way the observed increase of photoelectric response with frequency would be accounted for.

Photoelectric Cathodes

A large variety of photoelectric cathodes are used in phototubes, according to the purpose for which they are intended.

One of the most important considerations is the threshold for photoelectric response, which is given in terms of the work function expressed in electron volts by this equation

$$\lambda_0 \, (\text{Å}) = \frac{12{,}396}{\phi}. \tag{23.4}$$

In nearly all applications of photocells sensitivity to visible light is required. The visible range extends from about 4000 Å (violet) to 7600 Å (red), and this means that if a cathode is to respond to visible radiation at all, the work function must be no higher than 12,396/4000, or 3.1 eV. Among clean metals, only the alkali and alkaline-earth metals have work functions less than this. But if response to the *whole* visible spectrum is required, the condition is not satisfied by any of the clean metals, not even cesium, whose threshold is at about 6500 Å. Fortunately, several composite cathodes have work functions substantially lower than those of any of the clean metals, and some of these have found widespread application in phototubes.

Several different kinds of composite cathode are used in present-day phototubes. Among these, one that has found particularly widespread application is the cesium/silver-oxide cathode. Silver oxide, formed by surface oxidation of a plate of silver, is exposed to cesium vapor at about 200°C, whereupon the chemical reaction

$$2Cs + Ag_2O \rightarrow 2Ag + Cs_2O$$

takes place, and the cesium oxide then becomes saturated with cesium, holding some of this metal in solid solution, and, in addition, it holds to its surface a monolayer of adsorbed cesium. The response of this cathode not only covers the entire visible spectrum but extends well into the infrared, the threshold being at about 12,000 Å.

Cathode Rays

In this book electrons have been introduced by studying the phenomena of field emission, thermionic emission, and photoelectric emission. Historically, however, streams of electrons were first observed in the course of the study of gas discharges in the latter part of the nineteenth century. These were of the kind we now class as glow discharges (see Chapter 24). They were produced in partially evacuated glass tubes through the ends of which electrodes were sealed, and they were investigated, in the first instance phenomenologically, with particular interest attaching to the effect of the residual gas pressure on the appearance of the discharge. The electrodes were usually in the form of disks, situated one at either end of the tube and facing each other. A PD, generally of the order of some tens of thousands of volts, was applied between these, using for the purpose an induction coil or a Wimshurst machine.

If, using such a tube, a discharge is passed through different kinds of gas contained within it, it is found that at the higher pressures the color of the glow varies with the kind of gas, and that the discharge exhibits a characteristic structure, with the details of which we need not concern ourselves here. As the pressure is reduced, the various features of this structure increase their distance both from one another and also from the cathode, until, in the pressure range from about 10^{-2} to 10^{-3} mm, the former glow having a color characteristic of the gas has entirely disappeared and has been replaced by a general luminosity for whose proper observation the room should be darkened. This luminosity, which exhibits no

603

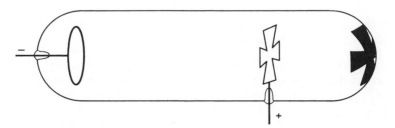

FIGURE 23.5   Shadow of obstacle in path of cathode-ray beam.

detailed structure, has a color that no longer depends on the nature of the gas but varies with the kind of glass of which the tube is constructed. Thus with lime-soda glass the color is yellow-green, whereas in the case of lead glass it is a gray-blue.

From such evidence the early workers in this field inferred that the luminosity observed in the pressure range 10^{-2} to 10^{-3} mm comes from the glass, not from the gas. The agency by which this was excited could, they felt, only be "rays" of some sort present within the tube and incident on the glass, perhaps rays emanating from one of the electrodes. When obstacles were placed within the tube, these were found to cast shadows on the glass—areas from which no luminosity could be observed. An example of such shadow formation is shown in Fig. 23.5; here a Maltese cross cut out of metal serves as anode, and its shadow is cast on the end of the glass tube remote from the cathode. Such shadows were observed regardless of whether the obstacles functioned as the anode (as in Fig. 23.5), were connected to an anode situated elsewhere in the tube, or were left electrically floating; but in all cases they were cast on the end of the tube remote from the cathode, never on the cathode end. From this it was inferred that the rays must have come from the cathode. Accordingly, they were called "cathode rays."

Two modifications in tube construction yielded further information of interest. These are indicated in Fig. 23.6. In part (a), where the anode was situated in a side branch of the tube, the positions of shadows cast by obstacles were found not to be affected appreciably by the change in the position of the anode. In the tube of part (b) the anode is brought up relatively close to the cathode and is provided in the center with a small aperture. In this case a narrow beam of cathode rays was found to pass through the aperture, producing a correspondingly small spot of luminosity on the glass wall beyond, at S. From these observations it was inferred that the cathode rays must travel along lines at least approximately at right angles to the surface of the cathode.

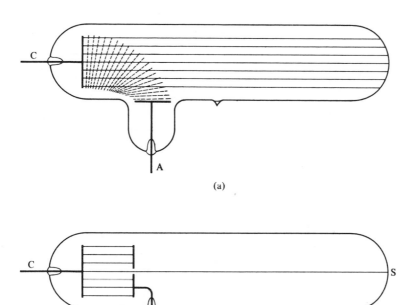

FIGURE 23.6   Special forms of cathode-ray tube.

The ejection of the cathode rays from this electrode could most naturally be explained by supposing that they consist of negatively charged particles, and this surmise was soon verified by collecting them through a hole in a metal box* mounted on a wire sealed through the glass wall. The box was found to become negatively charged. Corroborative evidence was obtained by establishing a magnetic field at right angles to the path of the rays and observing that this caused the spot S to be deflected in a direction such as would be expected for a stream of negatively charged particles (see Chapter 29). Yet further verification of the nature of the rays was obtained by deflecting the rays with an electrostatic instead of a magnetic field. This was done by mounting a pair of deflector plates inside the tube, one above and one below the beam of cathode rays, as indicated in Fig. 23.7, and applying a PD between these. This caused the spot to be deflected up or down, the direction of the deflection corresponding to that of the field as for negatively charged particles.

---

* Collection in a box avoids the difficulty of possible loss by reflection or because of secondary emission (see Chapter 24).

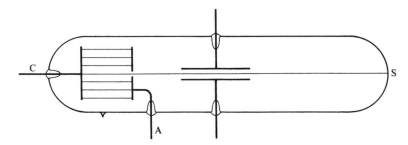

FIGURE 23.7   Primitive cathode-ray tube for producing narrow beam passing between deflector plates.

By combining electrostatic and magnetic deflections, and making the relevant measurements, J. J. Thomson, in 1897, succeeded in obtaining an approximate value for the ratio of charge to mass of the cathode-ray particles. The value he obtained agrees, within the limits of experimental uncertainty, with that now accepted for the ratio e/m of electrons, obtained by more reliable methods. Later, in the early years of the present century, the name "electron" came into general usage for the particles in question.

The smallness of the effect of changing the position of the anode from the end of the tube remote from the cathode to one side, as shown in Fig. 23.6(a), can readily be understood in terms of the system of equipotential surfaces between the electrodes. These are indicated by the dashed lines in the figure. In falling through these the electrons must acquire considerable momentum before arriving at a region where there is an appreciable lateral component of the field. Consequently, during their brief passage through this region they are only very slightly deflected in a direction toward the anode. In the region appreciably beyond the side tube in which the anode is situated the field is relatively weak, and so there can be no question here, either, of appreciable lateral deflection of the electrons.

The emission of electrons from the cathode, depending, as it does, on the presence of gas at a low pressure in the tube, is a gas-discharge phenomenon. This will be fully explained in Chapter 24.

It may be wondered how, in the arrangements shown in Figs. 23.6(b) and 23.7, the electrons can continue to impinge on the area S of the insulating glass wall, that is, how it is that by collecting these electrons this area does not acquire a negative charge sufficient to prevent any further electrons from reaching it. This is explained by the emission of secondary electrons from the glass. Secondary electron emission will be considered in Chapter 24.

The luminescence of the glass under electron bombardment is another feature calling for explanation. This is due to the "excitation" to higher-energy states of electrons in the glass by absorption of some of the kinetic energy of the bombarding electrons, followed by the conversion of the excitation energy to radiation.

X Rays

In the course of experiments with a cathode-ray tube Röntgen, in 1895, noticed that a screen impregnated with barium platino-cyanide (a phosphor), which was lying nearby, glowed brightly when the tube was in operation. Following up this accidental discovery, he found that the interposition of a sheet of black paper between the cathode-ray tube and the screen had no effect on the luminosity, but that it was diminished in intensity or stopped completely by the interposition of more substantial objects. He could only attribute this to some sort of rays emanating from the tube, and to these unknown rays he gave the name "x rays." Closer investigation showed that they were sent out in straight lines from the area on the glass wall of the tube where the cathode rays impinged.

Röntgen's discovery soon led to the development of the first primitive gas x-ray tube designed specifically for the production of a concentrated source of x rays. This differed from the cathode-ray tube with which Röntgen first observed x rays (1) in the substitution of a concave cathode for a plane one, this being found to produce the desired focusing of the electrons, and (2) in the use of a metal target on which these electrons were made to impinge, in place of the glass wall of the tube. With this tube Röntgen and others succeeded in obtaining shadowgraphs of bone structures and the like, not only on fluorescent screens, but also on photographic plates, for x rays were found to affect these also. Since those early days x-ray tubes have been changed and improved out of all recognition, and they are now in widespread use not only for diagnostic and therapeutic purposes in medicine but also in a number of industrial applications.

Gas x-ray tubes suffered from the disadvantage of gradual cleanup of gas during operation, necessitating periodical replenishment (for example, by diffusing in a little hydrogen through an attached palladium tube). Workers with x rays realized that it would be much better not to have to depend on a residual gas whose pressure has to be kept within narrow limits, replacing this by a good vacuum—something much more definite and permanent—and having an independent source of electrons.

607

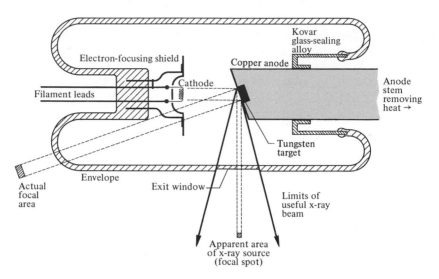

FIGURE 23.8   Modern x-ray tube.

A new kind of x-ray tube, in which this objective was realized, was developed in 1913 by W. D. Coolidge. In this tube a thermionic source of electrons replaced the earlier plate cathode in the "gas" x-ray tube. The establishment of the requisite high vacuum for the satisfactory operation of the thermionic cathode had by this time been rendered possible by the advances in vacuum techniques that had taken place since the discovery of x rays in 1895. All x-ray tubes now incorporate a thermionic cathode.

A modern x-ray tube is shown diagrammatically in Fig. 23.8, and the details of the focusing arrangement are indicated in section in a plane at right angles to that of Fig. 23.8 in Fig. 23.9. The cathode C is a short straight helix of tungsten wire, heated by a current in the same way as a lamp filament is heated; A is a massive copper anode, copper being the material chosen because of its high thermal conductivity. Behind the cathode is a focusing shield S which is usually connected internally to the negative end of the filament and so is given a potential slightly negative to that of the main

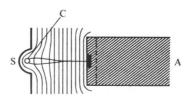

FIGURE 23.9   Focusing of x-ray beam by system of equipotentials.

emitting area of the cathode. The shield is so shaped and located as to give suitably curved equipotential surfaces outside it, in the direction toward the anode. The system of equipotentials is indicated in Fig. 23.9; these are analogous to contour lines on a map. Regarded as contour lines, this system would correspond to a downward-sloping valley at the top, merging into a downward-sloping plane surface. If a number of billiard balls were released at various points on the upper part of a surface having such a contour, they would roll down toward, and then along, the floor of the valley, and would continue in a narrow stream down the plane surface, finally arriving at the bottom in a concentrated stream. Similarly, in the x-ray tube as described, electrons emitted from the filament are guided by the system of equipotentials and concentrated into a narrow beam in the manner indicated in Fig. 23.9.

Actually the system of equipotentials has an action similar to that of a lens in optics, the equipotentials corresponding to refracting surfaces. They cast an image on the surface of the anode in the form of a short line focus, as indicated. With the arrangement shown, this image is "seen" by the object irradiated with x rays very obliquely to the surface of the anode, so that the *projected* area, as seen, is considerably foreshortened and appears as a small spot. This concentrated effective source permits shadowgraphs having reasonably sharp outlines to be obtained on fluorescent screens or photographic films. X rays are emitted in all directions from the focal spot, so for the protection of personnel it is usual either to provide an absorbing shield with a window around the apparatus or to construct the envelope of the tube of relatively thick glass, which absorbs a substantial proportion of the x radiation, except for a window of much thinner glass, which permits x rays to pass relatively freely to the outside in a selected direction.

Only a very small proportion of the kinetic energy of the electrons impinging on the anode is converted into x radiation*— almost the whole of it is dissipated as heat. In most modern tubes this causes the temperature of the focal spot to rise to a value well above the melting point of copper, and it is therefore necessary to provide a more refractory metal for the actual target. The metal used for this purpose is tungsten, whose melting point is 3650°K, and which has the incidental advantage that it gives a higher yield of x rays than do most metals. The piece of tungsten, of thickness

---

* The conversion efficiency is approximately proportional to ZV^2, where Z is the atomic number of the target metal and V the anode voltage. But even with a tungsten target ($Z = 74$) and $V = 100$ kV, the efficiency is less than 1 percent.

609

about 3 to 4 mm, is let into a depression in the massive copper anode, in which it is held firmly, and in consequence of thermal conduction the temperature of the inner face of the tungsten and the adjacent copper remains below the melting point of the latter. The heat is finally dissipated, largely by conduction along a massive copper stem that passes through the glass wall which is sealed to it. In tubes intended for operation at only moderate power, a few cooling fins are usually attached to the copper stem outside the tube, and the whole is immersed in a bath of oil. In the case of tubes operating at relatively high powers, such convective cooling is insufficient, and "forced" cooling must be resorted to. For this, the anode is provided with an internal tubular channel, through which the cooling oil is made to circulate.

The potential differences used for accelerating the electrons are commonly of the order of 20 to 100 kV. The currents vary considerably with the power of the tube. For relatively low-power tubes typical electron currents would be 1 or a few milliamps. High-power tubes may be operated at current strengths of 10 to 20 mA or, in certain cases, much more.

Reference has already been made to the fact that x rays consist of electromagnetic radiation of very short wavelengths. The evidence from which this has been inferred and certain other aspects of the physics of x rays will be considered in Chapters 39 and 40.

Thermionic Triode

This is an electronic device in widespread use both as an amplifier in radio reception and as an oscillator in radio transmitting systems (see Chapter 34). It differs from the diode, which we have already considered, in having a third electrode, a "grid," situated between the cathode and the anode. This electrode, which is generally constructed in the form of a wire wound on a frame, is normally operated within a range of potentials below the potential of the cathode, while the anode is given a positive potential, for example 100 to 200 V positive to the cathode. With a given anode potential, the more negative the potential of the grid, the smaller the electron current that is able to pass between the wires to the anode. The grid performs the function of a traffic policeman, regulating the flow of electrons passing from cathode to anode according to its potential. Being always negative to the cathode, the grid does not itself collect any electrons.

For simplicity, let us consider the electron-optical system of an

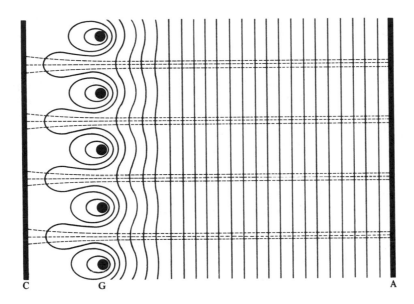

FIGURE 23.10 System of equipotentials in thermionic triode.

idealized form of triode, with plane cathode and anode surfaces, such as is shown in section in Fig. 23.10, C, G, and A representing the cathode, grid, and anode, respectively. In this the equipotentials are as represented by the thin lines.* These penetrate the spaces between the grid wires, approaching the areas of the cathode opposite these spaces more or less closely, according to the potential of the grid. When the penetration is deep, the field in the central regions of the cathode areas approached may be electron-accelerating all the way, right from the cathode surface, despite the existence of space charge. The areas giving saturated current vary with the degree of penetration, as determined by the grid potential. The electron current contributed by surrounding areas is space-charge-limited, again to a degree determined by the potential of the grid. Thus the total electron current, space-charge-limited plus saturated, increases or decreases according as the grid is less or more negative with respect to the cathode.

From the contributing areas of the cathode to somewhat beyond the grid the equipotentials correspond to contour lines of steep-sided, steeply downward-sloping valleys. Accordingly, the electron beams are brought to line foci on the anode, as indicated in the diagram.

---

*These are the equipotentials corresponding to a cold, nonemitting cathode. With the cathode hot, the system of equipotentials would be somewhat modified because of the space charge.

Modern Cathode-Ray Tube

In the cathode-ray tube shown in Fig. 23.7 a narrow beam of electrons is produced by screening off all except those that pass through a small aperture in the anode. This beam may be deflected up or down in the plane of the diagram by applying an appropriate field between the deflector plates. Obviously the provision of a second pair of deflector plates whose planes are perpendicular to those of the first pair would make it possible to deflect the electrons in a direction perpendicular to the plane of the diagram. In such a system the narrow beam of electrons would function as a sort of pencil, which may be guided by the fields applied between the pairs of deflector plates, and so caused to "write" any line or curve that may be desired on the far end of the tube wall. The visibility of this writing would be greatly improved by coating the inside wall of the tube in the target area with a suitable fluorescent material. A device of this kind would be very useful in such applications as tracing out wave forms (for example, those of sound waves) with the aid of appropriate auxiliary equipment, such as microphones. It would, however, be rather inefficient, because most of the electrons produced at the cathode are wasted; only a very small proportion pass through the aperture in the anode.

Two obvious improvements are indicated. One, corresponding to the case of the Coolidge x-ray tube, is the substitution of a thermionic source operating in a good vacuum for the one depending on the presence of residual gas. The other is an accelerating electrode system so designed that none of the electrons are wasted, all of them being converged, electron-optically, into a narrow beam by the provision of suitable electrostatic lenses. A system that produces a narrow beam of electrons in this way is known as an "electron gun."

An electron gun may be designed to produce either an electron beam of constant intensity or one whose intensity can be varied at will. Provision for intensity variation may be made by incorporating with the gun an additional electrode, a "modulating electrode." By regulating the electron current making up the beam that eventually impinges on the screen, the modulating electrode controls the brightness of the fluorescent spot appearing there. This electrode corresponds to the grid of a triode.

A design of electron gun as used in a modern cathode-ray tube, with two pairs of deflector plates following it, is shown in vertical section in Fig. 23.11. Its action will be understood with the aid of

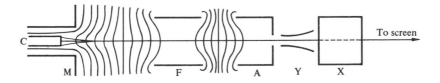

C

M   F   A   Y   X

To screen

FIGURE 23.11   Electrode system of modern cathode-ray tube, with equipotential surfaces.

the gravitational analogue previously referred to, the equipotentials, indicated by the thin lines, being thought of as contour lines.

The cathode C is of indirectly heated type, the heater (not shown) being inside. It consists of a hollow nickel cylinder whose closed end is coated with alkaline-earth oxides; it is from this coating that the electrons are emitted. The electrode M is the modulating electrode. Its potential is varied within a range the whole of which is negative with respect to the cathode. A typical instantaneous value might be −20 V. The next electrode, F, is the "focusing electrode." This is held at a potential some hundreds of volts positive to the cathode, and the configurations of this, the modulating electrode, and the cathode are such as to produce a system of equipotentials as shown. We see that in the region of the cathode we have, on the gravitational analogue, a steep-sided, downwardly sloping valley, which concentrates the electrons into a narrow beam. This is followed by a saddle. Although the saddle tends to spread the electrons passing over it outward, these have already gained such high momenta and are still converging sufficiently strongly that the only effect of the saddle is to make them converge less strongly, not to diverge them.

The field in the region of the surface of the cathode depends on the potential applied to the modulating electrode; this regulates the extent to which the equipotentials penetrate toward the cathode surface and the associated magnitude of the space charge. The more negative the modulating electrode is made, the less is the penetration and the smaller the proportion of emitted electrons able to surmount the space-charge potential barrier.

The anode A is held at a potential of some thousands of volts positive to the cathode. Between the focusing electrode and this is another system of equipotentials whose overall effect is to keep the electrons concentrated in a narrow beam as well as to accelerate them further.

613

Beyond the anode are a pair of deflector plates Y and X. If we think of Cartesian coordinates on the surface of the screen, then the X and Y pairs of deflector plates deflect the beam parallel to the x and y axis, respectively.

Besides being most useful and versatile instruments in the laboratory, cathode-ray tubes have come into widespread use for displaying television pictures. In television reception the focal spot is made to sweep the screen along a succession of parallel lines, covering the whole area of the picture 25 or 30 times per second.* This is done in synchronism with a corresponding succession of sweeps of the picture at the transmitting end, where variations of light intensity received from successive areas covered in the sweeping process are converted into variations in the amplitude of radio waves.

Problems

23.1   Calculate the order of magnitude of the thermionic emission from a cesium/silver-oxide photoelectric cathode at 20°C, given that the photoelectric threshold is at 12,000 Å, and assuming that the thermionic constant A for this cathode is of the order of 0.1 amp/cm²/deg².

23.2   The bright yellow line in the spectrum of sodium vapor has a wavelength of 5893 Å. How many photons per second are emitted from a sodium-vapor lamp whose power output in light of this wavelength is 10 W?

23.3   It is found that molecular oxygen, O_2, can be dissociated into its constituent atoms by irradiation with ultraviolet light of any wavelength from 1750 Å downward. Explain this phenomenon, and calculate the dissociation energy in electron volts. If dissociation is brought about by light of wavelength smaller than the threshold value, what else must happen in addition to dissociation?

23.4   A certain cathode is irradiated with filtered monochromatic ultraviolet radiation of wavelength 2537 Å from a mercury arc, and the potential of the anode relative to the cathode is gradually reduced from a substantial positive value downward. At first the plot of photoelectric current against anode voltage is horizontal. Then, at a certain value V_1 of the voltage it dips downward with further voltage reduction, and finally, at a lower voltage V_2, the current becomes zero. Explain this. Given that $V_1 - V_2 = 0.8$ V, and assuming space-charge effects to be negligible, calculate, in electron volts, the work function of the cathode.

23.5   A photoelectric cathode in the form of a plate and a parallel anode plate are mounted in a vacuum, and these are connected externally to the terminals of an electrostatic voltmeter. It is found that when the

---

* Twenty-five in most British Commonwealth countries, where the alternating-current mains frequency is 50 Hz, and 30 in the United States, where it is 60 Hz.

cathode is irradiated by a beam of monochromatic ultraviolet radiation of wavelength 2537 Å its potential relative to the anode rises by 2 V. The electrostatic voltmeter is now replaced by a microammeter, and this registers a current of 0.2 μA. Explain these effects, and obtain a minimum value for the ultraviolet power falling on the cathode. Why must the actual power be greater than the minimum value calculated?

23.6   When a certain photoelectric cathode is irradiated with filtered monochromatic radiation of wavelength 2537 Å from a mercury arc, a photoelectric emission of 5.2×10^{-9} A per microwatt of incident power is obtained. What is the average number of photons incident on the cathode for each electron released?

23.7   An electron, accelerated from rest through a PD of 200 kV, impinges on a metal target. If the whole of its kinetic energy is converted to the energy of a photon, what will be the wavelength of this photon?

23.8   The electrons impinging on the screen of a certain television tube have energies of 10,000 eV. What is the shortest x-ray wavelength these electrons can produce? Can you suggest a reason why the viewer comes to no harm from this x radiation?

24

DISCHARGES IN GASES

The first systematic investigations of electric discharges in gases, carried out in the nineteenth century, revealed a wealth of intriguing phenomena for which the physicists of that time were unable to find a scientific explanation. Viewed in retrospect, this situation is hardly surprising, for, as we now know, various processes by which electrons are liberated from metals, and the interactions of these electrons with molecules of the residual gas, play a key role in gas discharges, and electrons were not discovered until near the close of the nineteenth century. Moreover, many gas-discharge phenomena require an understanding of basic quantum principles for their explanation, and it was not until 1900 that the quantum theory was born.

Gas discharges are of the greatest importance in a variety of technological applications, whose development has been rendered possible only through an understanding of the physical processes on which they depend. But for the physicist these discharges have an interest quite apart from such applications. From the very beginning, gas discharges have held promise of providing the key to the long-standing problem of the constitution of molecules or, in those cases where the gases are monatomic, of atoms. This is because during much the greater part of their free paths the molecules are effectively uninfluenced by others, and so we have a correspondingly uncomplicated situation. And, as an accompaniment of the discharge, each gas emits a characteristic line spectrum. This spectrum, based as it must be on the particular constitutional system of the atom or molecule in question, should give us valuable information concerning this system, if only a way of interpreting it could be found.

Success in interpreting one of the simplest spectra, the spectrum of atomic hydrogen, was first achieved by Niels Bohr during the second decade of the present century. This was a direct application of the quantum theory as developed at that time to a hypothetical "planetary" system of the hydrogen atom, in which a single planetary electron describes one of a number of possible orbits about a relatively massive positively charged nucleus.

To understand why these characteristic spectra are emitted by gases carrying an electric discharge it is necessary to know by what mechanism the molecules are excited to optical emission by the discharge. This, in turn, involves an understanding of the broad features of the physical processes on which the passage of the discharge depends. It is to a consideration of these that the present chapter will be devoted. This will, incidentally, contribute to an important degree toward preparing the ground for such study of Bohr's theory of the hydrogen atom as the reader may undertake at a later stage.

Photoionization

In our discussion of Millikan's oil-drop experiment reference was made to the production of ions in air by x rays. These rays have been shown to consist of electromagnetic radiation of very short wavelengths; thus qualitatively the radiation is of the same nature as ordinary light. But because of the high frequency of x radiation its photons are much more energetic than are those of visible light. The absorption of these by the molecules of a gas render it conducting. We can account for this only by supposing that some of its molecules become split up into positively and negatively charged components. The fact that this occurs under irradiation by x rays, although not in response to illumination by ordinary light, can surely only mean that the photons of ordinary light are not energetic enough to effect the separation of the molecules into their components.

The most natural assumption concerning the nature of the pair of charged particles into which a molecule is converted by x radiation would be that the negative component is simply an electron, while the positive component, the **positive ion,** is what remains of the molecule when this electron has become detached from it. There is evidence that the primary products of the splitting up of the molecules are, indeed, electrons and positive ions of the kind suggested. However, in conditions where the electrons encounter neutral molecules before being absorbed by an electrode, they may, if the molecules are of the right kind, become attached to these, forming what are known as **negative ions.** Such attachment does, in fact, occur in atmospheric air irradiated with x rays. Evidence concerning the nature of the ions, of both kinds, includes measurement of the rates at which they drift through the gas in response to an applied field.

618   It is not to be imagined that an encounter between a slow electron and a neutral molecule necessarily leads to the formation of a negative ion—far from it. The probability of attachment is always very small indeed; the average number of encounters between an electron and a neutral molecule that is required before attachment occurs varies from about 4×10^4 for O_2 and H_2O to something of the order of 10^8 in the case of NH_3 and CO. Attachment of electrons to the molecules of the noble gases and to those of H_2 and N_2 apparently never occurs at all; negative ions of these gases have never been observed.

Under the conditions envisaged there must also be encounters between positive and negative ions or between positive ions and electrons, such encounters providing an opportunity for mutual neutralization of charge, or **recombination.** In air irradiated by a steady beam of x rays, a state of dynamic equilibrium is eventually reached where the rates of separation of electrons from molecules and of recombination just balance each other.

Photoionization due to high-frequency ultraviolet light and x rays received from the sun is responsible for rendering the upper atmosphere electrically conducting. Because of the presence of ions and electrons in the upper atmosphere, high above the troposphere, this region is known as the ionosphere.* The conducting properties of the ionosphere play a fundamentally important role in radio communication over long distances. These properties are by no means constant, but are subject to diurnal, seasonal, and other variations, representing a complex interplay of the processes of ionization and recombination.

Attachments of neutral molecules to electrons, and recombinations of oppositely charged particles, can obviously take place only if the electrons are given the opportunity of encountering neutral molecules, and oppositely charged particles of encountering one another. Such opportunities are not provided, except to a negligible minority of the charged particles, in a rarefied gas where the free paths are large and where almost all the electrons and positive ions are collected at electrodes before they have collided with even a single neutral molecule. Under such conditions the

---

* The ionosphere extends from a height of about 65 km to about 1000 km, beyond which it merges into the exosphere. It is separated from the troposphere by a region known as the stratosphere, in which we have, among other things, (1) the temperature inversion responsible for the acoustical skip distance or zone of silence discussed in Chapter 11, and (2) the formation, by photochemical action, of a layer of ozone from ordinary oxygen—the "ozone layer." No significant photoionization occurs in the stratosphere.

only charged particles that play any significant role are electrons and positive ions.

Ionization by Collision

In our discussion of the thermionic diode we dealt with the case where the vacuum is so good that any effects of collisions of electrons with residual gas molecules are negligible. Let us now consider the case where such a diode contains gas at a pressure of the order of 10^{-3} to 10^{-4} mm, the gas being chemically inert, so that it does not "poison" the cathode. At such a pressure, the mean free paths of both the gas molecules and the electrons are still long, so that only a very small proportion of the electrons collide with gas molecules in the interelectrode space. Nevertheless, the relatively small number of collisions that do occur produce important effects, greatly modifying the current–voltage characteristics of the tube in the space-charge-limited region. The nature of this modification for a particular value of the cathode temperature is shown in Fig. 24.1. The dashed curve is the characteristic that would be obtained at this cathode temperature under good vacuum conditions and corresponds to one of the curves in Fig. 22.5. When the gas is present this is changed to the characteristic shown by the solid curve.

From the fact that substantially the same saturated plate current is obtained regardless of whether or not the gas is present, it seems reasonable to infer that the gas does not affect the emission, and that such new carriers as may be created by the passage of the electrons through the gas do not make a significant contribution to

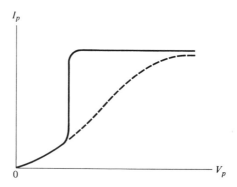

FIGURE 24.1 Effect of residual gas on current-voltage characteristic of thermionic diode.

the total number of carriers.* At the same time, the sharp rise in the space-charge-limited current, which occurs above a certain value of the plate potential, shows that at this point the space-charge potential barrier is drastically reduced, and, at a slightly higher plate potential, abolished altogether. There appears to be only one way in which this could occur—by the partial neutralization of the space charge—and this, in turn, can only be due to the presence of positive carriers, or ions. We must conclude that the impact of sufficiently energetic electrons with neutral molecules detaches electrons from the latter, and so produces not only these new free electrons but also positive ions.

It may be wondered how the positive ions, while making no significant contribution to the current, can have such an important effect on the space charge. The explanation lies in the great disparity in mass between positive ions and electrons. For example, an argon ion has 7.39×10^4 times the mass of an electron. Let us assume this to be "singly charged," that is, to have been formed from an argon atom by the detachment of one electron. Then if this has fallen from rest through a certain potential difference it must acquire the same kinetic energy as does an electron that has fallen through the same PD. But for equal kinetic energies the ratio of velocities must be equal to the square root of the inverse ratio of the masses; that is, the velocity of an argon ion would be only $1/\sqrt{7.39 \times 10^4}$, or $\frac{1}{272}$, that of an electron having the same energy. Actually, electrons and positive ions in any particular part of the interelectrode space will not in general have fallen through the same PD, but the ratio of velocities derived on the assumption that they have will still be right in order of magnitude. Because of the relative sluggishness of the positive ions, their lifetimes in the interelectrode space must be correspondingly long, and so they will have a quite disproportionately large effect on the space charge; their population density in the interelectrode space can be comparable with that of the electrons even though only one in a few hundred of the electrons passing from cathode to anode makes an ionizing collision with a gas molecule.

The phenomenon of ionization by collision as observed in a thermionic diode gives us a convenient means of determining, at least roughly, the energy needed to separate an electron from any particular kind of molecule. We have merely to locate the point

---

* The only alternative explanation, which we may dismiss at once as being merely "ad hoc" and altogether too far-fetched for serious consideration, would be that the presence of the gas does depress the emission but supplies new carriers just sufficient to compensate for the decrease.

on the space-charge-limited curve where the characteristic for a tube containing the relevant residual gas begins to rise above that for the same tube under good vacuum conditions. This point gives the anode voltage for which electrons that have almost reached the anode have acquired enough energy to ionize gas molecules. This anode voltage therefore approximates closely to the PD through which these electrons have fallen. The method is even applicable to electronegative gases, like oxygen, provided a tungsten cathode is used and operated at such a high temperature that the filament remains substantially clean despite the presence of the gas, adsorbed oxygen and tungsten oxide being distilled off as soon as they are formed.

The *energy* necessary for ionization is conveniently expressed in electron volts. Alternatively, one might specify the number of volts through which an electron would have to fall from rest to acquire this energy. Actually the latter is the more usual form in which the information is given, and the number of volts in question is called the **ionization potential** V_i of the particular kind of molecule.

The method of finding ionization potentials referred to above, although simple, does not lend itself to highly accurate determinations. This is due in part to the fact that there is only an extremely small probability that an electron having the requisite kinetic energy will ionize a molecule with which it collides unless this energy exceeds the ionization energy by a substantial margin. There are also other factors limiting the accuracy of the method. Fortunately, much more accurate methods are available, but a discussion of these here would take us too far.

The ionization potentials of a selection of gases and vapors are given in Table 24.1.

TABLE 24.1 Ionization Potentials

Gas or vapor	V_i (V)	Gas or vapor	V_i (V)
He	24.46	Hg	10.39
Ne	21.47	Pb	7.38
A	15.68	Mg	7.61
N_2	15.5	Ca	6.09
H_2	15.6	Ba	5.19
H	13.53	Li	5.36
CO_2	14.4	Na	5.12
CO	14.1	K	4.32
H_2O	12.56	Rb	4.16
O_2	12.5	Cs	3.87

It is interesting to calculate the wavelengths of radiation whose quanta correspond to these ionization potentials. These range from 504 Å for helium to 3190 Å for cesium. We see that the whole range is in the ultraviolet, extending to the far ultraviolet. X rays have wavelengths orders of magnitude shorter than these; consequently, irradiation by these must provide far more energy than is required for ionization. The surplus must appear in the form of kinetic energy of the positive ion and the electron. Because of momentum conservation, practically the whole of this kinetic energy must be possessed by the electron, whose mass is relatively so much smaller. In being brought to rest, this electron is able to ionize many molecules by collision. Thus we have the interesting situation that even in what is normally described as photoionization in the x-ray range of wavelengths, most of the ionization produced is actually due to collisions.

IONIZATION BY IMPACT OF POSITIVE IONS One would naturally expect that the kinetic energy of positive ions, as well as that of electrons, should be available for detaching electrons from neutral gas molecules by collision. This expectation is, in fact, borne out by experiment, although, because of their relatively large mass, ions are less efficient ionizers than electrons. This is because, not only does the law of conservation of energy have to be satisfied, but also the law of conservation of momentum. When the projectile producing ionization is an electron, almost the whole of its kinetic energy is available for conversion to energy of ionization without violation of the momentum-conservation law; the kinetic energy of the positive ion produced is negligible compared with that of the electron before impact. But if the projectile is a positive ion, the momentum-conservation law requires that its kinetic energy must be greater than the energy of ionization by a substantial amount if ionization is to result from the impact. By how much the ionization energy must be exceeded depends on the ratio of the masses of the positive-ion projectile and the new ion to be produced. As a simple calculation shows, if these masses are equal, the kinetic energy of the ion projectile must be at least twice the ionization potential; and if the projectile has twice the mass of the new ion to be produced, the energy of the projectile must be at least three times the ionization energy.

It must be clearly understood that what we have been discussing here is the minimum energy an ion must have to ionize a neutral molecule by collision. But whether an ion formed in an ordinary gas discharge is able to *acquire* this minimum energy by "falling'" downfield among the molecules of the gas is another question. This will be dealt with later.

Secondary Electron Emission

We have seen that electrons can be detached from gas molecules by making the requisite energy available to them in the form either of photons or of the kinetic energy of other electrons or (fractionally) of positive ions that collide with these molecules. This suggests that it might well be possible to extract electrons from solid metals, also, by bombarding them with electrons or with ions from an external source, as well as by the processes we have already studied: the application of strong electron-accelerating fields, heating, and irradiation. Emission of electrons from a metal that occurs in response to bombardment by "primary" electrons, or by any other particles, is known as **secondary electron emission.**

To investigate whether electron emission from a metal does, in fact, occur in response to electron bombardment, and to obtain quantitative information concerning it, it is necessary to direct a stream of electrons of known strength onto the surface of the metal in question and to provide for the permanent removal of all electrons emitted from the metal in consequence of this bombardment. If the primary electron current bombarding a metal plate is I, and the *net* rate of reception of negative charge by the plate (the "plate current") is I_p, then $I - I_p$ must be the rate of emission of secondary electrons plus reflection (if any) of primary electrons, and the ratio $(I - I_p)/I$ is the mean number of secondary electrons (plus reflected primaries) produced per primary. To avoid circumlocution we shall include reflected primaries with secondaries, classing them all as secondary electrons.

An experimental arrangement that is suitable for a quantitative investigation of secondary electron emission is represented schematically in Fig. 24.2, the electrodes C, F, A, and P of which are mounted within an evacuated container, not shown. The electrode C is an indirectly heated oxide cathode whose coating is confined to the flat right-hand end of this cylindrical electrode. Electrons emitted from this are focused by the focusing electrode F and the anode A into a narrow beam which passes right through these and impinges, without loss of electrons to either F or A, on the surface of the plate P. The strength of the primary electron current I is measured by the "cathode ammeter" A_c. Because the plate is positive to the cathode and none of the electrons are lost en route, they must all be incident on P; their momenta carry them on to P despite the retarding field to which they are subjected between A and P. This field, which is electron-retarding for the primary beam, is electron-accelerating from right to left and ensures that no secondary electrons emitted from P can return to it; they

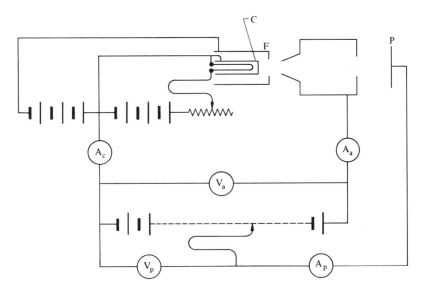

FIGURE 24.2 System for quantitative investigation of secondary electron emission phenomena.

must all be collected by the anode A. The anode ammeter A_a measures the secondary electron current I_a and the plate ammeter A_p measures the net electron current I_p collected by the plate. Clearly, if none of the primary electrons are lost to A, I_a should be equal to $I - I_p$. The provision of all three current-measuring instruments gives the possibility of checking that this is really so.

Under suitable conditions I_p is found to be negative; that is, the plate loses more electrons than it gains. The number of secondaries per primary, $(I - I_p)/I$, is then greater than 1.

The phenomenon of secondary electron emission is not confined to metals; nonmetals, and even insulators, will emit secondary electrons when bombarded by primaries of sufficient energy.

The number of secondary electrons emitted per primary, or the **secondary emission coefficient,** is found not only to depend on the energy of the bombarding primaries, but also, for a given primary energy, to increase with decreasing work function of the surface. In all cases it passes through a maximum for a primary energy of the order of 400 to 500 eV, but this is a fairly flat maximum; the coefficient does not change greatly between 300 and 1000 eV. At its maximum the coefficient has the value of about 1.2 for nickel, about 5 for an activated alkaline-earth oxide such as

625

the coating of a thermionic oxide cathode, and about 8 for a cesium/silver-oxide photoelectric cathode.

The falling off of the secondary emission coefficient with increasing primary energy beyond the optimum is presumably due to very fast electrons penetrating to considerable depths (on an atomic scale) below the surface. Electrons acquiring energy from primaries at too great a depth probably lose so much of this extra energy on their way to the surface that they are unable to penetrate the potential barrier located there.

The great bulk of secondary electrons have quite low energies; very few have energies in excess of 30 eV. The average is about 10 eV.

POTENTIAL OF FLOATING ELECTRODE If in a system of the kind shown in Fig. 24.2 the connection to the plate P is broken, the potential of this electrode will, in general, either rise or fall. If more than one secondary electron is emitted from P for every primary received by it, the potential will rise to a value such that some of the secondaries return to this electrode, enough to give a net loss of secondaries which just balances the rate of absorption of primaries. Owing to the relatively small energies with which secondaries are emitted, this stable floating potential will not be more than a few volts higher than V_a. If, on the other hand, less than one secondary is emitted per primary, the potential of P must fall. It will continue to fall, either to a lower potential for which the secondary emission coefficient has risen to the point where an average of one secondary is emitted per primary or, if no such lower potential exists, down to approximately the potential of the cathode. In any case, this near-zero potential will always be an alternative floating potential, even when the secondary emission coefficient for some range of higher potentials is greater than 1.

SECONDARY EMISSION DUE TO POSITIVE IONS Secondary emission is also found to occur in response to bombardment by positive ions. However, this process is very inefficient, as is the ionization of gas molecules by the impact of positive ions, except only when the bombarding ions have energies of many thousands of electron volts. Under these conditions the secondary emission coefficient is greater with positive ions than with electrons, several electrons being emitted per positive ion. This is probably due to the fact that positive ions do not normally penetrate the surface, and the energy becomes available at the surface rather than deep down. At relatively low bombarding energies, however, a very large number of positive ions are needed to liberate even a single electron.

626

Unlike the ionization of gas molecules by impact with positive ions, which is generally of minor significance, secondary emission under positive-ion bombardment has, as we shall see presently, consequences of the greatest importance, both where the bombarding ions have exceptionally high energies and also where these energies are quite low.

Processes Involved in Gas Discharges

We have already noted one rather special case of a gas discharge, in which the space-charge-limited characteristic of a thermionic diode was modified radically by the presence of gas at a pressure of the order of 10^{-4} mm. In this case primary electrons were provided by thermionic emission from the cathode, and we were concerned only with the modification of a vacuum characteristic ascribable to ionization of the gas.

We have a rather different situation in the case of the breakdown of the insulation of air which occurs at a high field strength, as in a spark or in a discharge from a point. Here the discharge has to establish and maintain itself in the absence of any initial electron emission from the cathode. How is this possible?

We can account for insulation breakdown in a gas only by supposing that somehow a proportion of the neutral molecules are split up into positive ions and electrons, and that these, streaming toward the electrodes under the action of the field, carry the current.

In our discussion of thunderstorms in Chapter 21 it was noted that ions are continually being created in the atmosphere by radioactivity and by cosmic rays. Here, as in other cases of ionization that we have considered, the primary process is the separation of one or more electrons from a neutral molecule. These electrons provide the trigger action by which more ions and electrons are created in a sufficiently strong field.

The breakdown field for air at atmospheric pressure is about 30,000 V/cm. It will be interesting to investigate what energy electrons accelerated by such a field acquire between successive collisions, and also in the course of a series of collisions. From evidence of various kinds it may be inferred that the mean free path of electrons in air at STP is roughly 8×10^{-5} cm, although actually it is not constant but increases with increasing electron velocity. In being accelerated from rest through this distance by a field of 30,000 V/cm an electron would acquire a kinetic energy of $8 \times 10^{-5} \times 30,000 = 2.4$ eV. On consulting Table 24.1 we see that this

627

is rather less than one fifth of the energy required to ionize oxygen, the most easily ionizable constituent of air, and less than one sixth that required for the ionization of nitrogen. An electron is therefore unlikely to ionize an air molecule on the first hit. It will have the requisite energy to do so only if it has an exceptionally long free path, some six times longer than the mean, or more than this. In probability theory it is shown that if N_0 electrons start from given points in a gas in which the mean free path is λ_e, the number N that will travel distances exceeding d before colliding with a molecule is given by the equation

$$N = N_0 e^{-d/\lambda_e}.$$

Hence if we may take 8×10^{-5} cm as the value of λ_e in air, the proportion N/N_0 of electrons that would have free paths equal to or greater than d, with d equal, say, to $6\lambda_e$, would be only 1 in 460. Owing to the low ionization efficiency of electrons that have not fallen through a PD substantially in excess of the ionization potential, probably well under 1 in 1000 electrons would ionize a molecule on the first hit.

Let us now consider the much larger number of collisions in which no ionization occurs. In such a collision two things may happen. If the kinetic energy of the electron is large enough, the molecule hit may absorb some or all of this and pass into a higher energy state, known as an "excited" state, immediately after which it will revert to its normal condition, or "ground" state, generally emitting a light quantum in the process, or, if the reversion to the ground state occurs stepwise, a series of light quanta. More usually, however, the electron will make an "elastic" collision with the molecule, in which both momentum and kinetic energy are conserved. In such a collision, owing to the relatively very small mass of the electron, this will not pass on an appreciable fraction of its kinetic energy to the molecule. Consequently, in most of its collisions with molecules, some nearly head-on, others glancing, an electron will retain substantially the whole of its kinetic energy until this has been built up to a relatively high value. After executing several free paths, and traveling toward the anode a distance of the order of 5×10^{-4} cm, an electron will generally have accumulated enough energy to make an ionizing collision with a molecule. Eventually—perhaps after traveling a little farther—it will do so, losing most or all of its energy in the process, and then it will start again.

The new electron created by the ionizing collision, as well as the original electron, will now produce additional electrons by collision, all these will produce yet more electrons, and so on; once the process is started there will be an enormous population buildup of both electrons and ions. There will, of course, be some recombination following ionization, and some electrons will become attached to neutral molecules to form negative ions. But these processes can only moderate the buildup; they cannot prevent its occurrence.

Let us now look into the question of ionization by collisions of positive (or negative) ions with neutral molecules.

The mean free path of a positive ion in air at STP is only about 6.3×10^{-6} cm, or only about $\frac{1}{12}$ or $\frac{1}{13}$ that of an electron. Accordingly, if an ion could, like an electron, accumulate energy in its drift along the direction of the field without appreciable loss on account of collisions, it would acquire a kinetic energy equal to the ionization energy after executing some 12 or 13 times as many free paths as are required in the case of an electron. Actually, as we have seen, a kinetic energy of a positive ion merely equal to the ionization energy is not enough; in a pure gas, in which the projectile and the target both have the same mass, at least twice this energy would be required. In view of this, and the distribution in direction among the free paths, perhaps a fair estimate of the number of free paths that would be required for accumulating sufficient energy for ionization would be of the order of 1000. But this estimate is based on the assumption that the ion suffers no appreciable loss of energy in making these collisions. This assumption is far from justified. In a direct hit on a stationary* target molecule of equal mass, an ion would transfer the whole of its kinetic energy to the uncharged target, and in an oblique collision it would generally lose a substantial proportion of its energy; only in a glancing collision would the loss be inappreciable. Consequently, an ion can never, or virtually never, accumulate enough energy in its drift downfield to ionize a neutral molecule; for all practical purposes we may regard positive ions as completely ineffective as ionizers in ordinary gas discharges. That this is so was actually first established experimentally, before its theoretical explanation was worked out.

Whereas the electrons and such negative ions as are formed must all drift toward the anode, the positive ions must all travel in the opposite direction, toward the cathode, constituting a double avalanche of charged particles. If no other process intervened the avalanche would soon expend itself, and the formation of a new avalanche would have to await a fresh triggering ionization by

---

* The thermal velocities of gas molecules are so small compared with the velocities produced in charged particles by the field that for the purposes of this discussion we may treat them as negligible.

stray radioactivity, cosmic rays, or other external agency. But a self-sustaining discharge is, in fact, not kept going simply by a succession of such triggerings; experiments with x rays in which these provide the trigger show that the discharge is substantially independent of the x-ray intensity. The essential mechanism for sustaining the discharge must therefore be something that is brought into being by the discharge itself.

What could this be? We know that the positive ions produced are incapable of making ionizing collisions and so liberating new electrons behind those in the avalanche, nearer the cathode. On the other hand, positive ions *can*, if sufficiently energetic, produce secondary electrons on impact with the cathode surface. Let us now look into this possibility.

For the work function of a metal surface exposed to air we may assume a value of about 6 eV, and if a positive ion incident on the surface has an energy in excess of this it will presumably have a finite chance of producing a secondary electron.

The minimum of 6 eV required for the production of a secondary electron is small compared with the 25 to 30 eV that would have to be built up before an ionizing collision with a gas molecule could take place. Although the accumulation of this latter energy by an ion is so fantastically improbable that for practical purposes we may consider that it never occurs, the attainment of an energy of 6 eV or slightly more, although extremely rare, may conceivably occur often enough to provide for maintenance of the discharge by the release of a secondary electron at the cathode surface. The situation is rendered substantially more favorable for such release by the modification of the field between the electrodes resulting from space charges when the discharge is established; the field is particularly strong in the neighborhood of the cathode ("cathode fall").

Although, despite the existence of this cathode fall favoring the accumulation of energy by positive ions, the probability of a particular ion releasing an electron must be exceedingly small, it must be remembered that each electron emitted from the cathode ultimately gives rise to an enormous number of positive ions. It is only necessary that one of these produce a secondary electron, replacing the original trigger electron, for the discharge to be kept going. A quantitative estimate of the probability of this occurring makes it appear that secondary electrons produced by positive ions could well be responsible for maintaining the discharge, or at least could be an important contributing factor.

Actually there is another mechanism for releasing electrons from the cathode that we must not overlook. We have already noted that molecules excited to a higher-energy state by electron impact normally emit photons in the process of returning to the ground state. The emission of photons must also result from recombination of electrons with ions. A substantial proportion of these photons must be sufficiently energetic to release electrons from the cathode. Recent experiments in which the details of the buildup of the discharge after triggering have been studied show that under a wide range of conditions this photoelectric emission from the cathode plays a very important, even a dominant, role.

We have so far confined our discussion to discharges occurring in air at atmospheric pressure. Qualitatively, the same kinds of processes must occur in other gases at other pressures. Actually, particular interest attaches to discharges occurring at relatively low pressures, of the order of millimeters, as in neon signs. In all cases where the mean free paths of the various particles remain small compared with the linear dimensions of the discharge space, the field E required for any particular condition in a given gas, for example for the initial trigger action or for the maintenance of the discharge, must be proportional to the pressure p. This is because the mean free path is inversely proportional to the pressure, and the energy built up in a free path is proportional to the product of its length and the field strength. Thus E/p is an important parameter in the theory of gas discharges.

ACTION OF POINTS In the use of points for spraying or collecting charge the field is not strong enough to produce ionization except in the immediate neighborhood of a point. There are two cases to consider: that in which the point is the cathode and surrounding objects function as the anode (or an electrode is specially provided to serve this function) and that in which the point is the anode.

If the point is the cathode, sufficient positive ions and photons are produced in its immediate neighborhood to maintain the requisite supply of secondary electrons and photoelectrons from the cathode surface. The stream of positive ions impinging on, and neutralized at, the point constitutes the process of the collection of charge, while the electrons plus negative ions repelled from the point simultaneously constitute the charge "sprayed" onto the anode. The impacts of these negatively charged particles with gas molecules en route to the anode impart to these a momentum in the direction away from the point, so producing an "electric wind."

631

Thus a row of points connected to one terminal of a Wimshurst or Van de Graaff machine is capable of blowing out a candle flame.

Corresponding phenomena are observed when the point is the anode. Here the field at the cathode, or at the surfaces functioning as cathode, is so weak that there is no question of secondary electrons being produced from this electrode. The electrons maintaining the discharge must therefore be produced by some other mechanism. If the cathode is not too remote, some of these might be emitted photoelectrically in the manner already discussed. In addition to any such photoelectric supply, the only other possibility appears to be photoionization of the gas at a distance from the anode. Some of the requisite photons could presumably be produced by recombinations between electrons and positive ions. Such photons would have the ionization energy of the gas or, if the electron combining with the ion has appreciable kinetic energy, perhaps a little more. These photons must be capable of ionizing both molecules in the ground state and excited molecules. In addition to photons produced in this way there might well be others of sufficient energy produced by the impact of electrons on the anode. We might think of the photons produced by such electron impacts as exceptionally soft x rays. In view of the relatively long free paths of electrons and the more favorable conditions for buildup of the energy of electrons in a series of collisions, fields of strength sufficient to produce positive ions having energies of 6 eV or more at a pointed cathode might well be expected also to give rise to electron impacts with the pointed anode producing photons of sufficient energy (around 15 eV) to ionize air molecules, particularly in view of the fact that the photon energies would be given by the sum of the kinetic energies of the impinging electrons and the work function of the anode surface.

A discharge from a point is known as a **corona discharge.** Such a discharge is typically accompanied by the conversion of some of the electrical energy, via microwinds, into an acoustic disturbance that can sometimes be heard as a high-pitched hiss. Owing to excitation of some of the molecules, previously referred to, a faint glow may also be visible in the dark in the region where the field is strongest.

GLOW DISCHARGE This is the kind of discharge that occurs in a neon sign, in which the pressure is of the order of a few millimeters and the current something like 10 to 100 mA. The supply of electrons at the cathode end of the tube is kept up by the processes we have already considered: secondary emission due to positive-ion bombardment of the cathode and photoelectric emission. Both of these are favored by the provision of a cathode of low work function.

As was mentioned in Chapter 23, the emission of electrons from the cathodes of the older types of "gas" cathode-ray tubes and x-ray tubes calls for explanation. We have in effect dealt with this in our discussion of secondary emission due to the impact of positive ions. We may regard the discharge in gas cathode-ray and x-ray tubes as an extreme case of a glow discharge. At pressures in the range 10^{-2} to 10^{-3} mm only a small proportion of the electrons emitted from the cathode make ionizing collisions with residual gas molecules on their way to the anode; the great majority travel straight across without passing sufficiently close to any gas molecule to ionize it. However, the positive ions created by such ionizing collisions as do occur impinge on the cathode with such high energies that each one releases a large number of electrons. In this way the electrons passing to the anode are replaced at a sufficient rate to keep the discharge going.

ARC DISCHARGE This is the same as a glow discharge except that there is a much more copious emission of electrons from the cathode, occurring either thermionically or otherwise. An arc carries a relatively higher current than a glow discharge, and the PD between the electrodes is generally less.

Hot-cathode sodium-vapor or mercury-vapor lamps, which come into this category, are widely used for street lighting. Hot-cathode discharge tubes containing mercury and/or other gases or vapors are also used for indoor illumination. The cathodes are of the oxide-coated type.

The inner walls of both cold-cathode glow-discharge tubes of the neon-sign type and hot-cathode tubes are commonly thinly coated with a material known as a "phosphor." Such tubes are classed as "fluorescent tubes." The role of the phosphor is to increase the quantity and improve the quality of the light output by absorbing ultraviolet light and reemitting some of the absorbed energy in the visible part of the spectrum. For example, cadmium borate gives a pink color, zinc silicate green, and calcium tungstate blue. White light may be obtained by mixing various materials in suitable proportions.

Mercury arcs are employed in the laboratory as a source of ultraviolet light, the tubes being constructed of fused silica, which is transparent to ultraviolet. In the simplest form of these, which is represented diagrammatically in Fig. 24.3, the discharge passes

633

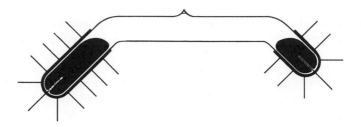

FIGURE 24.3 Mercury arc.

between two surfaces of liquid mercury. Presumably the electrons are emitted from the cathode photoelectrically.

Another well-known form of arc is the carbon arc. The discharge takes place between the ends of two carbon rods maintained at a PD of about 50 V. The arc is struck by bringing the ends of the rods momentarily into contact and then separating them by a short distance. The arc carries a current of the order of 10 to 100 A. The positive carbon develops a depression, or "crater," which becomes intensely hot, attaining a temperature of about 3600°K, and it is from this that most of the light is radiated. Although the temperature of the negative carbon is much lower than this, it may well be hot enough to give a substantial electron emission thermionically. Presumably there would also be an appreciable photoelectric emission from this surface. It is not known which effect is the more important.

In operating either a cold-cathode glow-discharge tube or an arc it is necessary to make suitable provision for limiting the current.

SPARK This is a discharge of very short duration which occurs when the breakdown field strength in the air or other gas is reached, but is not maintained, either because the circuit is such that the passage of the discharge substantially reduces the PD between the electrodes or because of a redistribution of the field due to space-charge effects. The passage of the spark heats the gas in its path to a very high temperature, and, as spectroscopic examination reveals, vapor derived from the electrode material is commonly present in the discharge path.

Problems

24.1 An electron makes a head-on elastic collision with a stationary molecule of nitrogen. What proportion of the kinetic energy of the electron is transferred to the nitrogen molecule?

24.2   According to an observer O_1, a charged particle of mass m_1, travelling with velocity v, collides with and just ionizes a stationary molecule of mass m_2, so that after the collision the projectile and target particles, including the separated electron, move off with a common velocity.

Show that the velocity v_c of the center of mass of the two particles relative to the observer O_1 is $[m_1/(m_1 + m_2)]v$. Consider another observer O_2, relative to whom the center of mass is stationary. Imagine that O_2 works out the problem of the conversion of the sum of the precollision kinetic energies as observed by him into the energy of ionization, E_i, of one of them. Hence, show that from the point of view of the observer O_1 the energy of the projectile which is just able to ionize the target particle is $[(m_1 + m_2)/m_2]E_i$.

Now check that when, as observed by O_1, the projectile has the energy $[(m_1 + m_2)/m_2]E_i$ prior to impact with the target particle, the kinetic energy lost on impact is indeed E_i.

Use this result for finding the minimum energies required for ionization by collision of (a) a hydrogen molecule by an electron, (b) a hydrogen molecule by an H^+ ion, (c) a hydrogen molecule by an H_2^+ ion, (d) a nitrogen molecule by an N_2^+ ion, and (e) a nitrogen molecule by an O_2^+ ion.

24.3   A photon may be shown to have a momentum of h/λ, where h is Planck's constant and λ is the wavelength (see Chapter 2 of Volume III). Find, in electron-volts, the smallest amount by which the energy of a photon impinging on a molecule of nitrogen would have to exceed the ionization energy of this molecule in order that photoionization might occur.

24.4   If a 10-eV electron collides "head on" with a stationary mercury atom, what will be the energy of the electron after impact (a) if the collision is elastic, and (b) if the mercury atom absorbs 5 eV of energy in internal electronic excitation?

DIRECT-CURRENT THEORY

In Chapter 22 a current in a metal conductor was identified with a flow along it of a stream of "free" internal electrons, and in Chapter 24 it was shown that in gas discharges the current consists of a two-way traffic stream, electrons flowing one way and positive ions the other. But this does not take us very far. We have yet to study the various effects of currents, not only for their intrinsic interest, but also because it is by their effects that currents are measured. Also, we shall find it necessary to develop certain new concepts to provide a basis for the quantitative discussion of currents and their generation.

In the present chapter we shall take a car battery, which is familiar to everyone, as a convenient standard device for producing reasonably large currents for experimental study, leaving for a later chapter a discussion of the theory of its action. Using this, we shall then consider as our main topics the principles of the measurement of current strength, an extension of the concept of potential difference and its measurement in current electricity, Ohm's law of proportionality between current strength and potential difference, systems of conductors, and the concept of electromotive force.

Measurement of Current Strength

Before discussing how a current-measuring instrument such as an ammeter may be used to *measure* currents, we shall consider it merely as a "black box" which *registers* the passage of a current through it. We shall assume no more than the existence of some unique relation between current strength and reading, so that we may take a steady deflection of the needle as indicating a corresponding constant, although unknown, magnitude of the current I passing through it. We shall now see how we may use this instrument to establish Faraday's first law of electrolysis, and so, with the aid of a voltameter, how to calibrate the ammeter in arbitrary units.

We first set up a circuit such as that shown in Fig. 25.1, in which A is an ammeter, B a battery, V an electrolytic cell, Rh a rheostat for regulating the current, and S a switch. To make the discussion more definite, let us, more specifically, suppose the

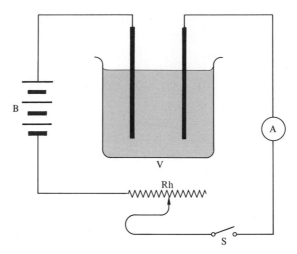

FIGURE 25.1   Cell for electrolysis of silver.

electrolytic cell to consist of a glass or other insulating vessel containing a solution of silver nitrate in which are immersed two electrodes of silver.

By weighing the electrodes before and after passing a current, we find that the current causes the weight of one to increase and that of the other to decrease, both by the same amount. The former is the cathode and the latter the anode.

The cathode having been identified, the switch is closed with the rheostat set to give a certain deflection of the pointer of the ammeter, which is noted, so that it may be reproduced on subsequent occasions. The switch is now opened again, and the cathode is taken out of the solution, washed, dried, and accurately weighed. The cathode is then reinserted and the switch is closed for a considerable length of time, after which it is opened again. This period of time is accurately measured, and during it the deflection of the pointer is kept constant at the previous value by making such adjustments to the rheostat as may be required from time to time. Finally, the cathode is again taken out of the solution, washed, dried, and weighed, and the increase in weight is noted.

This sequence of operations is repeated several times, with variations made in only one item, the electrolytic cell; cells of different sizes and shapes are used, operating at different temperatures, and containing solutions of different concentrations. In this way it is found that on each occasion the same mass of silver is deposited on the cathode by the standard quantity of electricity passed, or, to put it another way, a given current strength produces the same rate of deposition of silver, this being independent of the size of the cell or the electrodes, the distance separating the electrodes from each other, the concentration of the solution, the temperature, or any other factor.

We may now, if we wish, repeat the sequence of operations with the single cell replaced by N identical cells, provided with identical connecting wires, joined in parallel. The current I must now be divided between N branches, which, because they are all alike, must each carry the same current, I/N. And the mass of silver deposited in each cell is $1/N$ of that obtained with the single cell. This is so, regardless of the number N of cells used. This means that the mass deposited in any cell is proportional to the quantity of electricity that has passed through it.

Actually, this additional experimental work, involving as it does the preparation of a number of identical units, is not really necessary; we already have all the information we need. All sorts of shapes and sizes of electrolytic cells were used in the first series of experiments, so we know that the design of the cell has no effect on the result obtained. Hence the same result would have been obtained with cells of an ideal design ensuring a uniform distribution of the current I over the area A of the cathode, giving uniformity of current per unit area, i ($= I/A$), over the whole surface. Let us, then, imagine a number of such ideal cells with different cathode areas. Obviously, I being the same in all cases, i is inversely proportional to A. Also, the total rate of deposition of silver, R, was found to be the same in all cases, so the rate of deposition per unit area, r ($= R/A$), must be inversely proportional to A. Hence r and i must be proportional to one another:

$$r \propto i.$$

Consequently, the total rate R of deposition over any area A we wish to contemplate, being equal to rA, is proportional to the current I ($= iA$) passing into this area:

$$R \propto I.$$

This result is, in effect, the same as is stated in Faraday's first law of electrolysis: **The mass of any substance liberated in electrolysis is proportional to the quantity of electricity passing through the electrolyte.**

Having established Faraday's first law, we now have a reliable basis for calibrating any current-measuring instrument. For this purpose we use the circuit shown in Fig. 25.1, in which the elec-

639

trolytic cell is known more specifically, because of its application to the measurement of current, as a **voltameter.**

For calibration of the instrument the sequence of operations previously described is followed, but this time for several different steady deflections of the pointer. Because rates of deposition—calculated from mass increments and times—are proportional to currents, we obtain in this way the current values, in arbitrary units, corresponding to a number of different deflections.

UNIT OF CURRENT STRENGTH In Chapter 19 the coulomb was defined in terms of the force exerted between two point charges separated by a known distance in a vacuum, and it was later shown how a precision measurement of a charge in coulombs could be carried out, using an attracted-disk electrometer. In the same chapter the unit of current, the ampere, was referred to briefly, being, in effect, defined as that current which carries 1 C past any point in a second. Historically, however, the ampere came first and the coulomb afterward. Also, from a practical point of view it is more convenient to define the ampere independently and then to regard the coulomb as a derived unit. It is for this reason that the system of units in which we are working is the MKSA, not an MKSC (C for coulomb) system.

As we shall see in Chapter 29, it has been agreed to define the ampere in terms of the mechanical interaction between two currents. It has been established that the passage of 1 A, so defined, through a silver voltameter deposits silver on the cathode at the rate of 1.11800 mg/sec. For the purposes of the present chapter we may take this result as a basis for calibrating an ammeter, not merely in arbitrary units, but in amperes.

Moving-Coil Galvanometer

Practically all instruments for measuring direct currents depend for their action on the torque exerted on a current-carrying coil by a magnetic field to which the coil is exposed.* These instruments, besides being known as ammeters, are also sometimes called galvanometers. Although there is no essential difference between an ammeter and a galvanometer, there is a tendency to use the latter term rather than the former for particularly sensitive instruments, measuring small currents.

---

* It is assumed that the reader is already familiar with the elements of magnetostatics, including the concept of a magnetic field.

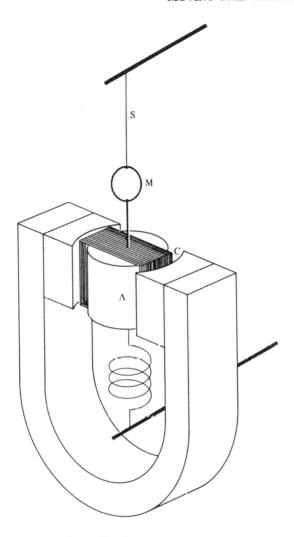

FIGURE 25.2 Moving-coil galvanometer.

Figure 25.2 shows a form of moving-coil galvanometer developed many years ago by A. d'Arsonval, and commonly known, after him, as the d'Arsonval galvanometer. This is also shown in plan in Fig. 25.3. The coil C and an attached mirror M are suspended by a fine metal wire or ribbon S, which serves as one of the current leads to the coil. It also provides the greater part of an elastic control, which is an essential feature of the instrument. The other current lead is taken out at the bottom via a weak helical spring, which also makes some contribution to the elastic control. The coil moves in the space between hollowed-out pole pieces attached to a horseshoe magnet and soft-iron cylindrical armature A.

641

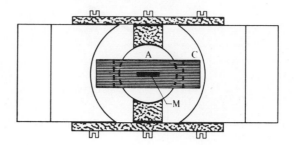

FIGURE 25.3 Moving-coil galvanometer as seen from above.

When a current is passed through the coil, this experiences a torque about a vertical axis and rotates. This rotation brings into operation an elastic restoring torque, and the coil finally comes to rest at an angle to its zero orientation such that the restoring torque just balances the deflecting torque. In the instrument shown in Fig. 25.2 this angle is read with the usual lamp, mirror, and scale arrangement. In other instruments a pointer is used. With the construction shown, the angle is found to be proportional to the current, at least approximately, as long as the vertical sides of the coil are well within the space between the hollowed-out pole pieces and the armature. The full quantitative theory of the instrument will readily be understood after reading Chapter 29.

All other forms of moving-coil galvanometer or ammeter represent variations on the design of Figs. 25.2 and 25.3. In some, the coil, again rotating about a vertical axis, is either mounted between a pair of pivots or supported on a single pivot above the coil. In other cases the coil rotates about a horizontal axis and is supported between a pair of pivots. In all pivoted instruments the elastic control is provided by a flat spiral spring or a pair of such springs. These instruments are limited in their sensitivity by the unavoidable frictional resistance at the pivots. Relatively sensitive instruments are therefore usually of the suspended-coil type shown in Fig. 25.2. The most sensitive galvanometers are capable of reading down to about 10^{-10} A.

BALLISTIC GALVANOMETER Not only does a galvanometer coil settle down to a steady deflection when a steady current is passed through it, but it is also set into rotational oscillations of different amplitudes in response to different current "pulses"— discharges of very short duration—passed through the coil while in its zero orientation. An example of such a pulse would be that given by the discharge of a capacitor through the coil. One might hope that the amplitudes of the oscillations would vary in some unique manner with the total quantity Q of electricity carried by the pulse, and a quantitative analysis shows that this is indeed the case. A galvanometer used in this way to measure charges Q sent through it is known as a **ballistic galvanometer.**

In the theory of the ballistic galvanometer, use is made of the fact that for any definite orientation of the coil, including the zero orientation with which we are particularly concerned here, the torque L which the magnetic field exerts on it (the "magnetic torque") is proportional to the current I passing through it. That this is so for the zero orientation may be shown experimentally by using a galvanometer of slightly modified construction, in which provision is made for always keeping the coil in its zero orientation by turning a torsion head at the top of the suspension, from which the latter is supported, through an appropriate angle. From Hooke's law we know that the angle turned through by the torsion head (the "angle of twist" of the suspension) is proportional to the torque elastically applied to the coil. Because the elastic torque balances the deflecting torque, the latter, also, must be proportional to the angle the head is turned through. Using such a modified galvanometer, we pass through it different currents, measured either directly with a voltameter or indirectly with another galvanometer, of conventional construction, which has been calibrated against a voltameter, and we also measure the corresponding angles through which the torsion head must be turned to keep the coil in its zero orientation. It is found in this way that the magnetic deflecting torque is proportional to the current. This is, of course, equally true for a galvanometer of conventional construction with the coil in its zero orientation; that is, it applies to the current and corresponding torque at every stage of the current pulse sent through the galvanometer.

Because the pulse is too short to allow any appreciable movement of the coil during its passage, the magnetic deflecting torque is, in effect, the *only* torque acting on the coil until after the pulse is over. Only then, when the coil has turned through an appreciable angle, does an elastic torque come into play. For *any* torque L applied to the coil, we have (see Chapter 7)

$$L = J \frac{d\omega}{dt},$$

where J^* is the moment of inertia of the coil and frame about its

---

* We use this symbol here, in place of the more usual I, to avoid confusion between moment of inertia and current.

axis of rotation and ω is its angular velocity. And, as the experiment just discussed establishes, we have, at every stage of the pulse,

$$L = KI,$$

where K is a constant. Equating the right-hand sides of these two equations and writing dq/dt in place of I, q representing the charge passed from the beginning of the pulse to the moment contemplated, we see that

$$J\frac{d\omega}{dt} = K\frac{dq}{dt}.$$

Integration of this at once gives the relation between the final angular velocity ω_f that has been built up at the end of the pulse and the total charge Q passed,

$$J\omega_f = KQ,$$

or

$$Q = \frac{J}{K}\omega_f.$$

This expression for Q is not yet in a useful form; none of the three quantities J, K, or ω_f are measurable by any simple means. We therefore try to express Q in terms of alternative quantities that are directly and simply observable. To this end we consider the behavior of the system from the end of the pulse onward. To simplify the discussion we assume, in the first instance, that there is no viscous resistance to the rotation of the coil or any other dissipative process generating heat; in other words, we assume that we are dealing with an ideal, conservative system. We shall see later how to allow for the fact that actually the system is not ideal in this sense.

Let the restoring torque exerted on the coil by the elastic control be k times the angle θ turned through, and let the coil rotate through a total angle θ_0 before momentarily coming to rest and reversing its direction of rotation. Then during the rotation through this angle the work done on the suspension, which is also the elastic potential energy E_p built up in it, is given by the equation

$$E_p = \int_0^{\theta_0} k\theta \, d\theta = \tfrac{1}{2}k\theta_0{}^2.$$

Under the assumed ideal conditions, this must have been derived from the kinetic energy developed by the current pulse; that is, $\tfrac{1}{2}k\theta_0{}^2$ must be equal to $\tfrac{1}{2}J\omega_f{}^2$. Hence for ω_f in the expression for Q we may substitute $\sqrt{k/J}\,\theta_0$, so obtaining the relation between Q

and θ_0,

$$Q = \frac{J\omega_f}{K} = \frac{\sqrt{kJ}}{K}\theta_0.$$

We may put this into a more useful form by writing $-k\theta$ for the torque exerted by the elastic control on the coil in the positive direction for θ, and equating this to $J(d\omega/dt)$, or $J(d^2\theta/dt^2)$, so obtaining the differential equation

$$\frac{d^2\theta}{dt^2} + \frac{k}{J}\theta = 0,$$

the most general solution of which is

$$\theta = A \sin\left(\sqrt{\frac{k}{J}}\,t + \phi\right),$$

where A and ϕ are constants of integration, the amplitude factor A having, in the case under discussion, the value θ_0. This is a sinusoidal variation of angle with time whose period T is $2\pi\sqrt{J/k}$. Hence for $\sqrt{kJ}$ in the expression above for Q we may substitute $Tk/2\pi$, so obtaining the somewhat more useful alternative expression

$$Q = \frac{T}{2\pi}\frac{k}{K}\theta_0, \qquad (25.1)$$

in which both T and θ_0 are directly observable quantities.

It is not difficult to see that the factor k/K may be identified with the calibration constant c of the galvanometer for steady-current use defined by the equation of simple proportionality between current and angular deflection

$$I = c\theta,$$

applying for small currents. For, when the coil is at rest and makes an angle θ with the zero orientation it is subjected to two equal and opposite torques—the magnetic deflecting torque KI and the elastic restoring torque $k\theta$—so that

$$I = \frac{k}{K}\theta.$$

As the reader will know, moving-coil galvanometers are normally designed in such a way that for any given current the magnetic torque is, at least roughly, independent of the orientation of the coil over a considerable range of θ. This would make the current simply proportional to θ over the whole working range. In practice this objective is not always achieved; the plot of I

645

against θ is sometimes curved. Fortunately, the theory of the ballistic galvanometer is in no way affected by this. Whether or not I is proportional to θ over the whole range, it must be proportional to θ *over a sufficiently restricted range near the zero;* this is because a short-enough section of any curve may be regarded as a straight line. Thus, regardless of the form of the calibration curve, we may define c as the slope of the tangent to the I–θ curve at the origin. With c so defined, we have, substituting for k/K in Eq. (25.1),

$$Q = \frac{cT}{2\pi}\,\theta_0. \qquad (25.2)$$

In practice, the moving parts of a galvanometer do not constitute a conservative system, as we have assumed; there is always some viscous and perhaps also other resistance to the movement of the coil, giving a certain degree of "damping." This means that the first angular excursion θ_1 of the coil, known as the "throw" of the galvanometer, is never quite as large as the value θ_0 that it would have attained in the absence of such damping. It is, however, a simple matter to calculate θ_0 from the observed throw and a few succeeding excursions on either side of the zero. Let these excursions be θ_1, θ_2, θ_3, ... It is found that these constitute a slowly converging geometrical progression; that is, that

$$\frac{\theta_1}{\theta_2} = \frac{\theta_2}{\theta_3} = \frac{\theta_3}{\theta_4} = \cdots = r.$$

This means that in the equation

$$\theta = A \sin\left(\sqrt{\frac{k}{J}}\,t + \phi\right)$$

describing the oscillation of the coil, A, instead of being a constant amplitude factor, diminishes exponentially with time, falling in each successive half-period by the factor $1/r$. Now, the interval that elapses between the passage of the current pulse and the first extreme excursion θ_1 is $T/4$, and in this quarter-period the full range of states of motion is represented, from maximum angular velocity to zero—the same range of states of motion is represented in this quarter-period as in any succeeding *half*-period. Therefore, because in each of these half-periods the amplitude is reduced by the factor $1/r$, in a quarter-period it must be reduced by the factor $1/\sqrt{r}$; only so can the reduction factors for the succeeding half-periods, each of which consists of two consecutive quarter-periods, be $1/r$. Thus we have

$$\theta_0 = \sqrt{r}\cdot\theta_1.$$

Hence, substituting for θ_0 in Eq. (25.2), we may express Q in terms of observed quantities by the equation

$$Q = \sqrt{r} \cdot \frac{cT}{2\pi} \, \theta_1.$$

It is easily seen that if the steady-current calibration is carried out in terms of the readings s on a circular scale in the usual lamp-and-scale or telescope-and-scale arrangement, and the slope of the I–s curve at the origin is κ, then

$$Q = \sqrt{r} \cdot \frac{\kappa T}{2\pi} \, s_1,$$

s_1 being the throw observed as a scale reading. If the scale, instead of being circular, is straight, the appropriate correction must of course be applied.

Potential Difference in Current Electricity

In electrostatics the potential difference $V_A - V_B$ between two points A and B was, in effect, defined as the quotient of the electrical energy W_{AB} that can be converted to mechanical energy corresponding to the conveyance of a small test charge from A to B and the magnitude q of this charge. In current electricity energy conversions are frequently between electrical energy and some form other than mechanical; for example, in the passage of a current along a wire the conversion is from electrical energy to heat, in the charging of a storage battery it is from electrical energy to chemical energy, and in its discharge it is from chemical energy to electrical. In none of these cases is the conversion measurable in terms of the performance of mechanical work, and so the electrostatic definition is not applicable. Although in all these cases it would be possible to *measure* the PD between two points with an electrostatic voltmeter, it is not the question of measurement that we are concerned with here. What we need in theoretical discussions of situations in current electricity is a less restrictive *definition* of PD than the electrostatic one; we need a generalized definition of which the electrostatic definition is a special case, and which at the same time will be directly applicable to all situations of electricity in motion.

Obviously the generalization we seek is the one expressed in mathematical form by the equation

$$V_A - V_B = \frac{dW}{dQ}, \tag{25.3}$$

where dW/dQ is the charge rate of conversion of electrical potential energy to *any* other form of energy.

It will now be convenient to express Eq. (25.3) in terms of the time rates of energy conversion and passage of charge past a point. This is done by writing the equation in the form

$$V_A - V_B = \frac{dW/dt}{dQ/dt} = \frac{P_{EO}}{I_{AB}},\qquad\textbf{(25.4)}$$

where P_{EO} is the rate of energy conversion from electrical energy (subscript E) to some other form of energy (subscript O), or power derived from electrical energy, while I_{AB} is the (conventional) current flowing from A to B.

Since we are working in the MKSA system of units, expressing P_{EO} in watts and I_{AB} in amperes, $V_A - V_B$ is, as in the electrostatic case considered in Chapter 19, in volts. Accordingly, multiplying both sides of Eq. (25.4) by the current, we may write

$$\text{volts} \times \text{amperes} = \text{watts.}$$

It should be noted that Eq. (25.4) is to be taken algebraically; either P_{EO} or I_{AB} can be negative as well as positive. Thus if energy is, in fact, being converted from electrical to another form, then, P_{EO} being positive, $V_A - V_B$ has the same sign as I_{AB}; the potential falls in the direction of current flow. But if other energy is being converted into electrical, so that P_{EO} is negative, $V_A - V_B$ must be opposite in sign to I_{AB}; the potential rises in the direction of current flow.

Ohm's Law

Corresponding to a given current flowing in a wire there is a definite rate of heat production between any two given points. This, divided by the current strength, is by definition the PD between the two points. Hence an investigation of the rate of heat production as a function of current strength must show how the PD between the points and the current are related quantitatively.

Although it is notoriously difficult to achieve high accuracy in calorimetry, it is nevertheless possible, by the exercise of sufficient ingenuity, to make reasonably accurate measurements. Such measurements would show that for a given wire the rate of heat production depends not only on the current but also on the temperature. At any *constant* temperature the rate of generation of heat is found to be proportional to the square of the current; that is,

$$P_{EO} = RI^2, \tag{25.5}$$

where R is a constant characteristic of the particular specimen of wire used and its temperature. Hence, dividing both sides by I, we obtain the relation

$$V = \frac{P_{EO}}{I} = RI \tag{25.6}$$

between the potential difference V and the current: **For a given conductor in a constant physical condition the current is proportional to the potential difference between the points of entry and exit.**

Experimental evidence leading to this law, known as **Ohm's law,** was first obtained by Georg Simon Ohm in 1826. Actually the experiments from which Ohm deduced the proportionality of current to potential difference were of quite a different nature from that considered above. These experiments are of some historical interest and reveal considerable ingenuity and insight on Ohm's part. However, since they have now been superseded by much more direct and accurate methods of establishing the law, we shall not concern ourselves here with the details of Ohm's work.

An obvious alternative method of determining potential differences in experiments leading to Ohm's law would be to measure them directly. For this purpose any suitable electrostatic voltmeter could be used, for example, either an absolute attracted-disk electrometer or some other electrostatic instrument that has been calibrated against such an electrometer.

The constant R in Eq. (25.6), which is equal to the ratio of the potential difference to the current, may be thought of physically as a measure of the difficulty of sending a current through the conductor. Accordingly, this ratio is known as the **resistance** of the conductor. The unit of resistance is called an **ohm.** This is the resistance of a conductor through which a current of N amperes would flow when a PD of N volts is applied across it, N being any number. The symbol for this unit is the Greek letter Ω.

It is instructive to consider the conditions that must apply within a conductor to give Ohm's law. Constancy of the physical condition of the conductor means, among other things, that the number n of carriers of charge per unit volume is constant. Let each of these carry a charge q. Then, under the action of the applied field E, each carrier will be acted on by a force Eq in the direction of the field, which will cause it to drift in the direction of the force.

649

This current can be a unique function of the field, as, in fact, it is observed to be, not increasing with time, only if the drift of the carriers is opposed by a mean retarding force per carrier equal to Eq. For the relation between this force and the mean drift velocity v let us make the simplest possible assumption—that it is one of proportionality—that is, that

$$Eq = kv,$$

where k is a constant characteristic of the conductor. We should then have a situation analogous to one in which a body moves with terminal velocity v through a resistive medium. For the current density J (current per unit area) we then have

$$J = nvq$$
$$= \frac{nq^2}{k} E.*$$

This proportionality between J and E obviously implies a corresponding proportionality between the current I flowing along a conductor and the potential difference V applied across its terminals.

The proportionality factor nq^2/k in the equation relating J with E is an important constant characteristic of the conducting medium. This is known as the **conductivity** σ of the medium.

In the case of metals the carriers of current are, of course, electrons. In electrolytes they are ions. The mean drift velocity v of any particular kind of carrier per unit field, v/E, is known as the **mobility** of this carrier.

OHMIC AND NONOHMIC CONDUCTORS Ohm's law applies not only to metals but also to several nonmetallic conductors. Among the latter are materials like silicon, germanium, galena, iron pyrites, and so on, which look like metals but are not usually classed as such. These are examples of certain materials known as **semiconductors**; they are sometimes also called, more specifically, **semimetals.** Among other semiconductors, which do *not* look like metals, are cuprous oxide, alkali halides that have been subjected to heat treatment in alkali vapor, and a large number of other materials.

All semiconductors fall into one or other of two categories, known as n-type and p-type semiconductors. In n-type semiconductors (n for negative), current is carried directly by free electrons, as in metals; the essential difference between a metal and

---

* If there is more than one kind of carrier, as in an electrolyte, this equation must be taken as referring to the *partial* current density, due to the kind of carrier under consideration.

an n-type semiconductor is that whereas in a metal the number of free conduction electrons per atom is independent of the temperature, in an n-type semiconductor the concentration of free electrons increases extremely rapidly with rising temperature. In p-type semiconductors there are no free electrons, and conduction occurs by an electron-replacement mechanism which from a certain point of view is equivalent to the movement of positive carriers, hence the name p-type—p for positive.

Many semiconductors, of both kinds, have the peculiarity that across the boundary of a metal electrode brought in contact with them a substantial potential step is developed when a current is passed, this varying with the nature of the semiconductor and the metal electrode, the current density, and the direction of the current. When carrying out experiments with these materials it is necessary to make allowance for such potential discontinuities; otherwise the results will be falsified. When the experiments are carried out in the proper way it is found that all semiconductors conform strictly to Ohm's law.

In testing electrolytes there is a similar difficulty with potential jumps at and near the electrodes. However, when the proper experimental precautions are taken it is found that with these materials, also, the current–voltage relation is ohmic, at least within the range of field strengths ordinarily applied.

Not all media through which a current can be passed are ohmic conductors. An outstanding example of one that is not is a gas; our discussion of discharges through gases in Chapter 24 shows that with these there can be no question of conformity to Ohm's law.

Charge and Discharge of Capacitor in an *RC* Circuit

In our discussion of capacitors in Chapter 21 no consideration was given to time effects in the charging and discharging processes. Let us now investigate how the state of charge of a capacitor varies with time in the case where the capacitor is connected in circuit via an ohmic conductor, or a "resistor." Such a circuit is generally referred to as an "*RC* circuit."

Let us consider first the charging process, in which a constant PD of magnitude V_0 is applied, through a conductor of resistance R, to a capacitor of capacitance C, as indicated in Fig. 25.4. This PD is dropped across the system in two steps, the drop V_R across the resistor being RI, while that across the capacitor, V_C, is Q/C.

651

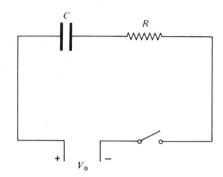

FIGURE 25.4 Arrangement for charging a capacitor through a resistor.

Thus

$$V_0 = V_C + V_R = V_C + RI,$$

or, since

$$I = \frac{dQ}{dt}$$

and

$$Q = CV_C,$$

$$V_0 = V_C + RC \frac{dV_C}{dt},$$

or

$$\frac{1}{V_0 - V_C} \frac{dV_C}{dt} = \frac{1}{RC},$$

which may also be written in the form

$$\frac{1}{V_0 - V_C} \frac{d(V_0 - V_C)}{dt} = -\frac{1}{RC}.$$

Integrating this and applying the condition that when $t = 0$ (the moment of switching on), $V_C = 0$, we obtain

$$\ln\left(\frac{V_0 - V_C}{V_0}\right) = -\frac{t}{RC},$$

or

$$1 - \frac{V_C}{V_0} = e^{-t/RC};$$

that is,

$$V_C = V_0(1 - e^{-t/RC}). \tag{25.7}$$

Similarly, for $Q \ (= CV_C)$ we have

$$Q = Q_0(1 - e^{-t/RC}), \tag{25.8}$$

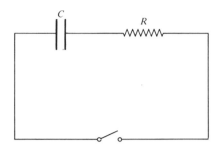

FIGURE 25.5 Arrangement for discharging a capacitor through a resistor.

where Q_0 is written for CV_0. And for I $[= C(dV_C/dt)]$ we have

$$I = \frac{V_0}{R} e^{-t/RC}. \tag{25.9}$$

These equations may also be written in the forms

$$V_C = V_0(1 - e^{-t/\tau}), \tag{25.10}$$
$$Q = Q_0(1 - e^{-t/\tau}), \tag{25.11}$$

and

$$I = \frac{V_0}{R} e^{-t/\tau}, \tag{25.12}$$

where τ is substituted for RC. This quantity, which has the dimensions of time, is an important property of the RC circuit; it is known as the **time constant** of the circuit.

For the discharge of this capacitor through the resistor (see Fig. 25.5) we have, again taking for the positive direction of I that in which the current actually flows,*

$$V_C = RI = -RC\frac{dV_C}{dt},$$

or

$$\frac{1}{V_C}\frac{dV_C}{dt} = -\frac{1}{RC}.$$

On integrating this, using the condition that when $t = 0$, $V_C = V_0$, we obtain the equation

$$V_C = V_0 e^{-t/RC} = V_0 e^{-t/\tau}, \tag{25.13}$$

and correspondingly

$$Q = Q_0 e^{-t/RC} = Q_0 e^{-t/\tau}. \tag{25.14}$$

---

* This will be the reverse of the previous direction.

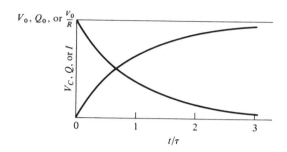

FIGURE 25.6   Curves for charge and discharge of a capacitor through a resistor.

Also, since

$$I = -\frac{dQ}{dt} = -C\frac{dV_c}{dt},$$

we must have

$$I = I_0 e^{-t/\tau}, \tag{25.15}$$

where I_0, the initial current, is equal to Q_0/RC; that is, since $Q_0/C = V_0$, to V_0/R. All these time variations are of one or other of two types, represented by the two curves of Fig. 25.6.

SMOOTHING OF PULSATING DIRECT CURRENT   In many situations where a steady supply of direct current is wanted, this is most conveniently obtainable by rectifying and smoothing an alternating supply, such as that available from a municipal power supply or from a transformer (see Chapter 34) fed from such a supply.

A current which, while unidirectional, pulsates in magnitude, may be obtained by operating the system shown diagrammatically in Fig. 25.7. The symbol on the left of this figure represents the alternating supply, while that on the top is the conventional representation of a rectifier. A rectifier is any device that allows current

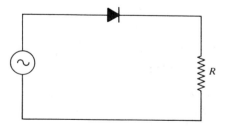

FIGURE 25.7   System for producing pulsating unidirectional current through a resistor.

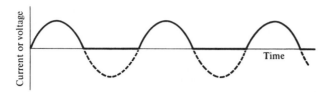

FIGURE 25.8 Time variation of unsmoothed half-wave-rectified current.

to flow through it in one direction (from left to right in the figure) without any significant associated voltage drop, while blocking it in the other direction. We have already considered one example of a rectifier, a thermionic diode. A number of other rectifiers are now available which are more convenient; in these the rectification occurs either at the boundary between a metal and a semiconductor or at a junction between an n-type and a p-type semiconductor.

In the absence of the rectifier, the alternating voltage applied to the resistive load would cause a current to flow whose time variation would be that represented by the combined solid and dashed curves of Fig. 25.8. The same combined curves would also represent, to some scale, the time variation of the voltage across the resistor. However, with the rectifier included in the system, as in Fig. 25.7, this suppresses every alternate half-cycle, giving a current through the load and a voltage across it, both of which vary with time in the manner shown by the solid curves and the intervening horizontal lines in Fig. 25.8. The dashed curves in this figure represent the half-cycles that have been suppressed.

The simplest way to smooth this pulsating dc, converting it into a reasonably steady current through the load, is to connect a capacitor of suitable capacitance across the latter, as indicated in Fig. 25.9. This has the effect of converting the I–t or V–t curve of Fig. 25.8 into one like that shown in Fig. 25.10. After the capacitor has been charged to its peak voltage, it cannot discharge back

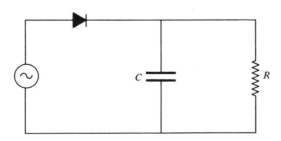

FIGURE 25.9 System incorporating capacitor for smoothing half-wave-rectified current.

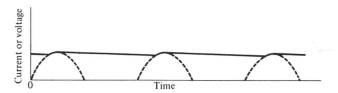

FIGURE 25.10 Time variation of smoothed half-wave-rectified current.

through the rectifier but only through the load, and, if the capacitor has a sufficiently large capacitance, the amount by which the PD across both it and the load falls before it "picks up" again just prior to the occurrence of the next peak is relatively small. Obviously, the curve from each peak to the following pickup point conforms to the exponential decay law considered under the heading of the discharge of a capacitor in an RC circuit. An approximate value for the fall of PD occurring in each cycle may be obtained by substituting T for t in the corresponding equation, where T is the period of the alternating supply.

Conductors in Series and in Parallel

We shall now investigate how the resistance R of a system of ohmic conductors joined either in series or in parallel may be expressed in terms of the separate resistances $R_1, R_2, R_3, \ldots, R_N$ of these.

CONDUCTORS IN SERIES   This case is shown in Fig. 25.11. Let the current flowing through the series of resistors be I. Then if the potentials at the points $A, B, C, D, \ldots, X$, and Y are V_A, $V_B, V_C, V_D, \ldots, V_X$, and V_Y, we must have

$$V_A - V_B = R_1 I,$$
$$V_B - V_C = R_2 I,$$
$$V_C - V_D = R_3 I,$$
$$\cdot$$
$$\cdot$$
$$\cdot$$
$$V_X - V_Y = R_N I.$$

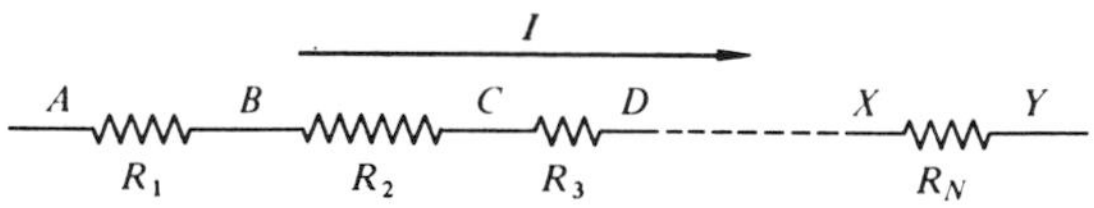

FIGURE 25.11   Resistors in series.

Hence, adding all these equations, we see that

$$V_A - V_Y = (R_1 + R_2 + R_3 + \cdots + R_N)I,$$

and so for the resistance R of the series of resistors we have

$$R = \frac{V_A - V_Y}{I} = R_1 + R_2 + R_3 + \cdots + R_N; \quad \textbf{(25.16)}$$

that is, the resistance of a number of conductors in series is the sum of the separate resistances.

CONDUCTORS IN PARALLEL   In this case the current I divides into the currents I_1, I_2, I_3, . . . , I_N, flowing in the separate conductors, as indicated in Fig. 25.12. The potential difference V between the entrance and exit ends may be expressed either in terms of the total current and the overall effective resistance R or in terms of the current flowing along any one of the separate branches and the corresponding resistance. Thus we have both

$$V = RI$$

and

$$V - R_1 I_1 - R_2 I_2 = R_3 I_3 = \cdots = R_N I_N.$$

But

$$I = I_1 + I_2 + I_3 + \cdots + I_N.$$

Substituting for these currents their values in terms of V and the corresponding resistances, we see that

$$\frac{V}{R} = \frac{V}{R_1} + \frac{V}{R_2} + \frac{V}{R_3} + \cdots + \frac{V}{R_N},$$

that is, that

$$\frac{1}{R} = \frac{1}{R_1} + \frac{1}{R_2} + \frac{1}{R_3} + \cdots + \frac{1}{R_N}. \quad \textbf{(25.17)}$$

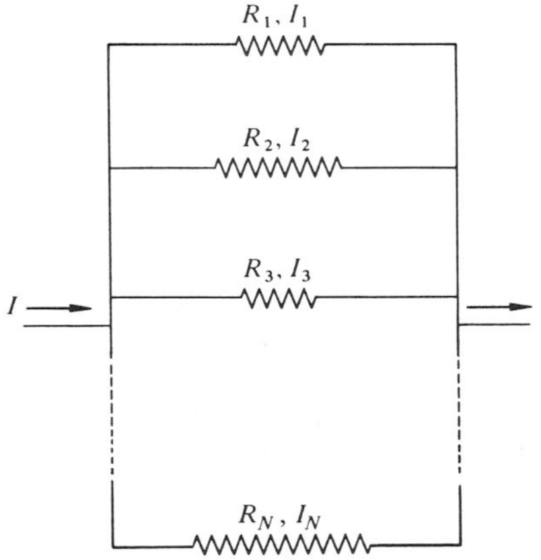

FIGURE 25.12   Resistors in parallel.

Thus the reciprocal of the overall effective resistance is equal to the sum of the reciprocals of the separate resistances.

RESISTIVITY AND CONDUCTIVITY From the formula for the resistance of a number of conductors in series it follows that the resistance of a uniform wire must be proportional to its length. We may see this as follows. A wire may be considered as made up of a number N of equal short elements of length (for example, millimeter lengths) joined end to end. Let each such element have a resistance r. Then from our formula we see that the resistance of N of these connected in series is equal to Nr. Different lengths l of wire may be thought of as made up of different numbers N of these elements, these numbers being proportional to the lengths. Hence the resistance of the wire, being proportional to N, is also proportional to l:

$$R \propto l.$$

Correspondingly, from the formula for the resistance of a number of conductors in parallel, it follows that the resistance of a wire of given length and given material is inversely proportional to its cross-sectional area A. To see this, it is convenient to consider wires of rectangular cross section. Such a wire may be thought of as made up of a number of equal wires, of the same length but of smaller cross section, packed together as indicated in section in Fig. 25.13. Let there be N' of these, and let each have a resistance r'. All these smaller wires are, in effect, connected in parallel. Hence, according to the formula,

$$\frac{1}{R} = \frac{N'}{r'},$$

or

$$R = \frac{r'}{N'}.$$

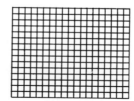

FIGURE 25.13

And since, for some standard size of the smaller wires, and so for a corresponding standard value of r', N' is proportional to the cross-sectional area A, we must have

$$R \propto \frac{1}{A}.$$

Since R is proportional both to l and to $1/A$, we may write

$$R = \rho \frac{l}{A},$$

where ρ is a property of the material of the conductor, known as its **resistivity.** It may be expressed in terms of the measured resistance, length, and cross-sectional area of the conductor by the formula

$$\rho = \frac{A}{l} R. \tag{25.18}$$

From this formula we see that ρ has the dimensions of a resistance multiplied by a length. Accordingly, the MKSA unit of resistivity is the **ohm-meter.**

The reciprocal of the resistivity, $1/\rho$, is readily identifiable with the quotient of the current density J and the field E, which we defined earlier in the chapter as the conductivity σ of the material. Thus, since, by definition, $R = V/I$,

$$\sigma = \frac{1}{\rho} = \frac{l}{A}\frac{1}{R} = \frac{l}{A}\frac{I}{V} = \frac{I/A}{V/l} = \frac{J}{E}. \tag{25.19}$$

The MKSA unit of conductivity is the ohm^{-1}/meter. This unit is also sometimes called the mho (ohm spelt backward) per meter.

Values of ρ for a selection of metals and alloys at 0° and at 100°C are given in Table 25.1, together with the mean temperature coefficient α between these temperatures, defined in terms of the resistivity ρ_0 at 0°C and the Celsius temperature Θ by the equation

$$\rho = \rho_0(1 + \alpha\Theta). \tag{25.20}$$

The variation of resistivity with temperature within this range is approximately linear.

Electromotive Force

An essential part of every circuit in which an electric current is maintained is a device in which electricity is "pumped" from a

TABLE 25.1 Resistivities of Various Metals and Alloys
(units: Ω-m $\times 10^{-8}$)

Metal or alloy	Temperature (°C)		Temperature coefficient over range 0–100°C (deg^{-1})
	0	100	
Ag	1.51	2.13	4.1×10^{-3}
Cu	1.56	2.24	4.3×10^{-3}
Au	2.04	2.84	4.0×10^{-3}
Al	2.48	3.55	4.5×10^{-3}
W	4.9	7.3	4.8×10^{-3}
Ni	6.14	10.33	6.8×10^{-3}
Fe	8.9	14.7	6.5×10^{-3}
Manganin (4% Ni, 12% Mn, 84% Cu)	48		$<10^{-5}$
Constantan (45% Ni, 55% Cu)	49		$<10^{-5}$
Nichrome (60% Ni, 15% Cr, 25% Fe)	112		1.6×10^{-4}

low to a high potential,* a device such as a battery or a dynamo. In performing this function the device continuously creates electrical energy, and this it can do only at the expense of some other form of energy, for example mechanical energy in the case of a dynamo deriving its power from a hydroelectric system, or chemical energy in the case of a battery.

As in the case of a mechanical pump (for example a water pump), so also here, the pumping action is never 100 percent efficient; there is always some loss of energy in the form of heat. Thus, in the case of a dynamo, some of the energy made available to it is lost by conversion to heat in the bearings, some heat is produced on account of the resistance of the windings through which current is made to flow, some is generated by the flow of eddy currents in the iron (see Chapter 32), and some is produced by sparking and other losses at the commutator brushes. There may also be other loss items.

---

* This reference to electricity and high and low potentials is intended to be taken algebraically, using the customary convention with regard to signs. What really happens may well be that *electrons* (*negative* carriers) are pumped from a low to a high *negative* potential, that is, from a high to a low conventional potential. This is *algebraically* equivalent to pumping positive electricity from a low to a high conventional potential.

We may divide these losses into two categories. In one of these we place purely mechanical losses, and, in addition, any heat that may be generated electrically, but *otherwise* than by current fed to the external circuit. In the other category we place such heat as may be generated within the device by current that does flow in the external circuit. The first category would include heat produced in the bearings, that produced by viscous losses in the surrounding air, heat produced by friction at the commutator brushes, and "iron losses," that is, heat generated by the flow of eddy currents and by hysteresis (see Chapter 31) in the iron. In the second category would be ohmic loss in the winding which carries the current that is fed to the external circuit, and heat generated by sparking at the commutator brushes.

Losses in the former of these categories are obviously of no interest in any consideration of the essential pumping action of the machine; they are merely an irrelevant complication, with which we need not concern ourselves further. All items belonging to this category, which merely act as a drag on the functioning of the machine, should accordingly be subtracted from the energy supplied to it. What remains we may then conveniently term the *net* energy supplied; it is only this that is available for conversion to electrical energy producing a current in the external circuit.

Losses in the latter category represent heat derived from some of the electrical energy that has already been produced by the machine. These are on the same footing as are conversions of electrical energy to other forms in the external circuit; from the point of view of the essential functioning of the machine it would be meaningless to distinguish between that part of the electrical energy generated which happens to be lost within the machine and that which is converted to other forms outside. Both these items should be lumped together as electrical energy that has been produced from some other form.

Because the generation of electrical energy from other energy does not involve the generation of heat, this must be an essentially reversible process in the thermodynamical sense. If some of the electrical energy produced by this process happens to be converted to heat within the generator, such conversion being irreversible, this does not affect the ideal reversibility of the conversion that preceded it—of other energy into electrical energy. It is therefore perfectly logical to regard the conversion of other energy into electrical energy as reversible in principle, despite the circumstance that some of the electrical energy produced is, in fact, wasted within the generator.

661

This reversibility of the essential operation of a dynamo means that it can also function as a motor, converting electrical energy into mechanical. Similarly, a storage battery, which converts chemical energy into electrical, eventually becomes discharged; and in the subsequent process of charging, electrical energy is converted to chemical. And in the case of the storage battery, as in that of the machine, there is an irreversible conversion of electrical energy to heat within the device whenever a current passes, associated with its internal resistance. In both cases this loss must be ignored in any theory of the essential process of operation of the device.

Let us divide the net nonelectrical energy ("other energy," O.E.) supplied to the generator—dynamo, storage battery, or other device, and converted to electrical energy (E.E.)—by the quantity Q of electricity that is made to flow past any particular point in the circuit. Then the quotient we obtain has the same dimensions as a potential difference, and must therefore be expressible in the same units, in volts. If we divide both numerator and denominator by the time taken, we have the quotient of the reversible power (O.E. $\rightarrow$ E.E.) to current strength. To this quotient we give the name **electromotive force,** emf. This is usually represented by the symbol E.*

This concept of electromotive force includes the idea of an associated direction, the direction of pumping. This is the direction in which the potential is increased by the pumping action. If, within the device functioning normally as a generator, current I_{AB} is pumped from terminal A to terminal B, then E has the direction from A to B. And if the rate at which other energy is being converted to electrical energy is denoted by the symbol P_{OE},† we may express our definition of electromotive force in the mathematical form

$$E = \frac{P_{OE}}{I_{AB}}. \tag{25.21}$$

In the situation envisaged, both P_{OE} and I_{AB} are positive, so E must

---

* While it is somewhat unfortunate that this is the same symbol as is used for electric field strength, it is unusual for the two quantities to appear together, and so there is seldom any danger of confusion. On the rare occasions where we have to deal with both an electric field strength and an electromotive force in the same discussion, the symbol used for one of these can be modified to distinguish it from the other; for example, electric field strength could be represented by the symbol E_f (f for "field").

† The subscript refers to the direction of the conversion, that is, from other energy to electrical energy.

also be positive. If we regard the direction from A to B within the device as the positive direction, we may interpret the sign of E as giving information concerning its direction; thus its positive sign indicates that its direction is from A to B.

Now let us suppose the device to be operating in reverse. The current now flows in the direction from B to A, so that I_{AB} is negative. Also, electrical energy is now being converted to other energy, and accordingly P_{OE} is also negative. Thus E in Eq. (25.21) retains its sign; the electromotive force has a unique direction, which is independent of whether the device is functioning as a generator or in reverse.

When the device is operating as a generator, the electromotive force has the same direction as the current, and the associated potential rise within the device is in this direction. The electromotive force in this case is often referred to as a "forward emf." On the other hand, when the device is operating in reverse, the direction of the electromotive force is opposite to that of the current, that is, there is an associated potential *fall* within the device in the direction of the current. In this case the electromotive force may be called a "back emf."

It does not follow from any of our discussion that the emf of a device, as defined, is necessarily independent of the current strength; it could conceivably vary with the latter. However, in practice it is found not to do so; in the case of a device operating under given conditions, such as a storage cell in a given chemical condition at a particular temperature or a dynamo rotating at a given speed, the emf is a constant characteristic of the device and its conditions of operation, in magnitude as well as in direction. Also, theoretical discussion of the mechanisms to which electromotive forces are due show that the emf should be independent of the current in the corresponding particular cases.

The question of the nature of what we have referred to as electrical energy needs a little elaboration. In electrostatics the only electrical energy with which we were concerned was potential energy. In current electricity the question arises of whether there might not be, in addition, an item corresponding to kinetic energy in mechanics, associated with the state of motion of the electricity. As we shall see later, there is, in fact, such an item. This might be called either kinetic electrical energy or magnetic energy; the latter is probably the better term. In the present context of defining electromotive force it is to be understood that we are including the magnetic item in what is referred to as electrical energy. In the

663

special case of the steady flow of current in a circuit the magnetic energy of the circuit is, obviously, constant; it is then only the potential energy that figures in the electromotive force.

It might be objected that the potential energy at every point of the circuit is also constant. Notwithstanding this, the argument is valid, for, as we shall see later, magnetic energy must always be considered for the circuit as a whole; it cannot be itemized over particular parts of the circuit. Potential energy, on the other hand, *can* be itemized, and it is precisely with such itemization of potential energy per unit charge in relation to the terminals of the device that we are concerned in our definition of electromotive force.

The term "electromotive force" is really a misnomer, but it is one that has become so thoroughly established in physical literature that it does not seem likely to be replaced by any more suitable term in the foreseeable future. Although in the operation of a generator a force *is* presumably somehow exerted—through a distance—on the carriers of charge, it is not the *force* per unit charge that defines the emf but the reversible *energy conversion* per unit charge.

The rise or fall of potential in the direction of the current within the device that corresponds to its emf is, in general, only one item in the total rise or fall. In addition there is always the ohmic drop of potential $R_0 I$ due to the internal resistance R_0 of the device associated with the passage of a current I in either direction. This must be combined algebraically with E to give the total rise or fall, as the case may be. Thus, according as the device is operating directly, as a generator, or in reverse, we have for the potential rise across the terminals

$$V = E - R_0 I, \qquad (25.22)$$

or, for the potential fall,

$$V = E + R_0 I, \qquad (25.23)$$

the rise or fall being reckoned in the direction of the current in both cases.

It follows from these equations that in the limiting case where no current is being drawn from, or sent through, the device, V is simply equal to E; that is, the open-circuit PD across the terminals is the same as the emf. Hence, what we really measure when we compare the open-circuit potential differences across devices having electromotive forces, for example with a potentiometer (see Chapter 26), is the ratio of these electromotive forces.

664

Cells in Series and in Parallel

Let us first consider a number of cells not necessarily all of the same kind, having electromotive forces E_1, E_2, E_3, . . . , and internal resistances R_1, R_2, R_3, . . . , connected in series and on open circuit. In such a system we have a succession of steps of potential E_1, E_2, E_3, . . . , and so the potential difference across the whole series must be given by the equation

$$V = E_1 + E_2 + E_3 + \cdots . \qquad (25.24)$$

If the terminals of the series are now connected to an ohmic system of resistance R, a current I will flow, determined by the condition that the total rise of potential over the whole series of cells is equal to the fall of potential across the external system; for if we travel once around the circuit we must come back to the potential with which we started. This condition is expressed by the equation

$$(E_1 - IR_1) + (E_2 - IR_2) + (E_3 - IR_3) + \cdots = IR,$$

or

$$E_1 + E_2 + E_3 + \cdots = I(R_1 + R_2 + R_3 + \cdots + R). \quad (25.25)$$

From this we see that the current is the sum of all the electromotive forces divided by the sum of all the resistances, internal and external.

Now let us consider a number, N, of cells, *all of the same kind*, each having an electromotive force E, the internal resistances of which are R_1, R_2, R_3, . . . , R_N, these cells being connected in parallel. Then on open circuit the PD across the whole system must obviously be the same as that across each cell, E. If now this system of cells is used to send a current through a resistor of resistance R, we must have the currents I_1, I_2, I_3, . . . , I_N in the separate cells combining to give the total current I through the resistor, according to the conditions

$$E - R_1I_1 = E - R_2I_2 = E - R_3I_3 = \cdots = E - R_NI_N,$$

or

$$R_1I_1 = R_2I_2 = R_3I_3 = \cdots = R_NI_N; \qquad (25.26)$$

$$I_1 + I_2 + I_3 + \cdots + I_N = I; \qquad (25.27)$$

and

$$E - R_1I_1 = RI. \qquad (25.28)$$

Let X be written for R_1I_1. Then, from Eq. (25.26), we have

$$I_1 = \frac{X}{R_1}, I_2 = \frac{X}{R_2}, I_3 = \frac{X}{R_3}, \cdots, I_N = \frac{X}{R_N};$$

665

and so, from Eq. (25.27),

$$\sum_{r=1}^{N} I_r = I = X\left(\frac{1}{R_1} + \frac{1}{R_2} + \frac{1}{R_3} + \cdots + \frac{1}{R_N}\right)$$
$$= R_1 I_1\left(\frac{1}{R_1} + \frac{1}{R_2} + \frac{1}{R_3} + \cdots + \frac{1}{R_N}\right).$$

Hence

$$R_1 I_1 = \frac{I}{1/R_1 + 1/R_2 + 1/R_3 + \cdots + 1/R_N}. \quad (25.29)$$

But also, from Eq. (25.28), we have

$$R_1 I_1 = E - RI. \quad (25.30)$$

Thus equating the right sides of Eqs. (25.29) and (25.30), we find that

$$E = I\left(R + \frac{1}{1/R_1 + 1/R_2 + 1/R_3 + \cdots + 1/R_N}\right). \quad (25.31)$$

We see that all the internal resistances combine in the same way as do those of a number of simple resistors connected in parallel.

Kirchhoff's Laws

Wheatstone's bridge and systems of resistors in parallel and cells in parallel, constituting part of a complex circuit, are examples of what are known as "networks" containing two or more meshes. Other examples will be considered in Chapter 26. The determination of the currents flowing in the various elements of such meshes depends on the use of one or the other, or both, of two general principles formulated by the German physicist G. H. Kirchhoff (1824–1887), which are usually referred to as Kirchhoff's laws. The first of these laws states, in effect, that there is no accumulation of electric charge at any junction of a network. The second is the statement that every point of the network has at any given moment a unique potential.

The formal statements of these laws are as follows:
1. The algebraic sum of the currents that meet at any junction of a network is zero.
2. In going around any complete mesh of a network, from any point back to the same point, the algebraic sum of all the increments of potential encountered on the way is zero.

These two laws suffice for the solution of any network problem, however complex, giving as many independent equations as there are unknowns.

The usefulness of Kirchhoff's laws is not confined to steady-current situations; subject to certain provisos they may also be applied to systems in which currents vary with time. Here it is essential that the round trip envisaged in the second law should (in imagination) be made in zero time; all potential rises and falls considered must be those characteristic of the same instant. This is an obvious requirement consequent on the time variation of the potentials of points in such systems. With regard to the first law, there is the complication that, in their functioning, circuits are not always as simple as they may appear on first sight. This applies particularly in the case of circuits carrying alternating currents of very high frequency. Thus an innocent-looking junction, with its adjacent wires, may well function as one "plate" of a capacitor, with the earth or some neighboring part of the circuit acting as the other electrode. It is clearly not permissible to equate the sum of all the currents feeding into such a junction to zero at all times. However, at relatively low frequencies such effects are quantitatively of no importance, and the first law may then be regarded as valid, just as it is for steady currents.

For the application of Kirchhoff's laws it is, of course, essential to have full knowledge of the relevant properties of all the elements of the circuit. Thus all resistances must be known, not only those of resistors but also those of cells, dynamos, motors, and so on. Again, it is necessary to know the electromotive forces, both in magnitude and in direction, of all devices having an emf, whether these electromotive forces are steady or (as in an alternator) vary with time. Also, if the network includes capacitors, the capacitances of these must be known.

The unknowns whose determination Kirchhoff's laws render possible are the currents flowing along the various elements of the network, or, alternatively, in the case of those elements that include capacitors, the charges on these; the corresponding currents are then the time variations of the charges.

In dealing with these unknowns algebraically it is not necessary to know in advance either the directions of the currents or, if capacitors are included, the polarities of the charges on these. For example, if for an element PQ of a mesh the current flowing from P to Q is assigned the symbol I_{PQ}, it is in no way implied that I_{PQ} is necessarily positive; the possibility is left open for it to be negative, that is, for the current actually to flow the other way. Correspond-

667

ingly, if the resistance of this element is R_{PQ}, the potential dropped from P to Q is, algebraically, $R_{PQ}I_{PQ}$, regardless of what sign I_{PQ} happens to have. Similarly, if the charge on an arbitrarily selected capacitor plate is assigned the symbol Q, the possibility remains open for Q to be negative; whether it is, in fact, positive or negative will be found in the eventual solution for Q. And, in the case where an alternator is included in the network, and the emf of the alternator in a certain direction is correctly represented as a function of time by the expression $E_0 \sin(\omega t + \phi)$, this is algebraically the potential rise in this direction corresponding to the emf at the time t, regardless of whether for this value of t the expression is positive or negative.

The following simple examples will serve to illustrate these principles. In the corresponding circuit diagrams the resistances shown are to be understood as including the internal resistances of any devices other than resistors, as well as the resistances of resistors, in the various mesh elements. The arrows shown alongside the devices having an emf point in the direction of this emf.

In Fig. 25.14 let the electromotive forces and resistances be as marked, and let us assign values I_1, I_2, and I_3 to the currents flowing in the resistors R_1, R_2, and R_3, respectively; these are to be understood as the currents (whether positive or negative) flowing in the directions of the arrows.

Kirchhoff's first law applied to either of the two junctions A or B gives us the equation

$$I_1 + I_2 + I_3 = 0. \tag{25.32}$$

Going around the left-hand mesh anticlockwise and adding algebraically all the corresponding potential rises, we find that according to the second law

$$E_1 - R_1 I_1 + E_3 I_3 = 0. \tag{25.33}$$

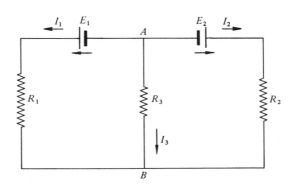

FIGURE 25.14

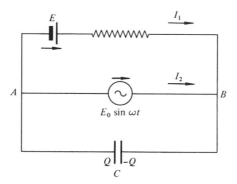

FIGURE 25.15

And going clockwise around the right-hand mesh gives us in similar fashion

$$E_2 - R_2I_2 + R_3I_3 = 0. \qquad (25.34)$$

Thus we have three equations from which we can solve for the three unknowns, I_1, I_2, and I_3.

Although our second example, the circuit diagram of which is given in Fig. 25.15, might properly be considered as a problem in alternating-current theory, we shall nevertheless consider it in the present chapter as an example of the application of Kirchhoff's laws.

Applying the first law to either of the two junctions A or B, we obtain the equation

$$I_1 + I_2 + \frac{dQ}{dt} = 0. \qquad (25.35)$$

Now applying the second law to the two meshes, we have, for the upper mesh,

$$-E_0 \sin \omega t + E - R_1I_1 + R_2I_2 = 0, \qquad (25.36)$$

and for the lower mesh

$$E_0 \sin \omega t - R_2I_2 + R_3 \frac{dQ}{dt} = 0. \qquad (25.37)$$

Thus again we have as many equations as there are unknowns, the latter being I_1, I_2, and Q.

Problems

25.1   The belt of a Van de Graaff generator is 60 cm wide and travels at a speed of 20 m/sec. One side delivers charge to the dome at a rate of 80 μA. What is the corresponding surface density of charge on the belt?

If the PD between the dome and the source of charge (at earth potential) is 4 MV, and the other side of the belt carries charge (of opposite polarity) away from the dome at a rate of 60 μA, what is the net power required to drive the generator?

25.2 A metal ring of diameter 20 cm is attached by insulating spokes to a shaft, given a charge of 0.5 μC, and rotated about its axis at 500 rps. What is the corresponding current? Would you expect this current to produce any heating of the ring? (Assume that the charge is stationary relative to the metal.)

25.3 Calculate the mean drift velocity of the electrons in a copper wire of 1 mm diameter which carries a current of 10 A, assuming that there is one free electron per atom and that the proportion of free electrons that are able to respond to an electric field is 1 in 1000.* What would be the corresponding strength of the electric field within the wire at a temperature of 20°C? Is the speed with which an electric signal can be sent along a wire related to the drift velocity of the electrons? Explain your answer.

25.4 A rheostat is marked "10 A, 25 Ω." What is the maximum permissible power dissipation indicated by this label? Would it be safe to connect the rheostat across 240-V supply mains? If it were so connected, what current would flow through it?

25.5 A resistor has a resistance of 10,000 Ω and a rating of 4 W. What is the maximum voltage that may be applied to its terminals, and what is the corresponding current?

25.6 The winding of a large electromagnet consists of copper tubing cooled by passing a stream of water through it. The resistance of the winding is 0.3 Ω, and it carries a current of 600 A. Given that the rate of flow of the water is 0.04 m³/min, calculate its temperature increase.

25.7 Calculate the potential differences $V_A - V_B$, $V_B - V_C$, $V_C - V_D$, $V_D - V_E$, and $V_E - V_A$ in the circuit shown (a) when the switch is open, and (b) when it is closed. Assume that the cells have zero internal resistance.

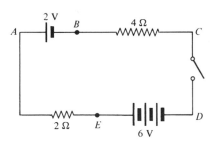

Figure for problem 25.7.

25.8 A high-resistance voltmeter connected to a battery on open circuit reads 6.15 V. When a 1-Ω resistor is connected across the terminals of the battery the reading falls to 5.95 V. Account for this, and calculate whatever quantities you can from the information provided.

---

* For an explanation of the apparent contradiction here, see Chapter 7 of Volume III.

25.9 Show that a given battery delivers maximum power to a resistor when the resistance of the latter is equal to the internal resistance of the battery.

25.10 A direct current of 100 mA is required in a part of a circuit, and the maximum fluctuation that can be tolerated is 1 mA. This is to be obtained from 50-Hz 240-V mains (peak voltage 339—see Chapter 33), in conjunction with a rectifier and capacitor as shown in Fig. 25.9. What must be the approximate values of R and C in this system?

25.11 A neon tube is connected in parallel with a capacitor in the system shown in the figure. It may be assumed that the tube does not conduct at all until the PD between its terminals reaches a value V_1, and that it then "strikes" and behaves as a perfect conductor, so that the capacitor discharges instantaneously to the stage where the voltage has fallen to V_2, when the discharge in the tube is extinguished, and the capacitor begins to recharge. Sketch a graph showing how the capacitor

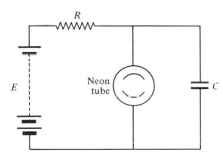

Figure for problem 25.11.

voltage varies with time, assuming that $V_1 < E$, and show that the neon tube flashes f times per second, where

$$f = \left[CR \ln \left(\frac{E - V_2}{E - V_1} \right) \right]^{-1}.$$

It is found that the behavior is rather erratic if E only very slightly exceeds V_1. Suggest a reason for this. (Assume the internal resistance of the battery to be included in R.)

25.12 In the figure, I_1, I_2, and I_3 are the currents that flow after the switch S is closed. Given that the emf of the cell is 2 V, calculate the

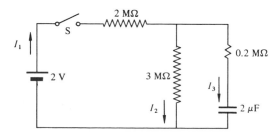

Figure for problem 25.12.

values of I_1, I_2, I_3, and the voltage V across the capacitor (a) immediately after the switch is closed, and (b) after a considerable time has
elapsed. The switch is then opened. Calculate the value of V 1 sec later.

25.13   When a certain PD is applied across three equal resistors
connected in series the power consumption is 15 W. The three resistors
are now connected in parallel across the same PD. What power will now be
consumed?

25.14   Two wires of resistances 20 Ω and 30 Ω are joined (a) in series,
and (b) in parallel, and connected in each case to a 110-V supply. Calculate
the rate of heat production, in calories per second, with each method of
connection.

25.15   You are given a cell of unknown emf and internal resistance,
a 10-Ω resistor, and a milliammeter of resistance 50 Ω. You find that on
connecting the milliammeter plus resistor to the cell the same current of
24 mA is registered, regardless of whether the resistor is connected in series
or in parallel with the milliammeter. What are the emf and internal resistance of the cell?

25.16   What is the resistance of the system shown in the figure
between the terminals A and B?

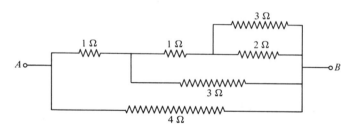

Figure for problem 25.16.

25.17   Find the resistance X between the terminals A and B of the
infinite array of resistors shown by making use of the fact that, if the points
C and D of the system shown on the left are connected to A and B, the
resistance between E and F will then be X. If $R_1 = 16,000$ Ω and $R_2 =
2500$ Ω, and if a potential difference of 100 V is maintained between the
terminals A and B, what is the power consumed in the first of the resistors
of resistance R_2?

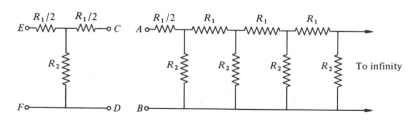

Figure for problem 25.17.

25.18   Twelve terminals, A, B, C, D, E, F, G, H, I, J, K, and L are attached to an insulating board as shown. Every terminal is connected to every other terminal by a wire of resistance R. The wires make electrical contact only with the terminals, not with one another. Find the resistance between any two terminals.

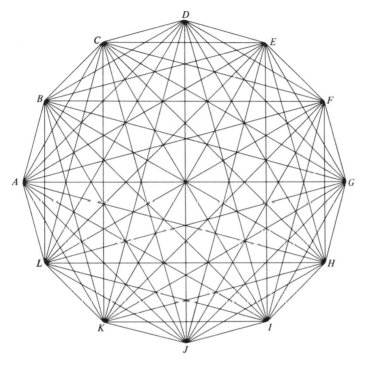

Figure for problem 25.18.

25.19   What are the resistances at 20°C between opposite faces of solid cubes of copper of sides 1 m and 1 cm?

25.20   A wire is drawn out to twice its original length. Assuming that the resistivity remains unchanged, calculate the factor by which its resistance has been increased.

25.21   An electrical strain gauge employs a wire in one leg of a bridge circuit bonded by insulating cement to a beam subjected to stress. The resulting strain deforms the wire, producing a proportional change in the bridge. Making the naive assumption that the volume of the wire does not change as it is stretched slightly and that the resistivity is not changed by the stress, calculate the gauge factor $(\delta R/R)/(\delta l/l)$, the ratio of the proportional change in resistance to the proportional change in length of the wire.

25.22   What is the resistance of 1 km of copper wire of diameter 1 mm at 20°C?

25.23   What length of manganin wire of diameter 0.325 mm would be required to wind a 10-Ω resistor?

673

25.24 Two wires at 20°C, one of aluminum and the other of copper, have the same length and the same resistance. What are the ratios of (a) their weights, and (b) their diameters? Point out the significance of your findings for long-distance power transmission.

25.25 Compare the resistivities of Nichrome (commonly used for electric radiators) at 20° and at 1000°C.

25.26 A motor connected to a 110-V dc supply develops 300 W of useful mechanical power and dissipates 15 W frictionally and otherwise. The current from the supply is 3 A. What electrical power is being supplied to the motor? What is the efficiency of the motor, and what is its resistance?

25.27 Discuss all potential rises and falls in the various parts of a circuit consisting of a dc generator of emf 240 V and internal resistance 3 Ω which delivers 15 A through leads of total resistance 1.5 Ω to a motor of resistance 2 Ω. Find the emf of the motor.

25.28 A steam turbine drives a dc generator at 3000 rpm, and the moment of inertia of the rotating machinery is 50 kg-m². At this speed the generator produces an emf of 400 V. At a certain moment the supply of steam is suddenly shut off. Given that the combined resistance of the generator and the circuit to which it is connected is 10 Ω, find how long it will take for the speed of the rotating system to fall to 600 rpm. Assume that the emf is directly proportional to the angular velocity of the generator rotor, and neglect frictional and viscous losses.

25.29 A battery maintains a PD of 6 V between the points B and D of the network shown in the figure. Find (a) the current delivered by the battery, and (b) the magnitude and direction of the current in the arm AC.

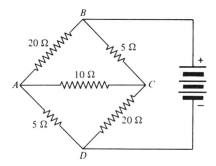

Figure for problem 25.29.

DIRECT-CURRENT INSTRUMENTATION

Our discussion of dc theory in Chapter 25 had necessarily to include some basic instrumentation, but much the greater part of dc instrumentation has yet to be dealt with. It is with the consideration of this that we shall be concerned in the present chapter.

Comparison of Resistances

It would in principle be possible to determine resistances on all occasions by measuring the current and the corresponding rate of generation of heat, or, alternatively, by measuring the current and the corresponding PD between the entrance and exit points, using for the latter some form of electrostatic voltmeter. The former procedure would be exceedingly tedious and time-consuming, and the latter, although less tedious, would also generally not be very convenient. Fortunately, such absolute determinations have only to be resorted to on relatively rare occasions; quite accurate and convenient methods are available for *comparing* the resistances of two conductors. By using one of these methods, we may determine the resistance of one conductor if that of the other is known.

WHEATSTONE'S BRIDGE The method most commonly used for comparing resistances is one employing a divided circuit with a "bridge" provided across the arms. This circuit, which is known, after its originator, as Wheatstone's bridge, is shown diagrammatically in Fig. 26.1. Current derived from a cell via a key K_1 and capable of being regulated by manipulating the rheostat Rh enters the network at the point A and leaves at B. Between A and B the current divides into two branches, ACB and ADB, each of which consists of two resistors connected in series. Of these, one, let us say that between the points A and C, of resistance R_{AC}, is the one whose resistance we wish to determine. R_{CB} is known, and the ratio R_{AD}/R_{DB} is known and variable at will.* Between C and D a galvanometer G is connected through a protective resistor of high resistance, which latter can be shorted out for increased sensitivity when this is required, by closing the switch S. It is usual also to

---

* How this ratio is known, and how it may be varied, we shall see later.

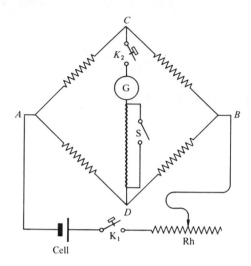

FIGURE 26.1 Wheatstone's bridge.

provide a key K_2 in the galvanometer arm, for further protection of the instrument. The ratio R_{AD}/R_{DB} is varied until, on first closing the key K_1 and then K_2, the current registered by the galvanometer is zero. The bridge is then said to be balanced, and, as will now be shown,

$$R_{AC} = R_{CB} \frac{R_{AD}}{R_{DB}}. \qquad (26.1)$$

Let the potentials of the points A, B, C, and D be denoted by the symbols V_A, V_B, V_C, and V_D, respectively. When no current passes through the galvanometer, V_C and V_D are necessarily equal; for, writing I_g for the current flowing from C to D through the galvanometer and R_g for the resistance of the galvanometer arm, we have, by Ohm's law,

$$V_C - V_D = R_g I_g,$$

and so, when I_g is zero, $V_C = V_D$. Let the currents flowing in the branches ACB and ADB when the bridge is balanced be I_1 and I_2, respectively. Then, again applying Ohm's law, we have

$$V_A - V_C = R_{AC} I_1$$

and

$$V_C - V_B = R_{CB} I_1,$$

and dividing the first equation by the second,

$$\frac{V_A - V_C}{V_C - V_B} = \frac{R_{AC}}{R_{CB}}.$$

676 Similarly, applying Ohm's law to the branch ADB, in which the current I_2 is flowing, we find that

$$\frac{V_A - V_D}{V_D - V_B} = \frac{R_{AD}}{R_{DB}}.$$

But, since $V_C = V_D$, the left sides of the last two equations are equal, and hence also the right sides; that is,

$$\frac{R_{AC}}{R_{CB}} = \frac{R_{AD}}{R_{DB}},$$

or

$$R_{AC} = R_{CB} \cdot \frac{R_{AD}}{R_{DB}}. \tag{26.1}$$

We may note that the form of this equation implies that it would still be valid if the battery and galvanometer were interchanged.

Different versions of Wheatstone's network employ different systems for the "ratio arms" AD and DB. Among these are the slide-wire bridge and the dial-box versions, which we shall consider presently. Before doing so, however, it will be convenient to deal with the general question of the establishment of a prescribed ratio of resistances.

Establishment of a Prescribed Ratio of Resistances Identical resistors must of necessity have equal resistances. Let us write r for the resistance of any one of these. Then if we have two sets of such resistors, one consisting of a number N_1 connected in series and the other of N_2, the resistances R_1 and R_2 of these sets must be $N_1 r$ and $N_2 r$, respectively, and so, without necessarily knowing the value of r, we may write

$$\frac{R_1}{R_2} = \frac{N_1}{N_2}.$$

We may confirm that the resistances of the separate units that we have endeavored to make identical in all respects really *are* equal by connecting first one and then another as one portion of a circuit (for example, an arm of a network) containing a battery and a galvanometer whose reading is sensitive to variations in the resistance of the unit connected. Then equality of the readings of the galvanometer on different occasions would mean equality of the resistances of the corresponding units connected. In practice one would normally use a Wheatstone bridge, connect first one unit and then another as one arm, and, having balanced the bridge with one, confirm that substitution of the other for this does not disturb the balance. If a disturbance *is* observed, appropriate adjustments to

677

one or other of the units may be made until, by trial and error, it is found that the interchanging of units no longer produces any effect. When this condition has been achieved the resistances of the units may be assumed to be accurately the same.

We may free ourselves of the restriction of having to use integral numbers of units each of resistance r by using the method of substitution just described for finding a length L of uniform wire that also has this resistance. Applying the same argument as before to equal short elements of length, for example millimeter lengths, of this wire, which are, in effect, connected in series, we readily see that any shorter length l of the wire must have the resistance $(l/L)r$. Thus an appropriate shorter length may be used to provide any desired fractional unit of resistance.

Slide-Wire Bridge This device, shown diagrammatically in Fig. 26.2, consists simply of a uniform wire of suitable material, usually 1 m long—hence the alternative name "meter bridge"—stretched on a board between two massive metal blocks provided with terminals, and a third electrode, moveable along a guide in a direction parallel to the wire, which may be brought into contact with this wire at any point along its length by depressing a spring. A scale is mounted just under or alongside the wire for convenience of measurement of the position where contact is made. The lettering corresponds to that of Fig. 26.1.

The metal blocks between which the wire is stretched and the attached terminals are assumed to have negligible resistance. The unknown and standard resistors are connected in circuit using short

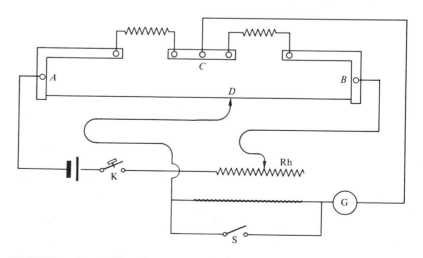

FIGURE 26.2 Slide-wire version of Wheatstone's bridge.

stout copper wires whose resistances are also regarded as negligible. The bridge is balanced by finding a point on the wire such that when contact is made with it by depressing the spring on the slider no current can be detected in the galvanometer arm of the circuit.

If we may assume the meter wire to be truly uniform, the resistances of the parts AD and DB must be in the same ratio as the corresponding lengths l_{AD} and l_{DB}; that is,

$$\frac{R_{AD}}{R_{DB}} = \frac{l_{AD}}{l_{DB}},$$

and so, substituting for R_{AD}/R_{DB} in Eq. (26.1), we have

$$R_{AC} = R_{CB} \frac{l_{AD}}{l_{DB}}.$$

This version of Wheatstone's bridge suffers from two main weaknesses. One of these is the assumption that the wire is uniform. Even if the wire *is* uniform to begin with, it tends to wear unevenly and become deformed with continued use. The other limitation is that of the accuracy of reading the exact point D on the wire for which a balance is obtained. Both these shortcomings are avoided in the dial-box version of Wheatstone's bridge, which we shall now consider.

Dial Box In this instrument no slide wire is incorporated, and so it does not deteriorate in its functioning as a result of wear. It uses a system of dial switches, each of which is, in effect, an ordinary dial version of a resistance box. An example of such a resistance box is shown diagrammatically in Fig. 26.3. In this there are 11

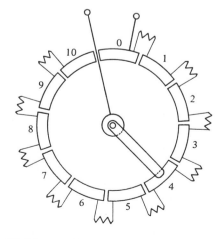

FIGURE 26.3 Dial-box system.

metal contact bars arranged in a circle, numbered 0, 1, 2, . . . , 10. About the center of this circle an arm can rotate, to make contact between the center and any one of the bars. Adjacent pairs of bars are connected by resistors as shown, all of equal resistance, so that by appropriate selection of the bar contacted by the rotating arm any desired number of units may be included between the mounting at the center and the bar labeled 0, up to 10. Wires connect these to terminals, as shown.

In the usual dial-box form of Wheatstone's bridge there are four of these dial switches arranged in series, in each of which there are 9 (not 10, as in Fig. 26.3) available resistors. In the first, second, third, and fourth of these switches the resistances of the resistors increase in steps of 1000, 100, 10, and 1 Ω, respectively; so by suitable positioning of the four arms any whole number of ohms may be selected, from 1 to 9999. These four dial switches in series constitute the arm CB shown in Fig. 26.1. The ratio R_{AD}/R_{DB} can be set by a single dial switch at any positive or negative power of 10 ranging from an upper limit of 10^3 to a lower limit of 10^{-3}. These dial switches, and a built-in dry cell and galvanometer, are connected in circuit according to the scheme of Fig. 26.1, constituting a convenient, portable, self-contained unit. The bridge is balanced, or as nearly balanced as possible, by adjusting the known resistance of the four dial switches in series constituting the arm CB to a suitable value for the ratio R_{AD}/R_{DB} used. This adjustment of the known resistance for a fixed ratio in the ratio arms is an interesting variant of the procedure used in the case of the meter bridge.

In using the bridge the most favorable ratio R_{AD}/R_{DB} to select is the one which, although not requiring a known resistance greater than the 9999 Ω available, nevertheless requires the highest possible value within this limit. This then contains the maximum number of significant figures and so gives the highest available accuracy.

The smallest step by which the known resistance can be changed is 1 Ω, so it will in general not be possible to adjust this to a value that gives an exact balance. The procedure adopted is therefore the following. First, the total resistance in the arm CB is adjusted to the nearest whole number of ohms on the too-low side of the correct value for a balance, and the corresponding deflection of the galvanometer on one side of the zero is noted. Then the resistance in this arm is increased by 1 Ω, bringing it to the too-high side, and the deflection of the galvanometer—this time on the other side of the zero—is again noted. From these two galvanometer readings it is possible by interpolation on the basis of simple proportion (valid where we are dealing with small quantities) to calculate the fraction of an ohm beyond the resistance adjusted for on the first occasion that would have given an exact balance.

POTENTIOMETER METHOD   A limit to accuracy inherent in the use of a Wheatstone bridge, even in the dial-box version, is the resistance of the leads by which the unknown resistance is connected in circuit. This difficulty is avoided in an alternative method of comparing resistances based on the use of a potentiometer. This will be considered later.

Insulation in Static and Current Electricity

In electrostatic experiments it is found that whereas such materials as polystyrene and polylite can normally be relied upon to give good insulation, glass functions as a satisfactory insulator only under conditions of low atmospheric humidity, and wood is quite hopeless as an insulator. Yet in equipment used in current electricity it is not at all unusual to use wood for insulation purposes; for example, a meter bridge is normally mounted on a wooden board, and it never occurs to anyone to question its efficacy as an insulator. At first sight this situation may seem rather puzzling, yet in reality there is no mystery about it.

In electrostatics the conductors we normally use have rather small capacitances, and unless these are mounted on particularly good insulators their charges will leak away before we have had time to perform experiments with them. For example, in carrying out Faraday's ice-pail experiments we might use a sphere of, say, 5 cm diameter. This would have a capacitance of 2.78×10^{-12} F; let us say, in round figures, 3×10^{-12} F. Let this be charged to a potential of V volts, so that it carries a charge of $3 \times 10^{-12} V$ coulomb. We might perhaps tolerate a leakage of 0.1 percent of this per second, but hardly much more. This would mean a current of $3 \times 10^{-15} V$ ampere, and the corresponding resistance, V/A, would be $3.3 \times 10^{14} \, \Omega$. From this typical example we see that in electrostatics we cannot usually tolerate resistances of insulators of a smaller order than $10^{14} \, \Omega$.

In current electricity, on the other hand, the voltages with which we are concerned are generally relatively small and the currents relatively large. As a rather extreme case, in the sense of requiring quite unusually good insulation, let us consider a current of 1 μA flowing in a conductor in response to a PD maintained at 100 V. Let us suppose that the accuracy of the experiment is such that we could not tolerate a leakage current of more than 0.1

681

percent of this, that is, 10^{-9} A. This current would flow in response to a PD of 100 V if the insulation resistance were 10^{11} Ω. Thus, even in this case, in which the conditions we have imagined are exceptionally onerous, we could get by with an insulation resistance of only $\frac{1}{3000}$ of what would be needed for the electrostatic experiment we considered above. In such a case we would, in fact, use somewhat better insulating material than wood. But in the great majority of cases, where we use rather smaller voltages and/or much larger currents, the factor between our minimum requirements of insulation resistance in electrostatics and in current electricity is correspondingly larger, and accordingly we find relatively poor insulators, like wood, perfectly satisfactory.

Voltmeters

It would in principle be possible to measure potential differences on all occasions with an electrostatic voltmeter. However, these instruments are somewhat inconvenient for general use, and a better practical alternative is to make these measurements with a different type of instrument, which we shall now consider.

The instrument to which we ordinarily refer when we speak of a voltmeter is, in essence, simply an ohmic conductor that automatically registers the current passing through it. In practice this means a galvanometer or ammeter having a suitable resistance. It is immediately obvious from Ohm's law stated in the form

$$V = RI$$

that we merely have to multiply current values by the constant factor R to obtain the corresponding potential differences between the terminals of the instrument. If, therefore, we know the value of R, and, in addition, the current I in terms of the angular deflection of the coil, we have, in effect, the calibration of the instrument in volts.

In practice it is desirable that the use of this instrument should disturb as little as possible what it is designed to measure; the voltmeter should not "bleed" the circuit appreciably. This requirement is met by providing a high resistance to limit the current. Hence the description of a voltmeter so often given—"a high-resistance galvanometer."

The usual form of construction of a voltmeter is not that of a galvanometer with a high-resistance moving coil, but rather one with an ordinary coil of fine copper wire connected in series with a resistor of suitably high resistance. This arrangement avoids excessive heating of the galvanometer coil. The series resistor, in which most of the resistance of the instrument as a whole resides, is usually constructed of manganin, which has a low temperature coefficient of resistivity. In this way it is contrived that the overall temperature coefficient of the instrument is kept small. An idea of the distribution of resistances between the galvanometer coil and the series resistor is given by the example of an actual instrument, a millivoltmeter reading to 120 mV full scale, of total resistance 500 Ω, made up of 10 Ω in the meter winding and 490 Ω in the manganin series resistor. Less-sensitive instruments would have an even greater proportion of their total resistance in the series resistor.

This system provides the possibility of using alternative series resistors with a particular galvanometer, giving different ranges. An instrument incorporating this feature is shown diagrammatically in Fig. 26.4. Each series resistor has a terminal for external connection marked with the full-scale voltage reading given by the instrument when this terminal is used.

The alternative series resistors, or "range multipliers," as they are also sometimes called, may be either built into the instrument, in the manner indicated in Fig. 26.4, or connected to it externally.

It must be remembered that what actuates the needle is really the *current* flowing through the instrument, and so the full-scale current must be the same whatever the series resistor used. This current is equal to V/R, where V is the voltage corresponding to the full-scale reading when the total (meter plus range multiplier) resistance is R. Thus the ratio V/R, or R/V, is independent of the range multiplier used. The value of R/V, stated as the number of "ohms per volt," is commonly inscribed on voltmeters. This is very useful information; it enables one to calculate the total resistance of the instrument corresponding to any of the alternative ranges,

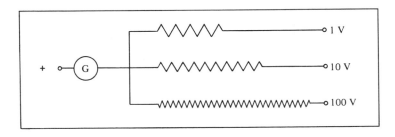

FIGURE 26.4 System of range multipliers in voltmeter.

simply by multiplying the number of ohms per volt by the full-scale voltage reading.

Ammeters

Ammeters, like voltmeters, may be designed for use over a number of alternative ranges. The same galvanometer is used in all cases, this taking only a small fraction of the total current, most of which is bypassed through a resistor connected in parallel with the galvanometer. A resistor connected in this way and serving this function is known as a **shunt.** By providing a selection of shunts, each having a different resistance, a corresponding selection of ranges is afforded.

The arrangement of a galvanometer G and shunt S is indicated diagrammatically in Fig. 26.5.

The potential V dropped across the system may be expressed either in terms of the current I_g flowing through the galvanometer and the galvanometer resistance R_g or in terms of the shunt current I_s and shunt resistance R_s. Thus

$$V = R_g I_g = R_s I_s,$$

whence

$$\frac{R_s}{R_g} = \frac{I_g}{I_s} = \frac{I_g}{I - I_g}.$$

For example, if the galvanometer is to take only $\frac{1}{100}$ of the total current, that is, if the sensitivity of the combination is to be only $\frac{1}{100}$ that of the galvanometer alone, we must have

$$\frac{R_s}{R_g} = \frac{1}{100 - 1} = \frac{1}{99}.$$

Owing to the parallel arrangement of galvanometer and shunt in an ammeter there is no possibility of making the scale insensitive

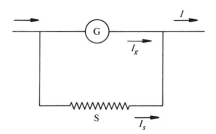

FIGURE 26.5 Galvanometer and shunt functioning as ammeter.

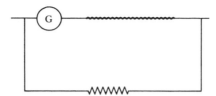

FIGURE 26.6 Ammeter consisting of galvanometer, series resistor, and shunt.

to variations of temperature if the simple system of Fig. 26.5 is adhered to. It might be thought that the scale could be kept independent of temperature variations by selecting a material for the shunt having approximately the same temperature coefficient of resistance as the galvanometer coil. This is not so, however; passage of current through the instrument will not in general raise the temperatures of the coil and shunt by the same amount, and unequal heating would have a significant effect on the ratio of R_g to R_s.

The only way to avoid this difficulty is to substitute for the simple galvanometer of Fig. 26.5 a complete voltmeter system, using for both the series resistor of the voltmeter component and the shunt an alloy such as manganin which has a small temperature coefficient of resistivity. We then have the system represented in Fig. 26.6. All good ammeters are of this form.

When dealing with this system the easiest way to calculate the shunt resistance required for any particular range is to treat the galvanometer and series resistor simply as a voltmeter, neglecting the current that passes through it. In general this procedure is justified because of the very large ratio of resistances in the galvanometer and shunt arms required in this system to give a reasonably low overall resistance.

Let us, for example, consider the millivoltmeter previously referred to, having a full-scale reading of 120 mV and a resistance of 500 Ω, converted to an ammeter reading to (1) 1.2 A, (2) 2.4 A, and (3) 24 A, by providing it with the appropriate shunts. We shall now calculate the resistances the shunts should have to give these ranges.

As a preliminary, let us work out the value of I_g, to see whether this really is negligible. Applying Ohm's law, we find that when the instrument is reading full scale this current is $120 \times 10^{-3}/500$ A, that is, 240 μA. This is less than could be detected as a movement of

685

the needle even on the lowest range, 1.2 A. Hence we need not concern ourselves with it.

The current I may accordingly be identified with I_s, and for R_s, therefore, we have

$$R_s = \frac{V}{I},$$

where V is the voltage dropped across the instrument when reading full scale, 0.12 V. Thus for the three ranges we have

$$R_s = \frac{0.12}{1.2, \ 2.4, \ \text{or} \ 24} \ \Omega,$$

that is,

$$R_s = 0.1 \ \Omega, \ 0.05 \ \Omega, \ \text{or} \ 0.005 \ \Omega.$$

It should be noted that, with I_g negligible, the particular value of the resistance of the voltmeter is of no consequence; it does not enter into the calculation of R_s. In other words, any particular shunt will give the same current scale on *all* voltmeters of the same rating, regardless of their resistances.

Ammeter–Voltmeter Combinations

It frequently happens that we wish to measure both the current passing through a device (for example, a resistive "load") and the PD across it. If we use an ammeter and a voltmeter for this purpose, there are two alternative ways of connecting these instruments in circuit—those shown in Fig. 26.7. In part (a) the PD across the load

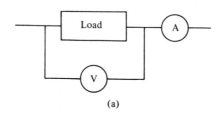

(a)

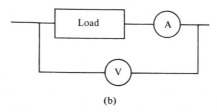

(b)

FIGURE 26.7 Alternative ammeter-voltmeter combinations.

is measured correctly, but the ammeter measures not merely the current through the load but also that taken by the voltmeter. On the other hand, in (b) the ammeter reads the correct current, but the PD measured by the voltmeter includes that dropped across the ammeter. In any particular case the error incurred with one system of connection will in general be less important than with the other; for example, the system (a) may give an ammeter reading that differs, relatively, by a smaller amount from the current through the load than does the voltmeter reading in (b) from the PD across the load; indeed, the error will, in many cases, be negligibly small. Just how important the error is with either system of connection may always be calculated, and, when necessary, it may be allowed for. However, not only are such corrections troublesome, but ammeters and voltmeters do not, in any case, lend themselves to highly accurate determinations of currents and voltages. The most accurate measurements are made with an instrument known as a **potentiometer,** which we shall now consider. The potentiometer is, incidentally, the standard instrument used for calibrating ammeters and voltmeters.

Potentiometer

This is an accurate voltage-measuring device which avoids entirely any bleeding of the circuit to which it is applied. It incorporates a standard cell, a system of resistors, and a sensitive galvanometer. Its high accuracy is due to (1) the reproducibility and reliability of the type of standard cell used, and (2) the fact that it is possible to construct a set of resistors the ratios of whose resistances can be found, and suitably adjusted, to a high degree of precision— for example, by using a Wheatstone bridge. The accuracy of the potentiometer is in no way affected by any uncertainty with regard to the calibration of the galvanometer, since, as in the comparison of resistances by the Wheatstone-bridge method, the only function of the galvanometer is to register the fact that no current is flowing in the galvanometer arm—the potentiometer method is a "null" method.

Although, as we shall see presently, the potentiometer can be put to a variety of uses, it is primarily an instrument for comparing potential differences. Let us, therefore, in the first instance, consider it from this point of view.

The principle of the potentiometer is shown in Fig. 26.8, which represents the case in which the potential differences to be compared are those across the terminals of two cells when "on open circuit"; that is, when they are not delivering any current. As we

687

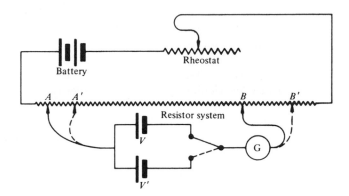

FIGURE 26.8 Potentiometer circuit diagram.

see from the figure, the only equipment needed for the comparison, apart from the two cells, is a battery, a rheostat, a resistor, a two-way switch, and a galvanometer. The battery would in practice generally be a lead storage battery. The rheostat is for convenience of adjustment of the current, but is not absolutely essential. The resistor must be such that adjustable portions of it can be tapped off, these having resistances that are known, although not necessarily in ohms; the units can be quite arbitrary. The two-way switch is for convenience of switching into the galvanometer arm of the circuit either one cell, having an open-circuit potential difference V across its terminals, or the other, the terminals of which have an open-circuit PD of V'.

The procedure is as follows. First, with the two-way switch connecting in circuit the upper of the two cells shown, a portion AB of the resistor, having a resistance R_{AB}, is tapped off such that no current flows in the galvanometer. Then, with the two-way switch connected to the lower cell, a portion $A'B'$ of the resistor, of resisrance $R_{A'B'}$, is tapped off such that the galvanometer current is again zero. Then, as we shall now show,

$$\frac{V'}{V} = \frac{R_{A'B'}}{R_{AB}}. \tag{26.2}$$

Let us consider the case of the former of these connections. With a current I flowing in the resistor, we have, by Ohm's law,

$$V_A - V_B = R_{AB}I. \tag{26.3}$$

The PD between the right-hand terminal of the cell and the point B must be $R_g I_g$, where R_g is the resistance of the galvanometer plus connections and I_g is the current flowing through the galvanometer. But the adjustment is for zero I_g. Hence B must be at the same potential as the right-hand terminal of the cell. Similarly, with no current flowing in this branch, A must be at the same potential as the left-hand terminal. For this condition, $V_A - V_B$ must be the same as V, the PD across the cell on open circuit, and so, substituting for $V_A - V_B$ in Eq. (26.3), we have

$$V = R_{AB}I.$$

Similarly,

$$V' = R_{A'B'}I.$$

Hence, provided that I is the same on both occasions,

$$\frac{V'}{V} = \frac{R_{A'B'}}{R_{AB}}. \tag{26.3}$$

The condition just referred to, that of constancy of the current I flowing through the resistor, will, in fact, be satisfied in the case of the steady operation of a battery of good quality delivering a current that is not too large. A storage battery connected in a circuit of constant resistance operates remarkably steadily over long periods of time. And the resistance of the circuit through which current is sent by the battery is, in fact, constant; for connection of part of the resistor to the galvanometer arm of the circuit can obviously have no effect when no current flows in this arm. That I is indeed constant can always be verified by testing for constancy of the resistances R_{AB} and $R_{A'B'}$ needed to give zero current in the galvanometer.

If one of the two cells is a standard cell, of known open-circuit PD, then obviously the PD of the other may also be found. Thus, if V refers to the standard cell, we have, from Eq. (26.3),

$$V' = \frac{R_{A'B'}}{R_{AB}} V.$$

WESTON STANDARD CELL The standard cell now universally used in potentiometer circuits is the Weston cell, the open-circuit PD between whose terminals is remarkably reproducible, and remains constant over an almost indefinite period of time as long as no appreciable current is ever allowed to pass through it. The PD also has an exceptionally low rate of variation with temperature.

This cell is shown in Fig. 26.9. Through the bottoms of two sides of a glass container of H form are sealed platinum wires; these constitute the terminals for connection to the circuit. On the positive side, the internal electrode, which makes contact with the

689

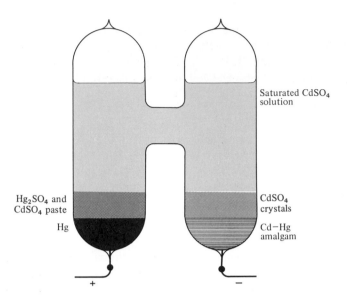

Saturated CdSO₄ solution

Hg₂SO₄ and CdSO₄ paste

Hg

CdSO₄ crystals

Cd−Hg amalgam

+   −

FIGURE 26.9 Weston standard cell.

corresponding platinum wire, consists of liquid mercury on top of which floats a paste consisting of a mixture of mercurous sulfate, Hg_2SO_4, which is an almost insoluble compound, and cadmium sulfate. On the negative side the electrode is an amalgam of cadmium in mercury of specified strength, 10 to 15 percent cadmium, upon which rests a layer of crystals of cadmium sulfate. A saturated solution of cadmium sulfate is then added, until it fills the tubes above the level of the cross piece, after which the tubes are sealed at the top. The function of the solid cadmium sulfate on both sides is to keep the solution adjacent to the electrodes saturated at all temperatures. The mercurous sulfate on the positive side keeps the mercury on which it floats free of cadmium, by operation of the chemical reaction

$$Cd + Hg_2SO_4 \rightarrow CdSO_4 + 2Hg.$$

The open-circuit potential difference between the terminals of a Weston cell made according to this formula has been found, by an accurate absolute method of determining potential differences, to be 1.01830 V at 20°C. The temperature variation of this PD has been determined with corresponding accuracy. Denoting the 20°C PD by V_{20}, the PD, V_θ, at any other temperature θ°C may be represented by the formula

$$V_\theta = V_{20} - 4.06 \times 10^{-5}(\theta - 20) - 9.5 \times 10^{-7}(\theta - 20)^2.$$

PRACTICAL FORMS OF POTENTIOMETER In a primitive form of the potentiometer the resistor component could be a meter bridge. The points A and A' could then conveniently always be one end of the wire—the zero end—while B and B' are the corresponding other points on the wire for which the galvanometer current is found to be zero. Then, resistance being taken as proportional to length of wire, we should have

$$\frac{V'}{V} = \frac{l_{A'B'}}{l_{AB}}.$$

However, as in the case of the Wheatstone bridge, a slide wire does not lend itself to work of high accuracy, and to use it—at least without any auxiliary equipment—in a potentiometer circuit would be to sacrifice unnecessarily most of the accuracy made possible by this system.

Despite this objection to the use of a simple slide wire, it is nevertheless possible to achieve reasonable accuracy with a resistor component consisting of a number of fixed resistors in series with a slide wire. The use of the fixed resistors in conjunction with the slide wire considerably "dilutes" the inaccuracy due to the latter. This arrangement, which is known after its originator as the **Crompton potentiometer,** is shown diagrammatically in Fig. 26.10. The slide wire is 1 m in length, and each of a large number of resistors connected in series with one another and with the slide wire is adjusted to have the same resistance as this wire. Then, for each of the two potential differences that are to be compared, the number of fixed resistors plus the length of slide wire is found for which there is no current in the galvanometer.

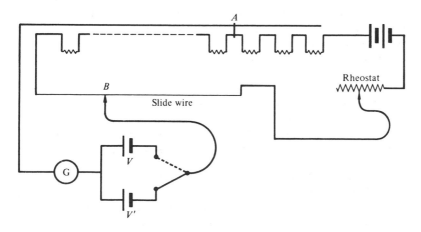

FIGURE 26.10 Crompton potentiometer.

If one of the two potential differences is that of a Weston standard cell on open circuit, the numerical value of the other PD may be found most conveniently by the following procedure, which makes the potentiometer "direct reading," avoiding the necessity for tedious calculation. Let us suppose the temperature to be 20°C, so that the open-circuit PD across the terminals of the standard cell is 1.01830 V. Then the first operation is to select the points A and B so as to include in the galvanometer arm of the circuit 10 fixed resistors and 18.30 cm of wire. The current I is then adjusted by manipulation of the rheostat so that with these preselected points A and B there is no current in the galvanometer when the two-way switch is connected to the standard cell. This is made possible by incorporating a suitable fine adjustment (not shown in the diagram) in the rheostat. It is easily seen that with this particular current I each fixed resistor drops 0.1 V and each centimeter of wire 0.001 V. The unknown PD is now switched into the circuit and, leaving the rheostat undisturbed, the new number of fixed resistors and length of wire are found that give a zero galvanometer reading. If this number of fixed resistors is n, and the length of wire is l cm, the value of the unknown PD must be $0.1n + 0.001l$ V. Since each millimeter of wire represents 10^{-4} V, and one might hope to find the point B to the nearest 0.1 mm, one should, ideally, be able to achieve, or at least approach, an accuracy to within 10^{-5} V. In actual practice the attainable accuracy would be less than this, largely because of the limitations deriving from the use of the slide wire.

From this it is only one step to the design of a potentiometer in which there is no slide wire at all, that is, in which only a resistance box containing fixed resistors is used. When this is done, interpolation between just-too-high and just-too-low values of resistances on the basis of galvanometer deflections becomes necessary, as with the dial-box version of Wheatstone's bridge. Although this is a little troublesome, it is a price well worth paying for the much higher accuracy attainable.

OTHER APPLICATIONS OF THE POTENTIOMETER   A potentiometer may be used not only for comparing potential differences but also for two other purposes: (1) for comparing resistances, and (2) for the precision measurement of a current.

To compare the resistances R_1 and R_2 of two resistors, a steady current may be passed through these, connected in series, and the corresponding potential differences V_1 and V_2 developed across them compared with a potentiometer, in the same way as the open-circuit potential differences across two cells. In the

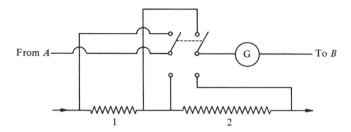

FIGURE 26.11 Use of double-throw switch in potentiometer circuit for comparison of resistances.

present case, however, a different switching arrangement would have to be employed; for example, a double-throw switch could be used, in the manner indicated in Fig. 26.11. With a given current, potential differences are proportional to resistances, so V_1/V_2 must be the same as R_1/R_2.

An outstanding advantage of the potentiometer method over the Wheatstone-bridge method of comparing resistances is that in the former the difficulty of the resistances of the leads is entirely avoided. That this is so is obvious from Fig. 26.11.

Currents may be measured much more accurately with a potentiometer than by any other method; indeed, the standard method for calibrating ammeters is that based on the use of a potentiometer. This depends on the availability of a standard resistor, of accurately known resistance R. The current I in question is passed through this and the ammeter to be calibrated, in series. Then the PD developed across the resistor is RI. This is measured with the potentiometer, and, R being known, I may be calculated.

Potential Divider

In radio jargon this is often referred to, erroneously, as a potentiometer. It consists simply of a resistor carrying a current, a portion of which may be tapped off in the manner indicated in Fig. 26.12. If the PD across the whole resistor, of resistance R_0, is V_0, then the corresponding potential difference V across a portion of this, of resistance R, must be $(R/R_0)V_0$.

This device, constructed in the form of a resistance box, is often referred to as a "volt box." This is provided with a number of alternative tappings, and can be used for the purpose of measuring a relatively large PD (V_0) maintained across its ends. A convenient

693

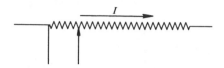

FIGURE 26.12   Potential divider.

fraction R/R_0 of this is measured with a potentiometer, and hence V_0 may be calculated.

Another application of this principle is the provision of the equivalent of a cell of very low voltage. Thus, a lead storage cell could be connected to the ends of a resistor of suitably high resistance and a small portion of this tapped off and connected to separate terminals. These would then provide a PD or supply a current in the same sort of way as would a cell of correspondingly low voltage.

Problems

26.1   A Wheatstone bridge is set up with three standard 100-Ω resistors for the known arms and a galvanometer of resistance R_g. The resistances of a number of resistors, all of which are close to 100 Ω, are calculated from the out-of-balance galvanometer currents observed when these are used, in turn, for the fourth arm. Taking the value of an unknown resistance as $100 + x$ Ω, where $x \ll 100$, and neglecting products of small quantities, use Kirchhoff's laws to obtain an expression for x in terms of the galvanometer current I_g, R_g, and the battery emf E. For what value of R_g is the power dissipated in the galvanometer a maximum?

26.2   Using Kirchhoff's laws, obtain an expression for the current I_g flowing through the galvanometer in a Wheatstone-bridge system in terms of the applied voltage V, the resistances of the four arms of the bridge (labeled as in the text), and the resistance R_g of the galvanometer, in the general case where the bridge is not balanced.

26.3   A voltmeter of resistance 26,000 Ω is employed to measure a large unknown resistance. When the instrument is connected across a dc supply it reads 140 V. When it is connected in series with the unknown resistance across the same supply the voltmeter reads 40 V. What is the magnitude of the unknown resistance?

26.4   An ammeter of negligible resistance is connected in series with a parallel combination consisting of a 2000-Ω voltmeter and a resistor of unknown resistance. If the voltmeter reads 120 V and the ammeter reads 0.60 A, what is the unknown resistance?

26.5   A millivoltmeter has a range of 100 mV, and its resistance is 480 Ω. Show how to convert it (a) to a voltmeter reading to 50 V, and (b) to an ammeter reading to 1.5 A.

26.6   A milliammeter reads up to 1 mA full-scale deflection and has a resistance of 50 Ω. How would you convert this (a) to a voltmeter reading to 10 V full scale, and (b) to an ammeter reading to 5 A full scale?

26.7 Is it permissible to use an ammeter of negligible resistance and a voltmeter carrying negligible current (instruments of usual type) to prove Ohm's law? Give reasons for your answer.

26.8 An ammeter having a resistance of 0.05 Ω and a voltmeter of resistance 10,000 Ω are used to measure the resistance of a conductor with the arrangements (a) and (b) shown in Fig. 26.7. The quotients of the voltmeter and ammeter readings are close to 4 Ω in both cases. What correction should be applied to the quotient actually obtained in each case to give the true resistance of the conductor?

26.9 A potentiometer has a resistor system of total resistance 400 Ω from which any resistance up to this value can be tapped off (see Fig. 26.8), and the voltage V of the standard cell is 1.0187 V. With the standard cell in circuit the current registered by the galvanometer is zero when the resistance tapped off is 83.13 Ω. When, instead, another cell, of emf V', is switched into the circuit, the resistance that has to be tapped off to make the galvanometer read zero is 119.20 Ω. (a) What is V'? (b) What is the current in the resistor system? (c) What is the maximum PD that could be measured with this potentiometer? (d) Without altering the setting of the rheostat, what single component of the potentiometer system would have to be changed, and in what way, to increase the maximum measurable PD? (e) Again without altering the rheostat setting, what single component would have to be changed, and how, to make it possible to measure small voltages with greater accuracy?

26.10 The PD across the terminals A and B of a cell as measured with a potentiometer is found to be V_1 on open circuit (switch S in the figure open) and V_2 when the cell is sending current through a resistor of resistance R (switch closed). What are (a) the emf of the cell, and (b) its internal resistance r?

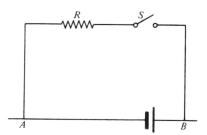

Figure for problem 26.10.

26.11 A potentiometer of 10,000 Ω total resistance is supplied with current by a 2-V storage cell. It is used to measure the PD (around 1.5 V) between two points in an electric circuit with a galvanometer as zero indicator which gives a detectable deflection for 1 μA of current. If the standard of PD (standard cell) has a voltage of 1.0184 $\pm$ 0.001 V, what is the proportional uncertainty in the result (apart from any uncertainty in the resistances of the potentiometer resistors)? Would the uncertainty be reduced if the galvanometer were replaced by an electrometer whose minimum detectable deflection is produced by 10 μV?

ELECTROCHEMISTRY

Electrochemistry may be regarded broadly as the "wet physics" counterpart of the study of discharges in gases. There are, however, certain important differences. Whereas in a gas discharge, ions are formed by collisions or by photoionization, in a solution of an electrolyte they are formed spontaneously, in accordance with certain statistical laws. And whereas in gas discharges we are concerned almost exclusively with positive ions and electrons, in electrochemistry we have to deal with positive and negative ions derived from molecules which, in solution, have a high probability of dissociation. The counterpart of the liberation of electrons from the cathode in a gas-discharge tube is the solution pressure of the metallic (positive) ions in an electrolytic or voltaic cell. This solution pressure is analogous to the vapor pressure of a liquid or that of electrons in thermionic emission; these are all governed by thermodynamical laws. There is a further link with thermodynamics in the theory of the reversible voltaic cell.

Electrolytic Processes

Electrolysis was first explained by the Swedish chemist Svante Arrhenius in 1887, by invoking the concept of ionic dissociation of the molecules. On this basis the broad quantitative features of the phenomenon, as set out in Faraday's laws of electrolysis, may be accounted for. We shall now consider a number of implications of Arrhenius's theory.

Let us, in the first instance, consider a particular electrolytic cell, one consisting of a copper sulfate solution in which are immersed two electrodes of copper. At the cathode, Cu^{2+} ions arriving there each combine with two of the free conduction electrons in the metal to form neutral copper atoms, which attach themselves to the electrode:

$$Cu^{2+} + 2e^- \rightarrow Cu.$$

On the other hand, at the surface of the anode the reverse of this takes place; copper atoms detach themselves from the metal, leaving two electrons behind and becoming free Cu^{2+} ions as they do so:

$$Cu \rightarrow Cu^{2+} + 2e^-.$$

But is it likely that the reaction should proceed only in one direction at the cathode and only in the other at the anode? Is there any conceivable reason why this should be so? Would it not be much more of the pattern of physical behavior that both reactions should occur simultaneously at each electrode, the applied field merely giving a quantitative preponderance of one over the other, according to its direction? If this were the case we should have a situation analogous to that existing in an evacuated tube provided with two hot metal electrodes both of which are emitting electrons thermionically and between which an electron current may be made to flow in either direction according to the field applied. This picture seems eminently reasonable, so let us adopt it as a working hypothesis. We imagine the metal to have a "solution pressure" of metal ions,* just as a liquid has a vapor pressure of neutral molecules or a thermionically emitting metal in a vacuum has a vapor pressure of electrons.

On this basis it should be possible to transfer copper from an anode to a cathode in the absence of any copper salt in solution, both plates simply being immersed in pure water. In practice, however, any such action is quantitatively many orders of magnitude less than that which takes place through a solution of copper sulfate. Why is this so?

The reason is not difficult to find. The charge-to-mass ratio for ions is very much smaller than for electrons, and for this reason alone their rate of movement under any given field conditions must be relatively very sluggish. Added to this circumstance is the fact that the ions have to travel through a resistive medium, a liquid, not through a vacuum. Hence, to give even an almost immeasurably small current of Cu^{2+} ions the population density of these ions in the water would have to be quite substantial, large enough to give a formidable positive space charge. In order that, notwithstanding this, the field should be ion-accelerating all the way across, from anode to cathode, it would be necessary to apply a fantastically large potential difference between these electrodes.

The essential role of the dissolved salt in an electrolytic cell is to render possible a high concentration of ions without the exis-

---

* The term "solution pressure" is introduced here without precise definition, merely to convey a qualitative idea. It will be defined quantitatively later in this chapter.

tence of an associated space charge, the mixture of positive and negative ions being electrically neutral.

If, with a PD applied between copper electrodes immersed in a solution of copper sulfate, the SO_4^{2-} ions remained in fixed positions, merely playing the passive role of neutralizing the space charge of the Cu^{2+} ions, the latter would travel in a steady stream from anode to cathode in response to the field, and the observed transfer of copper from one electrode to the other would be explained. Actually the situation is not so simple; the SO_4^{2-} ions move in the direction from cathode to anode concurrently with the movement of Cu^{2+} ions from anode to cathode.

If *all* the $CuSO_4$ molecules were dissociated into ions, and these moved, independently of one another, in response to the general field, the migration of the SO_4^{2-} ions toward the anode would quickly bring the electrolysis to a halt. There would be a piling up of these ions near the anode, but there would be no development of a negative space charge here, because a corresponding number of Cu^{2+} ions would be drawn out from the copper electrode, resulting merely in an increased concentration of the electrically neutral ion mixture. However, in the vicinity of the cathode there would be a depletion of the SO_4^{2-}-ion population, and after a time none of these ions would remain to neutralize the Cu^{2+}-ion space charge. In this region the situation would then be the same as that obtaining in a cell consisting of copper electrodes immersed in pure water, and for the reasons already discussed the electrolysis would be brought virtually to a halt.

Actually, current does continue to pass, and from this it must be inferred that SO_4^{2-} ions are somehow able to get back from the vicinity of the anode to that of the cathode; there must be a circulation of these ions. Obviously they cannot return as *isolated* ions against the field; the only way they can return is in association with positive ions, that is, as components of undissociated molecules, which are electrically neutral. If the dissociation of the dissolved salt is not complete, this will be provided for by the processes producing the dynamic equilibrium between dissociated and undissociated ion pairs. Any particular SO_4^{2-} ion will alternately form a temporary partnership with a Cu^{2+} ion that it encounters, forming a $CuSO_4$ molecule, and become free again. During its periods of freedom it must drift toward the anode in response to the field, but when it is attached to a Cu^{2+} ion it may, with its partner, travel in any direction. When, in consequence of the bias given to the directions of travel of the SO_4^{2-} ions during their periods of freedom, a certain concentration gradient of dissolved salt has been built

699

up, the traffic of SO_4^{2-} ions toward the anode will be balanced by a diffusion of the momentarily undissociated molecules toward the region where the solution is weakest, that is, toward the cathode. In this way an adequate concentration of dissolved salt is maintained near the cathode. Other processes that may contribute to this maintenance of concentration are convection and mechanical stirring.

Reason for Dissociation

In the dissociation of a salt, acid, or base into ions an activation energy is involved, and a state of dynamic equilibrium is attained between the number of undissociated molecules and the number of ions. The proportion of molecules that are dissociated is such as gives a balance between the rate of dissociation of neutral molecules and the rate of recombination of ions. Where the balance lies is determined by the activation energy involved and the temperature, according to statistical laws.* The statistical theory shows that not only does the proportion of molecules that are dissociated increase with increasing temperature, but also that at any given temperature anything which reduces the activation energy must increase this proportion.

The activation energy is the work that must be done to separate the charged constituents of a molecule against the Coulomb force of attraction between them. If we think of water as a continuous medium, we must expect the work of ionization to be much less in water than in a vacuum. This is because water has the very high dielectric constant of 80, and the Coulomb force at each stage of the separation must be reduced by this factor. Actually, of course, water is not a continuous medium; it consists of a complex of positive and negative particles separated from one another by distances of the same order of magnitude as those which separate the particles constituting the molecules of the salt, acid, or base. Thus the macroscopic dielectric constant cannot be used quantitatively to calculate the work of ionization. Nevertheless, the argument based on the dielectric constant does seem to have some sort of general validity; it is a fact that a solution of a given electrolyte of a given strength has a higher conductivity where the solvent has a high dielectric constant than where the dielectric constant is small. There are certain ionic substances which, besides dissolving in water, will also dissolve in methyl alcohol ($\kappa = 35$) and in liquid

---

* See Chapter 7 of Volume III.

ammonia ($\kappa = 22$), and some also in benzene ($\kappa = 2$). These exhibit conductivities diminishing in that order when these liquids are used as solvents, with the conductivity in benzene vanishing practically completely.

Electrolysis of Water

Water is an extremely poor conductor of electricity, and to electrolyze this compound at an appreciable rate it is necessary to "acidulate" it, for example by the addition of a little sulfuric acid, rendering it conducting. It now becomes dilute sulfuric acid, and the question arises whether it is still correct to refer to the electrolysis of this material as the electrolysis of *water;* for, after all, the ions to whose migration the release of hydrogen and oxygen at the electrodes is due are derived from H_2SO_4 molecules, not from H_2O. Despite full appreciation of this situation, the process is, in fact, always called the electrolysis of water, presumably because the products released at the electrodes are the chemical constituents of water.

It was formerly supposed that the ions derived from an H_2SO_4 molecule were two ionized atoms of hydrogen, H^+, and one $SO_4{}^{2-}$ ion. However, it is now believed that H^+ ions are not stable in contact with water molecules but attach themselves to these to form H_3O^+. Such ions are known as "hydronium" ions.

Let us suppose the electrodes to be platinum. The $SO_4{}^{2-}$ ions, arriving at the anode, take part in a chemical and electrical reaction with the water and the anode whose final outcome is, energetically and chemically, the same as if the $SO_4{}^{2-}$ ions gave up their charge to the anode, after which the SO_4 radicals replaced oxygen from H_2O, producing H_2SO_4 and liberating oxygen. The oxygen is deposited on the electrode and finally comes off in the form of bubbles. The ions arriving at the cathode are hydronium, and these combine with electrons to form water and neutral hydrogen atoms, which then unite in pairs to form molecular hydrogen:

$$2H_3O^+ + 2e^- \rightarrow 2H_2O + H_2.$$

The hydrogen is finally liberated from the cathode as bubbles of gas.

Pt–H₂SO₄–Pt Cell as a Generator

Let us now consider another aspect of the electrolysis of water in a Pt–H_2SO_4–Pt cell. If, before any gas has been deposited

701

on the electrodes, a small PD is applied between these, a correspondingly small current is observed to flow, but this soon stops. An increase in the PD restores the flow, but this again dies away to zero unless the PD exceeds about 1.7 V, in which case it is possible to maintain a steady current. Let the limit of PD below which no steady current can be maintained (about 1.7 V) be V_0, and let the somewhat greater PD actually applied be V. Then V_0 must be the algebraic sum of two steps in potential located at the regions of contact of the electrolyte with the electrodes. Only the remainder, $V - V_0$, is dropped within the electrolyte.

The case we are now considering differs from that of the Cu–CuSO$_4$–Cu cell in that here we have a net chemical change proceeding at a steady rate. Whereas in the electrolysis of copper sulfate with copper electrodes there is no change in the total amount of copper, of copper sulfate, or of water, in the present case water is steadily being separated into its constituents, hydrogen and oxygen. This cannot be done without the expenditure of power. It is the quotient of this power and the current strength that constitutes the greater part of the minimum PD, V_0, required for electrolysis.

Let us now calculate this quotient. The heat of formation of liquid water from hydrogen and oxygen at 15°C equals 69,000 cal/gram-molecule, or 2.89×10^5 J/gram-molecule. The quantity of electricity required to electrolyze 1 gram-molecule of water is $2 \times 96,488$, that is, 1.93×10^5 C. The former quantity divided by the latter, $(2.89 \times 10^5 \text{ J}) \div (1.93 \times 10^5 \text{ C})$, is 1.5 V.

The difference between this and the observed 1.7 V, although small, is nevertheless real. The quantity calculated from the energy conversion is known as the **reversible decomposition voltage.** The observed excess over and above this (0.2 V in the present case) is called the **overvoltage.** The excess is believed to be associated with the energy required for the formation and growth of bubbles. This was not taken into account in the above calculation.

If, at a stage where bubbling of gas from the surfaces of the electrodes has been established, these are disconnected from the supply and are joined to the terminals of a voltmeter, this will at first register a PD of about 1.5 V, but presently, as more and more electricity discharges through the voltmeter, this PD falls and finally becomes reduced to zero. This is accompanied by the disappearance of gas from the electrodes. Thus we have here a primitive form of reversible voltaic cell, and it is from the behavior of this that the reversible decomposition voltage derives its name. The reversible decomposition voltage is really the emf of the cell, and accordingly we shall henceforward use this term and denote it by the symbol E. It should be noted that the electrodes functioning as anode and cathode in electrolysis are still the anode and cathode, respectively, when the device is operating in reverse, as a generator, supplying current to an external circuit.

The energy required to form bubbles against surface-tension forces is not available in reverse. As the bubbles collapse this energy is simply converted to heat.

Just how the presence of oxygen and hydrogen on the electrodes produces an emf may be accounted for as follows. At the anode there is a certain tendency for the oxygen on the electrode, in association with water and with electrons derived from the electrode, to go into solution as hydroxyl ions, OH^-, according to the reaction

$$O_2 + 2H_2O + 4e^- \rightarrow 4OH^-,$$

followed by a further reaction of the hydroxyl ions with hydronium to form water:

$$OH^- + H_3O^+ \rightarrow 2H_2O.$$

This removal of H_3O^+ ions leaves an excess of SO_4^{2-} ions near the anode. The final result corresponds to a reversal of the events occurring at this electrode during electrolysis—the disappearance of the SO_4^{2-} ions arriving there, and of water, the formation of oxygen on the surface of the electrode, and the absorption of electrons at this surface.

The oxygen reacting with water and electrons to form hydroxyl ions presumably cannot come directly from the bubbles adhering to the anode, because the gas constituting these is electrically insulating. Almost certainly it must be derived from a monomolecular or monatomic layer of oxygen on the electrode surface between the bubbles. Such a layer might be replenished from the bubbles as it is used up in the reaction, by a two-dimensional flow of adsorbed oxygen atoms or molecules along the surface.

We may think of the reaction in which oxygen disappears from the surface of the anode as brought about by a solution pressure of the oxygen, although this term is not to be taken too literally as implying that the oxygen lost from the anode is to be found *as such* in the solution. The counterpart of this picture applying to the cathode would be a solution pressure of the hydrogen, tending to produce the reaction

$$H_2 + 2H_2O \rightarrow 2H_3O^+ + 2e^-.$$

703

The electrons liberated in this reaction are absorbed by the cathode. And, as we have seen, there is a corresponding extraction of electrons from the anode which are involved in the formation of hydroxyl ions from oxygen and water at this electrode. In this way electrons are transferred, through the solution, from the anode to the cathode. In the external circuit these electrons travel back, from the cathode to the anode.

The potential step at the anode is due to a local space charge of OH^- ions immediately adjacent to its surface in combination with the induced positive charge (corresponding to the electrons lost) on the surface of this electrode. If we idealize the negative space charge as a sheet of charge just outside the anode surface we have the equivalent of a charged capacitor, with the potential step located between the charges. Similarly, the potential step at the cathode is due to the combination of the local positive space charge of H_3O^+ ions close to its surface and the corresponding negative charge on the electrode. When the cell is functioning as a generator and delivering current to an external circuit these two potential steps upward in the direction of the current are balanced by the potential falls in the external circuit and through the electrolyte, the latter being due to the internal resistance of the cell. This potential fall within the cell from cathode to anode causes the ions to move in directions opposite to those in which they travel when the cell is being used for the electrolysis of water, the fall then being from anode to cathode.

Primary and Secondary Cells

Voltaic cells may be divided broadly into two categories, primary and secondary cells. The division is not very clear cut, as there are borderline cases. By a primary cell is meant one whose materials are used up finally during the cell's useful lifetime as a generator of current, in the sense that it is not practicable to reverse the chemical changes accompanying the delivery of current by sending current through the other way. A secondary cell, on the other hand, is one which, after being discharged, may conveniently be "recharged" by passing current through it in the opposite direction, the recharging current reversing the chemical changes that accompanied the discharge and restoring the status quo. The cells of car batteries belong to this category.

Actually the $Pt–H_4SO_4–Pt$ cell with which we introduced the subject of voltaic cells is in principle a secondary cell, although it is not a practical one. Also the Daniell cell, which will be considered later, is usually discussed theoretically as a reversible cell, although, as we shall see, it is in fact reversible in principle only if the quantities of electricity passed in either direction are infinitesimal. Even then the cell may be regarded as reversible only to the extent that inter-diffusion of the two electrolytes through the porous pot separating them is negligible. (Actually, such an idealization would be permissible in a thermodynamical discussion.) For these reasons the Daniell cell is used only as a primary cell. Practically the only secondary cells used at the present time are the lead storage cell, the Edison nickel–iron cell, and the cadmium cell. Of these, the lead storage cell is by far the most important.

Zn–H₂SO₄–Cu CELL This cell, which consists simply of electrodes of zinc and copper immersed in dilute sulfuric acid, is one of the earlier forms of primary cell to find at least some limited use in practice. Although this cell is now obsolete, it will be useful to discuss it because of the insight such a discussion will give us into the action of voltaic cells generally.

If we determined the PD between the electrodes of this cell on open circuit with a potentiometer, we should find the copper to be at a higher potential than the zinc; the zinc is the cathode and the copper the anode.

It is useful to express this result in terms of the solution pressures of the two electrode materials. According to the picture we advanced in developing the concept of solution pressure, the metal must finally be left at a potential that is negative relative to the electrolyte. Although we have no means of measuring separately the potential of either of the electrodes relative to the electrolyte, nevertheless, regardless of what these quantities may be, the potentiometer result does tell us, algebraically, that the amount by which the zinc is negative to the electrolyte exceeds that by which the copper is negative to it. If we take the negative potential developed relative to an electrolyte as a quantitative measure of what we have called the solution pressure of the electrode, we may express the potentiometer result by saying that the solution pressure of zinc is greater than that of copper. Similarly, if we deposited a quantity of hydrogen on a platinum or copper electrode by electrolysis, and then connected this and a zinc electrode immersed in the same electrolyte to a potentiometer circuit, we should find that the zinc is negative to the hydrogen-covered electrode. Accordingly, we must infer that the solution pressure of hydrogen is less than that of zinc.

With this background, let us now attempt to develop a detailed picture of the processes that occur in the Zn–H₂SO₄–Cu cell.

705

We shall first deal with the situation where the cell is on open circuit. At the surface of the zinc, Zn^{2+} ions are released into the electrolyte, leaving electrons behind. These ions set up a positive space charge adjacent to the electrode which ceases to build up further when the retarding field (retarding the emission of further ions) reaches a certain value. This gives a potential step of a definite magnitude characteristic of the zinc, making the zinc negative to the electrolyte. Correspondingly, a much weaker positive space charge of Cu^{2+} ions is formed adjacent to the anode, and so a smaller potential step, with the copper also negative to the solution, develops here. The difference between the two potential steps, equal to the PD between the electrodes, is the electromotive force of the cell, E.

If the copper and zinc are now connected externally through a resistor, current will flow through this, from the copper to the zinc, while within the cell this same current flows from zinc to copper. Hence, if the magnitude of the current is I and the internal resistance of the cell is R_0, the PD between the electrodes will be reduced by the amount $R_0 I$, while the potential drop in the external circuit, of resistance R, is RI. This situation is indicated in Fig. 27.1.

The field set up in the electrolyte drives both the SO_4^{2-} and the relatively few OH^- ions derived from the water toward the zinc cathode, while Zn^{2+} ions derived from the outer regions of the space charge surrounding the zinc electrode and H_3O^+ ions travel toward the copper anode. On arrival at the region of the space charge at the cathode, the SO_4^{2-} ions find Zn^{2+} partners in this space charge with which to form either $ZnSO_4$ molecules or dissociated ion pairs, and the Zn^{2+} ions so neutralized are immediately

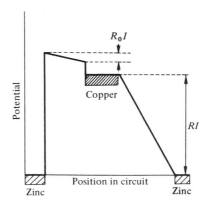

FIGURE 27.1 Variation of potential with position within $Zn-H_2SO_4-Cu$ cell.

replaced from the zinc electrode. At the same time the OH^- ions combine with H_3O^+ ions which are everywhere in copious supply, each pair of such combinations leaving an excess $SO_4{}^{2-}$ ion from the dissociated H_2SO_4. This $SO_4{}^{2-}$ ion then joins the main stream of these ions, whose effect at the cathode we have already discussed.

Of the Zn^{2+} and H_3O^+ ions traveling toward the copper anode, only the H_3O^+ are actually collected by this electrode, where, with electrons extracted from this, they form water and hydrogen, the latter of which is held on the surface of the copper. Any Zn^{2+} ions that reach the copper electrode are unable to remain there, because of the high solution pressure of these ions, which we have already noted. However, the hydrogen, whose solution pressure is much smaller, can and does remain. Instead of extracting electrons from the copper and depositing on this electrode as zinc, the Zn^{2+} ions take as partners $SO_4{}^{2-}$ ions derived from H_2SO_4 to form either neutral $ZnSO_4$ or the corresponding dissociated ion pairs. In so doing they leave a corresponding excess of H_3O^+ ions, which, like those we have already considered, extract electrons from the copper anode and leave hydrogen adhering to it.

In consequence of only H_3O^+ being collected by the anode, this electrode very soon becomes, in effect, hydrogen on copper, which is very like the hydrogen on platinum which we considered in the case of the $Pt–H_2SO_4–Pt$ cell. The potential step characteristic of hydrogen being less than that for zinc, current continues to flow, but the emf is now substantially smaller than at the beginning, before any hydrogen was collected on the copper. As the current continues to flow, hydrogen is steadily released from the anode in the form of bubbles and so is lost. There is no question of reversing the action of this cell by sending current through it the other way after this has happened.

The substantial increase in the potential step at the anode which is associated with the appearance of hydrogen at this electrode represents a corresponding "back emf." This phenomenon, which is a serious defect of this type of cell, is known as "polarization."

It is usual in cells having a zinc electrode to bring the whole surface of the zinc into contact with mercury before immersing it in the electrolyte, so producing a surface layer of amalgam. The purpose of this is to cover any inclusions of carbon or other impurities that might otherwise be exposed at the surface. In the absence of the amalgam covering, such exposed areas of impurity, in conjunction with the adjacent zinc in good electrical contact with them, would function as the electrodes of miniature voltaic cells

707

within the main cell, and the resulting "local action" would bring about serious wastage of both zinc and acid.

DANIELL CELL   Although the Daniell cell is now practically obsolete, it nevertheless has considerable scientific interest, illustrating important physical principles. It is for this reason that it is included in this discussion of cell types.

This cell, devised by J. F. Daniell in 1836, may be regarded as an elaboration of the simple cell just discussed which incorporates certain features whose purpose is to prevent polarization at the anode. It was without doubt the most satisfactory and efficient generator of electric current of its time, and it continued in fairly widespread use until comparatively recently. It is still used to some extent in physics teaching laboratories.

We have seen that the electrolyte of a simple cell, beginning as a solution of H_2SO_4, progressively changes to $ZnSO_4$. In the Daniell cell the electrolyte in contact with the zinc is normally made $ZnSO_4$ at the outset, although originally dilute H_2SO_4 was used. And the electrolyte in contact with the copper electrode is a strong solution of $CuSO_4$. This ensures that only Cu^{2+} ions are collected by the copper, and so there can be no polarization. Contact between the two electrolytes usually occurs through a porous earthenware pot, although in some forms, known as "gravity cells," the relatively less dense solution of $ZnSO_4$ simply floats on top of the saturated solution of $CuSO_4$. The more usual form of Daniell cell incorporating a porous pot is shown in Fig. 27.2.

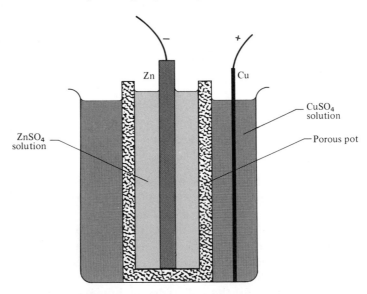

FIGURE 27.2   Daniell cell.

The action of the Daniell cell is quite simple. Zn^{2+} ions go into solution at the surface of the zinc, replacing others that pass through the porous pot into the copper sulfate solution. Here they simply remain, replacing Cu^{2+} ions that are deposited on the copper anode. The Zn^{2+} cannot compete with the Cu^{2+} ions for deposition on the anode, because of their relatively high solution pressure.

Owing to the deposition of the Cu^{2+} ions, the copper sulfate solution progressively becomes weaker as current is passed. To counter this, a ledge is sometimes provided just above the surface of the solution on which copper sulfate crystals are placed.

Although slight contamination of the copper sulfate solution by zinc sulfate can do no harm—this occurs in any case in the course of the cell's action—it is important to avoid any contamination of the zinc sulfate solution with copper sulfate. This may be prevented by making the level of the $ZnSO_4$ solution in the porous pot somewhat higher than that of the $CuSO_4$ solution outside.

As we have already noted, the Daniell cell is, in principle, a reversible cell. In practice, however, it would not be feasible to use it as a storage cell; the cell is theoretically reversible only for the passage of infinitesimal quantities of electricity. As we have just seen, the normal operation of the cell for an appreciable time causes the copper sulfate solution to be contaminated with zinc sulfate, and the mixing together of two materials is a thermodynamically irreversible process. Also, sending a current in the opposite direction must inevitably send Cu^{2+} ions to be deposited on the zinc. Only an extremely small contamination of the zinc with copper can be tolerated; the amount of copper present in the amalgam must not be sufficient to produce an appreciable reduction in the potential step at the zinc electrode. And, of course, any collection of copper by this electrode is, in its nature, irreversible.

The chemical change brought about in the operation of a Daniell cell is the replacement of copper in a copper sulfate solution by zinc. It will be interesting to calculate the emf of the cell from the heats of formation and solution of these substances. The heat of formation of a dilute solution of $ZnSO_4$ from Zn, SO_2, O_2, and water is 1.774×10^5 cal/gram-molecule, and that of a dilute solution of $CuSO_4$ from Cu, SO_2, O_2, and water is 1.273×10^5 cal/gram-molecule. The difference, giving the energy required for the replacement of zinc by copper in a dilute solution, is 5.01 cal/gram-molecule, or 2.10×10^5 J/gram-molecule. This must therefore also be the energy available from the replacement of copper by zinc. The corresponding quantity of electricity delivered is $2 \times 96,488 = 1.93 \times 10^5$ C/gram-molecule. Hence we should expect the emf

709

of the cell to be

$$E = \frac{2.10 \times 10^5}{1.93 \times 10^5} = 1.08 \text{ V}.$$

This calculation is for a high degree of dilution of both electrolytes. In actual Daniell cells neither solution is very dilute, and the copper sulfate solution is usually relatively much stronger than is the zinc sulfate solution. It is not difficult to see that the emf must be expected to be increased somewhat on this account. Measured values of this emf lie between 1.07 and 1.14 V, according to the concentrations.

Actually, neither this calculation nor that by which we calculate the emf of the Pt–H$_2$SO$_4$–Pt cell with anode and cathode covered with oxygen and hydrogen, respectively, is strictly valid theoretically. As we shall see later, when we apply the second law of thermodynamics to a reversible cell, the above calculations omit one item. It just so happens that this item is quantitatively quite small and does not affect the result appreciably in the cases we have considered.

LECLANCHÉ CELL This cell, which, like the Daniell cell, incorporates a zinc cathode and a porous pot separating the cathode from the anode, is shown in Fig. 27.3. The anode is a rod of carbon, and the electrolyte is a solution of ammonium chloride, NH$_4$Cl.

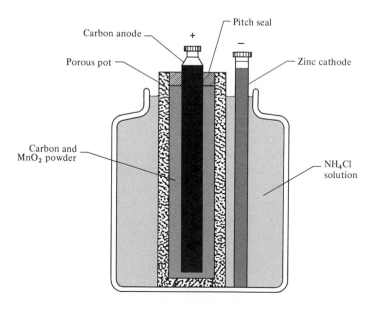

FIGURE 27.3 Leclanché cell.

In solution the salt molecules dissociate into NH_4^+ and Cl^- ions. At the surface of the zinc the solution pressure of this metal creates a potential step, just as in the case of the simple $Zn-H_2SO_4-Cu$ cell and of the Daniell cell. On the other hand, there is little or no tendency for the inert carbon anode to create ions, and so there is no important potential step here. When the electrodes are connected externally through a resistor, the field within the electrolyte drives the NH_4^+ and Zn^{2+} ions toward the anode and the Cl^- ions toward the cathode.

The Zn^{2+} ions that go into solution at the cathode cannot produce a deposit of zinc on the anode, owing to the high solution pressure of this metal. There can also be no continuous congregation of these ions in the region of this electrode, nor of Cl^- ions near the cathode. This is prevented by the encounters, everywhere within the electrolyte, between Zn^{2+} and Cl^- ions, a proportion of such encounters resulting in the formation of neutral $ZnCl_2$ molecules or quasi-molecules, with ensuing redistribution of these, just as in the case of Zn^{2+} and SO_4^{2-} ions in the simple $Zn-H_2SO_4-Cu$ cell. But in the Leclanché cell convection is less vigorous, and diffusion, always a slow process, is further retarded by the intervention of the porous pot between the electrodes. Perhaps it is because of this that the emf of a Leclanché cell falls appreciably when a continuous current of moderate strength is drawn from it, but recovers when the cell is rested.

Let us now consider what happens when the NH_4^+ ions arrive at the anode. These would, if allowed, react with water and collect electrons from this electrode according to the scheme

$$2NH_4^+ + 2H_2O + 2e^- \rightarrow 2NH_4OH + H_2,$$

liberating ammonia and hydrogen; and this hydrogen, depositing on the electrode, would soon set up a substantial back emf because of its solution pressure. But this is not allowed to happen. The carbon anode is surrounded by a tightly packed mixture of carbon and manganese dioxide in powder form, and the manganese dioxide functions as an efficient depolarizing agent, becoming reduced to the lower oxide, Mn_2O_3, with the production of water. The purpose of the carbon is to reduce the electrical resistance of the powder. The particles of carbon in the powder can, of course, take over from the carbon rod the function of effective anode.

The reducing action of the manganese dioxide poses the question of how a particle of this substance can interact with hydrogen when it is the particles of *carbon* to which the ammonium ions are drawn. One possibility would seem to be that hydrogen

711

is deposited on the surface of the carbon as a tenuous monatomic layer which behaves like a two-dimensional gas and spreads to the adjacent particles of manganese dioxide, and that these particles then react chemically with the hydrogen at a sufficiently early stage for the back emf built up to be inappreciable. Whether or not this is the right explanation, it is clear that more is involved in the process of depolarization than might at first appear.

DRY CELL   This cell, one form of which is shown in Fig. 27.4, is a portable, unspillable version of the Leclanché cell in which the zinc electrode is made in the form of a container, the porous pot is replaced by a canvas bag, and the electrolyte is a jelly of ammonium chloride and glycerine. The top is sealed with a layer of pitch or wax.

The emf of the Leclanché cell in both its forms is about 1.5 V.

LEAD STORAGE CELL   We have seen that if a current is passed for a time through a Pt–H_2SO_4–Pt cell, until its anode becomes charged with bubbles of oxygen and its cathode with hydrogen, it will function as a voltaic cell, supplying current in the reverse direction until the gases on its plates are used up. The main reason why it would not be feasible to put such a cell to practical use as a storage cell is that, owing to the limited amounts of gas it can hold without loss by bubbling to the surface, the length of time for which a current of any reasonable strength can be drawn from it is extremely short.

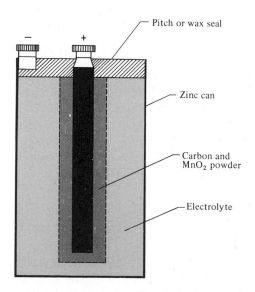

FIGURE 27.4 Dry cell.

The next step toward the development of the modern storage cell was the substitution by Planté, in 1859, of lead for platinum. Passage of current through this cell, instead of liberating oxygen at the anode, was found to oxidize it, forming a brown film of lead peroxide, PbO_2, at the surface of the electrode, hydrogen being simultaneously released at the cathode. On removal of the external source of emf and connection of the electrodes through a meter, a current was found to flow in the reverse direction, just as in the case of the $Pt–H_2SO_4–Pt$ cell.

It might be thought that the delivery of current by the cell must stop when such hydrogen as remains adhering to the cathode is all used up. However, this is not so; current continues to flow well beyond this stage. Actually, it would be surprising if current did *not* continue to flow, for at the stage when the hydrogen is used up the cell is still asymmetrical, the surface of the cathode being metallic lead while that of the anode is PbO_2. It is the emf corresponding to this asymmetry that is responsible for the continuing passage of current.

During discharge, PbO_2 on the anode is converted to $PbSO_4$, the lead changing from the quadrivalent to the divalent state. This involves the absorption of two electrons per atom of lead. These are taken from the metal of the electrode, in which they are replaced from the external circuit. To preserve electrical neutrality of the nonmetallic part of the anode this must be compensated for by the absorption of two hydronium ions from the electrolyte. The absorption of a molecule of H_2SO_4 is also required; when the cell is in action separate $SO_4{}^{2-}$ ions could not reach this electrode against the field. Thus for the complete anode reaction we have

$$PbO_2 + H_2SO_4 + 2H_3O^+ + 2e^- \rightarrow PbSO_4 + 4H_2O.$$

The lead sulfate remains on the anode, replacing the peroxide.

The cathode reaction is relatively simple. It consists of the absorption of an $SO_4{}^{2-}$ ion from the solution, with the formation of $PbSO_4$ and the release of two electrons:

$$Pb + SO_4{}^{2-} \rightarrow PbSO_4 + 2e^-.$$

This sulfate, also, adheres to the plate.

During charging, both these reactions are reversed.

The emf of a lead storage cell is about 2.0 V.

The total quantities of electricity passed during discharge ("capacities") of the earliest cells made by Planté were very small. He soon improved these cells by a process of "forming" the plates.

713

This consisted of passing a current backward and forward through the cell several times in succession, so eventually causing the chemical action to bite deep into the plates. After charging, the anode was left with a substantial layer of PbO_2, while on the cathode the lead was left, down to a corresponding depth, in a spongy form.

The next important improvement, giving cells substantially the same as those we use today, was introduced by Camille Faure, in 1881. Instead of subjecting the plates to an electrical forming process, Faure applied to their surfaces a paste of lead oxide with dilute sulfuric acid. This, after charging, besides giving the required PbO_2 on the anode, produced spongy lead on the cathode, both layers being appreciably deeper than those achieved by the forming process. At first Faure simply pasted the oxide onto the smooth surfaces of the electrodes, but he soon began slotting and perforating these to better hold the material. This idea has been developed over the years, and the weight of the metallic lead base has now been reduced to a minimum, this taking the form of a grid into the interstices of which the active material is pressed.

In our discussion of the chemical reactions accompanying charge and discharge we saw that while the cell is delivering current, sulfuric acid is used up, but that this reappears during the charging process. Hence in a fully charged cell the acid is more concentrated than in one that is approaching a state of discharge. Specific gravity increases with concentration, so measurement of the specific gravity with a hydrometer provides a convenient means of gauging the state of charge. The specific gravity of the acid in a fully charged cell should be about 1.28. During the course of the discharge this falls to about 1.16.

OTHER TYPES OF CELL   A number of types of voltaic cell other than those considered above, both primary and secondary, have been developed and used at various times, and some are at present in use for special purposes. However, a discussion of all these would be beyond the scope of this book.

Thermodynamical Theory of Reversible Cells

Just as the surface energy per unit area of a liquid is not quite the same thing as its surface tension, so also the emf of a reversible cell is not merely the amount of *chemical* energy reversibly converted to electrical per unit charge pumped, but the *total* reversible energy conversion per unit charge, such energy conversion possibly including a thermal item. Let us now investigate the effect such a thermal item must have according to the second law of thermodynamics.

Let us suppose that in working as a generator at a constant temperature the cell absorbs heat from its surroundings. This means that if the cell operates adiabatically, that is, the heat necessary to keep it at a constant temperature is withheld, its temperature must fall. Conversely, if the cell is made to operate adiabatically in reverse, its temperature must rise.

Let this cell now be taken through a reversible cycle as follows.

First, at the temperature T_1, corresponding to which its emf is E_1, the cell is allowed to pump a quantity of electricity Q through the electrolyte from cathode to anode.

Next, after the quantity Q of electricity has been pumped, the cell is thermally isolated from its surroundings while it continues to operate, now adiabatically, until its temperature has fallen to the very slightly lower value T_2.

The temperature is now kept constant at this value while a quantity Q' of electricity is forced through the electrolyte in the reverse direction, that is, from anode to cathode, such that a final adiabatic stage, with the cell still operating in reverse, will bring the cell back to its original chemical and thermal condition. This final adiabatic stage is then carried out.

This cycle of operations is indicated graphically in Fig. 27.5, where emf is plotted against quantity of electricity passed through the cell. The two horizontal lines are isothermals, and the two oblique lines are adiabatics. These and their intersections with one another define a cycle $ABCDA$ such as we have been discussing. It

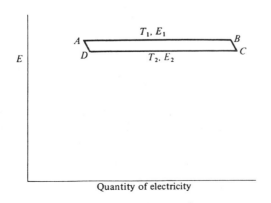

FIGURE 27.5 Reversible cycle in operation of ideal cell.

715

is clear from this diagram that, with given adiabatics, the closer the isothermals are to one another the more nearly must CD be equal in length to AB; that is, the more nearly must Q' be equal to Q. Let us suppose that the cycle under consideration is carried out under these conditions, which imply also that $T_1 - T_2$ is infinitesimally small.

Let W_c be the energy per unit quantity of electricity pumped which becomes available for conversion to electrical energy at the temperature T_1 as a result of chemical action, and let H^* be the quantity of heat (per unit charge) which is simultaneously absorbed by the cell from its surroundings. Then if E_1 is the emf of the cell at this temperature, we must have

$$E_1 = W_c + H. \tag{27.1}$$

The net electrical energy obtained from the cell during the cycle of operations is represented by the area $ABCD$. This is $(E_1 - E_2)Q$, where E_2 is the emf at the temperature T_2. The heat supplied at the higher temperature, T_2, is HQ. Thus the efficiency of the cell functioning as a heat engine is the former divided by the latter, $(E_1 - E_2)/H$. According to the second law of thermodynamics this must be equal to $(T_1 - T_2)/T_1$. Equating these two expressions, we have

$$\frac{E_1 - E_2}{H} = \frac{T_1 - T_2}{T_1},$$

or

$$\frac{H}{T_1} = \frac{E_1 - E_2}{T_1 - T_2},$$

the equivalent of which in calculus notation is

$$\frac{H}{T} = \frac{dE}{dT},$$

or

$$H = T \frac{dE}{dT}, \tag{27.2}$$

where T is now written in place of T_1. Eliminating H between Eqs. (27.1) and (27.2), and writing E in place of E_1, we obtain the equation

$$E = W_c + T \frac{dE}{dT}. \tag{27.3}$$

---

* This symbol is used here for heat, not Q, to avoid confusion with quantity of electricity.

The quantity W_c may be measured as a difference between heats of formation, or heats of formation plus solution, as in the examples of the Pt–H_2SO_4–Pt cell and the Daniell cell considered above.

Equation (27.3), which was first obtained by Helmholtz, is known, after him, as Helmholtz's equation.

In many cases the second term in the expression for E is quite appreciable, amounting to a few tenths of a volt. The reason why we obtained the correct value of the electromotive force of the Daniell cell by simply identifying E with W_c is that, as it just so happens, the temperature coefficient of the emf of the Daniell cell is practically zero, making $T(dE/dT)$ negligibly small.

In 1886 H. Jahn measured the temperature coefficients of the electromotive forces of a number of cells, together with the heats of chemical reaction, using an ice calorimeter. He found the electromotive forces of these cells calculated from Helmholtz's equation to be in very satisfactory agreement with the values determined experimentally.

Problems

27.1 An electrolyte whose ions have equal mobilities carries a current of 1 A; thus the positive ions carry 0.5 C of positive charge per second in one direction, and the negative ions carry 0.5 C of negative charge per second in the opposite direction. What is the total charge delivered to each electrode per second, 0.5 C or 1 C? Explain.

27.2 In an electrolytic cell containing a solution of copper sulfate the electrodes are horizontal sheets of copper, the upper one of which is the cathode and the lower one the anode. By this arrangement convection in the electrolyte is inhibited. After the passage of 1000 C through this cell in a time sufficiently short for the effects of diffusion to be negligible, it is found that 0.235 g of $CuSO_4$ has been lost from the solution in the neighborhood of the cathode, the solution near the anode having been correspondingly increased in strength. What is the ratio of the mobilities of the SO_4^{2-} and Cu^{2+} ions?

27.3 An ammeter, a rheostat, and a silver nitrate electrolytic cell are connected in series to a dc supply and the reading of the ammeter maintained at 4.00 A for 30 min. The weight of the cathode is found to have increased during this time by 7.98 g. What correction, if any, should be applied to the reading of the ammeter?

27.4 In a tank designed for the refining of copper by electrolysis, 20 kg of this metal is to be deposited on the cathode per hour. What current is required to effect this rate of deposition? If in this electrolysis a PD of 0.2 V is maintained between the electrodes, and the cost of electrical energy

717

is 1 cent per kilowatt-hour, what is the electrical-energy contribution to the cost of production of refined copper per kilogram?

27.5   In the production of hydrogen and oxygen by the electrolysis of dilute sulfuric acid the sum of the reversible decomposition voltage and the overvoltage is 1.7 V. A cell used for this purpose has an internal resistance of 0.1 Ω, and a power source applied to this cell has an emf of 3.0 V and an internal resistance of 0.08 Ω. What are the rates of generation of hydrogen and oxygen?

27.6   A current is passed between chemically inert electrodes through a solution of sodium chloride. What elements are released at the electrodes, and what further action, if any, occurs at either or both of these? What materials of commercial value are produced in this electrolysis?

27.7   A solution of sodium chloride containing 57.5 g of NaCl per liter is found to have a conductivity of 7.5 mhos/m at 18°C. What is the mean of the mobilities of the Na^+ and Cl^- ions?

27.8   In an earlier version of the Daniell cell dilute sulfuric acid was used as the electrolyte in contact with the zinc, not a solution of zinc sulfate, as now. What are the ions taking part in the action of such a cell, and what are their careers? How are conditions modified when the cell has been in action for some time?

27.9   How many grams of PbO_2 on the positive plate of a lead storage cell are converted to $PbSO_4$ for each ampere-hour of charge delivered, and what is the decrease in mass of the electrolyte for each ampere-hour?

28

THERMOELECTRICITY

The electromotive forces considered in Chapter 27 are derived from processes occurring in the region of contact of a metal with an electrolyte, and depend, at least primarily, on the conversion of chemical energy to electrical energy, or vice versa, although thermal energy also makes a minor contribution. We shall now consider certain all-metal systems in which electromotive forces are also developed, but in which these are derived entirely from conversions between heat and electrical energy.

Seebeck Effect

The first observation of a thermoelectric phenomenon was made in 1821 by Thomas Johann Seebeck incidentally to investigations he was carrying out on the magnetic effects of electric currents. He found that if a circuit including a galvanometer is made up of two different metals, say copper and iron, as indicated in Fig. 28.1, and the junctions 1 and 2 between these metals are maintained at different temperatures, the galvanometer needle shows a deflection. This deflection he found to be determined by both the nature of the two metals and the temperatures of the junctions.

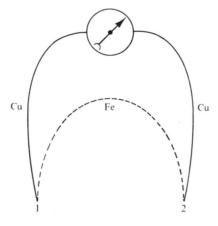

FIGURE 28.1 Thermocouple connected to galvanometer.

719

A pair of metals arranged in a circuit as in Fig. 28.1 is known as a **thermocouple.**

To account for the flow of current we have no choice but to postulate the existence of an electromotive force. Accordingly, for the proper investigation of the effect, a voltmeter should be used in place of the galvanometer, or, better, a potentiometer.

In the case we have just considered, the copper component of the thermocouple would normally be continued, via copper connecting wires, to the copper winding of the galvanometer, so that, in effect, we should have not only a single iron wire between one junction and the other but also a single copper wire. But this is a very special case, and we have now to consider the situation where the wires connecting the thermocouple to the device measuring the emf and the metal or metals constituting this device are not necessarily the same as either of the metals of the thermocouple. This more general case is represented in Fig. 28.2(a), where the thermocouple is constructed of two metals A and B, and the wires connecting the component B to the device measuring the emf are of metals C and D, which may be different from one another as well as from A and B. It will now be shown that if the portion of the circuit consisting of the measuring device and its connecting wires to the junctions of these with the ends of the metal B are at a uniform temperature, the circumstance that this device and its leads may consist of metals differing from A and B can have no effect on the algebraical sum of the electromotive forces, the "total emf," around the circuit. This is the "law of intermediate metals."

Let us consider the part of the circuit of Fig. 28.2(a) above a line PQ such that the whole of this is at a uniform temperature. There can be no resultant emf residing in this part of the circuit;

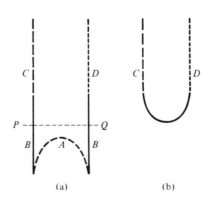

(a)   (b)

FIGURE 28.2

for, if there were, and its ends were joined, so giving the circuit shown in (b), this emf would cause a current to flow. Experiment shows, however, that no current ever does flow in an isolated all-metal circuit at a uniform temperature, no matter how complex this may be. We may note, incidentally, that if it did, this would be a violation of the second law of thermodynamics. It must therefore be inferred that the presence of intermediate metals, all at a uniform temperature, can have no effect on the total emf in a thermo-electric circuit.

Another general law, which is complementary to the law of intermediate metals, is that known as the "law of intermediate temperatures." According to this law the electromotive force E_{13} of a couple whose junctions are at the temperatures T_1 and T_3 is the sum of the electromotive forces E_{12} and E_{23} of two couples of the same metals the former of whose junctions are at the temperatures T_1 and T_2 and the latter at T_2 and T_3.

Using junction temperatures differing by only moderate amounts from room temperature, Seebeck experimented with a large number of metals and found that he was able to arrange them in a certain order such that when any two are used in a thermocouple the direction of the emf is from the metal at the hot junction occurring earlier to that occurring later in the series. Apart from certain alloys and metals of doubtful purity, this list is as follows: Bi-Ni-Co-Pd-Pt-U-Cu-Mn-Ti-Hg-Pb-Sn-Cr-Mo-Rh-Ir-Au-Ag-Zn-W-Cd-Fe-As-Sb-Te. The farther apart from one another the metals are in such a list the greater the emf is found to be for given junction temperatures. Thus the biggest effect would be obtained with bismuth and tellurium. Bismuth and antimony are found to be almost as good.

As we have already noted in Chapter 13, thermocouples have an important application in thermometry. In addition, "thermo-piles"—systems of a large number of thermocouples joined in series—are widely used in instruments known as radiometers for measuring radiation intensities.

Peltier Effect

The development of an emf in a thermocouple means that electrical energy is being generated from some other form, and the latter surely cannot be anything but heat. In other words, the thermocouple must function as a heat engine which, in principle at least, is reversible. The only reason why it is not truly reversible in

721

practice is that heat is conducted down temperature gradients in the constituent metal wires, and such conduction is, in its nature, irreversible. However, like friction in a mechanical heat engine, this is merely an incidental irreversible item, which is theoretically irrelevant, in an otherwise reversible process; and, like frictional loss, it can be allowed for in any quantitative thermodynamical discussion. It will be convenient to refer to an imaginary thermocouple in which there is no thermal conduction along its wires as an "ideal thermocouple."

Consideration of the action of an ideal thermocouple suggests that it may simply operate as a heat engine by taking in all its heat at the "hot" junction and rejecting heat—a lesser quantity—solely at the "cold" junction, converting the difference into electrical energy. Whether this is, in fact, the way it operates is a question we may leave open for the present. Meanwhile, this picture, accepted provisionally, suggests an experimental test for the reverse process—using an external source of emf to drive a current through the thermocouple in the reverse direction—and seeing if heat is liberated at the hot junction and absorbed at the cold junction.

An experiment very like this was tried in 1834 by Jean C. A. Peltier, a Parisian watchmaker who made a hobby of scientific experimentation. As we may see from Seebeck's list of metals, a copper–antimony thermocouple in action will produce a current in the direction from copper to antimony at the hot junction and from antimony to copper at the cold junction. In our supposed reversed heat engine we should therefore expect heat to be liberated at the junction where current is made to flow from antimony to copper and to be absorbed where it flows from copper to antimony. A little consideration will show that this must be so, regardless of which junction of the direct-acting heat engine was hot and which cold. It is therefore not necessary to take the temperature of the junction into account; a copper–antimony junction should be heated or cooled according as current is made to flow across it from antimony to copper or from copper to antimony. Peltier put this expectation to the test and at once found it to be verified. This heating or cooling of the junction between two metals according to the direction in which current is made to flow across it is known, after its discoverer, as the **Peltier effect.** On passing a current from antimony to copper Peltier observed a heating of 10 degrees, and on reversing the direction of the current he found the junction to be cooled by 5 degrees. He found the effects to be increased when bismuth was substituted for copper. Heinrich Lenz later succeeded in freezing water by the Peltier effect. A quite recent development has been the achievement of a temperature reduction of as much as 65 degrees by the passage of a current of a few amperes across a junction between bismuth telluride (a semiconductor) and copper. Correspondingly large effects are also produced at junctions between one semiconductor and another, if these are of the right kinds. Recently, a miniature refrigerator has been developed which depends on the passage of a current across a junction between two types of bismuth telluride known as n-type and p-type BiTe$_3$.

Application of Thermodynamics

Let us now return to the question hitherto left unresolved, of whether in the normal operation of a thermocouple heat is taken in entirely at the hot junction and rejected only at the cold junction. As Lord Kelvin (at that time William Thomson) pointed out in 1855, this question may be answered by a process of inductive reasoning involving an application of the second law of thermodynamics. The procedure is (1) to assume for the sake of argument that the Seebeck effect does ideally represent a reversible heat engine operating uniquely between two definite temperatures, (2) to find by thermodynamical reasoning how the emf given by such a heat engine should vary with these two temperatures, and then (3) to compare this prediction with what is actually observed.

Let us again consider the system of Fig. 28.1, and, simply as a matter of convenience, to aid us in our discussion, let us suppose that the thermocouple is made up of two particular metals; for example, copper and iron. With the temperature T_1 at junction 1 higher than T_2 at junction 2, let us suppose a current I to flow counterclockwise around the circuit. Then, as is easily seen by reference to Seebeck's list of metals and from our discussion of the Peltier effect, heat will be absorbed at junction 1 and given out at junction 2. Correspondingly there will be a forward Peltier emf, Π_1, at junction 1, while at junction 2 there will be a back emf, Π_2, so that the resultant forward emf is $\Pi_1 - \Pi_2$.

The rate of absorption of heat at the hot junction is $\Pi_1 I$, and the rate of evolution at the cold junction is $\Pi_2 I$. The ratio of these rates being the same as the corresponding ratio of quantities of heat absorbed and evolved, in a given time, we must have, by Carnot's principle,

$$\frac{\Pi_1 I}{\Pi_2 I} = \frac{\Pi_1}{\Pi_2} = \frac{T_1}{T_2},$$

or

$$\frac{\Pi_1 - \Pi_2}{\Pi_2} = \frac{T_1 - T_2}{T_2},$$

723

whence we obtain the expression for the resultant electromotive force,

$$E = \frac{\Pi_2}{T_2}(T_1 - T_2).$$

This means that with a given fixed temperature of the cold junction, and correspondingly a fixed value of Π_2, the thermoelectromotive force should simply be proportional to the amount by which the temperature of the hot junction exceeds that of the cold junction.

Experiment shows that this is far from being the case. In Fig. 28.3 the emf developed in a copper–iron thermocouple is shown as a function of the temperature of the hot junction, that of the cold junction being kept at 0°C. We see that while the emf at first increases with increasing temperature of the hot junction, it does so at a progressively decreasing rate, until at a certain value of this temperature the rate of increase falls to zero, and beyond this point the emf actually decreases as the temperature of the hot junction continues to increase. This behavior, which is typical of all thermocouples, shows that the assumption underlying our thermodynamical discussion, of heat intake and rejection only at the temperatures T_1 and T_2, respectively, is certainly not justified. In other words, the emf cannot be derived solely from processes occurring at the junctions.

Thomson Effect

From the failure of the simple thermodynamical theory that assumes Peltier electromotive forces to be the only ones responsible

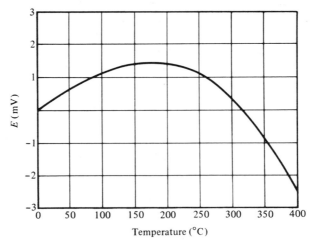

FIGURE 28.3 Variation of emf of a copper–iron thermocouple with temperature of "hot" junction.

for the Seebeck effect, Thomson inferred the existence of additional electromotive forces associated with the temperature gradients along the wires. The action of these electromotive forces, whose existence is now firmly established, is known as the **Thomson effect.**

Let us again refer to the circuit of Fig. 28.1, in which, in the interests of generality, we shall call the copper and iron the metals A and B, respectively. In this circuit, in addition to the Peltier contribution to the total emf $\Pi_1 - \Pi_2$, we must now postulate an additional Thomson emf. We may express this mathematically in terms of the emf in each metal per unit temperature difference, which is known as the **Thomson coefficient.** If we denote this quantity for the two metals by the symbols σ_A and σ_B, then for the total emf around the circuit we must have

$$E = \Pi_1 - \Pi_2 + \int_{T_2}^{T_1} \sigma_A \, dt - \int_{T_2}^{T_1} \sigma_B \, dt. \qquad (28.1)$$

As is implied by the last two terms of this equation, the Thomson coefficient σ of a metal is reckoned positive when the emf coincides in direction with that of rising temperature. If a current flows along a conductor in this direction, then, with σ positive, electrical energy is generated, and this must be at the expense of heat. The conductor must therefore be cooled. On the other hand, if the current flows down the temperature gradient, heat must be liberated, causing a rise in temperature of the conductor. In cases where σ is negative the heating and cooling effects are the reverse of these.

The Thomson effect may be demonstrated as follows. An iron rod is bent into the shape shown in Fig. 28.4, and while its two free ends A and C are surrounded by melting ice the loop B is heated with a bunsen burner. Each limb is provided with a thin insulating coating of a suitable material (for example, of asbestos), and on this a coil of wire having an appreciable positive temperature coefficient of resistance is closely wound. The whole is then thermally

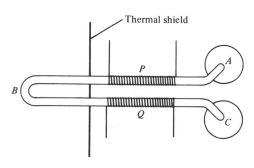

FIGURE 28.4 Arrangement for demonstrating Thomson effect.

725

insulated by being packed with asbestos wool. The coils P and Q are used as two adjacent arms of a Wheatstone bridge, and this is balanced. If a current of the order of 10 A is now passed along the rod from A to C, the balance of the bridge is found to be disturbed. The disturbance is in the direction indicating heating of the coil P and cooling of Q, or at least—having regard to the Joule heating of both limbs which must occur in any case—a rise in temperature of P above that of Q. On reversal of the direction of the current, the direction of the disturbance of the bridge is reversed also. From these observations it must be inferred that σ for iron is negative.

If copper is substituted for iron an effect may also be observed, but this time corresponding to a positive value of σ. The effect is much smaller with copper, because, owing to the high thermal conductivity of this metal, it is more difficult to maintain a sufficient temperature gradient, and also because the Thomson coefficient for copper is relatively small.

Most metals have Thomson coefficients of the order of a few microvolts per degree, and in the great majority of cases σ is positive. However, a few metals, among them iron, nickel, platinum, palladium, and bismuth, have negative Thomson coefficients. An intermediate case is lead; its Thomson coefficient is believed to be zero, or at least sufficiently nearly so to be taken as zero for all practical purposes.

While the method of demonstrating the Thomson effect just described might conceivably be made sufficiently quantitative to serve for actual measurements of σ, this has not in fact been done, because a more accurate, although less direct, method is available.

It may be shown by thermodynamical reasoning that the Peltier emf Π_{AB} at a junction between two metals A and B at a temperature T is given by the equation

$$\Pi_{AB} = T\frac{dE}{dT}, \tag{28.2}$$

where the other junction is kept at a fixed reference temperature (for example, that of melting ice) and E is the thermoelectric emf for the whole circuit, the positive direction for both Π_{AB} and E being taken as that from A to B at the junction in question. It is not difficult to see that dE/dT, although not E itself, is independent of the reference temperature. Thus the appropriate measurements enable us to determine Π_1 and Π_2 in Eq. (28.1) separately, and so to find the difference between the two mean values of the Thomson coefficients for the temperature interval in question. If this temperature interval is small this may be taken as the difference between σ_A and σ_B at a particular temperature: the mean between T_1 and T_2. This does not yet tell us the values of σ_A or σ_B separately, but if by some independent means we could find the value for one metal we could then determine the values for all the others. As stated above, the value of σ for lead is believed to be practically zero; thus the procedure we have considered for demonstrating the Thomson effect for iron and for copper would not show any effect at all if lead were used. On this basis, therefore, all the other Thomson coefficients can be found.

Thermoelectric Power

In assessing the sensitivity of a thermocouple to be used for thermometry the relevant quantity to consider is dE/dT. Also, as we have just seen, this quantity figures in the theoretical equation (28.2). For these and other reasons the rate of change of thermo-electromotive force with temperature is an important quantity, and accordingly it has been given a special name; it is known as the **thermoelectric power.**

Examination of the curve of Fig. 28.3 suggests that the thermoelectric power, given by the slope of this curve, might well vary linearly with temperature, or at least approximately so. Actually in this particular case the approximation to linearity is good only in the lower part of the temperature range, but for most thermocouples the variation of dE/dT with T approaches very closely indeed to linearity over a wide temperature range. This is shown in Fig. 28.5 for a number of metals (some possibly not very pure) used by Tait in 1873 in a thermocouple in conjunction with lead as a reference metal,[*] the direction around the circuit reckoned as positive being from lead to the other metal across the junction whose temperature is varied.

From plots such as these it is a simple matter to find the values of the constants a and b in the equation

$$\frac{dE}{d\Theta} = a + b\Theta \qquad (28.3)$$

characteristic of the various metals, the Celsius temperature scale now being used as a matter of convenience. Integration of this

---

[*] The difficulty presented by the relatively low melting point of lead (327°C) is circumvented by using a selected other, more refractory, metal as an experimental reference metal, whose behavior relative to a fictitious solid specimen of lead above its melting point is worked out by extrapolation.

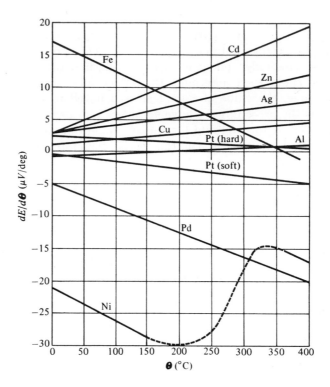

FIGURE 28.5 Variation of dE/dt with t for a number of metals used with lead as thermocouples.

equation yields the relation

$$E = a\Theta + \tfrac{1}{2}b\Theta^2, \tag{28.4}$$

between E and Θ, the constant of integration being zero for the case where the reference temperature is $0°C$. This is the equation of a parabola, and represents very closely, although not absolutely exactly, the variation of E with Θ in the great majority of cases.

Values of the constants a and b for a number of metals and alloys are given in Table 28.1.

The values of a and b in the equation of type (28.4) relating to a thermocouple made up of *any* two metals may be found by taking the differences of the values relative to lead. In using the table the following rule should be used: The direction around the circuit to be taken as positive is that across the junction at the temperature Θ from the first metal selected to the second. The values of a and b to be taken for the combination of the two metals are those given in the table for the first metal minus those for the second. The value of E then obtained relates to the positive direction

TABLE 28.1   Thermoelectric Constants (lead as reference metal)

Metal or alloy	a (μV/deg)	b (μV/deg^2)
Alumel	17.48	-0.00144
Aluminum	0.47	-0.003
Antimony	-35.58	-0.1450
Bismuth	43.69	0.4647
Chromel	-24.40	0.0
Constantan	38.105	0.0888
(60% Cu, 40% Ni)		
Copper	-1.34	-0.0094
Gold	-2.90	-0.0093
Iron	-17.15	0.048
Platinum	1.79	0.035
90% Pt–10% Rh	-6.413	-0.0064
Silver	-2.50	-0.012
Zinc	-3.0	0.01

as defined above. Let us, for example, consider the case of a copper–constantan thermocouple with $\Theta > 0°C$. Taking copper as the first metal and constantan as the second, we should then have $a = -39.44_5$ μV/deg and $b = -0.0982$ μV/deg^2, and accordingly E comes out negative. This means that the emf around the circuit in the direction from copper to constantan at the junction whose temperature is Θ is negative, that is, that the emf actually tends to drive current from constantan to copper at this junction.

From Eq. (28.3) we see that $dE/d\Theta = 0$ at the temperature Θ_n equal to $-a/b$. This temperature is known as the **neutral temperature** of the relevant thermocouple. Also, from Eq. (28.4), we see that $E = 0$, not only at 0°C, but also at the temperature Θ_i equal to $-2a/b$, which is known as the **inversion temperature.** At this temperature the emf changes direction.

Physical Explanations of Peltier and Thomson Effects

In our study of thermionic and photoelectric emission we saw that there is a potential step at the boundary between a metal and a vacuum, the magnitude of which varies from one metal to another. This suggests that there might well also be a potential step across a junction between two metals. Such a potential step, if it could exist, might account for the Peltier effect. However, the difficulty is to see how this could maintain itself, having regard to the fact that electrons can flow freely across the boundary in

729

either direction. According to all classical ideas such flow should immediately equalize the potentials on the two sides. It would seem, therefore, that classical physics is not competent to explain the Peltier effect. Actually a mechanism for this effect has been suggested, but this lies in the realm of quantum physics and is beyond the scope of this book.

The problem of explaining the Thomson effect looks, at first sight, somewhat more hopeful. The conducting electrons in a metal may be considered as an electron "gas" in a medium (the metal) which is, as a whole, electrically neutral. With increasing temperature these electrons must gain in kinetic energy, and in a metal wire along which there is a temperature gradient the relatively more energetic electrons at the hot end will tend to spread toward the cold end, so increasing the potential at the former and decreasing it at the latter until the electron-retarding field built up prevents any further net flow of electrons. The resulting situation is one of equilibrium. If the two ends of the wire are now connected through an external circuit in which there is no opposing emf, electrons will flow out of the cold end where the potential is low (negative potential high) and back into the hot end. The emf that drives the electrons around the circuit in this way is mathematically equivalent to one that drives positive electricity along the wire from the cold end to the hot end; that is, it corresponds to a positive Thomson coefficient.

In using up some of their kinetic energy to move against the field from the hot to the cold end of the wire, the electrons gain electrical energy at the same rate as they lose thermal energy. Thus the quantity of thermal energy lost per unit charge must be the product of the mean value of σ (henceforward to be called, for short, simply σ) and the fall in temperature. Hence we see that σ is simply the specific heat of the conducting electrons.

So far, so good. Now, however, we encounter difficulties. According to the law of equipartition of energy, which we considered in Chapter 15, the specific heat of the conducting electrons should be $\frac{3}{2}k$ per electron, the same as the specific heat per molecule of an ordinary gas at constant volume. But observed Thomson coefficients are of a much lower order of magnitude than this, and, as we have seen, some are even negative. Clearly, therefore, classical theory cannot account for the observed phenomena; again quantum theory has to be invoked for their explanation.

Actually this does not mean that the explanation of the Thomson effect advanced above is valueless. It still retains a certain qualitative validity, and it does show that σ is to be identified with the specific heat of the internal electrons. Quantum theory success-fully accounts for the low specific heats observed, including the negative specific heats found for certain metals.

Problems

28.1 A thermocouple is wanted for use as a thermometer at low temperatures, and a reasonably high sensitivity is required in a range extending down to as low a temperature as possible. Using the data of Table 28.1, find whether an antimony–bismuth or a copper–constantan thermocouple would be the more suitable. Give reasons for your answer.

28.2 For what reasons is a combination of platinum with a 90%–10% Pt–Rh alloy such a good thermocouple for the measurement of high temperatures? Plot the calibration curve for this thermocouple from $-200°$ to $+1600°C$.

28.3 If conducting electrons in a metal had the same energy distribu-tion as the molecules of an ordinary gas, what would be the magnitude of the Thomson coefficient?

28.4 The argument presented in the text which identifies σ with the specific heat of the conducting electrons makes no mention of either of the following facts: (1) The existence of thermal conduction means that elec-trons convey heat from the hot end of the wire to the cold end as they travel, in addition to transforming heat into electrical energy. (2) Most of the thermal energy carried by the electrons in traveling from the hot to the cold end is simply drained off by the cooling agency at the cold end; only a relatively small fraction is converted into electrical energy. In view of these complications, does the argument still hold? If so, explain why.

28.5 What are the values of the constants a and b in an equation of type (28.4) relating to a Cu–Fe thermocouple? Use these values to find the magnitude and direction of the emf for this thermocouple at 100°, 200°, 300°, and 400°C, the "cold" junction being at 0°C. Assuming the values found for the constants to apply below as well as above 0°C, what would you expect the emf to be at $-100°C$? What will be the neutral and inver-sion temperatures for this thermocouple?

SECTION VIII

MAGNETISM

29

MAGNETIC EFFECTS OF CURRENTS

Let us consider in conjunction with one another the following three items relating to magnets and electric currents: (1) In the study of magnetostatics one is concerned with the mechanical action between magnets. (2) The action of a moving-coil galvanometer depends on the force that a magnet exerts on a current-carrying wire in its neighborhood. (3) It is a simple matter to show that two parallel current-carrying wires exert forces on one another. Thus we have the situation that mutual forces are exerted (1) between magnets, (2) between a magnet and a current, and (3) between currents.

In accordance with our basic philosophical principle that things that *do* the same *are* the same, we must assume that a wire carrying a current and a magnet are, in some way, essentially the same; they must have some basic feature in common to which the exertion of these forces must be attributed. Logically, then, we should try either to explain currents in terms of magnets or magnets in terms of currents. Of these two alternatives it is the latter that is indicated rather than the former; for, as we know from atomic theory, a feature common to all atoms is the circulation of electrons in orbits about a central nucleus; that is, circulating electricity. But we know of nothing in the makeup of atoms to which we could attach the label "magnetism." It is always a sensible procedure to try to work from the known to the unknown, and we already have a considerable body of knowledge concerning electric currents and associated phenomena.

If we are to hope to succeed in explaining magnets in terms of currents it will obviously be necessary, as a first step, to examine in some detail the forces exerted by currents on one another— forces which, for convenience, we shall call "magnetic" forces. It is with the examination of these forces that we shall be concerned in this chapter and in Chapter 30. Then, when we have seen how these forces are determined by the current strengths and geometrical factors, we shall, in Chapter 31, endeavor to account for the phenomena exhibited by magnets in these terms.

In any particular system of current-carrying wires, each short length, or "element," of wire presumably acts on all the other

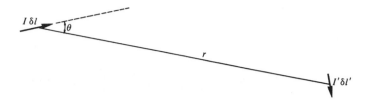

FIGURE 29.1 Two current elements whose mechanical interaction is to be investigated.

elements of the system. Our basic task is therefore to find the magnitude and direction of the force which one element, of length δl and carrying a current I, exerts on a second element, of length $\delta l'$ and carrying a current I'. Two such current elements, which we may conveniently refer to as $I\,\delta l$ and $I'\,\delta l'$, respectively, are indicated in Fig. 29.1. These could either be short sections of different circuits, or they could both belong to the same circuit.

Our investigation is best carried out in two stages, making use of the concept of a magnetic field produced by $I\,\delta l$ at the location of $I'\,\delta l'$. As a first step, one might investigate the direction and magnitude of this field, and, this having been found, one could then inquire into its effect on $I'\,\delta l'$. Actually, it is more convenient to proceed in the reverse order, because we cannot ascertain what field the element $I\,\delta l$ produces until we have *defined* a magnetic field quantitatively and established a method of measuring it. And such measurement is based on the action of the field on a current element or on a circuit built up of current elements.

Magnetic Field and Its Effects

Historically, the concept of a magnetic field derives from the observed tendency of a compass needle to set itself in a particular direction, the "direction of the field." The field may be visualized in the same kind of way as an electrostatic field, in terms of "lines of force," whose direction is everywhere that of the field. These lines may be plotted, step by step, with a compass needle. Figure 29.2 shows an axial section of such a system of lines of force, those produced by a current flowing in a pair of so-called Helmholtz coils. This consists of two similar coaxial circular coils of wire separated by a distance equal to the radius of either. When a current is passed through these connected in series, in the same direction in each, a substantially uniform magnetic field is produced throughout a considerable volume in the space between them.

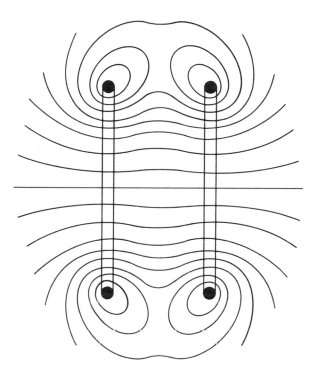

FIGURE 29.2 Magnetic lines of force associated with flow of current through a pair of Helmholtz coils.

Although it is useful to plot fields in this way, so obtaining an impression of their uniformity or otherwise in particular regions (such a procedure will later be justified), our present objective is to explain all magnetic effects in terms of currents alone, without reference to magnets. This may be done, as we shall now see, by invoking considerations of symmetry.

Let us, for example, decide to use a current-carrying circular coil of wire, or any coaxial system of such coils (for example, a pair of Helmholtz coils), to establish a magnetic field. Then, at any point on the axis of such a coil or system of coils, the magnetic field must necessarily be directed along this axis; this is demanded by the axial symmetry of the system. From considerations of symmetry we must also expect another current-carrying coil, of small dimensions, which we may refer to as a "search coil," located on this axis and free to rotate, to set itself coaxially with the coil or system producing the field; any other orientation would represent an arbitrary bias. Experiment confirms that such a search coil does indeed set itself in the orientation expected.

737

In conformity with the sense of the field as defined with a compass needle, we may define the sense of the field produced on the axis of a coil or system of coaxial coils as that which is related to the direction of current circulation in the latter in the same way as the directions of progression and turning of a right-handed screw are related.

It is found that the stable equilibrium orientation of a search coil located on the axis of the system producing the field is such that the current circulates around each in the same sense. Hence the right-hand screw relation holds also between the direction of the field and the circulation of current in the search coil. Accordingly, such a search coil may be used instead of a compass needle to plot a field, giving the lines of force everywhere in both direction and sense.

Not only may we use a coil system to produce a field at any point having a known direction, but we may also ensure that its magnitude is constant, even though we have not yet worked out how to define it quantitatively or to measure it. If we keep the current in the system producing the field constant, then the field itself, which depends on the current, must be constant also.

We shall presently investigate how the direction of the force exerted by a magnetic field on a current element is related to the directions of the current and the field and how this force varies in magnitude with the angle between these two directions, with the current strength, and with the length of the element. As a preliminary, however, it will be necessary to consider certain experimental complications, and to see how, in effect, we may eliminate these.

EXPERIMENTAL CONSIDERATIONS In some of the experiments now to be described the earth's magnetic field may be an embarrassment. Either the whole of this field, or its horizontal or vertical component, may be neutralized throughout a substantial volume by employing a pair of Helmholtz coils, suitably orientated and carrying an appropriate current. The test for such neutralization would be the absence of observable mechanical action on a search coil or compass needle placed within the region in question.

In addition to the Helmholtz coils used for neutralizing the earth's field or one of its components, another pair may be used for establishing a field in any desired direction, and of constant magnitude, in the manner already referred to.

Having established our field, we now wish to investigate its effect on a current element. But we cannot do this directly; we cannot isolate a current element from the circuit of which it forms a part. It will therefore be necessary to infer the direction and magnitude of the force on such an element from the mechanical effect produced on a whole circuit.

To simplify the investigation it will obviously be advantageous for the circuit experimented with to be of a simple form, for example rectangular. Accordingly, the inclusion in it of any large or geometrically complex device, such as a generator (for example, a battery), a rheostat, or an ammeter, is to be avoided. Yet the use of these devices is indispensable if we wish to maintain, regulate, and measure a current in the circuit. Thus we have the problem of arranging for what is to be *in effect* a simple, well-defined circuit, the current in which is maintained by a battery, regulated with a rheostat, measured with an ammeter, and supplied through leads, none of which make any contribution to the field whose mechanical effects are to be measured.

Let us consider the case where we wish to work with the small rectangle shown on the right in Fig. 29.3, whose sides we may regard as current elements. We might well have an appreciable disturbing field in the region of this rectangle due to the current in the battery, ammeter, rheostat, and associated wires enclosed by the dashed rectangle in the figure and labeled "supply." The existence of such a disturbing field could be tested for by changing the area enclosed by the supply or its orientation. Provided that the supply is sufficiently remote, such changes are not found to produce any measurable mechanical effect on the rectangle.

There remains the question of the possible effect of the leads connecting the supply with the experimental part of the circuit. We shall assume these to be close together, as indicated in the figure. Since the currents in the leads are the same, and flow in opposite directions, their effects at a distance will be expected to cancel each other out. That this is indeed the case may readily be confirmed by bringing a part of this "twin flex" close to a compass needle. It is

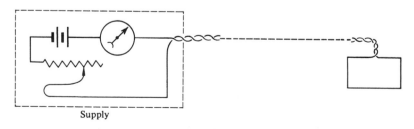

Supply

FIGURE 29.3 Circuit used for investigating action of magnetic field on current-carrying wire rectangle.

found that there is no effect on the needle, even with quite heavy currents. Thus the effective circuit with which we have to deal is simply the rectangle.

Not only does the current in the twin flex not produce any detectable field at an appreciable distance from itself, but, reciprocally, a magnetic field such as is produced by a magnet or by a current-carrying coil is found not to have any measurable effect on a current-carrying twin flex.

DIRECTION OF FORCE   The direction of the force exerted by a magnetic field on a current element is most conveniently found by exposing a current-carrying rectangle to the field under two conditions, in the first of which it is free to rotate about a central axis perpendicular to its plane, while in the second its axis of constraint is in the plane of the rectangle and parallel to one pair of opposite sides.

The first of these arrangements is shown in Fig. 29.4. With this it is found that the rectangle shows no tendency to rotate about its axis of constraint, regardless of the direction of the field; no torque is exerted about this axis. In the special case where the field is perpendicular to the plane of the rectangle this obviously means that whatever may be the forces acting on the four sides, there can be no components along these. For, if there were, all these components must, for consistency, be related to the direction of the current in the same way, and this would give rise to either a clockwise or an anticlockwise torque. In the general case where the field is not perpendicular to the plane of the rectangle the absence of components of force in the direction of the current must also be inferred, although in this case the inference is perhaps a little less obvious. Consideration of this more general case shows that in no self-consistent

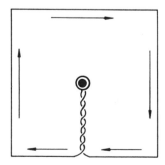

FIGURE 29.4   Current-carrying wire rectangle mounted on axle perpendicular to its plane and passing through its geometrical center.

scheme for the exertion of forces can the directions of the forces acting on the sides of the rectangle be other than at right angles to these sides. We must conclude that the only forces experienced by the sides of the rectangle are perpendicular to these: **The magnetic force exerted by a field on a current element is always at right angles to the element.**

Now let a rectangle be mounted so that its plane is vertical and it is free to rotate about a vertical axis in this plane, as shown in Fig. 29.5, and let it be exposed to a magnetic field.

From considerations of symmetry it is obvious that the lines of action of the forces which the field exerts on the top and bottom members of the rectangle must be through the axis of constraint; these forces therefore cannot contribute to any torque about this axis. Thus the only forces we need consider are those exerted on the vertical sides.

It is found that when the field is applied the rectangle comes to rest in such an orientation that the horizontal sides are at right

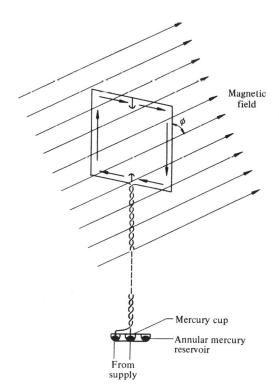

FIGURE 29.5 Current-carrying wire rectangle free to rotate about an axis in its plane, and exposed to a magnetic field.

angles to the direction of the field. Since in this equilibrium orientation no torque is acting about the axis of constraint, the forces on the vertical sides must be in the plane of the rectangle. We have already seen that the forces are perpendicular to the sides. The fact that they are in the plane of the rectangle therefore means that they are also parallel to the top and bottom sides, that is, at right angles to the field: **The force exerted on a current element is perpendicular both to the element and to the field.**

The forces on the vertical sides are obviously not affected by any rotation about the vertical axis of constraint, for these sides remain vertical in any such rotation. Hence the sense of the force on each of these, functioning as a current element, can readily be worked out from the particular orientation of the rectangle in which it is in *stable* equilibrium, or from the sense of the torque acting on it when it is in any orientation other than that of either stable or unstable equilibrium. The result may be recorded in the form of a suitable mnemonic, such as the following. The three quantities we are concerned with are the force F acting on a current element, the field B, and the current I. Arranging these in alphabetical order, we have B, F, I. This is one of three equivalent cyclic orders, the other two being F, I, B and I, B, F. We wish to find the sense of the force F in terms of I and B; so for this purpose we use the order F, I, B. Let arrows be constructed which point in the directions of I and B. Then not only is F at right angles to both of these, but its sense is such that, looking in the direction of F, we may go from I to B by a clockwise rotation of less than π.

VARIATION WITH ANGLE BETWEEN FIELD AND CURRENT ELEMENT   Let us now investigate how the force exerted by a field of a given magnitude on a given current element varies with the angle between the direction of the current and that of the field. This may be found with a rectangle such as that shown in Fig. 29.5 carrying a constant current and connected at the top via either a vertical wire capable of being twisted elastically, or some other form of spring, to a graduated torsion head. With the Helmholtz coils orientated to give field directions making different angles ϕ with the vertical, let the torsion head be rotated to twist the spring by the requisite amounts to turn the rectangle through a right angle from its equilibrium orientation so that its plane contains the direction of the field. If this is done it is found that the angle through which the spring has to be twisted is proportional to sin ϕ. Because of Hooke's law, the elastic torque applied by the spring is proportional to the angle of twist, and this elastic torque balances the magnetic torque. Hence the latter must also be proportional to sin ϕ. Since the arm of the couple is constant, being equal to the width of the rectangle, the force δF on each vertical side, whose length we shall later write as $\delta l'$, likewise must be proportional to $\sin \phi$:

$$\delta F \propto \sin \phi.$$

In the limiting case where ϕ is zero, δF is also zero; that is, **a magnetic field exerts no force on a current element having its own direction.**

The proportionality of δF to $\sin \phi$ means that B behaves like a vector, its horizontal component $B \sin \phi$ being fully effective in exerting forces on the vertical sides of the rectangle, while the vertical component, being parallel to these sides, has no effect: **A magnetic field is a vector quantity.** Accordingly, we shall henceforward represent it by the symbol **B** when we are considering its direction as well as its magnitude.

VARIATION WITH CURRENT With the rectangle having any constant orientation other than that in which the horizontal sides are perpendicular to the field, and with any constant value of the angle ϕ between the vertical sides and the field, let the current around the rectangle be varied. The torque exerted by the field on the rectangle, as measured by the angle through which the spring has to be twisted to maintain it in its constant orientation, is then found simply to be proportional to the current. This means that the force on each vertical side must also be proportional to the current, which we shall now designate by the symbol I' (the symbol I being reserved for another purpose):

$$\delta F \propto I'.$$

VARIATION WITH LENGTH Finally, let us consider how the force on a current element varies with its length. This could easily be found experimentally, for example by measuring torques as a function of the lengths of the vertical sides of the rectangle. Such measurements would show that, other things being equal, the force on a current element is simply proportional to its length:

$$\delta F \propto \delta l'.$$

That this must necessarily be the case may also be inferred from a simple theoretical argument without having to resort to any experimental test. Let the current element in question be considered as being made up of N shorter, equal "subelements" of some selected standard length, joined end to end. On each of these the same force must be exerted, and so the sum of these forces must be proportional to N, that is, to the length of the original element.

743

QUANTITATIVE DEFINITION OF **B** We have so far been able to assign a direction and sense to **B** but not a magnitude. However, we have found that for any definite, although unknown, value of **B,**

$$\delta F \propto I' \text{ for given values of } \delta l' \text{ and } \phi,$$
$$\delta F \propto \delta l' \text{ for given values of } I' \text{ and } \phi,$$

and

$$\delta F \propto \sin \phi \text{ for given values of } I' \text{ and } \delta l'.$$

All these are special cases of the more general statement that for given **B**

$$\delta F \propto I' \, \delta l' \sin \phi,$$

or

$$\frac{\delta F}{I' \, \delta l' \sin \phi} = \text{constant.}$$

The value of the constant must obviously depend on B, the magnitude of **B.** What could be more natural than to make this constant B itself, as a matter of definition? Accordingly, we now define B by the equation

$$B = \frac{\delta F}{I' \, \delta l' \sin \phi}. \tag{29.1}$$

We may also define B in words, using the special case where ϕ is a right angle: **The magnetic field B is the force exerted on a wire per unit length per unit current carried by it when the wire is at right angles to the field.**

Let us now return to our earlier finding concerning the direction and sense of the force exerted by a magnetic field on a current-carrying wire. By using vector notation we may incorporate this information with that concerning the magnitude of the force in a single equation. We do this in terms of what is known as the **vector product,** or **cross product,** of two vectors.

Let **A** and **B** be any two vectors and let their directions make an angle ϕ (less than π) with one another. Then we may define another vector **V** as that whose magnitude is $AB \sin \phi$, whose direction is at right angles to the directions of both **A** and **B**, and whose sense is such that looking in the direction of **V** we go from the direction of **A** to that of **B** by a clockwise rotation through the angle ϕ. All these specifications are included in the conventional representation of **V** as

$$\mathbf{V} = \mathbf{A} \times \mathbf{B,}$$

the right side of this equation being taken simply as an abbreviation for all that has been written above in words. **A** $\times$ **B** is called the vector product, or cross product, of **A** and **B**.

It is important to note that, according to our definition of the cross product,

$$\mathbf{B} \times \mathbf{A} = -\mathbf{A} \times \mathbf{B}.$$

The significance of the order in which the vectors are written in a cross product contrasts with the case of the dot or scalar product of two vectors or of the product of two scalar quantities, where the order is immaterial.

If we now consider $\mathbf{I}'$ as a vector of magnitude I' whose direction is that of the current in the wire, and $\mathbf{f}$ is written for the force *per unit length* that the field $\mathbf{B}$ exerts on this wire, it follows at once from Eq. (29.1) and from what we have found concerning the direction and sense of this force that

$$\mathbf{f} = \mathbf{I}' \times \mathbf{B}.$$

From the definition of B given in Eq. (29.1) we see that it may be expressed in N/A/m, or in kg/sec²/A. More usually, however, B is expressed in webers per square meter, the weber (Wb) being the MKSA unit of a quantity whose acquaintance we shall make in Chapter 32.

We have hitherto always referred to $\mathbf{B}$ as the "magnetic field." An alternative name commonly used for magnetic field is **magnetic induction.**

MEASUREMENT OF B Accurate measurement of B strictly in terms of its definition, that is, measurement of the force exerted on a short section of current-carrying wire constituting part of a circuit, would obviously not be practicable. It is therefore necessary to devise an alternative method of measurement that is free of the difficulties attaching to such a procedure.

We have, in effect, already considered this alternative method. This is the measurement of the torque exerted on a current-carrying coil whose plane contains the direction of the field.

Let us calculate the torque that a field $\mathbf{B}$ would exert on a rectangle of sides a and b about an axis parallel to the sides of length a when the current I' circulates around the rectangle and the field has the direction of the other two sides (Fig. 29.6). Such a field would, according to its definition, exert equal and opposite forces on the sides of length a equal to $BI'a$, the lines of action of these forces being separated by the distance b. Thus the torque L exerted on the rectangle would be $BI'ab$, or $BI'A$, where A is the area of the rectangle.

745

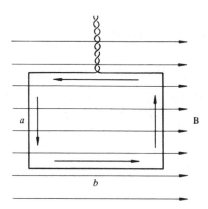

FIGURE 29.6 Current-carrying wire rectangle acted on by a magnetic field parallel to a pair of opposite sides.

It is not difficult to show that this result applies not only to the case of a rectangle but to a plane circuit of any geometrical form whose plane contains the direction of **B**. In the case of a coil of N turns, instead of a simple circuit in which the current goes around only once, the expression for the torque obviously becomes

$$L = NBI'A, \qquad (29.2)$$

so that for the quantity we wish to measure, B, we have

$$B = \frac{L}{NI'A}. \qquad (29.3)$$

This equation suggests an alternative definition of B which is equivalent to our earlier definition in terms of the force exerted on a current element. This is the torque per turn per unit current per unit area exerted on a current-carrying coil whose axis is at right angles to the field.

Torsion Magnetometer The most direct practical method of measuring a magnetic field based on the definition just stated is one in which provision is made for a torque to be applied elastically to a current-carrying coil. If the field is horizontal, or we are concerned only with the horizontal component of a nonhorizontal field, this torque could be produced by a twist in a wire from which the coil is suspended. The wire is twisted until the plane of the coil is parallel to the direction of the field, and then, from the twist and the elastic properties of the wire, the torque applied by the wire is calculated. This torque balances that exerted by the field, and so we have L. Substitution of this, together with the values of N, I', and A in Eq. (29.3), then gives us B.

Vibration Magnetometer From our earlier discussion of torques exerted on current-carrying circuits it is at once evident that Eq. (29.3) is a special case of the more general equation

$$B = \frac{L}{NI'A \sin \alpha}, \tag{29.4}$$

which applies where α is the angle between the axis of the coil and the direction of the field.

For values of α sufficiently small for $\sin \alpha$ to be identified with α itself, Eq. (29.4) may be written in the form

$$B = \frac{1}{NI'A} \left| \frac{L}{\alpha} \right|^{*};$$

that is, B is proportional to $|L/\alpha|$ by the factor $1/NI'A$. The value of $|L/\alpha|$ may be determined by suspending the current-carrying coil torsionlessly and allowing it to execute rotational oscillations of small amplitude under the magnetic "control" of the field. If the moment of inertia of the coil about the axis of oscillation is J,† the period of these oscillations is given by the equation

$$T = 2\pi \sqrt{\frac{J}{|L/\alpha|}},$$

whence we have

$$\left| \frac{L}{\alpha} \right| = \frac{4\pi^2 J}{T^2},$$

and so

$$B = \frac{1}{NI'A} \left| \frac{L}{\alpha} \right| = \frac{4\pi^2 J}{NI'AT^2}.$$

In practice it is difficult to avoid having a certain amount of elastic control associated with the current leads, this being superimposed on the magnetic control. It is a simple matter to allow for this, finding both the period T_1 with the current I' circulating through the coil and the longer period T_2 with the current switched off, and then using these two values of the period in a calculation of B.

Deflection Magnetometer Whereas the torsion and vibration magnetometers in the forms discussed above enable one to make an absolute determination of B, the deflection magnetometer is suitable

---

* We write $|L/\alpha|$ here rather than simply L/α because the latter is an essentially negative quantity.

† We use this symbol for moment of inertia here to avoid confusion with current, which is represented by the symbol I.

only for *comparing* a field B with a standard field B_0 whose value may or may not be known.

This being the case, there is, as we shall see, no particular point in using a current-carrying coil in the deflection magnetometer, although in principle this could be done; it is much more convenient to use a magnet.

The deflection magnetometer is used practically only for the comparison of horizontal fields, although a dip circle could be regarded as an exception to this rule; this compares the horizontal and vertical components of the earth's field. An example of the more usual application of the deflection magnetometer is the comparison of a field that either is or can be arranged to be horizontal with the horizontal component of the earth's field. Let us consider, in particular, this case.

A short bar magnet or compass needle is mounted on a pivot or suspended by a virtually torsionless fiber, and provision is made for measuring the angle through which it is deflected from the direction of the earth's field. It is arranged that the field B we wish to determine is not only horizontal but also at right angles to the earth's field, whose horizontal component may be denoted by the symbol B_0. Then the resultant field B_r will be as shown in Fig. 29.7, and the magnet will point in this direction. Since

$$B = B_r \sin \theta$$

and

$$B_0 = B_r \cos \theta,$$

we have

$$B = B_0 \tan \theta.$$

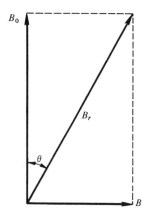

FIGURE 29.7 Vector parallelogram (= rectangle) of magnetic fields at right angles.

Use of Magnets in Torsion and Vibration Magnetometers If a magnet is suspended by a torsion fiber and caused to turn through an angle α from the direction of a magnetic field (or its horizontal component) by twisting the suspension, it is found that the twist required for this, and therefore also the torque applied elastically to the magnet, is proportional to sin α. In this respect a magnet behaves like a current-carrying coil. Also, if in a torsion or vibration magnetometer the current-carrying coil is replaced by a magnet, the ratio of the observed torques or frequencies corresponding to different magnetic fields is the same as when a coil is used. This justifies the use of a magnet in both these magnetometers in cases where we are concerned only with *comparisons* of fields. In such cases the use of a magnet has much to recommend it from a practical point of view, avoiding as it does all the complications due to current leads.

Field Produced by a Current Element

Our next task is to ascertain what field $\delta\mathbf{B}$ is produced at a distance r from a current element $I\ \delta\mathbf{l}$ at a point whose direction from the element makes with it an angle θ.* We shall need to determine not only the magnitude of this field but also its direction.

VARIATION WITH LENGTH OF CURRENT ELEMENT For given values of r, I, and θ, the field δB must obviously be proportional to δl. That this is so follows at once from reasoning similar to that from which we inferred that for a given field the force δF exerted on a current element having a given direction and carrying a given current I' is proportional to the length of the element $\delta l'$.

VARIATION WITH CURRENT Since the magnitude δB of the contribution $\delta\mathbf{B}$ of each element in a circuit carrying a current I must presumably vary with I according to the same law, this law must also apply to the magnitude B of the vector resultant $\mathbf{B}$ of all these contributions. Conversely, if we find experimentally the law according to which the field produced at any point by a complete circuit varies with the current carried by it, we may take it that this is also the law of variation with I of the δB due to any element $I\ \delta l$. We could, for example, investigate how the field B at the center of a current-carrying circular coil varies with the current. What we should find in this or any other case is that the former is

---

* The vector $\delta\mathbf{l}$ considered here may be defined as one having the magnitude δl and the same direction as the current.

simply proportional to the latter. It follows that the δB due to any current element is also proportional to I.

VARIATION WITH DISTANCE To find this, the simplest procedure is to determine the fields at the centers of a number of circular coils of different radii but all having the same number of turns and carrying the same current. Every element of such a coil is at the same distance r from the center and is at right angles to the corresponding radius. Clearly, the axial components δB_a of the fields produced at the center by these elements must all have the same direction and so are simply additive. Whether the current elements also produce field components in the plane of the coil is a question we cannot answer at this stage. Let us therefore avoid answering it by representing such components by the symbol δB_p and leaving it as an open question whether δB_p is finite or zero. In the latter case the field **B** at the center of the coil must obviously be axial, as we know it to be from considerations of symmetry. But, only a little less obviously, so it must also be in the former case. This is because the components δB_p due to equal current elements situated at opposite ends of any diameter must be equal and oppositely directed and so must cancel each other out, and all current elements in the coil are members of such pairs.

Since δB_a, δB_p, and their resultant δB are all the same kind of quantity, they all have the same dimensions. But this could hardly be true if they varied with r according to different laws. It is therefore a fair assumption that if we found the law applying to one of these quantities, for example, to δB_a, we should have found it for all three.

Let us assume an inverse power law for δB_a, which we may express as

$$\delta B_a = k\,\frac{I\,\delta l}{r^n},$$

where k and n are constants. Then for the whole coil, of N turns, we must have

$$B_a = k\,\frac{I}{r^n}\sum \delta l = 2\pi r N k\,\frac{I}{r^n} = 2\pi k N\,\frac{I}{r^{n-1}}.$$

Experimentally we find that

$$B_a \propto \frac{1}{r},$$

whence we may infer that an inverse-power law does indeed apply, with n equal to 2. Since not only δB_a, but also δB, must conform to this inverse-power law, we have

$$\delta B \propto \frac{1}{r^2}.$$

VARIATION WITH θ To ascertain how the magnetic field produced at a point O by a given current element at a given distance varies with θ we may carry out experiments with a constant short length of wire AB, as shown in Fig. 29.8, whose direction makes various angles with the line joining its center to O. In such experiments it is of course necessary to be satisfied that neither the supply system, PQ, nor the leads, PA and QB, make any contribution, or at least any *appreciable* contribution, to the field at O. With regard to the supply PQ, it follows from the way the field due to a current element varies with distance that if the supply is sufficiently remote the effect it produces at O must be negligible. To

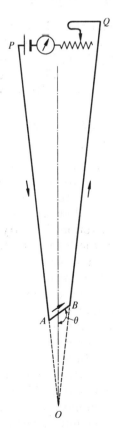

FIGURE 29.8   Arrangement for producing a magnetic field at a point O effectively due solely to current in short section of wire AB.

751

see this, we may regard the supply as the equivalent of a current element of length δl at a distance r from O. For a given current we have

$$\delta B \propto \frac{\delta l}{r^2}.$$

In the case under consideration δl is proportional to r. Hence δB is proportional to $1/r$, and so by increasing r we may make the field δB contributed by the supply as small as we wish. This leaves us with only the possible effects of the leads PA and QB to be considered.

There would appear to be a theoretical connection between our present problem and the result of our earlier series of experiments on the force δF produced by a given field on a current element $I'\,\delta l'$ inclined at an angle ϕ to the direction of the field, namely that

$$\delta F \propto I'\,\delta l'\,\sin\phi.$$

This means that δF is proportional to the product of I' and the component of $\delta \mathbf{l}'$ at right angles to the field. It seems reasonable to expect a certain reciprocity here, in the sense that only the component of $\delta \mathbf{l}$ at right angles to r should be effective in producing a magnetic field at O; in other words, that for given values of I, δl, and r, we should have

$$\delta B \propto \sin\theta.$$

In the system of Fig. 29.8 the currents in the leads PA and BQ flow directly toward and away from the point O. For none of the elements constituting these leads is the sine of the angle between it and the line joining it to O other than zero, so neither of these leads can contribute to the field at this point unless the expectation based on reciprocity considerations is faulty. If it *is* faulty, and the leads do have an effect on the field at O, the contribution of each lead must, from considerations of symmetry, be in its own direction. And since the directions of the two leads are not the same, these contributions must have a resultant in the plane of the diagram. But experimentally no component of the field in the plane of the diagram is observed; the field at O is accurately perpendicular to the plane of the diagram and is directed away from the reader, that is, into the paper. Also, in magnitude, the observed field is proportional to $\sin\theta$, θ being the angle between the direction of AB and that of the line joining its midpoint to O. This is strong, if perhaps not quite conclusive, evidence supporting our reasoning and the conclusion to which it leads. The only other possibility would be that the finite contribution of the leads in the plane of the diagram is exactly neutralized by a contribution in this plane made by the element AB, with the remaining contribution from AB, perpendicular to this plane, proportional to sin θ, and that the neutralization of the contributions in the plane of the diagram occurs for all values of the angle θ. It seems doubtful whether a mathematically self-consistent scheme giving this result could be found; and even if it could, the alternative explanation of the observed facts seems highly improbable. It is always good policy to try simple things first, so let us not further consider the improbable alternative. Actually, all experience in this field of physics is consistent with the proportionality of the contribution of any given element to sin θ. This is sufficient justification of our acceptance of this proportionality.

SENSE OF THE FIELD   This, which was referred to briefly for a specific case in the discussion above, conforms to the following simple general rule. Let us imagine any circle to be described about the current element as center. Then the direction of the field at any point on this circle is that of the corresponding tangent, and its sense is related to that of the current flow as seen from this point in the same way as the rotation of a right-handed screw is related to the direction of its progression.

AMPÈRE'S LAW   The separate variations of δB with I, δl, r, and θ found above may be combined in the single equation

$$\delta B = k \frac{I\,\delta l\,\sin\,\theta}{r^2}, \tag{29.5}$$

where k is a constant whose numerical value depends on the units in which the various quantities are expressed.

This law, although actually first proposed by Biot, is generally known as Ampère's law, after André Marie Ampère (1775–1836), who first discovered the mechanical interaction between current-carrying wires, and who independently derived the expression for δB from the results of a series of experiments with current-carrying wires having various configurations and orientations and situated at various distances from the point at which the field was determined.

The field **B** at a point due to a whole current-carrying circuit may be obtained from Eq. (29.5), in which the elements $\delta \mathbf{l}$ are made infinitesimal, by a process of vectorial integration over all the current elements constituting the circuit. The mathematical expression for **B** may be written in the form

$$\mathbf{B} = k \oint \frac{I\,d\mathbf{l}\,\sin\,\theta}{r^2}, \tag{29.6}$$

the symbol $\oint$ indicating integration around the whole circuit. It is to be understood that in performing this integration each $d\mathbf{B}$ must be given its appropriate direction and sense as discussed above.

753

Force between Current Elements

By writing δB in place of B in Eq. (29.1) to indicate that we are dealing with the case where the field is that due to a current *element*, and then eliminating δB between this equation and Eq. (29.5), we obtain for the force δF exerted by the current element $I \, \delta l$ on $I' \, \delta l'$ situated a distance r from it the equation

$$\delta F = k \frac{II' \, \delta l \, \delta l' \sin \theta \sin \phi}{r^2}. \tag{29.7}$$

The validity of this equation has been confirmed in all experimental tests to which it has been subjected.

There are certain peculiarities concerning the forces that current elements exert on one another which should be noted. As will be realized on reference to Fig. 29.1, these forces are not necessarily directed along the line joining the elements. Also, they are not, in general, *mutual* forces; there is no reciprocity in the sense that the force which one current element exerts on another is necessarily equal and opposite to that which the other exerts on the one. This will be evident on consideration of the two elements represented in Fig. 29.9. If we regard the left-hand current element of this system as acting on the right-hand one, θ and ϕ are both equal to $\pi/2$, and there is a force acting on the right-hand current element directed upward in the plane of the diagram. However, the right-hand element exerts no force at all on the current element on the left. Thus we have here a situation very different from that applying to the forces exerted between particles in mechanics.

This, implying, as it does, that current elements do not behave as particles, is of profound physical significance. It means that in the exertion of forces, current elements do not act independently, and that the assumption that they do so, on which Eq. (29.7) is based, is merely a convenient mathematical fiction, which, although useful as such, does not correspond to physical reality. It is a useful mathematical fiction, because, if Eq. (29.7), with δl and $\delta l'$ made infinitesimal, is used in a double integration to calculate the forces exerted on one another by a pair of *complete*

FIGURE 29.9   Two short current elements, one perpendicular to the line joining them, the other having a direction along this line.

circuits, the results always agree with the forces observed experimentally. *These* forces *are* equal and opposite and have a common line of action; also the calculated, as well as the observed, torques exerted on one another by such complete circuits are equal and opposite. Thus if the pair of circuits constitutes an isolated system, there is conservation of both linear and angular momentum.

Definition of the Ampere

The ampere was originally defined in terms of the force exerted by a circular current-carrying coil on a "unit magnetic pole" situated at its center. Fortunately, unlike the location of this pole, the ampere, in whose definition it was incorporated, in no way lacks precision, and so it has been possible to redefine this unit in more acceptable terms. It may be shown that the ampere as originally defined is identical with the unit of current that is required to balance Eq. (29.7) when δF is expressed in newtons; δl, $\delta l'$, and r are in meters; and k is given the numerical value 10^{-7}, the medium through which the action between the current elements occurs being a vacuum. This is the modern primary or fundamental definition of the ampere. The secondary definition given in Chapter 25 in terms of the electrolysis of silver in a silver voltameter depends on this.

A more formal definition, which is mathematically equivalent to the one given above and has been internationally agreed upon, is the following: The ampere is a current which, if maintained in each of two infinite straight parallel wires of infinitesimal cross section spaced 1 m apart in a vacuum, causes each to exert on the other a force of 2×10^{-7} N per meter of length.

This definition is, of course, unrealistic in the extreme; it would be unthinkable to attempt to measure a current in terms of it. Instead, the numerical value of k in Eqs. (29.5), (29.6), and (29.7), namely 10^{-7}, is used in practice to establish the ampere.

This is done by employing a so-called **current balance,** a form of balance by means of which the mechanical interaction between two current-carrying coil systems connected in series can be measured. In its original form the current balance consisted of a pair of coils having a common vertical axis, the lower member of which was fixed, while the upper coil was suspended from one arm of a beam balance. The electrical connections between the coils was such that when a current was passed it traveled the same way around each, causing the lower coil to exert an attractive force on

755

the upper. This force was not only "weighed" but was also calculated on the basis of Eq. (29.7) with k equal to 10^{-7}, in terms of the current I. Actually such a calculation gives the proportionality factor between the force and I^2 characteristic of the balance. Thus the magnitude of any particular current in amperes can be calculated from the observed force. Despite the considerable complexity of the calculation, with the force depending critically on the geometrical configuration of the coil system, it was found possible with this form of balance to determine currents to an accuracy of about 1 part in a million.

A much more elegant form of current balance, which, if only somebody had thought of it, would surely have been preferred, is shown in section in Fig. 29.10. This is really an electrodynamometer, other designs of which will be considered in Chapter 33. The two large coils are a pair of Helmholtz coils having a vertical axis. A much smaller coil is rigidly attached to the beam of a sensitive beam balance (not shown) in such a way that its axis is perpendicular to the central knife edge and is horizontal when the balance reads zero. The balance is so positioned that this coil lies in the central region between the Helmholtz coils where the field produced by a current passing through these is effectively uniform. When a current is passed through the Helmholtz coils and the small coil between them in series, a torque is exerted, via this small coil, on the balance, causing it to be deflected. Standard masses are now placed in the appropriate scale pan to restore the reading of the balance to zero. Since the restoring torque provided by these is equal and opposite to the magnetic deflecting torque, we have all the information required to calculate the current through the coils.

Let R be the effective mean radius and also the distance apart of the Helmholtz coils, while the effective mean radius of the central coil is r and the numbers of turns in each Helmholtz coil and in the central coil are N and n, respectively. Then a simple calculation shows that when a current of I amperes is passed through the coils,

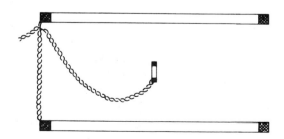

FIGURE 29.10   Coil system of a form of current balance.

the field B acting on the central coil is given by the equation

$$B = \frac{32\pi NI}{5\sqrt{5}\,R} \times 10^{-7}.$$

For the torque that this field produces on the central coil we have, therefore,

$$L = BnI \cdot \pi r^2 = \frac{32\pi^2 NnI^2 r^2}{5\sqrt{5}R} \times 10^{-7}.$$

Also

$$L = mgl,$$

where m is the mass placed on one scale pan to restore the zero reading and l is the length of the corresponding arm of the balance. Hence, equating the right-hand sides of the last two equations and rearranging, we have

$$I^2 = \frac{5\sqrt{5}}{32\pi^2}\frac{mglR}{Nnr^2} \times 10^7. \qquad (29.8)$$

In practice, various minor corrections would have to be made, for example for the slight variation of **B** over the area of the central coil and for any inequality between the effective radii of the Helmholtz coils and their distance apart.

By passing a current in series through a current balance and a silver voltameter and making the appropriate measurements, the ratio of the rate of deposition of silver to the current strength may be determined, and this may then be used for the calibration of working instruments, such as ordinary moving-coil ammeters.

Problems

29.1   A wire rectangle 8 cm by 12 cm carries a current of 3 A and is situated in a magnetic field of induction of 0.5 N/A/m. What must be the orientation of the rectangle relative to the field (a) to make the forces exerted on all four sides a maximum, and (b) to make the torque on the rectangle a maximum? What are the magnitudes of these maximum forces and of the maximum torque?

29.2   The rectangle of Problem 29.1 is so constrained that the only motion allowed it is rotation about a fixed axis. If it is so oriented that its plane contains the direction of **B** and its longer sides make an angle of 30° with this direction, what are the forces exerted on the four sides? What is the torque about the axis of constraint if this passes (a) through the midpoints of the shorter sides of the rectangle, and (b) through the midpoints of the longer sides? If, instead, the rectangle were so oriented that its shorter sides were perpendicular to the field, and the normal to its plane made an angle of 30° with the direction of **B**, what would be the forces

757

on the four sides and what would be the torque exerted about the two axes of constraint specified above?

29.3 A circular loop of wire of radius 5 cm carries a current of 10 A and is exposed to a magnetic field of 20 N/A/m in the direction of its axis. What is the resulting tension in the wire? (Neglect the effect of the field produced by the current in the loop itself.)

29.4 A current flows axially along a cylindrical pipe. Show that at all points within this pipe the resultant magnetic field due to this current is zero.

29.5 Show that the torque exerted by a field **B** on a plane circuit of any form to whose plane **B** is parallel is BIA, where I is the current and A is the area of the circuit.

29.6 When the axis of a wire loop of area A which carries a current I is at right angles to a magnetic field to which it is exposed, it experiences a torque L. How much work would have to be performed on this loop to turn it from its orientation of stable equilibrium in the field to its orientation of unstable equilibrium?

29.7 A galvanometer has a rectangular coil of 10 turns, and the effective dimensions of the part of the coil within the magnetic field are 4 cm by 2.5 cm. The magnetic field to which the coil is exposed is radial and has an effective mean magnitude of 40 N/m/A. The coil is suspended by a torsion wire that develops an elastic restoring torque of 4×10^{-8} N-m per degree of twist. If full-scale deflection is to be limited to 100°, what is the maximum current the galvanometer can measure without the use of a shunt?

29.8 Derive a formula for the magnetic field B in terms of the periods of vibration T_1 and T_2 of a vibration magnetometer subject to an elastic control. The magnetometer consists of a coil of N turns whose area and moment of inertia are A and J, respectively. T_1 is the period found when a current I is passed through the coil and T_2 is that characteristic of the elastic control only, the current being switched off.

29.9 (a) A long stiff uniform wire of length l is suspended at one end by a frictionless hinge of negligible moment of inertia, so that it can swing as a pendulum. Show that the period T of small oscillations is given by the equation

$$T = 2\pi \sqrt{\frac{2l}{3g}}.$$

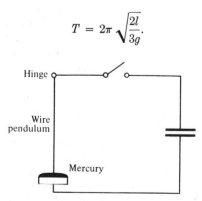

Figure for problem 29.9.

(b)   The wire referred to in part (a) is of length 50 cm and mass 25 g, and when at rest dips into a pool of mercury. It forms part of the circuit shown in the figure whose total resistance is 0.05 Ω. The capacitor has a capacitance of 10 μF and, with the switch open, is charged to 10 kV. Find the time required for the charge on the capacitor to fall to $\frac{1}{1000}$ of its initial value when the switch is closed.

(c)   If the wire is situated in a uniform horizontal magnetic field of 0.2 N/m/A directed parallel to the axis of the hinge, what is the amplitude of the oscillatory motion imparted to the wire when the capacitor is discharged? (Neglect any frictional or viscous resistance to motion, and assume that the amplitude is sufficiently small for the approximation $\cos \theta = 1 - \theta^2/2$ to apply.)

(d)   The arrangement described could be used for the measurement of charge in terms of the amplitude of the pendulum and other quantities. If the angular amplitude θ_m is produced by the discharge of a quantity Q of electricity, the charge sensitivity is θ_m/Q. Alternatively, the suspended wire could, in the absence of frictional forces, be used to measure small steady currents. If a steady current I produces a steady deflection θ_s (which must be sufficiently small for the lower end of the wire to remain in contact with the mercury), the current sensitivity is θ_s/I. Show that the charge sensitivitity is related to the current sensitivity by the equation

$$\frac{\theta_m}{Q} = \frac{2\pi}{T}\frac{\theta_s}{I}.$$

(Compare the corresponding relation for the ballistic galvanometer.)

29.10   A coil is to be wound on a form of fixed dimensions (see section shown in the figure), and a given potential difference is to be applied between the two ends of the winding. Find whether a greater magnetic field will be produced within the coil if fine wire is used rather than coarse, or vice versa, or whether it makes no difference either way. (Neglect the thickness of the insulation.)

29.11   A circular coil of 100 turns is wound in the form of a flat

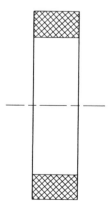

Figure for problem 29.10.

pancake of inner and outer radii 2 cm and 10 cm, respectively. What is the magnetic field at the center of this coil when it carries a current of 10 A?

29.12   A circular coil of 30 turns and radius 24 cm is mounted with its plane vertical and in the magnetic meridian at a place where the horizontal component of the earth's field is 2.5×10^{-5} N/m/A. What is the current in the coil if a small magnetic needle at the center is deflected 52°?

29.13   Show that the magnetic field at the center of a Helmholtz-coil system is given by the formula

$$B = \frac{32\pi N I}{5\sqrt{5R}} \times 10^{-7},$$

where N is the number of turns in each coil, R the coil radius, and I the current carried.

30

MAGNETIC FIELDS

Having established the basic laws of the exertion of magnetic forces in Chapter 29, we shall, in the present chapter, investigate the fields produced in a number of interesting special cases. We shall then see how a consideration of the results of these investigations leads to a theorem, Ampère's line-integral theorem, which may be regarded as the electromagnetic counterpart of Gauss's theorem in electrostatics, and which is of equally fundamental importance. Finally, after considering a theory of terrestrial magnetism, we shall establish a law for the force exerted by a magnetic field on a moving charged particle and see how this is applied in "magnetic-lens" systems for focusing beams of such particles.

Magnetic Permeability

Before considering the special cases just referred to, it will be useful, as a preliminary, to generalize our findings so far, to include the case where the medium through which the magnetic forces are exerted is something other than free space.

The quantity k in Eq. (29.5) is strictly constant, and has the numerical value of 10^{-7}, only for a vacuum. Although for most other media this equation still applies with sufficient accuracy for most purposes, with k having practically the same value as for a vacuum, there are some media for which k in this equation varies appreciably with I and geometrical factors, and has numerical values differing greatly from 10^{-7}.

In the MKSA system the quantity k is not actually used, but, instead, another quantity, μ, defined as $4\pi k$. This is known as the **magnetic permeability** of the medium. In terms of this, Eqs. (29.6) and (29.7) take the forms

$$B = \frac{\mu}{4\pi} \oint \frac{I \, dl \, \sin \theta}{r^2} \qquad (30.1)$$

and

$$\delta F = \frac{\mu}{4\pi} \frac{II' \, \delta l \, \delta l' \, \sin \theta \, \sin \phi}{r^2}, \qquad (30.2)$$

respectively.

For a vacuum the symbol μ in these equations should be replaced by μ_0, the numerical value of μ_0 being $4\pi \times 10^{-7}$.

The ratio of the permeability μ of any medium to μ_0, the permeability of empty space, is known as the **relative permeability** of the medium. Of course, this, like k, is not constant for all media, although, if we exclude exceptionally high values of B, it is for most.

By inspection of Eq. (30.2) we see that μ has the dimensions of a force divided by the square of a current. Accordingly, it may be expressed in N/A^2 or in $kg\text{-}m/s^2/A^2$. As we see from Eq. (30.1), it may also be expressed in $Wb/A/m$, and this is more usual.

Field Due to Infinitely Long Straight Wire

We shall now obtain an expression for the field B at the point O in Fig. 30.1, whose distance from an infinitely long thin wire carrying a current I is r. This will involve the use of Ampère's equation (29.5) and integration of the contributions of all current elements in the wire such as PQ. This integration is best carried out in terms of the angle θ, which, in the range 0 to π, includes the whole infinite wire.

It is readily seen, geometrically, that the distance of the element PQ from O is $r/\sin\theta$ and that its length is $(r/\sin^2\theta)\,\delta\theta$, also that θ may be identified with the θ in Eq. (29.5). Accordingly, for the contribution of this current element to the field at O, we have, writing $\mu/4\pi$ in place of k,

$$\delta B = \frac{\mu}{4\pi} I \frac{r}{\sin^2\theta} \, \delta\theta \cdot \sin\theta \cdot \frac{\sin^2\theta}{r^2} = \frac{\mu I}{4\pi r} \sin\theta \, \delta\theta.$$

All contributions $\delta\mathbf{B}$ have the same direction, at right angles to the plane of the diagram and into the paper; hence they are all

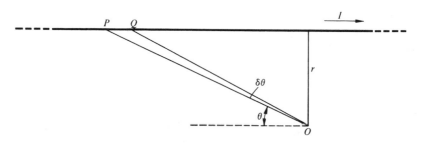

FIGURE 30.1

simply additive. For the total field B at O we have, therefore,

$$B = \frac{\mu I}{4\pi r} \int_0^\pi \sin \theta \, d\theta = \frac{\mu I}{2\pi r}. \qquad (30.3)$$

If, now, a current element $I' \, \delta l'$ is at O, and this is in the plane of the diagram, then the angle ϕ is $\pi/2$, and for the force δF acting on this element we have

$$\delta F = BI' \, \delta l' = \frac{\mu I I' \, \delta l'}{2\pi r},$$

and so for the force per unit length,

$$\frac{\delta F}{\delta l'} = \frac{\mu I I'}{2\pi r}.$$

If, in particular, $\delta l'$ is parallel to the infinite wire, being an element of one of two infinite parallel wires in a vacuum, and furthermore the distance r between the wires is 1 m and I and I' are both 1 A, then the force per unit length is $\mu_0/2\pi$, that is, 2×10^{-7} N/m. This corresponds to the formal definition of the ampere given in Chapter 29.

Field on Axis of Flat Circular Coil

The position of the point P on the axis may be expressed either in terms of the distance from the center of the coil or of the angle ψ shown in Fig. 30.2. The field is most conveniently found in terms of the latter.

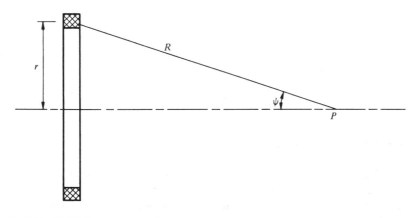

FIGURE 30.2

Each element of the coil is at right angles to the line joining it to P. Thus $\theta = \pi/2$, and so for the contribution which the element makes to the field we have

$$\delta B = \frac{\mu I}{4\pi R^2}\, \delta l,$$

with

$$R = \frac{r}{\sin \psi},$$

so that

$$\delta B = \frac{\mu I}{4\pi r^2} \sin^2 \psi\, \delta l.$$

The direction of $\delta \mathbf{B}$ is perpendicular to the line joining the element to P; hence it has both an axial component δB_a equal to $\delta B \sin \psi$ and a component at right angles to the axis equal to $\delta B \cos \psi$. Because the components perpendicular to the axis due to equal elements at opposite ends of a diameter are equal and oppositely directed and so cancel each other out, and all elements are members of such pairs, these components will make no contribution to the resultant field at P. Accordingly, we shall consider only the axial components, of magnitude

$$\delta B_a = \frac{\mu I}{4\pi r^2} \sin^3 \psi\, \delta l.$$

Such components, all having the same direction, are simply additive, so we have for the resultant field

$$B = \frac{\mu I}{4\pi r^2} \sin^3 \psi \sum \delta l.$$

For a coil of N turns $\Sigma\, \delta l$ is $2\pi r N$. The corresponding expression for the field is

$$B = \frac{\mu N I}{2r} \sin^3 \psi. \tag{30.4}$$

Note that the factor $\sin^3 \psi$ implies a falling away of B with increasing R according to an inverse-cube law.

Field Due to Current in Solenoid

The solenoid, of radius r and having n turns per unit length, is indicated diagrammatically in axial section in Fig. 30.3. We shall now calculate the field at a point O on the axis of the solenoid produced by a current I flowing in it. For this purpose we can use

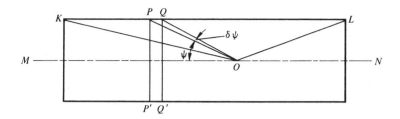

FIGURE 30.3

Eq. (30.4), considering the short section of the solenoid between the planes PP' and QQ' as the equivalent of a flat circular coil and integrating over the length of the solenoid.

If the number of turns between PP' and QQ' is δN, then the contribution δB that this section makes to the field at O is, as we see from Eq. (30.4),

$$\delta B = \frac{\mu I}{2r}\, \delta N \sin^3 \psi. \tag{30.5}$$

It is easily seen geometrically that PQ may be expressed in terms of the angles ψ and $\delta\psi$ shown in the figure by the equation

$$PQ = \frac{r}{\sin^2 \psi}\, \delta\psi.$$

The corresponding number of turns contained in this interval is $(nr/\sin^2 \psi)\, \delta\psi$. Hence, substituting this for δN in Eq. (30.5), we have

$$\delta B = \frac{\mu n I}{2} \sin \psi\, \delta\psi.$$

Integration of this between the angles MOK and MOL, which we shall write as ψ_1 and ψ_2, respectively, gives us for the total field at O,

$$B = \frac{\mu n I}{2} \int_{\psi_1}^{\psi_2} \sin \psi\, d\psi = \frac{\mu n I}{2} (\cos \psi_1 - \cos \psi_2). \tag{30.6}$$

In the limiting case of an infinitely long solenoid, for which $\psi_1 = 0$ and $\psi_2 = \pi$, $\cos \psi_1 - \cos \psi_2 = 2$, and so

$$B = \mu n I. \tag{30.7}$$

This will still be the magnitude of the field in a "long," even if not infinitely long, solenoid—long in the sense that $1 - \cos \psi_1$ and $1 + \cos \psi_2$ are both negligibly small.

It will be shown later that (1) the field well within the interior of a long solenoid is uniform over its whole cross section, and (2) outside the solenoid in regions remote from the ends it is zero.

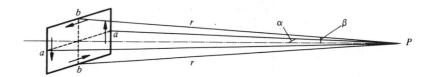

FIGURE 30.4

Field Due to Small Circuit or Coil

For simplicity let us consider a circuit of rectangular form and investigate the field (1) at a point on its "axis," that is, on a line passing through its center and perpendicular to its plane, and (2) at a point in its plane. After we have discussed these two special cases we shall see how to find the field at a point in *any* direction from the rectangle.

In Fig. 30.4 let P be the point on the axis of the current-carrying rectangle shown on the left at which we wish to find the field due to the rectangle.

We may regard the four sides of the rectangle as current elements. From considerations of symmetry it is evident that we need take into account only the axial components of the fields due to these elements; the components at right angles to the axis must cancel each other out.

The axial component of the field due to the current in each vertical side, of length a, is $(\mu I a \sin \alpha)/4\pi r^2$; while that in each horizontal side, of length b, contributes $(\mu I b \sin \beta)/4\pi r^2$.* Hence the total field is $(\mu I/2\pi r^2)(a \sin \alpha + b \sin \beta)$. But

$$\sin \alpha = \frac{b}{2r}, \qquad \sin \beta = \frac{a}{2r},$$

so that

$$a \sin \alpha + b \sin \beta = \frac{ab}{r} = \frac{A}{r},$$

where A is the area of the rectangle. Hence our expression for the field becomes simplified to

$$B = \frac{\mu I A}{2\pi r^3}. \tag{30.8}$$

This is the field for the "A (axial) position of Gauss," the case where the point P is on the axis of the circuit.

---

* We make no distinction here between the distances of the different sides of the rectangle from P.

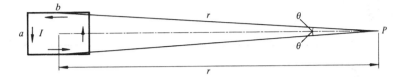

FIGURE 30.5

Now let us investigate the field for the "B (broadside) position of Gauss," where P is in the plane of the circuit. This case is shown in Fig. 30.5. Here the fields produced by all four sides of the rectangle are at right angles to the plane of the diagram.

The field due to the right-hand vertical side is $\mu I a/4\pi(r - b/2)^2$, or $(\mu I a/4\pi r^2)(1 - b/2r)^{-2}$, and is directed inward, while that due to the left-hand vertical side, which is directed outward, is

$$\frac{\mu I a}{4\pi(r + b/2)^2},$$

or

$$\left(\frac{\mu I a}{4\pi r^2}\right)\left(1 + \frac{b}{2r^2}\right)^{-2}.$$

The combined effect of these two sides is therefore a field of

$$\left(\frac{\mu I a}{4\pi r^2}\right)\left[\left(1 - \frac{b}{2r}\right)^{-2} - \left(1 + \frac{b}{2r}\right)^{-2}\right]$$

directed inward, and, $b/2r$ being small, this simplifies to $\mu I ab/2\pi r^3$, or $\mu I A/2\pi r^3$, where A is the area of the rectangle.

The field contributions due to the top and bottom sides of the rectangle are each equal to $(\mu I b \sin \theta)/4\pi r^2$, where θ is the angle shown, and these are both directed outward. Adding these and substituting $a/2r$ for $\sin \theta$, we obtain for the combined effect of these two sides $\mu I ab/4\pi r^3$, or $\mu I A/4\pi r^3$.

Subtracting this field due to the top and bottom sides from that due to the vertical sides, we obtain for the overall resultant field at P the value $\mu I A/4\pi r^3$:

$$B = \frac{\mu I A}{4\pi r^3}. \tag{30.9}$$

This is directed inward.

It should be noted that not only does the field at a distant point conform to an inverse-cube law for both A and B positions of Gauss, but also that in both cases this field is parallel to the axis of the rectangle.

767

FIGURE 30.6

It may readily be shown that Eqs. (30.8) and (30.9) are valid not only for rectangular circuits but for two-dimensional circuits of any geometrical form.

In the case of current-carrying *coils*, the right sides of Eqs. (30.8) and (30.9) must, of course, be multiplied by N, where N is the number of turns.

Let us now consider the problem of finding the direction and magnitude of the field produced by a current-carrying rectangle at a distant point that is neither on the axis of the rectangle nor in its plane, for example the point P in Fig. 30.6, which shows the rectangle in section. As may readily be seen with the aid of a model (which will greatly facilitate discussion of the three-dimensional geometry of the situation), the effect produced at P by the current flowing around this rectangle is the same as the combined effect of two rectangles, each carrying the same current, one of these being the projection of the original rectangle in a plane through its center O and perpendicular to OP, while the other is the projection in the plane at right angles to that of the diagram which contains the line OP. These two projections are indicated in Fig. 30.7. If ϕ is the angle OP makes with the axis of the rectangle, then the areas of the two projected rectangles are $A \cos \phi$ and $A \sin \phi$, respectively. The field at P due to the former of these, relative to which P is in the A position of Gauss, would be $(\mu I a \cos \phi)/2\pi r^3$, while the field due to the latter (B position of Gauss) would be $(\mu I A \sin \phi)/4\pi r^3$, r being

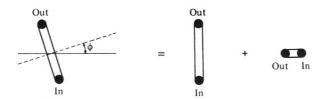

FIGURE 30.7 Components of current-carrying coil in two orientations at right angles determining magnetic fields at distant points.

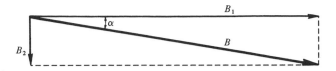

FIGURE 30.8

the distance OP. The former of these fields is directed along OP, while the direction of the latter is at right angles to this line. As is shown in Fig. 30.8, in which these fields are labeled B_1 and B_2, respectively, the resultant field B is therefore equal to $\sqrt{B_1{}^2 + B_2{}^2}$; that is,

$$B = \frac{\mu I A}{4\pi r^3} \sqrt{4 \cos^2 \phi + \sin^2 \phi}, \qquad (30.10)$$

and the direction of this field makes an angle α with OP such that

$$\tan \alpha = \frac{\sin \phi}{2 \cos \phi}. \qquad (30.11)$$

Just as for the A and B positions of Gauss, so also here, the validity of these results is not confined to the case where the circuit is rectangular; they are valid for all two-dimensional circuits, regardless of their geometrical form.

Magnetic Moment

In Eqs. (30.8), (30.9), and (30.10) the expressions for the field all contain the factor $\mu I A$, which, for free space as our medium, becomes $\mu_0 I A$. This quantity, which is a property of the circuit and the current carried by it, is a measure of the importance of the current-carrying circuit from two points of view. It is μ_0 times the torque exerted on the circuit per unit value of the field B when its axis is perpendicular to the field. Also, $\mu_0 I A$, in conjunction with the values of r and ϕ, determines the field produced by the current-carrying circuit at a distant point. These considerations suggest that we give this quantity a name. The name that has been chosen is the **magnetic moment** of the circuit. In the case of a coil of N turns the expression for the magnetic moment is, of course, $\mu_0 N I A$.

Some physicists define the magnetic moment as simply $N I A$, without the μ_0. Either definition is acceptable, but, of the two, $\mu_0 N I A$ appears to be somewhat more convenient for the discussion

769

of ferromagnetism to be undertaken in Chapter 31, and it is this definition, therefore, that is adopted here.

Field Due to Large Circuit

The procedure considered above for finding the field due to a current-carrying circuit applies only in the case of a circuit that is all in one plane, and for points at distances from it that are large compared with the linear dimensions of the circuit. It is not difficult to extend the procedure to cases where these restrictions do not apply.

We shall now consider *any* circuit, neither necessarily small compared with its distance from the point at which the field is to be determined nor necessarily two-dimensional. This circuit, with the direction of current flow around it, is indicated in Fig. 30.9.

Let us suppose the circuit to be subdivided in the manner shown, not by actual wires but simply in imagination, and let the linear dimensions of each "mesh" be small compared with its distance from the point at which we wish to determine the field. Then we may imagine the current to circulate around each individual mesh, in the manner indicated in the figure for a few of them, and, for each mesh, find the field it produces at the point in question. The resultant field at the point can then be found by vectorial composition. As we shall now see, the field determined in this

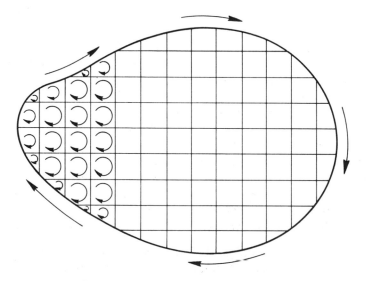

FIGURE 30.9 Equivalent mesh system of current-carrying circuit.

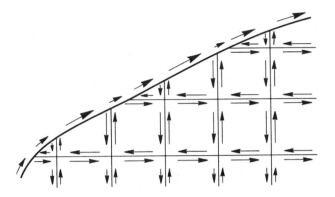

FIGURE 30.10

manner must necessarily be that due to the current flowing around the actual (not subdivided) circuit.

That this is indeed the case will be evident from Fig. 30.10, which shows a small part of the circuit on an enlarged scale. We see that there is mutual cancellation of the currents imagined to circulate around neighboring meshes for each element of each mesh except the elements constituting the actual circuit.

It may be noted here, parenthetically, that it is by imagining small two-dimensional nonrectangular circuits to be subdivided into rectangular meshes that the validity of Eqs. (30.8) through (30.11) for such circuits, as well as for rectangular ones, is most easily established.

Ampère's Line-Integral Theorem

When we have obtained a theoretical equation in physics it is often useful to manipulate it algebraically into other forms, for example by inverting it or turning it back to front, and to see if we can interpret physically the equation in its new form. This procedure often leads to important theoretical advances.

A case in point is the expression for the magnetic field in the vicinity of a long straight current-carrying wire,

$$B = \frac{\mu I}{2\pi r}. \tag{30.3}$$

Now $2\pi r$ is obviously the circumference of the circle centered on and perpendicular to the wire which we may construct through the

771

point. If we multiply both sides of this by $2\pi r$ we have

$$B \cdot 2\pi r = \mu I,$$

or, in terms of all the elements δs into which we may imagine the circumference of the circle to be divided,

$$B \cdot \sum \delta s = \mu I.$$

In this equation the symbol B may represent either the field at one point, as it does in Eq. (30.3), or, because the magnitude of the field is the same everywhere around the circle, it may equally well represent the field at the location of each element of path δs traversed as we go around the circle. Let us decide on the latter, and, in conformity with this, associate B definitely with δs by placing it on the other side of the summation sign:

$$\sum B \cdot \delta s = \mu I. \tag{30.12}$$

If we now imagine the circle to be described in a direction related to the current according to the right-hand screw relation, then the vector **B** has the same direction as δ**s** everywhere around the circle; that is, the angle α between them is zero, and so we may write, in place of Eq. (30.12),

$$\sum B \cos \alpha \cdot \delta s = \mu I.$$

The introduction of the angle α into the equation may seem hardly necessary, until we reflect that since **B** and δ**s** are both vectors we surely must expect the angle between them to have some relevance physically. Accordingly, it seems sensible to insert $\cos \alpha$ as a factor, at least experimentally.

Writing the last equation in integral form, we have

$$\oint B \cos \alpha \cdot ds = \mu I, \tag{30.13}$$

the symbol $\oint$ indicating integration once around the closed (here circular) path linking the current. The left side of Eq. (30.13) is known as the **line integral** of **B** around the closed path.

Although for the particular current-carrying wire and the closed path around it that we are here contemplating Eq. (30.13) tells us no more than does Eq. (30.3), it may well be that Eq. (30.13) has a *general* validity applying to *any* closed path linking *any* circuit. If so, this equation could prove as useful in solving problems relating to magnetic fields as we have found Gauss's theorem to be in solving electrostatic problems.

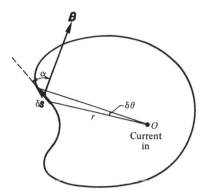

FIGURE 30.11

To obtain some indication of whether Eq. (30.13) is indeed valid generally, let us in the first instance test it for *any* closed path around the wire, as indicated in Fig. 30.11. Let the wire, which is perpendicular to the plane of the diagram and cuts this at the point O, carry a current I in the direction away from the reader. We shall consider a path that may take any three-dimensional course around this wire; it need not be confined to the plane of the diagram. If an element $\delta\mathbf{s}$ of this path, whose distance from the wire is r, makes an angle α with $\mathbf{B}$, then the component of this in the direction of $\mathbf{B}$ is $\delta s \cos \alpha$. Let $\delta\theta$ be the angle subtended by the projection of $\delta\mathbf{s}$ on the plane of the diagram at O. Then since this projection is also the component of $\delta\mathbf{s}$ in the direction of $\mathbf{B}$, we have

$$\delta s \cos \alpha = r \, \delta\theta,$$

and so

$$B \cos \alpha \cdot \delta s = \frac{\mu I}{2\pi r} \cdot r \, \delta\theta = \frac{\mu I}{2\pi} \, \delta\theta.$$

Integration of this gives us the same equation as applies to a circle constructed about the wire as axis,

$$\oint B \cos \alpha \cdot ds = \frac{\mu I}{2\pi} \int d\theta = \mu I. \tag{30.13}$$

In practice we could create a system approximating in its effects to an infinite straight current-carrying wire by joining the ends of a long straight wire by long leads to a supply, so that the far ends of the wire, the leads, and the supply are all so remote from the closed path considered that they have no appreciable effect on the field. Another alternative would be to set up a very large circuit in the form of a circle, and to consider a closed path linking it whose dimensions are small compared with those of the circuit;

773

the portion of the circuit contributing appreciably to the field could then be regarded as practically a straight line. In both of these cases the closed path is confined to a relatively small region near the path of the current. We shall now also test a case representing the other extreme, where the dimensions of the path considered are large compared with those of the circuit, and in which the region where it links the circuit is as far as possible from the current. We should have these conditions if the circuit were in the form of a circle, and, with the exception of the relatively remote portions of the closed path linking it, this were the axis of the circle, from a point at a great distance from it on one side to another at a great distance on the other. In this case the line integral for the whole closed path could, in effect, be identified with $\int_{-\infty}^{+\infty} B \, dx$, x being the coordinate of a point on the axis referred to the center of the circle as origin and α being zero at all points on the axis.

To evaluate this integral we may use the expression we obtained earlier for B in terms of the angle ψ shown in Fig. 30.12,

$$B = \frac{\mu I}{2r} \sin^3 \psi.$$

We have

$$x = r \cot \psi, \qquad dx = -\frac{r \, d\psi}{\sin^2 \psi},$$

and so

$$B \, dx = -\frac{\mu I}{2} \sin \psi \, d\psi,$$

whence

$$\int_{-\infty}^{+\infty} B \, dx = -\frac{\mu I}{2} \int_{\pi}^{0} \sin \psi \, d\psi = \frac{\mu I}{2} \int_{0}^{\pi} \sin \psi \, d\psi = \mu I.$$

The fact that in all cases considered we have obtained the same expression, μI, for the line integral strongly suggests that Eq. (30.13) applies to *any* closed path linking *any* current-carrying circuit, such as is indicated in Fig. 30.12. That Eq. (30.13) does indeed have general validity was first established by Ampère, and the corresponding theorem is known as **Ampère's line-integral theorem.** A proof of this will now be given.

In Fig. 30.13 let the upper closed curve represent the circuit carrying a current I in the direction indicated, and let $\delta \mathbf{l}$ be a short element of this circuit. Let P be a point on a closed path linking the circuit, at a distance r from $\delta \mathbf{l}$, and let us consider a small displacement $\delta \mathbf{s}$ of P along this path. Also let $\hat{\mathbf{r}}$ be a unit vector having the direction from $\delta \mathbf{l}$ to the point P, a dimensionless directed quantity

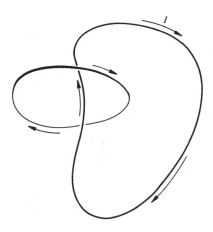

FIGURE 30.12 A closed path linking a circuit.

of magnitude unity. Then if $\delta\mathbf{B}$ is the vectorial contribution of the current element $I\,\delta\mathbf{l}$ to the magnetic field at P, we may either express this in magnitude by Eq. (29.5) and separately state its direction, or give both items of information by means of the vectorial equation

$$\delta\mathbf{B} = \frac{\mu I}{4\pi}\frac{\delta\mathbf{l}\times\hat{\mathbf{r}}}{r^2}. \tag{30.14}$$

Accordingly, we may express the contribution of the current element to the line integral in vectorial form by the equation

$$\delta\mathbf{B}\cdot\delta\mathbf{s} = \frac{\mu I}{4\pi r^2}(\delta\mathbf{l}\times\hat{\mathbf{r}})\cdot\delta\mathbf{s}. \tag{30.15}$$

If the right-hand side of this equation is transformed according to a certain theorem in vector algebra, its new form immediately suggests interpretation in terms of a solid angle. Accordingly, after

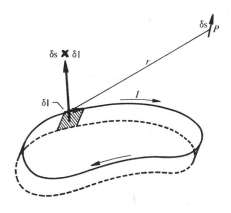

FIGURE 30.13

having satisfied ourselves of the validity of this theorem, we shall carry out the transformation.

The theorem in question is that if **A**, **B**, and **C** are any three vectors, then

$$(\mathbf{A} \times \mathbf{B}) \cdot \mathbf{C} = (\mathbf{B} \times \mathbf{C}) \cdot \mathbf{A} = (\mathbf{C} \times \mathbf{A}) \cdot \mathbf{B}.$$

That this is so may be seen at once by considering a parallelepiped whose sides are proportional to, and have the same directions as, these vectors. Clearly, each cross product then corresponds to the area of one of the sides associated with one of the directions normal to it, and the dot product of this and the remaining vector corresponds either to the volume of the parallelepiped, or minus this volume. It is easily seen that the cyclic order ensures that all three expressions correspond consistently to either one or the other. Accordingly, the validity of the theorem is confirmed.

Applying this theorem, we may now write Eq. (30.15) in the form

$$\delta\mathbf{B} \cdot \delta\mathbf{s} = \frac{\mu I}{4\pi r^2} (\delta\mathbf{s} \times \delta\mathbf{l}) \cdot \hat{\mathbf{r}}. \tag{30.16}$$

Before attempting to interpret this, let us refer again to Fig. 30.13, in which the dashed curve represents the circuit displaced by $-\delta\mathbf{s}$. Since all motion is relative, it is clearly permissible to consider P fixed instead of being displaced along $\delta\mathbf{s}$, while instead of keeping the circuit fixed we displace it by $-\delta\mathbf{s}$; the relative initial and final configurations of the circuit and the point P are the same in either case. It will suit our immediate purpose better to imagine the circuit to be moved rather than the point P.

In magnitude, the cross product $\delta\mathbf{s} \times \delta\mathbf{l}$ on the right side of Eq. (30.16) is the area of the shaded parallelogram shown in Fig. 30.13. Its direction is into the side which the reader sees and out the other side, and, of course, perpendicular to the plane of the parallelogram. The dot product of $\delta\mathbf{s} \times \delta\mathbf{l}$ and $\hat{\mathbf{r}}$ is the transverse projection of this area along $\hat{\mathbf{r}}$ as seen from P. And this, divided by r^2, is the solid angle $\delta\omega$ subtended by the parallelogram at P.

As seen from P, the current in the element $\delta\mathbf{l}$ will appear to the left or to the right of the parallelogram according as the angle between the directed normal to the latter and $\hat{\mathbf{r}}$ is smaller or greater than $\pi/2$. Hence, adopting the convention that $\delta\omega$ is to be counted as positive in the former case and negative in the latter, we have

$$\delta\mathbf{B} \cdot \delta\mathbf{s} = \frac{\mu I}{4\pi} \delta\omega,$$

or, if we write $(\delta B)_s$ for the component of $\delta \mathbf{B}$ in the direction of $\delta \mathbf{s}$,

$$(\delta B)_s \, \delta s = \frac{\mu I}{4\pi} \, \delta \omega.$$

From this, algebraical summation over the whole circuit gives us

$$\delta s \sum (\delta B)_s = \frac{\mu I}{4\pi} \sum \delta \omega,$$

or, since the sum of the components of a number of items in a certain direction is equal to the component of their resultant in the same direction,

$$B_s \, \delta s = \mathbf{B} \cdot \delta \mathbf{s} = \frac{\mu I}{4\pi} \sum \delta \omega. \qquad (30.17)$$

Now $\Sigma \, \delta \omega$ is the decrease, or minus the increase, in the total solid angle Ω subtended at P by the entire circuit associated with a displacement of P along $\delta \mathbf{s}$, Ω being reckoned positive if the current as seen from P appears to circulate anticlockwise, that is, if the circulation of the current about a line passing through the circuit toward P conforms to the right-hand screw relation. Hence, writing $\delta \Omega$ for this increment in Ω, Eq. (30.17) becomes

$$\mathbf{B} \cdot \delta \mathbf{s} = -\frac{\mu I}{4\pi} \, \delta \Omega. \qquad (30.18)$$

We must now carry out a summation around the entire closed path linking the circuit. This gives us

$$\oint \mathbf{B} \cdot d\mathbf{s} = -\frac{\mu I}{4\pi} \oint d\Omega. \qquad (30.19)$$

The evaluation of $\oint d\Omega$ presents some slight difficulty if we have to deal with a circuit that is not confined to a single plane. Let us therefore in the first instance consider a circuit that *is* confined to a plane, this being represented in section in Fig. 30.14. Even with this circuit there is a difficulty. As seen from a point in the plane of the circuit, does the current appear to circulate clockwise or anticlockwise? There is, of course, no definite answer to this. Let us therefore begin our journey around the closed path at a point P_1 slightly to the right of the plane of the circuit and end it at P_2, slightly to the left. From P_1 the current appears to circulate anticlockwise, and so the corresponding solid angle Ω_1 is, according to our convention, positive. Also its magnitude is almost 2π. As seen from P_2 the current appears to circulate clockwise, and so the solid angle Ω_2 for this point is negative. And its magnitude is slightly greater (algebraically) than -2π. Hence $\Omega_2 - \Omega_1$ is very close to -4π. By

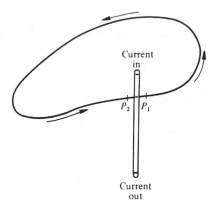

FIGURE 30.14   Closed path linking circuit in plane seen edge-on.

reducing the missed-out portion of the path P_2P_1 we can make its contribution to the line integral as small as we wish, and at the same time we make the increment in the solid angle approach progressively more closely to -4π. Clearly, therefore, we may take the increment in the solid angle corresponding to going around the entire closed path as -4π exactly. Accordingly, we have

$$\oint \mathbf{B} \cdot d\mathbf{s} = -\frac{\mu I}{4\pi} \int_{+2\pi}^{-2\pi} d\Omega = \mu I, \qquad (30.20)$$

which is Eq. (30.13) in vectorial form.

It is not difficult to see that for closed paths that do not link the circuit, such as those indicated in Fig. 30.15, $\oint d\Omega$, and hence the line integral, must be zero.

Let us now consider the case where the circuit is not confined to a single plane. Such a circuit may be replaced by a mesh system having the same boundary, around each mesh of which the current is imagined to circulate, and if the division is fine enough each mesh may be regarded as effectively plane. The closed path linking the circuit will pass through one of these meshes, but not through the others. To the mesh that the path does link the reasoning above applies in full, and since the current I is supposed to circulate around this mesh the contribution it makes to the line integral is μI. Since the path does not link any of the other meshes, these do not make any contribution. Thus the line integral for the whole circuit, as for a plane circuit, is μI.

APPLICATIONS OF THE LINE-INTEGRAL THEOREM   As a first application let us investigate the field at points within an infinitely long current-carrying solenoid not on the axis, and also at

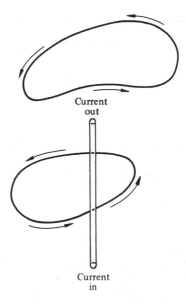

FIGURE 30.15 Closed paths not linking plane circuit.

points outside. We suppose the solenoid to carry a current I and to have n turns per unit length.

As a preliminary, we shall show that the field, wherever it is finite, must necessarily be parallel to the axis of the solenoid. Through any point either within the solenoid or outside, let a plane be constructed perpendicular to the axis. Then corresponding to every current element in the winding situated on one side of this plane there is a "mirror image," of the same length and at an equal distance, on the other side. Whereas the field contributions at the point selected due to these two elements have components parallel to the axis that reinforce one another, their radial components cancel each other out. There is also mutual cancellation of the components perpendicular both to the radius and to the axis. Hence whatever field there is anywhere is confined to the axial direction.

Consider, now, the line integral around a rectangular path such as the one shown in Fig. 30.16, where one side of the rectangle lies along the axis of the solenoid indicated diagrammatically in section. This rectangle does not link any current, so its line integral must be zero. This can be so only if the fields along the upper and lower (axial) sides are equal. Hence the field, whose magnitude on the axis we have found to be μnI, must be uniform over the whole internal cross section of the solenoid.

Next, let us consider the field at a point outside the solenoid. This we find by applying the line-integral theorem to a rectangle

779

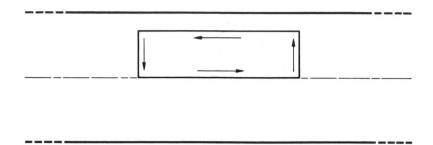

FIGURE 30.16   Rectangular closed path within infinite current-carrying solenoid, one side of which is along the axis of the solenoid.

such as that indicated in Fig. 30.17, whose length along the direction of the axis is l. Since this links nl turns, each carrying the current I, the line integral around this rectangle must be μnlI. But we know that the field inside the solenoid is everywhere of magnitude μnI and is directed parallel to the axis. Hence the contribution of the internal side of the rectangle parallel to the axis must be μnlI. But this, as we have just seen, is the value of the whole line integral. The external side of the rectangle can therefore make no contribution at all; the field outside the solenoid is zero.

As our next problem, let us investigate the field inside and outside a uniformly wound current-carrying toroid. A toroid may be imagined to be derived from a long solenoid that is curled around into a complete ring, as shown diagrammatically in two sectional views in Fig. 30.18, so that it no longer has any ends. Let the toroid have a total of N turns and carry a current I, and let the imaginary

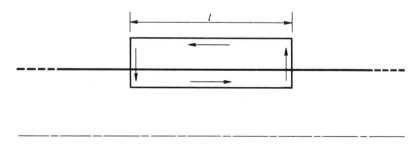

FIGURE 30.17   Rectangular closed path, one side of which is within an infinite current-carrying solenoid and parallel to its axis while the opposite side is outside the solenoid.

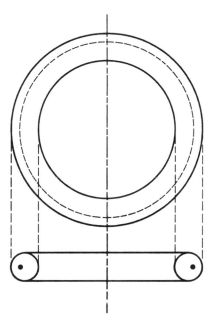

FIGURE 30.18 Circular closed path within a current-carrying toroidal winding, whose center lies on the axis of the toroid.

path to which the line integral is to be referred be a circle of radius r which is coaxial with the toroid. The dashed circle indicated in the figure is one such, which we may imagine to be wholly within the toroidal winding.

In this case it may readily be shown from considerations of symmetry that the field can nowhere on any such circle have any direction other than tangential.

Let the field on a circle wholly within the toroidal winding be of magnitude B. Then the line integral for this circle is $2\pi r B$. And since the circle links N turns each carrying a current I, the value of the line integral must, according to Ampère's theorem, be $\mu N I$. Equating these two expressions, we find that

$$ B = \mu \frac{N}{2\pi r} I = \mu n I, $$

where n is the number of turns per unit length around the toroid as counted at the distance r from the axis. Accordingly, if the radius of a turn is small compared with either the internal or the external radius of the toroid, the field inside is approximately uniform over the cross section of the winding, with the same value as for an infinitely long solenoid.

781

Note that the expression obtained above for the field is independent of the form of the cross section of the winding; for example, it would be the same for a ring of rectangular cross section for the same distance from the axis of the ring.

Since no path constructed externally to the winding encloses any current at all, the field at all points outside the winding must be zero, just as it is outside an infinitely long current-carrying solenoid.

These are two examples in which the line-integral theorem is used to ascertain magnetic fields. There is another application of rather a different kind which we may note here. This is one made by Maxwell in developing his theory of electromagnetic radiation.* Maxwell postulated the existence of a "displacement current" in a vacuum or a dielectric in which an electric field varies with time, and attributed to this displacement current the same magnetic effects as are exhibited by a current traveling along a wire. Then, combining mathematically Ampère's line-integral theorem in relation to such a displacement current with the law of electromagnetic induction (see Chapter 32), which deals with the electrical effects of varying magnetic fields, he obtained an equation which could only be interpreted as an electromagnetic wave equation. This is probably the most fundamentally important of all applications of Ampère's theorem.

Geomagnetism

The magnetic field at the surface of the earth varies in magnitude between about 3.5×10^{-5} N/A/m at the magnetic equator, where it is horizontal, to about 7×10^{-5} N/A/m at the magnetic poles, where it is vertical.

This field exhibits many irregularities, which are due partly to the uneven distribution of magnetic materials in the earth's crust and partly to currents flowing in the atmosphere. When these irregularities are smoothed out, the field we then have approximates what would be produced by an appropriate current flowing around a circular loop whose axis is inclined at an angle of about 11.5 degrees to the earth's axis of rotation.

The smoothed-out field has been analyzed into two components. One is the so-called "dipole field," a field that would be produced by a current-carrying loop whose dimensions are negligibly small compared with the diameter of the earth, the loop being so

*See Chapter 1 of Volume III.

positioned and orientated, and carrying such a current, as to give the best possible approximation to the earth's field. The other, the "nondipole field," is what must be superimposed on this to give the field actually observed. The motivation for carrying out this analysis in the first instance was probably that dipole fields, representing, as they do, a limiting case, have a particular interest of their own. Although there may be no obvious point in resolving the earth's field into dipole and nondipole components, this resolution has, in fact, turned out to be quite useful.

Magnetic records extending back over the last 400 years show that while the axis of the dipole field has not changed its orientation, at least during the past 200 years, the general pattern of the nondipole field is slowly rotating in a westward direction about the earth's geographic axis.

Fortunately, we do not have to rely solely on magnetic records extending over only a few centuries for our knowledge of the manner in which the earth's magnetic field has been changing with time. We now have information from what is known as "fossil magnetism," extending back to remote geological ages. Samples of sedimentary rock containing magnetic iron oxides are found to be magnetized in a certain direction, and there would appear to be little doubt that this direction must be that of the magnetic field at the locality in question at the time when the material was deposited. Similarly, ancient lava flows containing magnetic material reveal the direction of the magnetic field at the time when the lava cooled. In recent years there has been intensive investigation of fossil magnetism both in sedimentary rocks and in lava flows that occurred in past geological ages, and this has given us most interesting information concerning the changes that have occurred in the earth's field. Some of these changes have been dramatic, and there are even indications of reversals in the direction of the field having occurred at intervals of the order of millions of years. Another important result obtained from the study of fossil magnetism is that the constancy of the dipole axis observed during the last 200 years or so does not extend over periods of a higher order of magnitude. If the orientation of the earth's magnetic axis is averaged out over 10^4 to 10^6 years, the mean orientation coincides with the earth's geographical axis; on a geological time scale the changes in the field are statistically symmetrical about the earth's axis of spin.

If we are to hope to account for the earth's magnetism and its changes we must take into consideration the known physical condition of the interior of the earth as well as of its atmosphere. The atmosphere is known to make only a trivial contribution; the electric currents associated with thunderstorms and those flowing in the

783

outer parts of the atmosphere are known in order of magnitude, and these are totally inadequate to account for more than an insignificant proportion of the earth's field. We must therefore look below the surface of the earth, not above it, for the main cause of terrestrial magnetism.

The first to undertake a scientific study of magnetism, including terrestrial magnetism, was William Gilbert, physician to Queen Elizabeth I. As Gilbert wrote in his classical treatise *De Magnete*, he believed the earth to be a permanent magnet, in the same sense as a piece of steel or magnetite that has been exposed to a strong magnetic field is a permanent magnet. However, this theory is no longer tenable. It is now known that all magnetic materials lose their magnetic properties at high temperatures, and only a very thin outer shell of the earth's material, of the order of 20 km in thickness, is cool enough for it to be magnetic. The earth's field could be accounted for quantitatively in terms of magnetization of this shell only if the proportion of magnetic material within it were far greater than, in fact, it is. Even apart from this, it would hardly be possible to account for the *changes* that have been occurring in the earth's field on the basis of a permanent magnetic state of a solid portion of the earth.

There remains, then, only one possibility: The earth's magnetic field must be due to a current circulating in the deep interior of the earth. It has been estimated that if such a current were somehow brought into being and then left to itself without any electromotive force to maintain it, it would decay to insignificance after only a few tens of thousands of years. To account for its existence throughout geological time, therefore, it is necessary to find a mechanism that would be capable of continuously generating and maintaining it.

From the evidence of earthquake waves recorded at great distances from their point of origin, much information has been obtained concerning the earth's internal structure. Down to a depth of about 2900 km the earth is solid, but from here to about 5120 km there is a liquid region; and below this again, down to the center, there is an inner solid core, of radius about 1250 km. Both the liquid region and the inner solid core are believed to consist mainly of iron.

From various lines of evidence whose consideration here would take us too far, it is possible to form rough estimates of the electrical conductivity of different parts of the earth's interior. On the basis of these estimates it has been inferred that most of the current responsible for the earth's magnetic field circulates in the liquid region. And, from the slow westward drift of the nondipole component of this field observed during the last 200 years or so, it would appear that the part of the rotating current system that is mainly responsible for this field component rotates somewhat more slowly than does the outer solid mantle. Because this field component is nondipole it must be due predominantly to the currents flowing in the outer portions of the liquid region, and this, in turn, suggests that these likewise rotate more slowly than the mantle.

How can we account for such relative motion? In the absence of convection currents within the liquid and of any mechanical effects due to external bodies, the liquid must, because of viscosity, eventually settle down to a condition of uniform angular velocity which it shares with the mantle. It would be difficult to imagine any mechanism that would be capable of generating and maintaining a current within such a system, not to mention accounting for a relative angular motion of two components of an associated magnetic field or of either of these relative to the mantle. However, if there is convection within the liquid, this must necessarily lead to a variation of angular velocity with distance from the axis of rotation. Such convection could well result from loss of heat from the outer part of the liquid core by conduction through the mantle, and/or progressive growth of the inner solid core at the expense of the surrounding liquid, with the associated release of latent heat. It was formerly supposed that radioactivity in the deep interior of the earth might play an important role in promoting convection, but it is now believed that there is no significant radioactivity in the core.

If we started with a hypothetical liquid rotating at the same angular velocity throughout, and convection currents were then set up within it, the law of conservation of angular momentum would require that the angular velocity of a rising portion of liquid must decrease, whereas that of a falling portion must increase. We should therefore have the angular velocity within the liquid decreasing with increasing distance from the axis of rotation.

The radial movements characteristic of convection and the resulting variation of angular velocity with distance from the axis of rotation are both essential features of an ingenious hypothesis advanced some years ago by W. M. Elsasser and E. E. Bullard in an attempt to account for the earth's magnetic field. According to this hypothesis the liquid portion of the earth acts as a kind of self-exciting dynamo. The two features of the convection just referred to give conditions that, with a suitable electrical conductivity of the core, would make the functioning of such a dynamo possible.

785

As will be shown in Chapter 32, any motion of a conductor across a magnetic field produces an electromotive force which generates a corresponding current, if the conductor constitutes part of a closed circuit. It may be shown that a condition of zero current and associated zero magnetic field within a body of liquid moving in the way we have considered would be unstable; a magnetic field that might be set up within it, however small, would be built up by the movements within the liquid and the currents generated into one that is relatively large. The limit would be set by a balance between the generation of emf and the conversion of the energy of the circulating electricity into heat.

It has recently been shown that the electrical conductivity of the core is probably too small, and the convective dynamo engine proposed by Elsasser and Bullard correspondingly too inefficient thermodynamically, to account for the observed magnetic field of the earth. It is now believed that the dynamo is driven by a tendency for the core and mantle to precess at different rates. As was shown in Chapter 8, the ellipsoidal earth as a whole precesses under the combined influence of lunar and solar gravity. Other things being equal, we should expect a hollow body, such as the earth's mantle without the core, to precess faster than one, such as the core, which fills the inside of the mantle, each (mantle and core) being imagined as independent of the other. When all relevant factors are taken into consideration we must still expect the fictitious independent mantle and core to precess at different rates. Actually, of course, the two systems are not independent, and their interaction must greatly reduce the difference between their rates of precession. The problem is best considered from the point of view of the precessional torques exerted on mantle and core and the associated moments of inertia. On the basis of estimates of these quantities W. V. R. Malkus, in 1963, concluded that the precessional torques provide the driving agency for a satisfactory dynamo, using the earth's rotational energy.

Let us now inquire into the cause of the slow westward rotation of the nondipole component of the earth's field. This obviously means a corresponding difference between the angular velocities of the solid mantle, on whose surface the observations are made, and the nondipole field pattern. One factor that might be thought to determine the angular velocity of the mantle is the viscous interaction between this and the liquid immediately underlying it. If this were the only factor the angular velocities of these would have to be the same, absence of relative motion being the condition for zero torque exerted on the mantle. However, calculation based on the known orders of magnitude of the viscosity of the liquid and the electrical conductivity of the mantle shows that the viscous interaction is very small indeed compared with the electromagnetic interaction. The conducting mantle exposed to the rotating magnetic field is electromagnetically closely coupled—almost "locked"— to this field, in the sense that it is constrained to rotate at practically the same angular velocity as the field, with very little slip due to viscous interaction with the liquid below it. This coupling corresponds to that which causes the armature of an induction motor to rotate (see Chapter 32). It is due to an electromagnetic interaction between the rotating field and the mantle which, according to estimates based on observations of the variations in the length of the day, has a relaxation time constant of roughly 5 years.

The dipole component of the field, being the closest dipole approximation to the observed field, makes up the greater part of the latter; the nondipole component is merely a correction to the approximating dipole field. Hence it is substantially to the rotating dipole field that the earth's mantle is locked, and so it is that the nondipole component, which rotates faster than the dipole component, also has a higher angular velocity than the mantle.

Actually it is somewhat surprising that some slight eastward or westward drift of the dipole field is not also observed, for one would hardly expect the various parts of the electromagnetic coupling of the nondipole field with the mantle to cancel one another out. That this does, in fact, happen at the present time must be regarded as fortuitous, for, as we have already noted, the dipole field is by no means fixed relative to the mantle over long periods. In this connection it must be remembered that no part of the current system within the earth, not even the innermost part, is small enough to constitute a true magnetic dipole. Also, because we do not know the exact nature of the current distribution, the way in which the field has been resolved into dipole and nondipole components is somewhat arbitrary; it might with equal justification have been done in some other way.

Action of Magnetic Field on Moving Electrons

Since a current in a wire consists of a stream of electrons, the mechanical action of a magnetic field on a current-carrying wire must be due, basically, to the forces exerted on the electrons constituting the stream. Before considering this further, we shall investigate the action of a magnetic field on electrons moving in a vacuum.

787

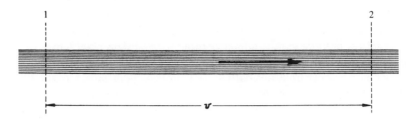

FIGURE 30.19   Beam of electrons.

Let us imagine we have a well-defined stream of electrons as indicated in Fig. 30.19, all traveling in a vacuum from left to right with velocity **v**, the cross-sectional area of the stream being A, and let these be exposed to a magnetic field **B** perpendicular to the plane of the diagram. Let the number of electrons per unit volume in this stream be n. Imagine two planes, 1 and 2, at right angles to the stream separated by a distance v numerically equal to the velocity of the electrons. Consider electrons which, at a certain instant of time, are in the plane 1. One second later, these electrons will have traveled a distance v to the right, and so will have arrived at the plane 2. And clearly, during this second, all electrons initially between the two planes will have crossed the plane 2. The volume occupied by these electrons is Av, their number is nAv, and the total charge carried by them is $nAve$. This, then, must be the current strength, and so the force per unit length of the stream exerted on the electrons must be $nAveB$. But the number of electrons contained in unit length of the stream, of volume A, is nA. Hence the force per electron is veB:

$$F = veB. \qquad (30.21)$$

We may note here that the vectorial equation corresponding to Eq. (30.21), which contains full information concerning the direction and sense of the force, as well as its magnitude, is

$$\mathbf{F} = e \cdot \mathbf{v} \times \mathbf{B},$$

the negative sign of e being taken into account, or otherwise

$$\mathbf{F} = |e| \cdot \mathbf{B} \times \mathbf{v}.$$

This force, exerted at right angles to the direction of motion of the electron, will obviously act as a centripetal force constraining it to travel in a circular path. To obtain the radius r of this circle we equate the force per electron, veB, to the expression for the centripetal force, mv^2/r, whence, solving for r, we find that

$$r = \frac{mv}{eB}. \qquad (30.22)$$

We see that the faster the electrons are moving transversely to a given field, the greater is the radius of the circular path they describe.

ELECTRONIC CHARGE-TO-MASS RATIO   In 1897 J. J. Thomson, working in the Cavendish Laboratory, Cambridge, made use of Eqs. (30.21) and (30.22) in a determination of the ratio of charge to mass of an electron. Although Thomson was not able to obtain anything better than a very rough value, reliable only to within about 40 to 50 percent, the method is nevertheless of considerable interest from a historical point of view.

Thomson's apparatus is shown diagrammatically in Fig. 30.20. C is the cathode of a "gas" cathode-ray tube, and A_1 and A_2 are two electrically connected anodes spaced some distance apart and provided with apertures through which a narrow beam of electrons emerges to pass between two deflector plates P_1 and P_2 on their way to a fluorescent screen at the end of the tube. By applying a potential difference between the deflector plates an electrostatic field may be produced which exerts a force either up or down on the electrons. Provision is also made for the application of a magnetic field at right angles to the plane of the diagram. Ideally this should be uniform and should be confined between the left-hand and right-hand edges of the deflector plates. Let us assume for the purposes of discussion that the magnetic field is so confined, and let us assume also that the electrostatic field between the plates is free of edge effects.

First, in the absence of either an electrostatic or a magnetic field, the position O of the fluorescent spot produced on the screen by the impinging electrons is noted. A magnetic field B is then applied and the new position O' of the spot observed. Finally, an electrostatic field E is applied between the plates such that the original position O of the spot on the screen is restored. Then from the deflection OO' produced by the magnetic field alone and the magnitudes of B and E the value of e/m may be calculated.

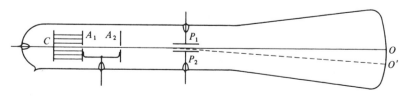

FIGURE 30.20   Thomson's apparatus for determining electronic charge-to-mass ratio.

Eq. (30.22) gives us the relation between e/m, the velocity v of the electrons, the magnetic field B, and the radius r of the circular path described by the electrons while traveling between the plates and subjected to the field B alone. Of these quantities, B is known, and r can readily be calculated from OO' and the dimensions of the tube. The remaining quantity v can be found from B and the electric field E required to bring the spot on the screen from O' back to O. When this additional field is applied, the forces veB and eE exerted on an electron in the magnetic and electric fields, respectively, balance each other, and accordingly v must be equal to E/B. So finally, substituting this for v in Eq. (30.22) and rearranging, we have e/m in terms of quantities all of which are known:

$$\frac{e}{m} = \frac{E}{rB^2},$$

(30.23)

The magnetic field used by Thomson was that due to a magnet, and such a field is neither uniform nor confined within a well-defined region. This, with other factors, severely limited the accuracy attainable. A modern version of the experiment, incorporating a thermionic source of electrons in a good vacuum, the establishment of a steady PD between cathode and anode, and the production of a uniform magnetic field confined within a solenoid running right through the tube, would probably give a value of e/m reliable to within about 1 percent.

MAGNETIC FOCUSING In Chapter 23 we considered electrostatic focusing of electrons in x-ray tubes and in cathode-ray tubes. Electrons in motion may also be focused by magnetic fields, and this is done in some designs of cathode-ray tube and in certain other electronic devices, notably in electron microscopes.

Magnetic focusing will best be understood if we consider first the time it takes for an electron to describe a complete circle when under the influence of a uniform magnetic field perpendicular to its direction of travel. This time, or period, T, is equal to the quotient of the circumference of the circle, $2\pi r$, and the velocity v; and from Eq. (30.22) we see at once that this is $2\pi m/eB$:

$$T = \frac{2\pi}{B}\frac{m}{e}.$$

(30.24)

Thus for a given field B the period is independent of the orbital velocity of the electrons.

This means that if several electrons are ejected from a point in different directions perpendicular to the magnetic field and with different velocities, they will all describe circles, of radii propor-

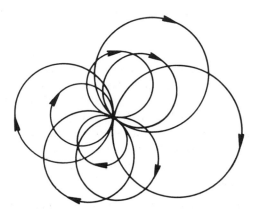

FIGURE 30.21   Circular paths of electrons projected in different directions and with different velocities at right angles to a uniform magnetic field.

tional to their velocities, and will all, after the same interval of time, arrive back at the starting point. This is indicated in Fig. 30.21, in which the magnetic field is perpendicular to the plane of the diagram and directed inward.

Now let us suppose that what we have just called the orbital velocity is only one component of an electron's total velocity; that is, that it possesses in addition an "axial" velocity component v_a in the direction of the field. The field does not exert any force on the electron due to its axial velocity, and so it will travel along a path representing a combination of a circle and a straight line at right angles to it; that is, it will describe a helical path. It is easily seen that the pitch p of the helix must be the product of the period T in the circular orbit and the axial velocity v_a:

$$p = T \cdot v_a = \frac{2\pi}{B} \frac{m}{e} v_a. \qquad (30.25)$$

Let us now consider Fig. 30.21 again, this time regarding it as the projection, in a plane perpendicular to the field, of the helices described by a number of electrons all originating from the same point—a "point object." If all these electrons have the same axial component of velocity, their helical orbits must all have the same pitch, and so they will all pass through another point situated on a line perpendicular to the plane of the diagram and passing through the point object, and distant p from it. This second point will therefore be an electron-optical point image of the original point

791

object. Further images will be formed at intervals p along the same line.

The image produced by a uniform magnetic field is at best merely a faithful replica of the object; it cannot, in the nature of things, be either larger than the object or smaller. To produce magnification or reduction in size a nonuniform field is necessary, in which the lines of force, while distributed symmetrically about an axis, are suitably curved in planes containing this axis. Such a system of lines of force constitutes what is known as a **magnetic lens,** just as a system of curved equipotential surfaces constitutes an electrostatic lens.

Although the theory of the action of a magnetic lens is less simple mathematically than is that of the formation of an image by a uniform field, it is quite similar in broad principle. In our study of the action of a uniform field we have seen that the field, acting on the component of the velocity of an electron perpendicular to the field, causes it to spiral in a path that periodically intersects the line of force passing through its point of origin. The axis of this helical path is a parallel line of force. Similarly, if the lines are curved, the electrons must broadly follow these lines in spiraling motions. And near the axis of the lens the paths of all the electrons passing through one point later intersect at another. Thus, if we call the first point an object, the second is its electron-optical image. With a magnetic lens of suitable form it can be contrived that all points within a restricted area on a plane perpendicular to the axis have images on another such plane. In this way it is possible to produce a magnified or reduced image of a two-dimensional object. A sectional view of a magnetic lens system that produces a magnified image is shown diagrammatically in Fig. 30.22. In this an electromagnet is used, the area in black representing iron and the cross-hatched area the winding, accommodated in an annular cavity inside the magnet. The lines of force constituting the lens are indicated by the full lines.

A magnetic lens, in addition to producing an enlarged or reduced image of the object, also rotates the image with respect to the object about the axis of the lens. This rotation may be through any angle. It is because of this that the lines labeled object and image in Fig. 30.22 are not provided with arrowheads as in the usual diagrams relating to ordinary optics. These lines are to be taken as simply representing the areas of the object and image, without indicating their relative orientation.

Electron Microscope Electron lenses find an important application in electron microscopy. The focusing in electron microscopes

Object

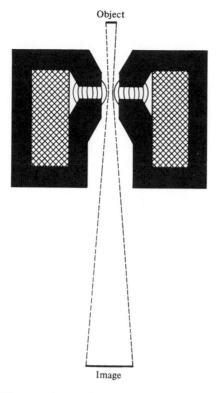

Image

FIGURE 30.22   Electron image formation by a magnetic lens.

can be provided for either with electrostatic or with electromagnetic lenses. For various reasons, better definition of the image is attainable with magnetic than with electrostatic lenses.

An electron microscope consists essentially of a series of lenses, usually two or three. The first lens produces a magnified image of the object. This image serves as an object for further magnification by the second lens, and, if there is a third lens, this still further magnifies the image produced by the second.

Figure 30.23 gives a diagrammatic representation of a two-stage electron microscope incorporating magnetic focusing. There are different types of microscopes; the one shown here is a "transmission microscope." After emission from a thermionic cathode at a high negative potential relative to earth, the electrons are first accelerated and at the same time electrostatically focused into a narrow beam by a specially shaped anode, which, with the metal casing of the microscope, is at earth potential. The object, in the form of a thin section, is placed in the narrow part of the beam, and the electrons emerging on the other side then pass in succession

793

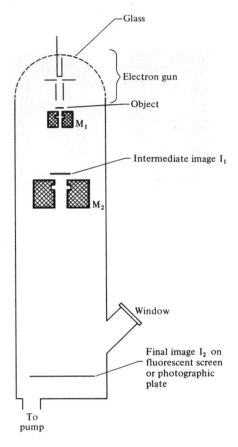

FIGURE 30.23 Two-stage electron microscope employing magnetic focusing.

through the magnetic lenses M_1 and M_2. M_1 produces the first magnified image I_1 of the object, and M_2 then produces a further magnified image I_2 of I_1. This final image is either rendered visible by the impact of the electrons on a fluorescent screen or is formed on a photographic plate or film.

The essential mechanism for the production of contrast in the image formed by a transmission microscope is differential scattering of electrons passing through the specimen. With variations of density and/or thickness from one region to another, different proportions of the electrons are scattered through appreciable angles out of their original directions. The lenses are able to focus correctly only those electrons emerging from any given point in the object whose transverse velocity components are relatively small; the peripheral regions of their fields are not suitable for bringing electrons with larger transverse velocities to the same focus, if, indeed, such electrons pass through the lenses at all. Hence electrons diverted by appreciable angles from their original directions cannot contribute to the image.

In Chapter 39 we shall see that, regardless of features of design or perfection of workmanship in the construction of an optical instrument, an ultimate limit to the definition of the image produced by it is set by the wave nature of the light used. The shorter the wavelength, the finer the detail obtainable; thus, apart from the effects of other factors, a somewhat more sharply defined image may be produced with blue light than with red.

The reader will perhaps be surprised to learn that this principle also applies in electron microscopy. Electrons are now known to have a wave aspect as well as being particles, the wavelength being given by the equation

$$\lambda = \frac{h}{\sqrt{2meV}}, \tag{30.26}$$

where h is Planck's constant, e the electronic charge, m the mass of the electron, and V the voltage through which the electron beam has been accelerated. If in this equation we substitute the values obtained for h, m, and e in Chapters 22 and 23 we obtain the numerical relationship

$$\lambda = \frac{12.2}{\sqrt{V}} \tag{30.27}$$

between λ in angstroms and V in volts. Actually this is not strictly correct except for rather low values of V. According to Einstein's theory of relativity, the mass of a particle increases as its velocity approaches that of light, and for electrons having the maximum energy of the range usually employed in electron microscopy, 60 keV (60,000 eV) the relativistic increase in mass amounts to about 12 percent. This would make the wavelength calculated from Eq. (30.27) too high by 6 percent. However, this is not a large error, so we may, for the purposes of the present discussion, take as a sufficiently good approximation for λ in all cases the value

$$\lambda = \frac{12}{\sqrt{V}} \text{ Å.} \tag{30.28}$$

The accelerating potentials used in electron microscopes usually lie between 20 and 60 kV. To these correspond, according to Eq. (30.28), a range of values of λ between 0.085 and 0.049 Å. These wavelengths are of the order of only 10^{-5} those used in optical microscopy, and, if we could base our calculation on this alone, it would mean that we should be able to produce a magnified image with an electron microscope having about 100,000 times the linear dimensions of an image of comparable definition obtained with an optical microscope. Thus, since the best optical microscopes produce useful magnifications of the order of 1000, those given by electron microscopes should be about 10^8. Unfortunately, however, other factors, unfavorable to the electron microscope, play an important part, and in practice useful magnifications obtainable with electron microscopes do not exceed about 100,000. But even this is very well worthwhile. At its best, an electron microscope can distinguish features in the specimen separated by only about 10 Å.

Problems

30.1   Two long parallel bars, each of which carries a current of 1000 A, are held with their axes 10 cm apart. What is the force between them per meter of length?

30.2   A pair of long straight wires whose axes are 2 mm apart connect a battery to a lamp. Given that the current feeding the lamp is 1 A, calculate the magnetic field B at a point in the plane of the wire axes whose mean distance from the wires is (a) 1 cm, (b) 3 cm, and (c) 10 cm.

30.3   Show that the magnetic field B at the center of a square loop of wire of side a which carries a current I is $2\sqrt{2}\mu_0 I/\pi a$.

30.4   (a)   A charge Q is uniformly distributed over a thin circular metal ring of radius a. Find the electric field intensity at a point on the axis of the ring and at a distance x from its center.

(b)   The ring is now made to rotate about its axis with angular velocity ω. Explain why we must expect the electric charge to rotate with it.

(c)   Show that when the ring rotates, the ratio of the electric field intensity E to the magnetic induction B at a point on the axis is given by the equation

$$\frac{E}{B} = \frac{x}{\mu_0 \epsilon_0 \omega a^2}.$$

30.5   Two equal, parallel, coaxial coils of radius r are separated by a distance l, and the same current I is sent in the same sense around each coil. Let x be the distance along the common axis from some agreed origin. Show that at the center of symmetry of the system not only is dB/dx zero for all values of l, but, if $l = r$, as in the Helmholtz-coil system, d^2B/dx^2 is also zero.

30.6   Show that the variation of the magnetic field due to a small current-carrying coil with distance and with direction from its axis is identical with that of the electric field due to an electric dipole, consisting of two equal and opposite charges separated by a short distance (see Problem 20.5).

30.7   A coaxial cable consists of a central insulated wire surrounded by a thin-walled tube, as shown in section in the figure. Current flows in one direction along the wire and returns along the surrounding tube. If the radius of the wire is a, the inner and outer radii of the tube are b_1 and b_2, respectively, and the current carried is I, how does the magnetic field vary with the distance r from the axis of the cable, from $r = 0$ to beyond $r = b_2$?

Figure for problem 30.7.

30.8 What is the magnetic field B, in magnitude and direction, on the two sides of an infinite plane sheet of current in which the current per unit transverse length is i?

30.9 In a simple model of a hydrogen atom an electron revolves in a circular orbit of radius 5.3×10^{-11} m about a proton, with the requisite centripetal force provided by the Coulomb attraction between these two particles. Considering this motion of the electron to constitute a current around the orbit, calculate the magnetic field produced at its center.

30.10 Assuming that, when the hydrogen atom as described in Problem 30.9 is exposed to a magnetic field $\mathbf{B}$ whose direction is normal to the plane of the orbit of the electron, the radius of the orbit does not change, show that the change δf in the frequency of the revolution of the electron resulting from exposure to this field is given by the equation

$$\delta f = \pm \frac{Be}{4\pi m},$$

where e and m are the electronic charge and mass, respectively.

30.11 A beam of electrons is bent into a circle of radius 2 cm by a magnetic field of 4.5×10^{-3} N/m/A. Calculate the velocity of the electrons.

30.12 How does the kinetic energy of a particle of given mass and charge that is traveling at right angles to a given magnetic field vary with the radius of curvature r of its path?

30.13 An electron is moving in a circular path at right angles to a uniform magnetic field. The magnitude of this field is now increased. Does this affect the linear velocity of the electron? If so, how?

30.14 An electron gun in an evacuated tube produces a narrow beam of electrons which passes between the turns of a long solenoid at right angles to and through its axis in the manner indicated in part (a) of the figure, where the beam, traveling at right angles to the plane of the diagram, is represented by a dot. Part (b) shows an axial view of the passage of the electrons from the gun, through the solenoid, and toward a fluorescent screen. If the solenoid has n turns per unit length, its radius is r, the distance of the screen from its axis is d, and the electrons, traveling with a

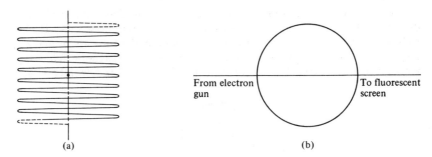

(a)   (b)

Figure for problem 30.14.

velocity v, have a charge e and mass m, by how much will the fluorescent spot on the screen be deflected when a switch is closed, sending a current I through the solenoid? (Assume that the angle through which the beam is deflected is small.) If the radius of the solenoid is halved, but the number of turns per unit length and the distance of the screen from its axis remain the same, by what factor must the current be increased to produce the same deflection on the screen?

30.15   Within the solenoid of Problem 30.14 is mounted a pair of deflector plates, symmetrically disposed about its axis, and the beam of electrons is directed between these, as shown in the figure. Deflections of the spot on the screen are produced (1) by applying a potential difference V between the plates only, (2) by passing a current I through the solenoid only, and (3) by applying the potential difference V and passing the current I simultaneously. Show that the deflection produced in case (3) is the algebraic sum of those produced in cases (1) and (2).

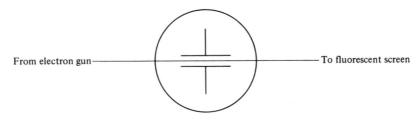

Figure for problem 30.15.

30.16   In the system of Problem 30.15, the solenoid, of diameter 10 cm, has three turns per centimeter, and the distance of the screen from its axis is 30 cm. The deflector plates are 5 cm long and spaced 1 cm apart. Given that a potential difference of 5 V applied between the deflector plates produces a deflection of 6.2 cm of the spot on the screen, and that the original (undeflected) position of the spot is restored by additionally passing a current of 141 mA in the appropriate direction through the solenoid, calculate (1) the velocity of the electrons, and (2) e/m. (Neglect edge effects.)

30.17   A cathode-ray tube is mounted near the center of, and coaxially with, a long solenoid which has n turns per meter and carries a current I. The potential difference between the cathode and anode of the cathode-ray tube is V. Show that an electron passing through a point on the axis of the tube at the far end of the anode or beyond, in a direction making a small angle θ with the axis, will move in a helical path, and, if the tube is sufficiently long, will return to the axis a distance x farther on, where

$$x = \frac{2\pi}{\mu_0 nI} \sqrt{2V} \sqrt{\frac{m}{e}},$$

m being the mass of the electron and e its charge. By making use of the fact that this expression does not involve θ, suggest a method of determining the ratio of charge to mass of the electron.

30.18 Show that when a beam of identical charged particles having a distribution of velocities enters a region where there is an electric and a magnetic field, both of which have the same direction at right angles to the incident beam, the pattern observed on a fluorescent screen or on a photographic plate normal to the screen is a parabolic curve. Show that for a different type of particle, having a different charge-to-mass ratio, a different parabolic curve is obtained. (This type of experiment led to the discovery of isotopes.)

30.19 Two infinite conducting plates are situated at $y = a$ and $y = -a$, so that their distance of separation is $2a$. They carry equal and opposite charges per unit area, so that there is an electric field E directed parallel to the y axis. In addition, in the space between the plates there is a uniform magnetic field $\mathbf{B}$ directed parallel to the z axis. At time $t = 0$, a particle of mass m carrying a charge e starts from rest at the origin, which is midway between the plates. Show that the equations of motion of this particle are

$$m \frac{dv_x}{dt} = Bev_y$$

and

$$m \frac{dv_y}{dt} = eE - Bev_x,$$

where v_x and v_y are the x and y components of its velocity, respectively. Hence show that the particle will not strike either plate if $E < B^2ea/2m$, and that, under these conditions, it will move in the x direction with an average velocity E/B.

30.20 (a) A wire of radius a has n effectively free conduction electrons per unit volume and carries a current I. Show that the mean drift velocity of the electrons is $I/\pi na^2e$.

(b) Use Ampère's line-integral theorem to find how the magnetic field within the wire varies with the distance r from the axis, assuming that the current density remains uniform over the cross-sectional area.

(c) This magnetic field results in the exertion of a force on the moving electrons which is balanced by an equal and opposite force arising from a radial electric field in the wire. Find the magnitude of this electric field as a function of r.

MAGNETIC PROPERTIES OF MATERIALS

We are now ready to see if we can explain magnets in terms of currents circulating within them.

A promising beginning in this direction was made as long ago as the early years of the nineteenth century by Ampère, who, observing that current-carrying solenoids behave like magnets, proposed a theory of magnetism according to which all magnets are simply systems of circulating electricity. In particular, he regarded each particle in a magnet as having circulating around it an "equatorial current," producing the effect of magnetic poles. So, in one sense, he anticipated, while in another sense he advanced beyond, the atomic theory of magnetism proposed by Weber and elaborated by Ewing, which we shall consider later. Ampère also ascribed terrestrial magnetism simply to currents flowing around the earth. In 1823 he published a paper giving a mathematical theory of magnetic phenomena, which Maxwell subsequently described as "perfect in form and unassailable in accuracy."

Unfortunately, Ampère's ideas, which were far in advance of his time, failed to gain general acceptance. The habit of thinking of magnetic phenomena in terms of "poles" was already too firmly established to be easily broken. It was perhaps not unnatural, in view of the very close correspondence found between the apparent interactions of these poles* and the interactions between positive and negative charges in electrostatics, to attribute to magnetic poles a basic significance which, in fact, they do not possess. The concept of a magnetic pole as something fundamental is now seen, in retrospect, as the expression of a great failure—a failure to think the problems presented by known electromagnetic laws through to their logical conclusion. This concept, and the confusion of thought

---

* As we now know, the mechanical interactions between magnets do not, strictly speaking, consist predominantly of forces exerted between the end regions, or poles, of the magnets, but occur between every part of one magnet and every part of the other. However, owing to the circumstance that there is mutual cancellation, or near-cancellation, of all the contributions to the force system except those due to the end regions of the magnets, the phenomena observed are the same *as if* practically only the end regions acted on one another.

of which it was born and which it tended to perpetuate, has clouded important issues in electromagnetic theory for well over a century.

Before attempting to explain the magnetic behavior of iron and other materials in terms of the interactions between currents, we shall first review the main facts that call for explanation. As we shall see, the theory of these phenomena in terms of currents will then follow naturally.

Magnetization and Demagnetization

Both the process of magnetizing a rod of steel by passing a current through a wire wound around it and the phenomena of induced magnetism are examples of iron or steel being made into a magnet by exposing it to a magnetic field due to some external agency. Let us now investigate more closely the "strength" of the magnet produced in this way in relation to the field to which the magnetization is to be attributed.

For this purpose it will obviously be desirable so to arrange matters that (1) every part of the specimen to be investigated is, as nearly as possible, subjected to the same field, and so is put into the same condition, and (2) we may assign a value to the applied field which is not affected by any response of the specimen to it. It will also be necessary for a satisfactory criterion of the state of magnetization of the material of the specimen to be found.

These requirements may be met by winding an insulated wire uniformly all the way around a ring or toroid constructed of the material whose magnetic properties are to be investigated. It was shown in Chapter 30 that within an *empty* toroidal winding that carries a current I the field B at a distance r from the axis of the toroid is $\mu_0 n I$, where n is the number of turns per unit length as counted around a circle of radius r centered on the axis. We also noted that if the radius of a turn is small compared with either the internal or the external radius of the toroid the variation of n with r is unimportant, so that the field inside the winding may be regarded as substantially uniform over its cross section, while outside the winding the field is zero.

If the toroidal winding is not empty but is filled with, let us say, iron, then the field produced by a given current carried by a winding of a given number of turns per unit length must presumably differ radically from $\mu_0 n I$. Thus we have the situation that, according to the filling, different effects are produced by the same cause. As this "cause" we may take the product nI, the effective circulating current per unit length of the winding. It is convenient to give this a name and assign it a symbol. The symbol used is H; that is, as a matter of definition we have

$$H = nI. \tag{31.1}$$

And the name that is given to this quantity is the **magnetizing force.** It is, of course, not really a force in the mechanical sense, any more than is an electromotive force. Nevertheless, this being well understood, the name may be regarded as suitable, conveying, as it does, the idea intended.

The MKSA unit of H is quite different from the unit of B. Since H is the effective current circulating around each unit of length of the winding, the unit is the **ampere per meter.**

In the older "absolute electromagnetic" system the unit of H is the oersted, and magnetizing forces are still expressed in oersteds in much of the current literature. It may be noted, therefore, for purposes of translation into MKSA units, that 1 oersted is equal to $1000/4\pi$ A/m.

It is not to be supposed that because B and H are measured in different units they are essentially different quantities, except in the strictly formal, dimensional sense. Although in fact this has not been done, H might equally well have been defined as $\mu_0 nI$, and in that case B and H would not even have had different dimensions. As it is, with μ_0 omitted from the definition of H, the dimensions of B and H must be different, for the simple reason that μ_0 is not a mere dimensionless number. Nevertheless, H is always *thought of*, like B, as a field, measured by the nI in a toroidal winding (or in a long solenoid) by which it either has been, or could be, produced. Like B, therefore, H will often be referred to as a field in the following pages. Although, purely as a matter of convenience, we have imagined this field, which we regard as the cause of what happens to an experimental sample of material, to be that produced by passing a current through a toroidal winding, other means could, in principle, be found. Thus, one could use a magnet, or a system of magnets, instead of this current, to produce H.

Obviously we cannot measure B within a ring of iron or any other solid material by any of the methods hitherto considered, because we cannot perform experiments within the body of the iron. There is, however, a way of circumventing this difficulty, based on the phenomenon of electromagnetic induction. We shall discuss this in Chapter 32. Meanwhile it will be necessary to take on trust the statement that the magnitude and direction of B within the iron *can* be determined.

803

HYSTERESIS AND SATURATION With such an arrangement, the core being of iron, let the current in the winding be taken, in steps, through a cycle between sufficiently high values in the two directions, the value of B being determined after each change in the current. The variation of B with H found in this way will have the character represented graphically in Fig. 31.1. This kind of curve is known as a **hysteresis curve.** Although all irons and steels exhibit hysteresis, the areas enclosed by the curves vary widely from one specimen to another. It may be shown that the area enclosed by a hysteresis curve is a measure of the energy dissipated as heat in taking the specimen through a cycle.

Another important feature of the B–H relationship is the phenomenon of **saturation,** the leveling out of the curve to near horizontality at sufficiently high positive and negative values of H. Since B is the sum of two contributions, namely, $\mu_0 H$ due to the current in the winding, and $B - \mu_0 H$ due to the magnetic state of the specimen, and we are really concerned only with the latter item, it becomes of interest to subtract $\mu_0 H$, that is, $\mu_0 n I$, from all the ordinates. If this is done, the lines joining the points so obtained are found to be truly horizontal at high values of H. This means that there is a certain maximum value which the contribution of the specimen cannot exceed.

In subtracting $\mu_0 H$ from B to obtain the contribution due to the specimen alone, we have made the tacit assumption that $\mu_0 H$ represents the contribution made by the current in the winding, despite the presence of intervening iron, that is, that this contribution is the same as if the medium were a vacuum. The fact that the

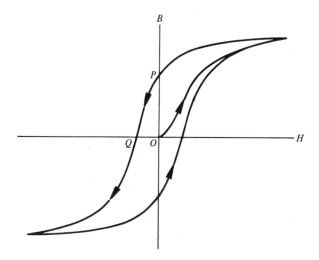

FIGURE 31.1 Magnetic hysteresis curve.

slope of the plot of $B - \mu_0 H$ against H becomes zero for high values of H suggests that this assumption is justified. The assumption is further supported by the consideration that all atoms, and hence all materials, consist almost entirely of empty space. Just as, because of this, there is really only one permittivity in electrostatics, the permittivity ϵ_0 of free space or a vacuum, so also here, there is only one true permeability μ_0, again that applying to a vacuum. The concept of media other than a vacuum, having other values of permittivity and permeability, ϵ and μ, respectively, is a mere fiction, serving no other purpose than that of convenience of mathematical treatment.

We have just referred to $B - \mu_0 H$ as the "contribution" toward the field B made by the specimen alone. Let us now consider just what this contribution means in terms of the magnetic condition of the material itself. In Chapter 30 we saw that a flat coil of N turns whose area is A and which carries a current I has a magnetic moment $\mu_0 NIA$, and that this is still the magnetic moment if the coil is extended axially into a solenoid. Let us imagine the toroidal winding considered above to be derived from such a flat coil of N turns, by first stretching it into a solenoid of length l and then curling it around into a toroid. At the stage before conversion into a toroid, where the winding is a long solenoid, its magnetic moment is $\mu_0 NIA$, and, because its volume is lA, its magnetic moment per unit volume is $\mu_0 nI$, or $\mu_0 H$, where n is the number of turns per unit length. Obviously the magnetic moment *per unit volume* cannot be affected by the final operation of curling the long solenoid into a toroid; this will still be $\mu_0 H$.

The two terms B and $\mu_0 H$ in the expression under consideration are necessarily the same kind of quantity, expressed in the same units, so it follows that B must be the magnetic moment per unit volume of the specimen plus current-carrying winding, and that therefore $B - \mu_0 H$ is the magnetic moment per unit volume of the material of the specimen. This is also called the **intensity of magnetization** M.

From the fact that the plot of M against H becomes horizontal at sufficiently high values of H, we see that for any given ferromagnetic material there is an upper limit M_s which the value of M cannot exceed.

RETENTIVITY AND COERCIVE FORCE From Fig. 31.1 it will be seen that on reducing H back to zero following saturation, a residual induction B_r corresponding to OP on the graph remains in the specimen. Since H is zero, this represents the intensity of

805

magnetization of the material which remains after saturation when no current is being circulated around the specimen. Accordingly, B_r is known as the **remanence,** or **retentivity,** of the material. It is a partial measure of the usefulness of the material for the construction of a permanent magnet.

In order completely to demagnetize the material after it has been saturated, that is, to reduce M to zero, a reverse magnetizing force must be applied which is slightly in excess of that corresponding to OQ. However, the change in B contributed by a material like iron is so overwhelmingly greater—except in the region of saturation —than is the corresponding change in $\mu_0 H$ that the reverse magnetizing force required to reduce M to zero may be taken as practically the same as that which reduces B to zero. This is H_c, the value of the reverse H which corresponds to OQ. H_c is known as the **coercive force,** or **coercivity,** of the material. It is a measure of the difficulty of demagnetizing a magnet made of the material.

A small value of H_c means a small area enclosed by the hysteresis loop, and a correspondingly small development of heat in the iron or other material per cycle. Smallness of H_c is therefore a practical requirement that must be met by any material to be used as the core of a transformer (see Chapter 34). On the other hand, a material that is to be used for the construction of a permanent magnet must have as large a value of H_c as possible, insofar as this does not conflict with the requirement of a high retentivity.

Ferromagnetic Materials

Materials whose general magnetic behavior is similar to that of iron and steel are known as **ferromagnetic.**

The only *elements* that are ferromagnetic at temperatures other than near 0°K are iron, cobalt, nickel, gadolinium, and dysprosium; although the two last-named exhibit ferromagnetic properties only below 16°C and below −168°C, respectively. To these must be added not only a number of alloys of which these are constituents but also some consisting entirely of metals that in the pure state are nonmagnetic. Among the latter are an interesting group of alloys whose ferromagnetic properties were discovered by Heusler in the opening years of this century and known, after him, as Heusler alloys. These are alloys of copper and manganese and one of the metals aluminum, tin, antimony, or bismuth. Several other nonferrous alloys having ferromagnetic properties have since been developed. Finally, certain oxides and other compounds have been found to be ferromagnetic;

examples of these are Fe_3O_4, Fe_2O_3, FeO_2, $FeTiO_3$, $Fe_4H_6O_9$, $FeCr_2O_4$, and $Fe_3Al_2Si_3O_{12}$. Certain intimate (for example, sintered) mixtures of such compounds have particularly interesting magnetic properties and have been put to practical use for special purposes.

Some of these materials are similar in their behavior to soft iron, having relatively low values of retentivity and coercive force, while others, having high values of these quantities, resemble magnet steel. Table 31.1 gives some typical examples of the former group and Table 31.2 of the latter.

Among the interesting points brought out by these tables, we may note the following:

1. The value of M_s for the 65%–35% Fe–Co alloy is substantially greater than that for pure iron, even though the value for pure cobalt is less than for pure iron.

2. Both the M_s of the nonferrous Heusler alloy listed in Table 31.1 and the B_r of manganese bismuthide included in Table 31.2 are comparable with the corresponding quantities for ferrous materials.

3. Although there is less than a factor of 10 between the highest and lowest values of M_s for all the materials listed in Table

TABLE 31.1   Magnetic Properties of Low-Coercivity Materials

Material	μ/μ_0 for $B = 2 \times 10^{-3}$ (N/A/m)	M_s (N/A/m)	Coercive force (A/m)
Purified iron (annealed)	5,000	2.15	4.0
Cobalt		1.76	950
Nickel		0.615	400
97% Fe, 3% Si	1,500	2.00	12
65% Fe, 35% Co		2.45	280
45 Permalloy: 54% Fe, 45% Ni, 0.3% Mn	4,000	1.60	5.6
Supermalloy: 15.7% Fe, 79 % Ni, 5% Mo, 0.3% Mn	100,000	0.80	0.15
Heusler alloy: 61% Cu, 26% Mn, 13% Al		0.48	560
Ferroxcube III (sintered powder): $MnFe_3O_4$ + $ZnFe_2O_4$	1,000	0.25	8.0

TABLE 31.2   Magnetic Properties of High-Coercivity Materials

Material	Percentage composition (remainder Fe)	Retentivity (N/A/m)	Coercive force (A/m)
Carbon steel	1 Mn, 0.9 C	1.0	4,000
36% Cobalt steel	36 Co, 5 W, 4 Cr, 0.7 C	0.95	19,000
Alnico I	12 Al, 20 Ni, 5 Co	1.05	35,000
Alnico VI	8 Al, 15 Ni, 24 Co, 3 Cu, 1 Ti	1.0	60,000
Alnico VII	6 Al, 18 Ni, 35 Co, 8 Ti	0.58	76,000
Hycomax	9 Al, 21 Ni, 20 Co, 2 Cu	0.95	64,000
Tyconal G	8 Al, 14 Ni, 24 Co, 3 Cu	1.3	48,000
Platinum–cobalt	77 Pt, 23 Co	0.59	210,000
Manganese bismuthide	100 MnBi	0.46	290,000
Vectolite	30 Fe_2O_3, 44 Fe_3O_4, 26 Co_2O_3	0.16	71,000

31.1, and of those of B_r for all the materials of Table 31.2, on the other hand the highest and lowest values of H_c shown in the two tables differ by a factor of well over a million.

From an examination of both tables it will be seen that the element manganese appears to play an important role. Besides entering into the composition of a number of ferrous alloys, it is an essential ingredient in all the Heusler alloys, and it is one of the metals of the intermetallic compound manganese bismuthide. Yet the presence of manganese does not always favor ferromagnetism. Although it does so when a few percent of this metal is added to iron, the 85%–15% Fe–Mn alloy is almost completely nonmagnetic.

It is interesting to note that the 75%–25% Fe–Ni alloy is also practically nonmagnetic, despite the outstanding ferromagnetism of both pure iron and pure nickel.

Weber–Ewing Theory of Magnetism

If a length of carbon-steel wire is heated to bright redness and then suddenly quenched in water, it is thereby rendered so brittle that it may easily be broken into small pieces. If such a wire is first magnetized and then broken into short lengths, it is found that each

FIGURE 31.2 Shorter magnets produced by breaking a long magnet transversely.

fragment has magnetic poles at its ends, and if the fragments are tested for polarity it is found that their north and south ends correspond to those of the original magnet, as indicated in Fig. 31.2. From this, W. Weber, supposing that such successive breakings might in principle be continued until finally the original magnet is separated into its ultimate molecules, argued that if this could be done each molecule would be found to be a magnet. Weber assumed the magnetism of the ultimate molecules to be an intrinsic property of the iron or other magnetic material, and postulated that while in a piece of unmagnetized material the molecular magnets are orientated at random, the application of a field causes the molecules to change their orientation, giving either partial alignment in the direction of the field or, under the influence of a sufficiently strong field, complete alignment, corresponding to magnetic saturation. These three conditions are indicated in parts (a), (b), and (c), respectively, of Fig. 31.3, the arrowheads representing the north poles of the molecular magnets.

It was not known in Weber's time that the atoms of a solid are arranged in a regular space lattice, so that it is meaningless to speak of a "molecule" of a metal. Today we should, instead, refer to the ultimate chemical particles as atoms, and Weber's "molecular theory" should accordingly be amended to "atomic theory."

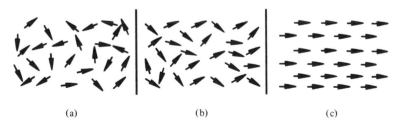

(a)   (b)   (c)

FIGURE 31.3 Weber's picture of orientations of ultimate molecular magnets
(a) in unmagnetized ferromagnetic material,
(b) in material magnetized below saturation point,
(c) in magnetically saturated material.

809

The details of the processes leading to partial and complete alignment in response to moderate and strong magnetizing forces, respectively, and the phenomenon of hysteresis, were investigated by J. A. Ewing, who showed that the whole cycle could be qualitatively accounted for by taking into consideration the combined effects of the externally applied field H and the interactions between neighboring elementary magnets.

Theory of Magnetic Domains

From a consideration of several facts it becomes evident that we cannot accept the idea, implicit in the Weber–Ewing theory, that the magnetic properties of a material are determined solely by whether its constituent individual atoms do or do not have a magnetic moment. Among the facts that it seems impossible to explain on such a basis are the following:

1. The superiority of an iron–cobalt alloy over pure iron in regard to the value of M_s.

2. The ferromagnetic behavior of alloys of nonferrous metals, for example the Heusler alloys and MnBi.

3. The nonmagnetic properties of ferrous alloys such as 85%–15% Fe–Mn and (at ordinary temperatures) 75%–25% Fe–Ni.

These facts suggest that, rather than the individual atoms, it must be whole groups of atoms that determine whether or not a material is ferromagnetic. Perhaps the determining factor is the way the atoms arrange themselves in such groups. With regard to item 1 we could ascribe magnetic moments to all iron and cobalt atoms, and assume that while in the pure metals their association with one another somehow prevents full alignment even in the strongest fields, on the other hand the arrangement of the atoms in an alloy of suitable composition is such as to favor a higher degree of alignment. Item 2 could be accounted for along similar lines. All atoms of at least one constituent of the nonferrous magnetic alloys, for example Mn, could be supposed to have magnetic moments, but while in the pure metal(s) these form closed groups, which do not have any resultant moment, the arrangements of the atoms within the alloys allow some appreciable degree of alignment. Finally, we might explain item 3 by supposing that the compositions of the nonmagnetic ferrous alloys are such as to produce closed groups giving practically zero external fields.

Further light is shed on the problem by a semiquantitative discussion of the magnetization and hysteresis curves for an assemblage of atom magnets based on the requirement that for equilibrium at each stage the total energy of the system shall be a minimum. This energy consists of two items: (1) that of the assemblage of atoms alone, each with its characteristic magnetic moment, this item depending on the degree of alignment of the magnetic axes, and (2) the energy of interaction of the resultant magnet with the applied magnetizing field. For each value of H the sum of these energies, considered as a function of the degree of alignment, can have either one or more than one minimum value, and correspondingly there can be one or more than one equilibrium magnetic moment, depending on whether H is large or small. Along these lines we may predict, at least roughly, the magnetization and hysteresis curves. It is found that for a given value of M_s the scale of H for these curves depends on the size of the magnetic units, the scale decreasing as the size of the units increases. If the units are assumed to be atoms, H_c comes out at something of the order of 500,000 A/m. This is many powers of 10 higher than the experimental values found for iron. On the same assumption of atomic units, the initial slope of the magnetization curve comes out at about 10^{-6}, whereas observed values are higher than this by factors of between 150 and 20,000.

From such comparisons it was inferred by Pierre Weiss, in 1907, that the units figuring in ferromagnetic processes must, in fact, be much larger than the individual atoms; only on this assumption could the relatively small values of H_c and the high values of the initial slope of the magnetizing curve observed for ferromagnetic materials be accounted for. These larger units, aligned groups of atoms, are now generally referred to as magnetic **domains.**

Bearing in mind the usefulness of iron filings in the investigation of large-scale magnetic fields, L. von Hàmos and P. A. Thiessen in 1931 conceived the idea that if the magnetic domains postulated by Weiss are large enough it might be possible to obtain visual evidence of their existence and of their response to a magnetic field with the aid of a sufficiently fine magnetic powder. Independently, F. Bitter brought a suspension of Fe_2O_3 whose particles had linear dimensions of the order of only 10^{-3} mm into contact with a polished surface of a magnetic material. Subsequent microscopic examination revealed that the particles had attached themselves to the specimen along lines enclosing small areas, which were, presumably, the exposed surfaces of magnetic domains. With changes in the applied field interesting changes were observed to occur in the positions of the boundaries between these areas and in the densities of the deposits along them. Work that has been carried out in the intervening years using this method has brought to light a wealth of information concerning the sizes of magnetic domains, their magnetic orientations, and the changes that occur in both of these in response to variations in the magnetic field.

It has been found in these investigations that magnetic domains vary widely in size, being much larger in single-crystal specimens than in polycrystalline material of small grain size. Their volumes range from about 10^{-3} mm^3 in polycrystalline material to as much as 10 mm^3 in single crystals, the corresponding range in the number of atoms comprising a domain thus being about 10^{17} to 10^{21}. In ordinary polycrystalline material the linear dimensions of the domains are normally somewhat less, although not of a smaller order of magnitude, than are those of the separate grains.

It appears that the magnetic axis has the same direction everywhere within a domain, and this presumably means that all the atomic moments in the domain are spontaneously aligned in this direction or close to it. Within each not-too-small grain the direction of alignment is found to vary from one domain to the next according to a geometrical pattern that is characteristic of the particular material, with a tendency for mutual cancellation of external magnetic effects; and in unmagnetized polycrystalline material the total distribution of the magnetic axes of the domains may be regarded as effectively random. Adjacent domains are separated by transition regions, known as "walls," about 1000 atoms thick, across which the orientations of the atomic moments change progressively.

When a field is applied, or an existing field is changed, there can of course be no question of the domains, closely interlocked as they are, adjusting their magnetic orientation by rotating as a body. There are, however, two alternative ways in which the magnetic orientation *can* change in response to changes in the field. All the atoms in a given domain may swing around together into a new alignment more closely corresponding to that of the field. This is known as "rotation." Alternatively, domains in which the atoms are already aligned in a direction approximating that of the field may grow in size at the expense of neighboring domains that are less favorably orientated. This process involves a displacement of the boundaries between neighboring domains, and accordingly it is generally referred to as "boundary displacement." Both rotation and boundary displacement are found to occur, boundary displacement preponderating at weak and moderate fields, while rotation is the main process by which the magnetic condition of the material approaches saturation when subjected to strong fields.

In material of such small grain size that the linear dimensions of the grains are only of the same order as the thickness of the walls separating adjacent domains, the process of boundary displacement clearly cannot occur except to a very limited extent, and so the magnetization of the material can change practically only by rotation, the process characteristic of strong applied fields. Hence very fine-grained material is particularly suitable for permanent magnets.

The fields involved in the hysteresis loop of a ferromagnetic material are those in the medium range, where changes in the magnetic condition are brought about mainly by boundary displacements. According to the present-day theory of magnetism, boundary displacements, other than those that occur at very weak fields, are thermodynamically irreversible in character, and it is on this account that hysteresis occurs. The enormous range in the values of H_c characteristic of different ferromagnetic materials has been accounted for theoretically in terms of structural differences and the presence or absence of lattice imperfections.

CIRCULATING ELECTRICITY IN ATOMS We know from modern atomic theory that the electrons constituting the negatively charged parts of atoms revolve in orbits about relatively massive positively charged nuclei, much in the same way as the planets of the solar system revolve in orbits about the sun. Corresponding to these orbital motions there must be magnetic moments, and in principle the latter might well combine vectorially to give a resultant other than zero for each atom.

Actually, although electrons do undoubtedly produce magnetic effects by virtue of their orbital motions, these motions cannot be held responsible for ferromagnetism. This is because the orbits are too firmly fixed with respect to the lattice structure of the material to be affected appreciably by a magnetic field. That the orbital motions of the electrons do indeed make no contribution to ferromagnetic behavior is known from certain gyromagnetic measurements whose discussion is beyond the scope of this book. The evidence of these measurements must be regarded as conclusive.

Not only are electron orbits unable to align themselves in response to an applied field, but, owing to the three-dimensional symmetry of the lattice, there also cannot be any resultant "orbital" magnetic moment producing the equivalent of a permanent magnet for the lattice as a whole.

O. W. Richardson in 1921, as well as a number of other physicists at about the same time, suggested that, like the planets in the solar system, electrons might conceivably spin about axes through their centers in addition to describing orbits about atomic nuclei. A few years later, in 1925, this idea was applied successfully

813

by Uhlenbeck and Goudsmit in explaining certain features of atomic spectra, notably the Zeeman effect—a splitting of spectral lines in strong magnetic fields. Since then the concept of spinning electrons has been confirmed by such a body of experimental evidence, as well as being demanded by theoretical considerations, that the existence of this spin must be regarded as firmly established. The magnetic moment produced by this spin is 1.1667×10^{-29} N-m^2/A. This elementary magnetic moment is known as the **Bohr magneton,** the name deriving from the magnetic moment of the smallest electron orbit according to the Bohr theory of the atom, to which that of the spinning electron is equal. It is interesting to note that the theoretical physicist Dirac has shown that the electron *must* have a spin producing a magnetic moment of this amount if it is to conform both to Einstein's principle of relativity and to the laws of quantum mechanics.

Since the orbital motions of electrons cannot be held responsible for ferromagnetic behavior, there is only one thing that can; it must be the electron spins.

The electrons belonging to any particular atom can produce an external magnetic effect as a result of their spins only if the separate spin magnetic moments, when combined vectorially, give a resultant other than zero. However, in solids, this is only rarely the case. The quantum theory of atomic structure requires that in all complete electron "shells" half the electrons must spin one way and half the opposite way, giving mutual cancellation of external effects. Similarly, for one reason or another, the valence electrons of a solid cannot produce a resultant magnetic moment. Thus, in nonmetallic solid elements the valence electrons play an essential role in providing the cohesive binding that holds the atoms in their lattice arrangements, and the spin moments of these electrons cancel each other out in pairs. In ionic crystals such as NaCl, valence electrons are transferred from one atom to another so as to form complete shells in both resulting ions, which, as we have seen, cannot give an external magnetic effect. And finally, for the valence electrons of metals also, there is mutual cancellation of spin moments. In the solid lattice, these electrons do not belong to particular atoms but constitute an internal electron "gas" whose existence is responsible for the electrical conductivity of the metal. As quantum theory shows, half the electrons constituting this gas must spin one way and half the opposite way.

There now remains only one possible way in which the spin moments of the electrons belonging to an atom can have a resultant other than zero; this can happen only when one or more of the internal shells of electrons is incomplete. Such incomplete shells occur in the transition elements and in the rare earths.

THEORY OF DOMAIN FORMATION We have seen that within each domain of a ferromagnetic material there is at least a high degree of alignment, if not perfect alignment, of the magnetic moments of the atoms, regardless of whether or not an external field is applied. This can only mean that the atoms exert forces on one another which bring about this condition. A quantum theory of these forces, known as "exchange forces," was worked out in 1928 by W. Heisenberg, and subsequently, in 1933, Bethe showed semiquantitatively under what conditions the exchange forces would bring about alignment of large assemblages of atoms. The theory indicates that not only the magnitude of the exchange forces, but also their nature, is determined by the distance of separation between adjacent atoms. Below a certain value of this distance the exchange forces are such as to tend to set the spin moments in opposite directions. As the distance is increased, these antialignment forces first decrease to zero, and then, with further increase in distance, the forces change their character, tending to align the magnetic moments parallel to one another. At a certain interatomic distance these forces pass through a maximum, beyond which they approach zero asymptotically as the distance is still further increased.

It appears that, among those pure metals whose atoms have resultant magnetic moments, only Fe, Co, Ni, Gd, and Dy have interatomic distances that are large enough to be favorable to domain formation. It appears furthermore that these forces are of sufficient magnitude actually to produce domains at ordinary temperatures, despite the randomizing tendency of thermal motions, only in Fe, Co, and Ni, and in addition, below 16°C, in Gd.

The nonmagnetic behavior of manganese, whose atoms have magnetic moments, is to be attributed to the distance between adjacent atoms in the manganese lattice being below the critical value for ferromagnetism, so that the exchange forces tend to set the atomic moments antiparallel. However, these same atoms can produce ferromagnetism in such alloys as the Heusler alloys and manganese bismuthide, because in these they are separated from one another by distances that are large enough for the exchange forces to set the magnetic moments parallel, with consequent domain formation.

Notwithstanding the explanations of magnetic behavior just given, it is not to be imagined that the volumetric concentration of atoms having magnetic moments is the only factor determining

815

whether or not a material is ferromagnetic. The intricacies of the magnetic behavior of alloys cannot be explained in these terms alone. A full theory would also have to take into account the association of the atoms of different kinds with one another in three-dimensional patterns of greater or less regularity and the effects of such association on the exchange forces.

Curie Temperature

In no ferromagnetic material above the absolute zero of temperature can the exchange forces bring about perfect alignment of the atomic magnetic moments; the tendency for them to do this is always countered to a greater or lesser extent by thermal agitation. As a result of this the effectiveness of the domains in contributing to the macroscopic intensity of magnetization progressively decreases as the temperature is raised from a low value, and the saturation intensity M_s is correspondingly less than the value M_0 characteristic of low temperatures. A theoretical discussion shows that the greater the number of atomic magnets that are shaken out of their proper alignment within any domain, the easier it is for the rest to follow. Consequently, with rising temperature, M_s/M_0 should fall below the value 1 at an increasing rate, until, when a certain limiting temperature is approached, the loss of magnetism becomes almost catastrophic; and beyond this temperature the material ceases to be ferromagnetic. This type of behavior, which was first observed experimentally by Pierre Curie (1859–1906), and which has now also been accounted for theoretically, is shown graphically in Fig. 31.4.

The upper limit of temperature beyond which a material is no longer ferromagnetic is known as the **Curie temperature,** or **Curie point.** The Curie point for iron is 770°C, for nickel it is 358°C, and for cobalt 1120°C. We have already noted the Curie points for gadolinium and dysprosium, 16°C and −168°C, respectively.

Paramagnetism

In most solids, whether or not the constituent atoms have a resultant magnetic moment due to unbalanced electron spins, the conditions for domain formation are not satisfied. Despite this, the orientation of the spin moments of such electrons as do not belong to closed shells are affected to *some* extent by an applied field, producing a certain degree of statistically preferred orientation.

816

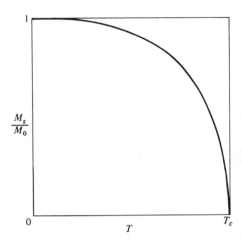

FIGURE 31.4 Ratio of saturation intensities of magnetization at temperatures T and $0°K$ as a function of T.

This has the effect of increasing B beyond the value $\mu_0 H$. Also, in gases and liquids whose molecules have a resultant magnetic moment, an applied field will produce some degree of alignment of the molecules. It is interesting to note that this could not occur but for collisions or other interactions between molecules; the magnetic axis of an isolated molecule would not tend to align itself with the field, but would merely precess about this direction like a top, without any change in the angle between the magnetic moment and the field. Under the conditions actually prevailing, interactions between molecules arising out of thermal agitation modify these gyrations in the sense of bringing the molecular magnetic axes into closer alignment with the field, despite the randomizing tendency of this same agitation.

In discussing the magnetic behavior of materials it is convenient to refer to a property known as their **magnetic susceptibility** χ. This is defined as the quotient of the intensity of magnetization M and the magnetizing force H that produces it:

$$\chi = \frac{M}{H}. \tag{31.2}$$

For nonferromagnetic materials this quotient is a perfectly definite quantity, not depending on whether H is being increased or decreased, and is constant for all but exceptionally large magnetizing forces. Nonferromagnetic materials for which χ is positive are classed as **paramagnetic.**

817

It will be recalled that in Chapter 30 μ was defined by the equation

$$B = \frac{\mu}{4\pi} \oint \frac{I \, dl \, \sin \theta}{r^2},$$

which, in the case where the material in question fills the interior of a long current-carrying solenoid, simplifies to

$$\mu = \frac{B}{nI},$$

or, since nI for a long solenoid has been defined in the present chapter as the magnetizing force H,

$$\mu = \frac{B}{H}. \tag{31.3}$$

Hence, with the intensity of magnetization defined as $B - \mu_0 H$, χ as defined above becomes $B/H - \mu_0$, or $\mu - \mu_0$. On the other hand, if the magnetic moment of a current-carrying coil were defined as NIA instead of $\mu_0 NIA$, then the corresponding expression for the intensity of magnetization would be $(B/\mu_0) - H$, or $(\mu H/\mu_0) - H$, and so the susceptibility defined as M/H would be $(\mu/\mu_0) - 1$, a dimensionless quantity. Many physicists prefer to define the susceptibility in this way. To convert values quoted by these to those appropriate to the definition used in this book, $\mu - \mu_0$, we have merely to multiply by μ_0, that is, by $4\pi \times 10^{-7}$. Thus, writing $\mu - \mu_0$ as χ, and denoting the dimensionless quantity, which we may call the "relative susceptibility," by the symbol χ_r, we have

$$\chi = 4\pi \times 10^{-7} \chi_r.$$

For paramagnetic materials the values of μ are so close to μ_0 that in all cases the difference, χ, is of a smaller order of magnitude than either. Whereas μ_0 has the value $4\pi \times 10^{-7}$ N/A^2, that is, 1.257×10^{-6} N/A^2, values of χ for solid and liquid paramagnetic materials are usually in the range 10^{-11} to 10^{-8} N/A^2, and for gases at ordinary temperatures and pressures χ is of the order of 10^{-12} N/A^2.

When a paramagnetic material is exposed to a field the resulting degree of alignment of the atoms or molecules is determined by a balance between the aligning tendency of the field and the randomizing tendency of the thermal motions. Hence we should expect the intensity of magnetization M produced by a given field to increase with decreasing temperature. For not-too-strong fields M should also increase with H, and theoretical considerations indicate that the relation between these should be one of proportionality. Curie found experimentally in 1895 that for a number of materials investigated by him the dependence of M on the absolute temperature T and the magnetizing force H is expressible by the formula

$$M = C \frac{H}{T}, \tag{31.4}$$

where C is a constant characteristic of the material. This is known as **Curie's law.**

This law implies inverse proportionality between χ and T. How well this holds is shown by the example of iron ammonium alum, for which χ is 1030 N/A at 15°K and 53 N/A at 290°K. The factor in temperature and the inverse factor in χ are here both very close to 19.4.

The intensity of magnetization obviously cannot increase without limit as H/T is increased, at sufficiently large values of this quantity the material must become saturated, just as does a ferromagnetic material at much lower field strengths. This is shown by Fig. 31.5, in which values of M/M_s obtained by W. E. Henry for chromium potassium alum are plotted against the corresponding values of H/T. The observations were made at four temperatures, 1.30°, 2.00°, 3.00°, and 4.21°K, and at a number of different values of the magnetizing force. Four kinds of "points" are marked in the graph, corresponding to the four temperatures. The curve, instead of simply being drawn through these points, is a quantum-

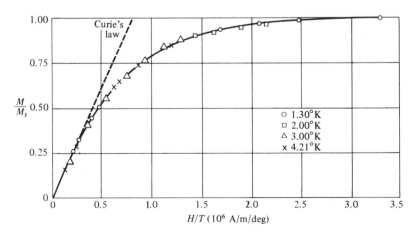

FIGURE 31.5 Quantum-theoretical curve of variation of M/M_s with H/T and corresponding experimental points obtained for the paramagnetic salt chrome potassium alum.

819

theoretical curve, and it will be noted how well it represents the experimental measurements.

We see that the curve corresponds to Curie's law up to an H/T of about 2.7×10^5 A/m/deg. This means that to detect any departure from Curie's law at room temperature a value of H in excess of 8×10^7 A/m would be required. To obtain 95 percent saturation at this temperature would require the extremely high value of 5.6×10^8 A/m.

Diamagnetism

The susceptibilities of some materials have negative values. This phenomenon is known as **diamagnetism.** An apparently successful theory of diamagnetism was worked out many years ago in which this was attributed to an electromagnetic induction effect that causes the orbiting electrons to rotate either faster or more slowly, according to the directions of the atomic magnetic moments relative to that of the applied field. However, this theory is not compatible with quantum mechanics and must therefore be abandoned. It has now been replaced by a quantum theory, which satisfactorily accounts for the phenomenon.

In contrast to the strong variation of χ with temperature in the case of paramagnetic materials, the χ of diamagnetic materials is practically temperature-independent.

The values of χ for a selection of diamagnetic materials are given in Table 31.3. Bismuth has the largest negative value of χ of any known substance.

Equivalence of Bar Magnet and Current-Carrying Solenoid

In Chapter 29 we saw that a short bar magnet and a small current-carrying coil are equivalent in regard to the manner in

TABLE 31.3 Susceptibilities of Diamagnetic Materials

Material	χ (N/A^2)	Material	χ (N/A^2)
Bismuth	-21×10^{-11}	Diamond	-2.8×10^{-11}
Antimony	-6×10^{-11}	Copper	-1.2×10^{-11}
Silver	-3.4×10^{-11}	Argon (STP)	-1.2×10^{-14}

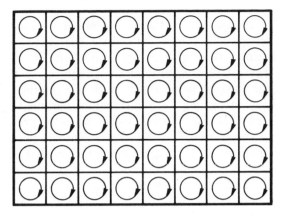

FIGURE 31.6 Idealized thin transverse slice of magnet with currents in microcircuits corresponding to magnetically aligned atoms.

which the torques exerted on them by a given magnetic field vary with the angles between their axes and the direction of the field. This is a special, limiting case of a much more general situation. It may be shown experimentally that there is also equivalence between a large bar magnet and a solenoid which has the same dimensions and carries an appropriate current. These are magnetically equivalent in all respects—not only in the torques exerted on them when orientated at various angles to a given external field but also in regard to the fields *produced by them* at all external points.*

Why this should be so may readily be seen. For simplicity let us imagine all the atomic moments in the magnet to be aligned in the direction of the axis, corresponding to the fully saturated condition, and let us suppose a thin slice of this, cut perpendicular to the axis, to contain these aligned atoms in a regularly spaced array, as indicated in Fig. 31.6. Each of these atoms produces an external magnetic field which would remain unaffected if we expanded the area of the equivalent circuit and diminished the current strength each by the same factor, so as to give the same product of area and current as before. Let us imagine this to be done, in such a way that the currents now circulate around the squares indicated. We would then have the situation depicted in Figs. 30.9 and 30.10.

---

* This is not strictly true for points relatively close to the ends of the magnet or solenoid, owing to the circumstance that the degree of alignment of the domains falls away somewhat toward the ends of the magnet. However, this does not affect the principle of the argument that follows.

821

Applying the argument previously presented in connection with these figures in reverse, we see that all the currents associated with the aligned atoms of Fig. 31.6 are equivalent in their external effects to a current flowing around the periphery of the slice, the material of the magnet being replaced by a vacuum. This applies to all the slices into which we may imagine the magnet to be cut. Finally, we see that the aligned atoms actually present may be replaced by a sheet of current flowing around the periphery of the magnet, the material of the magnet being removed. This equivalent sheet of current circulating around the surface of the magnet is known as the **Amperian current.** It may be approximated very closely by a current passing around a solenoid whose dimensions correspond to those of the magnet.

It is a simple matter to extend this reasoning to include the case where the atoms are not fully aligned and/or are not arranged in a perfectly regular pattern.

Electromagnets

Electromagnets consisting of an iron or iron-alloy armature provided with a winding are used extensively in industry, in household appliances, and in scientific work of various kinds. When current is passed through the winding the armature becomes a powerful magnet, but it reverts to its initial almost nonmagnetized state as soon as the current is switched off.

For practical reasons the winding is usually confined to a relatively small portion of the armature. When a current is passed through the winding the resulting Amperian current, acting in conjunction with the winding current, must provide a strong magnetizing force extending to all parts of the armature, including those parts which are remote from the winding. Thus the Amperian current propagates itself from the region of the winding, and it is for this reason that it is unnecessary to wind the whole of the armature.

Forces between Magnets

In the light of the present-day theory of magnetism it is clear that the forces exerted between magnets, whether "permanent" or temporary, must in effect be simply the forces between the elements of the Amperian currents.

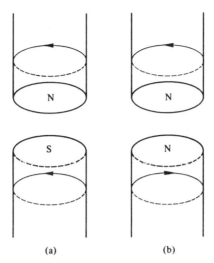

FIGURE 31.7 Forces between "poles" of magnets explained in terms of Amperian currents.

If a so-called north pole of one magnet is near the south pole of another, as indicated in Fig. 31.7(a) the Amperian currents circulate around these in the same direction and, since currents traveling in the same direction attract each other, the resultant mechanical effect must be one of attraction. On the other hand, if the two adjacent ends have the same polarity, as in Fig. 31.7(b), the Amperian currents, circulating in opposite directions, must give rise to a resultant mutual repulsive force. This is the true explanation of what is described in the older theory of magnetism as the attraction between unlike poles and the repulsion between like poles.

Calculation of these forces on the basis of Eq. (29.7) integrated vectorially over all the Amperian current elements yields the well-known inverse-square law of forces exerted between the end regions of magnets.

Problems

31.1 It is stated in the text that the area of the hysteresis loop represents the energy dissipated per cycle. Show that the quantity represented by this area has the dimensions of energy per unit volume.

31.2 The material of which a bar magnet is constructed has a coercivity of 19,000 A/m. The magnet is to be demagnetized by inserting it in a long solenoid having five turns per centimeter and passing a current through this. What should be the magnitude of the current?

31.3 The intensity of magnetization in saturated iron is 2.15 N/A/m. What is the corresponding magnetic moment contributed by each atom? (Density of iron = 7.9 g/cm³, atomic weight = 55.85.)

31.4 A permanent cylindrical steel magnet of diameter 5 mm and length 10 cm experiences a torque of 0.018 N-m when placed inside a long solenoid having one turn per millimeter and carrying a current of 10 A, with its axis at right angles to the axis of the solenoid. What is the magnetic moment of the magnet? What are the mean values of the intensity of magnetization and the Amperian current? (Neglect any effect of the earth's magnetic field on the torque experienced by the magnet.)

31.5 Find the value of the axial component of **B** at a point on the surface constituting one end of a long current-carrying solenoid. What is the magnetic flux* passing through a cross section of the solenoid remote from the ends? What proportion of this flux passes through each end? What would you expect to be the situation a short distance (say 1 or 2 diameters) in from an end?

In the case of a short solenoid, would (a) the magnitude of **B** at an end, and (b) the proportion of the flux that passes through the end be greater or less than for a long solenoid? Reasoning in terms of the correspondence between a current-carrying solenoid and a bar magnet, and flux in the latter, can you suggest a plausible explanation for the existence of magnetic "poles" and for the inverse-square law applying to these? (The ability to answer these questions depends on the realization of situations, rather than on calculation.)

31.6 What is the magnetic flux passing through the center of a uniformly magnetized cylindrical rod of steel 20 cm long and of 1 cm radius whose intensity of magnetization is 1 N/A/m? How much of this flux leaves the end of the magnet, and how much passes through the sides before reaching the end? What is the moment of this magnet?

31.7 What is the Amperian current of the magnet of Problem 31.6?

31.8 Do Amperian currents really exist? If they do, why do they not generate any heat?

---

* The magnetic flux passing through an element of area is defined as the product of this area and the component of **B** normal to it. The MKSA unit of flux is called the weber.

32

ELECTROMAGNETIC INDUCTION

In this chapter we shall consider the theory of a fundamentally important phenomenon in electromagnetism, **electromagnetic induction.** This is the development of an electromotive force in response to the motion of a conductor across a magnetic field or resulting from an interaction of a changing field with a fixed circuit. After having established the basic quantitative law of electromagnetic induction, we shall consider the operation of this law in a number of processes, such as the growth and decay of current in a circuit, the electromagnetic generation of alternating and direct currents, the action of electric motors, and the damping of a galvanometer. Then in the next two chapters we shall consider the role played by electromagnetic induction in certain aspects of alternating-current theory, instrumentation, and techniques.

Law of Electromagnetic Induction

Let us in the first instance consider the motion of a metal wire XY in a direction transverse to its length, as indicated in Fig. 32.1, this being supposed to occur in a uniform magnetic field **B** whose direction makes an angle θ with the normal to the plane defined by XY and the velocity **v** of the wire (the plane of the diagram). As we saw in Chapter 22, this wire contains both im-

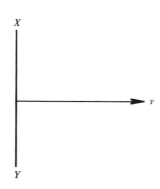

FIGURE 32.1 Length of metal wire traveling in a direction transverse to a magnetic field.

mobile positive charges and conduction electrons. Both the positive charges and the electrons, moving from left to right, constitute currents, and considering these in the same way as we did the stream of electrons represented in Fig. 30.19, we can calculate the force that the magnetic field exerts on each kind of charge. The positive charges are immobile, so the magnetic forces exerted on these cannot move them through the metal, and we need not consider them further. On the other hand, the electrons, which are mobile, *can* respond to the forces exerted on them by the magnetic field.

As we saw in Chapter 30, the component of **B** in the direction of **v** is without effect on the electrons. The component parallel to the wire produces a force at right angles to XY, with whose effects we shall not be concerned in the present discussion. Finally, the component $B \cos \theta$ perpendicular both to XY and to **v** produces a force along XY of magnitude $B \cos \theta \cdot e \cdot v$ on each electron, and it is the effect of this that we shall now consider.

Let the component $B \cos \theta$ be directed into the plane of the diagram. Then the force $B \cos \theta \cdot e \cdot v$ will be directed from X toward Y, and the electrons will move in this direction until an electrostatic field $E_f{}^*$ is set up which prevents any further flow of electrons. The condition for this is that

$$E_f e = B \cos \theta \cdot e \cdot v,$$

or

$$E_f = Bv \cos \theta.$$

It is easily seen that this (conventional) field must be directed downward, making X positive to Y. And since electric field strength is the same thing as negative potential gradient, the PD between X and Y must be $E_f l$, or $Bvl \cos \theta$, where l is the length of the wire. But vl is the rate of sweeping out area. Hence, writing this as dA/dt, and adopting the convention that dA/dt is positive for motion in the direction of v, we have

$$V_X - V_Y = B \frac{dA}{dt} \cos \theta,$$

or, with B and θ both constant,

$$V_X - V_Y = \frac{d}{dt} (BA \cos \theta).$$

---

* We use the symbol E_f for electrostatic field to avoid confusion between this quantity and the electromotive force E.

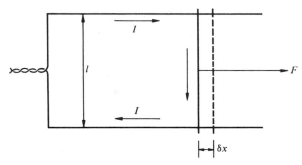

FIGURE 32.2 Wire sliding along a rectangular current-carrying wire frame whose plane is at right angles to a uniform magnetic field.

Let the wire now be used as a slider on a rectangular frame, in the manner indicated in Fig. 32.2, so forming part of a complete circuit. The PD developed between the contact points would now, in the absence of any other emf in the circuit, cause a conventional current to flow around in an anticlockwise direction, that is, in the direction opposite to that shown by the arrows. (We may ignore these arrows in the present discussion.) Thus we see that the wire moving across the magnetic field acts as a device developing an emf, producing a PD across its terminals when on open circuit and driving a current when the circuit is closed. It would appear, then, that we must attribute to the device an emf of magnitude equal to the open-circuit potential difference $d(BA \cos \theta)/dt$.

Let us now adopt a convention for the direction around the circuit which is to be taken as the positive direction. Let this be defined as that direction which is seen as clockwise when viewed in the direction of the component of **B** normal to the plane of the circuit. This makes the emf, E, which is anticlockwise, negative. Thus we have

$$E = -\frac{d}{dt} (BA \cos \theta), \qquad (32.1)$$

where A now stands for the area enclosed by the circuit.

That we do indeed have an emf of this magnitude, regardless of whether or not a current flows around the circuit, may be seen in another way. Let a current I flow around the circuit in the positive (clockwise) direction. (This current may be either positive or negative; in the former case an external source of emf would, of course, be needed to drive it.) Then the slider will experience a force having a component F in the direction of motion. This com-

827

ponent will be positive or negative according to the sign of I, and its magnitude will be the product of the length l of the slider between the contact points, the current I, and the component of the field perpendicular to the rectangle:

$$F = B \cos \theta \cdot Il.$$

On allowing the slider to move in response to this force through a distance δx, we have an amount of work $F \, \delta x$, or $B \cos \theta \cdot Il \, \delta x$, performed for us by conversion of electrical energy into another form. With this energy transformation is necessarily associated a back emf equal to the rate of energy conversion per unit current, so that if the movement occurs at uniform velocity in the time δt the forward emf E is given algebraically by the equation

$$E = -\frac{B \cos \theta \cdot Il(\delta x/\delta t)}{I}$$

$$= -B \cos \theta \, \frac{\delta A}{\delta t}$$

The condition of uniform velocity is automatically taken care of if δt is made infinitesimal, and our equation then becomes

$$E = -B \cos \theta \, \frac{dA}{dt},$$

or, with constant B and θ,

$$E = -\frac{d}{dt}(BA \cos \theta), \tag{32.1}$$

as before.

The factor $\cos \theta$ may be associated either with B or with A. In the former case the quantity in parentheses becomes $B_n A$, where B_n is the component of **B** normal to the plane of the circuit, and in the latter case it becomes BA_p, where A_p is the *projected* area of the circuit on a plane perpendicular to **B**. In terms of these quantities the last equation may be written

$$E = -\frac{d}{dt}(BA \cos \theta) = -\frac{d}{dt}(B_n A) = -\frac{d}{dt}(BA_p).$$

The quantity $BA \cos \theta$, or $B_n A$, or BA_p, is defined as the **magnetic flux** Φ that passes through the circuit. In terms of this we may write the law of electromagnetic induction in the form

$$E = -\frac{d\Phi}{dt}. \tag{32.2}$$

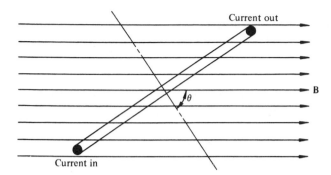

FIGURE 32.3 Plane circuit rotating about an axis in its plane which is perpendicular to a uniform magnetic field.

The unit of flux is the **weber** (Wb), corresponding to the definition of Φ as $BA \cos \theta$ when B is expressed in N/A/m and A in m².

The **flux density** is the flux Φ divided by the projected area $A \cos \theta$ through which it passes. It may therefore be identified with B_n. Accordingly, a unit for B alternative to the N/A/m is the weber per square meter (Wb/m²).*

An emf brought into being by a change in the flux linking a circuit is known as an **induced emf.**

Equation (32.2) has been derived on the assumption that (1) **B** is constant in magnitude and direction, and (2) the circuit is in a constant plane. Let us now inquire whether these restrictions are really necessary.

CIRCUIT ROTATING IN CONSTANT FIELD Let the circuit shown in section in Fig. 32.3 rotate about an axis perpendicular to the plane of the diagram. Let the direction along the normal to the plane of the circuit in which the current appears to circulate clockwise momentarily make an angle θ with the direction of the magnetic field **B**. Then, as we saw in Chapter 29, if the circuit is of area A and carries a current I, the field exerts on it a torque about an axis perpendicular to the plane of the diagram equal to $BIA \sin \theta$, the sense of this torque being anticlockwise. Hence the work that must be done *on* it to make it rotate (clockwise) through a small angle $\delta\theta$ is $BIA \sin \theta \cdot \delta\theta$. If this rotation occurs at uniform angular velocity during the time δt, the power P_{OE} is $BIA \sin \theta \cdot \delta\theta/\delta t$, and

---

* When we think of a magnetic field as a flux per unit area, we suppose the plane of this area to be at right angles to **B**, so that B_n and B are then the same.

for the induced emf during this time interval we must have

$$E = \frac{P_{OE}}{I} = BA \sin \theta \frac{\delta\theta}{\delta t}.$$

If we make δt infinitesimal, to avoid the necessity for specifying uniform angular velocity, this becomes

$$E = BA \sin \theta \frac{d\theta}{dt}$$

$$= -BA \frac{d}{dt}(\cos \theta).$$

Thus, in this case also (B and A constant),

$$E = -\frac{d}{dt}(BA \cos \theta) = -\frac{d\Phi}{dt}.$$

CHANGING MAGNETIC FIELD This case is best considered indirectly, by contemplating the last equation written in the form

$$E = -\frac{d}{dt}(B \cos \theta \cdot A),$$

which, for A constant, becomes

$$E = -A \frac{dB_n}{dt}.$$

Now, kinematically, rotation of the circuit in a constant field may be replaced by a rotating field of constant magnitude acting on a stationary circuit. But the only effective part of this field as far as the production of an emf is concerned is the component B_n. Hence the situation is the same as if dB_n/dt in the equation above were replaced simply by dB/dt, with **B** perpendicular to the plane of the circuit. We have, therefore, for the case of a circuit of constant area exposed to a changing field whose direction is always normal to the plane of the circuit,

$$E = -A \frac{dB}{dt} = -\frac{d\Phi}{dt}.$$

Since, as we have seen, Eq. (32.2) applies in all three of the special cases contemplated: (1) for varying A, (2) for varying θ, and (3) for varying B, the other two quantities remaining constant, it is indicated that it must surely apply quite generally. Thus we should expect it still to be valid where more than one of the quantities A, θ, and B changes at a time, or where the circuit is not all in one plane, or even where **B** is not uniform.

If N identical loops of wire are connected in series, so constituting a coil of N turns, the emf E induced in the coil as a whole will be N times $-d\Phi/dt$, or $-d(N\Phi)/dt$. Hence, writing for $N\Phi$, "the flux linked with the coil," the symbol Φ_l we have

$$E = -\frac{d\Phi_l}{dt}. \tag{32.3}$$

The phenomenon of electromagnetic induction was discovered independently, and almost simultaneously, by Henry in America and by Faraday in England, in 1830–1831. Faraday established experimentally all the information required for the setting up of a mathematical equation for the effect, but did not himself take this final step. It was not until 1845 that it was shown, by F. E. Neumann, that all Faraday's results could be summarized in the form of the single equation (32.3). In recognition of Faraday's inspired and painstaking work in providing the basis for Neumann's eventual mathematical formulation of electromagnetic induction, this equation has been named **Faraday's law.**

LENZ'S LAW   The law of electromagnetic induction is sometimes stated without reference to directions, this statement being that the induced emf in a circuit is equal to the rate of change of flux linked with it. This partial formulation requires supplementing with a further statement to make it complete. This supplementary statement, known as **Lenz's law,** is that the direction of an induced current (due to the induced emf) is such as to oppose the change of flux that produces it.

Thus there may be said to be either one or two laws of electromagnetic induction, according as the information is given in the form of Eq. (32.3) or as the partial verbal statement of Faraday's law supplemented by Lenz's law.

Self-Induction

Since with any current there is necessarily associated a magnetic field in the surrounding medium, there must be a certain flux linked with a circuit due to the current it carries. Consequently, if this current changes, so must the flux linked with the circuit, and this change of flux must induce an emf around the circuit. Such an emf, which, by Lenz's law, is always in the direction tending to oppose the change, is known as a **self-induced emf.**

831

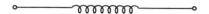

FIGURE 32.4   Conventional representation of an inductor.

Let the proportionality factor* between the flux Φ_l linked with a circuit and the current I flowing in it be L; that is, let

$$\Phi_l = LI.$$

Then for the self-induced emf we must have

$$E = -\frac{d\Phi_l}{dt} = -L\frac{dI}{dt}. \tag{32.4}$$

The quantity L is known as the **self-inductance** of the circuit. The unit of self-inductance in the MKSA system is that which gives an induced emf of 1 V for a rate of change of current of 1 A/sec. This unit is known as the **henry** (H), in honor of Joseph Henry, the discoverer of the phenomenon of self-induction.

Among the factors determining the self-inductance of a circuit are its geometrical form and area and the number of turns of wire around it. An additional factor of particular importance is the presence or absence of ferromagnetic material within the circuit. Thus the self-inductance of a helical winding filled with iron is—within the range of approximately linear variation of B with H—of the order of 1000 to 10,000 times that of the same winding when only air is present within it.

A circuit considered from the point of view of its self-inductance is known as an **inductor.** The conventional representation of an inductor is that shown in Fig. 32.4.

Energy of Current-Carrying Inductor

Let us consider an inductor connected to a steady external supply. Then the current in this will grow to some final value depending on its resistance R and the supply voltage.

At any moment the total fall of potential across the inductor is the sum of two items: the ohmic drop RI and $-E$, where E is the self-induced emf. Of these two items, the former is the rate of conversion from electrical energy to heat per unit current, and the latter

---

* Strictly speaking, Φ_l is proportional to I only in the absence of ferromagnetic material. However, even for an iron-cored circuit there is a good approximation to proportionality between the two quantities within a certain range of currents.

is the rate of conversion from electrical energy to some other form per unit current. Hence for the actual rate of conversion to this other form we have

$$\frac{dW}{dt} = -EI,$$

which, on substitution of the value for E according to Eq. (32.4), $-L(dI/dt)$, becomes

$$\frac{dW}{dt} = LI\frac{dI}{dt},$$

the integral of which is

$$W = \tfrac{1}{2}LI^2. \tag{32.5}$$

In view of the fact that in the final condition there is a magnetic field in the region of the circuit that was not there initially, we may call this energy W the **magnetic energy** of the current-carrying inductor.

We may note the analogy between this magnetic energy, due to electricity in motion, and the expressions for kinetic energy in mechanics, $\tfrac{1}{2}mv^2$ and $\tfrac{1}{2}I\omega^2$; I in the latter expression standing for the moment of inertia of the rotating body.

Growth and Decay of Current in an Inductive Circuit

If an external source of emf is connected to an inductive circuit, the current thereby caused to flow in the latter, with its associated energy, takes a finite time to become established. It also takes a finite time for the current to decay to zero when the source of emf is cut out. We shall now investigate these rates of growth and decay quantitatively.

Let us consider a circuit such as is shown in Fig. 32.5, having a resistance R and self-inductance L, provision being made for either

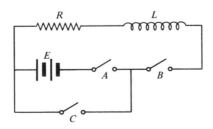

FIGURE 32.5 *R-L* circuit with provision for switching a battery either in or out of circuit.

connecting this to an external source of emf E or excluding the external emf from the circuit by appropriate operation of the three switches A, B, and C.

First, let us suppose that, with switch C open and A closed, B is also closed. From this moment on, the current builds up, the growth giving rise to an induced emf E_i given by the equation

$$E_i = -L\frac{dI}{dt}.$$

Accordingly, there is a net forward emf equal to $E - L(dI/dt)$, and this must be equal to the ohmic drop of potential RI associated with the resistance R of the circuit. Thus

$$E - L\frac{dI}{dt} = RI,$$

or

$$(E - RI)\,dt = L\,dI;$$

that is,

$$\left(I - \frac{E}{R}\right)dt = -\frac{L}{R}\,dI = -\frac{L}{R}\,d\left(I - \frac{E}{R}\right),$$

or

$$\frac{d[I - (E/R)]}{I - (E/R)} = -\frac{R}{L}\,dt.$$

Integrating this last equation, we have

$$\ln\left(I - \frac{E}{R}\right) = -\frac{R}{L}\,t + \text{const.},$$

and since, when $t = 0$, $I = 0$, the value of the constant must be $\ln(-E/R)$. Hence

$$\ln\left(1 - \frac{RI}{E}\right) = -\frac{R}{L}\,t,$$

or

$$I = \frac{E}{R}(1 - e^{-(R/L)t}). \tag{32.6}$$

This equation may also be expressed in the form

$$I = \frac{E}{R}(1 - e^{-t/\tau}), \tag{32.7}$$

where τ is written for the **time constant** L/R of the circuit. This is similar to the equation for the growth of charge of a capacitor, or of PD across its plates, in an RC circuit, which we discussed in Chapter 25.

Next, let us suppose that, after the lapse of sufficient time to allow practically the full current E/R to be established, the switch C is closed, so effectively eliminating the battery from the circuit. In practice, of course, the closing of switch C should immediately be followed by the opening of A, to avoid damaging the battery, but this last operation need not be considered further, as it has no effect on the decay of the current now to be investigated.

This case corresponds to the previous one without the external source of emf (that is, with zero E), so that

$$L\frac{dI}{dt} + RI = 0,$$

from which we obtain the equation

$$\ln I = -\frac{R}{L}t + \text{const.}$$

The value of the constant is found by applying the condition that when $t = 0$ (the moment when switch C is closed) the current has the value I_0 determined by the applied PD and the value of R according to Ohm's law. Thus

$$\ln\left(\frac{I}{I_0}\right) = -\frac{R}{L}t,$$

or

$$I = I_0 e^{-(R/L)t} = I_0 e^{-t/\tau}, \tag{32.8}$$

where the time constant τ again has the value L/R. The decay of current expressed by this equation follows the same kind of exponential law as applies to the decay of the charge of a capacitor when its plates are connected through a resistor.

Let us now consider a simple ERL circuit provided with only a single switch. The opening of this switch would give an extremely large value of $|dI/dt|$, and a correspondingly high E_i, which latter would be in the forward direction, tending to perpetuate the current. This E_i is often sufficient to cause a substantial spark to bridge the air gap at the moment of breaking the circuit, the length of the spark increasing with the rapidity of opening the switch.

Both the growth and the decay of current in an RL circuit can be demonstrated qualitatively in the following simple manner (see Fig. 32.6). First a winding consisting of a large number of turns on a hollow cylindrical frame of nonconducting material is connected to a battery via a pair of short stout pieces of wire mounted on insulating handles which the operator can either bring into contact to close the circuit or separate from one another to break it. The

835

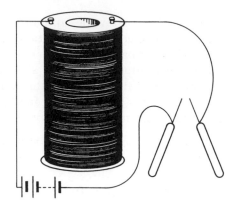

FIGURE 32.6 Apparatus for qualitative demonstration of self-inductance effects in "making" and "breaking" a current-
carrying circuit.

breaking of the circuit can be done either slowly or rapidly—the
latter by sliding one wire over the other at high speed until they are
separated. It will be observed that the spark obtained at "break"
is substantially longer in the latter case than in the former. If a
bundle of iron wires or closely packed strips of iron is now inserted
into the winding the same phenomena will be observed, but now the
sparks will be longer than formerly on both occasions, owing to the
greatly increased value of L. Finally, if the wires are flicked together
very rapidly, giving a correspondingly short interval between
"make" and "break," only a very feeble spark will be observed,
despite the rapidity of separation. The reason for this is that
insufficient time has been allowed for the current to build up. This
is an indirect way of showing that the establishment of the full
current in the circuit occupies an appreciable time.

Mutual Induction

Let us consider two circuits, 1 and 2, and let a current I_1 in the
first cause a flux $M_{12}I_1$ to be linked with the second, M_{12} being a
constant. Then if I_1 varies with time, such variation must give rise
to an induced emf E_2 in the second circuit such that

$$E_2 = -M_{12}\frac{dI_1}{dt}.$$

Correspondingly, if a current I_2 in the second circuit gives a flux
linkage of $M_{21}I_2$ with the first circuit, a change in this current must

produce an emf E_1 in the first circuit in accordance with the equation

$$E_1 = -M_{21} \frac{dI_1}{dt}.$$

It may be shown that M_{12} and M_{21} are necessarily equal to one another. We may therefore drop the subscripts from these symbols and write simply

$$E_2 = -M \frac{dI_1}{dt} \qquad (32.9)$$

and

$$E_1 = -M \frac{dI_2}{dt}. \qquad (32.10)$$

The constant M characteristic of the pair of circuits is known as their **mutual inductance.** The unit of mutual inductance is the same as that of self-inductance, the henry.

Clearly the assumption made in the foregoing discussion that M is constant will be justified only in the case where the two circuits have fixed configurations and positions relative to one another, and—strictly speaking—even then only in the absence of iron or other ferromagnetic material. However, even where some or all of the flux linkage occurs through iron, M will be very nearly constant over a certain range of the two currents. Obviously, other things being equal, M must be enormously greater where the flux linkage occurs through iron than where the medium is air or a vacuum.

Measurement of *B–H* Characteristic

In Chapter 31 it was stated that the value of B within an iron toroid could be measured inductively. We shall now see how this can be done.

This measurement depends basically on the induction of an emf in a winding linked with a toroidal ring of the iron or other material through which the flux is changing, and the measurement of the corresponding charge flowing around the circuit consisting of this winding and a ballistic galvanometer. The arrangement is indicated diagrammatically in Fig. 32.7. Current produced by a battery, regulated with a rheostat, and measured with an ammeter A is passed through a primary winding around the iron toroid, having a known number of turns n per unit length, so that the product of n and the current I_1 registered by the ammeter gives H.

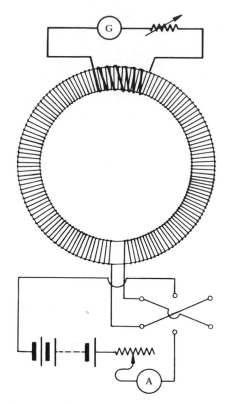

FIGURE 32.7 Apparatus for measuring B-H characteristic of ferromagnetic material.

Over a section of the toroid there is a secondary winding of N turns, and this is connected in circuit with a ballistic galvanometer G. Provision is made, for example by using a universal shunt (not shown in the figure), for the galvanometer circuit to have a high resistance; this makes the damping sufficiently small for the galvanometer to be used ballistically. A reversing switch is included in the primary circuit so that the current can be sent around the primary winding in either direction.

By appropriate adjustments of the rheostat and reversing switch the primary current I_1 is taken in steps around a cycle of "positive" and "negative" values having a range sufficient to produce saturation in the iron at the positive and negative maxima. At each step the current is changed sufficiently rapidly to satisfy the conditions for the ballistic use of the galvanometer. Then, as we shall see, the ballistic throws of the galvanometer are a measure of the corresponding changes in Φ_l linking the secondary, and so,

N and the cross-sectional area A of the iron being known,* of the changes in B.

For the charge δQ circulating around the secondary circuit, of resistance R, we have

$$\delta Q = \int I_2\, dt = \frac{1}{R} \int E_2\, dt = -\frac{1}{R} \int \frac{d\Phi_l}{dt}\, dt = -\frac{1}{R}\, \delta\Phi_l,$$

where I_2 and E_2 are the secondary current and emf, respectively, at any instant and $\delta\Phi_l$ is the change in flux linkage brought about by one of the abrupt changes in I_1. But

$$\Phi_l = NBA.$$

Hence

$$\delta B = \frac{\delta\Phi_l}{NA} = -\frac{R}{NA}\, \delta Q. \qquad \textbf{(32.11)}$$

Since δQ is given by the throw of the galvanometer and B is the algebraic sum of the quantities δB, beginning with the unmagnetized condition of the iron for which $B = 0$, the procedure outlined above gives us all the necessary information for constructing a complete B–H curve.

There is a curious feature about this system which should be noted. As we have seen, there is at no stage any field in the region outside the toroid. Yet the secondary winding, situated entirely in this field-free region, somehow responds to changes of field within the toroid. How can the secondary "know" what takes place inside the toroid?

To the best of the author's knowledge, nobody has yet been able to find an answer to this question.

Generators and Motors

The action of all dc and ac machines generating electric power depends on electromagnetic induction. And while the action of a motor could be described as based on the force that a magnetic field exerts on a current, the ensuing motion of the conductor brings into being an induced emf, so that in motors, too, electromagnetic induction is involved. We shall now consider the main features of the more usual types of both generators and motors.

---

* For simplicity we shall not make any distinction here between the cross-sectional area of the iron and the effective cross-sectional area of the primary winding.

EARTH INDUCTOR In establishing the law of electromagnetic induction one of the effects we considered was the emf induced in a circuit of constant area rotating about an axis in its plane when situated in a uniform magnetic field whose direction is at right angles to this axis. We found that for a rectangular circuit of area A this emf is given in terms of A and other quantities (see Fig. 32.3) by the equation

$$E = BA \sin \theta \, \frac{d\theta}{dt}.$$

It may readily be shown that the validity of this equation is not confined to the special case where the circuit is rectangular; it can have any form. And, if it has several turns instead of merely one, the emf is necessarily increased by a factor equal to the number of turns. Thus, quite generally, for a coil of area A having N turns,

$$E = NBA \sin \theta \, \frac{d\theta}{dt}. \tag{32.12}$$

This equation applies, *inter alia*, to the **earth inductor,** a device originally designed for finding the horizontal and vertical components of the earth's magnetic field. It consists simply of a coil that may be rotated about either a vertical or a horizontal diametral axis. If the terminals of the coil are connected to a ballistic galvanometer, and the coil is given appropriate rapid rotations through 180° about vertical and horizontal axes, the horizontal and vertical components of the earth's field may be calculated from the corresponding throws. By incorporating with the earth inductor a pair of "slip rings" with which brushes engage for continuous connection to an external circuit (see Fig. 32.8), this device may also be made to serve as a generator of alternating current varying sinusoidally with time. Let us, for example, consider the case where the coil is rotated about a vertical axis. It is then affected only by the horizontal component of the earth's field. If this is of magnitude B and the coil is rotated with uniform angular velocity ω, the induced emf is given by the equation

$$E = E_0 \sin \omega t, \tag{32.13}$$

where time is counted from one of the instants when $\theta = 0$, and

$$E_0 = \omega NBA. \tag{32.14}$$

ALTERNATING-CURRENT GENERATOR The electromotive forces obtainable with an earth inductor functioning as a generator are rather small. Of the four quantities ω, N, B, and A, whose product gives the peak voltage E_0, B is the one that is the most easily increased substantially. In practical ac generators this is pro-

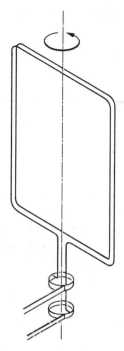

FIGURE 32.8   Earth inductor as alternating-current generator.

vided by a large magnet with hollowed-out poles facing each other. Between these the coil rotates. The coil, instead of simply being supported by a frame, is wound on a rotating soft-iron "armature" (see Fig. 32.9). The magnet between whose poles the armature with its winding rotates is known as the "field magnet." In the simplest

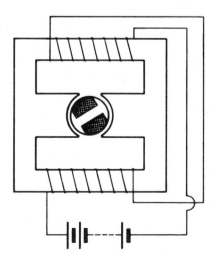

FIGURE 32.9   Practical alternating-current generator.

841

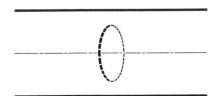

FIGURE 32.10 Eddy current in solid iron armature due to changing magnetic flux.

form of alternator, known as a "magneto," this is a permanent magnet, but much more usually it is an electromagnet through whose winding an appropriate steady current is passed. This current is usually supplied by an independent source, although in some cases it is obtained from the machine itself, rectified and smoothed (see Chapter 25) en route to the winding of the field magnet. Both the provision of a soft-iron armature and the use of an electromagnet to provide the field in which this armature rotates favor a large value of the maximum flux linked with the coil.

If the armature were of solid iron, changes of flux within it would induce currents not only in the winding but also around closed paths such as that indicated as a dashed circle (seen at an oblique angle) in Fig. 32.10. Such currents are known as "eddy currents." These not only extract energy from the machine, so reducing the efficiency of its operation, but in doing so they produce excessive heating of the iron. To reduce eddy-current losses the armature is built up of a number of thin sheets of iron whose planes contain the direction of the axis. These sheets are known as "laminations." They are provided with sufficient surface insulation before being packed together to substantially stop the flow of eddy currents. Despite the building up of the armature in this way, a certain amount of eddy-current loss still occurs, as a result of currents flowing around closed paths within the individual laminations, but such loss is no longer serious; it is of a smaller order than that which would occur in solid iron.

In the practical alternator as described, the flux through the coil does not vary strictly sinusoidally with time as the armature rotates with constant angular velocity, as it does in the case of the earth inductor. The same is therefore true of the emf induced. As may be shown by subjecting the "wave form" of the emf to a Fourier analysis (see Chapter 11), it can be represented by a series of terms of the form

$$E = E_0 \sin \omega t + E_1 \sin 3\omega t + E_2 \sin 5\omega t + \cdots ;$$

that is, it consists of a fundamental and a number of odd harmonics. Of these, the fundamental, of frequency equal to that with which the armature rotates, has an amplitude far greater than those of any of the harmonics; the latter are relatively unimportant.

DIRECT-CURRENT GENERATOR In its simplest form a dc generator is identical with an alternator in all respects other than the manner in which the ends of the coil are connected to the external circuit. The axle of the generator has mounted on it, not a pair of slip rings as in the case of an alternator, but a single "split ring," or "commutator." A pair of brushes engages with this at opposite ends of a diameter, in such a way that at the moment when the emf induced in the coil is about to change sign the halves of the commutator with which the brushes engage are interchanged. This arrangement is indicated diagrammatically in Fig. 32.11, in

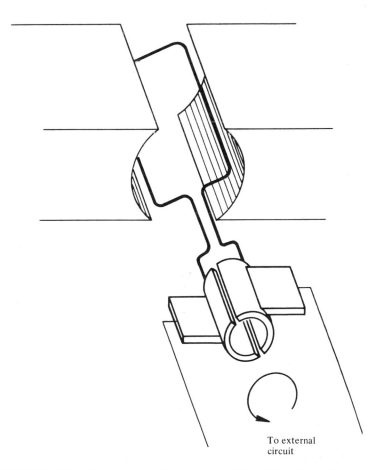

To external circuit

FIGURE 32.11 Generator producing a periodic unidirectional fluctuation of emf between zero and a maximum value.

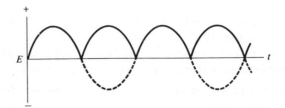

FIGURE 32.12 Variation of E with t produced by the machine shown in Fig. 32.11.

which, for simplicity, the iron armature has been omitted from the drawing. By this means, every alternate half-cycle of the "output" is reversed in polarity, and, as indicated in Fig. 32.12, a sinusoidally varying emf is converted into a fluctuating unidirectional emf.

In practical dc machines the fluctuations of emf are greatly reduced by changing the form of the iron armature and using a multisegmented commutator. Around the circumference of the armature are a number of equally spaced slots. Between diametrically opposite pairs of slots coils are wound, and these are connected in the manner now to be explained. Figure 32.13 is a diagrammatic

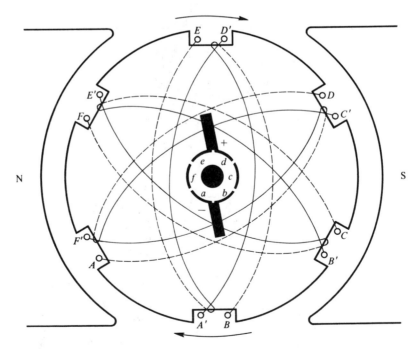

FIGURE 32.13 Practical direct-current generator designed to minimize "ripple."

representation of a machine whose armature has six slots and has wound on it six coils.* A' and B represent the two ends of the winding beginning and ending at one slot, B' and C the ends at the next, and so on, all the way around, as far as A. The ends A and A' of consecutive coils are both connected to the commutator segment a, B and B' to b, and so on. In this way all the coils are connected in series, constituting an endless winding, and the junctions between consecutive pairs are joined to commutator segments. Brushes engage with the commutator as indicated, and through these connection is made to the circuit supplied by the machine.

As the armature rotates, electromotive forces are developed in the coils that make F' positive to A and E' positive to F. Hence e is positive to a. Also C is positive to B' and D positive to C', so that d is positive to b. We see, then, that the endless winding is divided into two parts, from A' to E and from B to D', connected in parallel by the brushes, both parts developing an emf in the same sense. It should be noted incidentally that with the momentary positions of the coils shown, no emf is developed between A' and B or between E and D'. Hence there will be no sparking at the brushes.

Actually, with current being delivered by the machine, the current in the armature winding modifies the pattern of the magnetic flux in the armature, displacing the "neutral points" for the engagement of the brushes—the points at which sparking is a minimum—in the direction of rotation. Under working conditions, therefore, it is usual to advance the brushes in this sense.

It will be appreciated that with the usual much larger number of armature slots and corresponding coils than those shown in the diagram, the fluctuations of emf occurring during rotation will be very slight indeed; these are reduced to a mere ripple. When the machine is delivering current, the self-inductance of the armature winding tends to smooth this ripple still further; it is then barely perceptible.

In most dc machines the excitation of the field magnet is provided by current generated by the machine itself. There are two main systems, each of which has certain advantages and disadvantages. In one system the field winding is connected in series with the output, and in the other it is connected in parallel. The corresponding machines are known as series-wound and shunt-wound

---

* In practical machines the number is usually greater than this. The number of slots and coils shown in the figure is kept small to avoid undue complication of the diagram of connections.

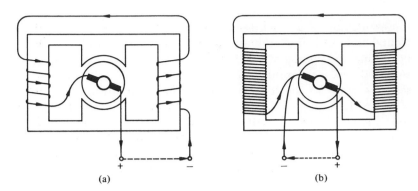

FIGURE 32.14　(a) Series-wound generator.
(b) Shunt-wound generator.

generators, respectively. These are shown in Fig. 32.14. In some machines, known as compound generators, some of the field winding is in series with the load and the remainder in parallel.

When the machine is at rest, the field magnet retains enough intensity of magnetization to give some current on starting. This current, or a portion of it, excites the magnet to a higher value of M, and in a very short time the full intensity is built up.

In all generators except magnetos and those whose field magnets are separately excited, it is solely because of the residual intensity of magnetization in the absence of current that the machine always delivers current in the same direction. If the direction of magnetization of the field magnet were reversed with the machine at rest, and the machine were then started, it would deliver current in the opposite direction.

The above discussion of generators applies only to relatively small and simple machines. Most practical dc generators are of more sophisticated design than that indicated in Fig. 32.13. Also, most large ac generators are designed with a more complex field structure than that shown in Fig. 32.9, the field magnet having not merely one pair of poles, but two or more pairs. This enables alternating current to be obtained with a lower speed of rotation of the armature than if there were only a single pair of poles.

An alternative name for a dc or an ac generator is a "dynamo."

ELECTRIC MOTORS　In a generator, nonelectrical energy is converted into electrical energy, and so the potential in the armature winding rises in the direction of current flow. In a motor, on the other hand, where the energy conversion is from electrical to some other form, the potential falls in the direction of the current.

All dc generators, and some ac generators, will function in reverse, that is, as motors.

In dc magnetos and separately excited dc machines operating as motors the current can be sent through the armature in either direction, each direction of current producing the corresponding direction of torque and consequent rotation. In series, shunt, or compound-wound dc machines the field magnet, being excited by current that is produced by the machine itself, necessarily reverses its polarity with reversal in the direction of current flow. Hence the torque is now independent of the direction of current flow, and so the direction of rotation when the machine operates as a motor must always be the reverse of that in which the machine rotates when it functions as a generator. This means that, in principle at least, the machine may also be run on ac.

In practice, the conditions under which a dc machine performs optimally as a motor differ somewhat from those under which it performs optimally as a generator, and corresponding adjustments have to be made. Hence most machines are designed specifically for either one kind of use or the other, rather than for both.

Although dc motors other than those with permanent or separately excited field magnets may be run on ac, this is not actually done except in the case of motors of relatively small size. In large machines designed as dc motors the self-inductance of both field and armature windings is considerable, and the effect of this would be seriously to reduce the efficiency of operation if these machines were run on ac. For this reason large ac motors are constructed quite differently from dc machines and are designed for use on ac only. The most common type of ac motor is the so-called "induction motor," whose armature currents, induced by a rotating field, flow entirely within the rotor unit; thus no commutator is required. Induction motors, and the supply systems designed for them, are of considerable interest. However, the detailed consideration of these here, as also of the great variety of dc machines, would take us too far; full discussion of these subjects would be more appropriate in a book on electrical engineering.

Damping of Galvanometer

If a current is passed through the coil of a moving-coil galvanometer, this is rotated from its zero-current orientation into a new orientation such that the deflecting torque arising from the action of the field on the current is just balanced by the elastic restoring torque. The production of this rotation corresponds closely to that which occurs in an electric motor due to the action of the field on the current in the armature winding and on the Amperian current in the armature.

The galvanometer coil does not usually just swing into its new orientation and stay there; at least under certain circuit conditions its moment of inertia carries it past its final orientation, it then turns back, again overshoots, and so it oscillates back and forth with steadily diminishing amplitude until finally the oscillation dies down to practically nothing and the coil comes to rest. It is of interest to study in some detail the electromagnetic processes accompanying this oscillation.

When the coil is stationary (for example, after it has attained its final orientation, or at the extremities of the preceding oscillation), the current I passing through it is determined simply by the externally applied emf E and the resistances of the external circuit, the coil, and the shunt (if any), in accordance with Kirchhoff's laws. However, when the coil is in motion an additional emf E' is induced in it because of its rotation in the magnetic field, corresponding to the emf induced in the armature winding of a dynamo. This induced emf, which is proportional to the angular velocity of rotation, produces in the winding a subsidiary current I' which is superimposed algebraically on I and which is equal to E'/R, where R is the resistance of the coil plus either the effective resistance of the external circuit and shunt connected in parallel, or simply the resistance of the external circuit if no shunt is used. According as the coil is rotating in the forward or in the backward direction, the direction of I' is opposite to or the same as that of I. In the former case, that of forward rotation, the supplementary torque resulting from I' is in the backward direction, whereas in the latter case it is in the forward direction. Thus we see that the torque due to I' is always in the direction opposing the rotation. The effect of this is to produce an electromagnetic damping of the oscillation additional to, and often greatly exceeding, the damping due to the viscous resistance of the air.

For any given angular velocity of the coil and the corresponding value of the induced electromotive force, I' is inversely proportional to R, so the opposing torque due to I' must also be inversely proportional to R. In the extreme case, approximated under certain experimental conditions, where R is infinite, I' is zero, and so there is no electromagnetic damping at all. The coil then oscillates for a long time before it eventually comes practically to rest. In this case the galvanometer is said to be "underdamped." On the other hand, below a certain value of R the coil is so heavily damped that no oscillations occur at all, there being merely an asymptotic approach to the final reading, which is never quite attained. The galvanometer is then said to be "overdamped," and the system is obviously quite unsuitable for quantitative work.

With any galvanometer there is a particular value of R for which the motion of the coil just ceases to be oscillatory. When R has this value the coil is said to be "critically damped." Critical damping is the condition most favorable for obtaining rapid readings. However, because it is difficult to distinguish critical damping from overdamping, it is sometimes preferred to use a value of R slightly in excess of that for critical damping, so giving a very slight, just perceptible oscillation. Under these conditions readings can be taken almost as rapidly as with critical damping.

UNIVERSAL SHUNT   We saw in Chapter 26 that by using different shunts with a galvanometer a corresponding number of different ranges may be covered with the same instrument. The foregoing discussion shows that it would not be possible to vary the sensitivity in this way and at the same time always to have near-critical damping. Fortunately, in cases where the resistance of the external circuit is large compared with the resistance of a shunt that would give near-critical damping, a way of escape from this difficulty is available. This is the use of a so-called **universal shunt,** or ring shunt, a shunt having the desired resistance from the point of view of damping which is used for a number of different sensitivities, these being secured by tapping off known fractions of the shunt for connection to the external circuit. The arrangement is indicated in Fig. 32.15.

Let R_g be the resistance of the galvanometer and R_s that of the shunt, and let a fraction $1/N$ of the shunt be tapped off for connection to the external circuit. Then the current divides into two

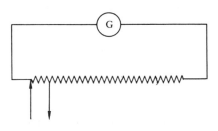

FIGURE 32.15   Galvanometer with universal shunt.

branches, having resistances $R_g + [(N - 1)/N]R_s$ and R_s/N. The current I_g in the galvanometer branch will be such that

$$I_g = I \frac{R_s/n}{R_g + [(N - 1)/N]R_s + (R_s/N)}$$
$$= \frac{1}{N} \frac{IR_s}{R_g + R_s},$$

or

$$I = N \frac{R_g + R_s}{R_s} I_g.$$

Now, $[(R_g + R_s)/R_s]I_g$ represents the scale of currents given by the instrument and shunt when connection to the external circuit is made at the ends of the shunt. The equation above shows that this scale may be multiplied by any desired factor N by making the connection, instead, to two points on the shunt the resistance between which is R_s/N.

Universal shunts are commercially available in the form of dial resistance boxes provided with three terminals, two of which are connected to the ends of the series of resistors, while the third terminal connects, via the movable arm, to the intermediate point selected.

PRACTICAL AMMETERS AND VOLTMETERS An alternative means for securing near-critical damping which avoids the necessity for working with a particular effective external resistance is one commonly used in practical ammeters and voltmeters. In these the coil is wound on a light aluminum or aluminum-alloy frame having such a resistance that the induced currents flowing around the frame due to its rotation in the magnetic field provide the requisite degree of damping. In these instruments the resistance of the coil is sufficiently high to ensure that the damping contributed by currents induced in this shall be relatively negligible, so that the instrument can never be overdamped, regardless of external circuit connections. By providing such instruments with the requisite shunts or series resistors, ammeters and voltmeters of different ranges can then be contrived in the usual manner without any complications with regard to damping.

Problems

32.1 A wire of length 10 cm and orientated in a direction perpendicular to a magnetic field of strength 0.1 N/A/m moves at a speed of 2 m/sec in a direction at right angles both to its own length and to the field. What PD is developed between the ends of the wire?

32.2 A car is traveling at 100 km/hr in a magnetic E–W direction where the vertical component of the earth's magnetic field is 5×10^{-5} Wb/m². What PD is developed between the sides of the car, 1.5 m apart? Could a passenger in the car use a voltmeter to measure this PD?

32.3 A horizontal wire of mass m and of negligible resistance rolls down two parallel metal rails, also of negligible resistance, which are inclined at an angle θ to the horizontal and whose distance apart is l. The rails are connected electrically by a resistor of resistance R, and a uniform magnetic field **B** is directed vertically upward. Show that (a) the wire eventually attains a constant velocity,

$$v_0 = \frac{mgR \sin \theta}{B^2 l^2 \cos^2 \theta},$$

and (b) at any stage where the velocity of the wire is v, the current that flows in the wire is

$$I = \frac{Blv \cos \theta}{R}.$$

(Neglect the self-inductance of the circuit.) Sketch a graph showing how the current increases with time, assuming that the wire starts from rest.

32.4 (a) Show that the displacement X from its mean position of the bob of a simple pendulum of length L executing small oscillations of amplitude X_0 may be expressed as a function of time t by the equation

$$X = X_0 \sin \sqrt{\frac{g}{L}} t.$$

(b) Find the corresponding expression for the displacement x from its mean position of a point on the suspension whose distance from the point of support is l.

(c) What is the potential difference, δV, developed between the ends of a short section of wire of length δl moving with velocity v at right angles to its length and to a magnetic field **B**?

(d) The pendulum bob in (a) is of metal and is supported by a thin flexible wire of negligible mass. Projecting below the bob is a short light wire whose lower end is continually immersed below the surface of a pool of mercury. A magnetic field **B** is applied at right angles to the plane of the oscillation. Neglecting the radius of the bob and the length of the short projecting wire in comparison with L, use the expressions obtained in (a), (b), and (c) to find how the potential difference V between the pendulum suspension point and the mercury varies (1) with t, and (2) with X, and sketch a graph showing these variations.

851

(e) If a resistor of fairly high resistance R is connected electrically between the suspension point and the mercury, what is the energy dissipated per oscillation? (Assume that viscous forces and the resistance of the remainder of the circuit are negligible.)

(f) By making use of the answer to (e) and of the fact that the energy of an oscillating simple pendulum is $mgX_0^2/2L$, show that the ratio of the amplitudes of successive oscillations is

$$\left[1 + \frac{\pi L^2 B^2}{2mR} \sqrt{\frac{L}{g}}\right]^{1/2}.$$

32.5 A rigid pendulum constructed of metal, whose period of oscillation is T, terminates in a wire dipping into a dish of mercury a distance l below the point of support. If the pendulum executes small oscillations of angular amplitude α at right angles to a uniform magnetic field **B**, show that a potential difference is set up between the support on which the knife-edges rest and the mercury which varies sinusoidally with time and whose peak value is $\pi \alpha l^2 B/T$.

32.6 A copper disk of radius 10 cm rotates at 1200 rpm about an axis through its center and perpendicular to its plane. If a uniform magnetic field of 1 Wb/m² is established in the axial direction, what potential difference will be developed between the axis of the disk and its rim? Could such a device as this be developed into an efficient dc generator?

32.7 A flexible wire of length 1 m and resistance 0.1 Ω is joined at the ends to form a continuous loop. This is initially circular in form and is orientated with its axis in the direction of the horizontal component of the earth's magnetic field, of magnitude 2×10^{-6} Wb/m². What is the total quantity of electricity that flows around the loop when it is pulled out straight, with negligible included area?

32.8 A coil of wire is rotated at constant angular velocity about an axis perpendicular to its plane and to a magnetic field to which it is exposed. Why must a greater torque be applied to maintain the rotation when the ends of the wire are joined than when they are not?

32.9 How does the mean rate at which heat is generated in the coil when the ends of the wire of Problem 32.8 are joined vary with the speed of rotation? (Assume that the wire has a constant resistance.)

32.10 A rectangular coil of wire of 50 turns, measuring 4 cm by 6 cm, rotates at 300 rpm about an axis passing through the midpoints of its shorter sides, in a region where the magnetic induction is at right angles to the axis of rotation and has the constant value of 0.3 Wb/m². Find the amplitude ("peak value") of the induced emf in the coil. What would be the effect of choosing some other line in the plane of the coil as the axis of rotation?

32.11 Show that Lenz's law may be considered to be a direct consequence of the principle of conservation of energy.

32.12 A current I is passed through a coil of N turns wound on a toroidal frame whose cross section is rectangular, as shown in the figure. Find (a) the value of B at any point in space, (b) the magnetic flux linking any turn of the toroid, and (c) the self-inductance of the coil.

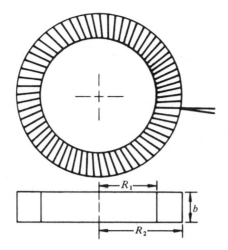

Figure for problem 32.12.

32.13   In the device shown schematically in the figure, a uniform copper disk of radius r is free to rotate about its axis. Electrical contact is made with the disk through a complete circular ring of carbon brushes X at its periphery. Another set of brushes Y makes contact with the axle, whose radius is negligible compared with r. The disk is mounted between the poles of a large magnet, and there is a substantially uniform magnetic field **B** normal to its surface. With the disk initially at rest, a battery of emf E is connected between X and Y, the set X being positive. Draw a diagram showing clearly the resulting forces exerted on the disk, and find an expression for its final angular velocity, assuming all frictional and viscous forces and the resistance of the circuit to be negligible. Hence show that the final kinetic energy of the disk is simply proportional to its mass m, regardless of what its radius might be. How will your conclusions be affected if the battery has a finite internal resistance? Explain briefly what will happen if, with the final angular velocity of the disk established, the battery is removed and X and Y are connected by (a) a resistor of negligible inductance, (b) an inductor of negligible resistance, and (c) a capacitor of capacitance C. In each case sketch graphs showing the way in which the angular velocity of the disk varies with time.

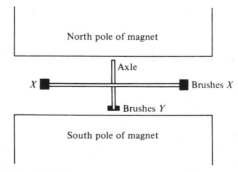

Figure for problem 32.13.

32.14   A uniformly wound solenoid of mean diameter 5 cm and length 1 m has 2000 turns of wire. What, approximately, is the self-inductance of this solenoid?

32.15   What is the magnetic energy of the solenoid of Problem 32.14 when it carries a current of 10 A? How is the magnetic energy per unit volume of the solenoid related to the magnetic field B within it?

32.16   What is the self-inductance per meter of a coaxial cable having inner and outer radii of 2 mm and 4 mm, respectively?

32.17   Does the self-inductance of the coil of a ballistic galvanometer play any role in determining the throw that occurs in response to a rapid change of flux through a part of the circuit in which the galvanometer is included?

32.18   An inductor has a self-inductance of 5 mH and a resistance of 3 Ω and is connected via a switch to a battery of negligible internal resistance. How long does it take after closing the switch for the current to reach 90 percent of its final value?

32.19   The current in an LR circuit including a battery and a switch attains 30 percent of its final steady value 1 sec after the switch is closed. What is the time constant of the circuit?

32.20   Show that the time constant of an LR circuit is equal to the time that would be required for the current to reach its final value if it continued to increase at its initial rate.

32.21   A stiff uniform wire PQ of length l hangs from a hinge at P (see the figure), and the end Q dips into a dish of mercury. P is connected via a switch to an inductor of self-inductance L, the other end of which is connected to the mercury pool. There is a uniform magnetic field $\mathbf{B}$ parallel to the axis of the hinge.

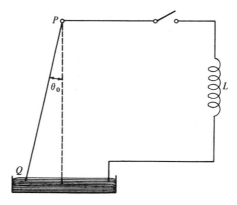

Figure for problem 32.21.

With the switch open, the wire can swing as a pendulum with period T_0. Consider now the case where the switch is closed and the wire released from rest at an angle of θ_0 to the vertical at the time $t = 0$. Find expressions for the angular displacement θ and the current I at any subsequent time, and sketch graphs showing the time variation of these quantities. (Neglect the resistances of all conductors and all frictional and viscous forces. Assume that θ_0 is a *small* angle.)

32.22   A long solenoid of 10 turns per centimeter has a cross-sectional area of 10 cm², and surrounding this in its central region is wound a coil of 100 turns. What is the mutual inductance between the two coils?

32.23   What is the mutual inductance between the rectangular circuit and the long straight wire shown in the figure?

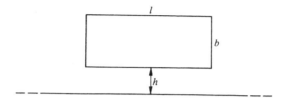

Figure for problem 32.23.

32.24   The long straight wire shown in the figure for Problem 32.23 carries a current I, and the rectangular circuit, of resistance R, is removed to a great distance from the wire. Show that this movement causes a charge Q to circulate around the loop, where

$$Q = \frac{\mu_0 I l}{2\pi R} \ln\left(1 + \frac{b}{h}\right).$$

(Neglect the self-inductance of the loop.)

32.25   A coil of 25 turns and cross-sectional area 4 cm² has a resistance of 5 Ω. It is mounted within, near the center of, and coaxially with, a long solenoid which has 10 turns per centimeter and carries a current of 2 A. The ends of the coil are connected to a ballistic galvanometer whose resistance is 35 Ω, and it is found that, when the solenoid current is suddenly reversed, the throw of the galvanometer is 30 cm as measured with a conventional lamp and scale. In a separate experiment it is found that the same throw is obtained when a certain capacitor is charged to 2.6 V and then discharged through the galvanometer. What is the capacitance of the capacitor?

32.26   A closely wound search coil, designed for quantitative exploration of magnetic fields, encloses an area of 4 cm² and has 150 turns whose total resistance is 60 Ω. This is connected to a ballistic galvanometer whose resistance is 30 Ω. When the coil is turned quickly from an orientation in which its axis is perpendicular to a uniform magnetic field into one in which its axis is parallel to the field, the galvanometer gives a throw that corresponds to the passage through it of 3.5×10^{-5} C. What is the magnitude of the field?

32.27   An iron-alloy cylinder of length 1 m and diameter 1 cm is bent into a closed toroid and is closely wound with insulated wire all the way around, the total number of turns being 1000. Linking the toroid and its winding is a coil of 100 turns which is connected via a resistor to a ballistic galvanometer. The total resistance of the galvanometer circuit is 2400 Ω. When a current of 1 A is suddenly established in the winding of the toroid, the galvanometer registers the passage of 1.6×10^{-2} C, and when this current is suddenly increased to 2 A a further 1.6×10^{-2} C is registered. Calculate (a) the relative permeability of the iron-alloy core, (b) the self-inductance of the winding on this core, and (c) the mutual inductance between this winding and the coil.

32.28   The frame of a galvanometer coil has dimensions 4 cm by 2 cm and is made of aluminum of resistance 0.1 Ω. If it is swinging with an angular velocity of 10°/sec and the effective mean radial flux density produced by the magnet is 0.2 Wb/m², what is the torque producing magnetic damping?

32.29   Two widely separated inductors of self-inductances L_1 and L_2 are connected in series. Show that their combined self-inductance is $L_1 + L_2$. Would the combined self-inductance still be $L_1 + L_2$ if they were close together? Explain.

32.30   As located in a circuit, two inductors, whose self-inductances are L_1 and L_2, have a mutual inductance M. Show that when the first of these carries a current I_1 and the second a current I_2, the magnetic energy of the system is W, where

$$W = \tfrac{1}{2}L_1 I_1{}^2 + \tfrac{1}{2}L_2 I_2{}^2 + MI_1 I_2.$$

33

AC CIRCUITS AND INSTRUMENTS

A scientist engaged in experimental work will need to make proper use of the facilities available to him if his work is to be effective. Among other things, it will be necessary for him to understand how his equipment functions, and to appreciate both its possibilities and its limitations. A large proportion of physical equipment utilizes electric current, and in many cases this is, for one reason or another, alternating rather than direct. Accordingly, the experimental scientist should not only have a good understanding of the various instruments used to measure alternating currents and voltages, and of the action of a transformer, but he should also be well versed in the theory of the properties of alternating-current circuits generally. These matters will be dealt with in the present chapter and in Chapter 34. These will be less concerned with the consideration of fundamental principles of physics than with providing the experimental physicist with an important "tool" of this trade.

PART I

PEAK, RMS, AND "MEAN" VALUES OF CURRENTS AND VOLTAGES

In many applications of electric currents we are concerned simply with the conversion of electrical energy to heat. This is, for example, the case in the operation of household electric radiators and cookers, as well as in innumerable industrial heating processes. Also in the operation of incandescent-filament lamps the temperature of the filament is determined by the rate of generation of heat from electrical energy. In all these cases the heating can be effected by using either direct or alternating current. In the latter case it becomes of interest to inquire what steady current or voltage would be equivalent in its heating effect to the alternating current or voltage actually used.

Let a voltage V that varies periodically with time be applied across a resistor of resistance R, and let the period of the voltage variation be T. At any moment the rate of generation of heat is

V^2/R, and, accordingly, the quantity of heat δQ^* generated during the short interval of time δt is $(V^2/R)\delta t$. For the total quantity of heat generated during a cycle we have, therefore,

$$Q = \frac{1}{R} \int_0^T V^2 \, dt,$$

and correspondingly the mean rate of generation of heat during a cycle, and therefore also during any number of cycles, is given by the equation

$$\frac{Q}{T} = \frac{1}{R} \frac{\int_0^T V^2 \, dt}{T}. \tag{33.1}$$

The second factor on the right side, being the limit of

$$\frac{\sum V^2 \, \delta t}{\sum \delta t},$$

may be interpreted as follows. Let the period T be divided into a very large number N of *equal* small intervals of time δt. We make the intervals equal because this facilitates interpretation of the expression—there can obviously be no objection to this. Under these conditions the expression may be written as

$$\frac{\sum V^2 \, \delta t}{N \, \delta t},$$

or, canceling out δt, as

$$\frac{\sum V^2}{N}.$$

Since there are N items in the numerator, this represents the mean value of V^2 over one period. Writing this as $\overline{V^2}$, Eq. (33.1) may be expressed in the form

$$\frac{Q}{T} = \frac{\overline{V^2}}{R}. \tag{33.2}$$

Let us now define a quantity V_r by the equation

$$V_r = \sqrt{\overline{V^2}}. \tag{33.3}$$

Then in terms of this quantity, known as the **root-mean-square**

---

* It is important not to confuse quantity of heat, represented in Chapters 14, 17, and 18 and also in the present discussion by the symbol Q, with quantity of electricity, for which the same symbol was used in Chapters 19, 20, and 21.

(rms) value of the voltage, Eq. (33.2) may be written in the form

$$\frac{Q}{T} = \frac{V_r^2}{R}.$$

(33.4)

Obviously V_r is the steady voltage that would produce a rate of heating equal to the average heating rate given by the periodically varying voltage.

Since the current I flowing in a pure resistor is proportional to the voltage V applied across it by the constant factor $1/R$, it is obvious that the mean rate of heating may, alternatively, be expressed by the equation

$$\frac{Q}{T} = RI_r^2,$$

(33.5)

where I_r is the root-mean-square value of I.

Relation between Peak and RMS Values for Sinusoidal Alternating Current

In the case where the voltage varies sinusoidally with time, according to the equation

$$V = V_0 \sin \omega t,$$

substitution for V in Eq. (33.1) gives

$$\frac{Q}{T} = \frac{V_0^2}{R} \frac{\int_0^T \sin^2 \omega t \, dt}{T}$$

$$= \frac{V_0^2}{R} \frac{\int_0^{\omega T} \sin^2 \omega t \, d(\omega t)}{\omega T},$$

or, since ωT is equal to 2π,

$$\frac{Q}{T} = \frac{V_0^2}{R} \frac{1}{2\pi} \int_0^{2T} \sin^2 x \, dx,$$

where x is written for the new variable, ωt. The integral on the right side of this equation is a well-known standard integral, whose value is π. Hence

$$\frac{Q}{T} = \frac{V_0^2}{2R} = \frac{(V_0/\sqrt{2})^2}{R}.$$

Comparing this with Eq. (33.4), we see that

$$V_r = \frac{V_0}{\sqrt{2}}.$$

(33.6)

859

Similarly,

$$I_r = \frac{I_0}{\sqrt{2}}, \qquad (33.7)$$

where the current amplitude, I_0, is equal to V_0/R.

In the specification of the voltage of an alternating supply it is always the rms voltage V_r, not the peak voltage V_0, that is referred to. Thus, when we say that our domestic alternating voltage is 110 volts, we mean that the rms voltage has this value. The corresponding value of the peak voltage would be $110\sqrt{2}$, or 155.5 V.

In the general case of a periodic variation of V that is not sinusoidal, the period being T, this can be analyzed into Fourier components according to an equation of the form

$$V = V_s + V_1 \sin(\omega t + \epsilon_1) + V_2 \sin(2\omega t + \epsilon_2) \\ + V_3 \sin(3\omega t + \epsilon_3) + \cdots, \qquad (33.8)$$

where ω is equal to $2\pi/T$, V_s is a steady non-time-varying component, V_1 corresponds to the V_0 of the previous discussion, and V_2, V_3, . . . are the amplitudes of the higher harmonics, while ϵ_1, ϵ_2, $\epsilon_{,3}$ . . . are the corresponding epochs. Some of these terms may have the value zero. For example, as we saw in Chapter 32, there is no steady component V_s in the output of an alternator, and also there are no even harmonics.

If both sides of Eq. (33.8) are squared and averages are taken over one or more periods, then, as a trigonometrical analysis would show, the means of all products of terms on the right side other than the squares of these terms would vanish, and hence

$$\overline{V^2} = V_s{}^2 + \tfrac{1}{2}V_1{}^2 + \tfrac{1}{2}V_2{}^2 + \tfrac{1}{2}V_3{}^2 + \cdots$$

Consequently,

$$V_r = \sqrt{V_s{}^2 + \tfrac{1}{2}(V_1{}^2 + V_2{}^2 + V_3{}^2 + \cdots)}. \qquad (33.9)$$

Relation between RMS and "Mean" Values for Sinusoidal Alternating Current

Although it is the rms value of voltage or current that determines the rate of generation of heat in a resistor through which the current is passed, the so-called "mean" value is also of interest in certain circumstances, and we shall therefore investigate the relation between this and the rms value in the case where the quantity varies sinusoidally with time. Of course, the true mean value $\overline{V}$ of a sinusoidally varying voltage, or the mean current $\overline{I}$, is algebraically zero. However, it is not the algebraic mean that we are concerned with here, but the mean of the *magnitude* of V or I, its sign being disregarded; that is, we are concerned with V_m, defined as the mean of $|V|$, or with I_m, the mean of $|I|$.

It becomes unnecessary to distinguish between $\overline{V}$ and V_m, or between $\overline{I}$ and I_m, if we consider only the period of time between two successive zero values of the quantity, such that during this half-period the quantity has only positive values. Accordingly, we may write

$$V_m = \frac{\int_0^{T/2} V\, dt}{T/2,},$$

with

$$V = V_0 \sin \omega t,$$

which, on substitution for V becomes

$$V_m = V_0 \frac{\int_0^{T/2} \sin \omega t\, dt}{T/2}.$$

Since dt is equal to $(1/\omega)\, d(\omega t)$, this may be written

$$V_m = \frac{2V_0}{\omega T} \int_0^{\omega T/2} \sin \omega t\, d(\omega t) = \frac{V_0}{\pi} \int_0^{\pi} \sin x\, dx$$

$$= -\frac{V_0}{\pi} [\cos x]_0^{\pi} = \frac{2}{\pi} V_0.$$

Thus

$$V_m = \frac{2\sqrt{2}}{\pi} \frac{V_0}{\sqrt{2}} = \frac{2\sqrt{2}}{\pi} V_r = 0.900 V_r. \qquad \textbf{(33.10)}$$

Correspondingly,

$$I_m = 0.900 I_r. \qquad \textbf{(33.11)}$$

PART II
AC MEASURING INSTRUMENTS

In the moving-coil type of ammeter or voltmeter the torque exerted on the coil is proportional to the current it carries and reverses its direction with the latter. Hence such an instrument cannot be used for the measurement of alternating currents or voltages unless the current is first rectified. Alternating-current instruments depending on current rectification are now used extensively. In addition, there are a number of other types of instruments whose registrations are functions of I^2 or V^2, and so depend simply on $|I|$ or $|V|$. Obviously, such instruments, not being affected by the sign of the current or voltage, must lend themselves to ac as well as to dc measurement.

Rectifier Systems

As we saw in Chapter 25, a single rectifier inserted in one of the lines connecting an alternating voltage supply with a resistive load suppresses every alternate half-cycle, giving a pulsating unidirectional current whose variation with time is of the kind shown graphically in Fig. 33.1. If the resistive load in question is the coil of a moving-coil galvanometer, this will be subjected to a pulsating unidirectional torque having a time variation of the same nature. However, except at very low frequencies there will be no appreciable time variation of the deflection produced; the moment of inertia of the coil will smooth out the reading to an effectively steady value.

Rectification corresponding to Fig. 33.1, produced by a single rectifier, is known as half-wave rectification. A better arrangement, giving full-wave rectification, is the one indicated in Fig. 33.2, in which four rectifiers are connected in such a way that, in whichever direction voltage is applied to the system, the resulting current flows in only one direction through the instrument. In effect, every alternate half-cycle, instead of being suppressed as it is in the half-wave rectifier system, is reversed in direction, and used, giving a current through the galvanometer whose time variation is of the kind shown in Fig. 33.3. This will not only double the deflection produced but will also reduce the oscillatory component of the deflection, permitting the use of the system at somewhat lower frequencies.

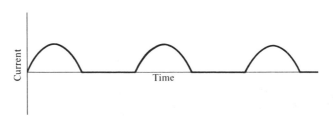

FIGURE 33.1   Current-time curve for half-wave rectification.

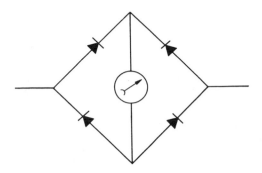

FIGURE 33.2 System giving full-wave rectification.

With either half-wave or full-wave rectification, such smoothing of the deflection as results from the moment of inertia of the coil can be improved to any degree desired by also smoothing the current, using the simple device of connecting a capacitor of suitable capacitance across the galvanometer (see Chapter 25).

Obviously, a half-wave rectifier could not be used directly in conjunction with a moving-coil type of dc ammeter for the measurement of an alternating current, for this would destroy the alternating character of the current it is designed to measure. Although, in principle, it *would* be possible to use the full-wave rectifying system shown in Fig. 33.3 incorporated in the main circuit, this is not normally done. The rectifiers used with instruments are not intended to carry currents greater than about 1 mA, and so currents of a larger order than this are measured indirectly, passing only a small "sample" of the current through the instrument. This could be done in various ways. The method normally adopted in practice is that indicated in Fig. 33.4. A step-up transformer (see Chapter 34) is used, whose primary winding, of a few turns only, carries the current to be measured, while the secondary is connected, via the rectifier (half- or full-wave) to the moving-coil instrument, provided, if required, with a series resistor. The rms secondary voltage is proportional to the rms primary current. The instrument in conjunction with the rectifier measures the mean rectified secondary cur-

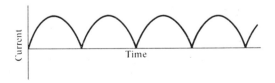

FIGURE 33.3 Current-time curve for full-wave rectification.

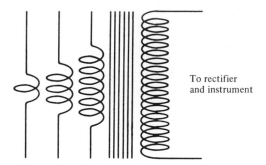

FIGURE 33.4 Step-up transformer with alternative primary windings, which, in conjunction with a rectifier system and a moving-coil galvanometer, gives the equivalent of an ac ammeter with alternative ranges.

rent, I_m, which, for a full-wave rectifier, would be equal to $0.900I_r$, where I_r is the current *passing through the instrument* (see the previous section). If a half-wave rectifier is used, the measured I_m is 0.450 of the I_r that *would* pass through the instrument in the absence of the rectifier. In either case, the relation of this I_r to the rms value of the primary current is known in terms of the circuit parameters; thus the latter is proportional by a known factor to the measured I_m. Different ranges are provided for by having alternative primary windings through which the current to be measured is passed.

Obviously the measurement of an alternating *voltage* does not require the use of a transformer; the dc measuring instrument is simply connected in series with the rectifier and a suitable series resistor.

In the discussion above we have tacitly assumed that the rectifier offers zero resistance to the current in the forward direction and infinite resistance in the backward direction. Such an ideal rectifier would have an I–V characteristic consisting of a vertical and a horizontal line passing through the origin, as shown in Fig. 33.5(a), the voltage being reckoned positive in the direction of easy flow. An equally good characteristic for a voltmeter system would be two lines meeting at the origin in the manner shown in Fig. 33.5(b). Here the reciprocal of the slope of the forward part of the characteristic corresponds to a resistance which merely has to be added to that of the instrument plus series resistor to give the effective overall resistance. However, actual I–V characteristics of semiconductor rectifiers are more like that shown in Fig. 33.5(c). A type of rectifier commonly used gives an intercept of the upper linear portion of

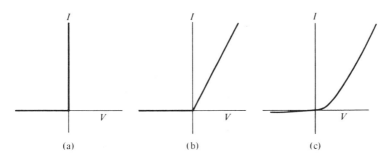

(a)   (b)   (c)

FIGURE 33.5

this on the voltage axis of about 0.3 V, and the maximum permissible voltage that may be applied across this rectifier without damaging it is about 1 V. Although this I–V relation is far from ohmic, it is nevertheless possible to achieve a close approximation to proportionality between current and voltage by "swamping" the potential fall across the rectifying contact with that of a resistor of high resistance connected in series with it; thus, if the PD across the resistor were of the order of 100 V, the mere 0.3 V of the intercept would be almost negligible. The maximum allowable current for most instrument rectifiers is about 1 mA; hence to drop 100 V across a resistor carrying this current would necessitate its having a resistance of 10,000 Ω. As long as the intercept voltage is negligibly small compared with that dropped across the resistor, I will be effectively proportional to V.

If relatively low alternating voltages are to be read with a rectifier-type instrument, these must first be stepped up to appropriately high values with a transformer and the stepped-up voltages then applied to the instrument.

It should be borne in mind that the use of a rectifier type of instrument is appropriate only in cases where the time variation of the current or voltage to be measured is sinusoidal. For any "wave form" other than sinusoidal the calibration involving the factor of 0.900 or 0.450 (as the case may be) between the mean and rms current values will no longer apply.

Electrodynamometer

This is an instrument in which the mechanical interaction between two current-carrying coils or coil systems is used as a measure of current, voltage, or power. One such instrument has already been considered, the current balance shown in Fig. 29.10.

865

Clearly, if ac instead of dc is sent through this, it will measure the mean torque acting on the small central coil, and since, as we have seen, the instantaneous torque is proportional to I^2, the mean torque must be proportional by the same factor to I^2, that is, to I_r^2. Hence, when carrying ac, the instrument measures I_r.

Like most absolute instruments, the current balance is somewhat troublesome to manipulate, and so other forms of electrodynamometer have been developed whose use is less time-consuming. In these, which have to be calibrated against a dc instrument, the feature of the design of Fig. 29.10 which ensures that the field in the region of the central coil shall be uniform need no longer be adhered to; this coil can, if desired, be relatively large. Also, it is more convenient to allow the coil to rotate against an elastic control such as a spiral spring than to measure the torque required to keep it stationary. Such an instrument is shown in Fig. 33.6. When no current passes, the axis of the central coil is orientated at an angle of about 120° to the axis of the fixed coils. The torque produced by the passage of current reduces this angle, and at full-scale reading it is about 60°.

As will be appreciated, the scale for this type of instrument used as an ammeter or a voltmeter is very uneven. Although the relation between θ and I_r^2 is roughly linear, equal increments of I_r produce much larger increments of θ at the upper end of the scale than at the lower end. This is a feature of all ac ammeters and voltmeters other than those of rectifier type.

The most important use of the electrodynamometer is as a wattmeter, for the measurement of power. For this there are two

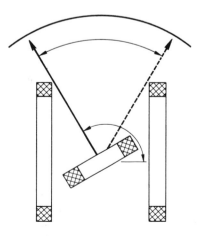

FIGURE 33.6 Electrodynamometer.

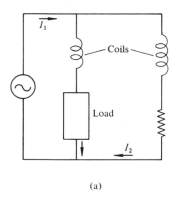
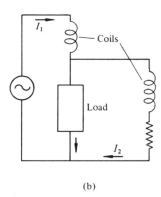

(a)   (b)

FIGURE 33.7   Circuits for use of electrodynamometer as a wattmeter.

alternative connections, as indicated diagrammatically in the two parts of Fig. 33.7. In these the pair of stationary coils is, for simplicity, represented as a single coil. The fixed and moving coils now carry different currents, I_1 and I_2, and the torque, besides being approximately proportional to sin θ (only approximately, because the field is not uniform), is also proportional to the product I_1I_2.

Both connections shown in the figure introduce a slight error, analogous to the case where power is measured in a dc system by using an ammeter and a voltmeter.

Neglecting in part (a) the voltage dropped across the first coil and in (b) the current carried by the second coil, we may say that the current through the load is I_1 and the voltage dropped across it is RI_2, R being the combined resistance of the second coil and the series resistor. Hence the instantaneous power is RI_1I_2, and the mean power $\bar{P}$ is $R\overline{I_1I_2}$. In a uniform field the mean torque exerted on the moving coil, and with it the angular deflection α, would be proportional to I_1I_2 sin θ. Thus $\bar{P}$ is approximately proportional to $\alpha/\sin\theta$ and hence roughly proportional to α.

The instrument can be calibrated using direct current. Different ranges may be provided for by appropriately varying R.

Moving-Iron Instruments

At not too high frequencies, where iron losses and self-inductance are not complicating factors of any importance, alternating currents may conveniently be measured with moving-iron instruments.

In broad principle, moving-iron instruments are similar to electrodynamometers. The essential difference is that while in the latter the mechanical interaction by which the current is measured is between the currents in two coils, in moving-iron instruments it is either (1) between the current in one coil and the Amperian current in a soft-iron armature, usually in the form of a vane, magnetized by the current in the coil, or (2) between the Amperian currents in two cylindrical armatures both of which are situated side by side within a coil conveying the current to be measured. In the latter case the cylinders repel one another. Examples of these two types of instrument are shown in Figs. 33.8 and 33.9.

In all moving-iron instruments it is necessary that the hysteresis loop for the iron used shall be narrow, and the enclosed area therefore small. A further requirement is that only that portion of the B–H characteristic shall be called into play where B is effectively proportional to H, so ensuring that in each configuration the Amperian current or currents shall be proportional by some factor to the current in the coil, throughout the whole current range. Only when these conditions are fulfilled will the instrument read true rms values of current irrespective of the wave form.

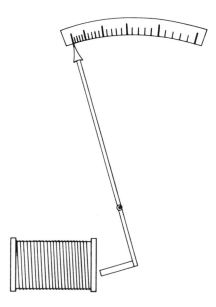

FIGURE 33.8 Moving-iron ammeter utilizing mechanical interaction between a current in a coil and the corresponding Amperian current produced in an iron armature.

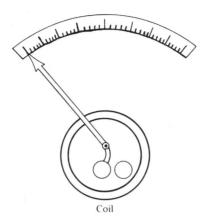

Coil

FIGURE 33.9   Moving-iron ammeter utilizing mechanical interaction between Amperian currents produced in two iron armatures.

Hot-Wire Instruments

In hot-wire instruments the currents to be measured, or some definite fraction of this current, is passed through a wire, and some physical effect of the resulting rise in temperature is made to produce a registration on a scale.

In modern instruments of this type, one junction of a thermocouple is welded to the center of the wire, and the thermocouple is connected to a dc ammeter of suitably high sensitivity whose scale is calibrated to read directly the current in the wire. For greater sensitivity, convective cooling is eliminated by enclosing the hot wire and the wires of the thermojunction within an evacuated bulb, and sealing the four leads—two "current" leads and two "instrument" leads—through the glass.

Owing to the smallness of the self-inductance and capacitance introduced into the circuit by a short, straight wire, instruments of this type have a particularly useful field of application in the measurement of radiofrequency currents and voltages.

Electrostatic Voltmeters

As we saw in Chapter 21, the force per unit area experienced by a charged surface on which the charge density is σ is $\sigma^2/2\epsilon_0$. In a capacitor of given configuration the charge density on each element of surface of either electrode is proportional to the potential

869

difference V between the electrodes. Hence the force on any particular portion of an electrode is proportional to V^2.

An electrostatic voltmeter may be regarded as a capacitor whose configuration varies with the orientation or position of its moving part—for example, in the case of a gold-leaf electroscope, with the deflection of the leaf. With any *particular* configuration the force on each element of area of the moving part is proportional to V^2, and hence, if V changes periodically with time, the mean force is proportional to $\overline{V^2}$. It is readily seen, therefore, that the deflection must be a unique function of V_r and that consequently, if an electrostatic voltmeter is calibrated using steady voltages, this calibration must be equally valid for the rms values of periodically varying voltages.

An electrostatic voltmeter may also be used for the determination of rms values of current, by measuring the V_r across a resistor of known resistance through which the alternating current is passed.

PART III
CURRENT–VOLTAGE RELATIONS IN AC CIRCUITS

RLC Circuit

Let us consider a portion of a circuit in which an inductor of self-inductance L and a capacitor of capacitance C are connected in series, the system also including a resistance R. This resistance may reside wholly in the inductor, or it may reside only partly in the inductor, the remainder being contributed by a separate resistor connected in series with the inductor and the capacitor. In either case it will be convenient, for purposes of mathematical discussion, to idealize the situation to one of spatial separation of the resistance, self-inductance, and capacitance, as indicated in Fig. 33.10.

Apart from the plates of the capacitor, there is nowhere any accumulation of charge, so the current I flowing at any moment along the system from left to right must be the same everywhere

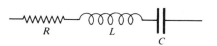

FIGURE 33.10 Series *RLC* system.

on the left of the capacitor, and also, because the capacitor *as a whole* does not accumulate charge, on the right of this.

Let us suppose that the current varies sinusoidally with time according to the equation

$$I = I_0 \sin \omega t. \tag{33.12}$$

Since

$$I = \frac{dQ}{dt},$$

$$\frac{dQ}{dt} = I_0 \sin \omega t,$$

integration of which gives us

$$Q = -\frac{I_0}{\omega} \cos \omega t + \text{const.},$$

and since there is no steady (non-time-varying) voltage across the capacitor the constant of integration must be zero. Thus

$$Q = -\frac{I_0}{\omega} \cos \omega t. \tag{33.13}$$

Let us now see how the potential drop V from left to right varies with time. This is made up, algebraically, of three parts: that dropped in the resistor, V_R; that dropped in the inductor, V_L; and that dropped across the capacitor, V_C.

For the potential drop across the resistor we have

$$V_R = RI = RI_0 \sin \omega t. \tag{33.14}$$

The emf E_L induced in the inductor, which, by definition, is the potential *rise* in this component, is given by the equation

$$E_L = -L \frac{dI}{dt},$$

and so for the voltage *fall*, V_L, we have

$$V_L = -E_L = L \frac{dI}{dt},$$

which, on differentiation of the expression (33.12) for I, becomes

$$V_L = \omega L I_0 \cos \omega t$$

$$= \omega L I_0 \sin \left(\omega t + \frac{\pi}{2} \right). \tag{33.15}$$

871

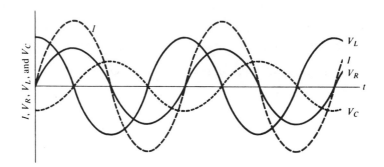

FIGURE 33.11 Time variations of sinusoidal alternating current in series RLC system and corresponding voltages dropped across the resistor, inductor, and capacitor.

For V_C we have, from Eq. (33.13),

$$V_C = \frac{Q}{C} = -\frac{I_0}{\omega C} \cos \omega t$$

$$= \frac{I_0}{\omega C} \sin \left(\omega t - \frac{\pi}{2}\right). \qquad (33.16)$$

Hence for the sum of V_R, V_L, and V_C, given by Eqs. (33.14), (33.15), and (33.16), respectively, we have

$$V = I_0 \left[R \sin \omega t + \omega L \sin \left(\omega t + \frac{\pi}{2}\right) + \frac{1}{\omega C} \sin \left(\omega t - \frac{\pi}{2}\right)\right]. \qquad (33.17)$$

The variations of the separate items V_R, V_L, and V_C with time, as well as that of I, are shown by the four curves of Fig. 33.11. We see that what I does at a particular time, for example pass through a positive or negative maximum, or through zero while rising or falling, V_R does simultaneously, while V_L does it a quarter of a period earlier and V_C a quarter of a period later. Thus V_R is in phase with I, V_L leads I by $T/4$, and V_C lags behind I by $T/4$. Or, in terms of phase angle instead of in terms of time, V_L leads I by $\pi/2$ and V_C lags behind I by $\pi/2$.

Use of Phasors in AC Theory

The quantities V_R, V_L, and V_C characteristic of any particular time t may be represented graphically, both in magnitude and in sign, in the manner indicated in parts (a), (b), and (c), respectively, of Fig. 33.12. In these, circles are drawn having radii of $I_0 R$, $I_0 \omega L$, and $I_0/\omega C$, and radii are constructed making angles with the horizontal direction from left to right of ωt, $\omega t + \pi/2$, and

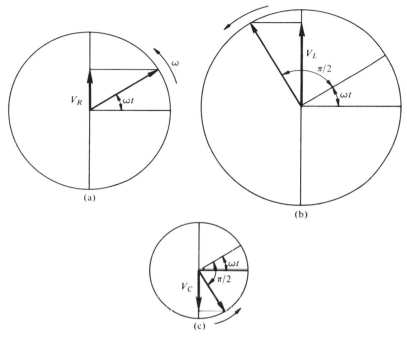

FIGURE 33.12 Graphical constructions giving V_R, V_L, and V_C at any specified time for RLC system carrying sinusoidal ac.

$\omega t - \pi/2$, respectively, these angles being reckoned as positive when swept out in the anticlockwise direction. Then, from Eqs. (33.14), (33.15), and (33.16), we see that the projections of these radii on vertical diameters represent V_R, V_L, and V_C, respectively, the signs of these quantities being positive or negative according as the projections appear above or below the centers of the circles. From these projections, with their proper signs indicated by arrows, the potential drop V across the whole system may then be obtained by the graphical construction shown in Fig. 33.13, which is self-explanatory.

The radii of the circles, provided with arrowheads to indicate the sense away from the center, are known as **phasors.**

A "streamlined" procedure for obtaining the same result is shown in Fig. 33.14, where, for clarity, the diagram is drawn to a larger scale. Here, in effect, the center of the second circle has been placed at the end of the first phasor and the center of the third circle at the end of the second phasor,* but the circles themselves

---

* The third phasor has been slightly displaced laterally to avoid losing its identity in the second.

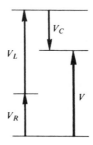

FIGURE 33.13 Geometrical representation of algebraical sum V of V_R, V_L, and V_C at a given time.

have been omitted. The beginning of the first phasor has then been joined by a straight line, labeled V_0, to the end of the third, and it is the projection of this on the vertical line that represents to scale the potential drop V across the whole system. That this is so will be realized by considering the sections of the vertical line defined by the feet of the perpendiculars drawn from the ends of the phasors, and comparing these sections with the lines labeled V_R, V_L, V_C, and V in Fig. 33.13.

The diagram of Fig. 33.14 gives the value of V at the time t. At a later time, t', the corresponding value of V would be given by a similar diagram in which each phasor has rotated through the further angle $\omega(t' - t)$, where ω is equal to 2π times the frequency f; that is,

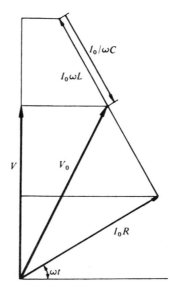

FIGURE 33.14 Phasor diagram, the projection of whose resultant on a vertical line gives V.

the whole diagram corresponding to the later time has rotated through this angle about the origin O. We see that all we have to consider is the rotation of a line representing V_0 to scale with angular velocity ω, and the projection of this on a vertical line. From this it is clear that V, like V_R, V_L, and V_C, varies sinusoidally with time, according to the equation

$$V = V_0 \sin (\omega t + \phi), \tag{33.18}$$

where ϕ is the angle marked in Fig. 33.15.

From the foregoing it is evident that in order to obtain the values of V_0 and ϕ in Eq. (33.18), it is merely necessary to construct the diagram shown in Fig. 33.15. This is, in effect, a vector diagram in which the components, of magnitude I_0R, $I_0\omega L$, and $I_0/\omega C$, are represented to scale by lines having their appropriate relative orientations, the resultant being of magnitude V_0. We see from Fig. 33.15 that the phase angle ϕ by which V leads I is given by the equation

$$\phi = \tan^{-1}\left(\frac{\omega L - 1/\omega C}{R}\right). \tag{33.19}$$

Although the graphical construction we have used here corresponds exactly to the polygon construction for combining vectors, we call the directed straight lines now under discussion phasors rather than vectors. This is because I_0R, $I_0\omega L$, $I_0/\omega C$, and V_0, to which the lengths of the sides of the polygon are proportional, are

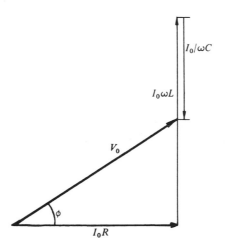

FIGURE 33.15 Phasor diagram of voltages drawn with usual orientation, in which the peak voltage I_0R dropped across the resistor is represented by a horizontal arrow pointing to the right.

not in fact the magnitudes of vectors. We are concerned merely with a vector-type construction for combining certain directed straight lines, with the object of obtaining the information sought. This includes the phase relation between the quantities represented by their projections as shown in Figs. 33.12 to 33.14; hence the name "phasors."

From Fig. 33.15 it is evident that

$$V_0 = ZI_0, \tag{33.20}$$

where Z is defined by the equation

$$Z = \sqrt{R^2 + \left(\omega L - \frac{1}{\omega C}\right)^2}. \tag{33.21}$$

Similarly, for the rms values of V and I, namely V_r and I_r, respectively, we must have

$$V_r = ZI_r. \tag{33.22}$$

If we compare this with the relation between PD and current in a dc circuit,

$$V = RI,$$

we see that the quantity Z plays the same role in determining the relation between V_r and I_r in an ac circuit as does R in determining the relation between V and I in a dc circuit. This quantity Z, which is the analogue of resistance in a dc circuit, is called the **impedance** of the circuit. The unit is the same as that of resistance, the ohm.

Equation (33.21) may be written in the form

$$Z = \sqrt{R^2 + X^2}, \tag{33.23}$$

where X, the **reactance** of the circuit, is the difference between ωL and $1/\omega C$. The former of these two items, ωL, is known as the **inductive reactance** X_L, and the latter, $1/\omega C$, is called the **capacitive reactance** X_C. Thus the reactance of the circuit is the difference between the inductive and capacitive reactances; that is,

$$X = X_L - X_C \tag{33.24}$$

or

$$X = X_C - X_L, \tag{33.25}$$

according as X_L or X_C is the greater.

In terms of reactances Eq. (33.19) may be written

$$\phi = \tan^{-1}\left(\frac{X_L - X_C}{R}\right). \tag{33.26}$$

Note the special cases of Eq. (33.21) applying to an RL and an RC circuit,

$$Z = \sqrt{R^2 + \omega^2 L^2} \tag{33.27}$$

and

$$Z = \sqrt{R^2 + \frac{1}{\omega^2 C^2}}, \tag{33.28}$$

respectively. We see that in the former case Z increases with frequency, while in the latter Z decreases as the frequency increases. Thus, for a given value of V_r applied to an RL circuit, I_r decreases with increasing frequency until, at very high frequencies, practically no current at all passes; the inductor acts as an efficient "choke." On the other hand, in an RC circuit, the current begins by being negligibly small at *low* frequencies but increases progressively as the frequency is increased, tending toward the limiting value of V_r/R.

In practice, a knowledge of Z and ϕ is all that is needed in many calculations concerning an RLC circuit, and to obtain this information it is usual to construct a phasor diagram similar to that of Fig. 33.15, but with the common factor I_0 omitted. In other words, the phasor components represented by the sides of the diagram are R, ωL, and $1/\omega C$, and the resultant is Z.

POWER FACTOR The power P developed in a circuit, that is, the rate at which electrical energy is converted, is, by the definition of V, expressed by the equation

$$P = IV. \tag{33.29}$$

In an ac circuit we are usually interested, not in instantaneous power, but in the mean power P over a prolonged period of time, expressed in terms of I_r and V_r, the rms values of I and V, respectively. This may be ascertained as follows.

Substituting for I and V in Eq. (33.29) their values as given in Eqs. (33.12) and (33.18), respectively, we have

$$P = I_0 V_0 \sin \omega t \sin (\omega t + \phi),$$

which, by trigonometrical manipulation, is easily shown to be the same as

$$P = I_0 V_0 [\sin^2 \omega t \cos \phi + \tfrac{1}{2} \sin (2\omega t) \sin \phi].$$

The mean value of the first term on the right side of this equation over a prolonged period is obviously the same as that for one complete cycle. This is

$$I_0 V_0 \cos \phi \, \frac{\int_0^T \sin^2 \omega t \, dT}{T},$$

877

which, as was shown in our discussion of rms values, is the same as $\frac{1}{2}I_0 V_0 \cos \phi$, or $I_r V_r \cos \phi$. On the other hand, it is obvious that the $\sin(2\omega t)$ factor in the second term must give zero, positive and negative values of this factor canceling each other out. Consequently, for the mean power we have

$$\bar{P} = I_r V_r \cos \phi. \tag{33.30}$$

The factor $\cos \phi$ by which the product of the rms values of current and voltage has to be multiplied to give the power is known as the **power factor.**

An interesting special case is that of a pure LC circuit, having no resistance. Here I and V are 90° out of phase, $\cos \phi$ is zero, and accordingly also $\bar{P}$. A current passing under these conditions, where no power is dissipated, is known as a "wattless current."

Resonance

From Eq. (33.21) we see that in an RLC circuit with given values of R, L, and C, the impedance Z is least when the inductive and capacitive reactances balance each other out, that is, when

$$\omega L = \frac{1}{\omega C},$$

or when

$$\omega = \frac{1}{\sqrt{LC}}. \tag{33.31}$$

To this value of ω corresponds the frequency

$$f_0 = \frac{1}{2\pi\sqrt{LC}} \tag{33.32}$$

characteristic of the circuit, known as its **resonant frequency.** It derives its name from the fact that for a given amplitude V_0 of alternating PD applied to the circuit whose frequency is varied, the current amplitude I_0, equal to V_0/Z, passes through a maximum at this frequency. Correspondingly, I_r, equal to V_r/Z, passes through a maximum.

It is a simple matter to calculate I_r as a function of f for fixed values of L and C and a series of values of R; and a corresponding series of "resonance curves" may then be drawn. These are similar to those shown in Fig. 11.7, representing acoustic resonance. In the electrical case, the curve with the highest maximum is that for the circuit having the smallest value of R.

On examination of such a series of curves we see that as we vary the frequency from f_0 on either side, the relative rate of falling away of I_r is greatest for the curve having the highest maximum and least for that whose maximum is lowest. In the former case the resonance is said to be "sharp," and in the latter "flat."

RADIO TRANSMISSION AND RECEPTION There is a method of setting an RLC circuit into oscillation alternative to the application of an alternating voltage to its terminals. Instead, the terminals may be connected together, and an alternating emf then induced in the circuit from an external source. This is the principle on which radio transmission systems operate.

A diagrammatic representation of a simple radio aerial system is shown in Fig. 33.16. The aerial, a "dipole aerial," consists simply of two conductors (rods or wires) of equal length placed in line with one another and separated by a short gap, the adjacent ends being connected to an inductor. This constitutes an RLC circuit whose capacitance is that between the two sections of the aerial and whose resistance resides in the inductor, the transmission line, and the aerial.

Energy is fed into the inductor connecting the two sections of the aerial by coupling this to another inductor which is energized from an oscillator system based on the use either of a thermionic triode (for high powers) or a transistor. The current in the aerial system may be adjusted to its maximum amplitude by suitable manipulation of the aerial parameters.

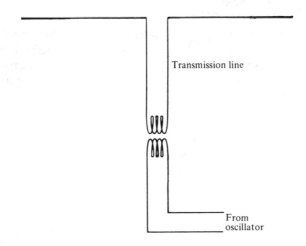

Transmission line

From oscillator

FIGURE 33.16 Essentials of radio transmission system employing a dipole aerial.

879

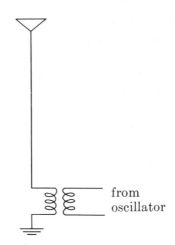

FIGURE 33.17   Alternative radio transmission system.

Another common type of aerial system is shown diagrammatically in Fig. 33.17. Here most of the capacitance is between the aerial and earth. This aerial is not as different from the dipole aerial as might be imagined. At each stage of its operation the electric field between the aerial and the earth is identical with what it would be if the earth were removed and replaced by the "electrostatic image" of the aerial in the earth's surface.* In this sense we may think of this aerial as the equivalent of a dipole aerial, except that the capacitance between the actual aerial and earth is larger than would be that between this aerial and its image.

An oscillating aerial system is something more than an *RLC* circuit in which energy loss occurs merely because of its resistance. In addition, some energy loss occurs radiatively; the electric and magnetic fields associated with the oscillation travel outward with the velocity of light, together constituting a train of electromagnetic radio waves.

Radio reception is, in essence, the reverse of transmission. The electromagnetic waves from the distant transmitter induce an alternating emf in the receiving aerial–inductor–earth system. This system can be tuned within a range of frequencies, for example by varying the capacitance of an air capacitor incorporated in it. When the receiver is tuned to the transmitter the amplitude of the current in the former attains its maximum amplitude. The

---

* It will be recalled, from Chapter 22, where we dealt with the image force acting on an electron escaping from a metal, that this electrostatic image is like an optical image in a plane mirror, with a charge of opposite sign to that of the "object."

aerial system is normally coupled magnetically to a tuned circuit, and the oscillating voltage induced in this is amplified by an electron-tube or transistor amplifier.

Problems

33.1 A rectangular coil of wire of 50 turns, of dimensions 4 cm by 6 cm, rotates at 300 rpm about an axis passing through the midpoints of its shorter sides, in a region where the magnetic induction is at right angles to the axis of rotation and has the constant value of 0.3 Wb/m². Find the peak and rms values of the induced electromotive force. What would be the effect of choosing some other line in the plane of the coil as axis of rotation?

33.2 The moving coil of an electrodynamometer used as an ammeter has a total angular range of movement of 60° between the zero and the final scale division, these being symmetrically disposed with respect to the fixed coils. The instrument reads 10 A full scale. Find the approximate positions of the 1-A, 2-A, 3-A, . . . , 8-A, and 9-A divisions.

33.3 A direct current of 100 mA is required in a part of a circuit, and the maximum fluctuation that can be tolerated is ±1 mA. This is to be obtained from 60-Hz 110-V mains (peak voltage 155.5) in conjunction with a single rectifier and capacitor. Sketch the circuit that must be used, and calculate the approximate values of the resistance R of the load and the smallest permissible capacitance C of the capacitor.

33.4 A rectifier gives a mean dc output of 300 V on which is superimposed a 60-Hz ripple component of 10-V peak and one of 120,000 Hz having a 1-V peak. A 10-μF capacitor is connected across its terminals. What currents will flow through the capacitor?

33.5 A 10-μF capacitor is connected across 110-V 60-Hz ac mains. Plot a graph showing the variation with time of the current starting from the zero of the *voltage* cycle, and calculate the rms current.

33.6 It is desired to pass a current of 1 mA through a capacitor without the potential difference across it exceeding 2 V (peak). Calculate the least values required for the capacitance for frequencies of (a) 60 Hz, and (b) 60 MHz.

33.7 A 4-μF capacitor is connected through an ammeter (reading rms current) to an alternator. The ammeter indicates a current of 0.1 A. Given that the frequency of the ac is 60 Hz, find the maximum amount of energy stored in the capacitor during a cycle.

33.8 A 100-μF capacitor is connected through a 30-Ω resistor to 110-V 60-Hz mains. Find (a) the reactance of the capacitor, (b) the rms current in the circuit, and (c) the difference in phase between the current and the voltage.

33.9 An electrical device of resistance 500 Ω requires a 150-V supply for satisfactory operation. If the mains supply is 240 V at 50 Hz, what capacitor would have to be connected in series with the device for its safe operation from the mains? Is this method of producing a voltage reduction satisfactory from the point of view of the supplier of the power as well as being economical from the point of view of the consumer?

33.10   An alternator rotating at 100 rps generates an emf of 50 V and is connected in series to a 10-μF capacitor and a 1-mH inductor. What is the rms current, and what is the phase relation between the current and the emf of the alternator?

33.11   A coil of resistance 100 Ω and self-inductance 0.1 H is connected through a 20-μF capacitor to the 110-V 60-Hz ac mains. Find (a) the current in the circuit and (b) the phase relation between the current and the supply voltage.

33.12   A coil of resistance 5 Ω and self-inductance 5 H is connected in series with a 5-μF capacitor. If an alternating current of 5 A and frequency 500 Hz passes through the combination, what are (a) the PD between its terminals, (b) the reactance of the combination, and (c) the average rate of generation of heat within the coil? At what frequency would the reactance be zero, and what would be the impedance of the combination and the PD across it at this frequency?

33.13   An ac ammeter is placed in series with a coil having a resistance of 5 Ω and self-inductance of 1 mH and a 20-μF capacitor. When the system is connected to a 1000-Hz supply the ammeter reads 2 A. Find, by constructing a phasor diagram, the rms voltage (a) across the coil, (b) across the capacitor, and (c) across the coil and capacitor together. Find also the power factor of the circuit.

33.14   A series circuit connected across a 200-V 60-Hz line consists of a capacitor of capacitive reactance 30 Ω, a noninductive 44-Ω resistor, and a coil of inductive reactance 90 Ω and resistance 36 Ω. Determine (a) the current in the circuit, (b) the PD across each unit, (c) the power factor of the circuit, (d) the power absorbed by the circuit, and (e) the capacitance of the capacitor.

33.15   An 800-Ω resistor, a 5-H inductor, and a 1.78-μF capacitor are joined in series. What are the impedances of this combination (a) at 50 Hz, and (b) at 500 Hz; and (c) what is the resonance frequency? If a PD of 50 V at 50 Hz is applied across the system, what are the voltages developed across the inductor and across the capacitor? If the resistance were changed to 1 Ω, what voltage would now be developed across the inductor?

33.16   A coil when connected to a 50-V 50-Hz supply passes 1 A. When connected to a 150-V 200-Hz supply it passes the same current. It is then connected to a certain dc supply and again passes 1 A. What is the voltage of the dc supply?

33.17   A 5-μF capacitor is charged to 100 V and then discharged through a coil of self-inductance 5 mH and resistance 0.1 Ω. Estimate the PD across the capacitor at the end of the first cycle.

33.18   Calculate the impedances of a 1-μF capacitor and of a 1-H inductor at 50, 500, and 5000 Hz. Repeat these calculations for a 0.1-μF capacitor and a 0.1-H inductor.

33.19   Two series *RLC* circuits having different values of resistance, self-inductance, and capacitance have the same resonant frequency f. What will be the resonant frequencies of systems in which these are connected (a) in series, and (b) in parallel?

33.20 A capacitor of capacitance C is charged to a PD of V_0, and by closing a switch at time $t = 0$ it is connected to a coil of self-inductance L and negligible resistance. Find expressions for the following quantities as functions of time t: (a) the current in the coil, (b) the charge on the capacitor, (c) the PD across the capacitor plates, (d) the energy stored in the capacitor, (e) the energy stored in the coil, and (f) the total energy stored in the circuit. Sketch graphs showing the time variation of all these quantities.

33.21 Assuming the coil of Problem 33.20 to have an appreciable resistance R, find a differential equation for the current I at time t. Try a solution to this equation of the form

$$I = I_0 e^{-\alpha t} \sin (\omega t + \phi),$$

and show that this is indeed a solution, provided that R is not too large. Find the upper limit of R for the validity of this equation. Find the value of α, and sketch graphs showing the variations with t of (a) the current, and (b) the PD between the plates of the capacitor.

33.22 An alternating voltage of constant amplitude but variable frequency is applied to a series RLC circuit, and at the resonant frequency f_0 a current of a certain amplitude flows in the circuit. At two other frequencies f' and f'', one of which is greater than f_0 while the other is less, a current of smaller amplitude flows, which is the same in each case. Show that f_0 is the geometric mean of f' and f''.

33.23 A high-frequency oscillator whose frequency can be varied is loosely coupled to an RLC circuit in such a way that an alternating emf is induced in this whose amplitude is independent of the frequency. With $R = 1.5 \, \Omega$, $L = 1.6 \, \mu\text{H}$, and $C = 2.7 \times 10^{-3} \, \mu\text{F}$, find what frequency of the oscillator will produce the maximum amplitude of current in the circuit. How will this amplitude be affected by a change in the frequency of the oscillator by (a) 1 percent, (b) 2 percent, (c) 4 percent, and (d) 10 percent? For a given percentage change in the frequency of the oscillator, will the amplitude of the induced current depend on whether the frequency has been increased or decreased?

33.24 The alternator shown in the figure generates an emf E of $E_0 \sin \omega t$. Express the voltages V_L, V_C, and V_R across the inductor, ca-

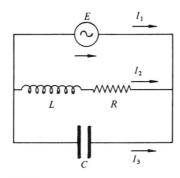

Figure for problem 33.24.

pacitor, and resistor, respectively, and the currents I_1, I_2, and I_3, as functions of time. Under what circumstances, if any, is I_1 zero?

33.25   Use a phasor diagram to show the relation between I_1 and E in the circuit shown in Fig. 33.18, and hence show that if C is varied, while L and R remain constant, the current will have a minimum value when $\omega^2 LC = 1 - CR^2/L$.

33.26   An ideal parallel resonant circuit having zero resistance is shown in the figure. Show from simple physical reasoning that the impedance between A and B is infinite for the resonant frequency $\omega/2\pi$, where $\omega^2 LC = 1$. (This property is often used to select a particular frequency from a mixture of signals, for example in tuning a radio receiver.)

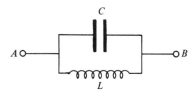

Figure for problem 33.26.

33.27   A simple method of comparing the capacitances of two capacitors is to make them into two adjacent arms of a bridge (de Sauty's bridge) of the kind shown in the figure, used in conjunction with an alternator and earphones. Show that the condition for no sound to be heard in the earphones is that $C_2/C_1 = R_1/R_2$. If the alternator and earphones were interchanged, would the equation still apply? If so, what does this equation mean, physically?

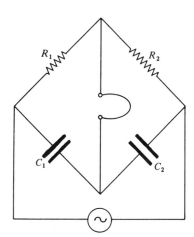

Figure for problem 33.27.

33.28   An alternating voltage $V = V_0 \sin \omega t$ is applied to the terminals A and B of the circuit shown. With the aid of a phasor diagram or otherwise, find an expression for the potential difference V_{PQ} between the points P and Q, and discuss the relation between V_{PQ} and V, as R is varied from 0 to infinity.

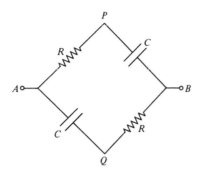

Figure for problem 33.28.

33.29   (a)   Show that, if no current passes through the galvanometer when the switch in the circuit shown is closed, then $L/C = R_1 R_2$.

(b)   Show that, if the galvanometer is replaced by an ac instrument and the battery is replaced by an alternating source of emf $E = E_0 \sin \omega t$, the condition for a balance is again $L/C = R_1 R_2$.

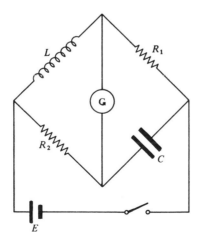

Figure for problem 33.29.

33.30   (a)   Given that E has an rms value of 100 V at 50 Hz, state what voltages and current would be indicated, respectively, by the voltmeters V_1 and V_2 and the ammeter A shown, if the capacitance C is 10^{-3} F.

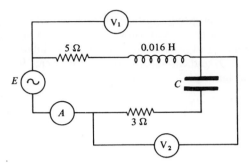

Figure for problem 33.30.

(Assume that A and all leads have zero resistance and that the internal resistances of V_1 and V_2 are infinite.)

(b)   Compute the power factor of the circuit.

(c)   Determine the phase relation between the voltages across the terminals of the instruments V_1 and V_2.

(d)   Calculate the new value of the capacitance C for which the ammeter would indicate maximum current. With this new value of C, what would be the phase relation between the current and the applied alternating emf?

34

ALTERNATING VOLTAGE
TRANSFORMATIONS

Electromagnetic induction is used for the transformation of alternating voltages in two main areas, power engineering and communication systems. We shall accordingly divide this chapter into the two corresponding parts, dealing with transformers and the application of the thermionic triode in ac circuitry, respectively. Notwithstanding the increasing use of transistors in communication systems, a discussion of these will not be included in Part II. This is because, at the level to which physical theory has been taken in this book, we do not have the necessary background to understand the action of these devices.

PART I
TRANSFORMERS

A transformer is a device, one form of which is shown in Fig. 34.1, whose function is to change an alternating voltage either from a relatively low to a higher value ("step-up transformer"), or from a higher to a lower value ("step-down transformer"). A transformer has two windings, as indicated: a primary, which is connected to the supply whose voltage is to be transformed, and a secondary, which delivers the stepped-up or stepped-down voltage, as the case may be. For a reason that will appear later, transformers operating at relatively low frequencies have an iron core. This is constructed of iron having a low hysteresis loss and is laminated to minimize eddy currents.

One of the most important applications of transformers is in the transmission of power over long distances. To avoid undue insulation difficulties in alternators these are not normally designed to generate voltages greater than 10 to 15 kV (10,000 to 15,000 V). But to restrict voltages to such relatively low values in long-distance power transmission would mean using correspondingly large currents in the transmission lines. With these would be associated serious ohmic losses, unless the transmission lines were of such heavy caliber that their cost would be prohibitive. The use of a step-up transformer at the generating station provides a way of

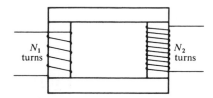

FIGURE 34.1 Transformer.

escape from this difficulty; for, as we shall see later, with the step up in voltage is associated a corresponding reduction in current, and the ohmic loss in the transmission lines is thereby kept within tolerable limits. Typical voltages used for transmission are 100 to 400 kV. Then, at the distant town or factory where the power is required, the voltage is stepped down by another transformer to a value that is reasonably safe for household or factory use.

Owing to unavoidable iron losses in iron-cored transformers and ohmic losses in the primary and secondary windings, the power delivered by a transformer is never quite as great as that fed in. Efficiencies of well-designed transformers are usually between 96 and 99 percent.

Even power dissipations as small as 1 to 4 percent would be sufficient to cause intolerable heating of the iron cores and windings of large transformers if something were not done to promote the cooling of these. The transformers used in power-distribution systems are cooled by immersing them in an oil bath and making provision for the oil to circulate, either by convection or otherwise. In this circulation the oil passes through cooling pipes.

An important property of a transformer is its **coupling coefficient.** This is defined as the ratio of the flux linked with each turn of the secondary, due to current in the primary, to that linked with each turn of the primary on account of this current. It is, in general, desirable to construct a transformer in such a way that its coupling coefficient shall approach as nearly as possible the ideal figure of 100 percent. In the design shown in Fig. 34.1 this is secured by the use of an iron core in the form of a "closed magnetic circuit." However, for the coupling coefficient to be high it is by no means essential that there should be an iron core. As we shall see presently, an air-cored transformer of suitable design may also have a coupling coefficient of practically 100 percent.

Ideal Transformer

An ideal transformer is one in which there are no ohmic or iron losses, and in which the coupling coefficient is 100 percent.

Although ohmic losses are never zero, it is possible to discuss theoretically a transformer that is otherwise ideal as if it were ideal in this respect also. This may be done by (1) subtracting the ohmic voltage drop in the primary from the supply voltage and considering the difference as the *effective* supply voltage, and (2) considering the resistance of the secondary as if it were part of the load. Where there is an iron core, the iron losses cannot similarly be eliminated; in this case the actual transformer must simply be regarded as a more or less close approximation to the ideal.

SELF- AND MUTUAL INDUCTANCES Of some interest in the theory of the ideal transformer are the self-inductances L_1 and L_2 of the primary and secondary windings and the mutual inductance M between these.

It will be instructive to consider, in the first instance, a particular design of ideal transformer, indicated in plan and section in Fig. 34.2, and to obtain expressions for the corresponding quantities L_1, L_2, and M. This transformer is in the form of a ring of rectangular cross section, and the primary and secondary are wound uniformly all the way around. We shall assume that the windings are close enough to avoid appreciable flux leakage, and we shall ignore their resistances. This transformer may be either iron-cored or air-cored;

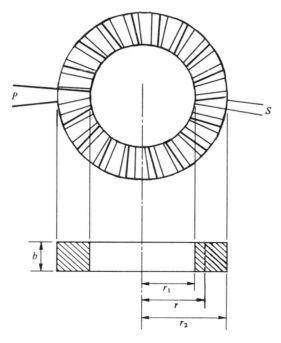

FIGURE 34.2 Air-cored or iron-cored transformer of ring form.

in either case the flux is confined entirely to the space within the windings, and so the coupling coefficient is 100 percent.

The inductances are readily found via the magnetic fields B at different distances r from the axis, and these fields, in turn, are most easily obtained by using Ampère's line-integral theorem derived in Chapter 30.

Let us consider as our closed path a circle of radius r and within the ring which is coaxial with the ring. Then, if the secondary is on open circuit, and the primary, of N_1 turns, carries a current I_1, the total current linked with the path is N_1I_1; $\cos \alpha$ everywhere along the path, of length $2\pi r$, is 1; and so we have

$$\oint B \cos \alpha \cdot dl = 2\pi r B = \mu N_1 I_1,$$

or

$$B = \frac{\mu N_1 I_1}{2\pi r}. \tag{34.1}$$

In this equation the positive direction of **B** within any part of the ring and that of I_1 around it are related according to the right-hand screw relation. In what follows this is to be taken as applying to all the quantities considered; some direction around the ring is arbitrarily defined as the positive direction for **B** and for magnetic flux, and this then also defines the directions to be taken as positive for the currents in the windings.

From Eq. (34.1) we may now calculate the flux Φ_1 due to the current in the primary linking each turn of this winding:

$$\Phi_1 = \int_{r_1}^{r_2} Bb \, dr = \frac{\mu N_1 I_1 b}{2\pi} \int_{r_1}^{r_2} \frac{dr}{r} = \frac{\mu N_1 I_1 b}{2\pi} \ln\left(\frac{r_2}{r_1}\right),$$

or

$$\Phi_1 = k N_1 I_1, \tag{34.2}$$

where k is written for $(\mu b/2\pi) \ln (r_2/r_1)$. For the total flux linkage we have, therefore,

$$\Phi_{l1} = N_1 \Phi_1 = k N_1{}^2 I_1.$$

Let us now rewrite this equation in the form

$$\frac{\Phi_{l1}}{I_1} = k N_1{}^2.$$

The left side of this will be recognized as the defining expression for the self-inductance L_1 of the primary. Accordingly, we have

890

$$L_1 = k N_1{}^2. \tag{34.3}$$

By corresponding reasoning it may be shown that the self-inductance of the secondary is given by the equation

$$L_2 = kN_2{}^2. \tag{34.4}$$

M may be found similarly. If a current I_1 flows in the primary, the secondary flux linkage Φ_{l21} due to this must be $N_2\Phi_1$, and so, substituting the value of Φ_1 as given by Eq. (34.2), we have

$$\Phi_{l21} = kN_1N_2I_1,$$

and hence for M, defined as this flux linkage per unit primary current, we obtain the expression

$$M = kN_1N_2. \tag{34.5}$$

While for the transformer of Fig. 34.2, k in Eqs. (34.2), (34.3), (34.4), and (34.5) has a particular value, $(\mu b/2\pi) \ln (r_2/r_1)$, these equations, with *some* value of k, must be equally valid for *any* ideal transformer. This follows from the fact that in any such transformer the flux in the core must be proportional by some factor k to the total effective rate of circulation of electricity around it. Obviously, if a current I_1 flows in the primary, of N_1 turns, the contribution this makes to the flux is proportional to both I_1 and N_1; that is, it is proportional to N_1I_1. Similarly, if a current I_2 flows in the secondary, of N_2 turns, the contribution this makes is proportional by the same factor to N_2I_2. And so, in the general case, where the currents I_1 and I_2 flow simultaneously, the flux Φ is proportional to $N_1I_1 + N_2I_2$; that is,

$$\Phi = k(N_1I_1 + N_2I_2), \tag{34.6}$$

the value of the constant k depending on the geometry of the transformer and the permeability of the core material. It is convenient to give the quantity $N_1I_1 + N_2I_2$ a special name and a symbol; we shall call it the **equivalent magnetizing current, I_e.**

From Eqs. (34.3), (34.4), and (34.5) we see that

$$\frac{L_1}{M} = \frac{M}{L_2} = \frac{N_1}{N_2}. \tag{34.7}$$

We shall have occasion to make use of these relations in what follows.

PRIMARY AND SECONDARY ELECTROMOTIVE FORCES Since the same flux links each turn of both primary and secondary, the same electromotive force E, equal to $-d\Phi/dt$, is induced in each of these turns. The primary may be regarded as N_1 turns connected in series and the secondary as N_2 of these turns. Hence the emf E_1 induced in the primary is N_1E, while E_2, the emf induced in the secondary, is N_2E. Accordingly, for the ratio of these electromotive forces we have

$$\frac{E_2}{E_1} = \frac{N_2}{N_1}. \tag{34.8}$$

In discussing the action of a transformer it is usual to speak in terms of the voltages V_1 and V_2 applied to (that is *dropped in*) the primary and the load, respectively. Since, by definition, the emf in any device or component is, algebraically speaking, the (reversible) potential *rise* in it, the potential *fall* is necessarily minus this emf. Hence V_1, the potential fall across the primary of the transformer, is equal to $-E_1$. On the other hand, the potential V_2 dropped across the load must be equal to the potential rise E_2 induced in the secondary winding; that is, V_2 is equal to E_2. Hence, substituting $-V_1$ for E_1 and V_2 for E_2 in Eq. (34.8), we obtain the voltage relation

$$\frac{V_2}{V_1} = -\frac{N_2}{N_1}; \tag{34.9}$$

We see that these voltages are in antiphase; that is, there is a phase shift of π of one relative to the other. And their magnitudes are in the ratio of the corresponding numbers of turns.

We shall now consider the actual values of E_1 and E_2. For E we have

$$E = -\frac{d\Phi}{dt} = -k\left(N_1\frac{dI_1}{dt} + N_2\frac{dI_2}{dt}\right),$$

and so

$$E_1 = N_1E = -k\left(N_1{}^2\frac{dI_1}{dt} + N_1N_2\frac{dI_2}{dt}\right)$$

and

$$E_2 = N_2E = -k\left(N_1N_2\frac{dI_1}{dt} + N_2{}^2\frac{dI_2}{dt}\right).$$

Substituting for $kN_1{}^2$, $kN_2{}^2$, and kN_1N_2 in the last two equations the corresponding symbols for the inductances according to Eqs. (34.3), (34.4), and (34.5), L_1, L_2, and M, respectively, these take the forms

$$E_1 = -L_1\frac{dI_1}{dt} - M\frac{dI_2}{dt} \tag{34.10}$$

and

$$E_2 = -M\frac{dI_1}{dt} - L_2\frac{dI_2}{dt}. \tag{34.11}$$

These are the expressions for the induced electromotive forces we should have anticipated on purely physical grounds from the definitions of the three inductances. We might almost have written these equations without derivation, regarding them as self-evident.

An alternative expression for E_2 may be obtained by eliminating E_1 between Eq. (34.8) and an equation found by applying Kirchhoff's second law to the primary circuit. Let the potential rise in the alternator of the primary circuit minus the ohmic drop in the alternator winding and leads be $E_0 \sin \omega t$. This must be the same as the potential fall V_1 in the primary winding of the transformer, or minus the potential rise E_1 in this winding. Thus

$$E_1 = -E_0 \sin \omega t.$$

Substitution of the right side of this equation for E_1 in (34.8) gives us for the emf in the secondary

$$E_2 = -\frac{N_2}{N_1} E_0 \sin \omega t. \tag{34.12}$$

CURRENT RELATIONS These are most conveniently ascertained by equating the expressions for E_2 in Eqs. (34.11) and (34.12). In this way we find that

$$M \frac{dI_1}{dt} + L_2 \frac{dI_2}{dt} = \frac{N_2}{N_1} E_0 \sin \omega t,$$

or

$$\frac{dI_1}{dt} = -\frac{L_2}{M} \frac{dI_2}{dt} + \frac{1}{M} \frac{N_2}{N_1} E_0 \sin \omega t,$$

or since, according to Eq. (34.7), L_2/M and N_2/MN_1 are equal to N_2/N_1 and $1/L_1$, respectively,

$$\frac{dI_1}{dt} = -\frac{N_2}{N_1} \frac{dI_2}{dt} + \frac{E_0}{L_1} \sin \omega t,$$

which, on being integrated, yields

$$I_1 = -\frac{N_2}{N_1} I_2 - \frac{E_0}{\omega L_1} \cos \omega t,$$

or

$$I_1 = -\frac{N_2}{N_1} I_2 + \frac{E_0}{\omega L_1} \sin \left(\omega t - \frac{\pi}{2} \right), \tag{34.13}$$

the constant of integration being zero.

It will be convenient to label the first term on the right side of this equation I_1' and the second I_1''.

From the defining equation for I_1',

$$I_1' = -\frac{N_2}{N_1} I_2, \tag{34.14}$$

we have

$$N_1 I_1' + N_2 I_2 = 0. \tag{34.15}$$

What does this mean physically? Obviously, since the component I_1' of the primary current flows in the primary winding, which has N_1 turns, and I_2 flows in the secondary, of N_2 turns, $N_1 I_1' + N_2 I_2$ is the contribution that I_1' and I_2 together make to the magnetizing current I_e. And, as we see from the last equation, this contribution is zero. Consequently, I_e must be due entirely to what remains of the current system after I_1' and I_2 have been crossed off. And this is I_1''. Thus, since the contribution this component makes is $N_1 I_1''$,

$$I_e = N_1 I_1'' = \frac{N_1}{\omega L_1} E_0 \sin\left(\omega t - \frac{\pi}{2}\right). \tag{34.16}$$

It will be observed that the expression for I_1'', namely

$$I_1'' = \frac{E_0}{\omega L_1} \sin\left(\omega t - \frac{\pi}{2}\right), \tag{34.17}$$

does not involve any item referring to the load; it is independent of the load. That this must be so is also evident physically. The emf induced in the primary is determined uniquely by the variations of flux in the core and must at all times balance the emf $E_0 \sin \omega t$ of the alternator. And since the latter is independent of the load, so must be the flux and its time variation. But, as we have seen, Φ is attributable solely to the component I_1'' of the primary current. Hence, finally, I_1'' must be independent of the load.

It is desirable that I_1'' shall be relatively small, so that there shall be no appreciable ohmic loss in the supply line associated with it. This means that the product ωL_1 should be large. At relatively low frequencies (for example, 50 or 60 Hz) this is provided for by the use of an iron core. But at radio frequencies, where the smallness of I_1'' is taken care of by the high value of ω, L_1 need not be large. An air core is therefore perfectly satisfactory, provided that the transformer is so constructed as to give virtually 100 percent coupling coefficient.

From the definition of I_1' we see that

$$\frac{I_2}{I_1'} = -\frac{N_1}{N_2}. \tag{34.18}$$

And since, with I_1'' small, I_1' approximates to I_1, we also have, to the corresponding degree of approximation,

$$\frac{I_2}{I_1} = -\frac{N_1}{N_2};$$
(34.19)

the currents, like the voltages, are in antiphase, and the ratio of $|I_2|$ to $|I_1|$ is the reciprocal of the ratio of the numbers of turns in the secondary and primary.

TIME VARIATION OF FLUX Since the flux Φ must necessarily vary with time in the same way as does the magnetizing current I_e to which it is due, this variation must be expressible by an equation of the form

$$\Phi = \Phi_0 \sin\left(\omega t - \frac{\pi}{2}\right),$$
(34.20)

where Φ_0 is a constant. The value of this constant may readily be ascertained by using the expression for E_1 obtained by applying Kirchhoff's second law to the primary circuit, namely,

$$E_1 = -E_0 \sin \omega t,$$

and equating this to the expression for the induced emf in terms of the variation of Φ. For this induced emf we have

$$E_1 = N_1 E = -N_1 \frac{d\Phi}{dt} = -\omega N_1 \Phi_0 \cos\left(\omega t - \frac{\pi}{2}\right)$$
$$= -\omega N_1 \Phi_0 \sin \omega t.$$

Equating the two expressions, we find that

$$\Phi_0 = \frac{E_0}{\omega N_1}.$$
(34.21)

PHASE RELATIONS We shall now consider the case where the secondary is connected to a series RLC load, as indicated in Fig. 34.3, and investigate the corresponding phase relations between the primary and secondary voltages and currents.

The phase relation between V_2 and I_2 may be seen at once from the equation itemizing the potential falls in the resistor, the inductor, and the capacitor,

$$V_2 = RI_2 + L\frac{dI_2}{dt} + \frac{Q}{C},$$
(34.22)

Q being written for the charge on the first capacitor plate encountered on going around the circuit in the positive direction. The corresponding phasor diagram, constructed for the case where the

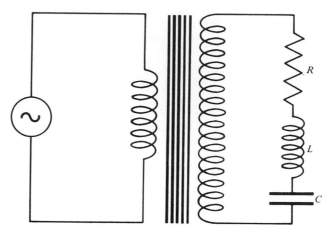

FIGURE 34.3   Transformer with series RLC load on secondary.

inductive reactance exceeds the capacitive, is shown in Fig. 34.4. In this the symbols for currents all have the extra subscript 0, to indicate current amplitudes, so that when multiplied by the relevant factors, R, ωL, $1/\omega C$, and Z_2, the products give the corresponding voltages; thus $Z_2 I_{20}$ is the same as the peak value V_{20} of the secondary voltage.

As this diagram shows, I_2 lags behind V_2 by the angle ϕ, where

$$\phi = \tan^{-1}\left(\frac{\omega L - 1/\omega C}{R}\right). \tag{34.23}$$

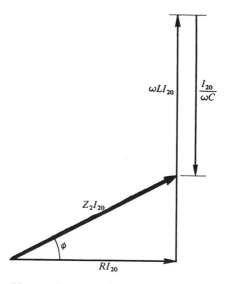

FIGURE 34.4   Phasor diagram of voltages for RLC load.

We may now construct another phasor diagram, Fig. 34.5, which gives not only V_2 and I_2, but also V_1, I_1', I_1'', I_1, I_e, and Φ in their correct relative phases. This diagram must be imagined to rotate anticlockwise with angular velocity ω, the instantaneous values of the various quantities being given by the relevant projections of the lines shown on a vertical line. Its orientation as shown in the figure corresponds to a time $(T/4) \pm nT$, where T is the period and n is a whole number. In obtaining the relative phases for the various quantities, account is taken of the fact that V_1 and V_2, and also I_1' and I_2, are in antiphase. And, in conformity with Eqs. (34.16) and (34.20), I_e and Φ are both shown with a phase lag of $\pi/2$ behind V_1.

PRIMARY AND SECONDARY POWERS Since, as shown by Eq. (34.16), I_1'' is in quadrature to V_1, I_1'' is a "wattless" current; it makes no contribution to the mean power delivered to the primary. We therefore need not consider I_1'' further.

Since not only V_1 and V_2, but also I_1' and I_2, are in antiphase with one another, it follows that I_1' lags in phase behind V_1 by the same angle ϕ as I_2 lags behind V_2. Hence the mean power delivered to the primary is $\frac{1}{2}I_{10}'V_{10}\cos\phi$ and that delivered to the secondary is $\frac{1}{2}I_{20}V_{20}\cos\phi$. But, as we have seen,

$$\left|\frac{I_2}{I_1'}\right| = \frac{I_{20}}{I_{10}'} = \frac{N_1}{N_2}$$

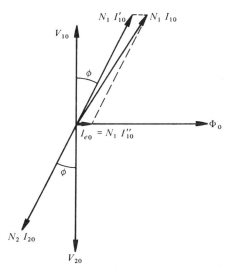

FIGURE 34.5 Phasor representation of a number of quantities involved in transformer theory.

and

$$\left| \frac{V_2}{V_1} \right| = \frac{V_{20}}{V_{10}} = \frac{N_2}{N_1},$$

so that $I_{20}V_{20}$ is equal to $I_{10}'V_{10}$. Hence the two powers are equal.

EQUIVALENT CIRCUIT Neglecting I_1'' in comparison with I_1', we may define the effective impedance Z_1 of the primary circuit as V_{10}/I_{10}. And the impedance Z_2 of the load is, of course, V_{20}/I_{20}. So, for the ratio of impedances we have

$$\frac{Z_1}{Z_2} = \frac{V_{10}/I_{10}}{V_{20}/I_{20}} = \frac{V_{10}}{V_{20}}\frac{I_{20}}{I_{10}} = \frac{N_1^2}{N_2^2},$$

that is,

$$Z_1 = \frac{Z_2}{T^2}, \tag{34.24}$$

where T is the "turns ratio," N_2/N_1.

Combining the definition of Z_1 with Eq. (34.24), we now have

$$I_{10} = \frac{V_{10}}{Z_1} = \frac{V_{10}}{Z_2/T^2} = \frac{V_{10}}{(1/T^2)\sqrt{R^2 + (\omega L - 1/\omega C)^2}}$$

$$= \frac{V_{10}}{\sqrt{(R/T^2)^2 + [(\omega L/T^2) - (1/\omega C T^2)]^2}}$$

If we write

$$\frac{R}{T^2} = R', \qquad \frac{L}{T^2} = L', \qquad CT^2 = C',$$

this equation for I_{10} can be expressed in the form

$$I_{10} = \frac{V_{10}}{\sqrt{R'^2 + (\omega L' - 1/\omega C')^2}}. \tag{34.25}$$

Thus it is as if we had the primary supply connected to a series *RLC* circuit having a resistance R', a self-inductance L', and a capacitance C', the quantities having the magnitudes defined above. An imaginary circuit having these values of the parameters is called the **equivalent circuit** of the loaded transformer.

Transformer Design

In the foregoing discussion the primary and secondary current and voltage relations have been considered in terms of the ratio of the numbers of turns, but it has not been indicated whether there is any upper or lower limit to the number of turns in, say, the primary. We shall now consider this question.

The equation

$$\Phi = \Phi_0 \sin\left(\omega t - \frac{\pi}{2}\right) \tag{34.20}$$

for the flux may be written in the form

$$\Phi = AB_0 \sin\left(\omega t - \frac{\pi}{2}\right),$$

where B_0 is the amplitude of the variation of B, and A is the cross-sectional area of the core, over the whole of which, for simplicity, we may consider B to be uniform, or sufficiently nearly so for the purpose of the present discussion. Hence

$$V_1 = -E_1 = N_1 \frac{d\Phi}{dt}$$

$$= \omega N_1 AB_0 \cos\left(\omega t - \frac{\pi}{2}\right)$$

$$= \omega N_1 AB_0 \sin \omega t,$$

whence we obtain for the peak value of the primary voltage

$$V_{10} = \omega N_1 AB_0. \tag{34.26}$$

If the transformer is iron-cored, the part of the B–H characteristic over which the core operates must be confined to the region below saturation where, to a reasonably close approximation, μ may be taken as constant. This condition sets an upper limit to B_0.

The values of V_{10}, ω, and B_0 being known in any particular case, Eq. (34.26) gives the appropriate value of $N_1 A$ for satisfactory operation. This is actually a *minimum* value; there would be no objection in principle to increasing $N_1 A$ beyond this. However, to do so would be wasteful of copper or iron, or of both. The question of what particular values of N_1 and A to use with the product of these equal to $V_{10}/\omega B_0$ is a matter of economics.

For transformer iron B_0 is generally of the order of 1 Wb/m^2.

Autotransformer

This is a type of transformer, used for either stepping up or stepping down voltages, in which part of the secondary winding functions also as the primary, or vice versa. As in the transformer with separate windings, so also here, the iron core is in the form of a closed magnetic circuit; hence the theory is essentially the same as that which we have already considered. The circuit diagram of a step-up autotransformer is shown in part (a) of Fig. 34.6, and that

899

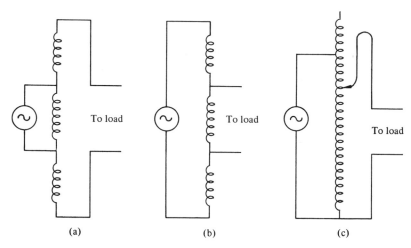

FIGURE 34.6   Types of autotransformers.

of a step-down transformer in (b). In part (c) a variable type of autotransformer is shown which can function either as a step-up or as a step-down device, according to what part of the winding is tapped off for the secondary.

Since the primary and secondary currents are in antiphase, it is permissible to make the turns common to the primary and secondary of relatively light-gauge wire; also there is, in any case, only one winding. For these reasons an autotransformer having given voltage and current ratings is less costly than a corresponding transformer with separate windings, but a disadvantage of the autotransformer is that there is no freedom of adjustment of the relative levels of potential of the primary and secondary circuits.

Induction Heating

In our discussion of generators and motors in Chapter 32 and in the discussion of transformers in the present chapter we have seen that if the armatures of these devices were constructed of solid iron there would be intolerable heating and waste of power as a result of eddy currents, and that therefore, to reduce eddy currents to a minimum, the armatures are built up of laminations. However, eddy currents are not always something to be avoided. They find an important application in a number of industrial processes and operations in which their heating effect is turned to good account. We shall consider induction heating in the present section, under the general heading of transformers, because the action of an induction heater really corresponds to that of a transformer—an air-cored step-down transformer of rather low coupling coefficient whose secondary consists of a single turn.

Since the rate of energy transfer from the coil functioning as the primary to the object that is to be heated increases with frequency, moderately high or very high frequencies are normally employed for induction heating. The frequencies used vary with the particular requirements; in most cases they range between 1000 and 500,000 Hz. The "work coil" usually consists of a few turns of heavy copper ribbon or tube mounted on an insulating, heat-resistant frame; ribbon is used where air cooling suffices, and the tubular form where it becomes necessary to circulate water through it to keep it cool. It may be either a helix or a flat spiral, whichever is better suited to the geometrical form of the object to be heated and the pattern of heat distribution required.

Induction heating is used for a wide variety of purposes, among which are the melting and refining of metals, welding, soldering and brazing, sintering, surface hardening, annealing, degassing of metal components of vacuum tubes, and gettering.

PART II
USE OF THERMIONIC TRIODE IN AC CIRCUITRY

We have already considered the electron optics of the thermionic triode in Chapter 23. In the various applications of the triode the cathode is almost always operated at a temperature sufficiently high to ensure a copious emission of electrons from it, an emission substantially in excess of the maximum electron current ever to be drawn to the anode. Under these conditions the electron current collected by the anode is a unique function of the potentials of anode and grid relative to that of the cathode, just as, in the corresponding case of the diode operating under space-charge-limited conditions, the thermionic current depends only on the potential of the anode.

Voltage Amplification

A circuit for a single stage of amplification is shown in Fig. 34.7.

The sole function of the grid is that of a traffic policeman; it must simply regulate the flow of electrons to the anode without itself collecting any of these. Accordingly, the mean grid potential

901

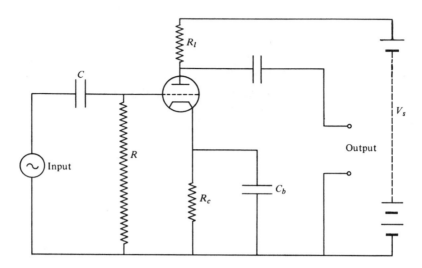

FIGURE 34.7 Circuit for use of thermionic triode to give voltage amplification.

must be such that even at its most positive it is still negative to the cathode. The requisite negative mean potential of the grid relative to the cathode is most conveniently provided for by arranging for the anode current to flow not merely through the tube and through the load resistor R_l (whose purpose will be considered later), but also through another resistor, the cathode-bias resistor R_c, connected in series with these. The grid is connected, via the grid resistor R, to the negative end of this.

The current carried by the cathode-bias resistor R_c, and with it the potential dropped across it, is made to remain effectively constant, despite the time variation of the anode current about its mean value, by the provision of the cathode bypass capacitor C_b. If the capacitance of this is large enough it will accommodate the charge variation associated with the alternating component of the anode current without any appreciable variation of the potential dropped across it and simultaneously across the resistor R_c. In other words, while the resistor carries almost exclusively the direct component of the anode current, the bypass capacitor, functioning as a smoothing capacitor, carries almost the whole of the alternating component.

If the input signal is derived from the secondary of a transformer, both the grid resistor R and the capacitor C, known as the grid-blocking capacitor, may be dispensed with; the terminals of the secondary may be connected directly to the negative end of R_c

and the grid. However, if a direct (nonalternating) voltage is associated with the alternating input voltage, the former must be isolated from the grid. This is done by inserting a blocking capacitor in series with the input and at the same time using a grid resistor, as shown. The function of the latter is to ensure that the mean potential of the grid shall be that of the negative end of R_c while allowing the grid potential to "swing" symmetrically above and below this in response to the input signal.

The blocking capacitor, of capacitance C, and the grid resistor, of resistance R, constitute a series RC circuit. Any direct voltage component of the input simply gives the capacitor a mean charge, on which is superimposed an alternating charge due to the alternating component of the input voltage. A corresponding alternating current is transmitted "through" the capacitor in the manner already considered in our discussion of RLC circuits.

The variations of potential directly affecting ("seen by") the grid are obviously those occurring across the resistor. It is desirable for these to be substantially the same as those of the input signal. This is achieved by making R large compared with $1/\omega C$. For any given pair of values of C and R this condition will be fulfilled for all frequencies above a certain limit; at lower frequencies the signal impressed on the grid will become relatively weaker. Thus for "audio" frequencies (frequencies within the audible range), typical values would be $R = 0.5$ MΩ (5×10^5 Ω) and $C = 0.1$ μF (10^{-7} F). With these values the ω for which the right-angled phasor triangle becomes an isosceles triangle ($1/\omega C = R$), giving grid potential fluctuations $1/\sqrt{2}$ those of the input signal, is 20. The corresponding frequency f is $10/\pi$, that is, just over 3 Hz. For the above condition to be satisfied f would have to be substantially greater than this. At the lower limit of audible frequencies ($f \approx 16$ Hz) the condition is sufficiently well satisfied for practical purposes; at higher frequencies it is satisfied even better.

Having discussed how to apply the input signal to the grid, let us now consider how this becomes magnified, or "amplified," by the operation of the tube. The potential V_p of the anode, or "plate," relative to that of the cathode, is $V_0 - R_l I_p$, where V_0 is the supply voltage V_s minus the cathode bias, R_l is the load resistance, and I_p is the plate current. Any increase in the grid potential V_g increases I_p, increasing the PD across the load resistor and correspondingly decreasing V_p. We see incidentally that the changes of V_p and V_g from their mean values are in antiphase. By a suitable choice of R_l the decrease in V_p may be made several times as large as the increase in V_g by which it is brought about. The

903

ratio of the change in V_p to the change in V_g to which it is due, signs being disregarded, is known as the **gain** of the amplifier.

Usual orders of magnitude of V_s and R_l are a few hundred volts and 100,000 Ω, respectively. Gains are usually somewhere within the range 10 to 20.

The output signal—the alternating component of the anode voltage—is communicated to the circuit in which it is required, via another blocking capacitor, after the manner indicated in Fig. 34.7. If a single stage of amplification is insufficient, the output signal may be fed to the grid of a triode in a further stage.

LC Oscillator

In the triode amplifier the input signal impressed on the grid is derived from an external source. If, instead of this being done, an arrangement is made for "feeding back" sufficient of the output to the grid at a suitable phase, the system will be self-maintained; it will function as an oscillator.

The question naturally arises of how the oscillation is started in the first instance. The answer to this lies in the disturbance created at the moment of switching on; this supplies the "trigger" that puts the system into operation.

In any practical oscillator system provision must be made for stability; the oscillations must not be allowed to build up to such an extent that the output signal becomes appreciably distorted. It would take us too far to consider here the details of how this stability is provided for.

Of the various feedback systems employed we shall here consider only one, that used in what is known as the *LC* oscillator. The basic features of this are shown in Fig. 34.8. This is similar to the system of Fig. 34.7, except that the *LC* loop, functioning as a "parallel *LC* circuit," takes the place of the resistive load of the system of Fig. 34.7.

From the point of view of the external circuit of which the *LC* loop forms a part, the inductor and capacitor are connected in parallel, but they may be considered to be in series within the loop itself. Any electrical impulse to which the loop is exposed will set it into oscillation, the alternating current surging back and forth within it having a frequency determined by the circuit parameters. In the absence of feedback, this oscillation would quickly die away. As it is, however, the oscillating current, flowing in the inductor,

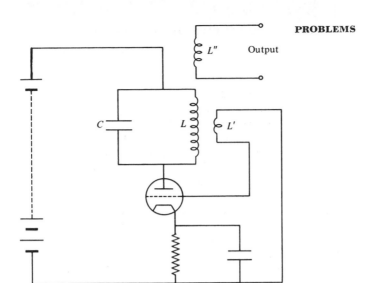

FIGURE 34.8 *LC* oscillator circuit employing thermionic triode.

induces an alternating emf of the same frequency in the inductor L' coupled with it, and this impresses a corresponding oscillating voltage on the grid. This, if in the correct phase, will not allow the current oscillation in the LC loop to die away; on the contrary, it will build it up to a substantial amplitude.

As is easily seen, the current within the RC loop is 90° out of phase with the voltage across either the inductor or the capacitor. This current, flowing in the inductive arm, induces in the inductor L' an emf that is 90° out of phase with the current. Thus, by suitably relating the directions of the two windings, it can be arranged that the signal applied to the grid is in antiphase to the ac component of the plate voltage, and this is what is required for the maintenance of the oscillation.

The output terminals are connected to an inductor L'', which, like L', is coupled to the inductive arm of the LC loop.

Problems

34.1   The secondary of a 1:2 step-up transformer is connected to a 200-Ω resistive load. What resistance connected across the primary voltage supply would carry the same current as flows in the primary of the transformer?

34.2   The power input to the primary of a transformer is 1 kW at 110 V, and the secondary feeds into a 6-Ω resistive load. A voltmeter connected across the load registers 76.6 V. What is the efficiency of the transformer, and what is the secondary current?

34.3   The primary of a transformer whose efficiency is 96 percent carries a current of 2 A when connected to 110-V mains with 100 mA flowing in a secondary resistive load. What is the voltage across the load?

34.4   A transformer stepping voltage down from 3300 V to 110 V has the usual "closed magnetic circuit" iron core. A voltmeter connected to the

905

ends of a wire linking this circuit reads 0.6 V. How many turns are there in the primary and secondary windings?

34.5   Design a step-up transformer to operate at 100 Hz on a primary voltage of 100 V, stepping this up to a secondary voltage of 1000 V, given that the induction must not exceed 1 Wb/m². Would this transformer operate satisfactorily at the same frequency for a primary voltage of (a) 50 V, or (b) 200 V? Would it be safe to use it off a 200-V 50-Hz supply? If it would not be suitable under any of these conditions, what would be the result of attempting to use it?

34.6   A transformer is designed for a primary voltage of 240 V, but is accidentally connected to 415-V mains. Sketch the wave form of the primary current and the secondary emf for the cases where the effective resistance in the primary circuit is (a) zero and (b) finite.

34.7   The secondary of a step-down transformer fed from 110-V 60-Hz mains develops an emf of 12 V and is connected to a load of resistance 2 Ω and self-inductance 20 mH. Neglecting losses in the transformer, find what currents flow in the primary and secondary windings and how they are related in phase to the corresponding voltages.

34.8   Three identical iron-cored transformers have their primary windings supplied by currents of $I_0 \sin \omega t$, $I_0 \sin (\omega t + 2\pi/3)$, and $I_0 \sin (\omega t + 4\pi/3)$, respectively. The secondary windings are connected in series as shown. For small values of I_0 it is found that the voltage between A and B is zero. Explain, with the aid of phasor diagrams if necessary, how this voltage will change as I_0 is increased. What will be its fundamental frequency?

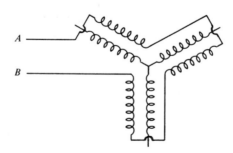

Figure for problem 34.8.

34.9   (a)   Show that the secondary current I_2 in the circuit depicted in the figure satisfies the equation

$$L' \frac{dI_2}{dt} + RI_2 + \frac{1}{C} \int I_2 \, dt = - \frac{N_2}{N_1} E_0 \sin \omega t,$$

where $L' = L_2(1 - k^2)$, k is the coefficient of coupling between primary and secondary, and N_2/N_1 is the turns ratio. (Assume that the resistances of the transformer windings are negligible.)

(b) Solve the differential equation for the case where $k = 1$ and find I_2 and I_1. Show the relations between the primary and secondary voltages and the currents I_1 and I_2 on a phasor diagram.

(c) If $k < 1$, what are the expressions for the power supplied to the primary and the power delivered by the secondary?

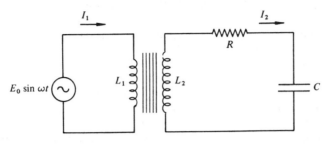

Figure for problem 34.9.

34.10 Show that with a given amplitude of the current oscillations in a coil used for induction heating, and with a given magnetic flux passing through the body to be heated per unit current in the coil, the rate of heating for a given temperature (and therefore resistivity) of the body is proportional to the square of the frequency.

34.11 How would you expect the "iron loss" occasioned by eddy currents in the core of a transformer to vary with the thickness of the lamination sheets? Explain your answer.

34.12 Explain how the wave form of the output of a triode amplifier becomes distorted if the amplitude of the input voltage is too large.

34.13 The anode of the triode of the first stage of a two-stage amplifier is connected to the grid of the second triode via a blocking capacitor, as shown, and a 0.5-MΩ resistor R connects this grid to earth. The amplifier is designed to operate between frequencies of 1 and 10 MHz.

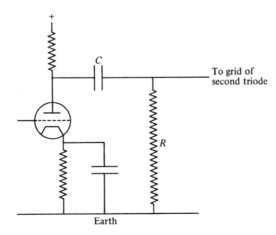

Figure for problem 34.13.

What minimum capacitance C of the blocking capacitor will be required if the voltage loss in this component is not to exceed 1 percent of the alternating voltage available anywhere in this frequency range?

34.14 In a circuit such as that of Fig. 34.7 the resistance of the resistor connected to the grid of the triode is 0.5 MΩ and the capacitance of the blocking capacitor is 0.1 μF. Between the input-signal terminals a resistor of relatively low resistance is connected, and across this a PD of 1 V is connected at time $t = 0$, disconnected at time $t = T/2$, reconnected at time T, disconnected at time $3T/2$, and so on, the intention being to impress a series of "square-wave-form" pulses on the grid. Investigate the magnitude and form of the successive pulses actually impressed on the grid in each of the three cases where the time interval between connection and disconnection is (a) $\frac{1}{100}$ sec, (b) $\frac{1}{30}$ sec, and (c) $\frac{1}{10}$ sec. Would the RC system in the grid circuit be suitable for the transmission of voltages varying sinusoidally with frequencies $1/T$ corresponding to these time intervals? If so, can you explain the difficulty with the square-wave pulses?

SECTION IX

LIGHT

THE PROPAGATION AND VELOCITY OF LIGHT

In our study of electric and magnetic phenomena we have from time to time had strong indications of a fundamental involvement of these phenomena in the propagation of light. The most important of these are the following:

1. In the mathematical expression for Coulomb's law,

$$F = k\,\frac{Q_1 Q_2}{r^2},$$

with F, r, Q_1, and Q_2 expressed in MKSA units, the numerical value of k is the product of a power of 10 (easily accounted for) and c^2, where c is the velocity of light.

2. It is surely no mere coincidence that while, on the one hand, many good insulators are transparent to visible light (and, as experiment shows, *all* are transparent in *some* region or regions of the visible, infrared, or ultraviolet spectrum), on the other hand the best conductors, metals, are opaque to all radiation.

3. Radio waves, which have the same velocity in a vacuum as light, and so are presumably essentially of the same nature, are generated electromagnetically.

4. Light, which exhibits interference, diffraction, and polarization phenomena (see Chapters 38, 39, and 40) must be propagated as a wave disturbance, for on no other basis would it be possible to account for these phenomena. And light travels freely through a vacuum. The only properties of a vacuum that have appeared in our study of physics are its capability of sustaining gravitational, electric, and magnetic fields. Hence the mechanism of the propagation of light is presumably to be explained in terms of one or more of these.*

---

* Actually, as we noted in Chapter 1, a successful electromagnetic wave theory of light was developed by Maxwell in the latter part of the nineteenth century. For a mathematical discussion of this theory, see Chapter 1 of Volume III.

In view of these indications that electromagnetic processes are involved in the propagation of light, it is logical that light should be the next general area of physics to be studied after electromagnetism.

The present chapter will be devoted, in the main, to the establishment of the laws of reflection and refraction to be used in the discussion of geometrical optics in the next two chapters. We shall also pass in review the most important experimental determinations of the velocity of light in a vacuum.

Concept of a Ray

In discussing optical phenomena it is often convenient to refer to "rays." This applies to all regions of the spectrum; thus we sometimes speak of rays in the radio range, and we quite commonly deal with rays of light, x rays, and gamma rays.

What, precisely, do we mean by a ray? The idea undoubtedly originated in connection with visible light, being derived from the commonly observed phenomenon of shadow formation. In a dusty or cloudy atmosphere we often get the impression that we can *see* rays, or, more strictly, a beam of light which we think of as a bundle of rays. This is due to the scattering of light by the suspended dust particles or water droplets, which thus become visible. We see the particles in the path of the light, but none in the shadow, and so we get the impression of seeing the whole illuminated volume.

The question now arises: How should we think of a ray in terms of the waves by which the electromagnetic disturbance is propagated? It would seem reasonable to define a ray as the line along which the energy of the radiation is propagated. And in an isotropic medium the energy must surely travel along the normals to the wave surfaces. This is obvious from considerations of symmetry. In such a medium, a parallel beam of rays corresponds to a progression of plane waves, while divergent and convergent pencils are associated with expanding and contracting spherical waves, respectively. We see that a ray is really an abstraction; notwithstanding certain appearances, we can never actually see or otherwise experience a ray of light or other electromagnetic radiation.

Huygens's Principle

Nearly 200 years before the formulation of the electromagnetic theory of light by Maxwell, a Dutch physicist, Christiaan Huygens (1629–1695), published a monograph (*Traité de la Lumière*, Paris, 1690) in which he proposed a mechanism for the propagation of

912

light waves. At this time it was not even generally accepted that light *is* propagated in the form of waves; Huygens appears to have been the first to put forward this idea, although without any real supporting evidence, while his contemporary, Newton, equally without evidence, maintained that light consists of a stream of high-speed particles. Not until 1801 was the Newton–Huygens controversy finally resolved, in favor of the latter, by Thomas Young (1773–1829), who for the first time succeeded in observing optical interference.

On the basis of his proposed mechanism of propagation Huygens worked out elementary theories of reflection and refraction which, with certain reservations, are still acceptable. The theory of refraction as formulated by Huygens has particular value in that it gives us a physical explanation of the constancy of the ratio of the sines of the angles of incidence and refraction for the passage of a ray of light across the boundary between two given media. But even more valuable has been the subsequent application of Huygens's ideas, by Fresnel (1788–1827) in 1814, to the theory of diffraction and those interference phenomena in which diffraction is involved.

In its original formulation Huygens's principle deals with the disturbances emanating from "wave surfaces," surfaces constituting the loci of points at all of which the disturbance is in the same phase. In the case of a disturbance generated at an infinitely distant point, the wave surfaces would be planes normal to the direction of propagation of the waves, while the wave surfaces for waves generated at a point a finite distance away would be spheres centered on this point.

According to Huygens, every point on a wave surface may be regarded as a fresh source of disturbance generating spherical "secondary wavelets." The envelope of all the secondary wavelets simultaneously generated and having the same radius r will be the new wave surface derived from the original one after a time t equal to r/v, where v is the velocity of propagation. This is indicated for the case of spherical waves in Fig. 35.1, where AB is an original wave surface generated at the point O, while CD, the envelope of all the secondary wavelets of the same radius generated at points on AB, is the new wave surface at a later instant of time.

In one sense Huygens's principle as formulated above may seem to us philosophically rather naive, in that it seeks to explain the propagation of a wave disturbance by invoking smaller waves, the secondary wavelets, and of course this is really no explanation at all. This aspect of Huygens's original picture, together with certain other questionable features, have been critically examined

913

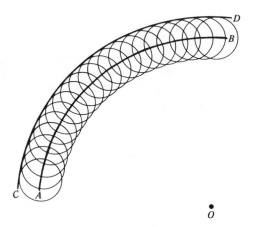

FIGURE 35.1   Huygens secondary wavelets generated at a wave surface, and their envelope in the forward direction.

by the nineteenth-century German physicist Kirchhoff, who developed a somewhat modified and much more sophisticated theory in which the secondary wavelets are placed on a sound quantitative footing, so that we need not now have any hesitation in using them.

An obvious generalization of the system of Fig. 35.1 is the one shown in Fig. 35.2, where *AB* represents *any* surface traversed by the waves traveling outward from their point of generation *O*, not necessarily a wave surface. From all points in *AB* as centers, spheres are constructed such that the radius of each sphere plus the distance of its center from *O* is the same. Then if Huygens's idea is correct, these spheres must be secondary wave surfaces all of which have the same phase, and their envelope, *CD*, should therefore be a wave surface of the train traveling outward from *O*. As we shall see later, Huygens himself in effect made use of this generalization of the original formulation of his theory in accounting for the geometrical laws of reflection and refraction.

From Fig. 35.2 we see at once that *CD* is part of a sphere with center *O*, as it should be. However, *EF* is also an envelope of the secondary wave surfaces, and we know that *EF* is *not* a wave surface of the train originating from *O*. To meet this difficulty, Huygens simply assumed that such "reflected" waves do not exist, that is, in effect, that the amplitude of the secondary wavelets in the backward direction is zero. If the idea of Huygens secondary wavelets is to be retained, it is obviously necessary to postulate the existence of an obliquity factor in the amplitude of the secondary wavelets whose value is a maximum in the forward direction and falls away

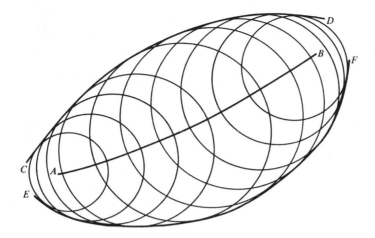

FIGURE 35.2   Secondary wavelets generated at an arbitrary surface and envelopes on two sides.

with increasing angle made with this direction, to become zero in the backward direction. It is not difficult to see that this obliquity factor must also vary with the state of polarization of the waves. In his rigorous analysis of the secondary wavelets Kirchhoff succeeded in obtaining full information concerning the obliquity factor.

In addition to the obliquity factor, there must also be a distance factor, making the amplitude of the secondary wavelets in any particular direction inversely proportional to their radius. This is demanded by energy considerations, for the energy flux density,* which may be shown to be proportional to the square of the amplitude, is also inversely proportional to the square of the radius.

We may next inquire what particular magic, if any, attaches to the *envelope* of a system of secondary wavelets. Although obviously the envelope in the forward direction of all the secondary wavelets *having the same phase* must give us a wave front, it is not always in this that we are interested. In the theory of interference and diffraction our concern is not with wave fronts but with the variation of the amplitude or intensity of the resultant disturbance due to all the secondary wavelets simultaneously arriving at a point,

---

* This is defined as the rate of flow of energy per unit area across a surface at right angles to the direction of propagation.

for example the focal plane of an eyepiece, with the position of this point. In combining the separate disturbances associated with all the secondary wavelets, with their different amplitudes and phases, essentially the same phasor technique is employed as in the combination of alternating currents or voltages studied in Chapter 33. We shall consider several examples of this kind of use of the secondary wavelets in Chapters 38 and 39.

In view of the very great importance of Huygens secondary wavelets, particularly in the theories of interference and diffraction, what appears at first sight to be the most disturbing result of Kirchhoff's analysis is that there is no self-consistent system of secondary wavelets whose combination corresponds to the known physical conditions everywhere, that is, at short distances from their points of origin as well as at relatively great distances. In other words, such wavelets do not really exist. Nevertheless, as Kirchhoff showed, there is *something*, corresponding to a mathematical expression, which does everywhere satisfy the physical conditions, and this "something" becomes indistinguishable from actual secondary wavelets at distances from the generating surface that are large compared with a wavelength. This something might accordingly be referred to as a system of pseudo-wavelets. However, to avoid circumlocution it is more usual to call these pseudo-wavelets simply wavelets. There can be no objection to this as long as we remember that we are actually dealing only with an approximation, and that there is no implication that wavelets, in the strict sense, exist. Since the approximation is acceptable only at "large" distances (large compared with a wavelength) from the generating surface, it is only for such distances that the wavelets should be used.

Mathematical analysis shows that the phase of each secondary wavelet lags behind that of the disturbance at the point of generation by $\pi/2$. However, this is compensated by the phase of the phasor resultant of all the wavelets arriving simultaneously at a point being $\pi/2$ ahead of that of the wavelet that is tangential to the main wave surface at this point. It is therefore as if neither phase difference existed. This mutual compensation justifies treating the envelope of a system of wavelets in phase with one another as a new wave front.

Laws of Reflection

In Fig. 35.3, let P be a point source of radiation, LOM a spherical wave surface centered on P, and AB a plane reflecting surface. If AB were just an *imaginary* surface within a continuous medium, we could construct Huygens secondary wavelets from it

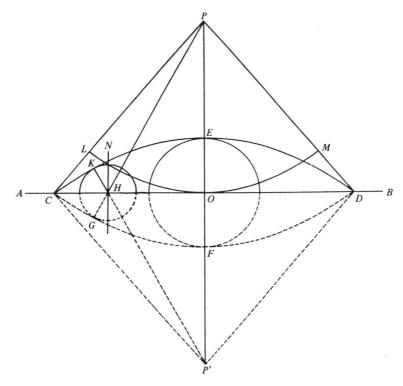

FIGURE 35.3 Huygens construction for derivation of laws of reflection.

as we did from the surface AB in Fig. 35.2, two representative examples of which are shown, centered on O and H. The envelope of these would be $CGFD$. But since AB is, in fact, a reflecting surface, the envelope to be constructed must instead be that which is tangential to the backward portions of the secondary wavelets; and this is $CKED$. This, then, is the reflected wave surface. From simple geometrical reasoning it is obvious that this is centered on P', which is on the normal to the surface of the mirror drawn through P and is the same distance behind the mirror as P is in front. P' is known as the "mirror image" of P. Corresponding to the incident ray PH is the reflected ray HK; and again from geometry it is clear that (1) PH, HK, and the normal to the mirror at H are coplanar, and (2) the angle of reflection, NHK, is equal to the angle of incidence, NHP. These are the two laws of reflection as found experimentally.

Laws of Refraction

Let AB in Fig. 35.4 represent a plane boundary between two isotropic media 1 and 2 in which electromagnetic radiation travels

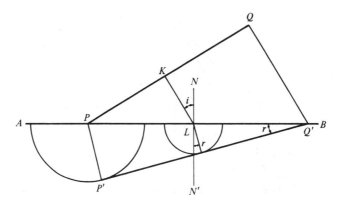

FIGURE 35.4 Huygens construction for derivation of laws of refraction.

with velocities v_1 and v_2, respectively. Let PQ be a plane wave surface advancing toward this boundary from medium 1 to medium 2, the direction of this advance being, of course, normal to PQ. Then in the time it takes for Q to reach Q' a secondary wavelet will have spread out from P in the second medium whose radius PP' satisfies the condition that

$$\frac{PP'}{QQ'} = \frac{v_2}{v_1}.$$

In this same time the point K on the original wave surface will have traveled to L and then from here a secondary wavelet of radius LM will have spread out in the second medium such that

$$\frac{LM}{SQ'} = \frac{v_2}{v_1}.$$

Thus we see that the envelope of all the secondary wavelets in the second medium is the plane wave surface $P'MQ'$, and this will travel in the direction normal to itself, that is, in the direction PP', or LM.

Regarding KL and LM as representative incident and refracted rays, respectively, we see that the angle of incidence, KLN, is equal to the angle SLQ', while the angle of refraction, MLN', is equal to the angle $MQ'L$. Calling these angles i and r, respectively, we have, therefore,

$$\frac{\sin i}{\sin r} = \frac{SQ'/LQ'}{LM/LQ'} = \frac{SQ'}{LM} = \frac{v_1}{v_2}. \tag{35.1}$$

The incident ray KL, the refracted ray LM, and the normal at the point of incidence NN' are all in the plane of the diagram.

The fact, established experimentally, that these three are coplanar is usually referred to as the first law of refraction.

From the equation above we see that the ratio of the sine of the angle of incidence to the sine of the angle of refraction, being equal to the ratio of velocities in the two media, v_1/v_2, is constant. The constancy of the ratio $\sin i/\sin r$, found experimentally, is what is usually called the second law of refraction. And the ratio of the sines of the angles of incidence and refraction, which accordingly has a particular significance for the two media, is given a special name; it is known as the **index of refraction,** or **refractive index,** of the second medium with respect to the first. This is usually represented by the symbol n_{12}:

$$\frac{\sin i}{\sin r} = n_{12}. \tag{35.2}$$

Equating the right sides of Eqs. (35.1) and (35.2), we have, according to Huygens's construction,

$$n_{12} = \frac{v_1}{v_2}. \tag{35.3}$$

We see that, instead of defining the refractive index as the ratio of the sine of the angle of incidence at a surface separating two media to the sine of the angle of refraction, we could, alternatively, define it as the ratio of the velocity of propagation of the waves in the first medium to that in the second. The latter is the more fundamental definition, and is to be preferred in theoretical discussions of refraction phenomena in nonisotropic media.

Let us now consider the case where a ray of light is incident on the boundary from the second medium and is refracted into the first medium, and let us denote the corresponding index of refraction, that of the first medium with respect to the second, by the symbol n_{21}. Then by corresponding reasoning we have

$$n_{21} = \frac{v_2}{v_1};$$

and so we see that

$$n_{21} = \frac{1}{n_{12}}. \tag{35.4}$$

The expectation to which Huygens's construction leads, that n_{21} should be equal to the reciprocal of n_{12}, is confirmed experimentally, and this experimental result is known as the third law of refraction. This is also sometimes referred to as the **reversibility law,** because it implies that if the ray incident in the second medium is ML, the corresponding refracted ray must be LK.

919

Finally, let us consider a number of different media 1, 2, 3, . . . , N, in which light travels with the velocities v_1, v_2, v_3, . . . , v_N, respectively. Then since

$$n_{12} = \frac{v_1}{v_2},$$

$$n_{23} = \frac{v_2}{v_3},$$

.

.

.

$$n_{N1} = \frac{v_N}{v_1},$$

we have, multiplying together all the left-hand sides and equating to the product of all the right-hand sides,

$$n_{12}n_{23} \cdot \cdot \cdot n_{N1} = 1. \tag{35.5}$$

That this is indeed the case may readily be established experimentally. This is the so-called fourth law of refraction, also known as the **cyclic law.**

Actually the third and fourth laws are not independent, for, writing the reversibility law in the form

$$n_{12}n_{21} = 1,$$

we see that this is really a special case of the cyclic law, applying where there are only two media.

In addition to the refractive index n_{12} for light traveling from a medium 1 to another medium 2, a useful concept is that of *the* refractive index n_1 of the first medium, or n_2 of the second medium, or, in general, n of any medium. This refers to the case where light travels from a *vacuum* into this medium. Applying the expression for refractive index in terms of velocities to this special case, we have

$$n = \frac{c}{v}, \tag{35.6}$$

where c is the velocity of light in a vacuum and v that in the medium in question.

From a knowledge of the refractive indices of different media we may at once work out the refractive index for the passage of light from any one medium to any other. Let the refractive indices of two media 1 and 2 be n_1 and n_2, respectively. Then for the passage of light from the former to the latter we have

$$n_{12} = \frac{v_1}{v_2} = \frac{c/v_2}{c/v_1} = \frac{n_2}{n_1}. \tag{35.7}$$

We may also obtain this result directly from the cyclic law (including the reversibility law) without invoking the theoretical relation between a refractive index and the corresponding ratio of velocities. For, using the subscript 0 for a vacuum, we have, by definition,

$$n_1 = n_{01}, \qquad n_2 = n_{02}.$$

The cyclic law applied to the case under consideration is

$$n_{12} \cdot n_{20} \cdot n_{01} = \frac{n_{12} \cdot n_{01}}{n_{02}} = 1;$$

that is,

$$n_{12} = \frac{n_{02}}{n_{01}} = \frac{n_2}{n_1}.$$

The refractive indices of a number of materials at 15°C for light of wavelength 5893 Å are given in Table 35.1.

TABLE 35.1

Material	Refractive index	Material	Refractive index
Air (atm. pr.)	1.000276	Turpentine	1.47
Water	1.33	Perspex	1.49–1.50
Teflon	1.35	Nylon	1.53
Alcohol	1.36	Rock salt	1.54
Calcium fluoride	1.43	Glasses	1.5–1.9
Fused silica	1.46	Diamond	2.42

No material has a unique refractive index, applying to light of all wavelengths. In general, the refractive index increases with decreasing wavelength; that is, in the visible range, as we traverse the spectrum from red to violet. For most solid and liquid materials the variation is of the order of 1 to 2 percent between the extremes of the visible range.

Total Internal Reflection

If a ray of light traveling in a medium of refractive index n_1 is incident on a boundary with a medium of lower refractive index n_2,

921

any light passing through must be refracted away from the normal according to the law

$$\frac{\sin i}{\sin r} = \frac{n_2}{n_1}. \tag{35.8}$$

Clearly, with n_2/n_1 less than 1, an upper limit for i must be reached where r is 90°, $\sin r$ then being equal to 1. For any greater angle of incidence there can be no refraction, and it is found that the incident ray is then reflected back into the first medium. Such reflection, occurring without any loss of energy, is known as **total internal reflection.** The angle i, which corresponds to the transition from refraction to total internal reflection, that is, that for which r is equal to 90°, is called the **critical angle.** Writing this as i_c, we have

$$i_c = \sin^{-1}\left(\frac{n_2}{n_1}\right). \tag{35.9}$$

Total internal reflection has many applications in optical instruments. Perhaps the most important is an isosceles right-angled prism. This may be used for turning the direction of travel of a ray of light incident on the prism either through 90° or through 180°, according to the way the light enters the prism. These two cases are shown in Fig. 35.5, which is self-explanatory. In both cases the light is incident on the reflecting face(s) of the prism at 45°. Let us suppose the prism is constructed of glass of refractive index 1.5. Then $n_2/n_1 = 1/n = 0.67$, and the corresponding value of i_c is 41.8°. Since this is less than 45°, the angle at which the reflection shown in Fig. 35.5 occurs, this reflection is total; *all* the light is reflected.

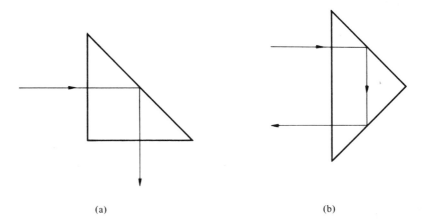

(a)   (b)

FIGURE 35.5   Two uses of total internal reflection in right-angled prisms.

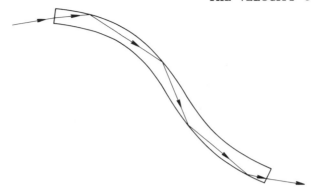

FIGURE 35.6 Action of a light pipe.

Another interesting application is a **light pipe.** This consists of a rod of glass or plastic material bent into an appropriate form corresponding to the path in which it is wished to guide the light, and usually cut off at right angles at both ends. This is shown in Fig. 35.6. Light fed into one end suffers a series of total internal reflections at high angles of incidence and finally arrives at the other end with its intensity diminished only by absorption along its path. Light pipes are useful in all situations where it is impracticable to place a source of light where the illumination is required and where a straight path for direct optical projection is not available. Such situations often arise; for example, in surgery.

Determinations of the Velocity of Light

The velocity of propagation of electromagnetic waves in a vacuum has been determined by a number of independent methods, all yielding values in close agreement with one another. We shall now pass in review the most important of these.

RÖMER'S METHOD Historically, the first determination of the velocity of light was that of the Danish astronomer Römer, in 1675, based on observations of the eclipses of Jupiter's satellites.

Jupiter's orbit about the sun is practically circular, and if a particular satellite may be assumed to describe a circular orbit about the planet, then this satellite may be thought of as a kind of clock; successive disappearances of the satellite into the shadow of the planet, or emergences from this shadow, should, to an imaginary observer on the sun,* mark out equal intervals of time—

---

* Actually such an observer would have the shadow behind his line of sight, and he would, more accurately, see disappearances behind, or emergences from, the apparent disk of the planet.

at least if the orbit of the satellite about the planet and that of the planet about the sun are in the same plane. For an observer on the earth, however, these successive eclipses would not *appear* to occur at regular intervals, because of the size of the earth's orbit about the sun and the finite velocity of light; during the period when the earth is increasing its distance from Jupiter the time intervals between *observed* eclipses would be lengthened, while during the period when this distance is decreasing they would be shortened. It is not difficult to see how, from the mean interval between successive observed eclipses, and the apparent irregularity, the time it takes for light to traverse a distance equal to the diameter of the earth's orbit can be calculated, and hence, if this distance is known, also the velocity of light in the vacuum of interplanetary space.

From a long series of observations of the times of emergence of the innermost of Jupiter's satellites from the shadow of the planet, Römer calculated the time for light to travel a distance equal to the diameter of the earth's orbit to be about 22 minutes. In Römer's time this distance was believed to be 276,000,000 km. The velocity of light calculated from these figures comes out to 2.1×10^8 m/sec.

As we now know, both the time and the distance used in this calculation were substantially in error. With the greater accuracy of astronomical observations and of timing made possible by present-day techniques, Römer's method would now be competitive with more modern methods for accuracy, were it not for some slight uncertainty concerning the scale of the solar system; this is the only significant remaining limitation to accuracy. Perhaps the most valuable feature of a modern determination of the apparent irregularities of satellite eclipses or other natural "clocks" in the planetary system would be that in conjunction with independent precision determinations of the velocity of light it would give us a reliable figure for the mean radius of the earth's orbit, the "astronomical unit," serving as a useful check on other methods for determining this quantity.

BRADLEY'S ABERRATION METHOD The next determination of the velocity of light, again astronomical, was made by Bradley in 1728. This method depends on observations of the apparent displacement, or "aberration," of fixed stars observed when the earth is traveling transversely to the direction of the star. Bradley, who was at the time professor of astronomy at Oxford, observed this shift in the apparent positions of the stars, always in the direction of motion of the earth in its orbit at the time in question, and realized the reason for this when, one day, he noticed a change in the apparent direction of the wind while sailing on the Thames. The two phenomena are, indeed, essentially of the same kind; it is simply a matter of relative motion—of the light waves in the astronomical case.

Since the determination of the velocity of light by Bradley's method depends on a knowledge of the speed of the earth in its orbit, and this, in turn, is expressible in terms of the length of the year and the radius of the earth's orbit, it is clear that measurements of stellar aberrations in conjunction with an independent determination of the velocity of light can, like the eclipses of satellites, serve as a basis for the calculation of the astronomical unit.

FIZEAU'S TOOTHED-WHEEL METHOD This was the first nonastronomical method used for determining the velocity of light, carried out by Fizeau in 1849. The principle of the method is shown in Fig. 35.7. Light from a concentrated source S is converged by a lens onto a surface of a half-silvered plate of glass G set at 45° to the axis of the beam, and the light reflected from this comes to a focus at F on the rim of a toothed wheel rotating at a high speed. (An axial view of the toothed wheel is shown separately on the right of the figure.) Let us suppose, in the first instance, that the wheel is stationary and that the focus is formed in the space between two teeth. Then the light diverging again beyond F is converted by the lens L_1 into a parallel beam, and after traversing a distance of several kilometers this is again brought to a focus by the lens L_2 on the surface of a concave mirror M whose center of curvature is at the center of the lens. The light reflected from M passes back through L_2 and L_1, is again brought to a focus at F, and the image formed here may be viewed through the half-silvered glass plate G and the lens L_3.

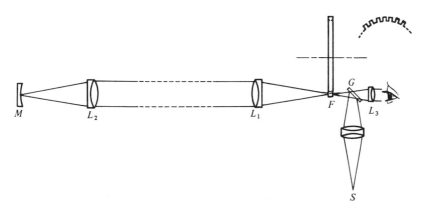

FIGURE 35.7 Fizeau's toothed-wheel method of determining the velocity of light.

Let us now suppose that the wheel is rotated at such a speed that light traveling from right to left and passing between two teeth is, on its return journey, brought to a focus at F on a tooth, which is now in the position previously occupied by a gap. In this case no light will be seen by the observer. Obviously there will be a series of speeds producing eclipses in this way, and by observing the lowest of these, or preferably several, and also counting the number of teeth on the wheel and measuring the distance FM, one obtains all the necessary information for calculating the velocity of light.

The reason for making the mirror M concave rather than plane is that, because of the finiteness of the size of the source S, there will be a corresponding finite area on the mirror M over which sets of converging rays from the lens L_2 are brought to point foci. With M concave, all these cones of light converging to points on M must be reflected as corresponding diverging cones, which will pass back through the lens L_2.

Working in Paris between stations located in Montmartre and Suresnes separated by somewhat more than 9 km, Fizeau obtained for the velocity of light the value 3.13×10^8 m/sec. The experimental arrangements were later modified and improved by Cornu and by Young and Forbes. The latter obtained as their final result 3.014×10^8 m/sec.

ROTATING-MIRROR METHODS An important advance on Fizeau's toothed-wheel method was one developed a few years later in which a rotating mirror was used. This was first proposed by Arago, pioneered experimentally by Fizeau and Foucault, at first working together, and finally used in 1862 by Foucault alone. In broad principle, the rotating mirror replaced Fizeau's toothed wheel, and Fizeau's eclipsing of the returning light by a tooth was replaced by a displacement of the final image of the source. Later, the Arago–Fizeau–Foucault method was substantially modified and improved by Michelson in America, whose work, beginning in 1878, continued well into the twentieth century. Michelson's arrangement is shown schematically in Fig. 35.8; the speed of rotation of the octagonal rotating mirror was found which produced an image in the focal plane of the telescope in the same position as when the mirror was stationary. From this speed and the double distance traversed by the light (70 km) the velocity of light could be calculated. Michelson's final result for the velocity in a vacuum,*

---

* This was obtained by multiplying the experimental value for the velocity in air by the refractive index of air, which, under Michelson's conditions, was 1.000225.

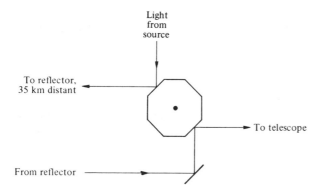

Light
from
source

To reflector,
35 km distant

To telescope

From reflector

FIGURE 35.8 Principle of Michelson's method of determining the velocity of light employing a rotating octagonal mirror.

published in 1927, was 299,796 ± 4 km/sec. He subsequently collaborated with Pease and Pearson, until his death in 1931, in a redetermination, using essentially the same method for light traveling in a long evacuated pipe, and the final value obtained in this work, published in 1932, was 299,774 ± 2 km/sec. Evidently, in at least one of these results, the estimate of the limits of error was too optimistic.

KERR-CELL METHOD In 1928 Karolus and Mittelstaedt developed a light-chopping method essentially similar to Fizeau's, in which, however, the "chopper" was electrical instead of mechanical. It was based on the use of a radiofrequency electric oscillator in conjunction with a Kerr cell, a vessel containing nitrobenzene, which, in common with certain other liquids, becomes doubly refracting under the action of an electric field. This effect, discovered by John Kerr in 1875, will be discussed in Chapter 40. In this method the light can be interrupted many millions of times per second. In 1941 Anderson, working with an improved version of Karolus and Mittelstaedt's apparatus, obtained for the velocity of light in a vacuum the value of 299,776 ± 6 km/sec.

More recent determinations by the Kerr-cell method in which earlier sources of systematic error were avoided have been made by E. Bergstrand in Norway (1950) and by I. C. C. Mackenzie in Scotlant (1954), who found the values 299,793.1 ± 0.3 and 299,792.4 ± 0.5 km/sec, respectively.

RADAR METHOD A precision determination of the velocity of electromagnetic waves may be made by timing the interval between the transmission of a pulse of radio waves and the reception at the transmitting station of the "echo" of this pulse reflected by an object at a known distance. One such determination was carried out by C. I. Aslakson in 1949 and yielded the value 299,792 ± 3.5 km/sec.

BEST PRESENT-DAY VALUE From a critical examination of all determinations of c, some by methods other than those considered here, R. T. Birge, in 1957, arrived at what he considered to be the "best" or most probable value. This was 299,792.4 ± 1.0 km/sec.

Problems

35.1 Show that the ray AP passing through A which is reflected at P by the plane mirror shown in the figure toward the point B is the one for which the composite path APB is a minimum.

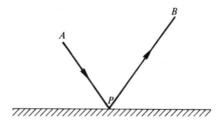

Figure for problem 35.1.

35.2 A mirror rotates about an axis in its plane with angular velocity ω, and during one half of each rotation reflects a parallel beam of light which is incident on it in a constant direction. What is the angular velocity of the reflected beam during these half-rotations?

35.3 A number of plane-parallel slabs of different transparent materials are placed together in contact, and a narrow parallel beam of light is incident on one side at an angle θ to the normal. The angle that the beam emerging from the other side of the system makes with the normal is found also to be θ. This is so for all values of the original angle θ. Show that this proves the validity of the cyclic law of refraction.

35.4 The refractive index of flint glass is 1.65, and that of water is 1.33, both relative to air. What are the refractive indices of flint glass relative to water, and of water relative to flint glass? Find the critical angle for light going from flint glass to water.

35.5 A semicylinder such as is shown in sectional view in the figure is constructed of glass of refractive index 1.65, and its flat horizontal upper surface supports a drop of liquid. For light directed radially toward this drop the critical angle θ_c for reflection at the glass–liquid interface is found to be 58°. What is the refractive index of the liquid?

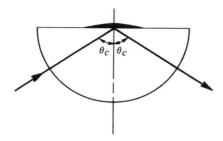

Figure for problem 35.5.

35.6 (a) What is the speed of light in a clear plastic material whose refractive index is 1.4?

(b) What is the refractive index of a material in which light travels at a speed of 1.8×10^8 m/sec?

MIRRORS AND PRISMS

In the field of optics, theoretical interest attaches mainly to interference, diffraction, polarization phenomena, and certain other aspects of physical optics. But in most experimental systems in physical optics the use of geometrical optical devices such as mirrors, prisms, lenses, and combinations of these in such optical instruments as telescopes and microscopes is involved. In view of this we shall study geometrical optics first; this will provide the reader with the necessary experimental background for the study of physical optics which is to follow.

Geometrical optics is based on the assumption, suggested by the casting of sharp shadows, that in a homogeneous medium light travels in straight lines, or along "rays." Actually, as we shall see when we come to study physical optics, the concept of an absolutely sharp shadow is an idealization; owing to the wave nature of light, shadows are never completely clear-cut. Nevertheless, the idea of rays of light, derived from this idealization, is a useful one.

We know from everyday observation that light rays are reflected at smooth surfaces, and that they may also be bent or refracted in passing across the boundary from one transparent medium to another. This reflection and refraction of the systems of rays originating from objects leads to the formation of images of these having definite locations, orientations, and sizes. In this chapter we shall first develop algebraical formulas relating to the images produced by mirrors, and then go on to consider the refraction of light by prisms. Then, in Chapter 37, we shall deal with the formation of images by lenses and optical instruments.

Real and Virtual Images and Objects

Before proceeding to discuss the formation of images specifically by mirrors, it will be useful to consider the nature of images generally, regardless of the means employed for their production.

Let a point A be either an original actual source of light or in effect a point light source, for example by virtue of diffuse reflection or scattering of light incident on a small area located at A. Such a point is known as a **point object.** Then another point B is said to be an **image** of A if all rays originating from A, or a proportion of them, subsequently either actually pass through B, or proceed *as if* they had passed through B without in fact having done so. In the former case the point B is said to be a **real image** of the point A, and in the latter it is called a **virtual image.**

Examples of the formation of real and virtual images are shown in Fig. 36.1. In part (a) a "pencil" of rays diverging from the point A is incident on a converging lens, after refraction through which the rays all pass through the point B; thus B is a real image of the point A. In (b), on the other hand, the rays are incident on a diverging lens, and this causes the rays, after refraction, to diverge still more. The emergent rays in this case all proceed along lines which, produced backward, intersect at B; they proceed as if they had come from B, but in fact no rays do pass through this point. Thus B is a virtual image of the object A.

As we shall see later, a lens does not necessarily have to be a diverging lens to produce a virtual image; under certain conditions a virtual image may also be produced by a converging lens.

It should be noted that any image produced by a lens on the side where the rays emerge is necessarily real, while, on the other hand, only virtual images can be produced on the side where the rays are incident. Thus, if we know on which side of a lens an image is produced, we also know whether this image is real or virtual.

Another example of the formation of a virtual image, this time by a plane mirror, is shown in parts (a) and (c) of Fig. 36.6. Here the rays reflected by the mirror all appear to have come from B but without in fact having passed through this point. B is therefore a virtual image of A.

Obviously, a real image formed by a mirror must necessarily be on the side of the mirror on which the rays are incident, while images formed behind the mirror are necessarily virtual.

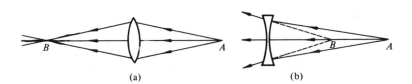

(a)   (b)

FIGURE 36.1   Production of real and virtual images of a point by means of lenses.

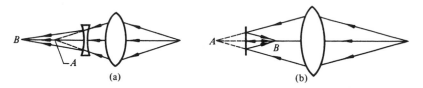

FIGURE 36.2 Production of a real image of a virtual point object (a) by a lens, and (b) by a mirror.

Not only images, but also objects, can be either real or virtual. The objects we have considered hitherto are all real; light rays actually diverge from these. On the other hand, if a convergent pencil of rays is incident on the surface of a lens or a mirror, converging toward a point on the other side, this point is said to be a **virtual object;** the incident rays do not actually pass through this point. Of course an incident pencil of rays that converges toward a point cannot be produced except by some converging lens or mirror other than the one we are considering. Examples of virtual objects A producing real images B are shown in parts (a) and (b) of Fig. 36.2, the former being produced by a lens and the latter by a mirror. In both cases the incident converging pencil of rays is shown as produced by a lens.

A virtual object must, by its nature, be on the side of the lens or mirror opposite that on which the rays are incident. As we shall see later, a virtual object may produce either a real or a virtual image.

Reflection by a Spherical Mirror

Let us, in the first instance, consider the three cases shown in Fig. 36.3. In each of these, A represents a real point object and B its image, the mirror having its reflecting surface on the right-hand side. The center of curvature of the mirror is at O.

Since AM is normal to the mirror surface, a ray incident at M must be reflected straight back along its path. Let us now consider another ray AP incident at some other point P. This will be reflected along PB, or along BP produced, where this line and AP make equal angles with OP. Then if the position of B, on the axis, were independent of that of P, this would be the point image of the object A.

Since in each case OP bisects the angle APB, either internally or externally,

$$\frac{OB}{OA} = \frac{BP}{PA}.$$

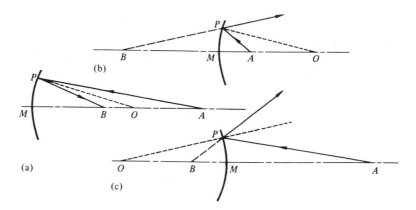

FIGURE 36.3   Three cases of image formation by spherical mirrors.

From this it follows that the point B is not independent of the position of P, that is, that a point object does not, in general, produce a point image with a spherical mirror. For, if B were independent of P, then, for another point of incidence P', at a different distance from M, we should have the corresponding equation

$$\frac{OB}{OA} = \frac{BP'}{P'A},$$

so that $BP'/P'A$ and BP/PA would be equal. But this, clearly, they are not. Hence a point object does not produce a point image.

Nevertheless, if the aperture (diameter of the circle coinciding with the edge of the mirror) is not too large, the ratio BP/PA for any position of P on the mirror will approximate reasonably closely to BM/MA, so that to this degree of approximation we have

$$\frac{OB}{OA} = \frac{BM}{MA}, \tag{36.1}$$

and since there is only one point B that satisfies this equation this will be an approximate point image of the object A for all rays reflected from the mirror; all such rays pass very close to B, or, in the case of a virtual image, appear to have come from a point close to B.

The failure of a point object to give an exact point image by reflection in a spherical mirror is an imperfection of the mirror known as **spherical aberration.**

Aplanatic Reflecting Surfaces

Only in one case does a spherical mirror produce an exact point image of a point object, namely, where A is coincident with the center of curvature O. B will then also be at this point. For this particular case, where there is no aberration, the mirror is said to be an **aplanatic reflecting surface.**

As we have already noted, a plane mirror always produces a point image of a point object; a plane mirror is an aplanatic surface quite generally. This is the only case where a mirror is aplanatic regardless of the position of the object. In all other cases there are only two object and image points (interchangeable) for which there is no aberration. These other cases are—apart from that of a spherical surface which we have just considered—those of an ellipsoid of revolution, a paraboloid, and a hyperboloid.

It is a well-known property of an ellipse that if the two foci A and B are joined to any point P on the ellipse (Fig. 36.4), then, XY being the tangent to the ellipse at this point, the angles APX and BPY are equal. This means that if the ellipse is rotated about its major axis and the surface thereby described, an ellipsoid of revolution, is a reflecting surface, a ray AP incident on this surface will be reflected along PB, so passing through B. This being true of all rays diverging from A and reflected by the mirror, B must be a point image of a point object situated at A. Similarly, A is a point image of B. With respect to its two foci, therefore, an ellipsoid of revolution is an aplanatic reflecting surface. This property of an ellipsoid is made use of in the Gregorian reflecting telescope, which, however, is now rarely used, and so is not discussed in this book.

A paraboloid is the limiting case of an ellipsoid of revolution in which one focus is at infinity. Accordingly, all rays originating at "the" focus (the one that is not at infinity) are reflected from a paraboloidal mirror as a parallel beam parallel to the axis. This property is turned to practical account in searchlights. Conversely,

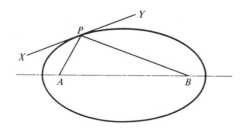

FIGURE 36.4 Ellipsoidal mirror as aplanatic surface for focal points.

935

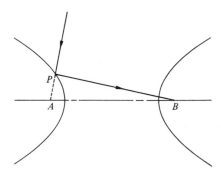

FIGURE 36.5 Hyperboloidal mirror as aplanatic surface for focal points.

all rays incident on the mirror in a direction parallel to the axis are reflected to form a point image at the focus. The main mirrors of reflecting telescopes used in astronomy are therefore made paraboloidal in form, so producing point images of stars.

A hyperbola is usually presented mathematically as an associated pair of curves such as those shown in Fig. 36.5, with two foci *A* and *B*. It may be shown that if these are joined by straight lines to any point *P* on one of the curves, for example the left-hand one, the angles these lines make with the tangent at this point are equal. Let us now suppose that this curve is rotated about its axis, so generating a hyperboloid of revolution, and let us consider a mirror of this form, with its reflecting surface on the convex side. Then, as the reader can at once verify for himself, the property of the hyperbola we have just considered means that a virtual point object situated at the near focus *A* of this curve will produce a real point image at the remote focus *B*; thus, with regard to the conjugate foci *A* and *B* this mirror is aplanatic. This property of a hyperboloidal mirror is made use of in the Cassegrainian telescope, to be discussed later.

Sign Convention for Algebraical Formulas

We are now ready to extend our discussion of reflection at the surface of a spherical mirror by developing an algebraical formula applying to this case and then showing how, by using this formula, it is possible to work out the positions of the approximate point images corresponding to all point objects on the axis of the mirror. Before embarking on this discussion, however, it will be convenient to consider the sign convention to be used, not only for the mirror formula, but also for formulas relating to image formation by refraction at a spherical surface of separation between two media having different refractive indices and by refraction through a thin lens.

There are various systems in use, but we shall here consider only one, the simple one-dimensional Cartesian system in which points on one side of an agreed origin are regarded as having positive coordinates, while the coordinates of points on the other side are taken as negative. Not only is this a system familiar to all who have used graphs, but it also has certain important advantages over other systems to recommend it, notably the essential simplicity of its application to complex problems involving the use of more than one mirror, refracting surface, and/or lens. It is the system normally used by workers in advanced geometrical optics.

Although this system, correctly presented, could hardly be simpler, it is regarded by some as difficult to understand. This is due to a confusion arising out of the common use of the terms "positive distance" and "negative distance." Distances are, by their nature, always positive. But Cartesian *coordinates* can have either sign. By calling coordinates by their right name all confusion is avoided.

In all the problems with which we shall be concerned use is made of an axis. This is commonly an axis of symmetry, such as the line passing through the center of a thin lens and perpendicular to its "plane." Usually, the point where this axis intersects the mirror, refracting surface, or lens is taken as the origin; points measured from this origin in one direction along the axis are taken as having positive coordinates and those measured in the other direction negative. For convenience, the system is usually drawn in such a way that its axis is horizontal, and in this case the coordinates of points on this axis are positive and negative x coordinates.

The simplest case of the use of coordinates to consider by way of illustration is that of reflection by a plane mirror, as shown in Fig. 36.6. In parts (a) and (b) the reflecting surface of the mirror is on the right, while in (c) and (d) it is on the left. In (a), A is a real point object and B is its virtual image, which, as we saw in Chapter 35, is as far behind the mirror as A is in front. Hence, denoting by u and v the coordinates (x coordinates) of object and image, respectively, relative to M as origin, we have

$$v = -u,$$

or

$$v + u = 0. \tag{36.2}$$

In (b) all rays are reversed, so that we have a virtual object at A and a real image at B, and clearly Eq. (36.2) applies as before. In

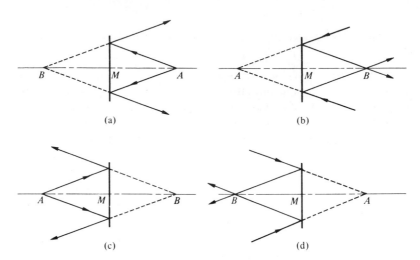

FIGURE 36.6 Positive and negative coordinates of point objects and images in various cases of reflection by a plane mirror.

(c), where the reflecting surface and real object A are on the left, B is a virtual object, and in (d), where all rays are reversed, object and image positions are interchanged, the object A being virtual and the image B real. Obviously, in (c) and (d) as well as in (a) and (b), Eq. (36.2) applies; this is the relevant equation in all cases where we are dealing with a plane mirror.

Spherical-Mirror Formula

Let us again consider the three cases shown in Fig. 36.3, and also Eq. (36.1), which we derived for the position of the approximate point image relative to that of a point object in these cases.

Let us, for example, express the quantities in Eq. (36.1) in terms of the coordinates u, v, and r for case (b). In this case,

$$OB = BM + MO = -v + r,$$
$$OA = MO - MA = r - u,$$
$$BM = -v,$$

and

$$MA = u.$$

Substituting these expressions in

$$\frac{OB}{OA} = \frac{BM}{MA}, \tag{36.1}$$

we have

$$\frac{-v + r}{r - u} = \frac{-v}{u},$$

whence, cross-multiplying, simplifying, and dividing throughout by uvr, we obtain the mirror equation

$$\frac{1}{v} + \frac{1}{u} = \frac{2}{r}. \tag{36.3}$$

It is readily seen that corresponding substitutions applied to the other two cases, (a) and (c), yield the same formula, (36.3).

If the rays are reversed in direction, object and image change places, so that what was formerly u becomes v, and vice versa. Obviously, making these substitutions can have no effect on the final formula obtained, this being symmetrical in u and v. It should be noted that in cases (b) and (c) reversal of the rays converts what was formerly a virtual image into a virtual object and what was a real object into a real image.

We obtain one more case by rotating the system (a) through 180° about a vertical axis through M but still having the reflecting surface on the right-hand side of the mirror. This gives us a convex mirror facing toward the right. The lines AP and BP are extended beyond P and the directions of the rays (necessarily on the right-hand side) are such that the incident ray travels from right to left and the reflected ray from left to right. We then have the system shown in Fig. 36.7. We see that here both object and image are virtual. All signs are now reversed, so that Eq. (36.3) is still valid. Reversal of the rays in Fig. 36.7 would still give us a virtual image of a virtual object, but now the object would be closer to the mirror than the center of curvature and the image farther away.

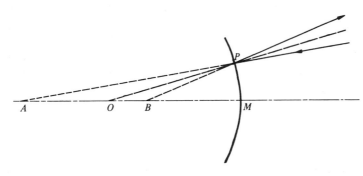

FIGURE 36.7   Virtual image of virtual point object due to reflection by convex mirror.

Finally, if in all these cases the diagrams are turned back to front, representing mirror orientations with reflecting surfaces on the left, and correspondingly light is incident from the left and reflected back toward the left, this merely has the effect of reversing all signs, and again Eq. (36.3) will apply.

An interesting special case is that in which the radius of curvature of the mirror is infinite, the mirror becoming plane. In this case Eq. (36.3) reduces to

$$\frac{1}{v} + \frac{1}{u} = 0,$$

or

$$v + u = 0, \tag{36.2}$$

the equation we have already derived for a plane mirror.

On comparing parts (a) and (b) of Fig. 36.3, in both of which light from a real object is incident on a concave mirror, we see that while in (a) the image, also, is real, in (b) it is virtual. This raises the question of what condition has to be satisfied in order that the image shall be real.

Obviously, for the image to be real, it must be on the same side of the mirror as its reflecting surface. This means that for a concave mirror v must have the same sign as r, while for a convex mirror the sign of v must be opposite to that of r. What sign v has in any given case may readily be ascertained by using the mirror formula (36.3).

FOCAL COORDINATE AND FOCAL LENGTH Consider now the special case where the rays incident on the mirror constitute a parallel beam, corresponding to an infinite value of $|u|$. The position of the image formed in this case is known as the **principal focus** of the mirror, and we may designate the coordinate of this point as the **focal coordinate** f. We see that by substituting ∞ or $-\infty$ for u in Eq. (36.3) the coordinate v of the image, which by definition is the focal coordinate f, becomes $r/2$. Thus we may write the mirror equation in the form

$$\frac{1}{v} + \frac{1}{u} = \frac{1}{f}, \tag{36.4}$$

with f equal to $r/2$.

From Eq. (36.4) we see that if either u or v is made infinite, the other of these quantities becomes f, or $r/2$. This means that not only (1) incident parallel light (light parallel to the axis) forms an image at the principal focus, but also (2) incident light diverging from or converging toward the principal focus, as the case may be, is reflected as a parallel beam.

A concave mirror has a positive focal coordinate if it faces in the direction taken as positive (for example, in the positive x direction), and a negative if it faces in the negative direction. For a convex mirror facing in the positive direction the focal coordinate is negative, and if it faces in the negative direction the focal coordinate is positive.

The plane perpendicular to the axis that contains the principal focus is known as the **focal plane** of the mirror.

It is convenient also to give $|f|$, or $|r|/2$, the distance between a mirror and its principal focus, a distinctive name. The term commonly used for this is the **focal length** of the mirror. This, being simply a distance, is by its nature always positive, regardless of whether the mirror is concave or convex, and whichever way it is facing.

In addition to the principal focus of a mirror, any pair of points A and B such that a point object located at one gives a point image at the other are known as **conjugate foci.**

Images of Extended Objects

Let A and B in Fig. 36.8 be conjugate foci on the axis of a mirror, and let AA' and BB' be the arcs of circles centered on O, the center of curvature of the mirror, such that $A'B'$, as well as AB, passes through O. Then $A'B'$ may be thought of as a new axis, cutting the mirror at the point P, with respect to which the mirror is placed somewhat asymmetrically, there being a greater area of surface on one side than on the other. It will be seen at once that

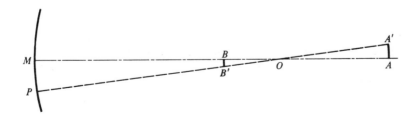

FIGURE 36.8

A' and B' must be conjugate foci on this new axis. Thus we may regard BB' as the image of AA'. The only effect of the asymmetrical disposition of the mirror about the axis $A'B'$ will be to make the spherical aberration at B' somewhat greater than at B, but if the aperture of the mirror is small this will not be a serious blemish. Also, if the angle MOP is small, the arcs AA' and BB' will approximate closely to straight lines perpendicular to the axis of symmetry OM; and it is usual to represent objects and images of finite size in this way, arrowheads being shown at A' and B', as in Fig. 36.9.

In Fig. 36.9, which corresponds to Fig. 36.3(a), the image is real and inverted. Similar constructions corresponding to parts (b) and (c) of Fig. 36.3 would show that in these cases, where the image is virtual, it is also erect. It is not difficult to see that the rule applies quite generally that where the object and its image are both of the same kind (both real or both virtual) the image is inverted, while if they are of opposite kinds the image is erect.

MAGNIFICATION   A partial definition of the **magnification** m of an image is the ratio of its linear dimensions to those of the object in a direction perpendicular to the axis. A full definition also includes reference to the relative orientations of image and object; if the image is erect (orientated similarly to the object) the magnification is said to be positive, while if it is inverted we call the magnification negative.

Considering the similar triangles $A'MA$ and $B'MB$ in Fig. 36.9, we see that

$$\frac{BB'}{AA'} = \frac{MB}{MA} = \frac{v}{u},$$

and that therefore the magnification, which is negative in this case, being equal to $-BB'/AA'$, may be expressed by the equation

$$m = -\frac{v}{u}. \tag{36.5}$$

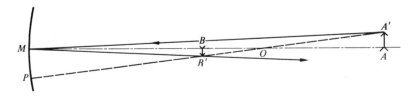

FIGURE 36.9   Inverted real image of real object produced by a concave mirror.

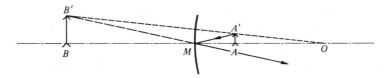

FIGURE 36.10   Erect virtual image of real object produced by a concave mirror.

It is easily seen that this equation applies to all cases. Let us, for example, consider the case shown in Fig. 36.10, which corresponds to Fig. 36.3(b). Here the virtual image BB' is erect, being similarly orientated to the object AA', and so according to our sign convention the magnification is positive. But, while u is positive, v is negative. Hence, considering the similar triangles $A'MA$ and $B'MB$, and applying the convention, we have, again,

$$m \; - \; - \; \frac{v}{u},$$

Refraction by a Prism

Let us now consider the passage of a ray through a prism of angle A, as shown in Fig. 36.11, the direction of the incident ray being perpendicular to the edge of the prism. Let θ be the angle of incidence on the first surface of the prism and ϕ the corresponding angle of refraction, while ϕ' is the angle of incidence inside the glass on the second surface and θ' the angle of refraction in the air outside. Then the total deviation D suffered by the ray in its passage through the prism is $(\theta - \phi) + (\theta' - \phi')$, or $(\theta + \theta') - (\phi + \phi')$. And, from geometry, we see that $\phi + \phi'$ must be equal to the angle of the prism, A.

A simple algebraical discussion shows that if ϕ were proportional to θ, and therefore ϕ' to θ' by the same factor, the deviation would

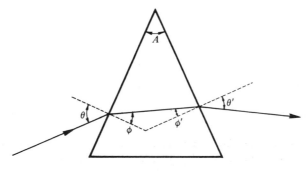

FIGURE 36.11   Reflection of light ray by prism.

be independent of θ. But, as we know, sin ϕ is proportional to sin θ, not ϕ to θ, and consequently D must vary with θ.

Now let us consider the case where the ray passes through the prism symmetrically, that is, where $\theta' = \theta$ ($= \theta_0$) and so $\phi' = \phi$, and let the corresponding deviation be D_0. Then if θ is made greater than θ_0, the deviation must be either increased or decreased; let us suppose for a moment that it is increased. But, as we see at once, if θ is increased beyond θ_0, θ' must be decreased below it. Since rays are reversible, this means that if an increase of θ increases D, a decrease of θ must also increase D; that is, the deviation must be a minimum for symmetrical passage of the ray through the prism. Correspondingly, if an increase of θ beyond θ_0 decreased D, the deviation would be a maximum for symmetrical passage of the ray. A more detailed discussion would show that it is actually a minimum, as is indicated by the graph of D against θ in Fig. 36.12.

Since at a maximum or a minimum the rate of change of a quantity is zero, D must be almost independent of θ over an appre-

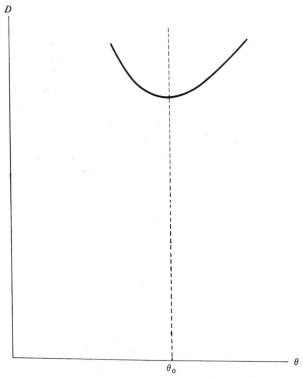

FIGURE 36.12   Deviation produced by a prism as function of angle of incidence of ray on first face of prism.

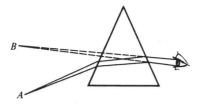

FIGURE 36.13   Image formation by a prism.

ciable range of values of θ in the region of θ_0. Hence all rays constituting a narrow monochromatic pencil incident on a prism whose angles of incidence are close to θ_0 will emerge from the prism with the angles they make with one another practically unchanged. If, therefore, this narrow pencil diverges from a source A on one side of the prism, and, after refraction through it, enters the eye of an observer on the other side, it will appear to have come from a point B at about the same distance from the prism, as shown in Fig. 36.13; B will be the image—a virtual image—of A.

Of course, a point image of A will be seen only if the light is monochromatic, or, alternatively, if the observer views it through a monochromatic filter. Otherwise the image seen will be drawn out into a spectrum.

Dispersive Power

We have just seen that in its passage through a prism a ray of heterogeneous light is broken up into a set of rays traveling in different directions, corresponding to the different wavelengths making up the original incident light. Such separation of wavelengths, whether brought about by passage through a prism or otherwise, is known as **dispersion.**

Dispersion brought about by refraction depends quantitatively on the variation of refractive index with wavelength, and is best discussed in terms of a property of a material known as its **dispersive power,** ω. This is designed to be a measure of the ability of a suitable device, for example a prism, to produce angular separation of different wavelengths per unit mean deviation. At the same time it is desirable that this shall be a *material* constant only; we do not want it to be dependent also on the angle of any particular prism or the orientation of such a prism with respect to the incident light.

945

We begin, by way of trial, by defining dispersive power in accordance with these ideas:

$$\omega = \frac{D_b - D_r}{D},$$

where D_b, D_r, and D are the deviations of rays of wavelengths λ_b, λ_r, and λ. Here λ is a convenient "standard" wavelength, selected, for example, out of the brightest part of the visible spectrum, and λ_b and λ_r are two other selected wavelengths, one on either side of λ. Thus λ might be the wavelength of sodium yellow, 5893 Å, λ_b might refer to the blue F line of the hydrogen spectrum, and λ_r might be the wavelength of the red C line, also of the hydrogen spectrum.

These deviations could be produced by a prism. But then, with ω defined as above, the value of this quantity would depend on the angle of the prism, and so not be a pure material constant. It would, however, become a material constant if the prism were thin, and all angles were small, with light passing symmetrically through the prism. For, in this case, the equation expressing the relation between the deviation for symmetrical passage of a ray of light through the prism, the angle of the prism, and the refractive index of its material, namely,

$$n = \frac{\sin\left[(A + D)/2\right]}{\sin\left(A/2\right)},$$

becomes simplified by substitution of the angles themselves for their sines to

$$n = \frac{A + D}{A} = 1 + \frac{D}{A},$$

or

$$D = (n - 1)A. \tag{36.6}$$

Thus each deviation would be proportional to the corresponding $n - 1$, and so for $(D_b - D_r)/D$ we should have

$$\frac{D_b - D_r}{D} = \frac{(n_b - 1) - (n_r - 1)}{n - 1};$$

that is,

$$\omega = \frac{n_b - n_r}{n - 1},$$

or, if we write δn for $n_b - n_r$,

$$\omega = \frac{\delta n}{n - 1}. \tag{36.7}$$

946   We adopt this, then, as our definition of dispersive power.

ACHROMATIC PRISM COMBINATION As a simple example
of the use of dispersive powers in the design of optical equipment,
let us consider the condition that has to be satisfied for a combination of two prisms constructed of different materials (for example,
kinds of glass) and oppositely orientated to produce deviation without
dispersion. For simplicity we shall suppose that the prisms are thin,
so that $(D_b - D_r)/D$ is, for each prism, independent of its angle.
Using subscripts 1 and 2 for the two prisms, we then have, from the
defining equation for ω,

$$D_{1b} - D_{1r} = \omega_1 D_1$$

and

$$D_{2b} - D_{2r} = \omega_2 D_2.$$

For deviation without dispersion the condition to be satisfied
is that the dispersions of the two oppositely orientated prisms shall
cancel each other out, that is, that

$$D_{1b} - D_{1r} = D_{2b} - D_{2r},$$

or

$$\omega_1 D_1 = \omega_2 D_2,$$

which, on substitution of the values of D_1 and D_2 in terms of prism
angles and the refractive indices for the standard wavelength as
expressed by Eq. (36.6), becomes

$$A_1(n_1 - 1)\omega_1 = A_2(n_2 - 1)\omega_2.$$

Thus the condition for achromatism is that

$$\frac{A_1}{A_2} = \frac{(n_2 - 1)\omega_2}{(n_1 - 1)\omega_1}. \tag{36.8}$$

Problems

36.1 A 5-cm plane mirror is just large enough to show the whole
image of a tree when held 30 cm from the eye of an observer whose distance from the tree is 100 m. What is the height of the tree?

36.2 What must be the minimum height of a plane mirror for a man
of height h to see a full-length image of himself? At what height above the
floor must the lower edge of this mirror be?

36.3 Two parallel-plane mirrors face each other separated by a
distance $4d$. An object is placed between them and at a distance d from one
of them. At what distances from the two mirror surfaces do successive
images of the object occur?

36.4 A concave mirror is set up 20 cm from a lamp filament and an
inverted image is formed which has three times the linear dimensions of the
object. What is the radius of curvature of the mirror? Where must the

object be placed in order that an image shall be produced which is the same size as before, but erect?

36.5 At what distance must an object be placed from a concave mirror of focal length $|f|$ to produce a magnification of m, where m is (a) positive, and (b) negative. In both cases state whether the image is real or virtual. Apply your results to the specific case where $|f| = 20$ cm.

36.6 A concave mirror has a radius of curvature R and is used to produce an image of an object having a magnification of (a) -2, and (b) $-\frac{1}{2}$. How far must the object be placed from the mirror in these two cases?

36.7 An object of height 4 cm is placed 15 cm in front of a convex mirror of radius of curvature 20 cm. Locate, by geometrical construction and by calculation, the position of the image, and find its magnification.

36.8 An object is (a) 40 cm, (b) 20 cm, (c) 5 cm in front of a convex mirror of radius of curvature 20 cm. Where is the image in each case, and what is the magnification?

36.9 A convergent pencil of light is directed toward a point 10 cm behind the mirror of Problem 36.8. Find, by geometrical construction and by calculation, at what point the reflected rays come to a focus.

36.10 Let x be the coordinate of a point on the axis of a spherical mirror. Find how $(dv/dx)/(du/dx)$ is related to the magnification m of an image. Interpret your result physically.

36.11 A concave mirror A of focal length 10 cm and a convex mirror B of focal length 25 cm have a common horizontal axis and face each other separated by a distance of 20 cm. An object of height 4 cm is placed at right angles to the axis and at a distance of 15 cm from A. Find the position, size, and nature of the image formed by reflection first at the surface of A and then at the surface of B. Is this image the same as, or different from, that formed by rays reflected first from B and then from A?

36.12 Light reflected from the inner surface of a filled teacup gives rise to a pattern on the top surface of the tea known as a caustic curve. Explain physically how a curve of this nature is formed. Outline the mathematical procedure for finding the equation of the caustic curve, and then work out the solution as far as you are able.

36.13 An achromatic prism combination is to be constructed of crown- and flint-glass components. The crown glass has refractive indices for light of wavelengths 6563 Å (red), 5893 Å (yellow), and 4861 Å (blue) of 1.5146, 1.5171, and 1.5233, and the corresponding refractive indices of the flint glass are 1.6224, 1.6272, and 1.6385, respectively. The crown-glass prism to be used has an angle of 10°. What must be the angle of the flint-glass prism?

LENSES AND OPTICAL INSTRUMENTS

Refraction at a Spherical Surface

Let us again suppose that we are dealing with monochromatic light, and let us investigate the formation of an image of a point object by refraction through a spherical surface of separation between two media of different refractive indices.

There are, in all, 16 independent possible cases. The light may travel from an optically less dense into a denser medium, or from a denser into a less dense.* Both the object and the image may be either real or virtual. And the refracting surface may be either concave or convex toward the incident light. Of all the possible cases, let us confine ourselves to the consideration of two only, shown in parts (a) and (b) of Fig. 37.1, which differ substantially from one another.

In both the cases selected the following applies:

$$\Delta APO: \quad \frac{\sin \alpha}{OA} = \frac{\sin \angle AOP}{PA},$$

$$\Delta BOP: \quad \frac{\sin \beta}{BO} = \frac{\sin \angle POB}{BP} = \frac{\sin \angle AOP}{BP};$$

whence

$$\frac{\sin \alpha}{\sin \beta} \cdot \frac{BO}{OA} = \frac{BP}{PA},$$

or, since $\sin \alpha / \sin \beta = n_{12}$,

$$n_{12} \frac{PA}{BP} = \frac{OA}{BO}. \tag{37.1}$$

From this it follows, as in the case of a spherical mirror, that B is not independent of the position of P, that is, that a point object does not produce a point image. Nevertheless, if only small angles are involved, spherical aberration is correspondingly small, and so we may write, as an acceptable approximation to Eq. (37.1),

$$n_{12} \frac{LA}{BL} = \frac{OA}{BO}. \tag{37.2}$$

---

* By an optically dense medium is meant one that has a high refractive index.

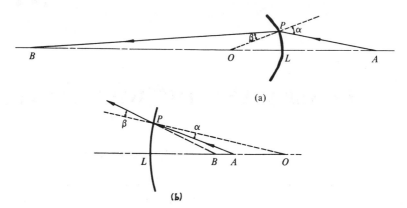

(a)

(b)

FIGURE 37.1 Two cases of refraction at a spherical boundary between two media.

Let us now express the distances in Eq. (37.2) in terms of the coordinates u, v, and r. Thus, in case (a),

$$LA = u,$$
$$BL = -v,$$
$$OA = OL + LA = -r + u,$$

and

$$BO = BL - OL = -v - (-r)$$
$$= -v + r,$$

so that Eq. (37.2) becomes

$$n_{12} \frac{u}{-v} = \frac{-r + u}{-v + r},$$

which simplifies to

$$\frac{n_{12}}{v} - \frac{1}{u} = (n_{12} - 1) \frac{1}{r}, \qquad (37.3)$$

or, because of Eq. (35.6),

$$\frac{n_2}{v} - \frac{n_1}{u} = (n_2 - n_1) \frac{1}{r}. \qquad (37.4)$$

Similarly, in case (b),

$$LA = u,$$
$$BL = v,$$
$$OA = LO - LA = r - u,$$

and

$$BO = LO - LB = r - v.$$

Equation (37.2) therefore becomes

$$n_{12} \frac{u}{v} = \frac{r - u}{r - v},$$

which, as before, simplifies to

$$\frac{n_{12}}{v} - \frac{1}{u} = (n_{12} - 1)\frac{1}{r}. \qquad (37.3)$$

Analysis along similar lines would show that also in all the other of the 16 possible cases the appropriate substitutions in Eq. (37.2) yield the same formulas, (37.3) and (37.4).

It is clearly to be understood that n_1 and n_2 in Eq. (37.4) stand for the refractive indices in the first and second media, respectively, traversed by the light. If, adhering to this convention, we reverse the direction of the light, we must interchange the subscripts 1 and 2 in the symbols representing the refractive indices of the two media.

On this basis, Eq. (37.4) is valid regardless of the direction of travel of the light; it applies for light traveling from left to right as well as for light traveling from right to left. That this is so may at once be seen as follows. In either of the cases considered let the rays be reversed in direction. Then what was formerly the image becomes the object, and vice versa. Also the symbol for refractive index previously having the subscript 2 must now be given the subscript 1, and vice versa. And, of course, the coordinate of the center of curvature of the refracting surface is not affected. To distinguish between the cases where the light travels from left to right from that where it travels from right to left, let the symbols used for the former case be primed. Then we have the following quantitative relations between the two sets of symbols:

$$v = u', \ u = v', \ n_2 = n_1', \ n_1 = n_2', \ r = r'.$$

Hence, substituting for the symbols in Eq. (37.4) their equivalents in terms of primed symbols, we have

$$\frac{n_1'}{u'} - \frac{n_2'}{v'} = (n_1' - n_2')\frac{1}{r'},$$

or, reversing the order of the terms on both sides,

$$\frac{n_2'}{v'} - \frac{n_1'}{u'} = (n_2' - n_1')\frac{1}{r'}.$$

We see that this is of precisely the same form as Eq. (37.4).

In the limiting case of a plane surface ($r = \infty$), Eqs. (37.3) and (37.4) reduce to

$$v = n_{12}u \qquad (37.5)$$

and its equivalent

$$n_1 v = n_2 u, \qquad (37.6)$$

respectively. Thus for an object under water, whose refractive index with respect to air, n_{aw}, is $\frac{4}{3}$ (approximately), n_{wa}, corresponding to n_{12} in Eq. (37.5), is $\frac{3}{4}$, and so $v = \frac{3}{4}u$; the apparent depth is three quarters of the real depth.

Thin Lens

In a thin lens, two refractions at spherical surfaces occur in close succession, and the corresponding thin-lens formula is most conveniently obtained by using Eq. (37.3) twice in succession. The image formed by the first refracting surface is regarded as the object for refraction at the second surface.

Lenses are most commonly used with air on either side. However, they can also be, and sometimes are, used in other media. We shall here consider the general case where the lens is immersed in *any* medium, not necessarily in air. The medium in which the lens is immersed we shall call medium 1 and that of which the lens is constructed medium 2, and we shall develop a thin-lens formula in terms of the geometry of the lens and the refractive index n_{12} for light traveling from medium 1 to medium 2.

Neglecting the thickness of the lens in comparison with the other distances with which we shall be concerned, and taking the location of the lens as origin, let us write w for the coordinate of the image formed by the first refraction, which serves as object for the second. And let r_1 be the coordinate of the center of curvature of the first refracting surface encountered by the light, and r_2 that of the second. Then, u and v being the coordinates of the original object and final image points, respectively, we have, from Eq. (37.3), for the first refraction

$$\frac{n_{12}}{w} - \frac{1}{u} = (n_{12} - 1)\frac{1}{r_1}, \tag{37.7}$$

and

$$\frac{n_{21}}{v} - \frac{1}{w} = (n_{21} - 1)\frac{1}{r_2}, \tag{37.8}$$

for the second. Let us now substitute $1/n_{12}$ for n_{21} in Eq. (37.8). This gives us

$$\frac{1}{n_{12}v} - \frac{1}{w} = \left(\frac{1}{n_{12}} - 1\right)\frac{1}{r_2},$$

which, on multiplication by n_{12} throughout, becomes

$$\frac{1}{v} - \frac{n_{12}}{w} = (1 - n_{12})\frac{1}{r_2}. \tag{37.9}$$

Finally, elimination of w between (37.7) and (37.9) by addition of these two equations gives us the thin-lens formula,

$$\frac{1}{v} - \frac{1}{u} = (n_{12} - 1)\left(\frac{1}{r_1} - \frac{1}{r_2}\right). \qquad (37.10)$$

Like Eq. (37.3) on which it is based, Eq. (37.10) is valid for either direction of travel of the light. In applying it, however, it must always be remembered that r_1 stands for the coordinate of the center of curvature of the first lens surface encountered by the light and r_2 for that of the second; and, of course, which is first and which second depends on the direction in which the light traverses the lens.

In the usual case where the lens is in air,* we may drop the subscripts from n_{12}, simply writing n for the refractive index of the glass or other material of which the lens is constructed. The equation then becomes, simply,

$$\frac{1}{v} - \frac{1}{u} = (n - 1)\left(\frac{1}{r_1} - \frac{1}{r_2}\right). \qquad (37.11)$$

As in the case of the spherical mirror, the locations of the points whose coordinates are u and v are known as conjugate foci. And the point where an incident set of paraxial rays (rays parallel to the axis) is brought to a focus, or the point from which, after refraction through the lens, the rays appear to have come, is called the principal focus. The focal coordinate f of this is obtained by giving u in the formula (37.10) the value of plus or minus infinity. Thus

$$\frac{1}{f} = (n_{12} - 1)\left(\frac{1}{r_1} - \frac{1}{r_2}\right). \qquad (37.12)$$

Accordingly, the lens formula is commonly written in the form

$$\frac{1}{v} - \frac{1}{u} = \frac{1}{f}. \qquad (37.13)$$

There is never any difficulty concerning the sign of f. It has the same sign as the right side of Eq. (37.12), in which r_1 and r_2 are the coordinates of the centers of curvature of the first and second surfaces encountered by the light. Alternatively, it may be ascertained very simply in any given case for the direction of the light used where it is known whether the lens is converging or diverging. If the lens is converging, incident parallel light must be brought

---

* As we see from the first item in Table 35.1, air is optically practically indistinguishable from a vacuum, so we may consider this case as that of a lens in a vacuum.

to a focus after passage through the lens, and so f is positive or negative according as the light traverses the lens in the positive or negative direction. On the other hand, if it is a diverging lens, the light, on emerging from it, will appear to have come from a point on the side where the light was incident, so that f will be positive or negative according as the light traverses the lens in the negative or in the positive direction.

The plane perpendicular to the axis that contains the principal focus of a lens is called its focal plane.

As in the case of a mirror, we designate $|f|$ as the focal length of the lens.

We have defined the principal focus as the point where an incident set of paraxial rays is brought to a focus, or the point from which, after refraction of such a set of rays through the lens, they appear to have come. This was for a particular direction of the light. Actually, since there are two possible such directions, there are, correspondingly, two principal foci. As is evident from the defining equation (37.12) for f, these are located at equal distances from, but on opposite sides of the lens, each functioning for one direction of the incident light.

Considering this in conjunction with the reversibility of rays, we see, furthermore, that for a given direction of the light a ray incident parallel to the axis (an incident "paraxial ray") is refracted through one principal focus, while an incident ray passing through the other principal focus is refracted parallel to the axis.

A ray passing through the center of the lens, where it is either thickest or thinnest, is obviously undeviated, the two lens surfaces here being parallel to one another.

EFFECT OF TILTING A LENS   A thin diametral slice cut from a lens such that the planes of the cuts are parallel to the axis may be regarded as a series of short sections of prisms, the angle of the prism changing from one member of the series to the next, as indicated in Fig. 37.2. In the case of a thin lens of small aperture the light passing through each such prism between conjugate foci on the axis must do so at least approximately symmetrically, corresponding to minimum deviation. Now let us imagine the lens to be tilted slightly about an axis through its center and perpendicular to the axis of the lens. This cannot affect appreciably the deviation of any incident ray, and since the shift in the portion of the lens producing this deviation is also small, practically the same image will be formed as before. There will also be no appreciable increase in spherical aberration.

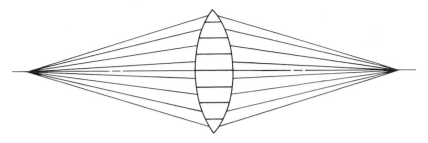

FIGURE 37.2 Action of lens considered as a series of prisms.

Since the rotation of the lens that we have envisaged involves a corresponding rotation of its axis, the point object and approximate point image under discussion are no longer on this axis. However, the line joining them still passes through the center of the lens, and the points are both still at the same distance from this center.

OBJECTS AND IMAGES OF FINITE SIZE   Let A and B in Fig. 37.3 be conjugate foci on the axis of a lens, and let AA' and BB' be arcs of circles centered on L, the center of the lens, such that $A'B'$, as well as AB, passes through this point. Then if these arcs are small relative to their radii, B' may be regarded as the image of A', the lens being tilted slightly with respect to the line $A'B'$. Thus BB' is the image of AA', and both approximate closely to straight lines perpendicular to AB. And, from geometry, it is clear that the magnification m is given with its correct sign by the ratio v/u. This is easily shown to be valid in all cases, regardless of whether the lens is converging or diverging or the object or image is real or virtual.

We see that the image is erect or virtual according as v and u have the same or opposite signs; that is, according as one is real and the other virtual, on the one hand, or both are real or both virtual, on the other.

POWER OF A LENS   The power k of a lens may be defined as $1/f$, corresponding to the case where the light traverses the lens

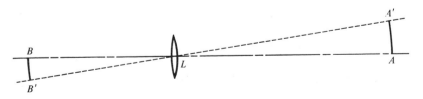

FIGURE 37.3 Real inverted image of real object formed by converging lens.

in the direction taken as positive, f being measured in meters. This definition makes the power positive for a converging lens and negative for a diverging. These are the signs used by opticians. The unit is the **diopter.**

Two Thin Lenses in Contact

These can be treated algebraically in the same way as we treated two adjacent refracting surfaces in deriving the thin-lens formula; that is, the image formed by the first lens traversed by the light is considered as functioning as the object for the second. Proceeding in this manner, we find that if f_1 and f_2 are the focal coordinates of the separate lenses for the direction in which the light travels through them, the object and image coordinates are given by the equation

$$\frac{1}{v} - \frac{1}{u} = \frac{1}{f_1} + \frac{1}{f_2};$$

(37.14)

that is, the combination behaves like a single thin lens of focal coordinate f such that

$$\frac{1}{f} = \frac{1}{f_1} + \frac{1}{f_2}.$$

(37.15)

This means that the power of the combination is equal to the sum of the separate powers of the lenses.

ACHROMATIC COMBINATION Let the powers of two thin lenses in contact in air, which are to function as an achromatic combination, be k_1 and k_2, and let ω_1 and ω_2 be the dispersive powers of the corresponding kinds of glass for the colors we wish to combine. Then for each of these colors

$$\frac{1}{v} - \frac{1}{u} = \frac{1}{f_1} + \frac{1}{f_2} = k_1 + k_2.$$

The condition for achromatism is that, for a given position of the object, v shall be the same for both colors, that is, that

$$\delta(k_1 + k_2) = 0.$$

(37.16)

To find an expression for the change δk in the power of a lens corresponding to a given change in wavelength, we must make use of the general expression for the power k. From the defining equation

$$k = \frac{1}{f} = (n - 1)\left(\frac{1}{r_1} - \frac{1}{r_2}\right)$$

for this it follows that

$$\delta k = \delta n \left(\frac{1}{r_1} - \frac{1}{r_2} \right).$$

But from Eq. (36.7) we see that δn is equal to $\omega(n - 1)$. Hence, making the corresponding substitution, we have

$$\delta k = \omega(n - 1) \left(\frac{1}{r_1} - \frac{1}{r_2} \right)$$

or

$$\delta k = \omega k. \tag{37.17}$$

On now substituting $\omega_1 k_1$ for δk_1 and $\omega_2 k_2$ for δk_2 in Eq. (37.16), we obtain the condition for achromatism,

$$\omega_1 k_1 + \omega_2 k_2 = 0,$$

or

$$\frac{k_1}{k_2} = - \frac{\omega_2}{\omega_1}. \tag{37.18}$$

We may note that the negative sign in this equation implies that one lens must be converging and the other diverging. That this must be so is also obvious physically.

Applications of Mirror and Lens Formulas

In addition to the applications we have already made of the mirror and lens formulas, the following will serve as examples of the great convenience and versatility of the algebraical method of solving optical problems.

PRESCRIPTION OF SPECTACLE LENSES FOR DEFECTS OF VISION A spectacle lens placed in front of a defective eye should in general cause an actual object, which the eye is unable to see properly, to be replaced by a virtual image of this object which, functioning as an object as far as the eye is concerned, it *can* see clearly. If the eye is short-sighted, the spectacle lens, to be effective, must produce an image of a distant object that is sufficiently close to the eye for clear vision. On the other hand, if the eye is long-sighted, the image produced by the lens must be farther away than the object. In an extreme case this might even have to be, in an optical sense, "beyond infinity"; that is, the lens might be required to produce converging light, converging toward a point behind the eye. In this case the lens no longer produces a virtual image; if the eye did not intervene the image produced would be real. But because the eye does intervene, its optical system has to deal with a virtual object (behind the eye), producing a real image of this on the retina.

957

Let us, for example, consider the case of an eye which, while able to cast a sharp image on the retina of all objects in the range of distances between 12 and 30 cm from the eye, is unable to see clearly any object beyond 30 cm. The problem here is to find a lens that will produce an image of the most distant object possible, one at infinity, which is 30 cm in front of the eye. This is found by using the lens formula and, taking the direction away from the eye as the positive direction, substituting ∞ for u and 30 cm for v. With these substitutions we have

$$\frac{1}{f} = \frac{1}{30} - \frac{1}{\infty},$$

from which we see at once that f must be 30 cm. The positive sign shows that the lens must be diverging.

We may now inquire what will be the shortest distance of an object that the eye will be able to see clearly through this lens. The answer to this question is obtained by substituting 12 cm for v and 30 cm for f in the lens formula. This gives us

$$\frac{1}{u} = \frac{1}{v} - \frac{1}{f} = \frac{1}{12} - \frac{1}{30},$$

from which we find that $u = 20$ cm; the least distance for distinct vision of an object viewed through the lens is 20 cm.

COMBINATION OF MIRROR AND LENS   Let us consider the following problem. A concave galvanometer mirror has a radius of curvature of 1 m, and this is to be used in the usual lamp-and-scale arrangement under two different conditions. In one the lamp and scale are both to be 50 cm from the mirror, while in the other they are to be 2 m away. It is required to find what lenses must be placed directly in front of the mirror in the two cases in order that the mirror–lens combination shall produce a sharp image of the lamp on the scale.

Let the focal coordinate of the required lens be f for light traveling in the direction toward the mirror. Then for light traveling away from the mirror its focal coordinate will be $-f$. Let the coordinate of the image formed by the lens on the first passage of the light through it be w. This will function as the object for reflection by the mirror, which latter will produce an image whose coordinate we shall call w'. Finally, this image will function as an object for refraction through the lens the second time.

Writing u and v for the coordinates of the original object and the final image, respectively, and r for that of the center of curvature of the mirror, we have for the first passage of the light through the lens

$$\frac{1}{w} - \frac{1}{u} = \frac{1}{f},$$

for reflection by the mirror

$$\frac{1}{w'} + \frac{1}{w} = \frac{2}{r},$$

for the second passage of the light through the lens

$$\frac{1}{v} - \frac{1}{w'} = -\frac{1}{f},$$

and, expressing the fact that the lamp and scale are at the same distance from the mirror–lens combination,

$$v = u.$$

Eliminating w, w', and v between these four equations, we find for f in terms of r and u,

$$\frac{1}{f} = \frac{1}{r} - \frac{1}{u}.$$

Finally, substituting for r its value, 1 m, and for u its two alternative values, (a) 0.5 m and (b) 2 m, we obtain for the corresponding values of f: (a) $f = -1$ m, and (b) $f = 2$ m. This means that for the 50-cm distance the lens must be converging and of focal length 1 m, while for the 2-m distance it must be diverging and of focal length 2 m.

Geometrical Constructions

It is always possible to find the image of a given object due to reflection at the surface of a mirror or refraction through a lens, either by an algebraical procedure, or geometrically, by an appropriate ray construction. In the algebraical method the mirror or lens formula may be used to find the position of the image, and then, by the additional application of the formula for the magnification, the entire image can be constructed, correct in size and orientation. However, it is often more convenient to obtain the image by a geometrical construction, and it is this that we shall now consider.

In the case of the mirror, rays incident on the mirror may be drawn from a selected point on the object and the point then found where the reflected rays intersect, either actually or virtually.

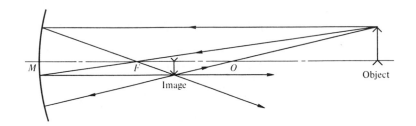

FIGURE 37.4 Geometrical construction for determination of image of extended object produced by a concave mirror.

Obviously two such incident and reflected rays are all that are needed to locate the image point. For the construction of these, application can be made of any two of the following three principles:

1. An incident ray passing through the center of curvature of the mirror is reflected directly back along the same path.

2. An incident paraxial ray is reflected through the principal focus.

3. An incident ray through the principal focus is reflected parallel to the axis.

An example of this procedure, in which all three rays are shown, is given Fig. 37.4.

In the case of a lens, also, any two of three rays may be drawn to locate an image point, according to the following principles:

1. An incident paraxial ray is refracted through the principal focus corresponding to the direction of incidence.

2. An incident ray through the other principal focus (through which paraxial rays traveling the other way would be refracted) is refracted parallel to the axis.

3. A ray incident at the optical center of the lens passes through undeviated.

An example is given in Fig. 37.5, again with all three rays shown.

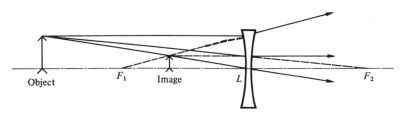

FIGURE 37.5 Geometrical construction for determination of image of extended object produced by a diverging lens.

Optical Instruments

The instruments we shall consider are the simple and compound microscopes, astronomical refracting and reflecting telescopes, and the prism spectrometer.

SIMPLE MICROSCOPE OR MAGNIFYING GLASS   We may, for simplicity, consider the eye as equivalent to a camera, with the retina replacing the plate or film. We may also, at least in the first instance, think of the magnifying glass as being so close to the eye that, in effect, it becomes part of the lens system of this equivalent camera. Because of the considerable reduction in the focal length of the system resulting from the addition of this extra lens, any object that is to cast a sharp image on the retina must be much closer to the eye than in the absence of this lens. A given object accordingly subtends a much larger angle at the center of the lens system and so produces an image on the retina that is larger in the same proportion than when the same object is viewed without the magnifying glass, hence the "magnification."

The normal eye sees most comfortably objects at a considerable distance. The rays from each point of such an object enter the eye as practically a parallel beam and are brought to a sharp focus on the retina with the eye relaxed. Ideally, then, a magnifying glass should be used in such a way as to simulate this condition: that is, the virtual image produced by the lens should be at infinity. This means that the lens should be held its focal distance $|f|$ in front of the object. When this is done, the size of the real image cast on the retina does not depend on whether the eye is held close to the lens or is some distance back. This will be realized from an inspection of Fig. 37.6(a), in which, for simplicity, the eye is shown as a "diagrammatic eye."

In the absence of the magnifying glass the largest *sharp* image would be cast on the retina with the object held the least distance for distinct vision D away, as shown in Fig. 37.6(b). Particular

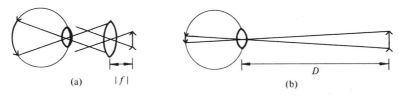

(a) $|f|$ (b) D

FIGURE 37.6   Basic principle of action of a magnifying glass shown for the special case where the object is at the principal focus of the lens.

961

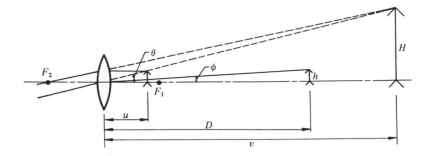

FIGURE 37.7 Principle of action of a magnifying glass for general case.

interest attaches to the ratio of sizes of images cast on the retina (1) with the magnifying glass under conditions of most comfortable viewing and (2) with the unaided eye accommodated to the maximum extent for close viewing. This ratio, which is sometimes loosely referred to as the "magnification," is more properly called the **magnifying power** (MP) of the lens used as stated.

It is possible to reduce the distance of the object from the lens slightly below $|f|$ and to compensate for the resulting divergence of the sets of rays entering the eye by an effort of accommodation, thickening the crystalline lens of the eye to obtain a sharp image on the retina. This image will be slightly larger than that produced under relaxed-eye conditions with the object held the distance $|f|$ from the lens. This more general case, which is shown in Fig. 37.7, will now be discussed, the lens being assumed to be close to the eye.

The magnifying power must clearly be the ratio of the angles subtended by the object at the lens or eye (1) when viewed through the lens and (2) when viewed under the most favorable conditions without the lens, that is, at the least distance for distinct vision. Thus

$$\text{MP} = \frac{\theta}{\phi} = \frac{H/v}{h/D} = \frac{h/u}{h/D} = \frac{D}{u}.$$

By using the lens formula we may express this in terms of D, v, and f instead of in terms of D and u:

$$\text{MP} = \frac{D}{v} - \frac{D}{f}, \qquad (37.19)$$

or, since f is negative, we may, if we prefer, express it in terms of positive quantities only:

$$\text{MP} = \frac{D}{|f|} + \frac{D}{v}. \qquad (37.20)$$

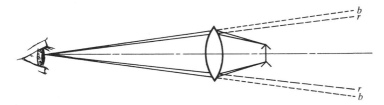

FIGURE 37.8 Chromatic aberration produced by magnifying glass held some distance from observer's eye.

The image may be seen clearly for all values of v between D (the smallest) and infinity, corresponding to a certain range of u below $|f|$, with $|f|$ as the upper limit. The corresponding values of the magnifying power will range between $1 + D/|f|$ and $D/|f|$, the latter being the value for relaxed-eye conditions of viewing.

To see whether it is really worthwhile to make the necessary effort of accommodation to obtain the larger value of $1 + D/|f|$, let us consider the example of a lens of focal length $|f| = 5$ cm. For this lens the range of values of magnifying power available to a person whose least distance for distinct vision is 25 cm would be between 6 and 5. Even for this relatively weak lens, the relaxed-eye magnifying power falls very little short of that for the eye accommodated to its maximum possible extent; the extra comfort gained by working with the eye relaxed is purchased at only a trifling cost in magnifying power.

Although the distance of the eye from the lens has no effect on the magnifying power, there is an important dependence of both the quality of the image seen and its extent (field of view) on this distance.

One important effect on the quality is shown in Fig. 37.8: the production of an image with colored edges. With the eye held a considerable distance from the lens the only rays from the outer parts of the object that enter the eye are those that have passed through the outer parts of the lens. Here the "equivalent prism" has a large angle, and there is a correspondingly large angular separation of rays of different wavelengths. Consequently, red and blue rays that originate from the same point in the object enter the eye in different directions, giving an image with a blue border. This is avoided by holding the eye close to the lens and so receiving only rays that have passed through its central region. Under these conditions there is no perceptible coloration of the image. At the same time other lens aberrations, such as image distortion, are avoided.

963

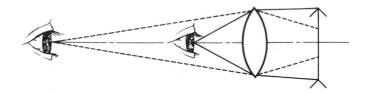

FIGURE 37.9 Progressive diminution of field of view with increasing recession of eye from lens.

A further advantage to be gained by working with the eye close to the lens is that this gives a large field of view; the area of an object in the focal plane contributing rays that enter the eye increases as the eye approaches the lens. That this is so will be evident from Fig. 37.9, which is self-explanatory.

COMPOUND MICROSCOPE A compound microscope differs from a simple microscope in that, instead of the object being viewed directly through a magnifying glass, an enlarged real image of the object is first produced, as in an enlarging camera, and a magnifying glass is then used to view this. A primitive form of this instrument is shown in Fig. 37.10. The first lens *O*, which produces the enlarged real image of the object, is known as the **objective,** and the second lens *E*, which is used as a magnifying glass, is called the **eyepiece.**

In practical compound microscopes both the objective and the eyepiece are lens *systems* rather than simple lenses. This is to reduce aberrations to a minimum. It is shown in more advanced optical studies that any sequence of lenses behaves, in certain respects, like a single thin lens. In the basic theory of the compound microscope we are concerned only with those features of the behavior of objectives and eyepieces that they have in common with single thin lenses. Hence Fig. 37.10, besides depicting a primitive form of the instrument, may also be regarded as a diagrammatic representation of actual, high-quality instruments employing more elaborate designs of objectives and eyepieces.

Whereas in low-power microscopes the objective is commonly made in the form of a simple cemented achromatic combination, in instruments of higher power the system is more complex, consisting of two cemented combinations plus, in some cases, single lenses, placed in such a sequence and with such spacings apart as to reduce aberrations to a minimum. The whole series is designed to produce a real magnified high-quality image of the object at a particular distance beyond the last lens. Four objective systems, with powers increasing from left to right, are shown in Fig. 37.11.

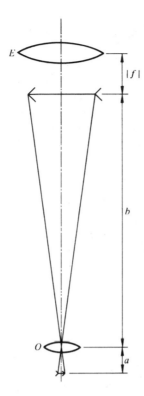

FIGURE 37.10 Principle of compound microscope.

The narrow cones of rays proceeding from the points of a real image formed by the objective of a microscope or a telescope are all roughly parallel to the axis. Because of this circumstance the conditions for optimum viewing are quite different from those applying to an actual object from which rays proceed in all directions. If a single lens were used as the eyepiece in one of these instruments, and the

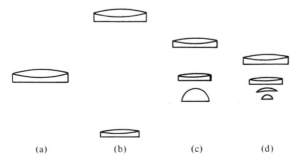

(a)   (b)   (c)   (d)

FIGURE 37.11 Various types of objective lens systems used in compound microscopes.

eye were held close to the lens, only rays from the central area of the real image would enter the eye. To obtain the maximum field of view the eye would have to be moved back from the lens a distance equal to its focal length; but then the rays received from the outer parts of the image, having passed through the peripheral parts of the lens, would give serious chromatic aberration. To achieve both maximum field of view and achromatism with the eye held in a single position it is necessary to use a compound eyepiece in a microscope or telescope. This consists of two lenses appropriately spaced apart from each other.

Referring to Fig. 37.10, we see that the magnification produced by the objective is b/a. Hence, using in conjunction with this the expression for the magnifying power of a simple microscope, we obtain for the overall magnifying power of the compound microscope the expression

$$\text{MP} = \frac{b}{a}\frac{D}{|f|},\tag{37.21}$$

where $|f|$ is the focal length of the eyepiece.

ASTRONOMICAL REFRACTING TELESCOPE Just as a compound microscope is the optical equivalent of an enlarging camera and a magnifying glass, arranged in series, so an astronomical refracting telescope is the equivalent of the more usual form of camera, designed to photograph relatively distant objects, plus a magnifying glass.

The objective of the telescope is a cemented achromatic lens combination, and the eyepiece is the same as is used in a microscope. As in the case of the microscope, we shall discuss the basic theory of the instrument by treating both lens systems as if they were simple thin lenses. This idealized system of the astronomical telescope is shown in Fig. 37.12.

The magnifying power is the ratio of the angular size of a distant object as seen through the telescope to that seen without.

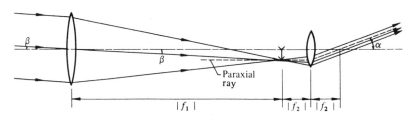

FIGURE 37.12 Principle of astronomical refracting telescope.

If the latter is the angle β shown in the figure, the former is α, and

$$MP = \frac{\alpha}{\beta}. \qquad (37.22)$$

If h is the linear size of the real image formed in the focal plane of the telescope, it is evident that α is equal to $h/|f_2|$ and β is $h/|f_1|$, where $|f_1|$ and $|f_2|$ are the focal lengths of the objective and eyepiece, respectively. Hence

$$MP = \frac{|f_1|}{|f_2|} = \frac{f_1}{f_2}. \qquad (37.23)$$

ASTRONOMICAL REFLECTING TELESCOPES We shall consider two forms of reflecting telescope: the Newtonian telescope, first constructed by Isaac Newton and still used, and a modification developed by Cassegrain. The latter has, for a given length, a higher magnifying power than the Newtonian, or, for a given magnifying power, it is shorter.

Newtonian Telescope This is shown in Fig. 37.13. It corresponds exactly to the astronomical refracting telescope except that the incident parallel rays from the distant object are converged toward a focus by a concave mirror instead of by a lens. The mirror is paraboloidal, this being the aplanatic form of reflecting surface for incident parallel rays. To avoid obstruction of the incoming light by the observer's head, the light is deflected through a right angle by reflection at the surface of a small plane mirror orientated at 45° to the axis, this being situated about the same distance in front of the focus of the objective mirror as this focus is from the side of the telescope tube. The real image is then viewed through an eyepiece whose axis is at right angles to the axis of the telescope.

If $|f_1|$ is the focal length of the paraboloidal mirror and $|f_2|$ that of the eyepiece, then for the magnifying power we have

$$MP = \frac{|f_1|}{|f_2|}.$$

Cassegrainian Telescope This telescope, shown in Fig. 37.14, employs two aplanatic surfaces: a paraboloidal main mirror, as in

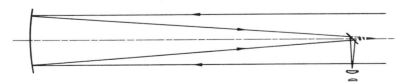

FIGURE 37.13 Newtonian reflecting telescope.

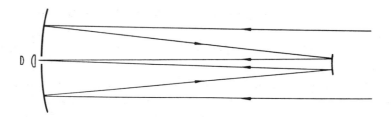

FIGURE 37.14   Cassegrainian reflecting telescope.

the Newtonian reflector, and a hyperboloidal mirror which intercepts the rays from the paraboloidal mirror before they come to a focus, the nearer focus of the hyperboloidal mirror coinciding with the focus of the paraboloidal mirror. The image formed by the paraboloidal mirror at its own focus thus serves as a virtual object, rays converging toward which are reflected by the hyperboloidal mirror to produce an enlarged image of this virtual object at its more remote focus. This final image is then viewed through an eyepiece situated just behind the main mirror, in which there is a small central hole to permit passage of the light.

The magnifying power of a Cassegrainian telescope is greater than that of an astronomical refracting or Newtonian telescope by the factor of the intermediate magnification given by the hyperboloidal mirror.

PRISM SPECTROMETER   Dispersion by a prism has already been noted, but without considering how the different wavelengths may be separated from one another in the most efficient manner with regard to economy of light and avoidance of overlapping. This is done by means of the prism spectrometer, which is shown diagrammatically in Fig. 37.15.

In the first place it is necessary to obtain a parallel beam to be directed onto the prism. Only if this is done will each ray of any given wavelength be deviated equally, so that the emergent beam is also parallel. If a convergent or divergent beam is passed through the prism the emergent system of rays will not be of a kind that can be focused to a point by any simple means.

This requirement means that we must have a concentrated source and that this be placed at the focal distance from an achromatic lens combination. Such a system is known as a **collimator.** In practice the source is usually a narrow slit illuminated from behind, and the slit and lens are mounted one at either end of a tube.

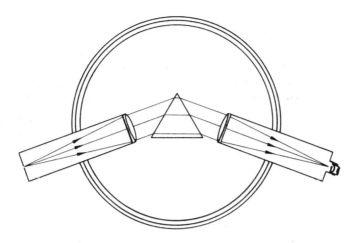

FIGURE 37.15 Prism spectrometer.

The prism converts a single beam of parallel rays of mixed wavelengths into sets of parallel rays, each set having its own wavelength and direction of travel. The telescope which follows, focused for infinity, then sorts out these sets, converting them into points in its focal plane at which the light is concentrated. In the usual arrangement where the source is an illuminated slit, the system of collimator, prism, and telescope produces in the focal plane of the telescope an array of colored images of the slit. The telescope is mounted so that it can be rotated about an axis passing through the prism and parallel to its "refracting edge." By suitably rotating the telescope, the observer is then able to make the image of the slit of any desired wavelength coincide with the cross wire. An arm attached to the telescope moves over a graduated circular scale, and, provided the instrument has been calibrated beforehand, this enables wavelengths to be read in terms of scale readings.

Calibration of the instrument requires that the telescope scale reading be related in some definite reproducible manner to wavelength. This it will be, for a given prism, only if the prism is orientated in a way that can be repeated with certainty. One possibility would be to set the prism for minimum deviation for each wavelength in turn. An alternative, somewhat less tedious procedure would be to set it for minimum deviation for a certain agreed wavelength only (for example, sodium yellow) and then to use it in this orientation for all wavelengths to be measured.

The definite arrangement of the prism on the spectrometer table having been decided on, it now only remains to calibrate it, that is, to determine the relation between the deviation produced

969

by this prism and the wavelength. This relation, when found, can conveniently be represented in the form of a graph, and for the accurate construction of this we need a sufficiently large number of points, suitably distributed. We therefore require a source giving a suitable line spectrum and some means for determining the wavelength of each line in turn. This is most conveniently done with a diffraction grating replacing the prism on the spectrometer table. The construction and theory of the diffraction grating will be discussed in Chapter 39.

Problems

37.1 A small object is on the bottom of a swimming pool, near one side, where the depth of water is 2 m. What is the apparent depth of the object as seen by an observer standing on the edge of the pool and looking straight down?

37.2 A parallel-sided slab of transparent material of thickness 1.25 cm has two fine scratches, one on each side. A microscope whose axis is perpendicular to these sides is focused first on the nearer scratch and then on the scratch on the other side of the slab. It is found that, from the first to the second setting, the distance the microscope has to be moved toward the slab is 7.96 mm. What is the refractive index of the material of the slab?

37.3 A block of flint glass, of refractive index 1.65 and of depth 5 cm, rests on the bottom of a beaker of water. The surface of the water is 10 cm above the top surface of the glass block. What is the apparent depth of a scratch on the inside of the bottom of the beaker below the surface of the water?

37.4 (a) A glass sphere in air has a refractive index of 1.5, and its radius is R. At what distance must a small object be placed from the surface of this sphere in order that its image shall be at the same distance from it on the other side of the sphere? (Assume that rays that would be refracted into the sphere with large angles of incidence are screened off. Do *not* treat the sphere as a thin lens.)

(b) Will the image be enlarged or reduced in size relative to the object, or will it be the same size? Will it be erect or inverted? Explain your answers.

(c) At what distance from the sphere would a point object have to be placed to produce a finally emergent beam of parallel rays?

37.5 A biconvex lens of glass whose refractive index is 1.5 has surfaces each of radius of curvature 10 cm. Find the position of the image of an object that is on the axis of the lens and 30 cm distant from it.

37.6 A diverging lens of focal length 20 cm is constructed of glass of refractive index 1.65 and has one plane surface. What is the radius of curvature of the curved surface?

37.7 A lens is required to form a virtual image with a magnification of 5 at a distance of 25 cm from the lens. If the glass selected for its construction has a refractive index of 1.60 for the frequency of the light to be used, and one surface is to be plane, what must be the radius of curvature of the other surface of the lens? Should this surface be convex or concave?

37.8 A lens held 30 cm from a miniature lamp produces a sharp image of the lamp on a screen situated 20 cm beyond the lens. The lamp is now moved 15 cm closer to the lens. How must the screen be moved in order again to receive a clear image? What will be the magnifications of the images in the two cases? What is the focal length of the lens?

37.9 The lens of a camera has a focal length of 6 cm and the camera is focused for very distant objects. Through what distance must the lens be moved, and in what direction, when the focus is readjusted for an object 2 m from the lens?

37.10 An object of height 4 cm is placed 100 cm in front of a converging lens of focal length 20 cm and perpendicular to the axis of the lens. What is the position, nature, and size of the image produced? What would be the nature of the image if the distance between the object and the lens were 10 cm?

37.11 The moon has a diameter of 3476 km and is at a distance of 384,400 km. What is the size of the image of the moon formed by a concave mirror of focal length 2 m?

37.12 Given that the angular diameter of the sun is 0°32′, calculate the focal length of a lens that will produce an image of the sun of diameter 2 cm.

37.13 Find the ratio of the focal lengths of a glass lens in water and in air, the refractive indices of the glass and water being 1.5 and 1.33, respectively.

37.14 An object and a screen are placed 1 m apart in a tank containing a liquid of refractive index 2. It is found that when a thin lens is placed in either of two positions between them a sharp image is produced on the screen, and that the magnification in one case is four times that in the other. Given that the lens is made of glass of refractive index 1.5, find its focal length in air. If the liquid were removed from the tank, would there still be two positions in which the lens could be placed to give a sharp image on the screen? Explain.

37.15 A thin biconvex glass lens is supported with its axis vertical on the surface of a pond. Its lower surface, of radius of curvature 20 cm, is in contact with the water, while its upper surface, of radius of curvature 50 cm, is exposed to the air. The refractive indices of the glass and the water are 1.5 and 1.33, respectively. An insect hovering above the lens and a fish below it see one another, each receiving sets of parallel rays derived from the other. What are the distances of the insect and the fish above and below the lens, respectively? What would be the effect, if any, of reversing the lens?

37.16 Consider the equation relating the positions of an object in a vacuum (or air) and its image in a medium of refractive index n, a single refraction of each ray occurring at a curved surface of separation between the two media. Show that if in this equation n is changed to -1, the equation becomes the *mirror* equation for light *reflected* at this same curved surface. Explain the physical significance of this.

37.17   Show that for two thin lenses in contact whose separate focal coordinates for a particular direction of travel of light passing through them are f_1 and f_2, the focal coordinate f of the combination is given by the formula

$$\frac{1}{f} = \frac{1}{f_1} + \frac{1}{f_2}.$$

37.18   A thin converging lens of focal length 20 cm is placed in contact with a thin diverging lens of 10-cm focal length. What is the focal length of the combination? Will it be converging or diverging?

37.19   What single lens is equivalent to a thin converging lens of focal length 6 cm in contact with a thin diverging lens of 10-cm focal length?

37.20   Will the focal length of a lens be greater or less for violet light than for red?

37.21   Design a cemented achromatic lens combination whose diverging and converging components are plano-concave and biconvex, respectively, using crown (*c*) and flint (*f*) glasses, and making the focal length 1.5 m. The following information is given:

$$n_c = 1.517, \quad n_f = 1.655, \quad \omega_c = 0.0168, \quad \omega_f = 0.0290.$$

37.22   A converging lens of focal length 10 cm and a diverging lens of 6-cm focal length are placed 20 cm apart, and an object is placed 15 cm from the converging lens on the side remote from the diverging lens. Find the position and nature of the final image.

37.23   A thin converging lens of focal length 10 cm and a thin diverging lens of 20-cm focal length have a common axis and are separated by a distance of 10 cm. An object is placed 20 cm from the converging lens on the side remote from the diverging lens, and an image is formed by passage of the light from the object through both lenses. What is the position and nature of this image, and what is its magnification? After having obtained your result, state, without carrying out any further calculation, whether (a) the position, (b) the nature, or (c) the magnification would be affected, and, if so, how, by interchanging the two lenses.

37.24   A person "sees himself" in either a plane or a concave spherical mirror. Sketch ray diagrams showing cases where the image of himself that he sees is (a) real, and (b) virtual. Explain the situation where, if the observer were not in the path of the reflected rays, a real image would be formed, but actually, with the observer present, this image becomes, for his eye, a virtual object. Where, relative to the observer, is this virtual object situated?

37.25   (a)   A man decides to test his eyesight with the aid of a small plane mirror. When he stands in front of the mirror he finds that he can see himself clearly, with his eyes relaxed, at a maximum distance of 1 m from the mirror. He then reduces his distance from the mirror and finds that with an effort of accommodation he can continue to see himself clearly down to a distance of 15 cm. What spectacle lenses should he wear in order to be able to see very distant objects clearly? What is his least distance for distinct vision when wearing these?

(b) An object is placed 60 cm in front of a concave mirror of radius of curvature 1 m. When not wearing spectacles what is the maximum distance the man may stand away from this mirror while still seeing clearly the image (of the object) formed by the mirror?

37.26 A person wishes to test his eyesight with the aid of a concave mirror of radius of curvature 1 m. He finds that with his eyes relaxed he can see himself clearly in the mirror when at a distance of 90 cm from it, and that with an effort of accommodation he can continue to see himself clearly as close as 70 cm from the mirror. What spectacle lenses does he need in order to see clearly objects at infinity with his eyes relaxed? When wearing these, what will be his least distance for distinct vision?

37.27 Show that a plane mirror in contact with a converging lens of focal length $|f|$ is equivalent to a concave spherical mirror of radius of curvature $|f|$.

37.28 If a biconvex thin lens is placed on a plane mirror, it is found that an object on the lens axis 20 cm from the lens coincides with its own image. If the mirror is removed and the lens is floated on a dish of mercury, coincidence of object and image occurs when the object is 13 cm from the lens. If the lens is turned over, this distance is found to be 10 cm. Find the refractive index of the glass of which the lens is made.

37.29 A thin biconvex lens has surfaces both of radius of curvature 15 cm and the glass of which it is constructed has a refractive index of 1.5. One of the lens surfaces is silvered, so that it acts as a mirror, and an object is placed 40 cm from the lens on the other side. Find the position of the image of this object.

37.30 A metal vessel has a spherical bottom, brightly polished, and is filled with water to a height of 50 cm. A miniature lamp is located 20 cm above the bottom of the vessel and is screened from view on its upper side. An observer looking down into the water judges the lamp to be 12 cm below the surface of the water. What is the radius of curvature of the bottom? (Take the refractive index of water to be $\frac{4}{3}$.)

37.31 An object of height 3 cm is placed 20 cm in front of a diverging lens of focal length 10 cm. Locate, by geometrical construction and by calculation, the position of the image, and find its magnification.

37.32 A converging lens of focal length 5 cm and diameter 2.5 cm is used as a magnifying glass under relaxed-eye conditions. Find, both by geometrical construction and by calculation, the extent of an object in the focal plane of the lens whose image will be seen by an eye (a) 5 cm, (b) 10 cm, and (c) 25 cm from the lens. Find at what distances from the center of the lens rays from a point 0.25 cm off axis pass to be received by the eye at these various distances.

37.33 A compound microscope of overall length 30 cm consists of two lenses of focal lengths 1 cm and 5 cm. What is the magnifying power of this microscope for a person whose least distance for distinct vision is 25 cm? How far from the objective is the object viewed?

37.34   An astronomical telescope has an objective of focal length 1 m and an eyepiece of 5-cm focal length and is adjusted for relaxed-eye viewing of very distant objects. This telescope is now used to view an object 5 m beyond the objective, again with the eye of the observer relaxed. How far must the eyepiece be pulled back for a sharp image to be cast on the retina, and what will be the magnifying power under these conditions?

37.35   A Ramsden compound eyepiece consists of two converging lenses of the same focal length $|f|$ spaced apart a distance d, where $d = \frac{2}{3}|f|$. Show that, for relaxed-eye viewing, the image viewed (formed by the microscope or telescope objective) should be $\frac{3}{8}d$ from the first lens.

37.36   A flint-glass prism of 60° angle is used in a spectrometer. What will be the angle between emergent beams of light of wavelengths 6563 Å and 4861 Å, at minimum deviation in each case? (For $\lambda = 6563$ Å, $n = 1.650$; and for $\lambda = 4861$ Å, $n = 1.669$.)

INTERFERENCE AND DIFFRACTION I

This chapter will be devoted essentially to the study of optical interference phenomena produced by the superposition of disturbances from two point sources or their equivalents in various forms. In Chapter 39 we shall deal with a number of phenomena due to the superposition of disturbances from more complex systems.

Interference Produced by Two Point Sources

There are a number of experimental arrangements whereby optical interference can be produced by waves traveling out from two effectively point sources. We shall now consider these in turn.

YOUNG'S FRINGES   The first to observe an interference pattern produced by light from two point sources simultaneously falling on a screen was Thomas Young in 1801. Young allowed light passing through a primary single pinhole to impinge on a pair of secondary pinholes and thence to fall on a screen, where he observed a regular pattern of alternate bright and dark bands. For some reason the term "fringes" has come into general use for such bands, and the pattern produced by Young is accordingly often referred to as "Young's fringes." A better arrangement for producing and observing these fringes, shown in Fig. 38.1, is one in which the pinholes are replaced by parallel slits and in which the screen is replaced by the focal plane of an eyepiece. Instead of the primary slit, a straight current-carrying incandescent tungsten filament in an evacuated bulb could be used, adjusted for parallelism with the secondary slits.

Coherence   It may be wondered why the source shown on the left of Fig. 38.1, together with the condensing lens and the primary slit, could not be dispensed with and the secondary slits simply replaced by two straight parallel incandescent filaments. The reason is that, with two independent sources, waves originating from them would not have any definite phase relation to one another. Although at any particular moment there must be *some* phase relation between the disturbance at a given point on one source and that at a point on the other, there is no reason why this

FIGURE 38.1 System for producing Young's fringes.

should either remain constant with time or be the same for any other pair of points; indeed, we should expect these phase relations to vary in a random fashion both with time and from one pair of points to another. Accordingly we cannot expect any definite interference pattern to be observable on a screen exposed to radiation from the two sources or at the focal plane of a lens placed in its path. Such independent sources are said to be **incoherent.**

In the system of Fig. 38.1, constancy of effective phase relations, or **coherence,** is secured by arranging that the interference is between Huygens secondary wavelets generated at points in the secondary slits from a single primary wave disturbance to which both are exposed.

Even with this provision made, the theory of the formation of a definite interference pattern due to light originating from an actual source, consisting of innumerable centers producing radiation with phases distributed at random, is far from simple. Another feature of actual systems that must be taken into account in any complete theory is the use of slits of finite length and having a finite width. These complicating factors will be discussed later. Meanwhile, let us idealize the situation, imagining we have a point source of light and supposing both primary and secondary slits to be replaced by pinholes of infinitesimal size.

Fringe Spacing Let S_1 and S_2 in Fig. 38.2 represent the pinholes to which the secondary slits have been idealized, receiving waves from a monochromatic source in phase with one another, and let AB be a screen or the focal plane of an eyepiece parallel to S_1S_2.

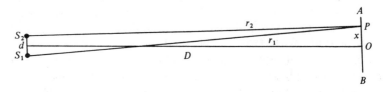

FIGURE 38.2

Let the distance between S_1 and S_2 be d and let the distance of the screen (or focal plane of the eyepiece) from S_1 and S_2 be D. Let O be the foot of the perpendicular from the midpoint of S_1S_2 to the screen, and, O being taken as the origin of a Cartesian coordinate system with the x axis parallel to S_1S_2 and the y axis perpendicular to this, let P be a point on the screen with coordinates x and y. Then the condition for reinforcement of the disturbances reaching P from S_1 and S_2, that is, for a maximum, is that $r_1 \sim r_2 = N\lambda$, where r_1 and r_2 are the distances S_1P and S_2P, respectively, λ is the wavelength of the light, and N is any whole number, including zero. And the condition for destructive interference at P, that is, for a minimum, is that $r_1 \sim r_2 = (N + \frac{1}{2})\lambda$; the disturbances arriving at P will then be in antiphase; that is, there will be a phase shift of π of one relative to the other.

Now,

$$r_1 = \sqrt{D^2 + \left(x + \frac{d}{2}\right)^2 + y^2}$$
$$= \sqrt{D^2 + \frac{d^2}{4} + x^2 + y^2 + xd}$$
$$= \sqrt{D'^2 + xd},$$

where D'^2 is written for $D^2 + (d^2/4) + x^2 + y^2$, and, since the transverse dimensions in Fig. 38.2 are grossly exaggerated and in practice both x and d are very small compared with D', we may use the algebra of small quantities, obtaining for r_1 the expression

$$r_1 = D'\left(1 + \frac{d}{2D'^2}x\right).$$

Similarly,

$$r_2 = D'\left(1 - \frac{d}{2D'^2}x\right),$$

and so

$$r_1 - r_2 = \frac{d}{D'}x.$$

Hence the condition for a maximum becomes

$$x = N\frac{D'\lambda}{d}, \qquad N = 0, \pm 1, \pm 2, \ldots,$$

while that for a minimum is that

$$x = (N + \tfrac{1}{2})\frac{D'\lambda}{d}.$$

977

FIGURE 38.3 Photograph of Young's fringes.

The number $|N|$ defines what is known as the "order" of a maximum. Accordingly, for the x coordinates x_N of the pair of maxima of any order $|N|$ other than zero we have

$$x_N = N \frac{D'\lambda}{d},$$ (38.1)

with N either positive or negative. Since for values of x and y with which we are concerned in practice the difference between D' and D is negligible, we may write, as a correspondingly good approximation to Eq. (38.1),

$$x_N = N \frac{D\lambda}{d}.$$ (38.2)

It will be instructive to work out the actual fringe spacing in a typical case. Let us suppose $d = 0.5$ mm, $D = 30$ cm, and sodium light is used. The yellow sodium lines are a close doublet, of wavelengths 5890 Å and 5896 Å; for the purposes of the present calculation we may use the figure 5.9×10^{-4} mm. Then for the corresponding fringe spacing we obtain the value of just over 0.35 mm.

From Eq. (38.2) we see that the maxima are equally spaced straight lines, the distance between consecutive maxima being $D\lambda/d$. A photograph of such a fringe system is shown in Fig. 38.3. By using an eyepiece provided with a cross wire in its focal plane which can be moved in the x direction by means of a micrometer screw, and measuring the distance between two maxima of counted difference of order, the wavelength of the light used may be determined with fair accuracy.

Use of the Term "Diffraction" If an interference pattern is produced by interfering Huygens secondary wavelets, as in the case of Young's fringes, the phenomenon may be referred to either as interference or as diffraction. The use of the latter term does not, however, always imply that the interfering waves are Huygens secondary wavelets; for example, when a narrow beam of x rays passes through a crystal, so-called diffracted beams will under suitable conditions produce a regular pattern of spots on a photographic plate or film, and this is known as a diffraction pattern. In this case the waves producing the pattern are not *Huygens* secondary wavelets, although they are secondary wavelets in another sense.

FRESNEL BIPRISM The proportion of light from the source that actually contributes to Young's fringes is necessarily rather small. Much brighter fringes are obtainable in certain alternative systems. The first of these is one devised by Fresnel, using a "biprism," in effect a pair of prisms of small angle placed base to base with refracting edges remote from each other. The arrangement is shown in Fig. 38.4. No significance is to be attached to the substitution here of an extended source backing the slit for the more distant concentrated source and condensing lens shown in Fig. 38.1; either system for illuminating the slit could be used in either case.

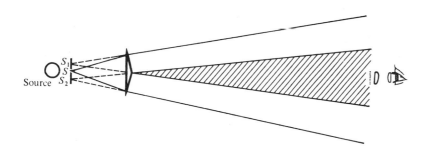

FIGURE 38.4 System for producing interference fringes with a biprism.

In the Fresnel-biprism system the effective sources are the virtual images S_1 and S_2 of the actual illuminated slit S, these being produced by the two components of the biprism. Because these are images of the same object, they are necessarily coherent. Interference fringes similar to those already discussed may be observed anywhere in the wedge-shaped region shown shaded.

FIGURE 38.5   Production of interference fringes with Fresnel mirrors.

The distance d between the images S_1 and S_2 may be measured by an indirect optical method which takes no account of the particular device by which the duplication of the source has been produced. A micrometer eyepiece is first mounted in a suitable position, facing the images of the slit. A converging lens whose focal length is less than one quarter of the distance between these images and the cross wire of the eyepiece is then mounted between them and the two positions found for which this produces a sharp pair of images of S_1 and S_2 in the focal plane of the eyepiece. With the lens in each of these positions the distance between the images is measured; let these distances be d' and d''. Then, as the reader may easily verify, d must be equal to $\sqrt{d'd''}$.

The lens may now be removed and the micrometer eyepiece used to measure the fringe spacing.

FRESNEL MIRRORS   Two closely spaced virtual images of a source may also be produced by means of a pair of mirrors, inclined at a very small angle to one another. This arrangement, which was also devised by Fresnel, is shown in Fig. 38.5, in which, for simplicity, the illuminated slit and its images are simply represented by dots. In observing the fringe system it is necessary to use a screen to prevent light from the actual source S from illuminating the region of the fringes. The region within which fringes may be observed is indicated by shading.

LLOYD'S SINGLE MIRROR   In this arrangement, shown in Fig. 38.6, the effective sources are the actual source S_1 and its image

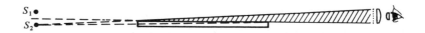

FIGURE 38.6   Production of interference fringes with Lloyd's single mirror.

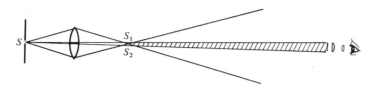

FIGURE 38.7 Production of interference fringes with Billet split lens.

S_2 in a mirror obtained by reflection at very oblique incidence. It may be shown, in an analysis based on Maxwell's electromagnetic theory of light, that on "external" reflection from the surface of a dielectric whose optical density is greater than that in which the light is incident (air in the present case) the light suffers phase reversal.* Owing to this, the center of the fringe system obtained with Lloyd's single mirror is a *dark* band, not a bright one as in the previous cases.

BILLET SPLIT LENS In this system, which is shown in Fig. 38.7, a converging lens is cut in half, and the two halves are then cemented together with a finite thickness of cement separating them. Each half produces its own real image of an illuminated slit S placed at a suitable distance behind. These real images, S_1 and S_2, now function as effective sources for the production of interference in the region indicated by shading.

Phasor-Diagram Method of Treatment

As in ac theory, so here, a very powerful method for investigating quantitative relations is the use of phasor diagrams.

Let us consider first a simple example, that of two plane-polarized wave trains passing through a point, producing electric and magnetic simple harmonic disturbances there such that both electric disturbances, and correspondingly both magnetic disturbances, are in the same direction. Either the electric or the magnetic component of each wave train may be represented mathematically by its wave equation in the manner discussed in Chapter 3; and, the distance of the point along the line of travel being substituted for x in this, the wave equation for the electric or the magnetic field, as the case may be, reduces to an equation of a simple harmonic

---

* Phase reversal occurs only under the conditions specified. There is no phase change on reflection at a boundary the medium beyond which has an optical density less than that of the medium in which the light is incident.

disturbance. Let the equations for the two separate disturbances with which we are here concerned be

$$s_1 = A_1 \sin(\omega t + \epsilon_1)$$

and

$$s_2 = A_2 \sin(\omega t + \epsilon_2),$$

where s_1 and s_2 each stand for the field intensity under consideration (either electric or magnetic), and let it be required to find the algebraic sum s of s_1 and s_2.

The geometrical procedure follows exactly the same lines as in the discussion of ac theory, except that now $\epsilon_2 - \epsilon_1$ may be *any* angle. As before, we construct the second circle of reference about the end of the first phasor as center and obtain the resultant of the corresponding phasor triangle by joining the center of the first circle to the end of the second phasor. The diagram, with circles omitted, is then as shown in Fig. 38.8, the projection s on a vertical line of the resultant, of length a, being the sum of the projections s_1 and s_2 of the components. We may now think of this whole phasor diagram as rotating anticlockwise with angular velocity ω, and so we see that the resultant disturbance s, like the component disturbances, is simple harmonic, being given by the equation

$$s = A \sin(\omega t + \epsilon),$$

where

$$A = \sqrt{A_1{}^2 + A_2{}^2 + 2A_1A_2 \cos(\epsilon_2 - \epsilon_1)}$$

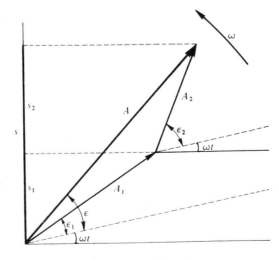

FIGURE 38.8 Principle of use of phasor diagrams for solving problems in optical interference.

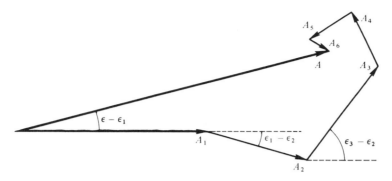

FIGURE 38.9 Phasor polygon for dealing with a finite number of finite contributions to a disturbance.

and

$$\epsilon = \tan^{-1}\left(\frac{A_1 \sin \epsilon_1 + A_2 \sin \epsilon_2}{A_1 \cos \epsilon_1 + A_2 \cos \epsilon_2}\right).$$

This reasoning may obviously be extended to apply to the general case where it is wished to compound more than two simple harmonic disturbances.

It should be noted that, by converse reasoning, we may infer that any simple harmonic disturbance can be regarded as made up of two or more simple harmonic disturbances in the same direction having any desired phase relations to one another and appropriate corresponding amplitudes.

As in ac theory, the amplitude A represented in Fig. 38.8 may be found by simply drawing the appropriate phasor triangle, without regard to orientation. By an extended construction of the same kind, we could include a third component, then a fourth, and so on, as indicated in Fig. 38.9. In some applications we have to deal with an infinite number of simple harmonic disturbances, each of whose amplitudes is infinitesimally small. The corresponding phasor diagram then becomes a curve, such as the one shown in Fig. 38.10, instead of a series of straight lines joined end to end; but the principle is the same, and the resultant amplitude is again represented by the length of the straight line joining the origin to the last point reached.

PHASOR TREATMENT OF INTERFERENCE PRODUCED BY TWO POINT SOURCES In our earlier discussion of interference produced by two point sources we were concerned only with finding the positions of maxima and minima. By using the phasor construction we may now also find how the intensity varies with position between a maximum and a minimum.

983

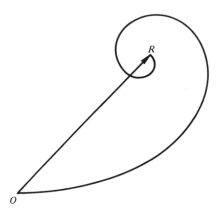

FIGURE 38.10   Phasor diagram for dealing with an infinite number of infinitesimal contributions.

Our phasor construction relates only to disturbances in the same direction; hence, strictly, we should use it only for plane-polarized waves polarized in the same plane. In the discussion to follow we shall therefore suppose that the waves are polarized accordingly. It will be shown later that the results so obtained apply also for so-called "unpolarized" light.

We have seen that the path difference $r_1 - r_2$ (Fig. 38.2) is, to a close approximation, equal to $(d/D)x$. This contains $(d/\lambda D)x$ wavelengths, and therefore corresponds to a phase difference ϕ of $(2\pi d/\lambda D)x$. Since in Fig. 38.2 S_2 is closer to P than is S_1, the disturbance contributed by the waves from S_2 to P, of amplitude A_2, will be ahead of that contributed by those from S_1, of amplitude A_1, by ϕ in phase, and so the relevant phasor diagram will be that shown in Fig. 38.11. Taking the amplitudes A_1 and A_2 as equal to one another and not varying appreciably with x and y over the ranges of these with which we are normally concerned, this diagram will have the forms shown in Fig. 38.12 corresponding to a series of values of ϕ.

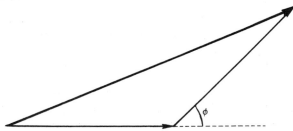

FIGURE 38.11   Phasor triangle relating to two coherent disturbances of equal amplitude differing in phase.

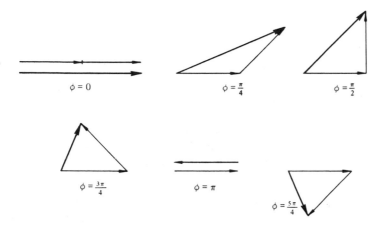

FIGURE 38.12 Series of phasor diagrams of the type shown in Fig. 38.11 giving resultant amplitudes for a number of phase differences between components.

It may be shown that in electromagnetic radiation the energy flux density at any point is proportional to E^2, or to B^2, or to EB.* Accordingly, the *mean* energy flux density is proportional to E_0^2, or to B_0^2, or to E_0B_0, where E_0 and B_0 are the amplitudes of the electric and magnetic fields, respectively. The resultant amplitude discussed above may be taken as referring to either of these, that is, either to E_0 or to B_0. Values of this amplitude, obtained in the series of phasor diagrams shown in Fig. 38.12, are plotted against ϕ in Fig. 38.13. The corresponding intensities (mean radiation flux densities) are obtained by squaring the amplitudes. These intensities, also, are plotted to an arbitrary scale in Fig. 38.13.

* See Chapter 1 of Volume III.

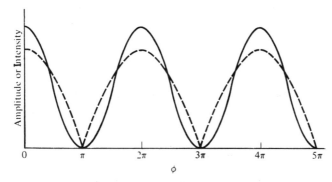

FIGURE 38.13 Graphs of amplitude and intensity against phase difference between components for idealized double-source type of interference.

Actual Light Sources

Practical light sources consist of innumerable radiating centers radiating with different states of polarization. At any external point exposed to these radiations the corresponding component disturbances (electric or magnetic) in any given direction will have phases distributed at random. Also, for any given total power radiated, there will be a wide range of amplitudes, distributed according to a certain probability law. We might well ask how such sources can possibly give rise to a regular succession of maxima and minima in an interference pattern.

STATES OF POLARIZATION   Let us first deal with the problem presented by the different states of polarization. We shall ultimately be concerned with the component periodic disturbances at an external point in particular directions, corresponding to plane-polarized components of the radiation. However, rather than attempt to analyze, say, an elliptically polarized wave train into plane-polarized components, it will be convenient to adopt the opposite procedure of compounding plane-polarized components in planes mutually at right angles. As far as a point in the path of the radiation is concerned, this means compounding simple harmonic disturbances at this point whose directions are at right angles to one another as well as perpendicular to the direction of propagation of the radiation.

Although it is electric or magnetic vectors that we wish to compound, our problem is basically the same as the composition of displacements, because the electric or magnetic vectors in question can be represented by such displacements. Let us, therefore, consider the case of two simultaneous simple harmonic motions of a particle in the x and y directions of a Cartesian coordinate system represented by the equations

$$x = A_1 \sin \omega t$$

and

$$y = A_2 \sin (\omega t + \phi).$$

On eliminating t between these equations we obtain as the equation of the path of the point giving the resultant disturbance,

$$\frac{x^2}{A_1{}^2} - \frac{2xy}{A_1 A_2} \cos \phi + \frac{y^2}{A_2{}^2} = \sin^2 \phi. \tag{38.16}$$

This is the equation of an ellipse. The major and minor axes of the ellipse are not in general parallel to the coordinate axes; they will be parallel to these only if ϕ is $\pm \pi/2$.

By considering the converse of the above procedure we see that any elliptical motion having the appropriate relation of velocity to position on the ellipse can be resolved into simple harmonic motions in any selected mutually perpendicular directions having certain amplitudes and phases. This applies, therefore, also to waves; any elliptically polarized wave train may be regarded as made up of two plane-polarized wave disturbances having planes of polarization in any selected pair of mutually perpendicular planes containing the direction of propagation. Elliptical polarization includes, as limiting cases, circular and plane polarization. Hence all the actual disturbances produced simultaneously at a point by the radiations from the different centers may be considered as made up of two sets of simple harmonic disturbances in directions at right angles, and in each set we have a random distribution of phases and a distribution of amplitudes conforming to some probability law. Our next task will be to see how to deal with each of these sets.

RESULTANT OF PLANE-POLARIZED COMPONENTS WITH DISTRIBUTED AMPLITUDES AND PHASES In principle the separate disturbances arriving at a point may be compounded by the phasor polygon method which we have already studied, the polygon having as many sides as there are radiating centers. The key to the determination of the resultant phasor is provided by a theorem in probability. It may be shown that if we have a large number of phasors whose magnitudes are distributed according to some definite probability law and whose directions are distributed at random, then the most probable magnitude of their resultant is proportional to the square root of the number of phasors compounded.

Incidentally, this provides a statistical explanation of the fact that the intensity is proportional to the number of radiating centers, as it must be from energy considerations; for the intensity is proportional to the square of the amplitude of the resultant simple harmonic disturbance.

We now see how it is that an actual light source behind the primary slit in the apparatus used to produce Young's fringes, or behind the slit of the biprism experiment, and so on, produces a definite wave disturbance emanating from each point in this slit, as if this point were an independent source. Each point in the slit may be considered as one that is affected by all the separate radiations passing through it from the various centers in the source, and to which our statistical phasor reasoning applies accordingly. From this point Huygens secondary wavelets are radiated corresponding to the resultant of all the disturbances reaching it. These secondary wavelets constitute the radiation that produces further Huygens secondary wavelets at the secondary slits in the production of Young's fringes, or which are refracted through the two halves of the biprism, and so on.

Effects of Finite Widths and Lengths of Slits

A complete geometrical discussion of the effects of the finite widths and lengths of slits (for example, those used in the production of Young's fringes), although not difficult, would be too lengthy to undertake here. It will, however, be useful to indicate the general lines of such a discussion. As we have just seen, each point in the primary slit may be regarded as the equivalent of an independent source. Let such a point be labeled α, and let points in the secondary slits situated along a line drawn perpendicular to the lengths of the slits be labeled β and γ, while P is a point in the plane where interference fringes are observed. Huygens wavelets will be generated at β and γ from disturbances originating at α, giving rise to the rays βP and γP. The relevant path difference, together with the elements of area surrounding the points α, β, and γ, where the Huygens secondary wavelets are generated, determines the disturbance finally produced at P. This path difference is $\alpha\beta P \sim \alpha\gamma P$. Now let us consider all possible combinations of such points in the planes of the three slits, for example, α, β, γ; α, β, γ'; α, β, γ''; . . . ; α, β', γ; α, β', γ'; α, β', γ''; . . . ; α, β'', γ; α, β'', γ'; α, β'', γ''; . . . ; α', β, γ; α', β, γ'; α', β, γ''; and so on. Obviously there cannot be exact equality of all path differences of the type $\alpha\beta P \sim \alpha\gamma P$. Nevertheless, if the variation among these path differences does not amount to more than a small fraction of a wavelength for all points P within a certain area, the nature of the interference pattern produced in this area will be essentially the same as if infinitesimally small apertures were used instead of slits. With apparatus of dimensions ordinarily used in these experiments this condition can be satisfied by slits a few millimeters in length and an appreciable fraction of a millimeter in width.

Obviously, the effective sources at the various points α, α', α'', . . . in the primary slit will not be coherent; neither, corresponding to any one of these, will the contributions reaching any particular point P from all the pairs of points in the secondary slits. Nevertheless, by reasoning corresponding to that already used for the incoherent radiating centers in the source, we see that there will be definite amplitudes of the disturbances at all points in the secondary slits and at P. And the *scale* of the final resultant amplitudes will vary from one point P to another in a definite manner, according to the variations of path difference characteristic of each such point. Along these lines we can account quantitatively for the maxima and minima, and for the intervening regions of intermediate intensity, in a completely satisfying manner.

Problems

38.1   A biprism is mounted at a suitable distance from an illuminated slit, and fringes are measured in the focal plane of an eyepiece, this plane being 120 cm from the slit. Their mean spacing is 0.345 mm. A converging lens is now introduced between the biprism and the focal plane of the eye-piece, and it is found that there are two positions of this for which it produces a sharp pair of images of the slit. In one case these images are 2.16 mm apart and in the other 1.50 mm. Find the wavelength of the light used.

38.2   In designing a Fresnel biprism to be constructed of glass of refractive index 1.5, it is planned to produce a pair of virtual images of a slit placed 5 cm behind the biprism whose distance between centers is 2 mm. What should be the angle of each prism? Using a biprism with components having this angle between faces and illuminating the slit with sodium yellow ($\lambda = 5893$ Å), what will be the fringe spacing in a plane 2 m beyond the biprism?

38.3   The spectrum of mercury contains two yellow lines fairly close together, their wavelengths being 5769.6 Å and 5790.6 Å. The slit of a Fresnel biprism system is illuminated with light from a mercury arc that has been sent through a yellow filter which transmits only these two lines of the mercury spectrum. What is the order N of the first interference maximum of the longer wavelength that coincides with a minimum of the shorter wavelength?

38.4   The two components of a Fresnel double-mirror system are set at an angle of 1° to each other, and a narrow slit illuminated by mono-chromatic light is situated 3 cm from their line of intersection. The focal plane of an eyepiece used for viewing the interference fringes produced by the reflected light is 1 m from this line of intersection, and in this plane the fringe spacing is observed to be 0.537 mm. What is the wavelength of the light?

38.5   The focal length of a Billet split lens is 12 cm and the separation between the lens halves is 0.4 mm. If the split lens is placed 30 cm from a narrow slit illuminated by sodium light ($\lambda = 5893$ Å), what will be the fringe spacing on a screen held 1 m from the lens?

38.6   Investigate whether the intensity curve in Fig. 38.13 has the form of a sine curve, as it appears to have.

38.7   One of the secondary slits of a system for producing Young's fringes is 50 percent wider than the other. What are the relative intensities at points on the screen receiving maximum illumination when one or the other slit, separately, is covered? What are the relative amplitudes con-tributed by the two slits when light passing through both produces an interference pattern? In the latter case, what is the ratio of the intensity of a maximum to that of an adjacent minimum?

38.8   Dipole aerials which oscillate in phase and with equal amplitudes are situated at S_1 and S_2 (see the figure), the aerials being orientated at right angles to the plane of the diagram and separated by a distance 2λ, where λ is the wavelength of the radiation. A quadrant AB of a circle centered on the point 0 midway between the centers of the aerials is drawn in a plane at right angles to the aerials, the radius of this circle being large compared with a wavelength (much larger than as shown in the figure), and a detector C is moved along the arc AB from A to B. Find for what angles θ ($= \angle COA$) the signals received at C have (a) maximum and (b) minimum values. Draw a graph showing how the intensity I of the received signal varies with θ.

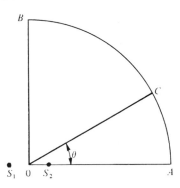

Figure for problem 38.8.

38.9   Two dipole aerials S_1 and S_2 (see the figure), which oscillate in phase and with the same amplitude, are situated at the points $(0, 0)$ and $(0, 4\lambda)$, where λ is the wavelength of the radiation emitted, both aerials being bisected by the plane of the diagram and orientated at right angles to it. Draw a graph showing how the intensity I of the signal received by a detector varies with x as the detector moves along the x axis from $(2\lambda, 0)$ to $(10\lambda, 0)$.

Figure for problem 38.9.

38.10   An electromagnetic wave train is traveling in the z direction of a Cartesian coordinate system. Given that

$$E_x = E_0 \sin \omega t$$

and

$$E_y = 2E_0 \sin (\omega t + \phi),$$

draw out the ellipses traced by the end of an arrow representing the electric-field vector for $\phi = 0$, $\phi = \pi/4$, and $\phi = \pi/2$.

38.11 In a system for producing Young's fringes, the lengths of the secondary slits and the width of the primary slit are negligible, while the widths of the secondary slits are a and the distance between their centers is d. For what order N of the maximum that would appear with secondary slits of negligible width will dI/dx in fact be zero, where I is the intensity of the light in the plane of the fringes viewed and x is the distance from the central maximum?

38.12 In a system for producing Young's fringes, the lengths of all three slits and the widths of the secondary slits are negligible. The width of the primary slit is a, its distance from the plane of the secondary slits is b, and the distance between the secondary slits is d ($\ll b$). Is there a certain value of a for which no fringe system will be visible? If so, explain this phenomenon, and calculate the value of a for disappearance of the fringes. If there is more than one value for disappearance, consider only the smallest.

38.13 In a system for producing Young's fringes, the widths of all three slits are negligible, the secondary slits are of length l and are a distance d apart, and, with the primary slit illuminated by light of wavelength λ, the fringes are viewed in a plane whose distance from the secondary slits is D. Find the distance y from the axis of symmetry in this plane at which the Nth-order maximum is no longer distinguishable from the adjacent minima.

38.14 In the case of Young's fringes obtained with slits of finite width, the contrast between maxima and minima deteriorates progressively with increasing distance x (see Fig. 38.2) from the center of the fringe system. There is a corresponding loss of contrast with increasing x in the case of the fringes produced with Lloyd's single mirror. There is, however, no such falling away of contrast with increasing x where the fringes are produced with Fresnel's double mirror, with a biprism, or with a Billet split lens. Explain why this is so.

INTERFERENCE AND DIFFRACTION II

Diffraction Grating

A diffraction grating is a system of a large number of parallel slits or apertures, all of equal width and uniformly spaced; and it normally has appreciable dimensions, the system of apertures occupying a width of something like 2 to 10 cm. Its function is to produce spectra, in the manner to be explained below. In the use of a grating it is desirable that the effective optical path shall increase by equal steps from one aperture to the next, all the way across. This condition can be secured only by arranging that parallel light from a collimator or a distant point object is incident on the grating and that the diffracted light is viewed through a telescope focused for infinity. In this respect the use of a grating resembles that of a prism in a spectrometer. Diffraction observed under these conditions, that is, of parallel light incident on and leaving the diffraction device, is known as **Fraunhofer diffraction.** On the other hand, diffraction in which either the incident light is not parallel or the emergent light is not received as a parallel beam, or both, is called **Fresnel diffraction.** An example of Fresnel diffraction is that in which Young's fringes are produced.

Let us begin our study by considering the system shown in Fig. 39.1, in which a very primitive diffraction grating is used, consisting of three slits only, and let us idealize the situation by supposing the slits to be very narrow compared with the distance d between them. Then in a direction making an angle θ with the direction of incidence the path difference between rays from successive slits is $d \sin \theta$, and the corresponding phase difference ϕ is $(2\pi d \sin \theta)/\lambda$. The phasor diagrams for a series of values of ϕ are shown in Fig. 39.2, and the corresponding intensities of the diffracted light obtained from the resultants of these and of phasor diagrams (not shown) for intermediate values of ϕ are plotted against ϕ in Fig. 39.3.

A corresponding series of phasor diagrams for four slits is shown in Fig. 39.4, and a graph of intensity against ϕ is shown in Fig. 39.5. Figure 39.6 shows the intensity graph for five slits.

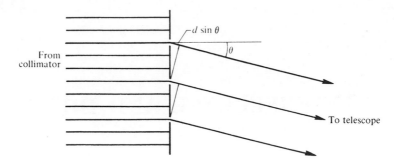

FIGURE 39.1 Fraunhofer diffraction of light by primitive three-slit diffraction grating.

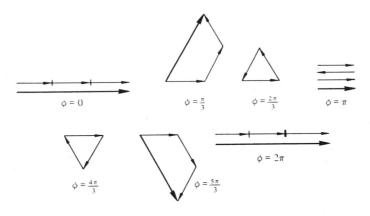

FIGURE 39.2 Series of phasor diagrams relating to a three-slit diffraction grating.

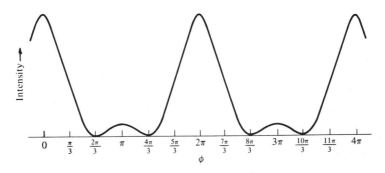

FIGURE 39.3 Graph of intensity against phase difference ϕ between contributions from successive slits of a three-slit diffraction grating.

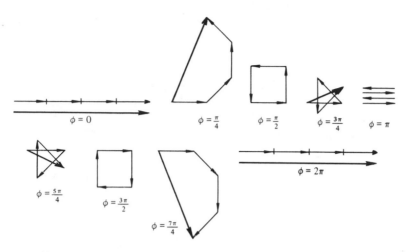

FIGURE 39.4 Series of phasor diagrams relating to a four-slit diffraction grating.

It will be observed that while for two slits (Fig. 38.12) there are main maxima only, in the case of three slits there is one subsidiary maximum between each pair of main maxima, with four slits the number of subsidiary maxima between main maxima is two, and with five slits it is three. It may be shown quite generally that in the diffracted light produced by a grating of N "lines" (apertures) there are $N - 2$ subsidiary maxima between each successive pair of main maxima. These main maxima occur at angles θ such that

$$d \sin \theta = N\lambda, \qquad (39.1)$$

where N is any whole number.

It should be noted that, since $\sin \theta$ cannot be greater than 1, only spectra of orders (values of N) below d/λ are observable.

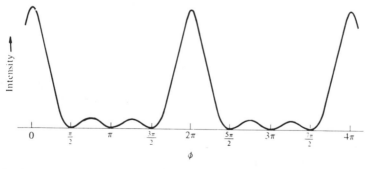

FIGURE 39.5 Graph of intensity against phase difference ϕ between contributions from successive slits in a four-slit diffraction grating.

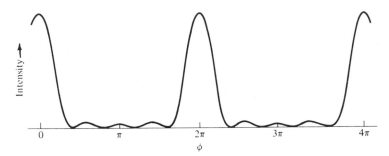

FIGURE 39.6 Graph of intensity against phase difference ϕ between contributions from successive slits in a five-slit diffraction grating.

From Figs. 38.13, 39.3, 39.5, and 39.6 it will be seen that—at least for numbers of slits up to five—the greater the number of slits the more the diffracted light is concentrated into the directions of the main maxima, these maxima becoming progressively narrower. This continues as the number of apertures is further increased, and it is for this reason that diffraction gratings used in practice have very large numbers of lines.

The phasor diagrams shown and the intensity graphs based on them are idealized in two senses: They are constructed as for very narrow slits, and it has been assumed that the Huygens secondary wavelets are equally strong in all directions. In practice the apertures must necessarily have a finite width, and on this account the lines in the phasor diagrams should be replaced by arcs of circles having the same lengths as these lines, the curvature of the arcs increasing with θ.* Also, the obliquity factor for the Huygens secondary wavelets referred to in Chapter 35 should be taken into account.

Diffraction gratings are made by ruling a very large number of fine equally spaced grooves with a diamond point, either on a plane glass surface or on polished metal, the former for the production of a transmission grating and the latter to produce a reflection grating. The theory of the action of reflection gratings, which have special applications, for example for use in the ultraviolet, is broadly similar to that of transmission gratings, and it will be sufficient here to consider only the latter. The grooves of a transmission grating scatter the light and so function as obstacles, while the spaces between the grooves transmit the light as would slits cut in an opaque screen. The lines are ruled with a special ruling machine. Gratings

---

* That this is so will be shown later, under the heading "Fraunhofer Diffraction by a Rectangular Aperture."

are commonly ruled with something of the order of 4000 to 6000 lines per centimeter.

It might be thought that in order to produce the equivalent of very narrow apertures a grating should be ruled rather heavily, leaving spaces between the grooves that are much narrower than the grooves themselves. Actually, the reverse is done; the diamond point is only allowed to press lightly on the surface, so as to produce narrow grooves with relatively wide spaces in between. Somewhat surprisingly, this gives the same effect, as far as the production of spectra is concerned, as if obstacles and apertures were interchanged. That this must be so follows from an important principle enunciated by Babinet, which is of significance not only in the theory of the diffraction grating but also in other cases of Fraunhofer diffraction.

Babinet's Principle

This states that if parallel light is incident on any system of obstacles and apertures the intensity of the (parallel) light diffracted in any direction other than that of incidence is the same as that of the light which would be diffracted in this direction if apertures and obstacles were interchanged.

Let A and B in Fig. 39.7 represent a complementary pair of apertures and obstacles. If we combine the apertures in A with

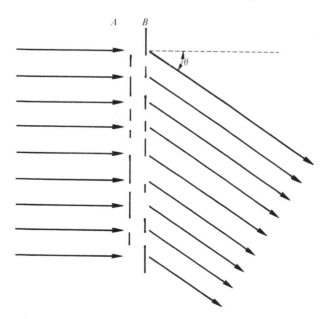

FIGURE 39.7 Complementary systems of obstacles and apertures considered in Babinet's theory.

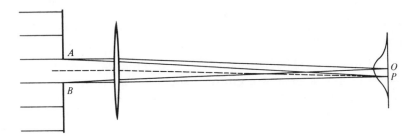

FIGURE 39.8 Fraunhofer diffraction by a single slit.

those in B we get an unimpeded light path and consequently no diffraction; A's apertures plus B's apertures give no light at any angle θ other than zero. This can only be so if the disturbance at the focus of the objective of a telescope set to receive light at the angle θ due to A's apertures is, at all times, equal in magnitude and opposite in phase to the disturbance due to B's apertures. And the equality is all that interests us here; we are not concerned with the phase relation. Hence Babinet's principle is proved.

Fraunhofer Diffraction by a Single Slit

Let monochromatic light from a collimator the width of whose slit is negligible impinge on a slit of width a, and let the light diffracted by this be viewed through a telescope on the other side, as indicated in Fig. 39.8. Then all the rays are brought to a focus at the central point O in the focal plane of the telescope. From the point of view of physical optics this means that all the corresponding disturbances arriving at this point are in phase and so produce a maximum; that is, that from the plane of the slit onward the number of wavelengths is the same for all the rays in their composite paths through air and glass. In other words, all the rays have traversed equal "optical paths."*

Now let us consider light arriving at a neighboring point P such that the line joining this point to the center of the telescope objective makes an angle θ with the direction of the light from the collimator, and let us consider in conjunction with this the plane BC indicated by the dashed line in Fig. 39.9, which is inclined at

---

* The optical path of a ray or sequence of rays is defined as the path in a vacuum that would contain the same number of wavelengths as are contained in the total path actually traversed.

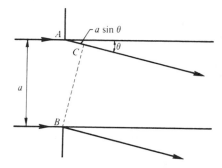

FIGURE 39.9

the angle θ^* to the plane of the slit AB. Then from the plane BC onward, through the telescope objective, to P, all the rays have equal optical paths. And from the collimator slit through the collimator lens as far as the plane AB, all optical paths are equal. But from the plane AB to BC the rays traverse extra paths proportional to their distance from B, the maximum extra path, $a \sin \theta$, being for the ray through A. The rays arriving at P therefore have a corresponding range of phases. Since the extra path traversed by the ray through A is $a \sin \theta$, the disturbance at P due to this ray lags in phase behind that due to the ray through B by $(2\pi a \sin \theta)/\lambda$.

If θ is such that $a \sin \theta$ is λ, then the phase difference between the disturbances due to the extreme rays is 2π. This case is of particular interest. Let us imagine the slit to be divided into a large even number, $2n$, of narrow strips of equal width, so that they all make contributions of the same amplitude to the disturbance at P, and let these be numbered consecutively 1, 2, 3, . . . , $2n$. Then the distances apart of strips 1 and $n + 1$, 2 and $n + 2$, . . . , n and $2n$ will all be $a/2$, the path differences of the corresponding rays will be $\lambda/2$, and the phase difference between the disturbances at P due to the members of each pair will be π; these disturbances will be in antiphase and will cancel each other out. Hence the whole slit will produce zero intensity at the point P.

The phasor diagram for $a \sin \theta = \lambda$ whose construction is based on the assumption that the phase of the disturbance contributed is uniform over the width of each strip—which of course it really is not—is a polygon of $2n$ equal sides, each making an angle of π/n with the prolongation of its predecessor. In the limit, when n is made infinitely large, this becomes a circle.

---

* This angle is exaggerated in Fig. 39.9 relative to that shown in Fig. 39.8.

Obviously for other values of θ the phasor diagram is a circular arc whose chord represents the resultant amplitude. When θ is zero (central maximum) this arc and chord both become straight lines, corresponding to an infinite radius. For an angle θ such that $a \sin \theta = \lambda/2$ this line becomes curled up into a semi-circle; for $a \sin \theta = \lambda$ it is, as already noted, a full circle; for $a \sin \theta = 3\lambda/2$ it is curled up into a circle and a half; for $a \sin \theta = 2\lambda$ it becomes two circles; and so on. For each complete number of circles the resultant is zero, and—to a very close approximation—for each odd number of half-circles beginning with three the resultant is a maximum. We see that the amplitudes for the central maximum and the subsidiary maxima on either side are in the ratio $\pi : \frac{2}{3} : \frac{2}{5} : \frac{2}{7} : \frac{2}{9} :$ $\cdots$, and, since the intensity is proportional to the square of the amplitude, the intensities are in the ratio

$$\frac{\pi^2}{4} : \frac{1}{9} : \frac{1}{25} : \frac{1}{49} : \ \cdots \ \doteq 1 : \frac{1}{22} : \frac{1}{62} : \frac{1}{121} : \ \cdots \cdot$$

Owing to the effect of obliquity on the Huygens secondary wavelets the intensities of successive maxima will actually tail off somewhat more rapidly than this. However, for small values of θ this effect will be negligible.

From the formula for a subsidiary maximum,

$$\sin \theta = (N + \tfrac{1}{2}) \frac{\lambda}{a}, \qquad N = \pm 1, \ \pm 2, \ \pm 3, \ \ldots , \qquad \textbf{(39.2)}$$

we see that, apart from the central maximum, the maxima are equally spaced on a scale of $\sin \theta$, or, for small angles, approximately on a scale of θ.

MISSING DIFFRACTION-GRATING SPECTRA An interesting effect of diffraction by a single slit is the absence of a diffraction-grating spectrum under certain conditions in directions where, according to the simple grating theory, a maximum should be produced.

Let the grating constant be d and let the width of each aperture be a (Fig. 39.10). Then the spectrum of order N will be of zero

FIGURE 39.10   Diffraction-grating obstacles and apertures.

intensity, that is, will be missing, if for the same value of θ for which

$$d \sin \theta = N\lambda,$$

we simultaneously have the condition for zero intensity received from each individual aperture,

$$a \sin \theta = M\lambda,$$

N and M both being whole numbers, that is, if

$$\frac{d}{a} = \frac{N}{M}. \tag{39.3}$$

If, for example, a were equal to $d/3$, the third, sixth, ninth, . . . orders would be missing.

In terms of phasors, the reason why, under the conditions discussed above, certain orders do not appear is that for the corresponding directions the elements of the phasor diagram are all curled up into one, two, or some other number of complete circles.

Note that the algebraical condition for missing spectra applies only for normal incidence. If the grating is used at other than normal incidence the conditions will be modified accordingly.

Fraunhofer Diffraction by a Circular Aperture

A single slit is, in effect, a rectangular aperture. Let us consider in particular the case where the rectangle is a square. If a circular aperture is substituted for this, as shown in Fig. 39.11(a), contributing strips of equal width no longer have equal areas, and so the elements of the relevant phasor diagram respresenting the contributions of the outer strips are shorter than those for the more central ones. Consequently, at the angle θ, equal to $\sin^{-1} (\lambda/a)$, for which the phasor diagram for a rectangular aperture would just close, as shown by the dashed circle in part (b) of the figure, that for a circular aperture, shown by the solid curve, does not yet close. To make it close, the angle must be increased beyond θ to a larger angle α, equal to $\sin^{-1} (1.22\lambda/a)$. The phasor diagram for this angle is shown in Fig. 39.12.

As the angle continues to increase, beyond α, the corresponding further curling up of the phasor diagram gives further maxima and minima in the resultants, and so in the intensity of the light seen in these directions. If the collimator slit is replaced by a pin-

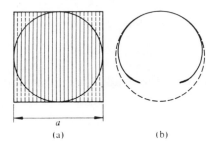

a

(a) (b)

FIGURE 39.11 (a) Square and circular apertures giving Fraunhofer diffraction.

(b) Corresponding phasor diagrams relating to these apertures. The broken circle, just closed, is the phasor diagram for the directions giving the first minima (zero) on either side of the main maximum in the case of the square aperture. The full curve is the phasor diagram for diffraction by the circular aperture in the same directions.

hole, or, alternatively, the light received by the circular aperture is that from a distant point source, such as a star, the diffraction pattern seen in the focal plane of the telescope will have circular symmetry and will consist of a circular bright area surrounded by diffraction rings, as shown in Fig. 39.13. This is what the image of a star viewed through a telescope looks like, even without the interposition of a screen with a circular aperture; for the tube of the telescope in one end of which the objective is mounted itself functions as such an aperture.

RESOLVING POWER OF A TELESCOPE By the resolving power of a telescope is meant its ability to show, as distinct from one another, objects whose angular separation is small. Quantitatively it may be defined as $1/\phi$, where ϕ is the least angular separation of objects that it will show as separate, or "resolve." It is found in practice that this angle may be identified with α, the angle between the direction of the undeviated light and the diffracted rays giving the first minimum.

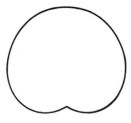

FIGURE 39.12 Phasor diagram for the first minimum (zero) in the diffraction produced by a circular aperture.

FIGURE 39.13 Fraunhofer diffraction pattern formed in the focal plane of a telescope receiving parallel light, for example from a star. (Fundamental Photographs.)

If we plot the intensity due to a single point source such as a star against angular position along a line passing through the center of the diffraction pattern shown in Fig. 39.13, we obtain one of the solid curves shown in Fig. 39.14; and if we do the same for another star of equal brightness and separated from the first by the angle α we obtain the other curve, identical with the first but shifted along the angle axis by α. If both stars are viewed together, the brightness distribution seen along the line in the focal plane of the telescope passing through both centers will be that corresponding to the curve obtained by adding the ordinates of the separate curves; this is the dashed curve in the figure. With the central maximum of each curve coinciding with the first minimum on one side of the other, as here, the resultant intensity-distribution curve has a slight dip at its center, and it is found that this condition represents the borderline of our ability to recognize what is seen as two separate

1003

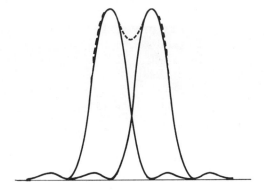

FIGURE 39.14 Intensity curves for the separate diffraction patterns produced in the focal plane of a telescope receiving light from stars of equal brightness whose angular separation is on the borderline of resolution, and their resultant.

objects. Thus for the resolving power (RP) as defined above we have

$$RP = \frac{1}{\phi} = \frac{1}{\alpha} = \frac{1}{\sin^{-1}(1.22\lambda/a)}, \tag{39.4}$$

or, because with λ/a as small as it is with all actual telescopes, $\sin^{-1}(1.22\lambda/a)$ may be identified with $1.22\lambda/a$ itself,

$$RP = \frac{1}{1.22}\frac{a}{\lambda}. \tag{39.5}$$

Note that for a given wavelength the resolving power is determined by a alone; it is independent of the focal length of the telescope. It works out that for the wavelength to which the eye is most sensitive, about 5.6×10^{-5} cm, and using a telescope of 20-cm aperture, it would be just possible to separate stars having an angular separation of 0.7 sec. With the largest telescope in the world, the 200-in. (5-m) telescope on Mount Palomar, California, the theoretical limit of resolution (ignoring the effect of irregularities of atmospheric refraction) is 0.028 sec.

Interference Due to Thin Films

Sufficiently thin films of all transparent materials show brilliant color effects when illuminated by an extended source of white light. Familiar examples are thin soap bubbles, films of glass obtained in glass-blowing operations, films of oil on water, and oxide films formed on polished metal surfaces.

Let I in Fig. 39.15 represent a ray incident on the film. From this there will be derived a series of "reflected" rays $R_1, R_2, R_3, \ldots$, and another of "transmitted" rays $T_1, T_2, T_3, \ldots$. In the case of a plane film of uniform thickness, the derived rays of each series must be parallel and will thus be capable of interfering when made to converge at the focus of a lens or lens system, for example on the retina of an observer's eye when this is relaxed (focused for infinity).

For a quantitative investigation of the interference it is necessary to find the values of the equivalent air-path differences between the rays R_1 and R_2, R_2 and R_3, and so on, and between T_1 and T_2, T_2 and T_3, and so on. By an equivalent air-path difference is meant the air path that would contain the same number of wavelengths as corresponds to the phase difference of the rays in question at the focus. In the case under consideration this focus will be on the retina of the observer's relaxed eye or in the focal plane of the objective of a telescope (and again on the retina), and the phase difference will be the same as that between points such as P_1 and P_2, P_2' and P_3, $\ldots$, Q_1 and Q_2, Q_2' and Q_3, $\ldots$, in the figure.

Let the angles of incidence and refraction be i and r, respectively, the thickness of the film t, the refractive index of the material of the film relative to air n, and let the equivalent air-path difference between the rays R_p and R_q, or between T_p and T_q, be denoted by δ_{pq}. Then, taking into account the shortening of the wavelength

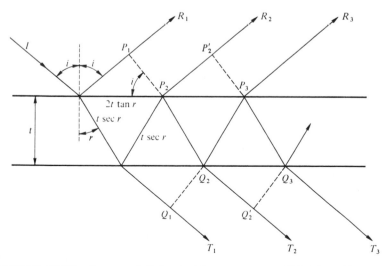

FIGURE 39.15 A ray of light incident on a parallel-sided film, and associated sets of reflected and transmitted rays.

within the film to $1/n$ of its air value and the phase reversal occurring on partial reflection of the incident ray to give the ray R_1,* we have for the rays R_1 and R_2

$$\delta_{12} = 2nt \sec r - 2t \tan r \sin i + \frac{\lambda}{2}$$

$$= \frac{2nt}{\cos r} (1 - \sin^2 r) + \frac{\lambda}{2}$$

$$= 2nt \cos r + \frac{\lambda}{2}. \tag{39.6}$$

Since no external reflection is involved in the derivation of the rays R_2, R_3, . . . , or of T_1, T_2, T_3, . . . from the incident ray, δ_{23}, δ_{34}, . . . for the reflected rays and δ_{12}, δ_{23}, δ_{34}, . . . for the transmitted rays are each equal simply to $2nt \cos r$.

Let us now investigate the relative amplitudes of the various reflected and transmitted rays. The ray R_1 is derived from the incident ray I simply by an external reflection; R_2 by a refraction, an internal reflection, and a final refraction; R_3 by a refraction, three internal reflections, and a final refraction; and so on. T_1 is derived from the incident ray by two refractions; T_2 by a refraction, two internal reflections, and a final refraction; T_3 by a refraction, four internal reflections, and a final refraction; and so on. In general the coefficient of transmission through a surface of discontinuity between two transparent media is nearly 1, while the coefficients of external or internal reflection are relatively small and not very different from one another. It is clear that under these conditions the amplitudes of R_1 and R_2 will be roughly the same, whereas those of the further reflected rays will be relatively very small, forming with the amplitude of R_2 a strongly convergent geometrical progression. Also, all the transmitted rays, T_1, T_2, T_3, . . . will have amplitudes constituting a convergent geometrical progression, the convergence being the same as for R_2, R_3, R_4, and so on. The amplitude of T_1, in whose derivation no reflection is involved, will of course be much greater than that of R_1.

The phasor diagrams for the cases where $2nt \cos r$ is equal to $N\lambda$, $(2N + 1)\lambda/2$, and $(2N + 1)\lambda/4$, N being a whole number (including 0), will be like those shown in Fig. 39.16.† The diagrams in group (a) refer to the reflected light and those in group (b) to the transmitted. By squaring the resultant amplitudes in the various cases it is a simple matter to obtain curves connecting the intensity of

---

* See Chapter 38 under the heading "Lloyd's Single Mirror."

† Strictly, every alternate diagram of the third type will be the mirror image of that shown, the imaginary mirror surface being horizontal.

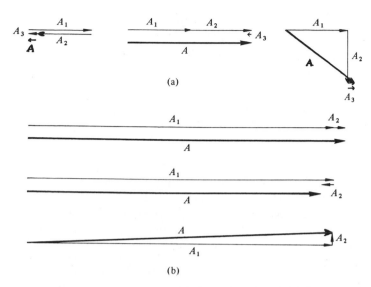

FIGURE 39.16 Phasor diagrams (a) for reflected and (b) for transmitted sets of rays.

the reflected and transmitted light with cos r. Such curves are shown in Fig. 39.17, curve (a) referring to the reflected and (b) to the transmitted light. We see that the contrast between maxima and minima is much poorer in the transmitted light than in the reflected. The two curves are complementary, as they must be from energy con-

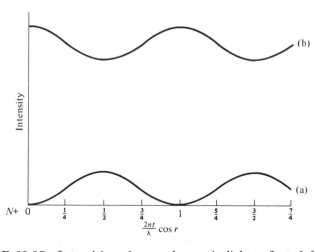

FIGURE 39.17 Intensities of monochromatic light reflected from and transmitted through a parallel-sided film as a function of cos r.

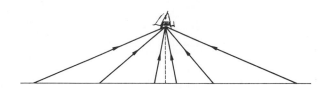

FIGURE 39.18   Directions in which interference maxima of light of a given wavelength reflected from a thin film reach an observer's eye.

siderations. When $2nt \cos r$ is equal to $N\lambda$ the intensity of the reflected light is a minimum and that of the transmitted a maximum, while when it is equal to $(N + \frac{1}{2})\lambda$ the intensity of the reflected light is a maximum and that of the transmitted a minimum.

It may be noted that Figs. 39.16 and 39.17 have been drawn correctly to scale to show the relative amplitudes and intensities of the reflected and transmitted light in the case of a film of glass of refractive index 1.5 in air, the light being incident nearly normally on the surface of the glass.

For both reflection and transmission the isochromatic lines, or lines of constant color, are concentric circles situated at infinity, centered about the perpendicular drawn from the eye to the film as axis, in the manner indicated in Fig. 39.18.

Let us now consider two numerical examples, for light reflected from a film of oil, of refractive index 1.5.

As our first example, let us take the case of a film of thickness 3000 Å, and let it be required to find what color it will appear when viewed at an angle of 60° to the normal. With this angle of incidence, $\sin i = \sqrt{3}/2$, $\sin r = (2/3)(\sqrt{3}/2) = 1/\sqrt{3}$, and $\cos r = \sqrt{2/3} = 0.816$.

As a general condition for a maximum for reflected light we have

$$2nt \cos r = (N - \tfrac{1}{2})\lambda, \qquad N = 1, 2, 3, \ldots ,^* \qquad (39.7)$$

so that those wavelengths will be strongly reflected for which

$$\lambda = \frac{4nt \cos r}{2N - 1}. \qquad (39.8)$$

In the case under consideration $4nt \cos r$ is equal to $4 \times \frac{3}{2} \times 3000 \times 0.816 = 14,700$ Å. Hence there will be reflected maxima

---

* It should be noted that this N is greater by 1 than the N used in the discussion of Fig. 39.16.

at 14,700 Å, 14,700/3 = 4900 Å, 14,700/5 = 2940 Å, ... Of these the only one in the visible range is 4900 Å; consequently, in this direction the film will appear blue-green.

Strictly, dispersion should be taken into consideration; n is never the same for all wavelengths, and it is therefore not permissible to treat $4nt \cos r$ as if it were a constant. This difficulty might have been avoided by putting the question around the other way and inquiring at what angle light of wavelength 4900 Å, for which the refractive index is 1.5, would be reflected most strongly from a film 3000 Å thick. We should then have found 60° for this angle.

Neglecting dispersion, let us now consider the case of a somewhat thicker film. Let $4nt \cos r$ for some selected angle of viewing be 60,000 Å, that is, about four times the value in the preceding problem. Then visible reflected maxima will be at wavelengths

$$\frac{60,000}{9}, \quad \frac{60,000}{11}, \quad \frac{60,000}{13}, \quad \frac{60,000}{15},$$

that is, at

6700 Å,	5500 Å,	4600 Å,	4000 Å,
red	yellow-green	blue	violet

and the resulting impression due to mixing these will be very nearly white; that is, no color will be observed. It is because of the crowding of the maxima over the visible part of the spectrum that thicker films do not ordinarily show any color.

BANDED SPECTRUM While the unaided eye cannot see colors reflected from thick films or plates, these colors can be seen quite easily with the aid of a focusing lens and a spectrometer, as indicated in Fig. 39.19. The lens is placed its focal distance in front of the collimator slit. Then from among all the sets of parallel rays reflected from the plate or film (and *all* rays may be regarded as members of sets of parallel rays), only those reflected at a certain angle will, after passage through the lens, enter the slit of the collimator. The spectrometer then sorts out the colors in this light in the usual way. What is seen in the focal plane of the telescope with this arrangement is known as a "banded spectrum."

BLOOMING OF LENSES An interesting application of the optical properties of thin films is the "blooming" of the surfaces of lenses in such optical instruments as telescopes and microscopes. Untreated glass surfaces reflect something of the order of 4 percent of light incident on them externally or internally at near-normal incidence, and in cases where there are a large number of lens

1009

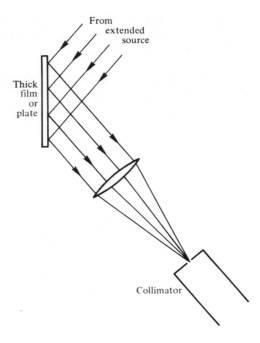

From
extended
source

Thick
film
or
plate

Collimator

FIGURE 39.19   Arrangement for observing a banded spectrum.

surfaces to be traversed the combined effect of such reflections is to produce a "light veil" over the whole field of view. When seen through this the object loses much of its contrast. The purpose of the blooming process is to reduce these reflections.

The process in question consists in the deposition on each lens surface, by distillation in a vacuum, of a thin film of a transparent material whose refractive index is intermediate between that of the glass and that of air. The thickness of the film should be such that for the part of the visible spectrum to which the eye is most sensitive the light reflected from the front face is in antiphase to that reflected from the back face. Under these conditions the two reflected radiations cancel each other out, at least partially, and there is a corresponding increase in the light transmitted. For complete cancellation the two reflected amplitudes should be equal, and it may be shown that this condition is satisfied if the refractive index of the film is equal to the geometric mean between that of air and glass. If we take the refractive index of the glass as 1.66, this means that ideally the refractive index of the film should be 1.29, while for a glass of refractive index 1.50 that of the film should be 1.22. Unfortunately, no suitable coating material has been found with such a low value of refractive index; the best that can be done is to use magnesium fluoride, MgF_2, whose refractive index is 1.38. Although this does not reduce the reflected light to zero, it does eliminate most of it.

We may note that there is a phase reversal at *both* reflections, so that the optical path difference between the two components of the reflected light, with cos r sufficiently nearly equal to 1, is simply $2nt$. Thus the optimum film thickness is such that

$$2nt = \frac{\lambda}{2},$$

whence, substituting 1.38 for n and 5600 Å for λ, we obtain for t the value

$$t = \frac{\lambda}{4n} = \frac{5600}{5.5} \doteq 1000 \text{ Å}.$$

WEDGE-SHAPED FILMS   The chief interest of wedge-shaped films is in connection with Newton's rings, which we shall study presently. In view of this, let us confine our investigation to the conditions under which Newton's rings are observed, one feature of which is that the film is an air film with glass on either side, while another is that the interference is seen in reflected light, with the incidence not far from normal. As acceptable approximations we may therefore write i for sin i and r for sin r. And, if n is taken as the refractive index of the glass relative to air, that of air relative to glass is $1/n$, and so n becomes r/i, not i/r.

As we readily see from Fig. 39.20, the angle at which the first two reflected rays converge is $2\alpha/n$. The distance apart of the points

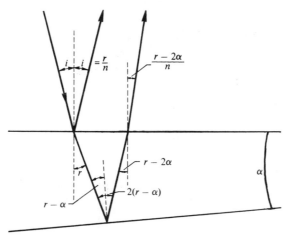

FIGURE 39.20   Rays arising from a single incident ray reflected from a wedge-shaped film.

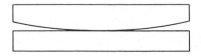

FIGURE 39.21 Plano-convex lens resting on a plane block of glass or other material, as used for the observation of Newton's rings.

from which these leave the upper surface of the film is about $2t(r - \alpha)$, where t is the mean thickness of the film in this region. Hence the distance from the film of the point where the two rays meet is $t(r - \alpha)n/\alpha$. This is of the same order of magnitude as t, except only where α is very small indeed and r is large compared with α. Even in this case, if the film is very thin a sharp image *can* be seen with the eye focused almost anywhere, including a point very close to the film. Hence, quite generally, the optimum condition for seeing interference in the light reflected from a wedge-shaped film is with the eye focused—as nearly as makes no difference—on the film itself.

NEWTON'S RINGS Let two glass surfaces, one convex and the other plane, be placed in contact as shown in Fig. 39.21. Then, under suitable conditions of illumination, a series of circular interference fringes may be observed, formed by light reflected and refracted at the two surfaces.

The experimental arrangement for viewing these fringes, known as **Newton's rings,** after Newton, who was the first to observe them, is shown in Fig. 39.22. Light from a monochromatic

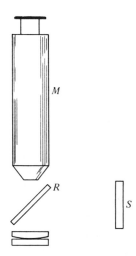

FIGURE 39.22 Experimental arrangement for observing Newton's rings.

source, such as a sodium flame or vapor lamp S, is reflected downward onto the lens and plate by a reflector R consisting of a plane sheet of glass. The interference fringes may then be observed through the reflector by means of a microscope M focused on the air film.

The air film between the adjacent glass surfaces may be regarded as a series of wedges. If the radius of curvature of the convex surface is R, then at a distance x from the point of contact the angle α of the wedge is equal to x/R, while the thickness t of the air film is such that

$$t(2R - t) = x^2,$$

or, to a very close approximation, t is equal to $x^2/2R$. For near-normal incidence, corresponding to which $\cos r$ is sufficiently nearly equal to 1, the condition for a maximum in the reflected light is that

$$2t = \frac{x^2}{R} = (N - \tfrac{1}{2})\lambda. \tag{39.9}$$

We see that the interference fringes, which are located at the air film, will consist of a series of rings whose center is dark. If a white-light source is used, then, since the scale of the ring system is different for each wavelength, a series of colored rings will be observed.

If the two glass surfaces are not in contact, as would be the case when they are separated by dust particles, the equation above must be replaced by

$$\frac{x^2}{R} + \delta = (N - \tfrac{1}{2})\lambda, \tag{39.10}$$

where δ is twice the least distance separating the surfaces.

Haidinger's Fringes

The system of Newton's rings is, as we have just seen, produced by using light at substantially normal incidence on a film of varying thickness. A closely similar pattern may be produced by a not-too-thin parallel-sided film or plate on which monochromatic light is incident at angles ranging from zero (normal incidence) upward. A suitable arrangement for viewing these rings, known as **Haidinger's fringes,** is shown in Fig. 39.23. The rings are at infinity. With a plate of any appreciable thickness it is possible to see these only if (1) monochromatic light of a high degree of homogeneity is used, and (2) the eye is properly relaxed for the reception of parallel light.

1013

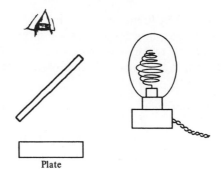

FIGURE 39.23 Experimental arrangement for observing Haidinger's fringes.

We have already obtained the equation for the maxima in our discussion of thin-film interference,

$$2nt \cos r_N = (N - \tfrac{1}{2})\lambda,$$

r_N now being written for the angle of refraction corresponding to the Nth-order maximum. We can convert this to an equation giving the angles θ_N (previously written i) to the normal at which maxima are seen as follows. From Fig. 39.24, in which t represents the thickness of the plate, we see that

$$\cos r_N = \frac{t}{\sqrt{t^2 + x_N{}^2}},$$

which, for $x_N{}^2/t^2 \ll 1$, may be written, with negligible error, as

$$\cos r_N = \left[1 + \left(\frac{x_N}{t}\right)^2 \right]^{-1/2} = 1 - \tfrac{1}{2}\left(\frac{x_N}{t}\right)^2.$$

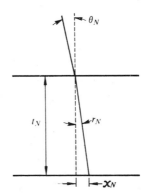

FIGURE 39.24

Substituting this in our equation for the series of maxima, we obtain, after simplification and substitution of θ_N for $n(x_N/t)$,

$$\theta_{N^2} = \frac{n\lambda}{t}\left[\left(\frac{2nt}{\lambda} + \frac{1}{2}\right) - N\right],$$

or, writing $N_0 + \Delta$ for $(2nt/\lambda) + \frac{1}{2}$, where N_0 is a whole number and Δ is a fraction,

$$\theta_{N^2} = \frac{n\lambda}{t}[(N_0 - N) + \Delta]. \qquad (39.11)$$

This is of the same form as the equation for Newton's rings; in both cases the radii (or angular radii) are proportional to the square roots of the natural numbers (including 0) all increased by a fraction. It is interesting to note that in the case of Haidinger's fringes the order of interference increases from the outside toward the center of the ring system.

Michelson's Interferometer

A system of rings corresponding essentially to Haidinger's is produced in an interferometer devised by Michelson, in which the interference is produced between what are in effect parallel surfaces separated by air, whose distance apart can be varied continuously. The basic features of this are shown in Fig. 39.25. The monochromatic source S, which is of appreciable size, constitutes, in conjunction with the lens L, the equivalent of an extended source. P is a plate of glass having plane and parallel surfaces, of which the side remote from the source is half-silvered. The system producing

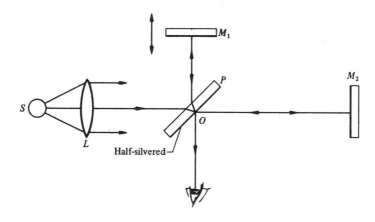

FIGURE 39.25   Basic features of Michelson's interferometer.

the interference rings is, in effect, the thick air film between the virtual image of the front-silvered mirror M_1 in the mirror P and the mirror M_2. A photograph of a ring system produced by a Michelson interferometer is shown in Fig. 39.26.

While the plate used for Haidinger's fringes has a definite thickness, the distance between the image of M_1 and M_2 may be varied at will by moving the mirror M_1 toward or away from P, as indicated by the double arrow. As M_1 is moved away from P, the existing rings expand outward and new ones grow out from the center. Since the number of new fringes appearing is equal to the number of half-wavelengths through which the mirror M_1 has been moved, this interferometer provides the possibility of counting the number of wavelengths in any given distance, for example, in principle, in the distance between the scratches at the two ends of the standard meter. This was referred to in Chapter 1. In practice

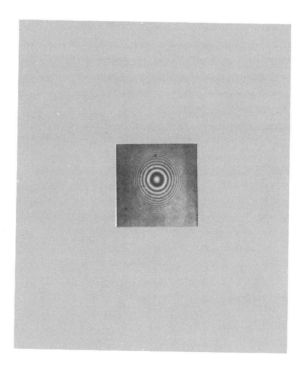

FIGURE 39.26 System of interference rings produced by a Michelson interferometer.

it is not necessary actually to count the entire number; an ingenious system has been devised whereby the correct number can be arrived at, which involves the counting of only a relatively small number of new rings.

In addition to being used for its original purpose of finding the number of wavelengths of a selected spectral line in a meter, this instrument has also been used by Michelson and Morley in a series of experiments designed to measure the velocity with which the earth moves through the "ether," the hypothetical medium in which light was once supposed to be propagated.*

X-Ray Spectroscopy

In one important method of determining x-ray wavelengths, use is made of the regular arrangement of the atoms in a crystal lattice. These act as scattering centers which, under x irradiation, send out secondary wavelets whose behavior is similar to that of the Huygens wavelets we have considered in other contexts. The scattered wavelets combine to produce maxima in various directions. Thus the crystal functions as a three-dimensional diffraction grating. And on the basis of a knowledge of the way in which the atoms are arranged in the lattice, and the scale of the structure, x-ray wavelengths may be calculated from the maxima observed under appropriate conditions.

Our first problem is to ascertain the type of lattice structure characteristic of the crystal we wish to use, and the scale.

The type of structure may be inferred from the Laue spots observed when a beam of continuous x radiation† is made to traverse the crystal and the diffracted beams are allowed to impinge on a photographic plate or film. A Laue-spot photograph obtained with a crystal of sodium chloride, in which the central spot is masked off, is shown in Fig. 39.27.

Let us suppose that by some means we knew the type of structure, for example that we knew it to be a simple cubic lattice. Then for x rays traversing the crystal perpendicularly to one of the faces of the cube we could predict the pattern of Laue spots that must be

---

* The negative results of these experiments played an important part in prompting Einstein to postulate his epoch-making theory of special relativity. For a discussion of this, see Chapter 2 of Volume III.

† By continuous x radiation is meant radiation consisting of a continuous range of wavelengths. This is the x-ray equivalent of white light in the optical region of the spectrum.

FIGURE 39.27 Laue-spot pattern formed by diffraction of x-rays by a crystal of sodium chloride. (Courtesy of Soo Kock Tan and J. Lawrence Katz, Laboratory for Crystallographic Research, Rensselaer Polytechnic Institute.)

produced, each spot being due to reinforcing rays of *some* wavelength scattered from all the atoms exposed to the radiation. Conversely, from the observed pattern we can work backward and infer the type of lattice that produced it.

The scale can now be worked out if we know how many atoms there are in a given volume. This information we have from the measured density of the material and the masses of the atoms concerned, the latter being known from the atomic weights, electrochemical data, and the electronic charge.

Thus equipped with complete qualitative and quantitative knowledge of the lattice structure, we may proceed to use the crystal for finding specific x-ray wavelengths, for example the wavelengths of the characteristic radiation—the "line spectrum" of the

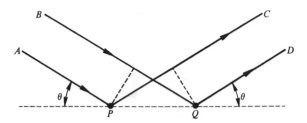

FIGURE 39.28

material of the target—which is superimposed on the continuous radiation.

In x-ray spectroscopy use is made of the reflection of x radiation, as of light from a mirror, from sets of planes of atoms, or, as they are called, "lattice planes."

First let us consider parallel rays AP and BQ incident on two adjacent atoms P and Q of a certain lattice plane, as shown in Fig. 39.28, together with the rays PC and QD scattered by these in the plane of the incident rays and in directions making the same angle θ with the lattice plane as do AP and BQ. Then, considering this as Fraunhofer diffraction, we see at once that the path difference between the composite rays APC and BQD is zero; the scattered rays will reinforce one another. This will be true of any angle θ, but only for certain such angles will there be *general* reinforcement from all the atoms in the whole assemblage of lattice planes parallel to the one considered. And only if there is this general reinforcement will there be a diffracted beam in the direction of the rays PC and QD.

The condition for general reinforcement may be seen from Fig. 39.29, in which the dashed lines represent two adjacent lattice

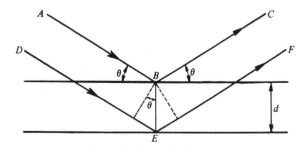

FIGURE 39.29 X-rays "reflected" at a given angle by adjacent net planes of a crystal.

planes separated by the distance d. We see that the path difference between ABC and DEF is $2d \sin \theta$. Hence the condition for general reinforcement of x radiation reflected from sets of parallel lattice planes is

$$2d \sin \theta = N\lambda, \tag{39.12}$$

where N is any whole number, the smallest value of θ at which reinforcement is obtained corresponding to first-order diffraction ($N = 1$). This equation, which was first obtained by W. H. and W. L. Bragg (father and son), is known as **Bragg's law.**

To find the wavelengths of an x-ray line spectrum a particular set of lattice planes in a crystal is selected to reflect the radiation, and the crystal is rotated so that the angle between the incident x-ray beam and the lattice planes varies continuously. At the same time a lead absorbing screen is used to screen off all diffracted radiation except that which has been produced by Bragg reflection from the lattice planes selected. This is done by appropriately synchronized movement of the screen, such that only the reflected radiation is allowed through the aperture. This radiation then passes on either to an ionization chamber, or, more usually, to a photographic plate or film. In this way a line spectrum of the radiation is obtained, and the wavelength of each line may be calculated from the corresponding angle θ and the lattice-plane spacing d, using Bragg's law.

As we have already noted, crystals not grown under very special conditions have defects of various kinds, for example vacant lattice sites. Consequently, the value of d corresponding to any particular set of lattice planes calculated from the density of a crystal and Avogadro's number is slightly in error, and so also is the value of λ based on this d, using Bragg's law.

Fortunately, we do not have to rely on crystal spectrometers for our knowledge of x-ray wavelengths. Since 1925 it has been possible to measure these by a more direct and more reliable method. This depends on the use of a ruled diffraction grating of reflection type, with the x-ray beam incident at a very small angle with the plane of the grating. It has been found that at very glancing angles x rays are totally reflected from metal surfaces. Ruled metal reflection gratings may accordingly be used for the measurement of x-ray wavelengths.

It should be noted, however, that despite the fact that the ruled-grating method is more direct in principle and gives more accurate absolute values of x-ray wavelengths, it is, in practice, rather troublesome to apply. For this reason the ruled-grating method is used only for the accurate determination of the wavelengths of a few "standard" spectral lines. This having been done, crystal spectrometers, which lend themselves more readily to the determination of relative values, are used to find the wavelengths of other lines in terms of those of the standards.

LINE AND CONTINUOUS X RADIATION   One of the earliest x-ray spectrograms, obtained by the Braggs using an ionization chamber for measuring the diffracted x rays, is shown in Fig. 39.30. Although the resolution is not good by modern standards, it serves well enough to show the basic principle of x-ray spectroscopy. The target of the x-ray tube was of platinum, and the diffracting system was a set of lattice planes of a crystal of rock salt. Apart from the maximum farthest to the right, two sets of three peaks each were obtained, as shown, and from analysis on the basis of Bragg's law these were found to be consistent with the assumption of first-order diffraction ($N = 1$) for the first group, on the left, and second-order diffraction ($N = 2$) of the same wavelengths for the second.

By employing modern techniques it is possible to obtain much sharper lines. Such line spectra are found to be characteristic of the material of the target, just as line spectra in the optical range of wavelengths are characteristic of the gases or vapors from which they are obtained. Whether or not all lines in a particular x-ray line spectrum are excited depends on the voltage at which the tube is operated; for the excitation of any line it is necessary that the kinetic energy of the electrons bombarding the target shall be at least as great as the energy quantum hf of the line in question, h being Planck's constant. If the voltage is not sufficient to excite the entire spectrum, only those lines will be excited for which eV is greater than or equal to hf; and if the tube voltage is insufficient

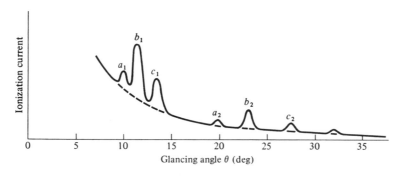

FIGURE 39.30   Early spectrogram of line and continuous x radiation produced at a platinum target and observed with a rotating crystal of rock salt and an ionization chamber.

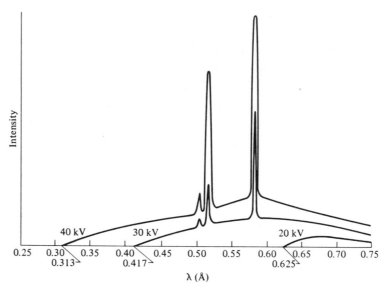

FIGURE 39.31 Spectrograms of x radiation produced at a molybdenum target with two different voltages of bombarding electrons.

to excite even the lowest-frequency line, only the continuous radiation will appear. Figure 39.31 shows schematically (not to scale) curves of intensity against wavelength in the shorter-wavelength region of the x-ray spectrum of molybdenum for two bombarding voltages. As will be seen, only the higher of these, 25 kV, excites the K_α and K_β lines of this element; for the lower voltage the short-wavelength limit of the radiation is longer than the wavelengths of these lines.

The background spectrum on which the characteristic spectrum is superimposed is also of considerable interest. This was originally given the name *Bremsstrahlung* by German writers, derived from *bremsen*, to brake, and *Strahlung*, radiation. It describes very well the process by which the continuous radiation is generated, namely, the stopping of the bombarding electrons and the conversion of part, or, in the limit, the whole of their kinetic energy into radiation. The original German name has since been adopted widely in English-language x-ray literature, being written without the capital as "bremsstrahlung."

The continuous radiation obtained with a tungsten target at four bombarding voltages is shown in Fig. 39.32. The chief point of interest concerning the continuous radiation is the short-wavelength limit of its range. The frequency f_0 corresponding to this limit is

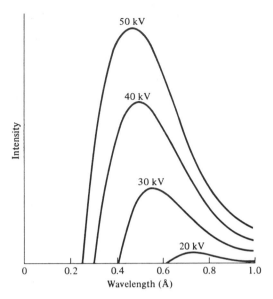

FIGURE 39.32 Continuous x radiation obtained with a tungsten target at four different bombarding voltages.

the same for all target materials and is related to the bombarding voltage V by the relation

$$hf_0 = eV. \tag{39.13}$$

Problems

39.1   Monochromatic light incident normally on a grating ruled with 5000 lines per centimeter gives its first main diffraction maximum on either side of the central maximum at an angle of 14°31′26″. Find the wavelength of the light. In how many higher orders, and in what directions, will light of this wavelength be diffracted? Would it be possible to obtain still higher orders, using oblique incidence? Explain.

39.2   Light from a mercury arc of wavelength 5461 Å is incident normally on a grating having 2000 lines per centimeter. What is the angular deviation of the third-order images on either side of the central maximum?

39.3   The spacing of the lines in a diffraction grating is not known. When this grating is set up in a spectrometer with normal incidence, the angle between the two second-order maxima obtained with sodium light ($\lambda = 5893$ Å) is found to be 150°. What is the number of lines per unit length of the grating?

39.4   The D lines of sodium, whose wavelengths are 5890 Å and 5896 Å, are measured with a diffraction grating that has 5000 lines per centimeter, the light being incident normally on the grating. Calculate the angular separations of the two D-line components in both the first-order and second-order spectra.

39.5   Show, by reasoning based on phasor diagrams, that in the spectrum of monochromatic radiation produced by a diffraction grating of N lines there are $N - 2$ subsidiary maxima between each pair of successive main maxima, and that if N is large the intensity of the first subsidiary maximum on either side of a main maximum is $4/9\pi^2$ of that of the latter.

39.6   A certain diffraction grating used with normal incidence is found to produce five images of the collimator slit (including the "straight-through" image) with the 4044-Å line of potassium, but only three with the 4067-Å line. What can you deduce about the grating? How many images would this grating produce (still with normal incidence) at 8095 Å in the near infrared?

39.7   A distant sodium street lamp seen through a woven nylon curtain appears to be accompanied by a series of "images" spaced 0.3° apart. What is the spacing of the nylon fibers? ($\lambda_{Na} = 5893$ Å.)

39.8   Parallel sodium light ($\lambda = 5893$ Å) falls normally on the plane of a slit of width 0.5 mm, and the diffracted light is viewed in a telescope of focal length 50 cm which is so directed that the cross wire coincides with the central maximum. Find the distances from the cross wire of (a) the first minimum, (b) the first subsidiary maximum, and (c) the second subsidiary maximum on either side.

39.9   If in Problem 39.8 the "parallel" light is derived from an illuminated slit at the focus of a lens of 20-cm focal length, below what limit must the width of this first slit be kept if the details of the diffraction pattern are not to be lost?

39.10   It is possible to calculate what should be the positions of points on a graph of intensity against angle for Fraunhofer diffraction by a single slit without having actually to construct phasor diagrams and measure their resultants. Show how this can be done.

39.11   An opaque screen with a rectangular aperture of sides a and $2a$ is placed symmetrically directly in front of the objective of a telescope of relatively large diameter and of focal length $|f|$, and parallel light of wavelength λ passes through the aperture along the direction of the telescope axis. Describe the diffraction pattern produced in the focal plane of the telescope and calculate the positions of the maxima of first, second, and third orders.

39.12   A diffraction grating is made by a photographic process so that alternate strips are clear and opaque. Find which spectra (if any) would be missing in the following cases: (a) equal widths of clear and dark, (b) clear spaces twice as wide as the dark, (c) dark spaces twice as wide as the clear, and (d) dark spaces three times as wide as the clear.

39.13   Parallel light falls normally on a grating consisting of five equally spaced slits cut in an opaque screen. The width of each slit is one third of the distance between the centers of adjacent slits; that is, $a = \frac{1}{3}d$. Construct a graph of intensity against angle for the Fraunhofer diffraction pattern over the range from the central maximum to the fourth-order main maximum on one side. (Although the graph need not be accurate in detail, the positions of all maxima and minima and the relative intensities of the maxima should be shown correctly. Assume that $a \gg \lambda$.)

39.14 Repeat Problem 39.13 for the case where the grating of that problem is replaced by an array of five parallel and coplanar strips of opaque material, each of which has a width of one third of the distance between the centers of adjacent strips.

39.15 The headlights of an approaching car, 1.1 m apart, are observed with a telescope whose objective has a diameter of 5 cm. In front of the telescope is a vertical slit whose width can be varied, and it is found that when the slit width is 1.2 cm the headlights are on the borderline of resolution. How far away is the car? (Assume an effective mean wavelength of 5600 Å.)

39.16 Assuming that the resolving power of the unaided eye is limited by diffraction alone, and that the diameter of the pupil at night is 6 mm, find the maximum distance at which the headlights of an approaching car, 1.1 m apart, can be seen as separate from one another. (Take $\lambda = 5600$ Å.)

39.17 Neglecting the effects of irregularities of refraction in the earth's atmosphere, calculate the linear separation of two objects on the moon's surface that could just be resolved by the 200-in. Mount Palomar telescope.

39.18 Beyond what distance will it no longer be possible to resolve the headlights of a car, 1.1 m apart, when these are viewed through a telescope of aperture 3 cm?

39.19 All except a narrow ring around the periphery of a telescope objective is masked off by an opaque disk coaxial with the telescope placed directly in front of it (see the figure).

(a) Does this produce a dark patch in the central region of the image of a distant object formed in the focal plane of the telescope? Explain your answer.

(b) How, if at all, will the presence of the disk in front of the telescope affect the resolving power? Explain your answer with the aid of phasor diagrams.

Figure for problem 39.19.

39.20 What minimum thickness of soap film is required for it to appear green by reflection of white light at normal incidence? If the film had 10 times this thickness, for what wavelengths in the visible range would there be reinforcement? (Take $\lambda = 5.5 \times 10^{-7}$ m for green light and $n = 1.33$ for the soap film.)

39.21 White light is directed toward a glass plate of thickness 2 mm and refractive index 1.5 at an angle of incidence of 45°. If there were no variation of refractive index with wavelength, how many maxima within the visible range (3900 Å to 7600 Å) would appear in the spectrum of the reflected light?

39.22 (a) In the normal experimental arrangement for producing a banded spectrum an extended white-light source is used. What would be the effect of replacing this by a monochromatic source?

(b) What would be the effect of retaining the monochromatic source and replacing the lens–spectrometer system by a telescope focused for infinity whose axis makes an angle θ with the normal to the plate? (Assume that a range of angles around θ is observable in the telescope.)

(c) There is a particular main feature of interest in what is seen in case (b). Assuming the interference formula for a parallel-sided film or plate, obtain a quantitative expression for this feature.

39.23 Find the phase difference between light reflected from the inner and outer surfaces of a deposit of magnesium fluoride ($n = 1.38$) 1000 Å thick on a lens surface (a) for violet light of wavelength 4000 Å, and (b) for red light of wavelength 7000 Å. (Neglect variation of n with λ.)

39.24 A fine wire rests on an optically flat surface of a slab of glass, and an optically flat surface of another glass slab is placed on top, as shown. When viewed normally with light of wavelength 5890 Å, interference fringes are observed, the distance between adjacent dark bands being 0.5 mm. If the wire is 10 cm from the line of contact of the surfaces, what is its diameter?

Figure for problem 39.24.

39.25 In the system of Newton's rings formed with sodium light ($\lambda = 5893$ Å) when a convex lens surface is brought into contact with a plane glass surface, it is found that the twelfth bright ring has a diameter of 11.4 mm. What is the radius of curvature of the lens surface?

39.26 In a system of Newton's rings formed with filtered green light from a mercury arc, it is found that the diameter of the fifteenth bright ring changes from 12.2 mm to 10.1 mm on introduction of a certain liquid between the lens and the plate. What is the refractive index of the liquid for this radiation?

39.27 A plano-convex lens is mounted a short distance below the lower, polished surface of a plane slab of material of thickness 0.25 cm supported by its upper surface. Newton's rings are observed from below with light of wavelength 5893 Å. When the slab of material is heated through 100°C, it expands toward the lens and the first bright ring of the original pattern moves out and occupies the position previously occupied by the twenty-seventh. Find the coefficient of linear expansion of the material.

40

POLARIZATION

Polarized and "Unpolarized" Light

In Chapter 38 we saw that the centers radiating in any actual source do so with different states of polarization, distributed at random. Hence so-called unpolarized light consists of a mixture of elliptically polarized components, presumably with circular and plane polarization represented as limiting cases. We saw also that any elliptically polarized component of the mixture may be resolved into plane-polarized waves, polarized in any two selected planes containing the direction of propagation and perpendicular to one another. Since this applies to each one of the elliptically polarized components, the whole mixture is resolvable into two corresponding sets of plane-polarized waves whose phases are distributed at random and whose amplitudes are distributed according to some probability law. And, as we have seen, each such set can be compounded into a single wave train. Owing to the random distribution of the states of polarization of the radiation emitted by the individual centers, the most probable amplitudes of the two resultant wave trains polarized in planes at right angles must necessarily be equal.

There are a number of devices whose actions on the two plane-polarized components polarized in a certain pair of mutually perpendicular planes differ from one another. In some cases the nature of this difference is such as to make it possible to select one plane-polarized component for further study or treatment while rejecting the other. Such devices are known as **polarizers.**

A polarizer will obviously deal differently with plane-polarized incident light according to its orientation with respect to the plane of polarization of this light. It will also deal differently with incident elliptically polarized light according to its orientation. Hence a polarizer may be used to give information concerning the state of polarization of the light. A polarizer used in this way is known as an **analyzer.**

Polarization by Reflection

If light is reflected obliquely from the surface of a transparent body, such as water, or a glass plate, then in general both the reflected and the transmitted beams are found to be partly polarized; in each the intensities of the plane-polarized components having, say, their magnetic vector in and at right angles to the plane of incidence* are different. At a particular angle of incidence, which has been called the "polarizing angle," the reflected light is completely plane-polarized, although the transmitted is, as at all other angles, only partly plane-polarized. This means that some of one plane-polarized component of the incident light is reflected but none of the other. The plane of incidence (and of reflection) in this case has been defined as the "plane of polarization" of the reflected light. It is now known that of the two vectors of the plane-polarized reflected light it is the *magnetic* vector that oscillates in this plane. Hence this definition is equivalent to identifying the plane of polarization with that in which the magnetic vector oscillates.

The angle of incidence θ which gives completely plane-polarized reflected light was found empirically by Brewster to be such that

$$\tan \theta = n, \tag{40.1}$$

n being the refractive index of the reflecting material. This law, known as **Brewster's law,** has since been derived theoretically on the basis of the electromagnetic theory of light, which also gives the result that the magnetic vector of the reflected light oscillates in what was defined above as the plane of polarization.

For the usual range of refractive indices covered by different kinds of glass the corresponding values of polarizing angles are in the region of 57° to 58°.

Light reflected obliquely from lakes or other bodies of water, from wet roads, or even from dry bitumen, is largely plane-polarized in a vertical plane, and much of the glare from these can therefore be eliminated by wearing glasses that reject light polarized in this plane.

Polarization by Scattering

Let us consider the scattering of light by particles whose dimensions are of the order of the wavelength of the light or less. We should expect the electrons in such particles to respond to the alternating fields of the electromagnetic waves and themselves vibrate correspondingly. The resulting system of vibrating electrons

---

* The plane of incidence is defined as the plane containing the incident ray and the normal to the surface at the point of incidence.

could then be regarded as an independent source of radiation, producing scattered light.

The observed scattering produced by small particles accords well with this picture. As in the case of acoustic scattering, the smaller the wavelength relative to a given particle, the greater is the scattering action. This means that there is preferential scattering for the shorter wavelengths. It is because of this that we have blue skies and red sunsets; the scattering particles in the atmosphere scatter blue light more than red. Also, because so much of the shorter-wavelength end of the visible spectrum is scattered out of the light reaching us directly from the sun, the appearance of this body as seen through the atmosphere suggests a temperature much lower than its actual temperature; it does not look as blue as it should.

Let a beam of unpolarized light be incident normally on the plane of the diagram at the point O, shown in Fig. 40.1, where we may imagine a scattering particle to be located. This beam is, as we have seen, equivalent to two plane-polarized components polarized in planes mutually at right angles and having equal amplitudes, whose relative phases change in a random fashion from one moment to the next. The electrons in the scattering particle will accordingly be set into independent vibrations in the directions of the corresponding electric vectors. The vibrations of the electrons in each direction will be equivalent to an oscillating electric dipole, generating electromagnetic waves in the same manner as does a dipole radio antenna. Let the two lines shown in the figure represent such oscillating dipoles. Then, because there is no component of

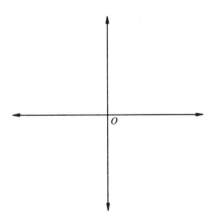

FIGURE 40.1 Equivalent independent vibrations of electrons of a particle in response to incidence of unpolarized light on it in a direction perpendicular to plane of diagram.

either electric vibration at right angles to the plane of the diagram, a scattered beam of light traveling in the plane of the diagram must necessarily be plane-polarized, with its electric vector vibrating in this plane. This is, in fact, observed in all cases where the scattering particles are small enough to scatter the light by the mechanism we have considered and not by specular reflection from their surfaces. Thus the light of the sky reaching us from a direction at right angles to that of the sun is found to be almost completely plane-polarized.

PRODUCTION OF POLARIZED X RAYS X radiation, like visible light, exhibits polarization phenomena, and this and other evidence leads us to infer that x rays must be propagated as electromagnetic waves. Polarized x rays were first produced, and their state of polarization demonstrated, by C. G. Barkla in 1906, using a scattering technique.

The experimental arrangement used by Barkla is shown in Fig. 40.2. A beam of x rays proceeding in the x direction of a Cartesian coordinate system (see the diagram on the right of the figure) is incident on a body A, for example a block of paraffin wax, from which scattered secondary x rays proceed in various directions, including the y direction, toward a second scattering body B. Tertiary rays scattered from B are now observed in the plane passing through B perpendicular to the y direction. It is found that these have their maximum intensity in the positive and negative x directions and are of zero intensity in the positive and negative z directions.

It would hardly be possible to account for this otherwise than on the assumption that x rays consist of electromagnetic waves. But on this basis the explanation is simple. The oscillating electric fields associated with the waves incident on A are all perpendicular to the x direction, and these induce corresponding oscillations in A's

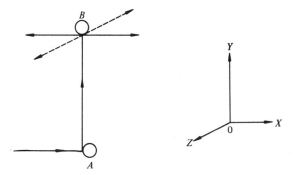

FIGURE 40.2   Polarization of x rays by double scattering.

electrons. Accordingly, in the secondary beam scattered toward B, the electric fields, being derived from the components of the electron oscillations in A that are perpendicular to the direction of propagation, are entirely in the z direction. B's electrons will, in turn, respond to these fields, vibrating only in the z direction. Hence radiation scattered from B will have its maximum intensity in directions perpendicular to the z direction, and there can be no tertiary radiation in the z direction.

Double Refraction

In our study of the propagation of light through such materials as glass and water we have been concerned with only one refractive index for any particular wavelength, the velocity of propagation not depending in any way on either the state of polarization of the light or its direction of travel. This is true also of cubic crystals; these, like liquids, are optically isotropic. But here isotropism ends; in noncubic crystals incident light is automatically broken up into two plane-polarized components polarized in planes at right angles to one another, and these travel in different directions and with different velocities. This phenomenon is known as **double refraction.**

A unit cell of the simplest type of noncubic crystal, known as a **uniaxial crystal,** is represented in Fig. 40.3. This is in the form of a rectangular parallelepiped with two sides equal to one another and the third different from these. The direction of this third side is known as that of the **optic axis** of the crystal. The corners of the cell are occupied by the fundamental units of the structure; these may be atoms, ions, or atomic or molecular aggregates. It will be convenient to think of the edges of the unit cell as parallel to the axes of a Cartesian coordinate system, the optic axis corresponding to the z direction while the directions of the edges perpendicular to these are those of the x and y axes.

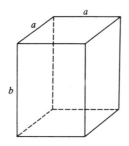

FIGURE 40.3   Unit cell of uniaxial crystal.

The electrons belonging to the atoms or ions are disturbed by the oscillating electric fields of electromagnetic radiation traversing the crystal, and, as we have already noted, this disturbance modifies the velocity of propagation of the waves; actually the velocity is always decreased. For a field in any given direction the disturbance of the electrons may be assumed to be proportional to the field strength; at least this must be so for fields that are not too strong.

Let us now consider fields in directions parallel to the x and y axes, that is, perpendicular to the z axis. Whatever may be the disturbance produced by a field in the x direction, a disturbance of the same magnitude must, by symmetry, be produced by an equal field in the y direction. And, since any field perpendicular to the z axis may be resolved into x and y components, each of which will produce its own disturbance, it appears likely that such a field will produce a disturbance proportional to the strength of the field by the same factor, whatever may be its direction; indeed, this *must* be the case if what we have called the "disturbance" may be treated vectorially. This expectation is borne out experimentally; it is found that just as cubic crystals are optically isotropic, so also the velocity of light plane-polarized with its electric vector perpendicular to the optic axis of a uniaxial crystal is—by the criterion of refractive index—independent of its direction of travel. A ray of light so polarized is known as an **ordinary ray.**

A uniaxial crystal is shown in side elevation and plan in Fig. 40.4,* and the line marked O represents an ordinary ray traveling within it. The dots indicate lines perpendicular to the plane of the diagram to show the direction of oscillation of the electric vector.

Next let us consider a ray such as that marked E in the figure, polarized in the plane at right angles to that of the corresponding ordinary ray, so that its electric vector oscillates in the direction of the transverse lines shown. Such a ray is known as an **extraordinary ray.** Since the oscillating electric vector of this ray has a z component, a component parallel to the optic axis, the disturbance of the electrons due to this vector will be proportional to the field by a different factor from that applying in the case of the ordinary ray, and hence the velocity of the extraordinary ray will also be different. Moreover, this velocity will vary with the angle that the direction of the extraordinary ray makes with the optic axis. In the

---

* Not all uniaxial crystals occur naturally in the tetragonal form shown in this figure. However, the discussion that follows applies in all cases; in this discussion reference is made only to the direction of the optic axis, not to the external form of the crystal.

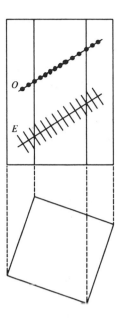

FIGURE 40.4 Directions of vibration of electric vector in ordinary and extraordinary rays within a uniaxial crystal.

limiting case where this angle is zero, that is, where the ray travels along the optic axis, the electric vector necessarily oscillates perpendicularly to this axis, as in the case of an ordinary ray, and so for this direction of travel the distinction between the ordinary and the extraordinary ray vanishes. On the other hand, when the extraordinary ray travels at right angles to the optic axis, its electric vector oscillates entirely along the z direction, and the difference between its velocity and that of an ordinary ray is then a maximum.

It may be shown that whereas for the ordinary radiation spreading out from a point source the wave fronts are spheres centered on the source, in the case of the corresponding extraordinary radiation they are ellipsoids of revolution whose axes of symmetry correspond to the optic axis. Such an ellipsoid may be either prolate or oblate; these two cases are shown in parts (a) and (b), respectively, of Fig. 40.5, the ordinary and extraordinary wave fronts being labeled O and E, respectively. An example of case (a) is calcite, while quartz is an example of case (b).

For ordinary rays the usual laws of refraction apply, as in the case of isotropic media. However, the behavior of extraordinary rays at surfaces of discontinuity is quite different. It is even possible for a normally incident ray to be refracted away from the normal,

1033

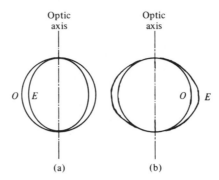

FIGURE 40.5 Ordinary and extraordinary wave fronts of radiation originating at a point within a uniaxial crystal for two cases.

as shown in Fig. 40.6. This figure shows how the envelope of the ellipsoidal Huygens secondary wavelets of the kind shown in Fig. 40.5 within the crystal travels in a direction intermediate between that of incidence and that of the optic axis. This direction of motion of the envelope is that of the refracted extraordinary ray.

If the incident ray shown in Fig. 40.6 is unpolarized, this will be broken up by the crystal into the two plane-polarized components corresponding to the ordinary and the extraordinary ray, and these will travel independently within the crystal as the rays marked O and E, respectively.

One of the best-known crystals showing double refraction is calcite, $CaCO_3$, also known as Iceland spar. This material cleaves in three directions, forming rhombs, all of whose edges meet at angles of either 78° or 102°. At two opposite corners all three angles

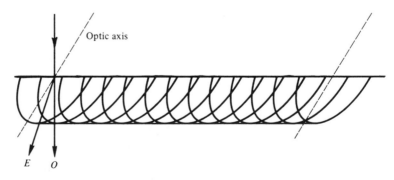

FIGURE 40.6 Refraction of extraordinary ray at surface of uniaxial crystal explained in terms of Huygens secondary wavelets.

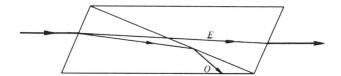

FIGURE 40.7 Production of emergent polarized light from incident unpolarized light by a Nicol prism.

are 102°, and the direction of the optic axis is that which makes equal angles with the edges that meet in one of these corners. From this it will be seen that the normal to any face of the rhomb will make a considerable angle with the optic axis. Correspondingly, there will be a substantial angular separation of the ordinary and extraordinary rays derived from an unpolarized normally incident ray. These rays, on finally emerging from the opposite face, will again travel normally to the surface, but along different paths, the ordinary ray along the prolongation of the original incident ray, and the extraordinary ray alongside it. Hence, if the crystal is now rotated about an axis normal to these faces, the emergent extraordinary ray will rotate about the ordinary ray.

Not all noncubic crystals are uniaxial; there are a number of others, known as biaxial crystals, whose internal structure and optical behavior are more complex. Consideration of these would be beyond the scope of this book. Suffice it to say that in biaxial crystals there are two directions in which light polarized in any plane will travel with the same velocity.

NICOL PRISM The difference in refraction between the ordinary and extraordinary rays has been turned to good account in the design of an ingenious device, known after its originator as the "Nicol prism," which functions as an efficient polarizer and analyzer. This is shown in Fig. 40.7. First two opposite ends of a cleaved crystal of Iceland spar are cut at a slight angle with the original faces and the new surfaces are polished. Then a cut is made right through the crystal in a certain direction,* with a thin wheel fed with diamond or carborundum powder, after which the surfaces are polished and reunited with Canada balsam cement. If the cut has been made at the correct angle with the optic axis, the ordinary ray, with its relatively large refractive index (1.66), will be totally internally reflected from the layer of Canada balsam cement

---

* Precise information concerning the directions to give the trimmed ends and the direction in which the cut through the crystal should be made can be obtained from most textbooks on optics.

($n = 1.54$), ultimately being absorbed by a coating of black paint applied to the relevant face of the prism. On the other hand, relatively little of the extraordinary ray is reflected; most of it is transmitted through the cement. Transmission is favored not only by the smaller effective refractive index* of this ray for the direction in which it travels, about 1.48, but also by its state of polarization, with the magnetic vector oscillating parallel to the surface of the cement.

Although some of the extraordinary ray is lost by reflections at the outer surfaces of the Nicol prism and at the two surfaces of the layer of cement, much the greater part of it is transmitted. The slight loss is relatively unimportant. What *is* important is that what is transmitted is completely plane-polarized; it is this that constitutes the great merit of the device.

Dichroism

Not only do the ordinary and extraordinary rays in uniaxial crystals travel with different velocities, but in some cases the absorption coefficients of these differ substantially. Such crystals are known as **dichroic,** and the phenomenon of unequal absorption which they exhibit is known as **dichroism.** As we should expect, the difference between the two absorption coefficients is greatest for light traveling at right angles to the optic axis.

A dichroic crystal that was formerly used widely as a polarizer and analyzer is the natural crystal tourmaline. Even in quite thin plates cut in planes containing the optic axis, of the order of thickness of only 2 mm, this absorbs virtually the whole of the ordinary ray, while transmitting the extraordinary ray relatively freely. Unfortunately, tourmaline has one rather serious defect: There is an appreciable amount of selective absorption even in the case of the extraordinary ray, and this causes the transmitted light to appear greenish or brownish. It is largely for this reason that tourmaline has now been superseded by other dichroic materials which are free of this defect.

Certain of these materials are used in the manufacture of a plastic film known as **Polaroid.** As now manufactured, this consists of polyvinyl alcohol, suitably treated. A sheet of this material is first stretched to line up its complex molecules in long parallel chains.

---

* As here used, the term "refractive index" must be considered to be defined as a ratio of velocities, not as a ratio of sines of angles.

It is then either impregnated with iodine or subjected to a special dehydrating process. In the iodine-impregnated material, known as H Polaroid, the iodine atoms arrange themselves in long strings in the direction of stretching, with a periodicity in this direction of about 3.1 Å. While this material absorbs one polarized component almost completely, its transmission of the other component, although good, is not perfect. At best, it transmits, as plane-polarized light, some 38 percent of the energy of unpolarized light incident on it between 4500 Å and 7500 Å. At the short-wavelength end of the visible spectrum it is less good, and at 3800 Å only slightly over 20 percent of the incident light is transmitted. The transmission improves somewhat as the extreme red end of the spectrum is approached, attaining a value of about 45 percent of 7600 Å. At wavelengths above 4400 Å the polarization of the transmitted light is virtually perfect, but below 4400 Å, down to 3800 Å, this is no longer so; at 4000 Å the light is only about 90 percent polarized.

The material resulting from dehydration of stretched polyvinyl alcohol is known as K Polaroid. This form of polaroid, although generally less good optically than H Polaroid, is more stable, withstanding intense illumination combined with relatively high temperatures without deterioration.

Both forms of polaroid film are mounted sandwich-fashion between glass or plastic plates for mechanical protection.

Production of Circularly and Elliptically Polarized Light

Doubly refracting material may be used to produce light in which the electric and magnetic vectors (always mutually at right angles) rotate, with either right-handed or left-handed rotation, the magnitude of each vector either being constant (circularly polarized light) or varying with direction (elliptically polarized light).

The material used for this purpose is always in the form of a very thin sheet, and, if it is uniaxial, its plane should contain the optic axis, so that on entering the sheet a normally incident ray is broken up into an ordinary and an extraordinary component, both traveling in the same direction (normal to the sheet) but with different velocities. With biaxial crystals the same conditions are satisfied if the normal to the sheet at a point where the two axes may be imagined to go through it is in the plane of these axes and bisects the angle between them. The most commonly used material is mica. This is biaxial, and, fortunately, cleaves in planes having the right orientation with respect to the optic axes.

1037

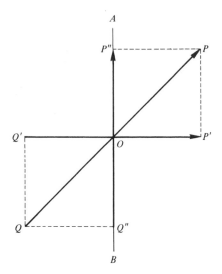

FIGURE 40.8 Production of equal ordinary and extraordinary polarized components within a uniaxial crystal from light incident normally on its surface (plane of diagram) in the case where the optic axis is parallel to this surface and the angle between the plane of polarization of the incident light and the optic axis is 45°.

In Fig. 40.8 let *AOB* be the direction in the plane of the crystal plate in which the electric vector of the extraordinary ray oscillates; in a uniaxial crystal this would be the direction of the optic axis, whereas in a cleaved sheet of mica it would be the intersection of the plate with the plane defined by the two optic axes. Let *OP* represent the amplitude of the oscillation of the electric vector of a plane-polarized ray incident normally on the plate, and let the plane *PQ* of this oscillation make an angle of 45° with *AB*. This ray is equivalent to two component plane-polarized rays whose electric vectors, having equal amplitudes corresponding to *OP'* and *OP''*, oscillate in phase with one another at right angles and parallel to *AB*, respectively. On entering the plate the former of these becomes the ordinary and the latter the extraordinary ray.

Because these rays travel through the plate with different velocities, there must be a progressive phase retardation of one with respect to the other, and after traversing a certain path this will amount to $\pi/2$. If this path is the thickness of the plate, the two rays, emerging with this phase difference, will recombine as a circularly polarized ray. A plate of this thickness is known as a **quarter-wave plate.**

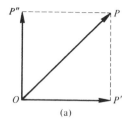
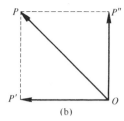

(a)   (b)

FIGURE 40.9 Ordinary and extraordinary components (a) of light incident normally on a half-wave plate and (b) of the emergent light, and corresponding planes of polarization of the light.

Clearly, a plate of twice this thickness, a "half-wave plate," will give emergent components derived from the ordinary and extraordinary rays having a phase difference of π, and these will recombine as a plane-polarized ray whose plane of polarization is at right angles to that of the incident ray. This is indicated in Fig. 40.9, parts (a) and (b) of which refer to the incident and emergent rays, respectively.

A three-quarter-wave plate, one of three times the thickness of a quarter-wave plate, will, like a quarter-wave plate, produce circularly polarized light, although this will not be of the same kind. If a quarter-wave plate produces right-handed circularly polarized light, that given by a three-quarter-wave plate will be left-handed, and vice versa.

It is easy to see that plates whose thicknesses are not whole-number multiples of that of a quarter-wave plate will produce elliptically polarized light. Also, all plate thicknesses other than whole-number multiples of that of a half-wave plate will produce elliptically polarized light if the plane of polarization of the incident light is inclined to AB in Fig. 40.8 at any angle other than 45°.

A given plate is a quarter-wave plate, half-wave plate, and so on, only for a particular wavelength; for other wavelengths, plates of different thicknesses must be used to produce the corresponding effects.

Photoelasticity

Certain noncrystalline and normally isotropic materials, such as glass and celluloid, exhibit double refraction when strained. The directions of the planes of polarization in these are parallel to the directions of the two principal stresses producing the strain.

This phenomenon is called **photoelasticity,** and advantage is taken of it to detect and study stresses in optical glass, lamps, radio tubes, and so on. An important engineering application is the study of stresses in models of various structures made of suitable materials, using polarized light for locating and measuring the stresses.

Transparent material that is strained and thereby rendered doubly refracting reveals its condition when properly examined in a "strain viewer," a type of polariscope. A simple strain viewer is shown diagrammatically in Fig. 40.10. It consists of a source of light, a diffusing screen, and two sheets of Polaroid between which the specimen is placed. The stress pattern exhibited by the specimen may either be viewed simply by eye, as indicated in the figure, or photographed. The Polaroids may be either set for extinction ("crossed") in the absence of the specimen or arranged with their optic axes parallel ("open Polaroids"). Stressed regions of the specimen convert the plane-polarized light incident on them into elliptically polarized light, and some of this (one plane-polarized component) emerges from the second Polaroid as plane-polarized light. In the case of crossed Polaroids this means a brightening of a previously uniformly dark field in the regions of stress, whereas with open Polaroids, the field, initially bright, is darkened in these regions, because the second Polaroid rejects one component of the elliptically polarized light. Of the two arrangements, that of crossed Polaroids is by far the more sensitive. For qualitative or semi-quantitative work a white-light source may be used, and the nature and degree of the stress judged from the colors seen; but for quantitative determinations of the stress it is better to use a monochromatic source, such as a sodium lamp. For maximum effect in any region of the specimen the directions of the two principal stresses in this region should be at 45° to the optic axes of the Polaroids. These directions normally vary from one region to another, so a full stress analysis can be carried out only if the specimen is viewed in different orientations.

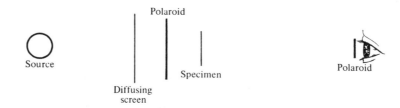

FIGURE 40.10   Simple strain viewer.

It is often desirable to eliminate the disturbing effects of refraction at the surfaces of the specimen under examination. This may be done by immersing the specimen in a rectangular tank with transparent sides filled with a liquid having approximately the same refractive index.

It is not always necessary for the stress to be actively applied during examination of the specimen. Cellophane, for example, exhibits double refraction due to a "frozen stress" derived from the process of manufacture, in which the material is rolled into sheets. These sheets behave like sections of a biaxial crystal, with the bisectrix perpendicular to the surface.

In the late 1940s a frozen-stress technique was developed for the three-dimensional analysis of stresses in engineering structures, depending on the use of stressed models of these constructed of certain kinds of resin, for example phenol–formaldehyde or glyptal resins. These models are subjected to the appropriate stresses while hot, allowed to cool under stress, and then released. It is found that after such treatment the material retains virtually the whole of the strain and the associated photoelastic behavior despite no longer being under stress. It may then be examined at leisure by cutting thin slices at various orientations and viewing these in a strain viewer in accordance with the ordinary two-dimensional procedure.

In Fig. 40.11 two examples of photoelastic stress patterns are shown.

Kerr Effects

It was found by J. Kerr in 1875 that a dielectric in which an electric field is established behaves like a uniaxial crystal, the direction of the optic axis being that of the field. This phenomenon is known as the **Kerr electrooptical effect.** The difference between the velocities of the ordinary and extraordinary rays, besides varying with the nature of the dielectric, is proportional to the square of the field strength. For a given field this difference is particularly large with nitrobenzene, and it is for this reason that this material is used in the Kerr-cell light chopper referred to in Chapter 35. In this, a cell consisting of a pair of parallel metal plates immersed in nitrobenzene is placed between a polarizer and an analyzer "crossed" so as to pass no light when the cell is not active. A beam of light is directed so that it passes through the cell parallel to the plates, and the polarizer is so orientated that this light enters the dielectric with its plane of polarization making an angle of 45° with the normal to the plates. This light then emerges

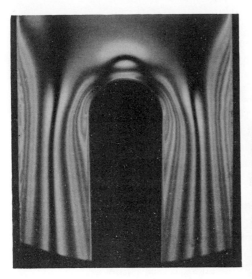

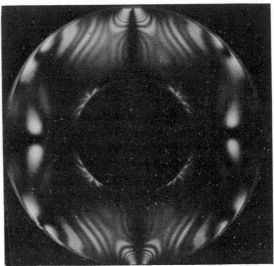

FIGURE 40.11 Photoelastic stress patterns.
(The Polaroid Company.)

either with its state of polarization unchanged or elliptically polarized, according as the field established between the plates is zero or finite. In the former case the analyzer completely suppresses the light, whereas in the latter it transmits a proportion of it.

In 1887 Kerr also discovered another effect, which we may refer to as the **Kerr magnetooptical effect.** He found that if plane-polarized light is reflected obliquely from the polished "pole" of a magnet it becomes elliptically polarized, unless the plane of polarization of the incident light either coincides with, or is perpendicular to, the plane of incidence.

Optical Rotation

Certain systems have the property of rotating the plane of polarization of plane-polarized light passing through them. We shall now consider some of these.

ROTATION BY UNIAXIAL CRYSTALS   Quartz, like calcite, is uniaxial, and exhibits phenomena of the same kind when illuminated by plane-polarized light incident at right angles to the optic axis. However, when light travels *along* the optic axis, quartz exhibits an important additional phenomenon not shown by calcite: a rotation of the plane of polarization. There are two kinds of quartz, right-handed and left-handed, rotating the plane of polarization clockwise and anticlockwise, respectively. Apart from this, the rotation is proportional to the thickness traversed, and the rotation per unit thickness is the same in both varieties. The rotation also varies with wavelength, being approximately inversely proportional to the square of the wavelength. It is thus much greater for blue than for red light. Its order of magnitude is 1 revolution per centimeter for blue light.

The rotatory power of quartz must be taken into consideration when using this material for lenses or prisms in optical apparatus with which polarized light is to be examined. The difficulty can be met by employing equal optical paths of right-handed and left-handed quartz.

The rotatory properties of quartz must be due to the arrangement of the atoms in the lattice; this may be inferred from the fact that fused silica exhibits no rotatory power.

Cinnabar has been found to behave similarly to quartz but to have a rotatory power about 15 times as great. As with quartz, some crystals are dextrorotatory, whereas others are levorotatory.

Certain other uniaxial crystals also exhibit this phenomenon.

OTHER ROTATORY MATERIALS   Certain materials rotate the plane of polarization equally for light traveling in all directions. Examples of crystalline materials in this category are sodium chlorate and sulfate of strychnia. The latter has the power of rotating the plane of polarization in solution as well as in the crystalline state. This suggests that we must seek the explanation of the rotation in the structure of the molecules.

Many liquids and vapors share with a solution of sulfate of strychnia this rotatory power, some being dextrorotatory and others levorotatory. Included among these are a number of organic liquids for example ordinary turpentine, and many substances in solution, such as sugars. It is found that the rotation produced by a given "active" substance in solution is simply proportional to the number of molecules traversed, and is independent of the strength of the solution or association in it with other, active or inactive, molecules.

However, in some cases the solvent, even if optically inactive, has some effect on the rotatory power.

In all cases the rotatory dispersion conforms roughly to the same law as applies to quartz.

To find a plausible explanation of optical rotation we merely have to imagine certain groups of atoms to have in their structure something corresponding to a helix, either right-handed or left-handed. In some cases this peculiarity of structure may be brought into being by the lattice forces and not persist when the crystalline structure is lost, although in other cases this is obviously not so. It is interesting to note that where the molecules of an active material are orientated at random, as in a liquid, their helical features would in general have "components" in the direction of travel of the light, and, if the molecules are all alike, these components must all be either right-handed or all left-handed, regardless of molecular orientation.

FARADAY EFFECT If an isotropic transparent medium is subjected to a magnetic field it acquires the property of rotating the plane of polarization of light traveling in a suitable direction through it. This effect is known, after its discoverer, as the **Faraday effect.** The light responds to that component of the field which is in the direction of travel of the light, and the maximum rotation is therefore obtained when the field and the direction of the light are made to coincide. The rotation per unit length of path is proportional to the field strength and reverses with reversal of the field. For a given direction of the field the rotation is clockwise with some materials and anticlockwise with others. The effect is most pronounced with media having a high refractive index, such as carbon disulfide or dense flint glass. As in other cases of rotation, rotatory dispersion occurs in the sense that the rotation increases with decrease of wavelength.

Problems

40.1 Prove that when a ray of light is incident on a refracting surface at the polarizing angle, the angle between the reflected and refracted rays is 90°.

40.2 When a parallel beam of light traveling in a liquid meets the surface of a glass plate of refractive index 1.52 at an angle of incidence of 48°7', the reflected light is found to be plane-polarized. What is the refractive index of the liquid?

40.3 Sunlight is reflected from the surface of a pond at such an angle that it is 60 percent (in intensity) polarized; that is, 60 percent is polarized in a particular plane, while the remaining 40 percent is "unpolarized." The reflected light is viewed through two Polaroid disks placed one behind the other. (a) Initially, these are adjusted to transmit as much light as possible. Then (b) the two disks are turned together, that is, with their axes kept parallel, until the intensity of the light is reduced to one half. Finally, (c) after the first disk has been restored to its original orientation, the second disk alone is turned until the intensity is reduced to one half of that for case (a). Calculate the angles of rotation in cases (b) and (c).

40.4 Calculate the percentage polarization of the light scattered by very small particles in a direction making an angle of 75° with the incident unpolarized beam. (Neglect any effects of multiple scattering.)

40.5 A beam of plane-polarized light is incident normally on a surface of a uniaxial crystal to which the optic axis is parallel, the angle between the plane of oscillation of the electric vector and the optic axis being 30°. What is the ratio of the amplitudes and the intensities of the ordinary and extraordinary rays within the crystal?

40.6 Two Polaroid disks are set up so that a beam of light passes through them in succession. One disk is fixed, the other rotatable about an axis perpendicular to its plane. This is first rotated until the light is extinguished. It is then turned (a) through 30°, and (b) through a further 30°. By what factor does the light intensity increase as the Polaroid axis is turned from orientation (a) to orientation (b)?

40.7 A series of three parallel Polaroid disks A, B, and C are placed between a source of light and an observer. At first A and B alone are set up, and arranged to give maximum transmission. Disk C is then added, and set relative to A and B to give zero transmission through the series. Finally, B is rotated about an axis perpendicular to the planes of the disks. How does the transmitted intensity vary with the angle θ through which B is turned?

40.8 A beam of unpolarized light is incident normally on a series of four Polaroid disks behind which is a photocell connected in circuit with a battery and a galvanometer. The orientations of the disks are adjusted until the galvanometer reading has its maximum value I_0. The first three disks are then rotated in the same sense about the direction of the beam as axis by 15°, 30°, and 45°, respectively. Through what angle, if any, must the remaining disk be rotated to give a maximum galvanometer reading, and what will this reading be?

40.9 A thin sheet of quartz whose plane contains the optic axis is placed between, and parallel to, two parallel Polaroid films with optic axes at right angles, and a beam of sodium light ($\lambda = 5893$ Å) is incident normally on these. Given that the refractive indices of quartz for the ordinary and extraordinary rays of this wavelength are 1.54425 and 1.55336, respectively, calculate what minimum thickness the sheet must have in order that the intensity of the finally emergent light shall be a maximum.

40.10 Calculate the minimum thickness of mica required to produce circularly polarized light from incident plane-polarized light for the wavelength 5893 Å. Assume that, for propagation at right angles to the cleavage planes, the refractive indices of the relevant plane-polarized components, each of which is polarized in one of the planes at 45° to the plane of polarization of the incident light, are 1.6050 and 1.6117. What would be the state of polarization of the emergent light if the thickness of the mica were (a) doubled, (b) trebled, (c) quadrupled?

40.11 A parallel beam of sodium light is incident on a pair of Polaroid disks set for extinction, and a sheet of mica, whose thickness is such that it functions as a quarter-wave plate for sodium light, is inserted between them. Describe and explain how the intensity of the emergent light varies as the sheet of mica is slowly rotated about an axis perpendicular to its plane. How many maxima and minima will occur in this light as the mica is rotated through 360°?

40.12 A half-wave plate of mica is inserted between a pair of open Polaroids, with the direction perpendicular to its optic axes making an angle of 45° with the optic axes of the Polaroids. Show that no light passes through the combination. One of the Polaroids is now rotated about an axis perpendicular to its plane. Show that when the angle through which it has been rotated is 90°, giving crossed Polaroids, the intensity of the light emerging from the system is a maximum.

40.13 How could you distinguish experimentally between unpolarized light and circularly polarized light?

40.14 A narrow parallel beam of plane-polarized light is incident normally on the face of a calcite rhomb, and the emergent light is directed on to a screen. Describe and explain the changes in what is seen on the screen as the rhomb is rotated about an axis coinciding with the direction of incidence. What will be found if (a) unpolarized and (b) circularly polarized light is used instead of plane-polarized?

40.15 Show that a plane-polarized wave train can be considered as the sum of two circularly polarized wave trains. In a certain medium, circularly polarized waves travel with velocities v_1 and v_2 for the clockwise and anticlockwise polarizations, respectively. Show that a plane-polarized wave train incident normally on a slab of this medium emerges with its plane of polarization rotated, and determine this rotation in terms of the thickness D through which the waves have traveled.

40.16 A closely wound solenoid is formed by winding a length of insulated wire from one end to the other of a cylinder of glass. In front of one end of this cylinder, with its winding, is placed a polarizer, and beyond the other end an analyzer, the polarizer and analyzer being crossed. A parallel beam of monochromatic light is directed through the system, along its axis, and at the same time an alternating current $I_0 \sin \omega t$ is passed through the winding. Obtain expressions for the variation of intensity of the emergent light with time, for the cases where the maximum angle θ_0 through which the plane of polarization of the light traveling through the glass is rotated is (a) small, and (b) finite, but less than 90°.

APPENDIX

GREEK ALPHABET

A	α	alpha	I	ι	iota	P	ρ	rho
B	β	beta	K	κ	kappa	Σ	σ	sigma
Γ	γ	gamma	Λ	λ	lambda	T	τ	tau
Δ	δ	delta	M	μ	mu	Υ	υ	upsilon
E	ϵ	epsilon	N	ν	nu	Φ	ϕ	phi
Z	ζ	zeta	Ξ	ξ	xi	X	χ	chi
H	η	eta	O	o	omicron	Ψ	ψ	psi
Θ	θ	theta	Π	π	pi	Ω	ω	omega

CONVERSION FACTORS

1 inch (in.) = 2.5400 centimeters (cm)

1 yard (yd) = 0.9144 meter (m)

1 mile (mi) = 1.6093 kilometers (km)

1 pound (lb) = 0.4536 kilogram (kg)

1 kilogram (kg) = 2.2046 pounds (lb)

0° Kelvin (°K) = -273.15° Celsius (°C)

1 calorie (cal) = 4.186 joules (J)

1 electron volt (eV) = 1.602×10^{-19} joule (J)

1 gauss (G) = 10^{-4} newton/ampere/meter (N/A/m)

1 oersted (Oe) = $1000/4\pi$ amperes/meter (A/m)

FUNDAMENTAL PHYSICAL CONSTANTS

Universal gravitational constant	$G = 6.673 \pm 0.003 \times 10^{-11}$ N-m^2/kg^2
Gravitational acceleration (standard)	$g = 980.616$ cm/sec^2
Atmosphere (standard)	1 atm $= 1.01323 \times 10^6$ dyn/cm^2
Universal gas constant	$R = 8.314 \times 10^7$ erg/g-mol/deg
	$= 8.314 \times 10^3$ J/kg-mol/deg
Stefan-Boltzmann constant	$\sigma = 5.6687 \times 10^{-8}$ W/m^2/deg^4
Wien's constant	$\lambda_m T = 2.898 \times 10^{-3}$ m-deg
Radiation constant	$c_1 = 3.7407 \times 10^{-16}$ W-m^2
Radiation constant	$c_2 = 1.4386$ m-deg
Permittivity constant	$\epsilon_0 = 8.85 \times 10^{-12}$ C^2/N/m^2
Common ratio containing permittivity constant	$\dfrac{1}{4\pi\epsilon_0} = 8.99 \times 10^9$ N-m^2/C^2
Electronic charge	$e = 1.6021 \times 10^{-19}$ C
Electronic mass	$m = 9.11 \times 10^{-31}$ kg
	$= 9.11 \times 10^{-28}$ g
Electronic charge-to-mass ratio	$e/m = 1.759 \times 10^{11}$ C/kg
	$= 1.759 \times 10^8$ C/g
Avogadro's number	$N = 6.0226 \times 10^{23}$ molecules/g-mol
Boltzmann's constant	$k = 1.3805 \times 10^{-23}$ J/deg
Planck's constant	$h = 6.6256 \times 10^{-34}$ J-sec
Permeability constant	$\mu_0 = 4\pi \times 10^{-7}$ N/A^2
Bohr magneton	$\mu_B = 1.1667 \times 10^{-29}$ N-m^2/A
Speed of light	$c = (2.997924 \pm 0.00001) \times 10^8$ m/sec

NATURAL TRIGONOMETRIC FUNCTIONS

sine

	.0	.1	.2	.3	.4	.5	.6	.7	.8	.9	
0°	.0000	.0017	.0035	.0052	.0070	.0087	.0105	.0122	.0140	.0157	.0175	89°
1°	.0175	.0192	.0209	.0227	.0244	.0262	.0279	.0297	.0314	.0332	.0349	88°
2°	.0349	.0366	.0384	.0401	.0419	.0436	.0454	.0471	.0488	.0506	.0523	87°
3°	.0523	.0541	.0558	.0576	.0593	.0610	.0628	.0645	.0663	.0680	.0698	86°
4°	.0698	.0715	.0732	.0750	.0767	.0785	.0802	.0819	.0837	.0854	.0872	85°
5°	.0872	.0889	.0906	.0924	.0941	.0958	.0976	.0993	.1011	.1028	.1045	84°
6°	.1045	.1063	.1080	.1097	.1115	.1132	.1149	.1167	.1184	.1201	.1219	83°
7°	.1219	.1236	.1253	.1271	.1288	.1305	.1323	.1340	.1357	.1374	.1392	82°
8°	.1392	.1409	.1426	.1444	.1461	.1478	.1495	.1513	.1530	.1547	.1564	81°
9°	.1564	.1582	.1599	.1616	.1633	.1650	.1668	.1685	.1702	.1719	.1736	80°
10°	.1736	.1754	.1771	.1788	.1805	.1822	.1840	.1857	.1874	.1891	.1908	79°
11°	.1908	.1925	.1942	.1959	.1977	.1994	.2011	.2028	.2045	.2062	.2079	78°
12°	.2079	.2096	.2113	.2130	.2147	.2164	.2181	.2198	.2115	.2233	.2250	77°
13°	.2250	.2267	.2284	.2300	.2317	.2334	.2351	.2368	.2385	.2402	.2419	76°
14°	.2419	.2436	.2453	.2470	.2487	.2504	.2521	.2538	.2554	.2571	.2588	75°
15°	.2588	.2605	.2622	.2639	.2656	.2672	.2689	.2706	.2723	.2740	.2756	74°
16°	.2756	.2773	.2790	.2807	.2823	.2840	.2857	.2874	.2890	.2907	.2924	73°
17°	.2924	.2940	.2957	.2974	.2990	.3007	.3024	.3040	.3057	.3074	.3090	72°
18°	.3090	.3107	.3123	.3140	.3156	.3173	.3190	.3206	.3223	.3239	.3256	71°
19°	.3256	.3272	.3289	.3305	.3322	.3338	.3355	.3371	.3387	.3404	.3420	70°
20°	.3420	.3437	.3453	.3469	.3486	.3502	.3518	.3535	.3551	.3567	.3584	69°
21°	.3584	.3600	.3616	.3633	.3649	.3665	.3681	.3697	.3714	.3730	.3746	68°
22°	.3746	.3762	.3778	.3795	.3811	.3827	.3843	.3859	.3875	.3891	.3907	67°
23°	.3907	.3923	.3939	.3955	.3971	.3987	.4003	.4019	.4035	.4051	.4067	66°
24°	.4067	.4083	.4099	.4115	.4131	.4147	.4163	.4179	.4195	.4210	.4226	65°
25°	.4226	.4242	.4258	.4274	.4289	.4305	.4321	.4337	.4352	.4368	.4384	64°
26°	.4384	.4399	.4415	.4431	.4446	.4462	.4478	.4493	.4509	.4524	.4540	63°
27°	.4540	.4555	.4571	.4586	.4602	.4617	.4633	.4648	.4664	.4679	.4695	62°
28°	.4695	.4710	.4726	.4741	.4756	.4772	.4787	.4802	.4818	.4833	.4848	61°
29°	.4848	.4863	.4879	.4894	.4909	.4924	.4939	.4955	.4970	.4985	.5000	60°
30°	.5000	.5015	.5030	.5045	.5060	.5075	.5090	.5105	.5120	.5135	.5150	59°
31°	.5150	.5165	.5180	.5195	.5210	.5225	.5240	.5255	.5270	.5284	.5299	58°
32°	.5299	.5314	.5329	.5344	.5358	.5373	.5388	.5402	.5417	.5432	.5446	57°
33°	.5446	.5461	.5476	.5490	.5505	.5519	.5534	.5548	.5563	.5577	.5592	56°
34°	.5592	.5606	.5621	.5635	.5650	.5664	.5678	.5693	.5707	.5721	.5736	55°
35°	.5736	.5750	.5764	.5779	.5793	.5807	.5821	.5835	.5850	.5864	.5878	54°
36°	.5878	.5892	.5906	.5920	.5934	.5948	.5962	.5976	.5990	.6004	.6018	53°
37°	.6018	.6032	.6046	.6060	.6074	.6088	.6101	.6115	.6129	.6143	.6157	52°
38°	.6157	.6170	.6184	.6198	.6211	.6225	.6239	.6252	.6266	.6280	.6293	51°
39°	.6293	.6307	.6320	.6334	.6347	.6361	.6374	.6388	.6401	.6414	.6428	50°
40°	.6428	.6441	.6455	.6468	.6481	.6494	.6508	.6521	.6534	.6547	.6561	49°
41°	.6561	.6574	.6587	.6600	.6613	.6626	.6639	.6652	.6665	.6678	.6691	48°
42°	.6691	.6704	.6717	.6730	.6743	.6756	.6769	.6782	.6794	.6807	.6820	47°
43°	.6820	.6833	.6845	.6858	.6871	.6884	.6896	.6909	.6921	.6934	.6947	46°
44°	.6947	.6959	.6972	.6984	.6997	.7009	.7022	.7034	.7046	.7059	.7071	45°
	.9	.8	.7	.6	.5	.4	.3	.2	.1	.0		

cosine

sine

	.0	.1	.2	.3	.4	.5	.6	.7	.8	.9	
45°	.7071	.7083	.7096	.7108	.7120	.7133	.7145	.7157	.7169	.7181	.7193	44°
46°	.7193	.7206	.7218	.7230	.7242	.7254	.7266	.7278	.7290	.7302	.7314	43°
47°	.7314	.7325	.7337	.7349	.7361	.7373	.7385	.7396	.7408	.7420	.7431	42°
48°	.7431	.7443	.7455	.7466	.7478	.7490	.7501	.7513	.7524	.7536	.7547	41°
49°	.7547	.7559	.7570	.7581	.7593	.7604	.7615	.7627	.7638	.7649	.7660	40°
50°	.7660	.7672	.7683	.7694	.7705	.7716	.7727	.7738	.7749	.7760	.7771	39°
51°	.7771	.7782	.7793	.7804	.7815	.7826	.7837	.7848	.7859	.7869	.7880	38°
52°	.7880	.7891	.7902	.7912	.7923	.7934	.7944	.7955	.7965	.7976	.7986	37°
53°	.7986	.7997	.8007	.8018	.8028	.8039	.8049	.8059	.8070	.8080	.8090	36°
54°	.8090	.8100	.8111	.8121	.8131	.8141	.8151	.8161	.8171	.8181	.8192	35°
55°	.8192	.8202	.8211	.8221	.8231	.8241	.8251	.8261	.8271	.8281	.8290	34°
56°	.8290	.8300	.8310	.8320	.8329	.8339	.8348	.8358	.8368	.8377	.8387	33°
57°	.8387	.8396	.8406	.8415	.8425	.8434	.8443	.8453	.8462	.8471	.8480	32°
58°	.8480	.8490	.8499	.8508	.8517	.8526	.8536	.8545	.8554	.8563	.8572	31°
59°	.8572	.8581	.8590	.8599	.8607	.8616	.8625	.8634	.8643	.8652	.8660	30°
60°	.8660	.8669	.8678	.8686	.8695	.8704	.8712	.8721	.8729	.8738	.8746	29°
61°	.8746	.8755	.8763	.8771	.8780	.8788	.8796	.8805	.8813	.8821	.8829	28°
62°	.8829	.8838	.8846	.8854	.8862	.8870	.8878	.8886	.8894	.8902	.8910	27°
63°	.8910	.8918	.8926	.8934	.8942	.8949	.8957	.8965	.8973	.8980	.8988	26°
64°	.8988	.8996	.9003	.9011	.9018	.9026	.9033	.9041	.9048	.9056	.9063	25°
65°	.9063	.9070	.9078	.9085	.9092	.9100	.9107	.9114	.9121	.9128	.9135	24°
66°	.9135	.9143	.9150	.9157	.9164	.9171	.9178	.9184	.9191	.9198	.9205	23°
67°	.9205	.9212	.9219	.9225	.9232	.9239	.9245	.9252	.9259	.9265	.9272	22°
68°	.9272	.9278	.9285	.9291	.9298	.9304	.9311	.9317	.9323	.9330	.9336	21°
69°	.9336	.9342	.9348	.9354	.9361	.9367	.9373	.9379	.9385	.9391	.9397	20°
70°	.9397	.9403	.9409	.9415	.9421	.9426	.9432	.9438	.9444	.9449	.9455	19°
71°	.9455	.9461	.9466	.9472	.9478	.9483	.9489	.9494	.9500	.9505	.9511	18°
72°	.9511	.9516	.9521	.9527	.9532	.9537	.9542	.9548	.9553	.9558	.9563	17°
73°	.9563	.9568	.9573	.9578	.9583	.9588	.9593	.9598	.9603	.9608	.9613	16°
74°	.9613	.9617	.9622	.9627	.9632	.9636	.9641	.9646	.9650	.9655	.9659	15°
75°	.9659	.9664	.9668	.9673	.9677	.9681	.9686	.9690	.9694	.9699	.9703	14°
76°	.9703	.9707	.9711	.9715	.9720	.9724	.9728	.9732	.9736	.9740	.9744	13°
77°	.9744	.9748	.9751	.9755	.9759	.9763	.9767	.9770	.9774	.9778	.9781	12°
78°	.9781	.9785	.9789	.9792	.9796	.9799	.9803	.9806	.9810	.9813	.9816	11°
79°	.9816	.9820	.9823	.9826	.9829	.9833	.9836	.9839	.9842	.9845	.9848	10°
80°	.9848	.9851	.9854	.9857	.9860	.9863	.9866	.9869	.9871	.9874	.9877	9°
81°	.9877	.9880	.9882	.9885	.9888	.9890	.9893	.9895	.9898	.9900	.9903	8°
82°	.9903	.9905	.9907	.9910	.9912	.9914	.9917	.9919	.9921	.9923	.9925	7°
83°	.9925	.9928	.9930	.9932	.9934	.9936	.9938	.9940	.9942	.9943	.9945	6°
84°	.9945	.9947	.9949	.9951	.9952	.9954	.9956	.9957	.9959	.9960	.9962	5°
85°	.9962	.9963	.9965	.9966	.9968	.9969	.9971	.9972	.9973	.9974	.9976	4°
86°	.9976	.9977	.9978	.9979	.9980	.9981	.9982	.9983	.9984	.9985	.9986	3°
87°	.9986	.9987	.9988	.9989	.9990	.9990	.9991	.9992	.9993	.9993	.9994	2°
88°	.9994	.9995	.9995	.9996	.9996	.9997	.9997	.9997	.9998	.9998	.9998	1°
89°	.9998	.9999	.9999	.9999	.9999	1.000	1.000	1.000	1.000	1.000	1.000	0°

	.9	.8	.7	.6	.5	.4	.3	.2	.1	.0	

cosine

tangent

	.0	.1	.2	.3	.4	.5	.6	.7	.8	.9	
0°	.0000	.0017	.0035	.0052	.0070	.0087	.0105	.0122	.0140	.0157	.0175	89°
1°	.0175	.0192	.0209	.0227	.0244	.0262	.0279	.0297	.0314	.0332	.0349	88°
2°	.0349	.0367	.0384	.0402	.0419	.0437	.0454	.0472	.0489	.0507	.0524	87°
3°	.0524	.0542	.0559	.0577	.0594	.0612	.0629	.0647	.0664	.0682	.0699	86°
4°	.0699	.0717	.0734	.0752	.0769	.0787	.0805	.0822	.0840	.0857	.0875	85°
5°	.0875	.0892	.0910	.0928	.0945	.0963	.0981	.0998	.1016	.1033	.1051	84°
6°	.1051	.1069	.1086	.1104	.1122	.1139	.1157	.1175	.1192	.1210	.1228	83°
7°	.1228	.1246	.1263	.1281	.1299	.1317	.1334	.1352	.1370	.1388	.1405	82°
8°	.1405	.1423	.1441	.1459	.1477	.1495	.1512	.1530	.1548	.1566	.1584	81°
9°	.1584	.1602	.1620	.1638	.1655	.1673	.1691	.1709	.1727	.1745	.1763	80°
10°	.1763	.1781	.1799	.1817	.1835	.1853	.1871	.1890	.1908	.1926	.1944	79°
11°	.1944	.1962	.1980	.1998	.2016	.2035	.2053	.2071	.2089	.2107	.2126	78°
12°	.2126	.2144	.2162	.2180	.2199	.2217	.2235	.2254	.2272	.2290	.2309	77°
13°	.2309	.2327	.2345	.2364	.2382	.2401	.2419	.2438	.2456	.2475	.2493	76°
14°	.2493	.2512	.2530	.2549	.2568	.2586	.2605	.2623	.2642	.2661	.2679	75°
15°	.2679	.2698	.2717	.2736	.2754	.2773	.2792	.2811	.2830	.2849	.2867	74°
16°	.2867	.2886	.2905	.2924	.2943	.2962	.2981	.3000	.3019	.3038	.3057	73°
17°	.3057	.3076	.3096	.3115	.3134	.3153	.3172	.3191	.3211	.3230	.3249	72°
18°	.3249	.3269	.3288	.3307	.3327	.3346	.3365	.3385	.3404	.3424	.3443	71°
19°	.3443	.3463	.3482	.3502	.3522	.3541	.3561	.3581	.3600	.3620	.3640	70°
20°	.3640	.3659	.3679	.3699	.3719	.3739	.3759	.3779	.3799	.3819	.3839	69°
21°	.3839	.3859	.3879	.3899	.3919	.3939	.3959	.3979	.4000	.4020	.4040	68°
22°	.4040	.4061	.4081	.4101	.4122	.4142	.4163	.4183	.4204	.4224	.4245	67°
23°	.4245	.4265	.4286	.4307	.4327	.4348	.4369	.4390	.4411	.4431	.4452	66°
24°	.4452	.4473	.4494	.4515	.4536	.4557	.4578	.4599	.4621	.4642	.4663	65°
25°	.4663	.4684	.4706	.4727	.4748	.4770	.4791	.4813	.4834	.4856	.4877	64°
26°	.4877	.4899	.4921	.4942	.4964	.4986	.5008	.5029	.5051	.5073	.5095	63°
27°	.5095	.5117	.5139	.5161	.5184	.5206	.5228	.5250	.5272	.5295	.5317	62°
28°	.5317	.5340	.5362	.5384	.5407	.5430	.5452	.5475	.5498	.5520	.5543	61°
29°	.5543	.5566	.5589	.5612	.5635	.5658	.5681	.5704	.5727	.5750	.5774	60°
30°	.5774	.5797	.5820	.5844	.5867	.5890	.5914	.5938	.5961	.5985	.6009	59°
31°	.6009	.6032	.6056	.6080	.6104	.6128	.6152	.6176	.6200	.6224	.6249	58°
32°	.6249	.6273	.6297	.6322	.6346	.6371	.6395	.6420	.6445	.6469	.6494	57°
33°	.6494	.6519	.6544	.6569	.6594	.6619	.6644	.6669	.6694	.6720	.6745	56°
34°	.6745	.6771	.6796	.6822	.6847	.6873	.6899	.6924	.6950	.6976	.7002	55°
35°	.7002	.7028	.7054	.7080	.7107	.7133	.7159	.7186	.7212	.7239	.7265	54°
36°	.7265	.7292	.7319	.7346	.7373	.7400	.7427	.7454	.7481	.7508	.7536	53°
37°	.7536	.7563	.7590	.7618	.7646	.7673	.7701	.7729	.7757	.7785	.7813	52°
38°	.7813	.7841	.7869	.7898	.7926	.7954	.7983	.8012	.8040	.8069	.8098	51°
39°	.8098	.8127	.8156	.8185	.8214	.8243	.8273	.8302	.8332	.8361	.8391	50°
40°	.8391	.8421	.8451	.8481	.8511	.8541	.8571	.8601	.8632	.8662	.8693	49°
41°	.8693	.8724	.8754	.8785	.8816	.8847	.8878	.8910	.8941	.8972	.9004	48°
42°	.9004	.9036	.9067	.9099	.9131	.9163	.9195	.9228	.9260	.9293	.9325	47°
43°	.9325	.9358	.9391	.9424	.9457	.9490	.9523	.9556	.9590	.9623	.9657	46°
44°	.9657	.9691	.9725	.9759	.9793	.9827	.9861	.9896	.9930	.9965	1.000	45°
	.9	.8	.7	.6	.5	.4	.3	.2	.1	.0		

cotangent

tangent

	.0	.1	.2	.3	.4	.5	.6	.7	.8	.9	
45°	1.000	1.003	1.007	1.011	1.014	1.018	1.021	1.025	1.028	1.032	1.036	**44°**
46°	1.036	1.039	1.043	1.046	1.050	1.054	1.057	1.061	1.065	1.069	1.072	**43°**
47°	1.072	1.076	1.080	1.084	1.087	1.091	1.095	1.099	1.103	1.107	1.111	**42°**
48°	1.111	1.115	1.118	1.122	1.126	1.130	1.134	1.138	1.142	1.146	1.150	**41°**
49°	1.150	1.154	1.159	1.163	1.167	1.171	1.175	1.179	1.183	1.188	1.192	**40°**
50°	1.192	1.196	1.200	1.205	1.209	1.213	1.217	1.222	1.226	1.230	1.235	**39°**
51°	1.235	1.239	1.244	1.248	1.253	1.257	1.262	1.266	1.271	1.275	1.280	**38°**
52°	1.280	1.285	1.289	1.294	1.299	1.303	1.308	1.313	1.317	1.322	1.327	**37°**
53°	1.327	1.332	1.337	1.342	1.347	1.351	1.356	1.361	1.366	1.371	1.376	**36°**
54°	1.376	1.381	1.387	1.392	1.397	1.402	1.407	1.412	1.418	1.423	1.428	**35°**
55°	1.428	1.433	1.439	1.444	1.450	1.455	1.460	1.466	1.471	1.477	1.483	**34°**
56°	1.483	1.488	1.494	1.499	1.505	1.511	1.517	1.522	1.528	1.534	1.540	**33°**
57°	1.540	1.546	1.552	1.558	1.564	1.570	1.576	1.582	1.588	1.594	1.600	**32°**
58°	1.600	1.607	1.613	1.619	1.625	1.632	1.638	1.645	1.651	1.658	1.664	**31°**
59°	1.664	1.671	1.678	1.684	1.691	1.698	1.704	1.711	1.718	1.725	1.732	**30°**
60°	1.732	1.739	1.746	1.753	1.760	1.767	1.775	1.782	1.789	1.797	1.804	**29°**
61°	1.804	1.811	1.819	1.827	1.834	1.842	1.849	1.857	1.865	1.873	1.881	**28°**
62°	1.881	1.889	1.897	1.905	1.913	1.921	1.929	1.937	1.946	1.954	1.963	**27°**
63°	1.963	1.971	1.980	1.988	1.997	2.006	2.014	2.023	2.032	2.041	2.050	**26°**
64°	2.050	2.059	2.069	2.078	2.087	2.097	2.106	2.116	2.125	2.135	2.145	**25°**
65°	2.145	2.154	2.164	2.174	2.184	2.194	2.204	2.215	2.225	2.236	2.246	**24°**
66°	2.246	2.257	2.267	2.278	2.289	2.300	2.311	2.322	2.333	2.344	2.356	**23°**
67°	2.356	2.367	2.379	2.391	2.402	2.414	2.426	2.438	2.450	2.463	2.475	**22°**
68°	2.475	2.488	2.500	2.513	2.526	2.539	2.552	2.565	2.578	2.592	2.605	**21°**
69°	2.605	2.619	2.633	2.646	2.660	2.675	2.689	2.703	2.718	2.733	2.747	**20°**
70°	2.747	2.762	2.778	2.793	2.808	2.824	2.840	2.856	2.872	2.888	2.904	**19°**
71°	2.904	2.921	2.937	2.954	2.971	2.989	3.006	3.024	3.042	3.060	3.078	**18°**
72°	3.078	3.096	3.115	3.133	3.152	3.172	3.191	3.211	3.230	3.251	3.271	**17°**
73°	3.271	3.291	3.312	3.333	3.354	3.376	3.398	3.420	3.442	3.465	3.487	**16°**
74°	3.487	3.511	3.534	3.558	3.582	3.606	3.630	3.655	3.681	3.706	3.732	**15°**
75°	3.732	3.758	3.785	3.812	3.839	3.867	3.895	3.923	3.952	3.981	4.011	**14°**
76°	4.011	4.041	4.071	4.102	4.134	4.165	4.198	4.230	4.264	4.297	4.331	**13°**
77°	4.331	4.366	4.402	4.437	4.474	4.511	4.548	4.586	4.625	4.665	4.705	**12°**
78°	4.705	4.745	4.787	4.829	4.872	4.915	4.959	5.005	5.050	5.097	5.145	**11°**
79°	5.145	5.193	5.242	5.292	5.343	5.396	5.449	5.503	5.558	5.614	5.671	**10°**
80°	5.671	5.730	5.789	5.850	5.912	5.976	6.041	6.107	6.174	6.243	6.314	**9°**
81°	6.314	6.386	6.460	6.535	6.612	6.691	6.772	6.855	6.940	7.026	7.115	**8°**
82°	7.115	7.207	7.300	7.396	7.495	7.596	7.700	7.806	7.916	8.028	8.144	**7°**
83°	8.144	8.264	8.386	8.513	8.643	8.777	8.915	9.058	9.205	9.357	9.514	**6°**
84°	9.514	9.677	9.845	10.02	10.20	10.39	10.58	10.78	10.99	11.20	11.43	**5°**
85°	11.43	11.66	11.91	12.16	12.43	12.71	13.00	13.30	13.62	13.95	14.30	**4°**
86°	14.30	14.67	15.06	15.46	15.89	16.35	16.83	17.34	17.89	18.46	19.08	**3°**
87°	19.08	19.74	20.45	21.20	22.02	22.90	23.86	24.90	26.03	27.27	28.64	**2°**
88°	28.64	30.14	31.82	33.69	35.80	38.19	40.92	44.07	47.74	52.08	57.29	**1°**
89°	57.29	63.66	71.62	81.85	95.49	114.6	143.2	191.0	286.5	573.0		**0°**
	.9	.8	.7	.6	.5	.4	.3	.2	.1	.0		

cotangent

COMMON LOGARITHMS

N	0	1	2	3	4	5	6	7	8	9	1	2	P.P. 3	4	5
10	0000	0043	0086	0128	0170	0212	0253	0294	0334	0374	4	8	12	17	21
11	0414	0453	0492	0531	0569	0607	0645	0682	0719	0755	4	8	11	15	19
12	0792	0828	0864	0899	0934	0969	1004	1038	1072	1106	3	7	10	14	17
13	1139	1173	1206	1239	1271	1303	1335	1367	1399	1430	3	6	10	13	16
14	1461	1492	1523	1553	1584	1614	1644	1673	1703	1732	3	6	9	12	15
15	1761	1790	1818	1847	1875	1903	1931	1959	1987	2014	3	6	8	11	14
16	2041	2068	2095	2122	2148	2175	2201	2227	2253	2279	3	5	8	11	13
17	2304	2330	2355	2380	2405	2430	2455	2480	2504	2529	2	5	7	10	12
18	2553	2577	2601	2625	2648	2672	2695	2718	2742	2765	2	5	7	9	12
19	2788	2810	2833	2856	2878	2900	2923	2945	2967	2989	2	4	7	9	11
20	3010	3032	3054	3075	3096	3118	3139	3160	3181	3201	2	4	6	8	11
21	3222	3243	3263	3284	3304	3324	3345	3365	3385	3404	2	4	6	8	10
22	3424	3444	3464	3483	3502	3522	3541	3560	3579	3598	2	4	6	8	10
23	3617	3636	3655	3674	3692	3711	3729	3747	3766	3784	2	4	5	7	9
24	3802	3820	3838	3856	3874	3892	3909	3927	3945	3962	2	4	5	7	9
25	3979	3997	4014	4031	4048	4065	4082	4099	4116	4133	2	3	5	7	9
26	4150	4166	4183	4200	4216	4232	4249	4265	4281	4298	2	3	5	7	8
27	4314	4330	4346	4362	4378	4393	4409	4425	4440	4456	2	3	5	6	8
28	4472	4487	4502	4518	4533	4548	4564	4579	4594	4609	2	3	5	6	8
29	4624	4639	4654	4669	4683	4698	4713	4728	4742	4757	1	3	4	6	7
30	4771	4786	4800	4814	4829	4843	4857	4871	4886	4900	1	3	4	6	7
31	4914	4928	4942	4955	4969	4983	4997	5011	5024	5038	1	3	4	6	7
32	5051	5065	5079	5092	5105	5119	5132	5145	5159	5172	1	3	4	5	7
33	5185	5198	5211	5224	5237	5250	5263	5276	5289	5302	1	3	4	5	6
34	5315	5328	5340	5353	5366	5378	5391	5403	5416	5428	1	3	4	5	6
35	5441	5453	5465	5478	5490	5502	5514	5527	5539	5551	1	2	4	5	6
36	5563	5575	5587	5599	5611	5623	5635	5647	5658	5670	1	2	4	5	6
37	5682	5694	5705	5717	5729	5740	5752	5763	5775	5786	1	2	3	5	6
38	5798	5809	5821	5832	5843	5855	5866	5877	5888	5899	1	2	3	5	6
39	5911	5922	5933	5944	5955	5966	5977	5988	5999	6010	1	2	3	4	6
40	6021	6031	6042	6053	6064	6075	6085	6096	6107	6117	1	2	3	4	5
41	6128	6138	6149	6160	6170	6180	6191	6201	6212	6222	1	2	3	4	5
42	6232	6243	6253	6263	6274	6284	6294	6304	6314	6325	1	2	3	4	5
43	6335	6345	6355	6365	6375	6385	6395	6405	6415	6425	1	2	3	4	5
44	6435	6444	6454	6464	6474	6484	6493	6503	6513	6522	1	2	3	4	5
45	6532	6542	6551	6561	6571	6580	6590	6599	6609	6618	1	2	3	4	5
46	6628	6637	6646	6656	6665	6675	6684	6693	6702	6712	1	2	3	4	5
47	6721	6730	6739	6749	6758	6767	6776	6785	6794	6803	1	2	3	4	5
48	6812	6821	6830	6839	6848	6857	6866	6875	6884	6893	1	2	3	4	4
49	6902	6911	6920	6928	6937	6946	6955	6964	6972	6981	1	2	3	4	4
50	6990	6998	7007	7016	7024	7033	7042	7050	7059	7067	1	2	3	3	4
51	7076	7084	7093	7101	7110	7118	7126	7135	7143	7152	1	2	3	3	4
52	7160	7168	7177	7185	7193	7202	7210	7218	7226	7235	1	2	2	3	4
53	7243	7251	7259	7267	7275	7284	7292	7300	7308	7316	1	2	2	3	4
54	7324	7332	7340	7348	7356	7364	7372	7380	7388	7396	1	2	2	3	4

Note: $\log_e 10 = 2.3026$ and $\log_{10} e = 0.43429$
$\log_e N = 2.3026 \cdot \log_{10} N$ and $\log_{10} e^x = 0.43429x$
$e = 2.71828\ldots$

N	0	1	2	3	4	5	6	7	8	9	1	2	P.P. 3	4	5
55	7404	7412	7419	7427	7435	7443	7451	7459	7466	7474	1	2	2	3	4
56	7482	7490	7497	7505	7513	7520	7528	7536	7543	7551	1	2	2	3	4
57	7559	7566	7574	7582	7589	7597	7604	7612	7619	7627	1	2	2	3	4
58	7634	7642	7649	7657	7664	7672	7679	7686	7694	7701	1	1	2	3	4
59	7709	7716	7723	7731	7738	7745	7752	7760	7767	7774	1	1	2	3	4
60	7782	7789	7796	7803	7810	7818	7825	7832	7839	7846	1	1	2	3	4
61	7853	7860	7868	7875	7882	7889	7896	7903	7910	7917	1	1	2	3	4
62	7924	7931	7938	7945	7952	7959	7966	7973	7980	7987	1	1	2	3	3
63	7993	8000	8007	8014	8021	8028	8035	8041	8048	8055	1	1	2	3	3
64	8062	8069	8075	8082	8089	8096	8102	8109	8116	8122	1	1	2	3	3
65	8129	8136	8142	8149	8156	8162	8169	8176	8182	8189	1	1	2	3	3
66	8195	8202	8209	8215	8222	8228	8235	8241	8248	8254	1	1	2	3	3
67	8261	8267	8274	8280	8287	8293	8299	8306	8312	8319	1	1	2	3	3
68	8325	8331	8338	8344	8351	8357	8363	8370	8376	8382	1	1	2	3	3
69	8388	8395	8401	8407	8414	8420	8426	8432	8439	8445	1	1	2	3	3
70	8451	8457	8463	8470	8476	8482	8488	8494	8500	8506	1	1	2	2	3
71	8513	8519	8525	8531	8537	8543	8549	8555	8561	8567	1	1	2	2	3
72	8573	8579	8585	8591	8597	8603	8609	8615	8621	8627	1	1	2	2	3
73	8633	8639	8645	8651	8657	8663	8669	8675	8681	8686	1	1	2	2	3
74	8692	8698	8704	8710	8716	8722	8727	8733	8739	8745	1	1	2	2	3
75	8751	8756	8762	8768	8774	8779	8785	8791	8797	8802	1	1	2	2	3
76	8808	8814	8820	8825	8831	8837	8842	8848	8854	8859	1	1	2	2	3
77	8865	8871	8876	8882	8887	8893	8899	8904	8910	8915	1	1	2	2	3
78	8921	8927	8932	8938	8943	8949	8954	8960	8965	8971	1	1	2	2	3
79	8976	8982	8987	8993	8998	9004	9009	9015	9020	9025	1	1	2	2	3
80	9031	9036	9042	9047	9053	9058	9063	9069	9074	9079	1	1	2	2	3
81	9085	9090	9096	9101	9106	9112	9117	9122	9128	9133	1	1	2	2	3
82	9138	9143	9149	9154	9159	9165	9170	9175	9180	9186	1	1	2	2	3
83	9191	9196	9201	9206	9212	9217	9222	9227	9232	9238	1	1	2	2	3
84	9243	9248	9253	9258	9263	9269	9274	9279	9284	9289	1	1	2	2	3
85	9294	9299	9304	9309	9315	9320	9325	9330	9335	9340	1	1	2	2	3
86	9345	9350	9355	9360	9365	9370	9375	9380	9385	9390	1	1	2	2	3
87	9395	9400	9405	9410	9415	9420	9425	9430	9435	9440	0	1	1	2	2
88	9445	9450	9455	9460	9465	9469	9474	9479	9484	9489	0	1	1	2	2
89	9494	9499	9504	9509	9513	9518	9523	9528	9533	9538	0	1	1	2	2
90	9542	9547	9552	9557	9562	9566	9571	9576	9581	9586	0	1	1	2	2
91	9590	9595	9600	9605	9609	9614	9619	9624	9628	9633	0	1	1	2	2
92	9638	9643	9647	9652	9657	9661	9666	9671	9675	9680	0	1	1	2	2
93	9685	9689	9694	9699	9703	9708	9713	9717	9722	9727	0	1	1	2	2
94	9731	9736	9741	9745	9750	9754	9759	9763	9768	9773	0	1	1	2	2
95	9777	9782	9786	9791	9795	9800	9805	9809	9814	9818	0	1	1	2	2
96	9823	9827	9832	9836	9841	9845	9850	9854	9859	9863	0	1	1	2	2
97	9868	9872	9877	9881	9886	9890	9894	9899	9903	9908	0	1	1	2	2
98	9912	9917	9921	9926	9930	9934	9939	9943	9948	9952	0	1	1	2	2
99	9956	9961	9965	9969	9974	9978	9983	9987	9991	9996	0	1	1	2	2

A-9

DEGREES OF ACCURACY

Estimation of Relative Error

We are generally less concerned with the absolute value of an error in measurement than with the proportion of the whole quantity that it represents. For example, if we were measuring the height of a telegraph pole we should hardly worry about an error as small as an inch. If, however, we were 1 inch in error in measuring the length of a lead pencil, this would be serious. In the latter case the error would represent a much larger proportion of the whole quantity than in the former; in other words, the *relative* error would be much greater.

If we gave for the measured height of the telegraph pole a figure such as 29 ft $2\frac{1}{2}$ in., it would be implied that the final $\frac{1}{2}$ in. was significant, that is, meant something, and that the true height was claimed to be nearer 29 ft $2\frac{1}{2}$ in. than 29 ft 2 in. or 29 ft 3 in., corresponding to an estimated possible error of about 0.07 percent. On the other hand, if we gave the height as 29 ft, we should imply that this was correct to the nearest foot, our estimate of the possible error being something like 6 inches, or 1 in 60. But "about 29 ft" would mean a larger possible error than this, perhaps 1 or 2 feet, that is, something of the order of 5 percent.

The implied accuracy of data given in decimal notation follows much the same lines. An example will make this clear. Suppose we measure the length of a brass cylinder and find it to be 7.04 in., using for measurement a scale in which each inch is divided into tenths. The 4 would represent an estimate of the fraction of a tenth of an inch, and in view of this and of the uncertainty of the coincidence of the other end with a division mark we might estimate the uncertainty of the measurement at about 0.01 in. The possible error is thus about 1 in 700. The figures 7, 0, and 4 are all significant, but the last of these may be in error by about 1. If the measurement had been somewhat less accurate, the length might have been given as 7.0 in., and then the 7 and 0 would both be significant, it being implied that the true length is probably somewhere between 6.9 and 7.1 in. But if we gave the thickness of a piece of metal foil as 0.003 in., the zeros would not be significant, in the present sense; the only significant figure from the point of view of relative accuracy would be the 3, and the implied uncertainty would amount to something like 30 percent. If the measured thickness were given as 0.0030 in., the zeros before the 3 would still not be significant, but the zero after the 3 would be, because, if it were not intended to have significance, there would be no point in writing it. It is, after all, simply a matter of common sense. The implication here would be that the true value probably lies between 0.0029 and 0.0031 in., the error thus possibly amounting to about 1 in 30. A neater way of giving this result would be to write it 3.0×10^{-3} in.

As a further example of zeros functioning as significant figures, let us consider the experimental value found for the velocity of light. In a recent precision determination of the velocity of light in a vacuum, the value of $299{,}792.4 \pm 0.4$ km/sec was obtained, and this is in good agreement with other recent determinations. Here the "plus or minus 0.4" represents an estimate of the limits of error. We seldom need to work to such extreme accuracy, however, and for most purposes all we need to remember is that the velocity is 3.00×10^5 km/sec, or 3.00×10^8 m/sec. In this the two final zeros are to be taken as significant.

Errors are of two kinds, random and systematic. A random error is one that is just as likely to be positive as negative, whereas a systematic error definitely tends to be one or the other, owing either to a faulty graduation or operation of the instrument or to some idiosyncrasy or faulty procedure on the part of the observer. Most measurements contain both a random and a systematic element of error, but if the instrument is of good quality and the observer is reasonably skilled and carries out his work carefully, the random element usually predominates.

If only a single measurement of a given quantity is carried out, the only basis for an estimate of possible error is a thorough appreciation of the various limitations to accuracy. However, the estimate finally arrived at will seldom be completely objective; it will usually be affected to some extent by the natural optimism or pessimism of the observer. Physical literature is full of examples of excessive optimism in the estimation of errors, and this is a tendency that should be severely curbed; it is a good principle always to err, if at all, on the side of conservatism.

In cases where systematic errors may safely be taken as small compared with random errors, the probable limit of error in the determination of a given quantity may be substantially reduced by taking the mean of a large number of measurements. In a sufficiently large number, positive and negative errors will occur roughly equally, and will thus to a considerable extent cancel each other out. Consequently, the mean will tend to be more accurate than most of the individual measurements.

Also, if a large number of determinations of the same quantity are made, it is no longer necessary to estimate the limits of error of individual measurements on the basis of experimental considerations alone; some information concerning these is now also obtainable from a scrutiny of the measurements themselves.

The following example will help to make these points clear. Let us suppose that the values obtained for nine individual readings of a certain quantity, all taken with equal care, are those entered in the first column of Table A.1. The mean of these is 12.622, and this may be taken as the most probable value of the quantity. This does not imply that it is the *true* value (which we have no means of knowing); it is merely the *best estimate of the true value* that we can make from the limited amount of information available. It is likely to be substantially nearer to the true value than are most of the individual readings.

TABLE A.1

Observations	Deviations from mean
12.62	0.002
12.63	0.008
12.64	0.018
12.63	0.008
12.60	0.022
12.61	0.012
12.60	0.022
12.64	0.018
12.63	0.008
Mean 12.622	Mean 0.013

The figures in the second column are the deviations of the individual readings from the mean, algebraic sign being disregarded. The largest of these is 0.022, and this may be taken as roughly the value of the largest error (deviation from the unknown *true* value) appearing among this sample of nine readings. Obviously this is not an acceptable estimate of the maximum possible error that may be incurred in any individual determination, for with a much larger number of observations, greater deviations would be likely to occur. Nevertheless, the largest deviation from the mean occurring in a limited number of readings does at least give us a lower limit for the maximum possible deviation from the true value; finding this is to some extent a safeguard against excessive optimism in our estimation of error.

In most cases consideration of experimental conditions allows us also to set a definite upper limit that no error can possibly exceed.

For example, if we are measuring a length of, say, something of the order of 10 cm with a scale marked in millimeters, we can usually be quite sure of not being in error by as much as $\frac{1}{2}$ mm; we can vouch for the maximum possible error being less than this.

We have, then, the situation that on the one hand a scrutiny of a series of readings of the same quantity gives us a lower limit for the maximum possible error, while, on the other hand, a consideration of experimental conditions enables us to assign an upper limit to this. The actual maximum error must lie somewhere between these two extremes, and from a consideration of both limits it is generally possible to form a reasonable estimate of the limits of error. For example, in the case of the figures given in Table A.1 we might, after due consideration, conservatively estimate the maximum possible error of any individual reading as, say, 0.04 or 0.05.

Relative Error of a Mean

If a large number of readings of the same quantity has been taken, it will naturally be desired to make use of this not merely as a safeguard against excessive optimism in estimating the limits of error of any one of them; one would also feel justified in assigning narrower limits of error to the mean than to the individual readings. Here, however, there is a difficulty. Strictly, it is not permissible to speak of a maximum possible error of the mean as substantially different from that of the separate readings, because, if we are very unlucky, the mean *could* be nearly as much in error as the least accurate individual reading. However, if the number of readings taken is large, this will be unlikely in the extreme, and it is indicated that we try to arrive at an estimate, not of the maximum possible error, but of the probable limits of error of the mean.

Some guidance in this is afforded by considering the related quantities, the probable error of any individual reading taken at random and that of the mean of these readings. Let these quantities be denoted by ϵ_p and e_p, respectively. Then statistical theory shows that, to a sufficiently close approximation,

$$e_p = \frac{\epsilon_p}{\sqrt{N}}, \tag{A.1}$$

where N is the number (assumed large) of readings taken. Not having any means of knowing the true value of the quantity, the best we can do is to identify the probable deviation of any individual reading from this, ϵ_p, with the average value of the deviations from the mean. With N large, this will be sufficiently nearly correct.

Thus, in the example above, ϵ_p may be taken as 0.013 and e_p as $0.013/\sqrt{9}$, or about 0.004.

If now we write ϵ_m for the estimated maximum possible error of individual readings and e_m for the maximum error likely to attach to the mean, it seems reasonable to assume that, for large values of N, e_m will be related to ϵ_m in the same way as e_p is to ϵ_p, that is, that

$$e_m = \frac{\epsilon_m}{\sqrt{N}}. \tag{A.2}$$

For example, if in the case we have been considering ϵ_m is assessed at between 0.04 and 0.05, e_m would be about 0.015. To make e_m less than ϵ_m by a factor of 10 would require as many as 100 readings to be taken.

It must be emphasized that we are not claiming that e_m is the *maximum possible* error of the mean but only that it is the maximum error *likely* to attach to the mean; it is our *estimate* of the limit of error. And in view of all the circumstances it will necessarily only be a somewhat rough estimate.

It might be supposed that the accuracy of a determination could be increased indefinitely, simply by increasing the number of measurements made. In practice this is not so. It must be remembered that our discussion refers to random errors only. Although it is usually possible to ensure that systematic errors are small compared with random ones, they can never be eliminated entirely. It is obviously useless to increase the number of observations to the point where the probable maximum random error of the mean becomes overshadowed by such systematic error as may still remain.

Although it is not profitable to make an excessive number of measurements, this number should, on the other hand, also not be too small. In the example considered the number was 9. If it were much less than this it would be quite unrealistic to apply a mathematical formula developed in probability theory to find the probable error of the mean and to arrive at an estimate of the maximum value of the error. If, for example, we had only three readings, there would be a 25 percent chance of either all being too high or all being too low. In such a case the deviation of each reading from the true value of the quantity would be much larger than that from the mean, and the probable error of the mean calculated by the procedure given above, and with it the probable maximum error, would be far too small.

Errors and Mistakes

It is important to distinguish between an error and a mistake. An error is the kind of inaccuracy we have been examining. A mistake, on the other hand, is due to a lapse on the part of the experimenter. For example, in counting scale divisions he may "jump" one division and so record a result as, say, 7.83 instead of 7.73. Or, in weighing, he may forget to include in the result one of the standard masses used or absentmindedly assign a wrong value to it. Because such lapses are normally of somewhat rare occurrence, they can often be detected and eliminated on inspection of a series of measurements of the same quantity. Thus, let us suppose that with a view to taking a mean, the following values have been obtained for a certain quantity: 6.432, 6.428, 6.428, 6.431, 6.329, 6.433, 6.427, 6.430, 6.429, and 6.432. From inspection it is obvious that the value 6.329 is due to a mistake, and should be discarded, the mean of the nine remaining values only being taken.

Error of Calculated Quantity in Terms of Errors of Separate Items

As an example of how errors in measurements limit the accuracy of results based on them, let us consider the determination of the density of a cylindrical body, for example a brass cylinder. Let the measured length be 17.88 cm, the mean of several measurements of the diameter 0.772 cm, and the mass 70.743 g. We may estimate 0.01 as the possible error of the length measurement, the corresponding relative error being 1 in 1800, or, in round figures, 1 in 2000. We cannot "split hairs" about errors or argue about the error of an error. After all, 0.01 cm is only a rough estimate of the possible error, which might easily be wrong by a factor of 2, and expressed as a proportional error 1 in 2000 is just as likely to be correct as 1 in 1800. Even the more conservative estimate of accuracy 1 in 1000 would do as well; according to one's degree of optimism one might make either estimate, with equal justification. Similarly, for the diameter we might estimate 1 in 1000 or 1 in 500—perhaps the former rather than the latter, in view of the fact that the figure given is the mean of several determinations. Finally, the mass would be accurate to within about 1 in 70,000.

If the measured length is too large and the diameter too small, or vice versa, each error will to some extent compensate for the other, and the calculated volume may be relatively more accurate than either of these measurements. But there is no means of knowing whether this will b′   and the two errors may equally well be additive. We may cater for this less favorable eventuality on the basis of the following general theorem, which applies to all cases of products or quotients (a quotient may be regarded as the product of one quantity and the reciprocal of another).

Let it be required to multiply together a number of quantities measured as P, Q, R, . . . , in which p is the possible *proportional* error in P, q that in Q, r that in R, and so on. This means that the true value of the first quantity lies between $P(1 + p)$ and $P(1 - p)$, and similarly for the other quantities. Making use of the approximate relation

$$(1 + p)(1 + q)(1 + r) \cdots = 1 + p + q + r + \cdots ,$$

or

$$(1 - p)(1 - q)(1 - r) \cdots = 1 - (p + q + r + \cdots),$$

where p, q, r, . . . are all small compared with 1, we see that the true value of the product must lie between

$$PQR \cdots [1 + (p + q + r + \cdots)]$$

and

$$PQR \cdots [1 - (p + q + r + \cdots)].$$

In other words, the possible proportional error of the product is the sum of those of the factors.

Applying this theorem to the present problem, we see that to obtain the possible proportional error ("error" for short) in the volume we must add together the error in the length measurement ($\frac{1}{1000}$ or $\frac{1}{2000}$) and twice that in the diameter (say $2 \times \frac{1}{1000}$). We double the error in the diameter, because in calculating the volume we have to *square* the diameter, so that any error in this measurement affects the result twice over. Thus the calculated volume will be accurate to, perhaps, 3 in 1000, or 1 in 300. A calculation of the volume such as would be made on the assumption of absolute accuracy of the data would be 8.36935 . . . cm³. The last few figures of this are, however, quite meaningless in view of the estimated limitation in the accuracy of the result—they cannot possibly be significant, and it is only misleading to write them. If we write 8.37 cm³, even the final 7 is not fully reliable, but should be retained for the purposes of further calculation.

The accuracy of the mass determination is far greater than can be made use of, and it is sheer waste of time to determine it to such high accuracy. A determination to the nearest centigram would still give an accuracy greater than that of the volume, but if the mass were determined only to the nearest tenth of a gram the accuracy would be comparable with that of the volume, and the quotient would therefore be less accurate than either. Hence, in order that the determination of the mass should not make the calculated density less accurate than it need be, while at the same time it is not carried to useless and time-wasting extremes of accuracy, the mass should be measured to the nearest centigram only, and its value, 70.74 g, then divided by the volume, 8.37 cm³. Of the alternatives for expressing the resulting value for the density, 8.452, 8.45, and 8.5 g/cm³, the first grossly exaggerates the accuracy of the result, the second slightly exaggerates it, and the third rather grossly understates it. The best choice is therefore to write the result as 8.45 g/cm³. The fact that we are somewhat uncertain of the last figure could be indicated by writing it as a subscript: 8.4_5 g/cm³. Or, if it is desired to indicate expressly that the result is accurate to about 1 in 300, we could write it in the form 8.45 ± 0.03 g/cm³.

In this problem we had occasion to estimate the accuracy of the square of the diameter of the cylinder, which entered into the calculation of its volume. Any power, whether integral or fractional, can be dealt with by simply applying the formula already derived,

$$\text{Proportional error} = p + q + r + \cdots \cdots \qquad \textbf{(A.3)}$$

Thus let $\mathbf{P}$ be the (unknown) true value of a quantity measured as P, $\mathbf{P}$ being estimated to lie between $P(1 + p)$ and $P(1 - p)$, where $p \ll 1$. Then the value of $\mathbf{P}^n$ must lie between $P^n(1 + np)$ and $P^n(1 - np)$; that is, the possible proportional error is simply np.

If a calculation involves additions or subtractions, then, of course, the *absolute* errors, not relative errors, must be added. For example, if $\mathbf{P}$ is estimated to lie between $P + a$ and $P - a$ and $\mathbf{Q}$ between $Q + b$ and $Q - b$, where $\mathbf{P}$ and $\mathbf{Q}$ are the true values and P and Q the measured values, respectively, then clearly $\mathbf{P} + \mathbf{Q}$ must lie between $(P + Q) + (a + b)$ and $(P + Q) - (a + b)$. Correspondingly, $\mathbf{P} - \mathbf{Q}$ must lie between $(P - Q) + (a + b)$ and $(P - Q) - (a + b)$. Note that if in the latter case P and Q are not very different from one another, the relative accuracy of $P - Q$ is much less than that of at least one of the quantities P and Q separately—less than that of either if a and b are of the same order of magnitude. For example,

$$(5.43 \pm 0.02) - (4.82 \pm 0.03) = 0.61 \pm 0.05.$$

We may summarize our discussion in the following two simple rules:

1. In multiplying or dividing, add *relative* errors.
2. In adding or subtracting, add *absolute* errors.

ANSWERS TO SELECTED PROBLEMS

1.1 The volume will be accurate to three significant figures:

$$V = \frac{4}{3}\pi r^3 = \frac{4}{3}(3.14)\left(\frac{4.37 \text{ cm}}{2}\right)^3 = 43.7 \text{ cm}^3.$$

A zero error of 0.10 cm in the caliper reading would mean an error of about 2.3 percent in the diameter and an error of about 3×2.3 percent $\simeq 7$ percent in the volume.

1.3 Since $x = 5.7 \pm 0.1 = 5.7(1 \pm \frac{1}{57})$ and $y = 4.2 \pm 0.1 = 4.2(1 \pm \frac{1}{42})$, the proportional error in x/y is calculated by

$$\frac{x(1 \pm \Delta x/x)}{y(1 \pm \Delta y/y)} \simeq \frac{x}{y}\left(1 \pm \frac{\Delta x}{x}\right)\left(1 \mp \frac{\Delta y}{y}\right)$$

$$\simeq \frac{x}{y}\left(1 \pm \frac{\Delta x}{x} \pm \frac{\Delta y}{y}\right)$$

$$\simeq \frac{5.7}{4.2}\left[1 \pm \left(\frac{1}{57} + \frac{1}{42}\right)\right]$$

$$\simeq 1.35[1 \pm (0.017 + 0.024)]$$

$$\simeq 1.35(1 \pm 0.04).$$

(a) The proportional error $= 0.04 = 4$ percent.
(b) The absolute error $= (0.04)(1.35) = \pm 0.05$.
The proportional error in $x - y$ is calculated by

$$(x \pm \Delta x) - (y \pm \Delta y) = x - y \pm (\Delta x + \Delta y)$$

$$= 5.7 - 4.2 \pm (0.1 + 0.1)$$

$$= 1.5 \pm 0.2.$$

(a) The proportional error $= 0.2/1.5 = 0.13 = 13$ percent.
(b) The absolute error $= \pm 0.2$.
We have $(x/y)/(x - y)$, where the proportional error in the numerator $= \pm 0.04$, and the proportional error in the denominator $= \pm 0.13$.
(a) The proportional error in the quotient $= 0.04 + 0.13 = \pm 0.17 = \pm 17$ percent. The quotient $= 1.35/1.5 = 0.90$.
(b) The absolute error in the quotient $= (0.17)(0.90) \simeq 0.15$.

1.5 $\rho = 0.7990 \pm 0.0002 \text{ g/cm}^3$.

1.7 The data can be linearized by changing to a new independent variable,

$$X = \frac{1}{x},$$

and a new dependent variable,

$$Y = \frac{1}{y}.$$

The original data satisfy

$$\frac{1}{x} + \frac{1}{y} = \frac{1}{10},$$

so the equation for the linearized form will be

$$Y = \tfrac{1}{10} - X.$$

1.9　If hT^2 is plotted as the new dependent variable instead of T, and h^2 as the new independent variable instead of h, it will be observed that the equation

$$hT^2 = \frac{4\pi^2}{g} k^2 + \frac{4\pi^2}{g} h^2$$

is now of the form $y = b + mx$, with the intercept $b = 4\pi^2 k^2/g$ and the slope $m = 4\pi^2/g$ and k and g can be determined from b and m.

1.11　Hint: Plot $\log p$ as the ordinate and h as the abscissa. The graph is linear and intercepts the $\log p$ axis at a value $\log p_0$ and has a negative slope $(-m)$:

$$\log p = -mh + \log p_0$$
$$\log p - \log p_0 = -mh$$
$$\log \frac{p}{p_0} = -mh$$
$$\frac{p}{p_0} = e^{-mh}$$
$$p = p_0 e^{-mh}.$$

The value of p_0 is obtained from the intercept and the value of m from the slope.

1.13　(a) Time^{-1}.
(b) Dimensionless.
Change the equation to read:

$$v = \omega r \cos \theta.$$

1.15　Hint: Substitute dimensions for variables as follows:

$$F = MLT^{-2},$$
$$\mu = ML^{-1},$$

and verify that the right side reduces to the dimensions of velocity, LT^{-1}.

Chapter 2

2.1　Given:

$$v = 40 \text{ miles/hr} = 58.6 \text{ ft/sec},$$
$$v_0 = 0 \text{ ft/sec},$$
$$t = 12 \text{ sec}.$$

Find a:

$$v = v_0 + at,$$
$$a = \frac{v - v_0}{t} = \frac{(58.6 - 0) \text{ ft/sec}}{12 \text{ sec}}$$
$$= 4.9 \text{ ft/sec}^2 \text{ (4.88 if we carry to three figures)}.$$

2.3　Given:

$$s = 5 \text{ m},$$
$$v_0 = 0 \text{ m/sec},$$
$$v = 150 \text{ m/sec}.$$

Find a (assumed to be uniform):

$$v^2 = v_0{}^2 + 2as$$

$$a = \frac{v^2 - v_0{}^2}{2s} = \frac{(150 \text{ m/sec})^2 - 0^2}{2(5 \text{ m})}$$

$$= 2250 \text{ m/sec.}$$

2.5 The bottle is *at rest* with respect to the water and the boat moves at constant speed relative to the water. Using the water as our initial frame of reference, we see that (time until turn around) equals (time back to bottle) $= \frac{1}{2}$ hr. Distance traveled by the bottle relative to shore = 4 km.

$$\text{speed of water} = \frac{4 \text{ km}}{1 \text{ hr}} = 4 \text{ km/hr.}$$

2.7 Given:

$$s_B = \text{distance traveled by } B,$$
$$s_A = \text{distance traveled by } A$$
$$= (s_B + 200),$$
$$a_A = \text{acceleration of } A,$$
$$a_B = \text{acceleration of } B,$$
$$v_{0A} = v_{0B} = 90 \text{ km/hr} = 25 \text{ m/sec}$$
$$= \text{initial velocity of } A \text{ and } B,$$
$$v_B = 99 \text{ km/hr} = 27.5 \text{ m/sec} = \text{velocity of } B \text{ at passing,}$$
$$v_A = \text{velocity of } A \text{ at passing,}$$
$$t = 20 \text{ sec.}$$

Find: $a_A; v_A; s_A.$
(1) First we find a_B:

$$v_B = v_{0B} + a_B t$$

$$a_B = \frac{v_B - v_{0B}}{t} = \frac{(27.5 - 25) \text{ m/sec}}{20 \text{ sec}} = 0.125 \text{ m/sec}^2.$$

(2) Next find s_B:

$$s_B = v_{0B}t + \frac{a_B t^2}{2}$$

$$= (25 \text{ m/sec})(20 \text{ sec}) + \frac{(0.125 \text{ m/sec}^2)(20 \text{ sec})^2}{2}$$

$$= 500 + 50 = 550 \text{ m.}$$

(3) Now find s_A:

$$s_A = s_B + 200 = 550 + 200$$
$$= 750 \text{ m.}$$

(4) Use previous results to find a_A:

$$s_A = v_{0A}t + \frac{a_A t^2}{2}$$

$$750 = (25)(20) + \frac{a_A (20)^2}{2}$$

$$a_A = \frac{(1500 - 1000) \text{ m}}{400 \text{ sec}^2}$$

$$= 1.25 \text{ m/sec}^2.$$

(5) Knowing a_A we can now find v_A:

$$v_A = v_{0A} + a_A t$$
$$= 25 \text{ m/sec} + (1.25 \text{ m/sec}^2)(20 \text{ sec})$$
$$= 75 \text{ m/sec}.$$

2.9   Let us call the upward direction positive. Then (see the figure):

$$v_{10} = \text{initial velocity of object 1} = +v_0,$$
$$v_{20} = \text{initial velocity of object 2} = -v_0,$$
$$a_1 = \text{acceleration of object 1} = -g,$$
$$a_2 = \text{acceleration of object 2} = -g.$$

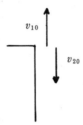

Find $s_1 - s_2$:

$$s_1 = v_{01}t + \frac{a_1 t^2}{2} = v_0 t - \frac{g t^2}{2}$$

$$s_2 = v_{02}t + \frac{a_2 t^2}{2} = -v_0 t - \frac{g t^2}{2}$$

$$s_1 - s_2 = v_0 t - (-v_0 t) - \frac{g t^2}{2} - \left(-\frac{g t^2}{2}\right)$$

$$= 2v_0 t.$$

2.11   (See the figure.)

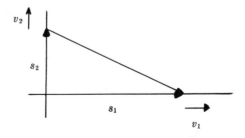

$$t = \text{travel time of car 2},$$

$$t + \frac{20}{60} = \left(t + \frac{1}{3}\right) \text{ hr} = \text{travel time of car 1}.$$

Find the magnitude of $\mathbf{s}$:

$$s_1 = v_1\left(t + \frac{1}{3}\right) = 60t + 20 \text{ km} = \text{distance of first car from origin},$$

$$s_2 = v_2 t = 100t = \text{distance of second car from origin}.$$

The magnitude of the separation between the cars is given by

$$s^2 = s_1{}^2 + s_2{}^2$$
$$= (60t + 20)^2 + (100t)^2$$
$$= 13600t^2 + 2400t + 400$$
$$= 20\sqrt{34t^2 + 6t} = 1.$$

2.13   (See the figure.)

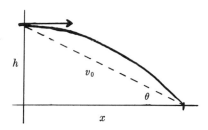

(1) $$h = \frac{gt^2}{2}.$$

(2) $$x = v_0 t.$$

Eliminate t between (1) and (2) by substitution from (2) into (1) to get

(3) $$h = \frac{g}{2}\frac{x^2}{v_0{}^2}.$$

θ is now obtained from

(4) $$\tan \theta = \frac{h}{x}.$$

by substitution from (3) to get

$$\tan \theta = \frac{gx}{2v_0{}^2},$$

$$\theta = \arctan \frac{gx}{2v_0{}^2}.$$

2.15   Take the positive y axis parallel to d and the positive x axis to the right

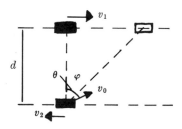

parallel to v_2, as in the figure. The velocity of the shell with respect to the water is given in component form by the pair of equations

(1) $$v_y = v_{0y} = v_0 \cos \theta,$$
(2) $$v_x = v_{0x} - v_2.$$

The transit time for the shell to cross d is obtained from

$$d = v_y t = v_{0y} t = v_0(\cos\theta)t$$

or

(3)
$$t = \frac{d}{v_0 \cos\theta}.$$

Ship 1 will travel a distance $s_1 = v_1 t$ or, substituting for t from (3),

(4)
$$s_1 = \frac{v_1 d}{v_0 \cos\theta}.$$

The angle φ of the actual horizontal projection of the path is given by

(5)
$$\tan\varphi = \frac{s_1}{d} = \frac{v_y}{v_x}.$$

Substituting for s_1 from (4) and v_x and v_y from (1) and (2) we get

$$\frac{v_1 d}{v_0 \cos\theta} = \frac{v_0 \cos\theta}{v_0 \sin\theta - v_2}.$$

Simplifying, and employing $\cos^2\theta = 1 - \sin^2\theta$, we have

$$v_1 v_0 d \sin\theta - v_1 v_2 d = v_0{}^2 \cos^2\theta$$

$$v_0{}^2(1 - \sin^2\theta) = v_1 v_0 d \sin\theta - v_1 v_2 d$$

$$1 - \sin^2\theta = v_1 d \sin\theta - \frac{v_1 v_2 d}{v_0{}^2}$$

$$\sin^2\theta + v_1 d \sin\theta - \frac{v_1 v_2 d}{v_0{}^2} + 1 = 0$$

$$\sin\theta = -\frac{v_1 d}{2} + \frac{1}{2}\sqrt{v_1{}^2 d^2 + 4(v_1 v_2 d/v_0{}^2 + 1)}.$$

2.17 (See the figure.)

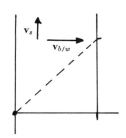

$\mathbf{v}_s$ = velocity of stream = 15 km/hr
= 4.16 m/sec relative to stream bed,
$\mathbf{v}_{b/w}$ = velocity of boat relative to stream.

The velocity of the boat relative to the stream bed is given by the vector sum

$$\mathbf{v} = \mathbf{v}_{b/w} + \mathbf{v}_s.$$

In component form this becomes the pair of equations:

(1) $\quad v_x = v_{b/wx} + v_{sx} = v_{b/w} + 0 = v_{b/w},$
(2) $\quad v_y = v_{b/wy} + v_{sy} = 0 + v_s = v_s,$

where x and y are perpendicular and parallel to the bank, respectively Distances across and downstream are given by the pair of equations

$$200 = v_x t = v_{b/w} t$$

and

$$400 = v_y t = v_s t = 4.16t,$$

from which elimination of t gives

$$v_{b/w} = 2.08 \text{ m/sec.}$$

The speed of the boat relative to the bed is given by

$$v^2 = v_x^2 + v_y^2 = v_{b/w}^2 = v_s^2$$
$$v^2 = 2.08^2 + 4.16^2$$
$$v = 4.65 \text{ m/sec.}$$

If the boat returns straight across with a new $v_{b/w}' = 4v_{b/w} = 4(2.08) =$ 8.3 m/sec, then the upstream components satisfy

$$v_{b/w} + v_{sy} = 0$$
$$v_{b/w} = -v_{sy} = -4.16 \text{ m/sec.}$$

The new heading of the boat will be such that $\sin \theta = v_{b/wy}/v_{b/w} =$ 4.16/8.3 = 0.5 or $\theta = 30°$ downstream.

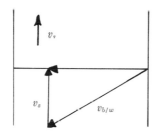

2.19 If the forward velocity of the axle in the figure is 30 m/sec, then the peripheral velocity due to rotation of the wheel around the axle must

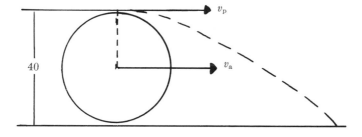

be added to that of the axle. At the top, the two velocities are parallel, giving the pebble a velocity $v = v_{\text{axle}} + v_{\text{pebble/axle}}$, or

$$v = 30 \text{ m/sec} + 30 \text{ m/sec} = 60 \text{ m/sec.}$$

The time of fall is obtained from

$$h = \frac{gt^2}{2}$$

$$t = \sqrt{\frac{2h}{g}} = \sqrt{\frac{2(40 \text{ cm})}{980 \text{ cm/sec}^2}} = \sqrt{0.0816 \text{ sec}^2} = 0.286 \text{ sec.}$$

In this time the pebble will advance a distance $s = 60$ m/sec $\times$ 0.286 sec $= 17.2$ m. The axle will have advanced only half as far, giving a net separation of 8.6 m.

2.21 At the top of the trajectory the entire acceleration of gravity acts normal to the path.

2.23 Since g enters in the denominator of the expressions for (a) time of flight, (b) range, and (c) maximum height, these values would be increased by a factor of $1/0.164 = 6.1$; that is, they would all be 6.1 times as large as the corresponding values on earth.

2.25

$$y_1 = h - \frac{gt^2}{2} = \text{height of stone 1 dropped from cliff,}$$

$$y_2 = v_0 t - \frac{gt^2}{2} = \text{height of stone 2 projected from ground.}$$

Stones pass where $y_1 = y_2$, hence

$$h - \frac{gt^2}{2} = v_0 t - \frac{gt^2}{2},$$

from which

$$t = \frac{h}{v_0}.$$

Replacing this value of t in the equation for y_2 gives

$$y_2 = v_0 \frac{h}{v_0} - \frac{g}{2} \frac{h^2}{v_0^2} = h \left(1 - \frac{gh}{2v_0^2} \right).$$

It is given that v_0 has the value v_1 would have at the bottom of the cliff.

$$v_0^2 = v_1^2 = 2gh,$$

which on substitution into the y_2 equation gives

$$y = h \left[1 - \frac{gh}{2(2gh)} \right] = h(1 - \tfrac{1}{4})$$

$$= \frac{3h}{4}.$$

The stones pass at a point three-quarters of the way up the cliff.

2.27

linear acceleration $= \dfrac{dv}{dt} = 3$ m/sec².

radius of drum $= 20$ cm $= 0.20$ m.

angular acceleration $= \dfrac{d\omega}{dt} = \left(\dfrac{1}{R} \right) \left(\dfrac{dv}{dt} \right) = \dfrac{3 \text{ m/sec}^2}{0.20 \text{ m}}$

$\qquad\qquad\qquad\qquad = 15$ m/sec².

2.29 The law $T^2 \propto R^3$ discovered by Kepler would be enough by itself to suggest that gravitational acceleration does not depend on the size of the accelerating mass, provided it is very small. Approximating the orbits as circles and using $a = R\omega^2 = 4\pi^2 R/T^2$, from which $T^2 = 4\pi^2 R/a$, substitution into Kepler's law gives

$$\frac{4\pi^2 R}{a} \propto R^3$$

or

$$a \propto \frac{1}{R^2}.$$

Since the masses of the planets vary irregularly without apparently affecting the acceleration, Galileo might have anticipated the results of his Pisa experiment. However, not only would he have had to recognize the gravitational nature of planetary accelerations, he would first have had to recognize that circular motion involves radial acceleration.

2.31 A synchronous orbit can be obtained only in the plane of the equator. The period of a complete revolution is $T = 24$ hr $= 8.64 \times 10^4$ sec. Let the radius of the orbit $= R$ and the radius of the earth $= R_0 = 6371$ km $= 6.371 \times 10^6$ m. The acceleration of gravity varies inversely as the square of the distance from the center of the earth so we may form the ratio

(1) $$\frac{g_R}{g_0} = \frac{1/R^2}{1/R_0^2} = \frac{R_0^2}{R^2}.$$

The centripetal acceleration required for a circular orbit of radius R is

(2) $$a_c = R\omega^2 = R\left(\frac{4\pi^2}{T^2}\right).$$

Substitute a_c from (2) for g_R from (1) to get

$$\frac{4\pi^2 R}{T^2} = g_0 \frac{R_0^2}{R^2},$$

from which we solve for R:

$$R = \sqrt[3]{\frac{T^2}{4\pi^2} g_0 R_0^2} = \sqrt[3]{\frac{(8.64 \times 10^4)^2}{4\pi^2} (9.8)(6.371 \times 10^6)^2}$$
$$= 4.22 \times 10^7 \text{ m} = 42{,}240 \text{ km}.$$

2.33 Let

(1) $$y = A \sin t.$$

Then

(2) $$v = \frac{dy}{dt} = \omega A \cos \omega t = V_{max} \cos \omega t.$$

For

(3) $$v = \tfrac{1}{2} V_{max}, \cos \omega t = \tfrac{1}{2},$$

and

$$\omega t = 60°.$$

Substitute (3) into (1) to get

$$y = A \sin 60° = 0.866 A.$$

2.35 Motion is harmonic with a period $T = 2\pi/20 = 0.314$ sec. Amplitude is 8.4 cm oscillating around $x = 6.5$ cm from $x = -1.9$ cm to $x = +14.9$

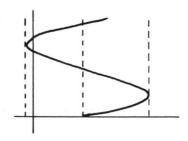

cm (see the figure). Maximum velocity is

$$\frac{dx}{dt}\Big|_{max} = \frac{d}{dt}\,(6.5 + 8.4 \sin 20t)\Big|_{20t=n\pi}$$
$$= 168 \text{ cm/sec}^2.$$

2.37 Total distance = 32 cm. A = amplitude = 32/4 = 8 cm. Hence,

$$a_{max} = A\omega^2 = 12 \text{ cm/sec}^2$$
$$\omega^2 = \frac{a_{max}}{A} = \frac{12 \text{ cm/sec}^2}{8 \text{ cm}} = 1.5 \text{ sec}^{-2}$$
$$\omega = 1.225 \text{ sec}^{-1}.$$

(a) Frequency of oscillation is

$$f = \frac{\omega}{2\pi} = \frac{1.225}{2\pi} = 0.195 \text{ sec}^{-1}.$$

(b) Maximum velocity is

$$v_{max} = A\omega = 8(1.225) = 9.78 \text{ cm/sec}.$$

The equation of motion is
$$y = 8 \sin 1.225t.$$

2.39 Motion is elliptical with the center at $x = 15$. The major axis is hori-

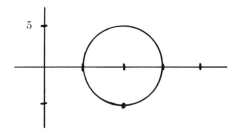

zontal with half-width 10 cm and the minor axis is vertical with half-width 5 cm. Motion is counterclockwise (see the figure).

$$f = \frac{15}{2\pi} = 2.39 \text{ sec}^{-1}.$$
$$v_{x\,max} = 150 \text{ cm/sec}; \qquad v_{y\,max} = 75 \text{ cm/sec}.$$
$$a_{x\,max} = 2250 \text{ cm/sec}^2; \qquad a_{y\,max} = 1125 \text{ cm/sec}^2.$$
$$v_x = 150 \cos 15t; \qquad v_y = -75 \cos\left(15t - \frac{\pi}{2}\right).$$

Chapter 3

3.1 The variables y, v, and a are sinusoidal with v phase shifted by $\pi/2$ and a phase shifted by π relative to y.

3.3 Given $f = 1$ Hz, $v = 10$ m/sec.

$$f\lambda = v$$

$$\lambda = \frac{v}{f} = \frac{10 \text{ m/sec}}{1 \text{ sec}^{-1}} = 10 \text{ m}.$$

(a) Points separated by 1 m differ in phase by $2\pi/10 = \pi/5$ rad.

(b) Points separated by 10 m differ in phase by 2π rad.

3.7 Given $y = \sin(10t - x)$. Comparing the given equation with

$$y = A \sin\left[2\pi f\left(t - \frac{x}{v}\right)\right]$$

shows that

$$A = 1$$
$$2\pi f = 10$$
$$\frac{2\pi f}{v} = 1.$$

Combining the last two equations gives the propagation velocity

$$v = 10 \text{ m/sec}.$$

The transverse velocity is given by

$$v_y = \frac{\partial y}{\partial t} = 10 \cos(10t - x)$$

$$v_{y_{\text{mid}}} = 10 \text{ m/sec}.$$

The ratio is

$$\frac{v_{y_{\text{mid}}}}{v_{\text{prop}}} = \frac{10}{10} = 1.$$

3.9 Given $v = 10$ m/sec, $\lambda = 1$ m, and $A = 5$ cm $= 0.05$ m.

$$f\lambda = v$$

$$f = \frac{v}{\lambda} = \frac{10 \text{ m/sec}}{1 \text{ m}} = 10 \text{ sec}^{-1} = \frac{1}{T}$$

$$y = A \sin\left(\frac{2\pi}{T} t - \frac{2\pi}{\lambda} x\right)$$

$$= 0.05 \sin(20\pi t - 2\pi x)$$

$$= 0.05 \sin[2\pi(10t - x)].$$

3.13 A phase shift of π is required. This can be accomplished either by the transformation $t' = t + T/2$ or by $x' = x + \lambda/2$.

3.15 The wave form moves in the positive sense in a counterclockwise direction.

3.17

$$x = A \sin\left(2\pi ft - \frac{2\pi x}{\lambda}\right)$$

$$A = 10^{-5} \text{ mm} = 10^{-8} \text{ m}$$

$$f = 500 \text{ Hz}$$

$$v = 350 \text{ m/sec}$$

$$\lambda = \frac{v}{f} = \frac{350 \text{ m/sec}}{500 \text{ m/sec}} = 0.70 \text{ m.}$$

$$x = 10^{-8} \sin\left(1000\pi t - \frac{2\pi x}{0.70}\right)$$

$$\frac{\partial x}{\partial t} = 1000\pi(10^{-8})\cos\left(1000\pi t - \frac{2\pi x}{0.70}\right)$$

$$\frac{\partial x}{\partial t_{\max}} = 10^{-5}\pi \text{ m/sec}$$

$$\frac{v}{\partial x/\partial t_{\max}} = \frac{350 \text{ m/sec}}{10^{-5}\pi \text{ m/sec}} = 1.1 \times 10^7.$$

The propagation velocity is about 10^7 times greater than the maximum particle velocity. The amplitude to wavelength ratio is

$$\frac{10^{-8}}{350} \simeq 3 \times 10^{-11}.$$

3.19 Given the approach frequency $f' = f_0/(1 - v/V)$, the receding frequency $f'' = f_0/(1 + v/V)$, and the ratio

$$\frac{f'}{f''} = \frac{f_0/(1 - v/V)}{f_0/(1 + v/V)} = 1.1226.$$

Solving for v:

$$\frac{1 + v}{V} = 1.1226\left(\frac{1 - v}{V}\right)$$

$$\frac{2.1226v}{V} = 0.1226$$

$$v = \frac{0.1226}{2.1226} V$$

$$= \frac{0.1226}{2.1226} (340)$$

$$= 19.5 \text{ m/sec.}$$

3.23

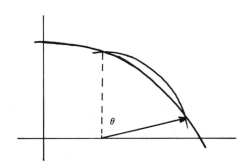

The main point is that $\theta \propto x$ at a given instant t in time. Thus, $y = R \sin(kx + \varphi)$, which is the same as the result of Problem 3.22 if $k = 2\pi/\lambda$.

3.25 A water wave is composed of the vector resultant of two such waves 90° out of phase.

3.27 The fundamental mode is given as $\lambda = 2l = 140$ cm and the frequency is given as $f = 250$ vibrations/sec. (a) $v = f\lambda = 250(140) = 35{,}000$ cm/sec.

3.29 By superposition, the standing wave equation is

$$y_T = y_1 + y_1$$
$$= A \sin 2\pi \left(ft - \frac{x}{\lambda} \right) + A \sin 2\pi \left(ft + \frac{x}{\lambda} \right)$$
$$= 2A \sin(2\pi ft)\cos \left(\frac{2\pi x}{\lambda} \right).$$

Nodes:

$$\cos \frac{2\pi x}{\lambda} = 0$$
$$\frac{2\pi x}{\lambda} - (2n - 1)\frac{\pi}{2} - \text{odd multiples of } \frac{\pi}{2}$$
$$x = (2n - 1)\frac{\lambda}{4}.$$

Antinodes:

$$\cos \frac{2\pi x}{\lambda} = 1$$
$$\frac{2\pi x}{\lambda} = n\pi$$
$$x = \frac{n\lambda}{2}.$$

3.31 (a) Maximum at center.
(b) All odd harmonics (antinodes in center).

3.33

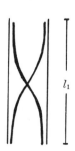

Given: $l_1 = 84$ cm, $l_2 = 85$ cm, $\lambda_1 = 2l_1 = 2(84) = 168$ cm, $\lambda_2 = 2l_2 = 2(85) = 170$ cm, and $f_1 - f_2 = $ beat frequency $= 12$ beats/5 sec $=$

2.4/sec.

$$\frac{v}{\lambda_1} - \frac{v}{\lambda_2} = 2.4$$

$$v\left(\frac{1}{168} - \frac{1}{170}\right) = 2.4$$

$$v\left[\frac{2}{(168)(170)}\right] = 2.4$$

$$v = (1.2)(168)(170) = 34{,}200 \text{ cm/sec.}$$

3.35 The reflected wave is Doppler shifted to $f' = f/(1 + v/V)$ because the source recedes from the reflector. The direct wave is Doppler shifted to $f'' = f/(1 - v/V)$. Beat frequency:

$$f'' - f' = \frac{f}{1 - v/V} - \frac{f}{1 + v/V}$$

$$= \frac{2fv/V}{1 - (v/V)^2}.$$

Given $v = 1$ m/sec, $V = 340$ m/sec, and $f'' - f' = 5$/sec.

$$5 = \frac{f(2)(\frac{1}{340})}{1 - (\frac{1}{340})^2}$$

$$f = 850[1 - (\tfrac{1}{340})^2]$$

$$\simeq 850 \text{ Hz.}$$

3.37
$$y = A \sin 2\pi\left(ft - \frac{x}{\lambda}\right)$$

$$\frac{\partial y}{\partial t} = A(2\pi f)\cos 2\pi\left(ft - \frac{x}{\lambda}\right)$$

(1)
$$\frac{\partial^2 y}{\partial t^2} = -A(4\pi^2 f)\sin 2\pi\left(ft - \frac{x}{\lambda}\right)$$

$$\frac{\partial y}{\partial x} = -A\left(\frac{2\pi}{\lambda}\right)\cos 2\pi\left(ft - \frac{x}{\lambda}\right)$$

(2)
$$\frac{\partial^2 y}{\partial x} = -A\left(\frac{4\pi^2}{\lambda^2}\right)\sin 2\pi\left(ft - \frac{x}{\lambda}\right)$$

$$v = f\lambda$$

(3)
$$v^2 = f^2\lambda^2.$$

Multiply (2) by (3):

$$v^2\frac{\partial^2 y}{\partial x^2} = f^2\lambda^2\left(-A\frac{4\pi^2}{\lambda^2}\right)\sin 2\pi\left(ft - \frac{x}{\lambda}\right)$$

$$= -A4\pi^2 f^2 \sin 2\pi\left(ft - \frac{x}{\lambda}\right)$$

$$= \frac{\partial^2 y}{\partial t^2}.$$

3.39

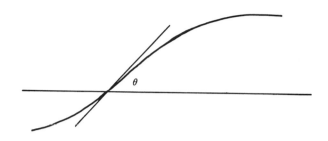

$$\tan \theta = \frac{\partial y}{\partial x}$$

$$y = f(x - vt)$$

$$v_T = \frac{\partial y}{\partial t} = -vf'(x - vt) - \text{transverse velocity.}$$

$$v = \text{propagation velocity}$$

$$-\frac{v_T}{v} = -\frac{[-vf'(x - vt)]}{v} = f'(x - vt)$$

$$\tan \theta = \frac{y}{x} = f'(x - vt).$$

Therefore,

$$\tan \theta = -\frac{v_T}{v}.$$

Chapter 4

4.1

At point a:

T_a

mg

Mg

$$\sum F_a = 0$$

$$T_a - mg - Mg = 0$$

$$= mg + Mg$$
$$= (10^{-1} \times 9.8) + (10 \times 9.8)$$
$$= 98.98 \simeq 99.0 \text{ N.}$$

At point c:

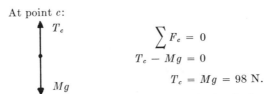

$$\sum F_c = 0$$
$$T_c - Mg = 0$$
$$T_c = Mg = 98 \text{ N.}$$

At point b (assuming uniform density of string):

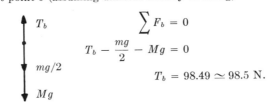

$$\sum F_b = 0$$
$$T_b - \frac{mg}{2} - Mg = 0$$
$$T_b = 98.49 \simeq 98.5 \text{ N.}$$

4.3

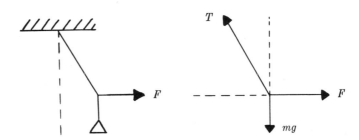

$$\sum F_{\text{hor}} = 0$$
$$-T \sin 30° + F = 0$$
(1) $$\qquad F = 0.5T.$$

$$\sum F_{\text{vert}} = 0$$
$$T \cos 30° - 9.8 = 0$$
(2) $$\qquad T = \frac{9.8}{\cos 30°} = 11.2 \text{ N.}$$

Substitute from (2) into (1):

$$F = 0.5(11.2) = 5.6 \text{ N.}$$

4.5 The two equal and opposite forces are applied to two different objects, the horse and the cart. To produce equilibrium, two equal and opposite forces would have to be applied to the same object. For the horse to accelerate it is only necessary for the horizontal force applied by the road in the forward direction to exceed the retarding force which the wagon applies to the horse. Similarly, the acceleration of the wagon results because the horizontal force applied by the horse to the wagon exceeds the retarding force of friction.

4.7 Given:
$$v_1 = 0,$$
$$v_2 = 100 \text{ km/hr} = 27.8 \text{ m/sec,}$$
$$s = 200 \text{ m.}$$

Find $\mu_s = f/n$:

$$v_2{}^2 = v_1{}^2 + 2as$$

$$a = \frac{(27.8)^2 - 0^2}{2(200)} = 1.93 \text{ m/sec}^2$$

$$\left. \begin{array}{l} F = Ma \\ N = Mg \end{array} \right\} \qquad \frac{F}{N} = \frac{Ma}{Mg} = \frac{a}{g} = \frac{1.93}{9.8} = 0.20.$$

4.9

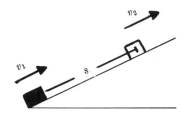

(a) Find s:

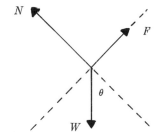

$$\sum F_x = ma_x$$

$$-f - W \sin \theta = ma_x$$

$$v_2 = v_1 + 2as$$

$$a_x = \frac{v_2{}^2 - v_1{}^2}{2s} = \frac{-20^2}{2s}.$$

Also, substitute

$$f = \mu N = \mu W \cos \theta$$

and

$$m = \frac{W}{g}.$$

$$-\mu W \cos \theta - W \sin \theta = \frac{W}{g} a_x = \frac{W}{g} \left(\frac{-20^2}{2s} \right)$$

$$-0.2 \frac{150}{170} - \frac{80}{170} = \frac{-20^2}{2gs}$$

$$-\frac{110}{170} = \frac{-400}{2(9.8)s}$$

$$s = \frac{400(170)}{110(2)(9.8)} = 31.5 \text{ m.}$$

(b) Total time to return to starting point:

t_1 = time up:

$$\bar{v}t_1 = s$$

$$t_1 = \frac{s}{v}$$

$$\bar{v} = \frac{v_1 + v_2}{2}$$

$$t_1 = \frac{31.5}{(20 + 0)/2} = 3.15 \text{ sec.}$$

t_2 = time down:

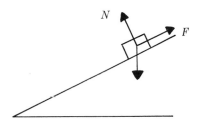

Take the positive sense of x down the plane:

$$\sum F_x = ma_x$$

$$W \sin \theta - \mu W \cos \theta = \frac{W}{g} a_x$$

$$\frac{80}{170} - 0.2 \frac{150}{170} = \frac{a_x}{9.8}$$

$$a_x = \frac{50}{170} (9.8) = 2.94 \text{ m/sec.}$$

Next use $s = \frac{1}{2}a_x t_2^2$:

$$t_2 = \frac{2s}{a} = \frac{2(31.5)}{2.94}$$

$$= 4.63 \text{ sec.}$$

Total time:

$$t = t_1 + t_2 = 3.15 + 4.63 = 7.78 \text{ sec.}$$

(c) $v_{\text{bottom}} = 2v_{\text{av}} = 2\frac{s}{t_2} = \frac{2(31.5)}{4.63}$

$$= 13.6 \text{ m/sec.}$$

4.11 The tension must be the same at all points in the rope. The increase in tension causing monkey A to accelerate will be immediately transmitted to monkey B, imparting an equal acceleration to him. The two monkeys will maintain equal attitudes.

4.13

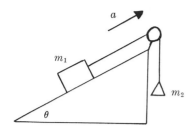
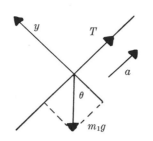
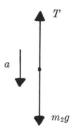

$$\sum F_x = m_1 a$$

(1)
$$T - m_1 g \sin \theta = m_1 a.$$

$$\sum F = m_2 a$$

(2)
$$m_2 g - T = m_2 a.$$

Add (1) to (2) to eliminate T:

$$m_2 g - m_1 g \sin \theta = (m_1 + m_2)a$$
$$a = \frac{(m_2 - m_1 \sin \theta)}{(m_1 + m_2)} g.$$

Substituting a into (1) or (2) yields

$$T = \frac{m_2(1 - \sin \theta)}{1 + m_2/m_1} g.$$

4.15   Number the masses given in the figure A, B, C (from left to right).

$$m_A = m_B = m_C = 1 \text{ kg (mass)},$$
$$F = 1 \text{ kgf}.$$

Acceleration:

$$F = (m_A + m_B + m_C)a$$
$$a - \tfrac{1}{3}g = 3.27 \text{ m/sec}^2.$$

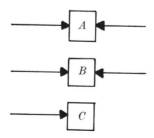

Block A:

$$\sum F = m_A a$$
$$F - F_{B \to A} = m_A a$$
$$F_{B \to A} = F - m_A a$$
$$= 1 \text{ kgf} - \tfrac{1}{3} \text{ kgf}$$
$$= \tfrac{2}{3} \text{ kgf}.$$

A-37

Block C: Similarly,

$$F_{B \to C} = \tfrac{1}{3} \text{ kgf.}$$

4.17 The answer depends on how abruptly the first car stops. If it smashes into an exceedingly massive object and stops instantaneously, then the lag distance is clearly unsafe because the rule of thumb increases the lag linearly with v, since $s = vl/15$, where l is the length of a car; but the stopping distance, even ignoring reaction time, goes as the square of v, since $s = v^2/2a$.

If, however, the first car is assumed to stop abruptly but both cars are capable of the same deceleration, then the actual deceleration distance will be the same for both and the lag distance will be eaten up only by the reaction time of the second driver. This should increase only linearly with v, since $s = vt_R$.

4.23

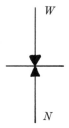

(a) Constant velocity of ascension:

$$\sum F = 0$$
$$N - W = 0$$
$$N = W.$$

Scale reading: 80 kgf.

(b) $a = 1 \text{ m/sec}^2$ upward.

$$\sum F = ma$$
$$N - W = \frac{W}{g} a$$
$$N = W \left(1 + \frac{a}{g} \right)$$
$$= 80 \left(1 + \frac{1}{9.8} \right)$$
$$= 80 \left(\frac{10.8}{9.8} \right)$$

$$\simeq 88 \text{ kgf} = \text{scale reading.}$$

(c) $a = 2$ m/sec² downward.

$$\sum F = ma$$

$$W - N = \frac{Wa}{g}$$

$$= W\left(1 - \frac{a}{g}\right) = 80\left(1 - \frac{2}{9.8}\right)$$

$$= 80\left(\frac{7.8}{9.8}\right)$$

$$= 63.6 \text{ kgf.}$$

(d) $N = 0$.

4.25

$$\sum p_x = 0,$$
$$V = \text{recoil velocity of gun,}$$
$$v = \text{velocity of shell,}$$
$$M = \text{mass of gun,}$$
$$m = \text{mass of shell.}$$
$$-MV + mv = 0 \qquad v = -V = v_{\text{muzzle}} = -V + 300$$
$$-1000V + 5(-V + 300) = 0$$
$$-1005V = 1500 = 0$$
$$V = \frac{1500}{1005} \simeq 1.5 \text{ m/sec}$$
$$s = 50 \text{ cm} = 0.50 \text{ m}$$

$$F - ma \qquad v^2 = v_0^2 - 2as$$
$$F = 1000(2.25) \qquad a = \frac{1.5^2}{2(0.5)} = 2.25 \text{ m/sec}^2$$
$$F = 2250 \text{ N.}$$

4.29 The molecules of heated gases collide with the back of the combustion chamber, transmitting their separate impulses to the chamber, and escape through the throat of the nozzle.

Chapter 5

5.1 (a) At the top:

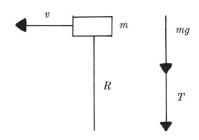

$$\sum F = ma$$

$$T + mg = m\frac{v^2}{R} \quad \text{(downward positive)}$$

$$T = m\left(\frac{v^2}{R} - g\right)$$

$$= m\left(\frac{2^2}{1} - 9.8\right)$$

$$= -5.8 \text{ N}.$$

In order to maintain the motion at the top of the circle, the string would have to exert an upward force; therefore the mass will not reach the top of the circle.

(b) At the bottom:

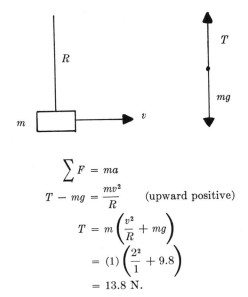

$$\sum F = ma$$

$$T - mg = \frac{mv^2}{R} \quad \text{(upward positive)}$$

$$T = m\left(\frac{v^2}{R} + mg\right)$$

$$= (1)\left(\frac{2^2}{1} + 9.8\right)$$

$$= 13.8 \text{ N}.$$

(c) At the side:

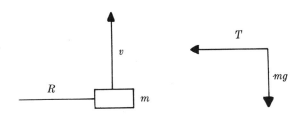

$$\sum F = ma$$

$$T = m\frac{v^2}{R}$$

$$T = (1)\left(\frac{2^2}{1}\right) = 4 \text{ N.}$$

5.3

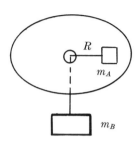

(a) Upper limit:

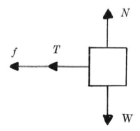

$$T + f = m_A a$$
$$m_B g + \mu m_A g = m_A R \omega^2$$
$$\omega^2 = \frac{(m_B + \mu m_A)g}{m_A R} = \frac{[0.6 + (0.4)(0.5)]9.8}{(0.5)(0.2)} = 78.4$$
$$f = \frac{1}{2\pi}\sqrt{78.6} = 1.41 \text{ rev/sec.}$$

(b) Lower limit:

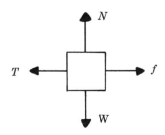

$$T - f = m_A R \omega^2$$
$$\omega^2 = \frac{T - f}{m_A R} = \frac{m_B g - \mu m_A g}{m_A R}$$
$$\omega^2 = \frac{[0.6 - (0.4)(0.5)]9.8}{(0.5)(0.2)} = 39.2$$
$$f = \frac{1}{2\pi} \sqrt{39.2} = 0.98 \text{ rev/sec} \simeq 1.0 \text{ rev/sec.}$$

5.5

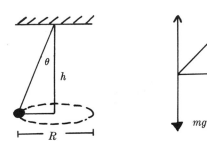

$$T \cos \theta - mg = 0$$
$$T = \frac{mg}{\cos \theta}$$
$$F_c = T \sin \theta = m R \omega^2$$
$$\frac{mg}{\cos \theta} \sin \theta = m R \omega^2$$
$$\omega = \sqrt{\frac{g \tan \theta}{R}}$$
$$\tan \theta = \frac{R}{h}$$
$$\omega = \sqrt{\frac{gR/h}{R}} = \sqrt{\frac{g}{h}}$$
$$T = 2\pi \sqrt{\frac{h}{g}}$$

5.7

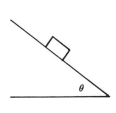
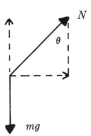

$$R = 200 \text{ m}$$
$$v = 60 \text{ km/hr} = 16.7 \text{ m/sec}$$
$$\sum F_x = ma$$
$$N \sin \theta = m \frac{v^2}{R}$$
$$\sum F_y = 0$$
$$N \cos \theta = mg$$
$$N = \frac{mg}{\cos \theta}$$
$$\frac{mg}{\cos \theta} \sin \theta = \frac{mv^2}{R}$$
$$\tan \theta = \frac{v^2}{Rg}$$
$$= \frac{16.7^2}{(200)(9.8)}$$
$$= 0.142$$
$$\theta = 8.1°.$$

If $v = 100 \text{ km/hr} = 27.8 \text{ m/sec}$, the required centripetal force is

$$F_c = m \frac{v^2}{R} = \frac{m(27.8)^2}{200} = 3.87m.$$

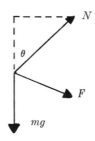

$$N \sin \theta + f \cos \theta = 3.87m$$
$$mg \tan \theta + f \cos \theta = 3.87m$$
$$f = \frac{3.87m - mg(0.142)}{0.986}$$
$$= \frac{(3.87 - 1.39)m}{0.986}$$
$$f = \frac{1.48m}{0.986} = 1.5m$$
$$\frac{f}{N} = \frac{1.5m}{mg \tan \theta} = 1.08$$

5.9

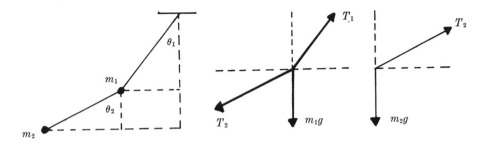

m_1:

$$\sum F_y = 0$$

(1) $\qquad T_1 \cos \theta_1 - T_2 \cos \theta_2 - m_1 g = 0$

$$\sum F_x = ma$$

(2) $\qquad T_1 \sin \theta_1 - T_2 \sin \theta_2 = m r_1 \omega^2.$

m_2:

$$\sum F_y = 0$$

(3) $\qquad T_2 \cos \theta_2 - m_2 g = 0$

$$\sum F_x = ma$$

(4) $\qquad T_2 \sin \theta_2 = m_2 r_2 \omega^2.$

Substitute from (3) into (4):

$$\frac{m_2 g}{\cos \theta_2} \sin \theta_2 = m_2 r_2 \omega^2$$

(5) $\qquad g \tan \theta_2 = r_2 \omega^2.$

Substitute from (3) into (1):

$$T_1 \cos \theta_1 - \frac{m_2 g}{\cos \theta_2} \cos \theta_2 - m_1 g = 0$$

$$T_1 \cos \theta_1 - (m_1 + m_2)g = 0$$

(6) $\qquad T_1 = \frac{(m_1 + m_2)g}{\cos \theta_1}.$

Substitute from (6) into (2):

(7) $\qquad (m_1 + m_2)g \frac{\sin \theta_1}{\cos \theta_1} - \frac{m_2 g}{\cos \theta_2} \sin \theta_2 = m_1 r_1 \omega^2.$

Substitute from (5) into (7) for ω:

$$\omega^2 = \frac{g \tan \theta_2}{r_2}$$

$$(m_1 + m_2)g \tan \theta_1 - m_2 g \tan \theta_2 = \frac{m_1 r_1 g \tan \theta_2}{r_2}$$

$$(m_1 + m_2)g \tan \theta_1 - \left(m_2 + m_1 \frac{r_1}{r_2}\right) g \tan \theta_2 = 0$$

$$(m_1 + m_2) \tan \theta_1 = \left(m_2 + m_1 \frac{r_1}{r_2}\right) \tan \theta_2,$$

where

$$\tan \theta_1 = \frac{r_1}{(r_1{}^2 + l^2/4)^{1/2}}$$

$$\tan \theta_2 = \frac{(r_2 - r_1)}{[(r_2 - r_1)^2 + l^2/4]^{1/2}}.$$

5.11

$$g_0 = 161 \text{ cm/sec}^2$$

$$r_0 = 1740 \text{ km} = 1.740 \times 10^8 \text{ cm}$$

$$v = \frac{dr}{dt}$$

$$a = \frac{dv}{dr}$$

$$a = v \frac{dv}{dt}$$

$$v \, dv = a \, dr$$

$$\frac{a_r}{g_0} = \frac{1/r^2}{1/r_0} = \frac{r_0{}^2}{r^2}$$

$$a_r = \frac{g_0 r_0{}^2}{r^2}.$$

To get v at the surface, taking the outward direction positive,

$$\int_{v_0}^0 v \, dv = \int_{r=r_0}^{r=\infty} a \, dr = -g_0 r_0{}^2 \int_{r_0}^{\infty} \frac{dr}{r^2}$$

$$\frac{v^2}{2} \Big|_{v_0}^0 = g_0 r_0{}^2 \frac{1}{r} \Big|_{r_0}^{\infty}$$

$$0 - \frac{v^2}{2} = g_0 r_0{}^2 \left(\frac{1}{\infty} - \frac{1}{r_0} \right)$$

$$v = \sqrt{\frac{2 g_0 r_0{}^2}{r_0}} = \sqrt{2 g_0 r_0}$$

$$= \sqrt{2(1.61 \times 10^2)(1.74 \times 10^8)} = 2.37 \times 10^5 \text{ cm/sec}$$

$$= 2.37 \text{ km/sec.}$$

5.17 The period is less, since the restoring force increases with θ faster than θ ($\sin \theta = \theta^2/2 + \cdots$).

5.21

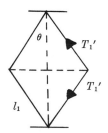

The restoring force at x is

$$T_1 \sin \theta + T_1' \sin \theta$$

but T_1 and T_1' have increased by Hooke's law to

$$T_1 = kl_1 = \frac{kl_0}{\cos \theta} = \frac{T_0}{\cos \theta}.$$

Thus the restoring force is

$$T_0 \frac{\sin \theta}{\cos \theta} + T_0' \frac{\sin \theta}{\cos \theta} = (T_0 + T_0') \tan \theta$$

$$= (T_0 + T_0') \frac{x}{l_0}$$

$$= cx \qquad (i.e., \text{ proportional to displacement}).$$

Hence the motion is harmonic.

Chapter 6

6.1

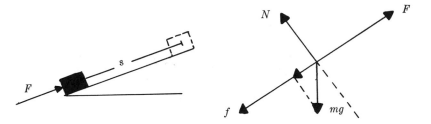

$$\text{Work} = F \cdot s$$
$$F = f + mg \sin \theta$$
$$f = \mu N = \mu W \cos \theta$$
$$F = \mu mg \cos \theta + mg \sin \theta$$
$$\text{Work} = (\mu mg \cos \theta + mg \sin \theta)s$$
$$= (0.3 \cos 20° + \sin 20°)(100)(9.8)(10)$$
$$= (0.282 + 0.342)(9.8 \times 10^3)$$
$$= (0.624)(9.8)(10^3)$$
$$= 6.1 \times 10^3 \text{ J.}$$

6.3

$$\Delta E_p = mgh_{cg} = mg\left(\frac{l}{2} - \frac{l}{2}\cos 60°\right)$$
$$= (1 \text{ kg})(9.8 \text{ m/sec}^2)[\tfrac{1}{2} - \tfrac{1}{2}(0.5)] \text{ m}$$
$$= 2.45 \text{ J}.$$

6.7

$$E_k = \frac{1000}{9.8}(80 \times 10^3)^2 \simeq 6.3 \times 10^{11} \text{ kg-m}.$$

6.9 Work done by friction $= \Delta E_k$.

$$fs = \tfrac{1}{2}MV$$
$$\mu Ws = \tfrac{1}{2}MV^2$$
$$\mu Mgs = \tfrac{1}{2}MV^2$$
$$s = \frac{v^2}{2\mu g}.$$

6.11

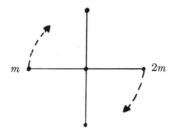

$$\Delta E_k = -\Delta E_p$$
$$\frac{1}{2}mv^2 + \frac{1}{2}(2m)v^2 = -mg\frac{l}{2} = 2mg\frac{l}{2}$$
$$\frac{3}{2}mv^2 = mg\frac{l}{2}$$
$$v = \sqrt{\frac{gl}{3}}$$
$$\omega = \frac{v}{R} = \frac{\sqrt{gl/3}}{l/2}$$
$$= \frac{2}{l}\sqrt{\frac{gl}{3}} = \sqrt{\frac{4g}{3l}}.$$

6.13

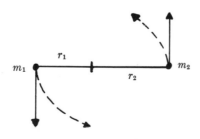

$$r_1 = \left(\frac{m_2}{m_1 + m_2}\right) r$$

$$r_2 = \left(\frac{m_1}{m_1 + m_2}\right) r.$$

Angular velocities are the same.

$$W_{r \to \infty} = \frac{Gm_1m_2}{r}.$$

6.15

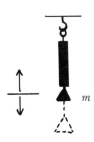

$$a_{\max} = 4\pi^2 f^2 A = 4\pi^2 (2)^2 (8)$$
$$= 1260 \text{ cm/sec}^2.$$
$$v_{\max} = 2\pi f A = 2(2)(8)$$
$$\simeq 100 \text{ cm/sec.}$$
$$E_{k_0} = \tfrac{1}{2}mv_0^2 = \tfrac{1}{2}(10)(100)^2 = 50{,}000 \text{ ergs.}$$

$$f = \frac{1}{2\pi}\sqrt{\frac{k}{m}}$$

$$k = 4\pi^2 f^2 m$$
$$= 1580 \text{ dynes/cm.}$$
$$ky = mg$$
$$y = \frac{mg}{k} = \frac{10(980)}{1580} = 6.2 \text{ cm.}$$

6.17

$$V_{\text{recoil}} \simeq 1.5 \text{ m/sec.}$$
$$E_{k_{\text{gun}}} = \tfrac{1}{2}MV^2 = \tfrac{1}{2}(4)(1.5)^2 = 4.5 \text{ J.}$$
$$E_{k_{\text{bullet}}} = \tfrac{1}{2}mv^2 = \tfrac{1}{2}(2 \times 10\tfrac{1}{2}{}^2)(300)^2$$
$$= 900 \text{ J.}$$

6.19

$$U = \text{final velocity of He,}$$
$$u = \text{final velocity of } (H_2^+),$$

$$\sum p_{\text{before}} = \sum p_{\text{after}}.$$

$$mv + 0 = mu + 2mU$$

(1)
$$v - u = 2U.$$

$$\sum E_{k_{\text{before}}} = E_{k_{\text{after}}}$$

$$\frac{mv^2}{2} + 0 = \frac{mu^2}{2} + \frac{(2m)U^2}{2}$$

$$v^2 - u^2 = 2U^2$$

(2) $$(v - u)(v + u) = 2U^2.$$

Divide (2) by (1):

(3) $$v + u = U.$$

Add (1) and (3):

$$2v = 3U$$
$$U = \tfrac{2}{3}v$$
$$u = -\tfrac{1}{3}v$$

The ionization would not take place because the "lost" kinetic energy would leave both particles at rest with a net momentum of zero. Since the original momentum was not zero, conservation of momentum would be violated.

6.21

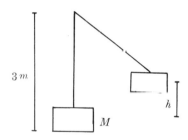

(1) $\qquad mv = (M + m)V \qquad$ (conservation of momentum).

(2) $\dfrac{(M + m)V^2}{2} = (M + m)gh \qquad$ (conservation of mechanical energy after collision).

(3) $\qquad\qquad V = \sqrt{2gh}.$

Substitute (3) into (1):

$$mv = (M + m)\sqrt{2gh}$$
$$v = \left(\frac{M + m}{m}\right)\sqrt{2gh}.$$

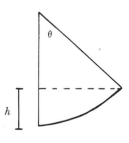

$$h = (1 - \cos\theta)$$
$$= 3(0.0088) = 0.0264 \text{ m}.$$

$$v = \left(\frac{1.005}{5 \times 10^{-3}}\right) \sqrt{2(9.8)(2.64 \times 10^{-2})}$$

$$= 144 \text{ m/sec.}$$

$$E_{k_{\text{before}}} = \tfrac{1}{2}mv^2 = \tfrac{1}{2}(5 \times 10^{-3})(144)^2$$

$$= 50 \text{ J.}$$

$$E_{k_{\text{after}}} = E_{p_{\text{top of arc}}} = (M + m)gh = (1.005)(9.8)(2.64 \times 10^{-2})$$

$$= 0.26 \text{ J.}$$

The lost mechanical energy would be dissipated thermally during the collision.

Chapter 7

7.1   Let power $= L\omega$.

$$10 \times 10^3 = \Gamma 600 \left(\frac{2\pi}{60}\right)$$

$$\Gamma = \frac{10^3}{2\pi} = 159 \text{ N-m.}$$

7.3

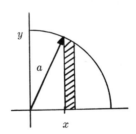

$$x^2 + y^2 = a^2$$
$$y = (a^2 - x^2)^{1/2}$$
$$dm = \sigma y \, dx,$$

where σ = area mass density. The moment of dm around a circular plate $= x \, dm = x\sigma y \, dx$.

$$\bar{X} = \frac{\displaystyle\int x \, dm}{\displaystyle\int dm} = \frac{\displaystyle\int_0^a x(a^2 - x^2)^{1/2}\sigma \, dx}{\displaystyle\int_0^a \sigma(a^2 - x^2)^{1/2} \, dx}$$

$$= \frac{-(\sigma/3)(a^2 - x^2)^{3/2}\Big|_0^a}{+\sigma\dfrac{a^2\pi}{4}} = \frac{4}{3}\frac{a^3}{\pi a^2} = \frac{4a}{3\pi}.$$

$$\bar{Y} = \frac{4a}{3\pi}.$$

7.5

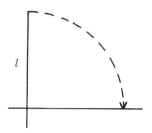

$$\Delta E_p = -\Delta E_k$$
$$mgh_{cg} = \tfrac{1}{2}I\omega^2$$
$$mg(5) = \tfrac{1}{2}m\frac{l^2}{3}\omega^2$$
$$\omega^2 = \frac{30(9.8)}{100}$$
$$\frac{v^2}{l^2} = \frac{30(9.8)}{100}$$
$$v^2 = \frac{30(9.8)}{100}100$$
$$v = \sqrt{30(9.8)} = 17.1 \text{ m/sec.}$$

7.7

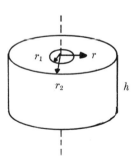

For a thin annulus of thickness dr, $dI = \rho h(2\pi r)\, dr\, (r^2)$. For the entire cylinder,

$$I = \int dI = \int_{r_1}^{r_2} 2\pi\rho h r^3\, dr$$
$$= 2\pi\rho h \frac{r^4}{4}\bigg|_{r_1}^{r_2} = \frac{\pi\rho h}{2}(r_2{}^4 - r_1{}^4).$$
$$M = \pi\rho h(r_2{}^2 - r_1{}^2),$$
$$I = \frac{M(r_2{}^2 - r_1{}^2)}{2}.$$

7.9

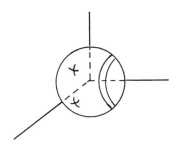

Slicing the sphere into discs and using the result that, for each disc,

$$dI = \tfrac{1}{2} \, dm \; y^2 = \tfrac{1}{2} \rho \pi y^2 \; dx \; y^2,$$

we have for the sphere

$$I = \int dI = \int \tfrac{1}{2} \rho \pi y^4 \, dx$$
$$x^2 + y^2 = a^2$$
$$y^2 = a^2 - x^2$$
$$y^4 = (a^2 - x^2)^2$$
$$I = \int_{-a}^{+a} \rho \pi (a^2 - x^2)^2 \, dx = \tfrac{8}{15} \rho \pi a^5$$
$$M = \tfrac{4}{3} \rho \pi a^3$$
$$I = \tfrac{2}{5} M a^2.$$

7.11

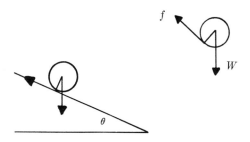

For the limiting case,
$$f = \mu N = \mu w \cos \theta,$$
$$L_{c_0} = fr = \mu w r \cos \theta.$$

7.13   Refer to the solution to Problem 7.7.

$$I = M \frac{(r_2^2 - r_1^2)}{2} = M k^2,$$

$$k = \sqrt{\frac{r_2^2 - r_1^2}{2}}.$$

7.15

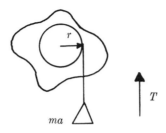

$$\sum F = ma$$
$$mg - T = ma$$
$$T = m(g - a)$$
$$\alpha = \frac{a}{r}$$
$$T = I\alpha = I\frac{u}{r}$$
$$m(g - a)r = I\frac{a}{r}$$
$$I = \frac{mr^2(g - a)}{a}$$
$$= mr^2 \left(\frac{g}{a - 1} \right).$$

7.17

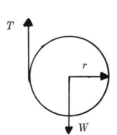

$$\frac{1}{2} I\omega^2 + \frac{mv^2}{2} = mgh$$
$$\frac{1}{2} \left(\frac{mr^2}{2} \right) \frac{v^2}{r^2} + \frac{mv^2}{2} = mgh$$
$$v^2 = \frac{4}{3} gh.$$

But
$$v^2 = 2ah.$$

Hence,
$$a = \tfrac{2}{3}g.$$

7.21 (a)
$$\omega_2 = (r_1/r_2)^2 \omega_1.$$
(b)
$$m\omega_1{}^2 r_1{}^2(1 - r_1{}^2/r_2{}^2).$$

7.23
$$\delta R = \frac{1}{2}\frac{R}{T}\,\delta T = 36.8 \text{ m.}$$

7.25
$$a = 2F/m; \ \ddot{\theta} = \dot{\omega} = 12F/ml.$$

7.27
$$\omega = \sqrt{3g/l}.$$

7.29 (a)
$$F = \tfrac{40}{7}\text{ kg.}$$
(b)
$$F = \tfrac{40}{3}\text{ kg.}$$

7.31 (a)
$$F_0 = 5mg - mg\cos^2\theta\left(\frac{1}{2} + \frac{4h}{l}\right).$$

(b)
$$F_h = mg\sin\theta\cos\theta\left(\frac{1}{2} + \frac{2h}{l}\right).$$

(c)
$$h = \frac{3\sin^2\theta + 7}{16\sin\theta\cos\theta}\,l.$$

(d)
$$\theta \simeq \cos^{-1}\sqrt{\tfrac{2}{3}}.$$

Chapter 8

8.1

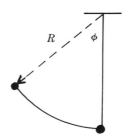

For $r \ll R$ and for small ϕ:

$$L = \frac{\partial E_k{}^+}{\partial\theta},$$

where θ is the angle turned through by the ball.

$$E_{k_{\text{rot}}} = \tfrac{1}{2}I\omega^2$$
$$E_{k_{\text{rot}}} + E_{k_{\text{trans}}} = mgh \simeq mgR(1 - \cos\phi)$$
$$\frac{1}{2}\left(\frac{2}{5}mr^2\right)\omega^2 + \frac{mv^2}{2} = mgR(1 - \cos\phi)$$
$$\omega = \frac{v}{r}$$
$$\frac{1}{2}\left(\frac{2}{5}r^2\right)\frac{v^2}{r^2} + \frac{v^2}{2} = gR(1 - \cos\phi)$$

$$\phi = \frac{r}{R}\theta$$

$$\frac{7}{10}v^2 = gR\left(1 - \cos\frac{r}{R}\theta\right)$$

$$\frac{v^2}{r^2} = \omega^2 = \frac{10}{7}\frac{gR}{r^2}\left(1 - \cos\frac{r}{R}\theta\right)$$

$$E_{k_{rot}} = \frac{I\omega^2}{2} = \frac{1}{2}\left(\frac{2}{5}mr\right)\frac{10}{7}\frac{gR}{r^2}\left(1 - \cos\frac{r}{R}\theta\right)$$

$$= \frac{2mgR}{7}\left(1 - \cos\frac{r}{R}\theta\right)$$

$$L = \frac{\partial E_{k_{rot}}}{\partial\theta}$$

$$= \frac{2}{7}mgR\left(+\frac{r}{R}\sin\frac{r}{R}\theta\right)$$

$$\simeq \frac{2}{7}mgr\frac{r}{R}\theta$$

$$= \frac{2}{7}mg\frac{r^2}{R}\theta$$

$$\frac{2}{5}mr^2\ddot{\theta} + \frac{2}{7}mg\frac{r^2}{R}\theta = 0$$

$$\ddot{\theta} + \frac{5g}{7R}\theta = 0$$

$$T = 2\pi\sqrt{\frac{7}{5}\frac{R}{g}}.$$

8.3

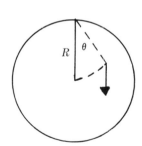

$$L = -mg(\sin\theta)R \simeq -mgR\theta$$
$$-mgR\theta = (I_0 + mR^2)\ddot{\theta} = 2mR^2\ddot{\theta}$$
$$2mR^2\ddot{\theta} + mgR\theta = 0$$

$$\ddot{\theta} + \frac{g}{2R}\theta = 0$$

$$\omega^2 = \frac{g}{2R}$$

$$T = 2\pi\sqrt{\frac{2R}{g}}$$

$$T = 2\pi\sqrt{\frac{(2 + 3\pi)R}{3\pi g}}$$

8.5

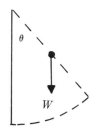

$$L = -W(\sin\theta)\frac{l}{2} \simeq -W\theta\frac{l}{2} \qquad \text{for small } \theta$$

$$= I\ddot{\theta}.$$

$$-mg\theta\frac{l}{2} - \left(I_0 + \frac{ml^2}{4}\right)\ddot{\theta} = 0$$

$$\frac{ml^2}{3}\ddot{\theta} + mg\theta\frac{l}{2} = 0$$

$$\ddot{\theta} + \frac{3g}{2}\theta = 0.$$

$$\omega^2 = \frac{3g}{2l}.$$

$$\frac{4\pi^2}{T^2} = \frac{3g}{2l}$$

$$T = 2\pi\sqrt{\frac{2}{3}\frac{l}{g}}.$$

A point mass located at $l' = \frac{2}{3}l$ would not alter the period of the system.

8.7

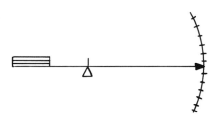

Sensitivity:

$$L = mg(12 \text{ cm})$$
$$= 10^{-2}(980)(12)$$
$$= 11.8 \text{ dyne cm.}$$

$$\theta = \frac{3.8(0.12)}{25} = 1.83 \times 10^{-2}.$$

$$\frac{L}{\theta} = \frac{11.8}{1.83 \times 10^{-2}} = 6.43 \times 10^{+2} \text{ dyne cm/rad.}$$

$$T = 15 \text{ sec}$$

$$= 2\pi \sqrt{\frac{I}{L/\theta}}$$

$$T^2 = 4\pi^2 \frac{I}{L/\theta}$$

$$I = \left(\frac{L}{\theta}\right)\frac{T^2}{4\pi^2} = (6.43 \times 10^2)\left(\frac{15^2}{4\pi^2}\right)$$

$$= 3620 \text{ gm cm}^2.$$

8.9

$$T_2 = 2\pi \sqrt{\frac{I_2}{L/\theta}}$$

$$T_1 = 2\pi \sqrt{\frac{I_1}{L/\theta}}$$

$$\frac{T_2}{\pi} = \sqrt{\frac{I_2}{I_1}}$$

$$I_2 = I_1 + \frac{ml^2}{12}$$

$$\frac{T_2}{T_1} = \sqrt{\frac{I_1 + ml^2/12}{I_1}} = \sqrt{1 + \frac{ml^2}{12I_1}}$$

$$\left(\frac{T_2}{T_1}\right)^2 = 1 + \frac{ml^2}{12I_1}$$

$$\frac{ml^2}{12I_1} = \left(\frac{T_2}{T_1}\right)^2 - 1$$

$$\frac{12I_1}{ml^2} = \frac{1}{(T_2/T_1)^2 - 1}$$

$$I_1 = \frac{ml^2}{12}\left[\frac{1}{(T_2/T_1)^2 - 1}\right] = \frac{20(12)^2}{12}\frac{1}{(8/5)^2 - 1} = \frac{240}{1.56}$$

$$I_1 = 154 \text{ gm cm}^2.$$

$$T_1 = 2\pi \sqrt{\frac{I_1}{L/\theta}}$$

$$T_1^2 = \frac{4\pi^2 I_1}{L/\theta}$$

$$\frac{L}{\theta} = \frac{4\pi^2 I_1}{T_1^2}$$

$$= \frac{4\pi^2(154)}{5^2}$$

$$\frac{L}{\theta} = 240 \text{ dyne cm.}$$

8.11

$$I = 50 \text{ kg} - m^2$$
$$\omega = 100\rho s = 200\pi \text{ rad/sec}$$
$$\Omega = 90°/5 \text{ sec} = \frac{\pi/2 \text{ rad}}{5 \text{ sec}} = \frac{\pi}{10} \text{ rad/sec}$$
$$L = I\omega\Omega = 50(200\pi)\left(\frac{\pi}{10}\right) = 1000\pi^2$$
$$= 9.86 \times 10^3 \text{ Nm.}$$

Chapter 9

9.1 $\rho = 1.66 \times 10^{-4} \text{ g/cm}^3 = 1.66 \times 10^{-1} \text{ kg/m}^3.$

$V = 100 \text{ m}^3.$
$W = \rho g V = 1.66 \times 10^{-1} \text{ kg/m}^3 \times 9.8 \text{ m/sec}^2 \times 100 \text{ m}^3$
$\quad = 1.63 \times 10^{+3} \text{ N.}$

9.3

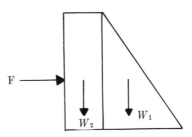

W_1 = weight per meter of rectangular slab
$\quad = \rho g A_1 = 2.3 \times 10^3 \times 9.8 \times 2 \times 10$
$\quad = 4.69 \times 10^5 \text{ N/m.}$

The line of action is vertical and 7 m from 0.

W_2 = weight per meter of triangular section
$\quad = \rho g A_2 = 2.3 \times 10^3 \times 9.8 \times \frac{(10 \times 6)}{2}$
$\quad = 6.66 \times 10^5 \text{ N/m.}$

The line of action of W_2 is $\frac{2}{3}(6)$ m from 0. The total force per unit length due to water is

$$F = p_a h = p_{\underline{bottom}} \atop 2 \ (10)$$

$$= \frac{pgh}{2}(10) = \frac{9.8 \times 10^3 \times 10 \times 10}{2}$$

$$= 4.9 \times 10^5 \text{ N/m}.$$

The line of action of F is one-third the distance from the base, directed horizontally to the left.

Overturning moment per unit length:

$$M_{\text{water}} = 4.9 \times 10^5 \text{ N/m} \times \tfrac{10}{3} \text{ m}$$
$$= 1.63 \times 10^6 \text{ (Nm)/m clockwise.}$$

Moment of wall:

$$M = 4.69 \times 10^5 \times 7 + 6.66 \times 10^5 \times 4$$
$$= 6.94 \times 10^6 \text{ (Nm)/m counterclockwise.}$$

9.5 No change.

9.7 The new "buoyant" force on the balloon will include a component in the direction of the acceleration; hence the balloon will lean *into* the direction of acceleration.

9.9 If a man were submerged in a fluid with a specific gravity equal to his own, he would seem "weightless" at any acceleration, in the sense that his capsule walls would not need to exert a force on him because the required force would be provided by "bouyancy." However, the hydrostatic pressure exerted on the man by the fluid would increase with the acceleration of the system.

9.11

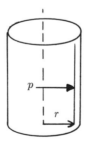

$$\sum F = (dm)a$$
$$(p + dp - p)A = (\rho A\ dr)r\omega^2$$
$$dp = \rho r\ dr\ \omega^2$$

(1)
$$\frac{dp}{dr} = \rho r\omega^2.$$

The static pressure is $p = \rho g y$, from which

(2)
$$dp = \rho g\ dy.$$

Substituting from (2) into (1),

$$\frac{g\,dy}{\omega^2} = r\,dr.$$

Integrating and taking $y_0 = 0$ at the center,

$$\frac{g}{\omega^2}\,y = \frac{r^2}{2}$$

$$y = \frac{r^2\omega^2}{2g}.$$

9.13   B.F. $= 6000\text{ kg} \times 9.8\text{ m/sec}^2 = 5.88 \times 10^4$ N. The volume of water displaced $= Ah = A(0.15)$ m. The weight of water displaced $= \rho g\,(\text{vol}) = 10^3 \times 9.8 \times A \times 0.15.$

$$5.88 \times 10^4 = 9.8 \times 10^3 \times 0.15 \times A$$

$$A = \frac{5.88 \times 10^4}{9.8 \times 10^3 \times 0.15} = 1.4 \times 10^1$$

$$= 14\text{ m}^2.$$

9.15

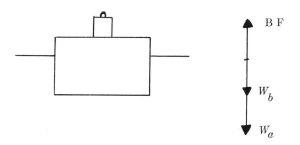

$$\rho_{\text{block}} = 0.7\text{ g/cm}^3 = 700\text{ kg/m}^3$$

$$\sum F = 0$$

$$\rho_{\text{water}}g(A \times 0.09) - \rho_{\text{block}}g(A \times 0.1) - mg = 0$$
$$m = 10^{-2}(\rho_{\text{water}} \times 0.09 - \rho_{\text{block}} \times 0.1)$$
$$= 10^{-2}(10^3 \times 0.09 - 700 \times 0.1)$$
$$= 0.2\text{ kg} = 200\text{ g}.$$

9.17

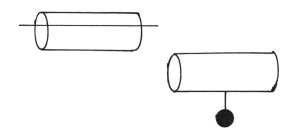

Floating log:

$$\text{B.F.} = \text{weight of log.}$$
$$\rho_{\text{water}}g(0.8V) = 400(9.8)$$
(1) $$V = 0.50 \text{ m}^3.$$

Submerged log:

$$\text{B.F.} = \text{weight of log} + \text{tension in string.}$$
(2) $$\rho_{\text{water}}gV = 400g + T.$$

Submerged iron:

$$T + \text{B.F.} = \text{weight of iron.}$$
(3) $$T = (\rho_{\text{iron}} - \rho_{\text{water}})gv.$$

Substitute (3) into (2):

$$\rho_{\text{water}}gV = 400g + (\rho_{\text{iron}} - \rho_{\text{water}})gv$$
$$0.50 \times 10^3 = 400 + (7.8 \times 10^3 - 10^3)v$$
$$v = \frac{100}{6.8 \times 10^3} = 14.7 \times 10^{-3} \text{ m}^3.$$

Mass of iron:

$$m = \rho_{\text{iron}}v = 7.8 \times 10^3 \times 14.7 \times 10^{-3}$$
$$= 115 \text{ kg.}$$

9.19

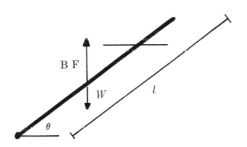

$$\sum M = 0 \text{ around } 0$$

$$(\text{B.F.})\left(1 - \frac{1}{n}\right)\frac{l}{2}\cos\theta - W\frac{l}{2}\cos\theta = 0$$

$$\rho_{\text{water}}gA\left(1 - \frac{1}{n}\right)l\left(1 - \frac{1}{n}\right)\frac{l}{2}\cos\theta - \rho_{\text{rod}}gA\frac{l}{2}\cos\theta = 0$$

$$\frac{\rho_{\text{water}}(1 - 1/n)^2}{2} - \rho_{\text{rod}} = 0$$

$$\rho_{\text{rod}} = \frac{10^3}{2}\left(1 - \frac{1}{n}\right)^2.$$

9.21 As the balloon rises, the external pressure reduces and the volume of gas in the bag increases, keeping the relative density of gas to air constant and the lift constant. Once the gas has expanded to the maximum volume of the bag, no further increase can take place without loss of gas and loss of lift. That height for which the pressure causes the gas to expand to the maximum volume of the balloon is the altitude ceiling. The submarine will sink when its average density exceeds that of the surrounding water. Since the water is nearly incompressible, its density will not change with depth, and the submarine will continue to sink indefinitely.

9.23

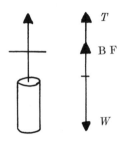

$$47.324 + 1.29 \times 10^{-3}V - 2.70V = 0$$
$$V(2.70 - 1.29 \times 10^{-3}) = 47.324.$$

Note that the second term in parentheses will make a correction in the third decimal place but the first term is given only to the second decimal. There is no point, therefore, in carrying the second term. Furthermore, there is no point in carrying more than three significant figures on the right side because the left side is only good to three figures.

$$V(2.70) = 47.3$$
$$V = \frac{47.3}{2.70}.$$

In the liquid, rounding to three figures:

$$32.2 + \rho_{liq}V - \rho_{Al}V = 0$$
$$32.2 + \rho_{liq}\left(\frac{47.3}{2.70}\right) - 2.70\left(\frac{47.3}{2.70}\right) = 0$$
$$\rho_{liq} = 15.1\left(\frac{2.70}{47.3}\right) = 0.822 \text{ g/cm}^3.$$

9.25

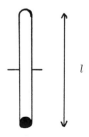

For any initial h, $l > h > 0$. Harmonic motion will result. At equilibrium:

$$\text{B.F.} = mg$$

$$\rho_l g A \frac{l}{2} = mg$$

$$\rho_l = \frac{2m}{Al}.$$

When accelerating, let $x =$ distance of the center $(l/2)$ from the surface of the water.

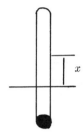

$$\sum F = ma$$

$$mg - \frac{2mg}{l} x - mg = ma$$

$$x + \frac{2g}{l} x = 0.$$

This is the differential equation of harmonic motion, with

$$\omega^2 = \frac{2g}{l}$$

$$2\pi f = \sqrt{\frac{2g}{l}}$$

$$f = \frac{1}{2\pi} \sqrt{\frac{2g}{l}}.$$

Amplitude = maximum value of x:

$$X = h - \frac{l}{2}.$$

9.27   (a) The effective weight density of both block and water will remain unchanged because the system is not accelerating. The block will continue half submerged.
(b) The block will remain half submerged because buoyant force will increase as $\rho_{\text{water}}(g + a)$. This force provides the force in excess of the weight of the block required to accelerate the block. This is $\rho_{\text{block}}(g + a)$. Since both vary as $(g + a)$ the fraction submerged will remain constant.
(c) Same as (b).

(d), (e) The block will neither be submerged nor will it float. Water will not even continue to assume the shape of the container.

9.29

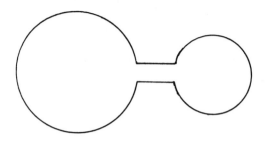

First expansion:

$$p_1 v_1 = p_2 v_2$$
$$p_1(4) = p_2(4 + \tfrac{4}{9})$$
$$p_2 = p_1 \frac{4}{4} + \frac{4}{9} = p_1 \frac{1}{\frac{10}{9}}$$
$$p_2 = \tfrac{9}{10} p_1.$$

After n repetitions:

$$p_n = (\tfrac{9}{10})^n p_1$$
$$\left(\frac{9}{10}\right)^n = \frac{p_n}{p_1} = 10^{-4}$$
$$n \log 0.9 = -4$$
$$n = 86.$$

9.31 Employ Stokes's formula [Eq. (9.9)]:

$$v = \frac{2g}{9\eta} r^2 (\rho - \delta) \qquad \delta \ll \rho$$
$$\simeq \frac{2gr}{9\eta} \rho.$$

(a)
$$v = \frac{2(980)(5 \times 10^{-3})^2(1)}{9(1.8 \times 10^{-4})}$$
$$r = 0.05 \text{ mm}$$
$$= 0.005 \text{ cm}$$
$$= 5 \times 10^{-3} \text{ cm}$$
$$\rho = 1.0 \text{ gm/cm}^3$$
$$\eta = 1.8 \times 10^{-4}$$
$$v = 3.02 \times 10^{+1} \text{ cm/sec}$$
$$v = 30.2 \text{ cm/sec.}$$

(b) $\qquad v = 483 \text{ cm/sec.}$

(c) $\qquad v = 12000 \text{ cm/sec.}$

(d)
$$v = \frac{2(980)(6 \times 10^{-1})^2(0.917)}{9(1.8 \times 10^{-4})}$$
$$= 4 \times 10^5 \text{ cm/sec.}$$

9.33 (a)

$$v \simeq \frac{10 \text{ cm}}{1 \text{ hr}} = \frac{10 \text{ cm}}{3600 \text{ sec}}$$

$$= \frac{2gr^2}{9\eta} (\rho - \delta)$$

$$\frac{10}{3600} = \frac{2(980r^2)}{9 \times 10^{-2}} (4 - 1)$$

$$r^2 = \frac{(9 \times 10^{-2})(10)}{3600(3)(2)(980)}$$

$$r = 2.0 \times 10^{-4} \text{ cm.}$$

(b)

$$v = 4.0 \times 10^{-5} \text{ cm.}$$

9.35

$$q = \bar{v}A = \bar{v}(\pi R^4) = \frac{(P_1 - P_2)\pi R^4}{8\eta l}$$

$$v = 200 \text{ cm/sec}$$
$$l = 5 \text{ m} = 500 \text{ cm}$$
$$R = 5 \times 10^{-1} \text{ cm}$$

$$(P_1 - P_2) = \frac{8\eta l \bar{v}}{R^2}$$

$$= \frac{8(0.011)(500)(200)}{(5 \times 10^{-1})^2}$$

$$= \frac{8(0.011)(5 \times 10^2)(2 \times 10^2)}{25 \times 10^{-2}}$$

$$= 3.52 \times 10^4 \text{ dynes/cm}^2$$

$$\rho gh = 3.52 \times 10^4 \text{ dynes/cm}^2$$

$$h = \frac{3.52 \times 10^4 \text{ dynes/cm}^2}{1 \text{ g/cm}^3 \times 980 \text{ cm/sec}^2} = 3.59 \times 10^1$$

$$= 35.9 \text{ cm.}$$

9.37

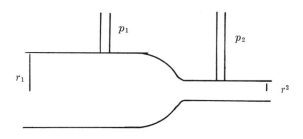

Applying Bernoulli's equation:

$$\frac{v_1^2}{2g} + \frac{p_1}{\rho g} = \frac{v_2^2}{2g} + \frac{p_2}{\rho g}$$

(1)

$$\frac{2(p_1 - p_2)}{\rho} = v_2^2 - v_1^2.$$

A-65

Next, employing the continuity equation:

$$Q_1 = Q_2$$
$$v_1 A_1 = v_2 A_2$$

(2)
$$v_2 = v_1 \frac{A_1}{A_2} = v_1 \frac{\pi r_1^2}{\pi r_2^2} = v_1 \frac{r_1^2}{r_2^2}.$$

Substitute into (1):

$$\frac{2(p_1 - p_2)}{\rho} = v_1^2 \frac{r_1^4}{r_2^4} - v_1^2 = v_1^2 \left(\frac{r_1^4}{r_2^4} - 1 \right)$$

(3)
$$v_1 = \sqrt{\frac{2(p_1 - p_2)}{\rho(r_1^4/r_2^4 - 1)}}$$

Replacing (3) in the continuity equation gives

(4)
$$Q = Av = \pi r_1^2 v_1 = \pi r_1^2 \sqrt{\frac{2(p_1 - p_2)}{\rho(r_1^4/r_2^4 - 1)}}$$

$$= \pi (15)^2 \sqrt{\frac{2(1.33 \times 10^5)}{1(81 - 1)}}$$

$$= 4.08 \times 10^4 \text{ cm/sec} = 4.08 \times 10^1 \text{ liters/sec.}$$

9.39

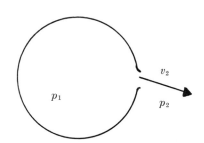

$$\frac{v_1^2}{2g} + \frac{p_1}{\rho g} = \frac{v_2^2}{2g} + \frac{p_2}{\rho g}$$

$$0 + 2 \frac{(p_1 - p_2)}{\rho} = v^2$$

$$p_1 - p_2 = 2.026 \times 10^6 \text{ dynes/cm}^2$$

$$v_2 = \sqrt{\frac{2(2.026 \times 10^6)}{0.83}}$$

$$= 2.21 \times 10^3 \text{ cm/sec.}$$

9.41

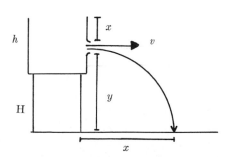

Range:

(**1**)
$$X = vt$$
$$y = \frac{gt^2}{2}$$
$$t = \sqrt{\frac{2y}{g}}.$$

Substitute into range Eq. (1):

(**2**)
$$X = v\sqrt{\frac{2y}{g}}.$$

From Bernoulli's equation:

$$v = \sqrt{2gx}.$$

Substitute into Eq. (2):

$$X = \sqrt{(2gx)\frac{2y}{g}}$$
$$y = H + h - x$$
$$X = \sqrt{4x(H + h - x)}$$
$$\frac{dX}{dx} = \frac{H + h - 2x}{[x(H + h - x)]^{1/2}} = 0$$
$$H + h - 2x = 0$$
$$x = \frac{H + h}{2} = \text{value of } x \text{ for maximum range.}$$

Chapter 10

10.1 $K = -V\,dp/dV$, where $K = 5 \times 10^{11}$ dyn/cm².

$$\frac{dp}{p} = \frac{-dV}{V} = 10^{-3},$$

$$dp = -K\frac{dV}{V} = 5 \times 10^{11} \times 10^{-3} \text{ dyn/cm}^2 = 5 \times 10^8 \text{ dyn/cm}^2.$$

The total pressure that the lead has to be subjected to is

$$P = P_0 + dp = (1.01 \times 10^6 + 5 \times 10^8) \text{ dyn/cm}^2$$
$$= 5.01 \times 10^8 \text{ dyn/cm}^2$$
$$\frac{W}{V} = p\left|\frac{dV}{V}\right| = 5.01 \times 10^8 \text{ dyn/cm}^2 \times 10^{-3} = 5.01 \times 10^{-5} \text{ erg/cm}^3.$$

10.3
$$W = \frac{1}{2}\frac{EAl^2}{L} = \frac{1}{2}\frac{(2 \times 10^{12})\pi(0.765 \times 10^{-1})^2(5 \times 10^{-1})^2}{10^3}$$
$$= 10^9\pi(7.65)^2(5)^2(10^{-6}) = (5 \times 7.65)^2\pi(10^3)$$
$$= 146\pi(10^3)$$
$$= 465 \times 10^3 \text{ erg.}$$

10.5 From the equation $k = 2k_1k_2/(k_1 + k_2)$, $k_i = E_iA/L_i$. Therefore,

$$k = \frac{2E_1E_2A}{E_1L_2 + E_2L_1} = \frac{2(10^{11})(2 \times 10^{11})A}{10^{11}(1.5) + 2 \times 10^{11}(2.5)} = \frac{\pi}{6.5} \times 10^5 \text{ N/m}$$

and

$$F = -kx.$$

Therefore,

$$|F| = |hx| = \frac{\pi}{6.5} \times 10^5 \times 2 \times 10^{-3} = \frac{2\pi}{6.5} \times 10^2 \text{ N}.$$

The force F required to increase the length of composite wire by 2 mm is

$$F = \frac{2\pi}{6.5} \times 10^2 \text{ N} = 96.6 \text{ N}.$$

Since

$$k_1 = \frac{E_1A}{L_1} = \frac{10^5\pi}{4 \times 2.5}$$

$$k_2 = \frac{E_2A}{L_2} = \frac{\pi}{2 \times 1.5} \times 10^5,$$

$$l_1 = \frac{4 \times 2.5}{6.5} \times 10^{-3} \text{ m} = 1.53 \text{ mm},$$

$$l_2 = \frac{2 \times 1.5}{6.5} \text{ mm} = 0.46 \text{ mm},$$

where l_1 is the increase in length of brass and l_2 is the increase in length of steel.

10.7

$$dm = \rho A \, dx$$
$$dF(x) = dm \, \omega^2 x$$
$$dF(x) = d\omega^2\rho A x \, dx$$
$$F(x) = \int_x^l \omega^2\rho A x \, dx = \frac{\omega^2\rho A}{2} (L^2 - x^2).$$

We had the force acting on the element dx, so the elongation of element dx was

$$\Delta = \frac{F(x) \, dx}{AE} = \frac{\omega^2\rho}{2E} (L^2 - x^2) \, dx.$$

The elongation of wire with the length x is

$$y(x) = \int_0^x dy = \int_0^x \frac{\omega^2\rho}{2E} (L^2 - x^2) \, dx = \frac{\omega^2\rho}{2E} L^2 x - \frac{\omega^2\rho}{6} x^3 \Big|_0^x$$

$$= \frac{\omega^2\rho}{2} \left(L^2 - \frac{x^2}{3}\right) x.$$

10.11 (a) The dimensions of torque are ML^2/T^2, where M = mass, L = length, and T = time.
(b) The dimensions of rigidity are M/LT^2.

10.13 The shear stress of the hollow cylinder $= dF/2\pi r \, dr$. The strain $= r\theta/l$. From the definition of shear modulus we obtain

$$dF = \frac{2\pi G\theta}{L} r^3 \, dr.$$

10.15

$$L = \frac{\pi G a^4 \theta}{2l} = \frac{\pi (1.7 \times 10^{11})(0.1 \times 10^{-3})^4 2\pi}{2 \times 50 \times 10^{-2}}$$

$$= 2\pi^2(1.7) \times 10^{11} \times 10^{-16} = 3.4\pi^2(10^{-5}) = 3.36 \times 10^{-4} \text{ N-m.}$$

The torque of the rod:

$$d\tau = x^2 \ddot{\theta}\, dm$$

$$\tau_{\text{total}} = 2\rho\, dr = \tfrac{2}{3}\rho\ddot{\theta}r^3 = \frac{M}{3}\ddot{\theta}r^2, \qquad \text{where } 2\rho r = M,$$

$$= \frac{100}{3} \times 10^2 \ddot{\theta} = \frac{10^4}{3}.$$

Then

$$\frac{10^4}{3}\ddot{\theta} + \frac{1}{2}\frac{\pi(1.7) \times 10^{12} \times (10^{-2})^4}{50}\theta = 0$$

$$\ddot{\theta} + \frac{\pi}{2}\frac{3 \times 1.7 \times 10^{8-8}}{50}\theta = 0$$

$$\ddot{\theta} + \frac{\pi}{2}\frac{5.1}{50}\theta = 0.$$

Therefore,

$$\omega = \sqrt{\frac{\pi}{2}(0.102)} = \sqrt{0.051\pi} = \sqrt{5.1\pi}(10^{-1})$$

$$T = 2\pi\sqrt{\frac{1}{5.1\pi}} \times 10^1 = 20\pi\sqrt{\frac{1}{5.1\pi}} = 20\pi\sqrt{6.25} \times 10^{-1} = 2\pi\sqrt{6.25}$$

$$= 15.7 \text{ sec.}$$

10.19   The friction force is $mg \sin \theta = mg \sin 15°$.

$$ma - mg \sin 30° - mg \sin 15°$$
$$a = g(\sin 30° - \sin 15°) = g(\tfrac{1}{2} - 0.258)$$
$$= 9.8 \times (0.5 - 0.258) = 9.8 \times 0.242 = 2.37 \text{ m/sec}^2.$$

Chapter 11

11.1

$$\lambda = 2l = 240 \text{ cm} = 2.40 \text{ m}$$

$$\mu = \frac{m}{l} = \frac{56 \text{ g}}{120 \text{ cm}} = 4.66 \times 10^{-2} \text{ kg/m}$$

$$F = 15 \text{ kgf} = 15 \times 9.8 \text{ N} = 147 \text{ N.}$$

$$v = \sqrt{\frac{F}{\mu}} = \sqrt{\frac{147 \text{ N}}{4.66 \times 10^{-2} \text{ kg/m}}}$$

$$= 5.61 \times 10 = 56.1 \text{ m/sec.}$$
$$f\lambda = v$$

$$f = \frac{v}{\lambda} = \frac{56.1 \text{ m/sec}}{2.40} = 23.2 \text{ sec}^{-1}.$$

11.3

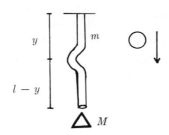

(a) With $M = 0$, the velocity of the impulse is

$$v_i = \sqrt{\frac{F}{\mu}},$$

where

$$F = \mu g(l - y).$$

Hence

$$v_i = \sqrt{\frac{\mu g(l - y)}{\mu}}$$

(1) $$v_i = \sqrt{g(l - y)},$$

(2) $$a_i = \frac{dv_i}{dt} = \frac{dv_i}{dy}\frac{dy}{dt} = \frac{-g^{1/2}}{2(l - y)^{1/2}}\frac{dy}{dt}.$$

Also,

(3) $$a_i = \frac{dv_i}{dy}v_i = \frac{-g^{1/2}}{2(l - y)^{1/2}}[g(l - y)]^{1/2} = \frac{-g^{3/2}}{2}.$$

Substituting (3) into (2):

$$\frac{-g^{3/2}}{2} = \frac{-g^{1/2}}{2(l - y)^{1/2}}\frac{dy}{dt}$$

$$g\,dt = \frac{dy}{(l - y)^{1/2}}$$

$$gt = 2(l - y)^{1/2}\Big|_0^y$$

(4) $$gt = 2[(l - y)^{1/2} - l^{1/2}]$$

relates time and distance of fall for the pulse. For free fall, the corresponding relation is

$$\frac{gt^2}{2} = y$$

(5) $$t = \left(\frac{y}{2g}\right)^{1/2}.$$

To get the crossover value of y, eliminate t between (4) and (5):

$$\left(\frac{y}{2g}\right)^{1/2} = \frac{2}{g}[(l - y)^{1/2} - l^{1/2}]$$

$$g^{1/2}y^{1/2} = 2r_2[(l - y)^{1/2} - l^{1/2}]$$

$$g^{1/2}[l - (l - y)]^{1/2} = 2r_2[(l - y)^{1/2} - l^{1/2}]$$

Let $l - y = h$:

$$g^{1/2}(l - h)^{1/2} = 2r_2(h^{1/2} - l^{1/2})$$
$$g(l - h) = 8(h - 2h^{1/2}l^{1/2} + l)$$
$$(g - 8)l - (g + 8)h = -16h^{1/2}l^{1/2}$$
$$(g - 8)^2l^2 - 2(g^2 - 8^2)lh + (y + 8)^2h^2 = 16^2hl$$
$$(g + 8)^2h^2 - 2(g + 192)lh + (y - 8)^2l^2 = 0$$
$$h = \frac{2(g + 192)l \pm \sqrt{4(g + 192)l^2 - 4(g^2 - 8^2)^2l^2}}{2}.$$

11.5

$$v = \text{velocity}$$
$$\mu = \frac{\text{mass}}{\text{length}}$$
$$y = A \sin \frac{2\pi}{\lambda}\left(x - \sqrt{\frac{F}{\mu}}\,t\right)$$
$$\frac{\partial y}{\partial t} = \frac{-2\pi A}{\lambda}\sqrt{\frac{F}{\mu}}\cos\left(x - \sqrt{\frac{F}{\mu}}\,t\right)$$
$$(v_y)_{\text{max}} = \frac{2\pi A}{\lambda}\sqrt{\frac{F}{\mu}}$$
$$v_{\text{max}}^2 = \frac{4\pi^2 A^2}{\lambda^2}\frac{F}{\mu}$$
$$\frac{E}{l} = \frac{(E_k)_{\text{max}}}{l} = \frac{\mu v^2}{2} = \frac{2\pi^2 A^2 F}{\lambda^2}.$$

11.7

$$v = \sqrt{\frac{\gamma RT}{M}}$$
$$\frac{dv}{dT} = \left(\frac{\gamma R}{M}\right)^{1/2}\frac{1}{2}T^{-1/2} = \frac{1}{2}\sqrt{\frac{\gamma R}{MT}} = \frac{1}{2}\left(\frac{\gamma RT}{M}\right)^{1/2}T^{-1}$$
$$= \frac{1}{2}\frac{v}{T} = \frac{1}{2}\frac{343.4}{293}$$
$$\frac{dv}{dT} = 0.585 \text{ m/sec/}^\circ\text{K.}$$

11.9 The vocal and chest cavities resonate at certain frequencies generated by the vocal cords. When these cavities are filled with helium, the resonant frequencies are higher because the velocity of sound in helium is higher than it is in air.

11.11 Length $\simeq$ duration of sound $\times$ velocity of sound.

11.15

$$v = \sqrt{\frac{K}{\rho}}$$
$$K = \rho v^2 = (10^3 \text{ kg/m}^3)(1450 \text{ m/sec})^2 = 2.1 \times 10^9 \text{ N/m}^2.$$

11.17

$$v = Cg^a\lambda^b$$
$$LT^{-1} = (LT^{-2})^aL^b = L^{b+a}T^{-2a}$$
$$a + b = 1$$
$$2a = 1$$

A-71

$$a = \tfrac{1}{2}$$
$$\tfrac{1}{2} + b = 1$$
$$b = \tfrac{1}{2}$$
$$v = Cg^{1/2}\lambda^{1/2} = C\sqrt{g\lambda}$$
$$7.00 \text{ m/sec} = C\sqrt{(9.8 \text{ m/sec}^2)(31.5 \text{ m})}$$
$$C = \frac{7.00}{(9.8 \times 31.5)^{1/2}} = 0.399.$$

Chapter 12

12.1
$$T \simeq \sqrt{\gamma^3 \rho / \sigma}.$$

12.3 $\Delta p = 4\sigma/r$. Total pressure $P = P_0 + \Delta p$, where P_0 is atmospheric pressure.

$$\Delta p(d = 250 \text{ Å}) = \frac{4(50)}{125 \times 10^{-8}} = 1.6 \times 10^8 \text{ dyn/cm}^2.$$

Therefore,

$$P(d = 250 \text{ Å}) = (1.6 \times 10^8 + 1.01 \times 10^6) \text{ dyn/cm}^2$$
$$= 1.61 \times 10^8 \text{ dyn/cm}^2$$

$$\Delta p(d = 10^3 \text{ Å}) = \frac{4(50)}{10^{-5}} = 200 \times 10^5 = 2 \times 10^7 \text{ dyn/cm}^2$$

$$P(d = 10^3 \text{ Å}) = 2.1 \times 10^7 \text{ dyn/cm}^2$$

$$\Delta p(d = 4 \times 10^3) = \frac{50}{10^{-5}} = 5 \times 10^6$$

$$P(d = 4 \times 10^3) = 6.01 \times 10^6 \text{ dyn/cm}^2.$$

12.5 From the equation on p. 317, $\sigma = r \, \delta p/2$.

$$\delta p = 3.8 \times 1 \times 980 - 3.2 \times 0.79 \times 980$$
$$= (3.8 - 2.528) \, 980 \text{ dyn/cm}^2 = 1.272 \times 980 \text{ dyn/cm}^2$$

Therefore,

$$\sigma = \frac{3.9 \times 10^{-2} \times 1.272 \times 980}{2} = 3.9 \times 4.9 \times 1.272 = 24.4 \text{ dyn/cm}.$$

12.7 $F = \sigma l = 72.8 \times 4 \times 10^{-1} = 7.28 \times 4$, so the pressure due to surface tension is

$$p_s = \frac{F}{A} = \frac{7.28 \times 4}{(\tfrac{1}{10})^2} = 7.28 \times 4 \times 10^2 \text{ dyn/cm}^2.$$

Assume we can depress below the free surface by h. Then the pressure due to water is
$$p_h = \rho gh = 980h \text{ dyn/cm}^2.$$

Therefore, $p_s = p_h$, and we obtain

$$h = \frac{1}{980} (7.28 \times 4) \times 10^2 = \frac{7.28 \times 4}{9.8} = 2.97 \text{ cm}.$$

12.9 Since the contact angle of water-glass equals zero,

$$y = \frac{2\sigma \cos \theta}{\rho g r} = \frac{2 \times 72.8}{980(\frac{5}{2})10^{-2}} = \frac{2 \times 72.8}{490(5)10^{-2}} = \frac{2 \times 7.28}{4.9(5)} \times 10 = 5.95 \text{ cm.}$$

The water will rise to 5.95 cm in the tube at 20°C.

12.11

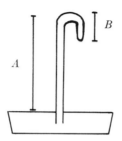

It would not work. As shown in the figure, the tube on side A is longer than the tube on side B, so the pressure due to the liquid on side A is greater than on side B, and the atmospheric pressure on both sides balances in addition to the surface tension produced by the liquid, which forms the hemispherical droplet in the outlet of tube B.

12.13

$$h = \frac{2\sigma \cos \theta}{\rho g r} = \frac{2(0.04 \times 10^5/10^2) \cos 14°}{1.1 \times 9.8 \times 3} = \frac{2(4 \times 9.7)}{1.1 \times 9.8 \times 3} = 2.4 \text{ cm.}$$

12.17 The surface tension of water is

$$\sigma = \frac{F}{l} = \frac{3.97 \times 980}{2\pi(r_1 + r_2)} = \frac{3.97 \times 980}{2\pi(4.2 + 4.4)} = 71.9 \text{ dyn/cm.}$$

(a) If the edge of the ring is 1 mm higher, it must add the weight of water drawn up by surface tension. The weight of water is

$$W = \rho v = 1 \times \pi[(4.4)^2 - (4.2)^2](10^{-1})$$
$$= \pi(1.72)10^{-1} = 0.54 \text{ g,}$$

so the weight 0.56 g must add to the other scale pan.
(b) If the edge of the ring is 1 mm lower than the horizontal surface, from Archimedes' principle the weight of 0.54 g must be subtracted from 3.97 g.

Chapter 13

13.1 −40°C.

13.3 (a) 61°C. (b) 251°C. (c) −67°C.

13.5 880°C, 1930°F.

13.7

$$\frac{V_0}{T_0} = \frac{V_1}{T_1}$$
$$V_1 = V_0 + V',$$

where V' is the increment in V_0 at temperature T_1. The actual increment is at temperature T_0, hence

$$\frac{V'}{T_1} = \frac{V}{T_0}$$

$$V' = \frac{T_1}{T_0} V$$

$$V_1 = V_0 + \frac{T_1}{T_0} V$$

$$\frac{V_0}{T_0} = \frac{V_0 + (T_1/T_0)V}{T_1}$$

$$T_1 = \frac{V_0}{V_0 - V} T_0 = \frac{T_0}{1 - V/V_0}$$

$$= \frac{273}{1 - V/V_0}.$$

13.9   The Kelvin scale is more likely to be relevant theoretically because it starts at absolute zero, which is not an arbitrary point.

13.11   Even if the mercury column is 76 cm long, giving a $\Delta p = 1$ atm $\simeq 10^6$ dyn/cm^2, this would yield

$$\frac{\Delta V}{V} = \frac{10^6 \text{ dyn/cm}^2}{2.6 \times 10^{11} \text{ dyn/cm}^2} \simeq 4 \times 10^{-6},$$

corresponding to a temperature error of

$$\Delta T = \frac{4 \times 10^{-6}}{1.8 \times 10^{-4}} \simeq 2 \times 10^{-2}°\text{C},$$

which would be negligible.

13.13   The wavelength will vary because of Doppler broadening.

Chapter 14

14.1

$$\Delta l = l_0 \alpha \, \Delta T = (100)(17 \times 10^{-6})(70)$$
$$= 1.2 \times 10^{-1}$$
$$= 0.12 \text{ m}.$$

14.3

$$\beta = 3\alpha$$
$$V = V_0(1 + 3\alpha \, \Delta T)$$
$$\rho = \frac{m}{V} \frac{m}{V_0(1 + 3\alpha \, \Delta T)} \frac{\rho_0}{1 + 3\alpha \, \Delta T}$$
$$= \frac{19.30}{1 + 3(1.43 \times 10^{-4})60}$$
$$= 18.7 \text{ gm/cm}^3.$$

14.5

$$T = 2\pi \sqrt{\frac{l}{g}}$$

$$\frac{T}{T_0} = \sqrt{\frac{l}{l_0}} = \sqrt{\frac{l_0(1 + \alpha \, \Delta T)}{l_0}} = \sqrt{(1 + \alpha \, \Delta T)}$$

$$\Delta T = -10°C$$

$$\frac{T}{T_0} \simeq 1 - \tfrac{1}{2}(1.9 \times 10^{-5})(10) = 1 - 0.95 \times 10^{-4}.$$

The period is shortened by 0.0095 percent. The clock will gain about 8 sec/day.

14.7

$$\frac{\Delta l}{l} = \alpha \, \Delta T = \frac{1}{y}\frac{F}{A}$$

$$1.1 \times 10^{-5} \, \Delta T = \frac{1}{2.1 \times 10^{12}} \, (2000 \times 10^3 \times 980) \, \text{dyn/cm}^2$$

$$\Delta T = \frac{2 \times 10^6 \times 980}{2.1 \times 10^{10} \times 1.1 \times 10^{-5}}$$

$$= 0.83 \times 10^2$$

$$= 83°C \text{ drop.}$$

14.9

$$200C(36 - 22.4) + 400(0.093)(36 - 22.4)$$

$$= 1000(0.093)(22.4 - 0)$$

$$C = \frac{93 \times 22.4 - 400(0.093)(13.6)}{200 \times 13.6}$$

$$= 0.58 \text{ cal/gm-°C.}$$

14.11

$$\frac{\Delta\theta}{\Delta t} = \frac{[100(0.093) + 500](\theta_1 - \theta_2)}{t} = \frac{[100(0.093) + 400C](\theta_1 - \theta_2)}{0.52t}$$

$$C = 0.64 \text{ cal/gm-°C.}$$

14.13

$$12.9\pi \left(\frac{10^{-1}}{2}\right)^2 = \frac{m}{0.917} - \frac{m}{1.00}, \qquad m = \text{mass of ice melted,}$$

$$12.9\pi \left(\frac{10^{-2}}{4}\right) = m\left(\frac{1}{0.917} - 1\right)$$

$$m = 1.11 \text{ g ice melted.}$$

$$12C(15) = 1.11(79.6)$$

$$C = \frac{1.11(79.6)}{12(15)} = 0.49.$$

14.15

$$100(0.48)(200 - T) = 20(70) + 20(539.5) + 20(0.48)(T - 100)$$

$$T = 193°C$$

All 20 g of water will be vaporized.

14.17

$$10(0.48)(50) + 10(539.5) + 10(100 - T)$$
$$= 60(0.53)(20) + 60(79.0) + 60T + [200 + 100(0.093)](T - 20)$$
$$T = 19.5°C.$$

14.19 Since the heat transferred through the compound slab must be the same,

$$\frac{d\theta}{dx} = \frac{1}{-KA}\frac{d\theta}{dt} \propto \frac{1}{K};$$

that is, the temperature gradients are inversely proportional to their thermal conductivities.

14.21 Since

$$A = 4\pi r^2,$$

$$d\theta = -\frac{d\theta/dt}{4\pi K}\frac{dr}{r^2}.$$

Therefore,

$$\theta_1 - \theta_2 = \frac{d\theta/dt}{4\pi K}\frac{1}{r_1} - \frac{1}{r_2}.$$

14.23 Using the result of Problem 14.22,

$$\frac{dq}{dt} = \frac{2\pi K(\theta_1 - \theta_2)}{\ln r_2/r_1} = \frac{2(3.14)(0.11)(120 - 115)}{\ln 1.5/2}$$
$$= -12.4 \text{ cal/cm-sec},$$

since the specific heat for steam at atmospheric pressure and between 100°C and 200°C is 0.48 cal/g-deg. Thus, the temperature decrease per unit length $\Delta\theta/dx$ is related to dq/dt by

$$\frac{\Delta x}{\Delta\theta}(2 \text{ g/sec})\ 0.48 \text{ cal/g-deg} = \frac{dq}{dt} \text{ cal/cm-sec}.$$

Therefore,

$$\frac{\Delta x}{\Delta\theta} = \frac{dq/dt}{2(0.48)} = -12.8°C/cm.$$

Chapter 15

15.1 $V = 11127.09 \text{ cm}^3 = 11.127 \text{ liter}.$

15.3 $P = 3.043 \times 10^7 \text{ dyn/cm}^2.$

15.5 $P = 601.33 \text{ mm Hg}.$

15.7 $l = \frac{mgV}{P_0A^2}; \quad T = \frac{2\pi}{A}\sqrt{\frac{mV}{P_0}}.$

15.9   Use $p = \frac{\rho}{M}RT.$

15.11   2 Mg.

15.13 $\sqrt{\bar{v}^2} = 6.1 \times 10^5 \text{ cm/sec; no.}$

15.15 $\qquad$ 9.78×10^9 molecules.

15.17 $\qquad$ 1.64×10^3 J.

Chapter 16

16.1 $p_{O_2+H_2O} = 756$ mm Hg at $T = 25°C = 298°K$. $V = 1$ liter. Find m_{O_2}.

p_{vap}:

$$\log p \simeq A - \frac{B}{T}$$

$$\log 17.51 = A - \frac{B}{293} \qquad \text{(Table 16.2)}$$

$$\log 55.13 = A - \frac{B}{313} \qquad \text{(Table 16.2)}$$

$$\log 55.13 - \log 17.51 = \frac{-B}{313} + \frac{B}{293}$$

$$\log \frac{55.13}{17.51} = B\left(\frac{1}{293} - \frac{1}{313}\right)$$

$$B = \frac{10^4}{2.2} \log 3.15.$$

$$\log p_{25} - \log 17.51 = B\left(\frac{1}{293} - \frac{1}{298}\right) = B(6 \times 10^{-5})$$

$$\frac{\log p_{25}}{17.51} - \left(\frac{10^4}{2.2} \log 3.15\right)(6 \times 10^{-5})$$

$$p_{25} = 17.51(3.15)^{0.27} = 24 \text{ mm Hg.}$$

Corrected pressure for O_2:

$$p = 756 - 24 = 732 \text{ mm Hg.}$$

Find V at STP:

$$\frac{p_0 V_0}{T_0} = \frac{pV}{T}$$

$$V_0 = \frac{p}{p_0} \frac{T_0}{T} V = \frac{732}{760}\left(\frac{273}{298}\right) 1 \text{ liter}$$

$$= 0.992 \text{ liter.}$$

Number of molecules O_2:

$$n = \frac{0.882 \text{ liter}}{22.4 \text{ liter/mol at STP}} = 3.93 \times 10^{-3} \text{ g-mol.}$$

Mass of O_2:

$$m = 3.93 \times 10^{-3} \text{ g-mol} \times 32 \text{ g/g-mol}$$
$$= 1.26 \times 10^{-1} = 0.126 \text{ g } O_2.$$

16.3 $p_T = 760$ mm Hg, $T = 20°C = 293°K$, $p_{vap} = 17.51$ mm Hg (Table 16.2), and $p_{air} = 760 - 17.51 = 742.5$ mm Hg. For dry air (using equivalent weight $\simeq 29$):

$$\rho = 1.20 \text{ g/liter} = 1.20 \text{ kg/m}^3.$$

For water vapor at 293°K:

$$p = 17.51 \text{ mm Hg.}$$

The ideal gas law yields

$$\rho_{H_2O} = \frac{m}{V} = 0.017 \text{ g/liter} \simeq 0.017 \text{ kg/m}^3,$$

and

$$\rho_{air} = 1.18 \text{ g/liter} = 1.18 \text{ kg/m}^3$$
$$\rho_{tot} = 1.18 + 0.017 = 1.25 \text{ g/liter} = 1.25 \text{ kg/m}^3.$$

16.5 At $V = 50/2.5 = 20$ liters $= 20 \times 10^3$ cm³, $\rho_{vap}/\rho_{sat} = 1.0 = 100\%$ humidity. $p_{vap} = 17.5$ mm Hg. Treat the vapor as an ideal gas to compute mass.

$$PV = \frac{m}{M} RT$$

$$m = \frac{PVM}{RT} = \frac{17.5}{760} \frac{1.013 \times 10^6 (20 \times 10^3)18}{(8.3 \times 10^7)293}$$

$$= 3.46 \times 10^{-1} \text{ g.}$$

Compression of the volume to 10 liters will condense

$$\tfrac{1}{2}m = 1.7 \times 10^{-1} = 0.17 \text{ g H}_2\text{O.}$$

16.7 (a) 2.1×10^{-6} mm Hg. (b) 3.4×10^4 mm Hg.

16.9 $l \simeq 4\sigma N^{2/3}V^{2/3}$, where σ is the surface tension, N is Avagadro's number, and V is the mean volume per volume, which is much larger than the diameter of the molecule.

Chapter 17

17.1 The change in mechanical potential energy equals the change in internal energy of the water:

$$\frac{mgh}{J} = mC\,\Delta T$$

$$\Delta T = \frac{gh}{JC} = \frac{9.80(300)}{4.18 \times 10^3} = 7.04 \times 10^{-1} = 0.704°C.$$

17.3

$$\tfrac{1}{2}E_{k_{bullet}} = \Delta U_{bullet}$$

$$\frac{\tfrac{1}{2}(\tfrac{1}{2}mv^2)}{J} = mc\,\Delta T$$

$$\Delta T = \frac{v^2}{4JC} = \frac{(150)^2}{4(4.18)(10^3)} = 1.35°C.$$

17.5

$$m = pV = 1.025 \times 10^3 \text{ kg}$$
$$\Delta U = mC\,\Delta T = (1.025 \times 10^3)(10^3)(1)°C$$
$$= 1.025 \times 10^6 \text{ cal.}$$
$$\Delta E = J\,\Delta U = 4.18(1.025 \times 10^6) = 4.29 \times 10^6 \text{ J.}$$
$$t = \frac{E}{p} = \frac{4.29 \times 10^6 \text{ J}}{10^6 \text{ J/sec}} = 4.29 \text{ sec.}$$

17.7 Assuming power delivered to both liquids is the same,

$$P = \left(\frac{J\ \Delta U}{\Delta t}\right)_{\text{water}} = \left(\frac{J\ \Delta U}{\Delta t}\right)_{\text{unknown}}, \qquad \text{where } \Delta t = \text{time (sec)},$$

$$\frac{Jm_{\text{H}_2\text{O}}C_{\text{H}_2\text{O}}\ \Delta T_{\text{H}_2\text{O}}}{\Delta t_{\text{H}_2\text{O}}} = \frac{Jm_xC_x\ \Delta T_x}{\Delta t_x}$$

$$\Delta T_x = \Delta T_{\text{H}_2\text{O}} = 60°\text{C}.$$

$$C_x = \frac{\Delta t_x}{\Delta t_{\text{H}_2\text{O}}}\ \frac{m_{\text{H}_2\text{O}}C_{\text{H}_2\text{O}}}{m_x} = \frac{1.8}{5}\left(\frac{1000}{800}\right) = 0.45 \text{ cal/g-°C}.$$

17.9

$\delta W = 0$ (rigid container)

$\delta Q = 0$ (no heat transferred across the boundaries of the container)

$\delta Q = \delta U + \delta W$

$\delta U = \delta Q - \delta W = 0.$

17.11 $P_2 = P_1 = 760$ mm Hg $= 1.016 \times 10$, $V_2 = 2V_1$, $m = 20$ g argon, $T_1 = 15°\text{C} = 288°\text{K}$.

$$\frac{P_1V_1}{T_1} = \frac{P_2V_2}{T_2}$$

$$T_2 = \frac{P_2}{P_1}\frac{V_2}{V_1}T_1 = \frac{2V_1}{V_1}\,288°\text{K} - 576°\text{K}$$

$$P_1V_1 = nRT_1$$

$$V_1 = \frac{m}{M}\frac{RT_1}{P_1} = \frac{20}{40}\frac{(8.32 \times 10^7)(288)}{1.016 \times 10^6} = 11{,}800 \text{ cm}^3$$

$$\Delta V = V_2 - V_1 = 2V_1 - V_1 = V_1 = 11{,}800 \text{ cm}^3$$

$$W = P\,\Delta V = 1.016 \times 10^6 \text{ dyn/cm}^2\ (11{,}800 \text{ cm}^3)$$

$$= 1.2 \times 10^{10} \text{ erg.}$$

$$\Delta T = T_2 - T_1 = 2T_1 - T_1 = T_1 = 288°\text{K}$$

$$Q = nC_p\,\Delta T = n\gamma C_v\,\Delta T$$

$$= (\tfrac{1}{2}\text{ mole})(\tfrac{5}{3} \times 3)\,\frac{\text{cal}}{\text{mole-°K}} \times 288°\text{K}$$

$$Q = 720 \text{ cal.}$$

$$\Delta U = Q - W = 720 \text{ cal} - 1.2 \times 10^{10} \text{ erg}/4.18 \times 10^7 \text{ erg/cal}$$

$$= 720 - 287$$

$$\Delta U = 433 \text{ cal.}$$

Alternatively,

$$\Delta U = nC_v\,\Delta T = \tfrac{1}{2}\text{ mole}\left(\frac{3 \text{ cal}}{\text{mole-°K}}\right)288°\text{K} = 433 \text{ cal.}$$

17.13 $T_2 = T_1$, $V_1 = 1$ liter, $P_1 = 5$ atm, $V_2 = 2$ liters.

$$W = \int P\,dV = \int \frac{nRT}{V}\,dv = nRT\ln\frac{V_2}{V_1} = P_1V_1\ln\frac{V_2}{V_1}$$

$$= 1 \text{ liter} \times 5 \text{ atm} \ln\left(\frac{2}{1}\right)$$

$$= 3.46 \text{ liter-atm.}$$

17.17 $T_1 = 0°C = 273°K$, $P_1 = 760$ mm Hg, $P_2 = 2P_1 = 2(760$ mm Hg$) = 1520$ mm Hg. Because the process is "sudden," assume $Q = 0$, that is, adiabatic, so

$$P_1{}^{1-\gamma}T_1{}^\gamma = P_2{}^{1-\gamma}T_2{}^\gamma$$
$$P_1{}^{1-\gamma}(273)^\gamma = (2P_1)^{1-\gamma}T_2{}^\gamma$$
$$T_2{}^\gamma = (\tfrac{1}{2})^{1-\gamma}\,273^\gamma$$
$$= 2^{\gamma-1}(273)^\gamma$$
$$T_2 = 2^{\frac{\gamma-1}{\gamma}}(273)$$
$$= 2^{\frac{1.4-1}{1.4}}(273)$$
$$= 2^{0.4/1.4}(273)$$
$$= 334°K.$$

17.21

$$Q = 0$$
$$W = -10 \text{ J}$$
$$\Delta U = Q - W = 0 - (-10) = 10 \text{ J}.$$

17.25 $T = 152.7°C.$

17.27 Prove it from the reversible process.

Chapter 18

18.1

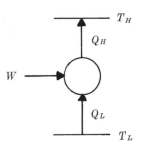

$$m = 1000 \text{ kg} = 10^6 \text{ g}, \quad T_H = 273 + 40 = 313°K, \quad T_L = 263°K$$
$$Q_L = mc_{H_2O}(20 - 0) + mh_f + mc_{100}[0 - (-10)]$$
$$= 10^6(20) + 10^6(80) + 10^6(0.5)(10)$$
$$= 1.05 \times 10^8 \text{ cal.}$$

$$W_{ideal} = Q_H - Q_L = Q_L\left(\frac{T_H}{T_L} - 1\right)$$
$$= 1.05 \times 10^8(\tfrac{313}{263} - 1)$$
$$= 2 \times 10^7 \text{ cal}$$
$$= (2 \times 10^7 \times 4.18) \text{ J}$$
$$= 8.36 \times 10^7 \text{ J.}$$

$$W_{actual} = \frac{W_{ideal}}{0.70} = \frac{8.36 \times 10^7 \text{ J}}{0.70}$$

$$P_{actual} = \frac{W_{actual}}{t} = \frac{8.36 \times 10^7 \text{ J}}{0.70 \times (24 \times 3600) \text{ sec}}$$

$$P = 1.38 \times 10^3 = 1380 \text{ W.}$$

18.3 $T = 327.4°C = 600.4°K$, $\Delta p = 1$ atm $= 1.016 \times 10^5$ N/m$^2 = 1.016 \times 10^6$ dyn/cm^2, $\rho_S = 11.005$ g/cm^3, $\rho_L = 10.645$ g/cm^3, $L = 5.47$ cal/g.

$$\Delta T = \frac{-(V_S - V_L)T}{L}\,\Delta p$$

$$V_S = \frac{1}{\rho_S} = \frac{1}{11.005 \text{ g/cm}^3} = 9.07 \times 10^{-2} \text{ cm}^3/\text{g}$$

$$V_L = \frac{1}{\rho_L} = \frac{1}{10.645 \text{ g/cm}^3} = 9.4 \times 10^{-2} \text{ cm}^3/\text{g}$$

$$\Delta T = \frac{(9.07 - 9.40) \times 10^{-2} \text{ cm}^3/\text{g} \times 600°K \times 1.01 \times 10^6 \text{ dyn/cm}}{5.47 \text{ cal/g} \times 4.18 \times 10^7 \text{ erg/cal}}$$

$$= \frac{(0.37) \times 10^{-2} \times 6 \times 10^2 \times 1.01 \times 10^6}{5.47 \times 4.18 \times 10^7}$$

$$= 9.7 \times 10^{-2}°K$$
$$= 0.097°K, \quad \text{elevation in melting point.}$$

18.5 Interchange $-P$ to $\mathfrak{F}$ and V to l in the text to get the Carnot engine for wire extension.

Volume II
Chapter 19

19.1 9×10^{-5} N.

19.3
$$(Q_1/Q_2) = \tfrac{1}{2}(-3 \pm \sqrt{5}).$$

(Hint: Use conservation of charge and the fact that the spheres are identical.)

19.5
$$Q = \sqrt{\frac{3a^3 mgl\pi\epsilon_0}{\sqrt{3l^2 - a^2}}}\ \text{C.}$$

(Hint: The suspension point is directly over the center of the triangle.)

19.7 $\phi = 67°$. The force is away from the center of the triangle and toward B.

19.11
$$F = \sqrt{\frac{Qq}{\pi^3\epsilon_0 l^3}}\,\frac{1}{\text{sec}}, \quad \text{where } l \text{ is the distance between } q.$$

(Hint: Approximate the forces by using the binomial expansion.)

19.13
$$F = \sqrt{\frac{qQ}{16\pi^3\epsilon_0 r^3 m}}.$$

If the ring had a charge of $2Q$ we would have had the same answer as in Problem 19.12. This is due to the analogy of opposite sides of the circle corresponding to the results in Problem 19.12, where the Q of that problem is expressed by $Q/2\pi r$ in this problem. But we integrate only around the semicircle.

Chapter 20

20.1

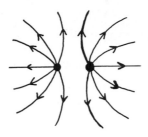

At distances large compared to those between the charges this system resembles a point charge of $+2q$ at the midpoint of the line joining the two charges. If these were negative charges the direction of the lines would be reversed.

20.3 We use an argument based on symmetry. If we choose an arbitrary point inside the cylinder and an arbitrary differential area on the cylinder we then define a cone whose base is this area and whose vertex is this point. We then look at a cone whose axis is in the opposite direction with the same vertex. Since the area of the bases of these cones is proportional to the square of the distance and is also proportional to the charge, and Coulomb's law is inversely proportional to the square of the distance, we find that the contributions from these differential areas cancel out. Thus, summing over the cylinder we get a null result.

20.5

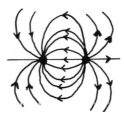

(Hint: Use the binomial expansion in the first part of the problem.)

20.7

$$E = \frac{1}{4\pi\epsilon_0} \frac{qr}{(a^2 + r^2)^{3/2}},$$

where a is the radius of the loop and r is the distance along the axis from the center of the loop.

$$E_{\text{sphere}} = 0 \text{ inside,}$$

$$E_{\text{sphere}} = \frac{Q}{4\pi\epsilon_0 r^2} \text{ outside,}$$

where r is the distance from the center of the sphere. The gravitational field inside the sphere should be equal to a point mass at the center whose mass is that enclosed by a sphere concentric with the original sphere but whose radius is from the center to the point in question.

20.9 $0.5\ \mu C$.

20.11 The electrostatic field inside the hollow and in the interior of the conducting material is zero. If a closed Gaussian surface is constructed in the interior of the conductor the integral $\int En\,dA$ is zero. Thus, the total charge enclosed must be zero and the total charge on the inner surface must be zero. If there is a patch of positive charge and negative charge on the inner surface there must be a potential difference between these two points. But $V_1 - V_2 = -\int \mathbf{E}\,dl$. For a path inside the conductor this potential difference is zero. Therefore, no charged patches can exist.

20.13

Inner surface cylinder I: $\sigma = 0$

Outer surface cylinder I: $\sigma_1 = \dfrac{\lambda_1}{2\pi R_1}$

Inner surface cylinder II: $\upsilon = \dfrac{-\lambda_1}{2\pi R_2}$

Outer surface cylinder II: $\sigma_2 = \dfrac{\lambda_1 + \lambda_2}{2\pi R_2}$

(a) $E = 0$.
(b) $E = \sigma_1 R_1/\epsilon_0 r$.
(c) $E = (\sigma_1 R_1 + \sigma_2 R_2)/\epsilon_0 r$, where $\sigma_2 = \lambda_2/2\pi R_2$.

20.15 (a) $\rho(r) = 2\epsilon_0 E/r$.
       (b) $V(r) = -Er$.

20.17 -0.5 N.

20.19 The potential must be positive with respect to a zero potential at infinity. The work that an outside agent would have to do on a positive test charge to bring it in from infinity must be positive. Thus, $V_{AB} = W_{AB}/q$ must be positive.

20.21

(a) $V = \dfrac{Q_1}{4\pi\epsilon_0 R_1} + \dfrac{Q_2}{4\pi\epsilon_0 R_2}$.

(b) $V = \dfrac{Q_1}{4\pi\epsilon_0 r} + \dfrac{Q_2}{4\pi\epsilon_0 R_2}$.

(c) $V = \dfrac{Q_1 + Q_2}{4\pi\epsilon_0 r}$.

20.23

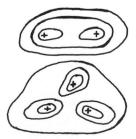

Chapter 21

21.1 6×10^6 V.

21.3 $V_{\text{final}} = 2V_{\text{initial}}$.

21.5 2.3×10^{-3} C/m².

21.7 0.10 nF.

21.9 Five different ways: three capacitors in series to each other or in parallel to each other, two in series parallel to the third one. Respectively, we get 6 μF, 90 μF, $66\frac{2}{3}$ μF$(10 + 20\|60)$, $28\frac{4}{7}$ μF$(10 + 60\|20)$ and 25 μF$(20 + 60\|10)$.

21.11

$$Q_1 = V \left\{ \frac{2 + C_1[(1/C_2) + (1/C_3)]}{(1/C_1) + (1/C_2) + (1/C_3)} \right\}$$

$$Q_2 = V \left\{ \frac{C_2[(1/C_3) + (1/C_1)]}{(1/C_1) + (1/C_2) + (1/C_3)} \right\}$$

$$Q_3 = V \left\{ \frac{C_3[(1/C_2) + (1/C_1)]}{(1/C_1) + (1/C_2) + (1/C_3)} \right\}.$$

$$V_1 = V \left[\frac{(2/C_1) + (1/C_2) + (1/C_3)}{(1/C_1) + (1/C_2) + (1/C_3)} \right]$$

$$V_2 = V \left[\frac{(1/C_3) + (1/C_1)}{(1/C_1) + (1/C_2) + (1/C_3)} \right]$$

$$V_3 = V \left[\frac{(1/C_2) + (1/C_1)}{(1/C_1) + (1/C_2) + (1/C_3)} \right].$$

(Hint: Use conservation of charge and the fact that the sum of the potential drops around a circuit is zero as well as the method of determinants to solve the simultaneous linear equations.)

21.13 9.0×10^{-2} C; 135 J.

21.15 5.31×10^{-4} J. (a) 8.85×10^{-5} J. (b) 3.19×10^{-3} J.; 1200 V.
(a) Energy is lost in the wires connected to the battery.
(b) Work is done on the system by removing the slab due to the induced charge on the surface of the dielectric.

21.17

(a) $C_0 \dfrac{1}{1 - f(1 - 1/\kappa)}$

(b) $V_0 \left[1 - f\left(1 - \dfrac{1}{\kappa} \right) \right]$

(c) $W_0 \left[1 - f\left(1 - \dfrac{1}{\kappa} \right) \right]$

If $\kappa \to 1$ we get the result for an air capacitor. If $\kappa \to \infty$ we get the result of the previous problem.

21.19

$$\frac{C_2(C_1C_4 + C_1C_5 + C_4C_5 - C_3C_5) + C_3(C_1C_2 + C_1C_4 + C_1C_5 + C_4C_5)}{C_1C_2 + C_1C_4 + C_1C_5 + C_2C_3 + C_2C_4 + C_3C_4 + C_4C_5 - C_3C_5}.$$

21.21   673 J.

21.23

(b) $\dfrac{\gamma}{\alpha\epsilon_0}\left(1 - \dfrac{x}{a}\right)$.

21.25

$$Q^2 = 8\pi^2\epsilon_0 r^2[3pr\,dr + 8\sigma\,dr].$$

The bubble expands due to the repulsion of the charges on the surface. (Hint: Treat the bubble as a spherical capacitor and find the change in its stored electrical energy. Solve for the initial pressure of the gas in terms of σ, p, and r.)

21.27 1.2×10^{-8} C; 860 V.

Chapter 22

22.5   (a) (1) $dE/dx = -ne/\epsilon_0$.
         (2) $i = nev$.
         (3) $m\,dv/dt = eE$.
     (b) $x = (ei/6m\epsilon_0)t^3$.

22.7   Plot I vs. $V^{3/2}$. The slope is

22.9

$$\sqrt{\dfrac{e}{m}}\,\dfrac{8\sqrt{2}}{9}\,\pi\epsilon_0\,\dfrac{7.62\text{ cm}}{1.27\text{ cm}}.$$

$$v = \sqrt{\dfrac{2eV_0}{m}}\left(\dfrac{x}{d}\right)^{2/3}.$$

$$t = 3\sqrt{\dfrac{m}{2eV_0}}\,d,$$

$$= 1.43 \times 10^{-9}\text{ sec.}$$

22.11 1.51×10^{-19} C.

22.13 6.97×10^{8} electrons.

22.15 1.794×10^{14}.

22.17 8.61×10^{-2} eV.

Chapter 23

23.1 10^{-15} A/cm^2.

23.3   7.083 eV. The atoms will have a kinetic energy equal to the excess energy of the photons.

23.5 0.4×10^{-6} W. The cathode losing electrons at a fixed rate causes its potential to rise. The actual power is larger due to the probability of a photon striking an electron and of the electron escaping being less than one.

23.7   0.0620 Å.

Chapter 24

24.1   0.0156%.

24.3 0.92×10^{-8} eV.

Chapter 25

25.1 0.67×10^{-5} C/m²; 560 W.

25.3   0.95 m/sec; 0.22 V/m. No, the electric signal establishes itself at velocities close to the speed of light. This situation is analogous to the flow of water through a pipe and the speed at which the pressure wave passes through the water in the pipe.

25.5   200 V; 0.02 A.

25.7   (a) $+2$ V, 0 V, $+4$ V, -6 V, 0 V.
      (b) $+2$ V, $+\frac{8}{3}$ V, 0 V, -6 V, $+\frac{4}{3}$ V.

25.9   Hint: Solve for the power in terms of the resistance of the battery, its electromotive force, and the unknown resistance. Then maximize the power with respect to the unknown resistance.

25.11   The assumption that we have an instantaneous discharge is no longer valid if E is only slightly larger than V_1, that is, due to the time interval between flashes being so short.

25.13   45 W.

25.15 $E = 1.49$ V; $R_B = 2$ Ω.

25.17
$$X = \tfrac{1}{2} \sqrt{R_1{}^2 + 4R_1 R_2}$$
$$P = 0.338 \text{ W.}$$

25.19 1.69×10^{-8} Ω; 1.69×10^{-6} Ω.

25.21   2.

25.23   1.73 m.

25.25 $\rho(20°C)/\rho(1000°C) = 0.889$.

25.27
$$\text{External voltage of generator} = 195 \text{ V;}$$
$$\text{PD across leads} = 22.5 \text{ V;}$$
$$\text{PD due to resistance of motor} = 30 \text{ V;}$$
$$E \text{ of motor} = 142.5 \text{ V.}$$

25.29   Hint: Use the method of determinants to solve the three simultaneous linear equations for the currents.
      (a)   0.15 A.
      (b)   0.05 A; A to C.

Chapter 26

26.1
$$x = \frac{-200 I_g(R_g + 100)}{I_g(R_g + 150) + E/2}$$
$$R_g = 100 \text{ Ω.}$$

26.3 $65 \times 10^3 \, \Omega.$

26.5   (a) Add 239,520 Ω in series; (b) place 0.067 Ω in parallel.

26.7   No, the galvanometer as used as a voltmeter depends on Ohm's law.

26.11   Six parts per thousand; yes.

Chapter 27

27.1   1 C.

27.3   3.97 A.

27.5   0.270 g H_2/h; 2.16 g O_2/h.

27.7 2.47×10^{11} m/sec N.

27.9   4.47 g/h of PbO_2; 3.67 g/h of H_2SO_4.

Chapter 28

28.1   Copper-constantan is better than antimony-bismuth. They have neutral temperatures of $-400°C$ and $-128°C$, respectively.

28.3 1.30×10^{-4} V/unit charge °K.

28.5

$$a = 15.81 \; \mu V/\deg;$$
$$b = -5.74 \times 10^{-2} \, \mu V/\deg^2.$$

1294 μV, 2588 μV, 2060 μV, 1732 μV.

$$-1868 \; \mu V.$$
$$\theta_n = 275°C; \qquad \theta_i = 550°C.$$

The emf is in the direction of Cu to Fe.

Chapter 29

29.1   (a) Rectangle should be perpendicular to the field.
1.2 N (8-cm side); 1.8 N (12-cm side).
(b) Rectangle should be parallel to the field.
0.144 N-m.

29.3   10 N.

29.5   Hint: Break plane up into small rectangles with a pair of sides parallel to **B**. Then calculate the resultant torque due to this equivalent system.

29.7   10 μA.

29.9   (b) 0.5 nsec (1 nsec = 10^{-9} sec).
(c) $\theta = (QB/m)(3l/2g)^{1/2} = 12.7°$

(Hint: Use impulse approximation for charge passing through circuit to find initial angular velocity.)

29.11 1.24×10^{-4} N/A/m.

(Hint: Treat as if 100 coils were concentric and approximate sum by an integral.)

Chapter 30

30.1   2 N/m.

30.7   The field B is tangent to the radius of the cylinder.
For $r \leq a$, $B = \mu_0 I r / 2\pi a^2$;
For $a \leq r \leq b_1$, $B = \mu_0 I / 2\pi r$;
For $b_1 \leq r \leq b_2$, $B = (\mu_0 I / 2\pi r)[1 - (r^2 - b_1^2)(b_2^2 - b_1^2)]$;
For $b_2 \leq r$, $B = 0$.

30.9   17 Wb/m².

30.11 16×10^6 m/sec.

30.13   The magnitude of the linear velocity remains the same but the radius of curvature decreases.

30.15   (a) $(eVL/mv^2l)d$, where l is the distance between the plates and L is the length of the plates.
(c) Hint: Note that for small deflections the magnetic force is approximately in the same direction as the electric force.

Chapter 31

31.3 2.01×10^{-23} A-m².

31.5 $\mu_0 n I / 2$; $\mu_0 n I A$; $\frac{1}{2}$.
(a) Less.
(b) About the same.

31.7 1.59×10^5 A.

Chapter 32

32.1 2×10^{-2} V.

32.3 $I = (mg \sin \theta / Bl \cos \theta)\{\exp [(-B^2 l^2 \cos^2 \theta / mR)t] - 1\}$

32.5   Hint: Use the procedures used in Problem 32.4, except solve in terms of α and T.

32.7 1.6×10^{-7} C.

32.9 $(B^2 A^2 / 2R)\omega^2$.

32.11   Hint: Choose a simple system and look at the mechanical work done on the system to produce the emf. Then consider the Joule heating and what happens if the current is in the opposite direction of Lenz's law.

32.13 $\omega = 2E/Br^2$ and K.E. $= \pi m E^2 / B^2$, where m is the mass per unit area. There will be no change in the above results if the battery has an internal resistance as they are null current situations. $\omega = \dfrac{2E}{Br^2} e^{-(B^2t/2mR\pi)}$, where R is the resistance—dissipation of energy; $\omega = \dfrac{2E}{Br^2} \cos \dfrac{E}{L} t$, where L is the inductance—exchange of energy; $\omega = \dfrac{2E}{Br^2} \dfrac{1}{[(cB^2/\pi m) + 1]^{1/2}}$, where C is the capacitance—null current.

32.15   0.5 J; $(\frac{1}{2})(B^2/\mu_0)$.

32.17   Yes, the back emf is proportional to dI/dt.

32.19   1.96 sec.

32.21 $\theta = A \cos rt + B$, where $A = \dfrac{mgl}{(B^2l^4/2L) + mgl} \theta_0$ and

$$B = \dfrac{B^2l/2L}{B^2l^4/2L + mgl} \theta_0;$$

$I = (Bl^2/2L)(\theta - \theta_0)$; and $\Omega = [(3g/2l) + (3B^2l^2/4mL)]^{1/2}$.

32.23 $(l\mu_0/2\pi) \ln (1 + b/h)$.

32.25   0.48 μF.

32.27 $10^5 \mu_0$; 0.1 mH (L of coil without core) and 10 H (with core); 1 H.

32.29   No, because mutual inductance would become a consideration.

Chapter 33

33.1   0.226 V; 0.160 V.

33.3   1,555 Ω; 1.65 μF.

33.5 $0.587 \sin (120\pi t + \frac{1}{2}\pi)$ A; 0.415 A.

33.7 1.74×10^{-2} J.

33.9   5.1 μF.
It is economical from both points of view as the energy stored in the capacitor on the charging part of the cycle is returned to the power company on the discharging part of the cycle. The consumer calculated his operating voltage as an rms value and thus took into account his power consumption.

33.11   0.968 A (rms)
The current lags behind the voltage by 6°.

33.13   (a) 16.1 V.
(b) 15.9 V.
(c) 10.5 V.
(d) 0.948.

33.15   (a) 828.
(b) 15,500.
(c) 16.9 Hz.
(d) $V_L = 94.7$ V; $V_C = 108$ V.
(e) $V_L = 368$ V

33.17 99 V.

33.19 (a) f.
       (b) f.

33.21 $L(d^2q/dt^2) + R(dq/dt) + (1/C)q = 0$; $\alpha = R/2L$;

$$\omega = [(1/LC) - (R/2L)^2]^{1/2};$$

$R < 2(L/C)^{1/2}$; $\varphi = \tan^{-1}\{1/[(4L/R^2C) - 1]^{1/2}\}$; $q = V_0Ce^{-\alpha t}\cos\omega t$;

and $I_0 = -V_0/L$.

33.23 2.42×10^6 Hz; $0.974I_{max}$ (1%); $0.91I_{max}$ (2%); $0.735I_{max}$ (4%); and $0.257I_{max}$ (10%).

For frequencies above the resonant one, a slighter decrease in current is recorded than for frequencies below resonance.

33.25 $I_1 = (E_0/Z)\sin(\omega t + \varphi)$;
       $\varphi = \tan^{-1}\{\omega[(CR) + (L/R)(C\omega^2L - 1)]\}$;
       $Z = \{1/[R^2 + (\omega L)^2]\}(\omega^2\{C[R^2 + (\omega L)^2] - L\}^2 + R^2)^{1/2}$

33.27 Yes.

Chapter 34

34.1 50 Ω.

34.3 2,110 V.

34.5 N.A. = 0.160 turns/m².
      (a) Yes.
      (b) No.
      (c) No.
      If the transformer were run in parts (b) and (c), it would burn out due to hysteresis loops set up in the iron core due to alternating currents in the flux links.

34.7 $I_1 = 0.183$ A $(I_1' \approx I_1)$; $I_2 = 1.68$ A; and I_1 and I_2 lag behind V_1 and V_2, respectively, by 75.1°.

34.9 (a) Before approximations are made, right-hand side of the equality should be $-k(N_2/N_1)E_0\sin\omega t$.

(b) $I_2 = -\dfrac{N_2}{N_1}\dfrac{E_0}{[k^2 + (1/\omega C)^2]^{1/2}}\sin(\omega t + \varphi)$;

$\varphi = \tan^{-1}(-1/R\omega C)$;

$I_1 \approx +\dfrac{E_0(N_1/N_2)^2}{[R^2 + (1/\omega C)^2]}\sin(\omega t + \varphi)$.

(Complete solution contains the added term $(-E_0/\omega L_1)\cos\omega t$.)
I_2 and I_1 are in antiphase and lag behind their respective voltages.
(c) Power delivered equals power | power supplied,

$$\dfrac{E_0^2/R}{(1/T^2)\{R^2 + [\omega L' - (1/\omega C)]^2\}}.$$

34.11   It would decrease with thinner laminas until magnetic domains start getting distorted.

34.13 $1.83 \ \mu\mu F$.

Chapter 35

35.1

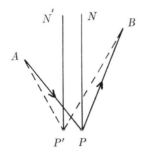

Show that $AP + PB \angle AP' + P'B$.

35.3   Hint: Use the third law or law of reversibility of refraction.

35.5 $n_{\text{liquid, air}} = n_{\text{glass, air}} \sin \theta_2 = 1.65 \sin 58° = (1.65)(0.848) = 1.36$.

Chapter 36

36.1   Height of tree $= (5/30)x = x/6$, where x is the distance between the man and the tree.

36.3   Let M_1 and M_2 represent the mirrors. Measure distances from M_1. Successive images occur at $d, 3d, 5d, 7d, 9d, 11d, 13d, 15d, 17d, \ldots$. Thus the nth image occurs at a distance $s_n = d(2n - 1)$, where $n = 1, 2, 3, \ldots$.

36.5   (a) $\dfrac{2}{r} = \dfrac{1}{u} + \dfrac{1}{v} = \dfrac{1}{|f|} = \dfrac{1}{u}\left(1 + \dfrac{u}{v}\right) = \dfrac{1}{u}\left(1 - \dfrac{1}{m}\right)$. Therefore $u = |f|$ $[1 + (1/m)]$, when m is negative and the image is real.
(b) Since $m \equiv -v/u$; and for positive m, v should be negative, a virtual image results. Therefore

$$\frac{1}{f} = \frac{1}{u}\left(1 + \frac{u}{v}\right) = \frac{1}{u}\left[1 - \left(-\frac{u}{v}\right)\right] = \frac{1}{u}\left(1 - \frac{1}{m}\right)$$

and $u = |f| [1 - (1/m)]$, when m is positive and the image is virtual. For $|f| = 20$, substitute in the equation and do the necessary evaluation for the given m.

36.7

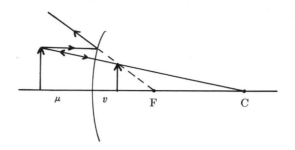

$(2/r) = (1/u) + (1/v)$, where u is positive and r and v are negative. Therefore

$$-\frac{2}{20} = \frac{1}{15} + \frac{1}{-v}$$

and

$$\frac{1}{v} = \frac{1}{15} + \frac{1}{10} = \frac{(2+3)}{30} = \frac{1}{6}$$

Thus, $v = -6$ cm.
Magnification $= m = -v/u = -(-6 \text{ cm})/15 \text{ cm} = \frac{2}{3}$.

36.9

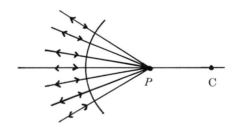

Since $r = 20$ cm and $2/r = 1/f$, $1/f = 2/20 = 1/10$, so $f = 10$. Therefore focal point and point P are the same. Clearly, in this case all the rays of the pencil make normal incidence to the surface of the mirror, and hence they are reflected back in the same line and opposite direction. They come to focus at the focal point and form a virtual point.

36.11

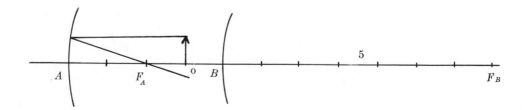

Here $AB = 20$ cm; $AF_A = 10$ cm; $BF_B = 25$ cm; and $AO = 15$ cm. Note that point B is the center of curvature for A, that is, $B \equiv C_A$. First reflection at A:

$$\frac{1}{v} = -\frac{1}{u} + \frac{2}{r} = -\frac{1}{15} + \frac{2}{20} = \frac{1}{10} - \frac{1}{15} = \frac{1}{30}$$

Thus, $v = 30$ cm from A. To find size, note first that

$$m = -u/v = -hv/hu = -30/15 = -hv/4.$$

Then $h_0 = (2)(4) = 8$ cm, and it is an inverted image. Then, the reflection at B

$$\frac{2}{r_B} = \frac{1}{u_B} + \frac{1}{v_B} \quad \text{and} \quad \frac{1}{v_B} = \frac{1}{r_B} + \frac{1}{u_B}$$

Then

$$\frac{1}{v_B} = \frac{1}{25} + \frac{1}{10} = \frac{7}{50} \quad \text{and} \quad v_B = \frac{50}{7} = 7.1 \text{ cm from } B$$

$$m_B = -\frac{h_{v_B}}{h_{u_B}} = -\frac{v_B}{u_B} = -\frac{7.1}{10} = +\frac{h_{\theta_B}}{8}$$

and

$$h_{v_B} = -\frac{7.1}{10} \times 8 = -5.68 \text{ cm.}$$

Following the same procedure but starting from B and then from A, the position of the image is found to be 16.1 cm from A. Clearly, the image is at a different position.

36.13   Given $\omega = (n_b - n_r)/(n_y - 1)$. The condition for achromatism is

$$\frac{A_1}{A_2} = \frac{(n^{(2)} - 1)}{(n^{(1)} - 1)} \cdot \frac{\omega_2}{\omega_1} - \frac{n_b^{(2)} - n_r^{(2)}}{n_b^{(1)} - n_r^{(1)}}.$$

Therefore $A_1 = (0.0087/0.0161) \times 10° = 0.0870/0.0161 = 5.45°$.

Chapter 37

37.1   1.5 m.

37.3   10.6 cm.

37.5   15 cm, on the opposite side.

37.7   3.75 cm, concave.

37.9   The lens must be moved to a position 0.2 cm farther away from the film.

37.11   1.81 cm.

37.13   3.85.

37.15   Initial position of lens: 54 cm, 72 cm.
Reversed position: 35 cm, 47 cm.

37.15   Hint: Obey the prescribed sign convention.

37.19   A converging lens with $|f| = 15$ cm.

37.21 Planoconcave radii: ∞, 56 cm.
Biconvex radii: 56 cm, 152 cm.

37.23 Final image is real, inverted, has magnification -2, and is 20 cm from diverging lens on the side remote from converging lens.
After interchanging lenses the final image is real, inverted, has magnification $-\frac{1}{2}$, and is 20 cm from converging lens on side remote from diverging lens.

37.25 (a) Diverging lenses with $|f| = 1$ m.
Least distance for distinct vision is 17.7 cm.
(b) 4 m.

37.27 Hint: There are two refractions and one reflection to be considered.

37.29 4.15 cm in front of lens.

37.31 6.67 cm from lens on same side as object with magnification of $\frac{1}{3}$.

37.33 Object is 1.04 cm in front of objective.
Magnifying power is -120.

37.35 Hint: For relaxed-eye viewing the light comes out of the eyepiece in parallel rays.

Chapter 38

38.5 1.26 mm.

38.9 The distance from the source at $(0, 4\lambda)$ to the point x is $(x^2 + 16\lambda^2)^{1/2}$. Thus the path difference is $(x^2 + 16\lambda^2)^{1/2} - x$. The positions of the maxima are given by $(x^2 + 16\lambda^2)^{1/2} - x = N\lambda$, and those of the minima by $(x^2 + 16\lambda^2)^{1/2} - x = (N + \frac{1}{2})\lambda$.

38.11 $N = m(\alpha/a)$, where $m = 1, 2, 3, \ldots$.

Chapter 39

39.1 For $\lambda = 5016$ Å, $N = 0$, $\theta = 0°$; $N = 1$, $\theta = 14°31'26''$; $N = 2$, $\theta = 30°6'$; and $N = 3$, $\theta = 48°48'$.

39.3 8200.

39.7 $a = $ spacing of fibers $= 1.12 \times 10^{-3}$ cm.

39.9 0.47 mm.

39.11 The central maximum is a rectangle of sides a and $2a$. There are secondary maxima in both the horizontal and vertical directions; these secondary maxima are also rectangles. In one direction (either horizontal or vertical), these maxima are located at a distance $(n + \frac{1}{2})|f|(\lambda/a)$ from the center and $n = 1, 2, 3, \ldots$. In the other direction, the distance from the center of the maxima is $(n + \frac{1}{2})|f|(\lambda/2a)$.

39.17   51.2 m.

39.19   (a) No, by Babinet's principle the diffraction pattern produced is the same as that for an objective with a narrow opaque ring around the edge. Thus there will be a *bright* spot in the center.
(b) The resolving power is proportional to the size of the aperture, (as explained above) is reduced.

39.21   6700.

39.23   (a) 248°.
(b) 142°.

39.25   480 cm.

39.27 $3.06 \times 10^{-5}/°C$.

Chapter 40

40.1   Hint: Use Brewster's law.

40.3 $I/I_0 = \frac{1}{4} = \cos^2 \theta$ implies that $\cos \theta = \frac{1}{2} = 0.5$ and thus $\theta = 90° - 30° = 60°$.

40.5   The wave component whose vibrations are parallel to the optical axis (the E-wave) can be represented as it emerges from the plate by $E_y = (E_0 \cos \theta) \sin \omega t = (E_0 \cos 30°) \sin \omega t$, where E_0 is the amplitude of the plane-polarized, incident wave. Let the y-axis coincide with the optical axis. The wave component whose vibrations are at right angles to the optic axis (the O-wave) can be represented by $E_x = (E_0 \sin \theta) \sin (\omega t) = (E_0 \sin 30°) \sin \omega t$. Then, $E_y/E_x = \cot \theta$ or (amplitude of O-wave)/(amplitude of E-wave) $= E_x/E_y = \tan \theta$. Finally, the amplitude ratio $= \tan \theta = \tan 30° = 0.5774$ and the intensity ratio $= \tan^2 \theta = 0.332$.

40.7

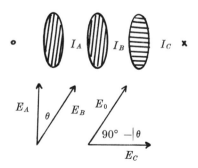

$E_B = E_A \cos \theta$ and $I_B = I_A \cos^2 \theta$.
$E_C = E_B \cos (90° - \theta) = E_B \sin \theta$, then $I_C = I_B \sin^2 \theta$.
$I_C = I_A \cos^2 \theta \sin^2 \theta$. Here, I_C is the transmitted intensity. Finally,
$I_{\text{transmitted}} = I_{\text{incident}} \cdot \cos^2 \theta \sin^2 \theta$

40.9 $x = \lambda(N_e - N_o)/n_e - n_o = 0.5893 \times 10^{-4}/0.00911 \times 10^{-2} = 0.319 \times 10^{-2}$ cm.
Thickness $= 0.0032$ cm $= 0.032$ mm.

40.11 Intensity $I = \frac{1}{2}E^2 \sin^2 2\varphi$, where φ is the angle between the polarizer optic axis and one of the directions of vibration in the mica. Minima occur at $\varphi = 0$, $\frac{1}{2}\pi$, π, $\frac{3}{2}\pi$, 2π. Relative maxima occur at $\varphi = \frac{1}{4}\pi$, $\frac{3}{4}\pi$, $\frac{5}{4}\pi$, ... The amplitude of the light incident on the mica plate is E.

40.13 Insert a quarter-wave plate. If the beam is circularly polarized the two components will have a phase difference of $\pi/2$ between them. The quarter-wave plate will introduce a further phase difference of $\pm\pi/2$ so that the emerging light will have a difference of either zero or π. In either case, the light will now be plane-polarized and can be made to suffer complete extinction by rotating a polarizer in its path. For unpolarized light the extinction is not possible.

40.15 Hint: Consider a Cartesian system of coordinates with the x-axis in the direction of propagation, the y-axis parallel to the optical vector of one wave, and the z-axis parallel to the optical vector of the other wave. The optical vectors of the two waves may be represented by expressions of the type

$$E_y = A_y \cos [\omega(t - x/v) + \varphi_1]$$
$$E_z = A_z \cos [\omega(t - x/v) + \varphi_2]$$

and then use the principle of superposition.

INDEX

An italic *n* after a page number indicates a footnote, and italic numbers refer to illustrations. (This index is complete for both volumes I and II.)

aberration of light 924–925
absolute thermodynamic scale
    465–466
abstractions 4
acceleration 17, 28, 29–32, 35
    centripetal 39
    constant 29–31
    for given force, variation of 97–98
    due to gravity, *see* gravitational
        acceleration
    and mass 98
    mean 38–39
    measurement of 32–34
    in SHM 46–47
    units of 17
    variation with force 95, 101
accuracy
    degrees of 11
    and error A-11–A-12, A-15,
        A-16–A-18
    of a mean A-12–A-13
    in measurement A-16–A-18
    need for 9–10
    and significant figures A-11–A-12,
        A-16
achromatic combination
    of prisms 947
    of thin lenses 956–957
achromatism 947, 956–957
acoustics (*see also* sound) 79
action of points (electrical) 631–632
action and reaction 105–106
activation energy 411
    in dissociation 700–701
adiabatic bulk modulus 436
adiabatic lapse rate 441
adiabatic volume change 434–438
    in Carnot's cycle 445–448
adiabatics 716
aerial, radio *879–881*
    dipole 879
alternating current
    sinusoidal 859–861

alternating current, sinusoidal, rms
        and "mean" values for 860–861
    rms and peak values for 859–860
alternating-current instruments
        861–870
    hot-wire 869
    moving-iron 867–*869*
    rectifier-type 862–*865*
alternating-current theory 872–878
    use of phasors in 872–878
alternating currents and voltages
        857–861
    values of
        peak, rms, and "mean" 857–859
amebas, dimensions of 19–20
ammeter 637–640, *684–686*
    calibration of 637–640
    electrodynamometer-type (ac)
        865–*867*
    hot-wire (ac) 869
    moving-iron (ac) 867–868, *869*
    rectifier-type (ac) 862–863, 864
    temperature-insensitive form of 685
ammeter-voltmeter combinations
    686–687
ammonia clock 14–15
Amontons, G. 269
ampere (unit of current) 495, 496*n*
    definition of 755–757
    per meter 803
    in terms of electrolysis 640–641
Ampère, André Marie 7, 753, 801
    equation 753, 762, 763
    theory of magnetism 801
Ampère's law 753
Ampère's line-integral theorem
        771–782
    applications of 778–782
    proof of 774–776
Amperian current 822
    in moving-iron ac instruments
        868, 869
amplification of voltage 901–904

amplifier 901, *902*, 903–904
  gain of 904
amplitude
  of simple harmonic motion (SHM) 45
  of simple harmonic disturbances 52
  of a wave 60
analyzer, for polarized light 1027
angle *16*, 166
  of contact 314
  of incidence
    polarizing 1028
  as natural unit 16
  solid, defined 511*n*
angular acceleration, as a vector 169–170
angular displacement *167*–168
angular momentum 163
  definition of 163
  as a vector 171
angular quantities 166
angular velocity 39, *168*–169
  as a vector 168–169
annealing 266
anode 575
  in electrolysis *638*–639
  of thermionic triode 610–611
  in thermionic tube 575
antinodes, nodes 75, 78
aplanatic reflecting surface *935*–936
Arago, D. F. J. 926–927
arc
  carbon 634
  mercury 633–*634*
arc discharge 633–634
Archimedean upthrust *222*–224, 231
Archimedes' principle *222*–224
  and buoyancy correction 223–224
  proof of 222–223
  statement of 222
area, units of 17
Arrhenius, Svante 697
astronomical telescope, *see* telescope
astronomy
  paraboloidal mirrors in 936
atmosphere 220–221
  effective height of 220–221
  pressure and temperature variations in 438–441
  temperature variation in 291–292
    determination of 292
atmospheric precipitation 416–421
atmospheric pressure 213, 220–222

atmospheric pressure, reason for 220–221
  variation with altitude 221–222
atmospheric temperatures,
  determination of 291–292
atomic clock 14–15
  ammonia 14–15
  cesium 15
atomic heat *365*–*366*
atomic masses 588
atomic sizes 588–589
atoms
  circulating electricity in 813–815
  magnetic properties of 813
  model of 7–8
  "planetary" theory of 617
  sizes of 588–589
attracted-disk electrometer 555–557
audibility 289–292
  effects of temperature gradient on 289–291
  effects of wind on 289–291
  zones of 291–292
auto transformer 899–*900*
Avogadro's law 376, 378
Avogadro's number 390–391, 586–587
axis of constraint 185–187

Babinet's principle *997*–998
ballistic galvanometer 642–647
  theory of 643–647
Barkla, C. G. 1030
barometer, mercury 219, *220*
battery (in electric circuits) 660, 712–714
battery, storage, *see* lead storage cell
beat, acoustical 79–80
beat frequency 80
Bergstrand, E. 927
Bernoulli, Daniel 236
Bernoulli's equation 238
Bernoulli's theorem 230, 236–240
  proof of 236–238
Berthelot, P. 336
Bethe, Hans 815
B–H characteristic 804–805, 837–839
  measurement of 837–839
biaxial crystal 1035
Billet split lens *981*
biprism *979*–980
Birge, R. T. 928
Bitter, F. 811

black-body radiation 348, 351–352,
    600n, 601
  Planck's formula for 351–352
  pressure of 348
total intensity of 348
"black-box" concept 637
blooming of lenses 1009–1011
Bohr magneton 17, 814
Bohr, Niels 7, 8, 617
boiling 421
  conditions for 421
Boltzmann's constant 386, 587
boundary layer 231
Bowden, F. P. 270, 271, 273
Boyle, Robert 225
Boyle's law 4, 10, 225–226, 336,
    375–376
Bradley, James 924–925
Bragg, W. H. and W. L. 1020, 1021
Bragg's law 1020
Brahe, Tycho 121
bremsstrahlung 1022–1023
  short-wavelength limit of 1022–
    1023
Brewster's law 1028
Brown, Robert 457
Brownian motion 456–457
brushes, in generator 840, 843, 845
bubble 311–313
  pressure within 312–313
  soap 313
bubbling method 316–317
bulk modulus of elasticity
  adiabatic 227, 283, 436
  definition of 226
  isothermal 226, 227
    of a gas 226, 227
    of a liquid 227
  of solids 249–250
bulk waves 287, 288
Bullard, E. E. 785–786
bumping (of boiling water) 421
buoyancy correction 223–224

calcite, double refraction of 1034–
    1035
calibration, of thermometers 342
Callendar, H. L. 339
Callendar's gas thermometer 339
calorie 360, 361n, 362
  value of 394, 396–397
calorimeter
  in finding specific heats 360–361

calorimeter, continuous-flow 428–430
  ice, Bunsen's 360–361
calorimetry 360–362, 428–430,
    648–649
capacitance
  of coaxial-cylinder capacitor
    546–547
  of concentric-sphere capacitor 544
  definition of 543
  of isolated sphere 545
  of parallel-plate capacitor 546
  units of 544
capacitor 542–547
  blocking 902–903
  bypass 902–904
  charge and discharge of 651–654
  coaxial cylinder, capacitance of
    546–547
  concentric-sphere 544–546
    capacitance of 544
  energy of charged 551–552
  in parallel and in series 550–551
  parallel-plate, capacitance of 546
capacity, see capacitance
capillarity 313–316, 320
  theory of 314–316
car on curved road 120–121, 179–
    180, 186
Carnot, Sadi 445
Carnot's cycle
  for ice–water system 472–473
  and ideal-heat engine efficiency
    434, 447–448
  for a perfect gas 445–448
  as reversible process 460
  for surface-tension heat engine
    473–476
  utilizing surface tension 474–476
Carnot's principle 460–463, 465, 466
Cartesian coordinates
  in geometrical optics 936–938
Cassegrainian telescope 936
cathode
  in electrolysis 638–639
  oxide (thermionic) 579–580
  photoelectric 598–599, 602–603
  of thermionic triode 610–611
  in thermionic tube 575
cathode rays 603–606
  detection of 604
  as electrons 605
cathode-ray tube
  applications of 614
  and charge-to-mass ratio 789–790

cathode-ray tube, gas 603–604
  and glow discharge 632–633
  modern 612–614
  principles of 603–*605, 606*–607
Cavendish, Henry 196
  determination of *G* 196–197
Cavendish–Boys method 196–*197*
cell (electrolytic)
  construction of 689–*690*
  Daniell 704–705, *708*–710
    reversibility of 709–710
  gravity 708
  lead storage, *see* lead storage cell
  Leclanché *710*–712
  primary 704, 705–712
  Pt–H₂SO₄–Pt 701–704
    emf of 702–703
    as generator 701–704
  reversible 714–717
    thermodynamical theory of
      714–717
  secondary 704, 705, 712–714
  in series and in parallel 665–666
  standard 689–*690*
  voltaic 702–703, 704–714
    field in electrolyte of 706–707
    local action in 707–708
    polarization in 707
  Weston 689–*690*
  Zn–H₂SO₄–Cu 705–708
Celsius scale of temperature 331,
  332–333
  ideal-gas 335–337
  and Kelvin scale 465
  standard 334–337
center of gravity 174–176
center of mass 173–180
  calculation of 176–*177*
  motion of 178
centimeter 13
centrifugal force 119–121
centrifuge, action of 121
centripetal acceleration 39
cesium clock 15
CGS system of units 100, 152, 495,
  496
charge (electric)
  absolute electromagnetic unit of
    496
  and discharge, of capacitor
    651–*654*
  induced 482–483
  location of 483–484, 515

charge (electric), MKSA unit of 495–
  497
  mobility of
    and field emission 565–566
    in metals 563–567
    and surface potential barrier
      565, 566–*567*
  positive; negative 486–487, 489
  resinous (negative) 486–487, 489
  spraying of, from points 564–*565*
  subdivision of 493–494
  surface density and field 518
  test for 490–491
  in thundercloud 539–541
  transfer of 484
  vitreous (positive) 486–487, 489
charged surface
  force on 552–555
charge-to-mass ratio, electronic
  582–583, *789*–790
charges
  classification of 484–489
  equal magnitude of 487–489
  forces between *485–487*
Charles's law 376–378
choke 877
circuit, *RC* 651
circular motion, uniform 37
Clapeyron's equation 466, 468, 471
Claude process 444
Clausius-Clapeyron equation 466–469
Clément and Desormes (determination of γ) 437
closed path, and line-integral
  theorem 772–*778*
cloud chamber, Wilson 305–306
clouds 416–418
  formation of 416–418
coercive force (coercivity) 806
coherence; incoherence 975–976
cold-cathode tubes 633
collimator 968
Collins liquefier 444
collisions
  elastic 381
  frequency *389*–391
  ionization by 620–623, 624,
    626–627, 628–630
  non-ionizing 628–629
commutator 843
compensated gas thermometer 339
compressibility 227
  values of 250

condensation
    spontaneous 416
    on various nuclei 415–416
condensation nuclei 415–416
    for ice 418
    in troposphere 416–417
        concentration of 416–417
condenser, *see* capacitor
conduction 366–369
conduction (electrical), nature of
    489
conductivity
    electrical 368–369, 650, 658–659
        values of 368
    thermal 367–369, 391–393
        mechanism of 367–369
        theory of, for a gas 391–393
        values of 368
conductors (electric) 483–484
    hollow 515–517
    ohmic; nonohmic 650 651
    in parallel *657*–658
    pointed 505–507
        collection of charge by *506–507*
        field near 505–507, *529*
        spraying of charge from *506–507*
    in series *656*–657
conical pendulum, *see under*
    pendulum
conjugate foci
    of lens 953
    of mirror 941
conservation of energy 147–151,
        426–430
    and heat 426–430
    law of 428
conservation of momentum, *see under*
    momentum
conservative force 150–151
convection 366, 369
    currents 369
conversion factors A–2
Coolidge, W. D. 608
corona discharge 534, 632
coulomb 495, 640
    definition of 495–496
Coulomb, Charles 269, 494
    and laws of friction 269
Coulomb's law
    and force on charged surface
        554–555
    formulation of 493–494, 497
    and intervention of metal 507–*509*
    and velocity of light 911

Coulomb torsion balance *487*, 494
couple 165–166, 200
    arm of a 166
    moment of a 165–*166*
coupling coefficient 888
critical angle 922
critical damping 849
critical data (table) 422
critical pressure 422
critical temperature 422
Crompton potentiometer 691
cross product (*see also* vector
    product) 744
cryophorus *407*
crystal
    biaxial 1035
    cubic 258–260
    dichroic 1036
    net planes in *1019*
    Nicol prism *1035*–1036
    perfect 265–266
    single *259*
    structure of 258–*261*
    uniaxial, refraction by *1031*–1036
crystal grain *260*
crystal lattice (*see also* space lattice)
    analysis by x-ray diffraction
        1017–*1018*
    as diffraction grating 1017–*1018*
    distorted *261*–266
    structure of 258–*261*
current
    carriers of 650
    convection 369
    direct
        smoothing of pulsating 654–
            656
    electric, as fundamental quantity
        12, 100
    thermionic, *see* thermionic current
    wattless 878
current balance 755
current elements 735–*736*
    field produced by 749–753
    unit of 640
Curie, Pierre 816, 819
Curie temperature (Curie point)
    816
Curie's law 819, 820
curvature, of liquid surface *413*–416
cyclic law, of refraction 920
cylinder, uniformly charged 518–
    519, 520

Dalton's law 413
damping, critical 849
Daniell cell 704–705, *708*–710, 717
　chemical reactions in 709–710
Daniell, J. F. 708
D'Arsonval, A. 641
Debye, P. J. W. 402
decapoise 231
decomposition voltage, reversible
　702–703
deformations
　elastic 249–256, 260–262
　friction 249, 266–275
　plastic 249, 256–266
　　diagrams of *261, 262, 263*
　　industrial applications of 266
　　structural changes in 258–266
degradation of energy 451–452
degrees of freedom
　excitation of 402
　of gas molecules 397–398
　　equipartition of energy among
　　397
　　and ratio of specific heats
　　397–398
density 211–213
　definition of 211
　of heavenly bodies 213
　of intergalactic space 213
　of nuclear matter 213
　relative 211–212
　units of 211
　of various materials *213*
　of water 212
depolarization 711
derived quantities 16–17
determination of *g* 32–34
dial box *679*–681
diamagnetism 820
dichroism 1036–1037
dielectric 547–550
　permittivity of 549–550
dielectric constant 547–550
　and Coulomb's law 548–549
　measurement of 548
　values of 549
diesel engines, efficiency of 464
diffraction
　Fraunhofer 993, *994, 998–1004*
　Fresnel 993
　x-ray 1017–1023
diffraction grating 993–997
　description of 993, 996–997
　effect of number of slits in 996

diffraction grating, four- and five-slit
　993, *995–996*
　and missing spectra 1000–1001
　phasor diagrams for *994, 995,
　996, 1002*
　ruling of apertures in 996–997
　three-slit 993, *994*, 995
　transmission; reflection 996–997
　two-slit *976, 985*, 995
diffusion
　gaseous 379
　　Graham's law of 389
dimensions 17–20
　of amebas 19–20
　of force 103
　uses of 18–20
diode, thermionic 572, *574, 620*,
　621
diopter 956
dipole aerial *879*
dipole field; nondipole field (earth's)
　782–783, 786–787
Dirac, P. A. M. 814
　pulsating, smoothing of 654–*656*
discharges in gases 617–634
　arc 633–634
　corona 632
　glow 632–633
　mechanism sustaining
　　630, 631
　processes involved in 627–634
　spark 633
　spectra produced by 617
　triggering of 629–630
dislocation
　edge *263*
　screw 263
dispersion 945–947
dispersion, rotatory 1043, 1044
dispersive power 945–947
displacement 28, 33–35
displacement current 782
dissipative forces 150
dissociation, in electrolyte
　700–701
disturbances, simple harmonic, *see*
　simple harmonic disturbances
domains, magnetic, *see* magnetic
　domains
dome, of Van de Graaff generator
　534, *535*
　and potential of 537, 544
Doppler effect 69–71
　equation for 70–71

dot product 139, 140–141
  definition of 140–141
  mathematical representation of
    140–141
double refraction
  by biaxial crystal 1035
  by calcite (Iceland spar) 1034–1035
  explanation of 1031–1034
  and photoelasticity 1039–1040
drop, liquid 308–309, *311*–313,
    539–540
  in cloud 312
  excess pressure within 310–312
  supercooled, splintering of 539–540
dry cell *712*
dry ice 420
Dulong and Petit's law 6, 364–366,
    401–402
  failure at low temperatures 402
  theory of 401–402
dynamics 27
dynamo
  energy loss in 660–662
  as generator 661–662, 663, 846
  as motor 662, 663
dyne 102–103

e (electronic charge) 582n, 583, 585
earth
  convection within 785–786
  internal structure of 784–785
  magnetic field of
    changes in 783
    cause of 784–787
  mass of 198
  rotation of 14
    effect on g *40*–42
    variation in rate of 14
earth inductor, as alternator 840,
    841
eddy currents *842*
Einstein, Albert 8, 15, 105, 145,
    600n, 795, 1017n
  influence of Michelson's experi-
    ments on 1017n
  theory of relativity 8, 15, 145,
    795, 1017n
elastic deformation, *see under*
  deformations
elastic limit 250
elastic moduli, *see* moduli of elasticity
elastic potential energy, *see under*
  energy

elasticity 249–256
elasticity, bulk modulus of, *see* bulk
  modulus of elasticity
electric current, as fundamental
  quantity 12, 100
electric discharge, in gases, *see*
  discharges in gases
electric field
  concept of 501–507
  definition of 501
  energy density of 553–554
  image 569–570
  polarization of 569–570
  unit of 502
electric motor
  ac; dc 847
  as reversed generator 846–847
electricity
  circulation of 538
  and friction 482
  in metals 563–568
  MKSA unit of charge 495–496
  positive and negative 486–487,
    489
  resinous and vitreous 486–489
electrification by friction 481–483,
    484, *485*–486
electrode
  in electrolysis 637–639
  floating 626
  modulating 612, *613*
electrodynamometer 865–*867*
  as ammeter; voltmeter 865–*866*
  as wattmeter 866–*867*
electrolysis 637–639
  of copper sulfate 697–700
  first law of 639
  laws of 697
  processes involved in 697–700
  of water 701
electrolyte, dissociation in 700–701
electrolytic cell 637–*638*
electromagnet 822
electromagnetic induction
  and generators; motors 839–847
  law of 7, 825–831
    equation 828
electromagnetic radiation
  availability of energy quanta 601–
    602
  theory of 7–8, 782
electromagnetic theory 7–8
electromagnetic unit, absolute 496

electromagnetic waves 69, 370–371,
    911–912
  equations for 69
  as light 911–912
electromagnetism, study of 7–8
electrometer
  attracted-disk 555–557
    applications of 557
  Braun's 542, *543*
  quadrant (Kelvin's) 542
electromotive force (emf)
  back 663, 707
  of Daniell cell 709–710
  definition of 662
  forward 663
  induced 829
  of lead storage cell 713
  of Leclanché cell 712
  and Peltier effect 722, 723–724
  of Pt–H$_2$SO$_4$–Pt cell 702–703
  primary; secondary 891–893
  of reversible cell 717
  self-induced 832
  theory of 659–664
  of thermocouple 720, 723–*724*
electronic emission, secondary, *see*
    secondary emission
electrons
  as cathode-ray particles 605–606
  charge of 17, 582*n*, 583, 585,
      586–588
    deductions from information
        586–588
  charge-to-mass ratio 582–583
  as current-carriers in metal 368–
      369, 568
  emission, *see* discharges in gases
  glow discharges of 603
  latent heat of vaporization of 576
  mass of 586
  ratio of charge to mass of *789*–790
  secondary emission of 606,
      624–627
  spin of 813–815
electron beam 612
electron gun 612–*613*
electron microscope 790, 792–796
electron optics 530, 610–611
  of thermionic triode 610–611
electron volt 587
electroscope 489–*491*, *508*, 541*n*,
    542
  gold-leaf 541*n*, 542
  and test for charge 490–491

electrostatics
  fundamental law of 494
  study of 6–7
electrostatic shielding 507–508
electrostatic voltmeter
  ac use of 870
  as capacitor 870
elements, current 735–*736*
ellipsoid of revolution 935, 1033
  of extraordinary rays 1033
Elsasser, W. M. 785
emf, *see* electromotive force
emission
  electron, *see* discharges in gases
  field 564–*565*
  photoelectric, *see* photoelectric
      emission
  secondary 606, 624–627
  thermionic 570–582
energy
  of activation 411
  concept of 137
  conservation of 147–151, 426–
      427
  conversion 147–150, 461, 664
    in electrochemistry 714–717
    and heat 426
  definition of 141
  degradation of 451–452
  elastic potential 142, 143–144
  electrical 663–664
  equipartition of, among gas
      molecules 397
  forms of 141–145, 425
  gravitational potential 141,
      142–143
  of heat 425–426
  internal 428, 430–431
  kinetic, *see* kinetic energy
  magnetic 663–664, 833
  mechanical 141–145
  potential 147–148, 524, 663–664
    and PD 524
  surface 308–310, 473–476
  total 425
  transfer of 369–371
  transport of 391
  units of 152
  zero-point 428
energy conversion, in thermo-
    electricity 719–731
energy loss, in direct-current theory
    660–662
energy quantum, *see under* quantum

energy transformations, in Daniell cell 709–710
equilibrium 185–*186*
  definition of 185
  thermal 330
equinoxes, precession of 204–205
equipartition of energy, theorem of 398, 401
  and Dulong and Petit's law 401–402
equipotential surface 526–530, 608–613
  definition of 526
  in modern cathode-ray tube 612–*613*
  in modern x-ray tubes *608*–609
  in thermionic triodes 610–*611*
equivalent circuit, of transformers 898
equivalent magnetizing current 891
erg 152
error A-11–A-18
  absolute and relative A-11–A-12
  and accuracy of measurement A-16–A-18
  of calculated quantity A-16–A-17
  estimation of A-11, A-12–A-14
  maximum possible A-13–A-14, A-15
  of a mean A-14–A-15
  and mistake A-16
  probable limit of A-12–A-13, A-14–A-15
  of product or quotient A-17–A-18
  proportional A-17
  random A-12
  and significant figures A-11–A-12
  of a sum or difference A-18
  systematic A-12
Esclagnon, E. 286
ether 1017
evaporation
  cooling by 442
  of electrons (thermionic emission) 576–*577*
Ewing, J. A., and theory of magnetism 801, 808–810
exchange forces 815
excitation, of gas molecules 628
expansion (of a gas)
  adiabatic 434
expansion, thermal, *see also* thermal expansion
  isothermal 433–434
  unresisted 432, 433, 459, 460

expansion chamber 305–306
expansivity
  of ideal gas 340–341
  linear 357
    values of 358, *359*
  volume 357
extended body, forces on 172–173, 178
extraordinary ray, behavior of 1032, 1033–*1034*
eyepiece
  compound, of microscope 964, *965*, 966
  simple and compound compared *965*–966

Fahrenheit scale of temperature 331, *332*
farad 544
Faraday effect 1044
Faraday, Michael 7, 517, 637, 639, 697, 831
  and electromagnetic induction 831
  ice-pail experiments 517
  law 831
  laws of electrolysis 637, 639, 697
    first 637, 639
Faure, Camille 714
du Fay, Charles, and electric charges 486
feedback 904–905
ferromagnetism 806–816
ferromagnetic materials 806–808
  of high coercivity 808
  of low coercivity 807
field, electric
  in atmosphere 541
  concept of 501–507
    definition of 501
  energy density of 553–554
field, electrostatic
  and potential gradient 528
  representation by equipotential surfaces 529–530
field emission 564–*565*
field magnet 841, 845–846, 847
film
  colors of thin 1008–1009
  interference due to thin 1004–1013
  liquid, stability of 309, 318–319
  wedge-shaped *1011*–1012
first law of thermodynamics 427

A-105

fixed points
  in thermometer calibration 342
Fizeau, A. H. L., and velocity of
  light 7, *925–926*
flotation of minerals *321–322*
flow, fluid, *see* fluid flow
flow, plastic, *see* plastic flow
fluid flow
  along a pipe 232, 233, 236–*237*
    Poiseuille's formula for 233–234
  laminar *229*
  tube of *230*, 236–238
  types of 228, *229*, *230*, *237*
  velocity of exit 239–240
  velocity gradients in 228–230, 236
fluorescent tubes 633
flux
  electric 514
  magnetic 828–829
    unit of 829
flux density 514, 518, 519
flux linkage 831
focal coordinate
  of lens 953
  of mirror 940–941
focal length
  of lens 954
  of lens combination 956–957
  of mirror 940–941
focal plane
  of lens 954
  of mirror 941
focus, principal
  of lens 953, 954
  of mirror 941
foot 13
foot-pound 152
foot-poundal 152
force 89–113
  absolute units of 100–103
  and acceleration 92, 95–98, 101
  active; passive 91
  centrifugal 119–121
    and car *120*–121
  centripetal 119, *120*
  on a charged surface 552–555
  complete definition of 103–104,
    105
  concept of 145–147
  conservative 150–151
  dimensions of 103
  dissipative 150
  as distance rate of increase of
    kinetic energy 146

force, effect of 93
  frictional 91
  gravitational 91, 92*n*, 93, 101
    units of 101
  idea of 91
  image, on electron 569–570
  magnetic, *see* magnetic force
  and mass, laws of 111–113
    according to Mach 113
  measurement of 94, 101–103
    as weight 101–103
  qualitative definition of 93
  reactional 93, 105–106
  as time rate of change of momen-
    tum 104–105, 146, 383
  unit of 100, 102–103
  as a vector 94–*95*, 101, 104
  viscous 91
  as weight 92–93, 103
    relation to mass 103
fossil magnetism 783
Fourier, J. B. 298
Fourier's theorem 298
FPS system of units 100, 152
frames of reference 89–91
  inertial 90–91
    definition of 91
Franklin, Benjamin 486, 489
Fraunhofer diffraction 993,
    994, 998–1004
  by circular aperture 1001–*1004*
  by square aperture 1001, *1002*
free fall 31
freezing 405–406
freezing mixtures 405–406
freezing point 405, 411–412, 471–473
  effect of pressure on 471–473
frequency (acoustic)
  beat 80
  fundamental 79
  harmonic 79
  relation to pitch 294–295
  of simple harmonic disturbances
    52–53
  standards of 294–295
  unit of 17
Fresnel, Augustin Jean, and diffraction of light 913
Fresnel biprism *979*–980
Fresnel diffraction 993
Fresnel mirrors *980*
friction
  coefficient of 268
    variations in 269–270

friction, and deformations of solids 249
  electrification by 481–483, 484, 485–486
  heat of 459–460
  kinetic 267–268, 274
    dissipation of energy in 267
    localized heating in 274
    uses of 267
  laws of 269, 271
  laws of, revised 271
  mechanism of 273
  polishing by 274
  rolling 271–272
    theory of 275
  static 266–268
    limiting 268
    uses of 267
  sliding 267, 274
  and surface contaminations 272–273
  theory of 272–275
  true contact-area determination 270–271
fringes, Haidinger's 1013–1015
frozen-stress technique 1043
froth 318, 321
fundamental constants of physics A–3
fundamental quantities of physics 12–13, 15, 100, A–3
  electric current as 12, 100
fusion, latent heat of 404

G, see under gravitation
g, see gravitational acceleration
gain, of amplifier 904
Galileo Galilei 31
galvanometer 343–344
  ballistic 642–647
    damping of 646
  D'Arsonval 641, 642
    damping of 847–850
    and practical instruments 850
  high-resistance 682–684
  moving-coil 640–647, 862, 863–865
  oscillation of 848–849
  and shunt 684–685
gamma (ratio of specific heats) 436–438
  determination of 437–438
  value for dry air 438

gas
  adiabatic volume change in 434–438
  behavior of 382
  discharges in, see discharges in gases
  equation of state for 375–378
  ideal, see ideal gas
  insulation breakdown in 627
  ionization of, see ionization of gas
  isothermal volume change of 433–434
  kinetic theory of 10–11, 375–398
  molecular heat of 394, 396–397
    at constant volume 394
  molecular picture of 378–381
  molecules of 378–381, 387–398
    collision frequency 389–391
    degrees of freedom 397–398
    effective diameter 391
    mean free path 381, 389–391
    mean velocity of 388–389
    rms velocity 387–389
    velocity distribution among 380–381
  pressure exerted by 382–385
    mechanism of 381
  pressure–volume relation for 224–226
  ratio of specific heats and molecular degrees of freedom 397–398
  temperature 385–386
    and rms velocity 388–389
  unresisted expansion of 394–395, 432, 459, 460
gas constant R 376, 378
  value of 378
gases, liquefaction of, see liquefaction of gases
Gassendi, P. 285
Gauss, Carl Friedrich 509
  positions of (axial, broadside) 766–767, 768, 769
Gaussian surface 515–519, 520
  box-shaped, and field 519–520
Gauss's theorem 509–523, 544
  applications 514–520
  and concentric-sphere capacitor 544
  mathematical formulation of 510, 514
  proof of 511–514
  statement of 514

Gauss's theorem, validity of 520
Gay-Lussac 377–378
generator
   ac 840–843
      armature 841, 842
         laminated 842
      and circulation of electricity 846
   dc *843*–846
      armature system of *844*–846
      compound 846, 847
      series-wound 845–*846*, 847
      shunt-wound 845–*846*, 847
   field magnet of 841, 845–846, 847
   Van de Graff 533–538
geomagnetism 782–787
   cause of 783–787
   changes with time 783–784
   dynamo theory of 785–786
   and precession 786–787
geometrical constructions, *see*
   geometrical optics
geometrical optics
   constructions in 959–*960*
   distances, coordinates in 936–*938*
   and light rays 912, 931
   sign convention in 936–938
getter 572*n*
Gilbert, William 784
glass, as a fluid 235
glass working 319–320
glow discharge 603, 632–633
gold-leaf electroscope 541*n*, 542
   as voltmeter 541–542
Goudsmit, S. 814
Graham's law of diffusion 389
grain
   boundaries 260–261
   crystalline *260*
   structure, in a metal *260*
gram 99, 100, 103
gram-calorie 362
gram-centimeter 152
graphs, function of 11–12
gravitation 123
   constant of (*G*) 123
   law of 123
   and planets 123–125
gravitational acceleration (*g*)
   "absolute" 40, 42
   definition of 31–32
   effect of earth's rotation on *40*–42
   as gravitational field 125
   of planets and satellites 42–44
   standard 32

gravitational acceleration (*g*), universal law of 44
   variation with distance 42
   variation with latitude 41
gravitational attraction 101, 151
gravitational constant (*G*) 196–198
   determination of 196–*197*, 198
   value of 197, 198
gravitational field 124
gravitational potential energy 141,
   142–143
gravity
   center of 174–176
   specific 211
Greek alphabet A–1
Gregorian telescope 935
grid
   function of (thermionic) 901–902
   of thermionic triode 610–*611*
Guericke's law 375
gyration, radius of 185
gyrocompass *202*
gyroscopic action 198–205
   applications and consequences 202
   illustration *200*–201
   theory of 198–202
   and uniform circular motion *200*

Haidinger's fringes 1013–1015
   and Michelson interferometer
      1015, 1016
hail 416, *419*
   formation of *419*
hailstone *419*
   structure of *419*
   trajectory of *419*
half-wave plate *1039*
Hallwachs's experiment 594–595
von Hàmos, L. 811
harmonic analyzers 299
harmonic (frequency) 79, 300
harmonic motion, simple, *see* simple
   harmonic motion
heat
   and conservation of energy
      426–430
   as form of energy 425–426, 428
   and internal energy 428, 430–431
   latent, *see* latent heat
   mechanical equivalent of 394–397,
      427, 430
      calculation of 396
      value of 427

heat, nature of 358–359
  in solids 402–405
    and atomic vibrations 402
  specific, *see* specific heat
  transfer 359, 366–371
  unit of 359–360
heat content, and total energy
  425–426
heat death 451
heat engine
  efficiency of 461, 462–464
  ideal 445–448
  practical 463–464
  and second law of thermo-
    dynamics 452–454
  and surface tension 473, *474–476*
Hebb, T. C. 286
Heisenberg, W. 815
Helmholtz coils 736, *737*, 738, 756
von Helmholtz, Hermann 428, 716
  717
  equation for emf of cell 716, 717
henry 832, 837
Henry, Joseph, and electromagnetic
  induction 831, 832
hertz 17
Hertz, Heinrich 594
Heusler alloys 806, 807, 808, 810,
  815
Heyl and Chrzanowski 197
Hirn, G. A. 426, 427
  and mechanical equivalent of
    heat 427
hollow conductor *see under*
  conductor
Hooke, Robert 251
Hookean stress-strain relationship
  256–257, 262
Hooke's law 49, 129, 251
horsepower 152
hot-cathode tubes 633
hot-wire ac instruments 869
Huygens, Christian 912–916
  light propagation, theory of
    912–916
  reflection, theory of 916–917
  refraction, theory of 918–919
Huygens secondary wavelets 913–
  916, 976, 978–979, 996, *1034*
  and diffraction grating 996
  and double refraction *1034*
  envelope of *914, 915*–916
hydraulic press 216
hydrodynamics 228–240

hydronium 701
hydrostatics 211–227
hypothesis, scientific 4
hysteresis 275, 804, 810, 813,
  837–839
  elastic 275
  measurement of 837–839
hysteresis curve *804*, 813

ice
  and Clapeyron's equation *472–*
    473
  vapor pressure of, in troposphere
    440–441
ice calorimeter, Bunsen's *360*–361
Iceland spar
  double refraction of 1034–1035
ice nuclei 418
ice-pail experiments 517
ice point 341
icing, of aircraft 418
ideal gas 335–337
  as base of temperature scale
    335–337
  expansivity of 340–341
image
  magnification of, *see*
    magnification
  point, *see* point image
  real 931–*933*, 939–943
  virtual 931–933, *939–943*
image field 569–570
image force 569–570
  on electron 569–570
image formation
  of extended object
    geometrical constructions for
      959–*960*
  by lens, of extended object
    955–956, 960
  by mirror, of extended object
    941–943
  by plane mirror 917
  by spherical mirror 933–*934*
impedance 876
incidence
  angle of 1028
  plane of 1028*n*
incoherence 976
india rubber *251*
  stress–strain relation for 251
induced charge 482–483, 493–494
induced emf 829

inductance
    of ideal transformer 889–891
    mutual 836–837
induction, electromagnetic, *see*
    electromagnetic induction
induction
    electrostatic 482–483, 491–*492*,
        493–494
        explanation of 491–*492*
        and lines of force 507
    mutual 836–837
induction heating 900–901
induction motor 847
inductive circuit
    growth, decay of current in
        833–836
    time constant of 834
inductive reasoning 4
inductor 832
    current-carrying energy of
        832–833
    earth, as alternator 840, *841*
inertia 15, 162
    moment of, defined 162
inertial frames of reference 90–91
    definition of 91
infrasaturation current (thermionic)
    574, 580–581
    variation with anode voltage 581
insects
    and surface tension 320
instrument conversions with volt-
    meter and ammeter 685–686
insulation
    in current electricity 681–682
    in static electricity 681
insulators 483–484
    see also dielectric
interference (optical) 71–72
    light sources producing 986
    due to thin films 1004–1013
    from two point sources 975–981
    due to wedge-shaped films *1012*
interference fringe 975, *976–978*,
    979, *980*
    spacing 976–978
interference maxima 978
interferometer, Michelson's *1015–*
    1017
"intermediate metals, law of" 720
"intermediate temperatures, law
    of" 721
internal energy 428, 430–431

inverse-square law (Coulomb's) 494
    test of 520–523
inversion temperature 433
    thermoelectric 729
ionization of gas
    by positive-ion impact 623
    by radiation 619–620
    triggering of 629–630
ionization potential 622–623
    values of 622
ionosphere 619*n*
ions
    in electrolyte 698–700
    gaseous 618–620
        negative 618–620
        positive 618–620
    as nuclei of condensation 415–416
    positive
        effect on thermionic current
            621–622
        ionization by 623
"iron losses" 661
isothermals 422, 716
    for carbon dioxide 422
    for liquid–vapor system 422
isothermal volume change 433–434,
    445–448
    in Carnot's cycle 445–448
    work done in 434
isotropism 1031
ionization chamber 305–306
ionization of gas 618–620
    and collisions with neutral mole-
        cules 629–630
    by electron impact 620–623

Jaeger's method of measuring sur-
    face tension *316*–317
Jahn, H. 717
joule 152
Joule, James 395, 426, 431–432
    and gas expansion experiment 395
    and mechanical equivalent of heat
        394
Joule's churn *426*–427
Joule-Kelvin effect 444
Joule-Kelvin experiment 431–433
junctions
    temperature variation of 721, 722,
        723–*724*
Jupiter, satellites of 923–924

Kapitza, P., and liquefaction of
    helium 444

Karolus and Mittelstaedt, and
    Kerr-cell method 927
Kater's pendulum *128*–129, 195–196
Kelvin (William Thomson, Lord
    Kelvin) 341, 377, 432, 433,
    465, 533, 723
    "water dropper" experiment 533
Kelvin scale of temperature 331,
    332, 336–337, 340–341, 465
Kepler, Johannes 121
    laws of planetary motion 121–*122*,
    123
Kerr cell
    and velocity of light 927
Kerr effects
    electro-optical 1041
    magneto-optical 1042
Kerr, John 927, 1041
kilocalorie 362
kilogram 99, 100, 103
    international 99
kilogram-meter 152
kilowatt 152
kinematics 27
kinetic energy
    in classical and relativistic
        mechanics 142, 144–145
    of a particle 144–145
    of rotating body 161–162
kinetic theory of gases 10–11
Kirchhoff, Gustav 666, 914, 915–916
    and Huygens wavelets 914,
    915–916
    network laws 666–669
Kirchhoff's laws 666–669
    application of 667–669
    statement of 666
Kundt's apparatus *287*–288

laminations, of ac armature 842
lamp
    gas-discharge types 633–*634*
        arc 633–634
        mercury-vapor; sodium vapor
        633–634
        neon 632–633
lapse rate, adiabatic 441
latent heat
    definition of 359, 363
    of fusion
        defined 363
        reason for 404
        values of 364

latent heat, of vaporization
    defined 363
    of electrons 576
    internal and external 469, *470*,
    471
    reason for 406
    values of 364
    variation with temperature
    468–471
lattice, space, *see* space lattice
von Laue, M. 258
Laue spots 258, 402, 1017–*1018*
law, scientific 3–4
lead storage cell 712–714
    capacity of 713–714
    chemical reactions in 713–714
    development of 713, 714
    specific gravity of acid in 714
Leclanché cell *710*–712
    chemical reactions in 711–712
    dry *712*
length 12, 13
    units of 13
lens
    aberration of *963–964*
    blooming of 1009–1011
    combination 956–957
        achromatic 956–957
        focal length of 956
        power of 956
    conjugate foci of 953
    converging; diverging 953–954,
        955–956, *960*
    focal coordinate of 953
    focal length of 954
    focal plane of 954
    formula for thin 952–953
        applications of 957–959
    image formation by 932–933,
        952–956
        of extended object *955*–*956*,
        960
    and mirror combination 958–959
    and mirror formulas, applications
        of 957–959
    power of 955–956
    principal focus of 953, 954
    spectacle, prescription of 957–
        958
    split, for producing interference
        981
    effect of tilting 954–955
Lenz, Heinrich F. E. 7, 722
Lenz's law 831

A-111

light
  circularly polarized 1027, 1037–
    1039
    production of 1037–*1039*
  diffraction of 913
  as electromagnetic radiation
    911–912
  electromagnetic theory of 911–912
  elliptically polarized 1027, 1033
    production of 1037–*1039*
  monochromatic 596
  plane-polarized 1027–1032, 1034,
    1036–1037, 1042–1044
  polarized 1027–1044
  scattering of
    effect in atmosphere 1029, 1030
  sensitivity of photocells to
    602–603
  sources of 986–988
    incoherent centers in 986
  unpolarized 1027, *1029*
  velocity of, *see* velocity of light
  wave nature of 911–912, 913–916
light pipe *923*
lightning stroke, charge of 541
Linde process 442–*443*, 444
line integral 772
line-integral theorem, Ampère's
  771–782
lines of force
  electric 502–507
    areal density of 503–505
    field strength of 503–505
    and electrostatic induction 507
    and equipotential surface 526–
      527, 529
  magnetic 736
liquefaction of gases 441–444
  by Claude process 444
  by isothermal compression
    441–442
  by Linde process 442–444
liquid drops, *see* drop, liquid
liquid films, stability of 318
liquid surfaces 305–322
  curved 312–313
    pressures on both sides 312–*313*
    effect on vapor pressure
      413–416
  energy of 308–310
  molecular picture of 306
  nature of 306
  thermal effects accompanying
    areal increase 309

liquid surfaces, and vapor pressure
  305
Lloyd's single mirror *980*–981
logarithms, table of common
  A-8–A-10
longitudinal waves 67–68
lubrication 235, 275

Mach, Ernst, and laws of force and
  mass 113
Mackenzie, I. C. C. 927
macroworlds and microworlds
  456–458
magnetic circuit 888
magnetic domains 810–813
  boundary displacements of
    812–813
  formation of 815–816
  nature of 812–813
  role in hysteresis 813
  rotation in 812–813
  sizes of 812
  transition regions between 812
magnetic energy 663–664, 833
magnetic field 736–749
  concept of 736–737
  due to current
    on axis of flat coil *763*–764
    in large multidimensional
      circuit 770–*771*
    in meshes 770–*771*
    in solenoid 764–*765*, 778–*780*
    in small rectangular circuit
      766–769
    in straight wire *762*–763
  definition of 744
  determination of 778–782
  direction of 736, *737*–738, 740–743
  of earth, neutralization of 738
  between Helmholtz coils 736–*737*
  measurement of 745–749
  effect on moving electrons
    787–789
  produced by current element
    749–753
  in toroid 780–782
  torque on coil 745–746
  as vector quantity 743
magnetic flux 828–829
  unit of 829
magnetic focusing
  image formation by 792
  theory of 790–792

magnetic force
  between current elements 735–
    736, *739*–745, 754–755
    variation with angle 742–743
    variation with current 743
    variation with length of element
      743
  direction of
    variation between field and
      current element 740–742
magnetic induction 745
magnetic lens 792, *793*
magnetic moment 769
magnetic permeability
  relative 762
  of free space 761–762
magnetic poles, concept of 801–802
magnetic susceptibility 817, 818, 820
magnetism
  domain theory of 810–816
  fossil 783
  terrestrial, *see* geomagnetism
  theory of 801
magnetization 802–806
  intensity of 805
magnetizing current, equivalent 891
magnetizing force 803
  unit of 803
magneto 842, 846
magnetometer
  deflection 747–748
  torsion 746
  vibration 747
magneton
  Bohr 17, 814
magnets
  Amperian currents in 822–823
  bar 820–822
  field 841, 845–846, 847
  forces between 822–823
  in certain magnetometers 749
magnification 942–943
  by spherical mirror *942–943*
magnifying glass, *see* microscope,
  simple
magnifying power
  of compound microscope 966
  of simple microscope *962–963*
Malkus, W. V. R. 786
manganese bismuthide 807, 808,
  810, 815
manometer *218*–221, 337, 339
  barometer, mercury *220–221*
  U-tube *218–219*, 220

mantle
  earth's 786–787
many-body system 151
mass
  and acceleration 98
  of atom 588
  center of 173–180
  conformity to mathematical laws
    99
  definition of 15, 89, 98–99, 101
  of electron 586
    value of 586
  equations for 101, 102, 104, 105
  and force, laws of 111–113
  as fundamental quantity 12
  as inertia 15
  as "quantity of matter" 112
  relation to weight 99, 100
  units of 99–100
  variation with velocity 105
masses, standard 223–224
mathematics, role in physics 9, 99
Maxwell, James Clerk 7, 398, 401,
  455, 782, 911*n*
  and equipartition of energy
    theorem 398, 401
  and wave theory of light 911*n*
Maxwell's demon 455–456
Mayer, Robert 428
McLeod gauge 340
mean, deviations from A-13
mean free path 381
  of electron in gas 627–628
  of gas molecules 389–391
  of positive ion in gas 629
  theory of 389–391
  values of 391
measure, units of
  of capacitance 544
  of current strength 640
  of density 211
  of electricity 495–496
  of electronic energy 587
  of force 102–103
    absolute 100
    gravitational 100
  of length 13, 16–17
  of magnetic flux 829
  of magnetizing force 803
  of mass and force 99–103
  of mutual inductance 837
  of potential 526
  of power 152
  of pressure 212*n*, 213

measure, units of, of derived quantities 16–17
of resistance 649
of resistivity 659
of self-inductance 832
of thermal conductivity 367
of time 13–15, 16
of viscosity 231
of work 152
measurement
accuracy in 9–10
errors in A-11–A-18
mechanical energy *see under* energy
mechanical equivalent of heat 394–397, 427, 430
calculation of 394
mechanics
classical 105
relativistic 105
statistical 458
study of 5–6
melting 402–405
theory of 402–404
melting point 404
variation with pressure *404*–405
mercury thermometer *see under* thermometer
Mersenne, M. 285
meshes, in networks 666
relation to magnetic field 770–*771*
metal
crystalline structure of 258, 259–*260*, 261
deformation of 260–263
impurities in 265–266
mobility of charge in 563–567
meter 13
standard 13
and Michelson interferometer 1016–1017
metrology 9–10
mica, as biaxial crystal 1037–1038
Michelson, Albert A. 926–927
and interferometer 1015
role in relativity theory 1017*n*
Michelson's interferometer *1015*–1017
and standard meter 1016–1017
microscope
compound 964–966
eyepiece 964, *965*, 966
magnification of 966
objective 964–965
practical 964, *965*

microscope, electron 790, 792–796
optical 796
simple *961*–964
aberration of *963*–*964*
magnifying power of *962*–963
transmission 793–795
microworld 456–458
Millikan, R. A. 563, 583, 585, 594
and electron emission 594
and oil-drop experiment *583*–585
millivoltmeter, and control of temperature coefficient 683, 685–686
minerals
flotation of *321*
mirror
aplanatic *935*–*936*
concave *942*, *943*
concave; convex 940–941, *942*, *943*, *960*
ellipsoidal *935*
hyperboloidal 935–*936*
and image formation 932, *933*–*934*
paraboloidal 935–936
in telescopes 936
plane *933*, 935, 937–*938*
formula for 937–938
and image formation 917
spherical
aberration of 934
conjugate foci of 941
focal length of 941
focal plane of 941
formula for 938–940
image formation by 933–*934*
principle focus of 940–941
mistake (and error) A-16
MKSA system of units 100, 152, 495, 496, 526, 544
mobility, of carriers of current 650
moduli of elasticity
bulk, *see under* bulk modulus of elasticity
relation among three kinds of *252*, 255–256
shear 249, 252, 255
table of 255
Young's 249, 250–251, 252
molecular heat, of a gas, *see under* gas
molecules
collisions between 379–381
forces between 306–307

molecules, in a gas 378–381, 382
    degrees of freedom of 397–398, 402
  and gas pressure 382–385
  in liquid surfaces 306–308
    energy of *307*–308
  velocities of, in a gas 387–389
moment
  of a couple 165–*166*, 178
  of a force 159, *166*
    theoretical derivation of 163, *164*
  of inertia 162, 171, 180–185
    definition of 162
    values of 180–185
  "law" of 159–160
moments, "law" of 163–164
momentum 104–105
  angular 163
  conservation of 110–111
  definition of 104–105
  rate of change of 108–111
motion
  laws of 90–91
    Newton's 112–113
  science of 27, 89
  simple harmonic, *see* simple harmonic motion
  in a straight line 28–31
    equations of 29–31
  uniform circular, *see* uniform circular motion
motor, electric, *see* electric motor
moving-coil galvanometer 640–647
moving-iron ac instruments 867–*869*
mutual inductance 836–837
  unit of 837

neon-sign tubes 633
net planes *1019*
network 666
neutral temperature (of a thermocouple) 729
Neumann, F. E. 7, 831
newton (unit of force) 102–103
Newton, Isaac
  and law of gravitation 123–124, 151
  laws of motion 90–91, 112–113
  light propagation, theory of 913
Newton's rings 1011, *1012*–1013
Nicol prism *1035*–1036
nodes, antinodes 75, 78
noise 300

note (of sound) *see* pitch (of sound)
nutation 205

object (in optics)
  point 791, 931–*933*, 937, *938*, 939–943
  real 933, 939–*943*
  virtual *933*, *939*–943
objective (microscope), system of lenses in 964, *965*
oersted 803
ohm 649
Ohm, Georg Simon 649
ohm-meter 659
Ohm's law
  establishment of 648–649
  statement of 649
  theory of 649–650
oil-drop experiment, Millikan's *583*–*586*
optical instruments 961–970
optical rotation 1042–1044
  by liquids and vapors 1043–1044
  materials producing 1043
  by uniaxial crystals 1043
optic axis 1031
optics, geometrical, *see* geometrical optics
ordinary ray, wave surface of 1032–*1034*
oscillating mass on spring
  and conservation of energy 148, 151
  and determination of g 129–132
  elastic potential energy of 143–144
  period of 129–130
oscillation
  forced 295–296
  rotational 193–198
    general theory of 193–194
oscillator, *LC*, feedback in 904–*905*
overtones 300
overvoltage 702
oxide cathode 579–580
oxide emitter, *see* oxide cathode
ozone layer 619n

parallel axes, theorem of *183*–185
paramagnetism 816–820
paraxial ray 954
particle 27–*28*, 144–145
  kinetic energy of 144–145

particle theory, *see* quantum theory
Pascal, Blaise 216
Pascal's principle, applications of 216
PD, see potential difference
Peltier effect 721–724, 729–730
    explanation of 729–730
    and thermodynamics 723–724
Peltier, Jean C. A. 722
Pelton wheel *108*–109
pencil (of rays) 913, 932, 933
pendulum
    compound *195*–196
    conical 49–*50*, 97
    and conservation of energy 147–148, 151
    forces exerted on *126*
    Kater's *128*–129, 195–196
    and reactional force *106*–108
    rigid–body *128*–129, *195*–196
    simple 50–52, 95–98, 125–129
        and force–acceleration relation 95–98
        period of oscillation of 127–128
        theory of 125–129
perfect gas, *see* ideal gas
period
    of simple harmonic motion 47–48
    of wave disturbances 61
permanent set 256
permeability, magnetic 761–762, 805
permittivity 497, 549–550
    of a dielectric 549–550
    of free space 549–550
Perrin, J. 586
phase angle
    in *RLC* circuit 875
    of simple harmonic motion 48
    of wave disturbances 61
phase difference 49
phase reversal, on reflection *74*
phasor diagrams *874–875*, 877, 981–*983*, *984–985*, *994*, *995*, *996*, *1002*, 1006, *1007*
    for diffraction gratings *994*, *995*, *996*, *1002*
    for interference by thin films 1006, *1007*
phasors
    in ac theory, use of 872–878
    in physical optics, use of 981–*983*
    and two-point-source interference 983–985
φ, *see* phase angle

phosphor 607, 633
photoelasticity 1039–1041
    and stress analysis *1040*–1041, *1042*
photoelectric cathode 598–599, 602–603
photoelectric cell 595, *596*
photoelectric emission 593–*597*, 598–599, 601
    and values of thermionic work function 597–599
    variation with frequency 596
photoelectric threshold 595–596, *598*, 599–*600*
photoelectrons 594
photoionization 618–620
    by radiation 619
    by ultraviolet light 619
    by x rays 618–619
photons 601, 631
    role in electron emission 631
phototube 595, *596*
physics, historic development of 5–9
pitch (of sound)
    effect of intensity 294
    relation to frequency, determination of *293*–294
Planck, Max 7, 351, 402, 600n
Planck's constant 600
Planck's quantum of action 17
Planck's radiation formula 351–352
Planté, G. 713
planetary motion, Kepler's laws of 121–123
plastic deformation *256*
plastic deformation, *see under* deformation
plastic flow
    structural changes in 258–266
    and work hardening 256–258
"plate voltage" 572
Plimpton and Lawton's experiment 520–*521*, 523
point charge *504*, *511*, 513–514
point image *938*, 939–943, 949, 952–960
point object 791, 931–*933*, 937, *938*, 939–943
pointed conductor *529*
points, action of, in electrostatics 507, 631–632
poise (unit of measure) 231
Poiseuille, J. L. M. 231
Poiseuille's formula 233–234

polarity, explanation of *823*
polarization
  by light scattering 1028–1031
  of mechanically propagated waves
    64
  plane of 65, 1028*n*
  by reflection of light 1027–1028
  role in interference 986–987
  in voltaic cell 707
  of x rays *1030*–1031
polarization field 569–570
polarized light, *see under* light
polarizer 1027
polarizing angle 1028
polaroid film 1036–1037, 1040–1041
polishing, frictional (theory of) 274
potential barrier
  explanation of 566–*567*, 568–570
  at metal surface 565, 566–*567*
  in thermionic diode 570–571, 621
  and thermionic emission 578–581
potential difference (PD) 524–526,
    539, 647–649, 662–664, 687–689
  comparison of (with potentiom-
    eter) 687–689
  in current electricity 647–648
  and electromotive force 662–663,
    664
  in electrostatics 647
  and ice–water relationship 539
potential divider 693–*694*
potential (electric) 523–530, 626
  definition of 523–524
  floating 626
  unit of 526
potential energy 147–148, 524,
  663–664
potential steps, at electrodes 704,
  705–707
potentiometer
  applications of 692–*693*
  in comparison of resistances 681,
    687–693
  Crompton *691*
  direct-reading 692
  practical 691–692
  principle of 687–*688*
  resistance-box 692
  slide-wire 691
  in thermometry 344
  and Wheatstone bridge 691, 692,
    693
pound 99, 100, 103
  relation to kilogram 99

poundal 102–103
power
  definition of 152
  of lens 955–956
  in rotational dynamics 163
  units of 152
power factor 877–878
Poynting, J. H., and determination
  of *G* 196
precession 198, 200–205, 786–787
  of earth's mantle and core
    786–787
  of equinoxes 204–205
  secondary 201–202
  of top 203–204
precipitation, atmospheric 416–421
pressure
  atmospheric 212*n*, 220–222
  critical 422
  definition of 212*n*, 213
  exerted by gas molecules
    "box" theory of 382–385
    mechanism of 381, 382–385
  measurement of 217, 218–221
  partial, Dalton's law of 413
  standard 220
  theory of 213–217
  units of 212*n*, 213, 218–219
  variation with depth *214*–215
pressure gauge (manometer) *218*
pressure gradient 121
pressure–volume relation
  for gases 224–226
  of liquids and solids 226
Priestly, Joseph 520
principle, scientific 3
prism
  combination, achromatic 947
  image formation by *945*
  refraction by *943–945*
    minimum deviation in 943–*944*
prism spectrometer
  calibration of 969–970
  collimator in 968
projectiles, flight of 37
pyrometer
  optical *352*–353
  radiation 347–348, *349*
  total-radiation, Féry's *349*–351,
    353

quanta, radiation 7, 601–602
  energy of 601–602

quantities
  derived 16–18
    units of measure of 16–17
  fundamental 12–13, 15, 100
quantum
  of energy 600
  Planck's 17
quantum mechanics 7–8
quantum theory 402, 597–602, 814
  and electron spin 814
  and excitation of degrees of
    freedom 402
  relation to radiation laws 600n
quarter-wave plate 1038–1039

radar method, and velocity of light
  927–928
radian 16
radiation
  black-body 600n, 601
  electromagnetic 7–8, 601–602, 782
    energy of quanta 601–602
    theory of 782
  of heat 348, 366, 369–371
  ionization by 618–620
  theory of 7–8, 782
radiation formula
  Planck's 351–352
  Wien's 352, 353
radio
  aerial *879*–881
  reception 880–881
  transmission *879–880*
  waves, identification of 594
radius of gyration 185
rain 416, 420–421
  cold 416, 420–421
    formation of 420
  electrically charged 540–541
  production of, by seeding 420–421
  warm 416, 420–421
range multiplier *683*
ray
  concept of 912
  of light 912, 931, 937–938, 939,
    943–945, 946–947, 1005,
    1032–1035
    passing through prism *943*–945,
      946–947
  ordinary; extraordinary 1032–1035
reactance, inductive and capacitive
  876
recombination, in ionized gas 619

rectification, of alternating current
  654–656
  full-wave 862–*864*
  half-wave *862*–864
rectifier 654–*655*
rectifier systems 862–*865*
reference, frames of 89–91
reflection 72–80, 292–293, 916–917,
  921–923
  Huygens's theory of 916–917
  phase change accompanying 73,
    292
  of sound 292
  theory of 916–917
  total internal 921–923
    critical angle for 921–922
refraction
  cyclic law of 920
  double 1031–1036
    explanation of 1031–*1034*
    by uniaxial crystal *1031*–1035
  Huygens's theory of *918*–919
  index of 919
  laws of 917–921
  by prism *943–945*
    minimum deviation in 943–*944*
  reversibility law of 919
  of sound 290
  at spherical surface 949–952
  by uniaxial crystal *1031*–1035
refractive index 919
  and Nicol prism 1038n
  values of 921
refrigerator, as reversed heat engine
  442, 454
Regnault, H. V. 286
relative density 211–212
relative permeability 762
relativistic mechanics 105
relativity, special (theory of) 8, 15,
  145, 795, 1017n
remanence 805–806
resinous electricity 486–489
resistance
  comparison of (of two conductors)
    675–681
  dial-box *679*–681
  unit of 649
resistivity *658*–659
  unit of (MKSA) 659
  values of 659, 660
resistor 651, *652, 653, 654*, 901–904
  cathode-bias 902–903
  grid *902*–904

resolution of vectors 36–37
resolving power, of telescope
    1002–1004
resonance 878–881
    acoustic 295–298
        observation of 296–*298*
    of ac circuit 878
    curves 878
    effect of damping on 297–*298*
    frequency 878
    "sharp"; "flat" 297–298, 879
resonator, acoustic *296*–297
rest mass 105, 145*n*
    defined 145*n*
retentivity 805–806
reversibility
    of Daniell cell 709
    law of 919
reversible cell, thermodynamical
    theory of 714–717
reversible decomposition voltage
    702–703
reversible and irreversible processes
    458–460
    definition of 458
Richardson, O. W. 578
    and electron spin 813–814
Richardson's equation 578, 587,
    588
    and work function ϕ 587, 588
Richardson line 578
rigid body 159, 193
rigidity 251–252, 255
RLC circuit 833–836, *870–872, 873*
    current–voltage relations in
        870–872, 873
rms, *see* root mean square (rms)
    value
rod waves 287–288
Römer, Olaf 923, 924
Römer's method 923–924
Röntgen, W. K. 438, 607
root mean square (rms) value 387,
    858–859
    of voltage 858–859
rotation
    of earth 14, *40*–42
        effect on *g 40*–42
    optical, *see* optical rotation
    of planets and satellites 43–44
rotational oscillation 193–198
    general theory of 193–194
rotational quantities as vectors
    166–171

rotatory dispersion 1043, 1044
"roughness factor" 575

sailing ship, exertion of force on
    109–*110*
saturation
    magnetic 804
    of thermionic current 573–574
scalar 34
scalar product, *see* dot product
scale, temperature, *see* temperature
    scale
scattering
    of light 1028–1031
    of x rays 1030–1031
Seebeck effect 343, 344, 719–721
Seebeck, Thomas Johann 719, 721
search coil, and magnetic field
    737–738
second, definition of 14
second law of thermodynamics
    452–453
secondary emission
    due to electron impact 606,
        624–627
    measurement of 624–627
    and positive-ion impact 606,
        626–627
secondary emission coefficient
    625–626
    values of 625–626
seeding, of clouds 420–421
seismic waves 287
self-induced emf 831–832
self-inductance
    unit of 832
semiconductors
    n-type; *p*-type 650–651
    and Peltier effect 722–723
semimetals 650–651
shear modulus 255
    see also rigidity
shearing strain 254
shearing stress 252–254, *255*
shear waves 287
shielding, electrostatic 507–508
ship, sailing 109–*110*
SHM, *see* simple harmonic motion
sidereal day 14, 41
    variation of 14
shunt *684, 849*
    ring *849*
    universal *849*

sign convention in geometrical optics 936–938
significant figures (in measurement) A-11–A-12
simple harmonic disturbances 49, 52–53, 79–80, 194
  and acoustic beats 79–80
  amplitude of 52
  definition of 52, 53
  frequency of 52–53
simple harmonic motion (SHM)
  acceleration in 46–47
  amplitude of 45
  definition of *45*, 48
  and mass on spring 130
  phase of 48
  phase angle of 48
  and pendulum 97, 126–127
  period of 47–48
  velocity in 45–46
  of waves 61
simple pendulum, *see under* pendulum
siphon, theory of *217*
slide-wire bridge *678*–679
sliding friction (kinetic) 267, 274
slip planes 263–*264*
  observation of *264*
slip rings, in alternator 840
slit, illuminated
  effect of varying dimensions of 988–989
  single, diffraction by 998–1001
  two or more 985, 993–997
snow 416, 418–419
  formation of 418–419
snowflakes 418–419, 420
soap bubble 313
sodium-vapor lamp 633
solenoid 820–822
  and magnetic field 778–*780*
  magnetic field on axis of 764–*765*
solid
  atomic structure of 258–260
  and heat 402–405
solution pressure 698
  of copper and zinc 705, 709
  of hydrogen and oxygen 703–704, 705
Sommerfeld, A. 7, 8
sound
  audibility, factors affecting 289–292
  zones of 291–292

sound, Doppler effect 69–71
  frequency 69–71, 294–295
  and pitch 69, 293–295
  standards of 294–295
  musical, and noise 300
  propagation in solids 287–288
    by bulk waves 287, 288
    by rod waves 287–288
    by shear waves 287
    velocity of 286–288
  quality of 298–300
    factors determining 298
  reflection of 292–293
    phase change accompanying 292
  refraction 290
  resonance, *see* resonance, acoustic
  scattering of 292–293
  velocities of
    in air 285–286
    in a gas 388
    in liquids and solids 286–288
    in various media (table) 288
  wave nature of 285
sound waves, *see under* waves
space charge
  of ions in electrolyte 668–669, 706–707
  in thermionic tube 580–581
    electron transit time in 580–581
  space-charge-limited current 580, 581
  see also infrasaturation current (thermionic)
  and positive ions 621–622
space lattice (*see also* crystal lattice)
  atomic structure of 258–261
  cubic 258–*259*, 260, 1017–*1018*
  dislocations in *261*, *262*, *263*–266
    elastic 260–*261*
  forms of
    body-centered-cubic 259
    face-centered-cubic 259–260
  impurities in 265–266
  and Laue spots for KCl 402–*403*
  perfect 265–266
  and sizes of atoms 588–589
  slip planes in 263–*264*
  in x-ray spectroscopy 1017–1021
spark 634
specific gravity 211, 714
specific heat
  constant-pressure 362–363
    values of 363
  constant-volume 362

specific heat, definition of 362
  determination of 428–430
  temperature variation of 362,
    365–366
  and Thomson coefficients 730–731
  of water, values of 361
specific inductive capacity, *see*
  dielectric constant
spectra
  missing, in single-slit diffraction
    1000–1001
  observable orders of 995
  x-ray *1018*–1019
spectrometer
  crystal 1020, 1021
  diffraction-grating 975
  prism, *see* prism spectrometer
    968–970
spectrum
  banded 1009, *1010*
  line 1018, 1021
  production by gas discharge 617
sphere, uniformly charged 517–*518*,
  520–523
spherical aberration 934
spring, oscillating mass on 129–*130*
square-wave form 293
standard cell 689–*690*
standard masses 223–224
standard temperature and pressure
  (STP) 220
standing waves, *see under* waves
star
  fixed, aberration of 924–925
  as point source *1003–1004*
statics 185–187
statistical mechanics 458
steam engine, reciprocating 463–464
steam point 341
Stefan-Boltzmann law 349
Steiner's theorem 183
Stokes, Sir George 231, 417
Stokes's formula 231–232, 417
STP, *see* standard temperature and
  pressure
strain
  shearing 254
  tensile 250–251
strain viewer *1040*
stratosphere 438
streamline 228
stress
  frozen 1041
  shearing 252–254, *255*

stress, tangential 252–254
  tensile 250, 252
stress–strain relationship 250–*251*,
  256–258, 262–263
stretched string, waves along
  theory of *279*–281
  velocity of 281
superposition 71–80
  principle of 71
surface density, of charge 518–519
surface energy 308–310, 473–476
surface, of liquid, *see* liquid surfaces
surface molecules 306
surface tension
  applications and consequences
    305, 319–322
  definition of 309–*310*
  and liquid-film stability 318–319
  measurement of *316*–317
  and surface energy 473–476
  temperature variation of 475–476
  theory of *310*
  values of 317

Tabor, D. 273
telescope
  Cassegrainian reflecting 936,
    967–968
  Newtonian reflecting *967*, 968
  reflecting 936, *967–968*
    Gregorian 935
    magnifying power of 967, 968
  refracting *966*
    magnifying power of 966–967
  resolving power of 1002–1004
television 614
temperature
  "brightness" 353
  critical 422
  definition of 329, 376
  of gas 385–386
    meaning of 385–386
    and rms velocity 388–389
  inversion 433
  thermodynamic scale of 464–466
  variation with altitude 438–441
temperature scale 329, 330–*331*
  absolute thermodynamic 341,
    465–466
  Celsius 331*n*, *332–333*, *334*–337,
    341
  electrical resistance 333–*334*
  Fahrenheit 331, *332*

temperature scale, fixed points in
330–331
Kelvin 331, *332*, 340–341
and mercury thermometer 331,
333, *334*
variation with property on which
based 333–334
tempering 265
tension and extension (of a spring)
130–132
terminal velocity
of cloud droplets 417–418
Stokes's formula for 231–232
terrestrial magnetism, *see*
geomagnetism
theory, scientific
physical 4–5, 8–9
role in unification 4–5
thermal equilibrium 330
thermal expansion 357–358
linear 357
volume 357
thermal phenomena, study of 6
thermionic current
and emission 573–574
infrasaturation 574, 580–581
variation with anode voltage 581
space-charge-limited 580–581
variation with anode voltage 581
saturated 573
variation with anode voltage 573,
574
thermionic diode 572, *574*, *620*, 621
thermionic emission 570–582
applications of 581–582
as evaporation of electrons 570–
574, 576–*577*
observation of *571*–573
and surface potential barrier
578–581
temperature variation of 574–577
thermionic triode 610–*611*, 901,
902–905
thermocouple 343, *344*, *719*, 720–
721, 722
ideal, action of 722
thermodynamic efficiency 447–448,
461
of Carnot heat engine 447–448
thermodynamic temperature scale,
absolute 465–466
thermodynamics
first law of 427, 451
and energy-conservation law 427

thermodynamics, and Peltier effect
723–724
of reversible cell 715–717
second law of 336–337, 341, 451,
452–453
evidence for truth of 455
and heat-engine efficiency 461
statement of 452–454
statistical nature of 457–458
and temperature scale 336–337,
341
violation of, in microworld
457–458
thermoelectric constants, values of
729
thermoelectric effect 343, 344
thermoelectric power 727–729
thermojunctions 343
thermometer
calibration of 342
carbon 347
compensated gas 339
definition of 330
electrical resistance 333–*334*
gas *337*–340
calibration, use in 342, 349–350
constant-pressure *338*–340
constant-volume *337*–338
gas fillings 339
temperature limits of 339
high-temperature, *see* pyrometer
347
liquid-in-glass 344, *345*, 346
alcohol 346
n-pentane 346
low-temperature 347
mercury-in-glass 331, 333–*334*,
335, 344–345
defects of 344–345, 377–378
platinum resistance 343
semiconductor 347
and temperature, scales of
330–334
thermocouple 343, *344*
vapor-pressure 346
thermopile 721
Thiessen, P. A. 811
Thomson coefficient 725
Thomson effect
demonstration of *725*–726
explanation of 724–725, 729–731
Thomson, J. J. 606, 789, 790
determination of e/m *789*, 790
thundercloud *540*–541

thunderstorm cell 538–541
  energy of 541
time 12, 13–15
  concept of 13
  measurement of 13–15
time constant
  of *RC* circuit 653
  of *RL* circuit 834
top, precession of *203*–204
toroid and magnetic field 780–782
torque
  and ballistic galvanometer 643–645
  concept of 159
  on current-carrying coil 745–746
  definition of 160, 161–163
  expressions for 162–163
  and gyroscopic action 199–202
  and law of moments 163–164
  as a vector *170*–171
Torricelli, Evangelista 220
Torricellian vacuum 220
torsion balance, Coulomb *487*, 494
total internal reflection
  in right-angle prisms *922*
tourmaline, dichroism of 1036
Traité de la Lumière 912
transformer
  auto- 899–*900*
  coupling coefficient 888
  design of 898–899
  efficiency of 887–888
  ideal 888, *889*–898
    air-cored 895
    current relations for 893–895
    equivalent circuit of 898
    flux variation in 895
    inductance (self- and mutual)
      of 889–891
    iron-cored 894–895
    phase relations in 895–897
    phasor diagrams for 895,
      896–*897*
    primary, secondary powers in
      897
    theory of 888–891
    turns ratio of 897–898
    voltage relation for 891
  step-down 887–*888*
  step-up 863–*864*, 887–*888*
  uses of 887–888
translation 27
transverse waves *62*, 64–67
trigonometric functions, table of
  A-4–A-7

triode, thermionic 610–*611*, 901,
  902–*905*
  anode of 610–611
  applications of 901
  cathode of 610–*611*
  electron-optical system of
    610–*611*
  grid of 610–*611*
  and voltage amplification 901,
    902
triple point of H_2O 341, 411–412
troposphere
  height of 438
  temperature variation in 438–441
  and weather phenomena 416–*417*
true value, best estimate of
  A-13–A-14
tube of flow *230*, 236–238
tungsten 576, 578–579
  thoriated 578–579
tuning fork 294–295, 296–297
turbine, efficiency of 464
turns ratio (of transformers)
  897–898
two-body system *106*, 107–108,
  150–151

Uhlenbeck, G. E. 814
uniform circular motion 37, 200
  and gyroscopic action 200
units
  arbitrary and natural 17
  of measure, *see* measure, units of
units, systems of 100, 102–103, 152
  CGS 100, 102–103, 152, 495, 496
  FPS 100, 102–103, 152
  MKSA 100, 102–103, 152, 495,
    496, 526, 544
unpolarized light 1027, *1029*
upthrust, Archimedean, *see* Archimedean upthrust *222*–224
U-tube manometer *218*, *219*, 220

Van de Graaff generator 533–538,
  545, 564, 565
  and field emission 564, 565
  maximum dome potential of 545
  pressurized 537
  priming of 537
  self-exciting *536*–537
Van de Graaff, R. J. 533, 535

A-123

van der Waals, J. D. 10–11,
    386–387
  equation for gases 386–387, 425
vapor
  saturated 407–408
  supersaturated 412, 416
  unsaturated 412
vapor pressure
  equation for 471
  of liquid 408
  and liquid-surface curvature
    413–416
  measurement of 408
  of solid 408
  temperature variation of (for
    various materials) 408–409,
    410
    mathematical representation of
      410
  and triple point 411–412
  in troposphere 438–441
  variation with temperature 469–471
vaporization
  latent heat of 363, 364, 406, 466,
    468, 469–471, 576
    of electrons 576
  theory of 406–407
vector
  components 36
  composition of 37–39
  definition of 34, 35–36
  parallelogram 34
  polygon 34–*35*
  representation of 139, 140–141
  resolution of 36–37
  triangle 34
  unit 774–775
vector algebra 36
vector product 744
velocity
  angular 39
  constant 29
  distribution of, among gas
    molecules *380*–381
  escape 388
  of light, *see separate listing*
  of mechanically propagated waves
    64
  as a quantity 28–31, 34–35
  terminal 231–232, 417–418
  units of 17
velocity gradient
  longitudinal 228, 230, 236
  transverse 228–229, 231

velocity of light 17, 923–928
  determination of 923–928
    Bradley's method 924–925
    Fizeau's method *925*–926
    Foucault's method 926
    Kerr-cell method 927
    Michelson's method 926–*927*
    radar method 927–928
    rotating-mirror methods
      926–*927*
  value of 928
Venturi meter *239*
da Vinci, Leonardo 269
viscosity
  definition of 230–231
  measurement of 233
  theory of 235–236
    for a gas *393*–394
  units of 231
  values of 234
  variation with temperature 234
viscous resistance 231
vitreous electricity 486–489
volume, units of 17
volt 526
volt box 693–*694*
Volta, Alessandro 526
voltage amplification 901–904
voltage, reversible decomposition
  702–703
voltameter, silver (in current
  measurement) 640–641
voltmeter
  Braun's electrostatic 542, *543*
  electrostatic 541–542, *543*,
    869–870
    ac use of 870
    as capacitor 870
    electrodynamometer-type (ac)
      865–867
    hot-wire (ac) 869
    practical, operation of 682–684
    principle of 682
    range multipliers in 683
    rectifier-type (ac) 862–863, 864

water
  bubbles and droplets 415–416
  and Clapeyron's equation
    472–473
  density of 212
  electrolysis of 701
  supercooled 418

water droplets 419
water vapor 415–416, 417, 418, 420–421
waterproof fabrics 321
watt 152
wattless current 878
wattmeter 866–*867*
waves
  amplitude 60
  circularly polarized *65*–66
  electromagnetic, *see* electro-
    magnetic waves
  elliptically polarized 66–67
  in fluid 281–283
    theory of 281–283
  in a gas 283–286
    velocity of 283–285
  kinds of 59
  light 69
  longitudinal 67–68
  in material medium 285 286
    identification with sound
      285–286
  "matter" 59
  motion of 59–80
  plane 69
  plane-polarized 64–65
  radio, identification of 594
  reflection of, *see under* reflection
  sine *60*
  sound 67, 68, 69–71
    equation for 69
  spherical, equation for 69
  standing 72–*75*, 76–79, 300
    compared to sound wave 300
    equation for 76
  along strings 59, *60*–67, 75–79, *279*–281
  transverse *62*, 64–67
  velocity of 64
  water 59
wave disturbances
  phase angle of 61
  period of 61
wave equations
  for sine waves 61, 62–64
    forms of 62
  for sound waves 68, 69
  of vibrating string 75–79
wave form 298–300
  of alternating current 842–843
  analysis of 298–299
    and sound quality 299
  synthesis of 299–300

wavelength 60, 70–71
  and electron microscopy 795–796
wave mechanics 7–8
wave propagation 279–283
  in a gas 283–285
    theory of 283–285
    velocity of 283–285
  in fluid 281–283
    nature of 281–*282*
    velocity of 283
  mechanical, defined 59
  along stretched string *279*–281
    velocity of 281
wave train 60, *62*
weber 829
Weber, Wilhelm, and theory of
  magnetism 801, 808–810
Weber–Ewing theory of magnetism
  808–810
weight 92–93
  and buoyancy-correction factor
    223–224
  and mass 99, 100, 103
Weiss, Pierre, and magnetic domains
  811
Weston cell 689–*690*
  see also standard cell
wetting and nonwetting 314, 321
Wheatstone's bridge 675–677, 687
  dial-box *679*–681
  slide-wire *678*–679
white dwarfs, density of 213
Wiedemann-Franz law 369*n*
Wiedemann-Franz law 7
Wien, W. 352
Wimshurst machine 564*n*, 565
wind 229, 289–291
  effects on audibility 289–291
Wood's metal 274
work
  coil 901
  concept of 137
  definition of 137–139
  as scalar product 139
  units of 152
work function 578
  values of 579
work function ϕ, values of 588
work hardening 256–258

x radiation
  characteristics 1021–*1023*

A-125

x radiation, continuous 1017n, *1021–1023*
  short-wavelength limit of 1022–1023
  line *1021–1022*
x-ray diffraction
  by crystals 1017–1020
  by ruled grating 1020–1023
x rays 607–610
  diffraction of 1017–1021
  discovery of 607
  as electromagnetic radiation 610
  electromagnetic nature of 1030–1031
  photoionization by 618–619
  polarization of *1030*–1031
  scattering of *1030*-1031
x-ray spectroscopy 1017–1023
x-ray tube 607–610

x-ray tube, electron-optical system of *608*, 609–610
  gas 607–608
  and glow discharge
  modern thermionic *608*–610

yard, legal 13
Young's fringes 975, *976*, *978*
Young's modulus of elasticity 249, 250–251, 252
Young, Thomas 975
  wave nature of light 913

zero, choice of 524
zones of audibility 291–292
zone of silence 291

Editorial processing by Science Bookcrafters, Inc.